Earthquake Engineering
Sixth Canadian Conference

Génie Sismique
Sixième Conférence Canadienne

Countless lives have been saved as a result of recent strides in earthquake engineering and related sciences. This trend has been furthered by the work of the Canadian National Committee on Earthquake Engineering which has, over the past twenty years, provided specialists with a forum for exploring new approaches to the problem. Engineers, scientists, researchers, geologists, seismologists, and other professionals have shared research and experience at the committee's conferences. The sixth of these, held in June 1991, is documented in this volume.

Three keynote papers provide the overall focuses for the volume. Each deals with one of the three major areas in the field: structures, in a paper on design developments in high-rise design and construction in Japan; geotechnical engineering, in a discussion of the effects of site conditions on ground motions; and seismology, in an account of the development of phased strong-motion time-histories for structures with multiple supports.

Shorter papers fall into three broad areas: response analysis and design of structural components; the interaction of seismicity, mitigation, soil response, and soil structure; and seismic codes and structures.

This conference along with other similar events throughout the world has contributed significantly towards understanding various phenomena needed for building safe, reliable, and economical structures that can meet the challenges presented by the forces of nature.

S.A. Sheikh and S.M. Uzumeri are faculty members of the Department of Civil Engineering, University of Toronto.

PROCEEDINGS / COMPTES RENDUS
SIXTH CANADIAN CONFERENCE ON EARTHQUAKE ENGINEERING
SIXIÈME CONFÉRENCE CANADIENNE SUR LE GÉNIE SISMIQUE
12–14 JUNE 1991 / TORONTO / 12–14 JUIN 1991

EARTHQUAKE ENGINEERING

Sixth Canadian Conference

GÉNIE SISMIQUE

Sixième Conférence Canadienne

Shamim A. Sheikh and S.M. Uzumeri

Editors / Éditeurs

Published with / Publié avec
University of Toronto Press
Toronto   Buffalo   London

© University of Toronto Press 1991
  Toronto Buffalo London
  Printed in Canada
  Reprinted in 2018

  ISBN 0-8020-5942-2

  ISBN 978-1-4875-8192-3 (paper)

The texts of the various papers in this volume were prepared for reproduction under the supervision of the papers' authors. The opinions expressed are those of the authors.

Les textes des articles figurant dans cette publication ont préparés sous la surveillance des auteurs concernés. Les opinions exprimées sont celles des auteurs respectifs.

Canadian Cataloguing in Publication Data

Canadian Conference on Earthquake Engineering (6th: 1991: Toronto, Ont.)
    Proceedings: Sixth Canadian Conference on Earthquake Engineering, 12–14 June 1991

"Under the auspices of the Canadian National Committee on Earthquake Engineering."
Title page and introductory material in French and English; papers in English only.
ISBN 0-8020-5942-2
ISBN 978-1-4875-8192-3 (paper)

1. Earthquake engineering – Congresses.
I. Sheikh, Shamim Ahmed, 1950–   . II. Uzumeri, S.M.
III. Canadian National Committee for Earthquake Engineering.
IV. Title.

TA654.6.C36 1991   624.1'762   C91-093932-2

Preface

The Sixth Canadian Conference on Earthquake Engineering starts the third decade of activities which initiated with the first conference in 1971 on the campus of the University of British Columbia. Since then this conference, under the auspices of the Canadian National Committee on Earthquake Engineering (CANCEE), has provided a unique forum for thousands of earthquake engineering experts that includes engineers, scientists, researchers, geologists, seismologists, and other professionals.

This conference, along with the other similar events throughout the world, has contributed significantly towards understanding phenomena needed for building safe, reliable, and economical structures. The behaviour, during recent severe earthquakes, of the structures designed and built using the recent techniques and code provisions is a testimony to our success in meeting the challenges offered by nature.

The growing interest in the conference is evident from the fact that a total of 179 abstracts were received from across Canada and seventeen other countries for consideration for inclusion in the three-day conference program. Of these, 138 abstracts were accepted and the authors were requested to submit full-length papers for consideration, with the condition that each paper would be presented at the conference by one of the authors. A total of 100 papers were submitted, of which 95 were accepted for inclusion in the Conference Proceedings and presentation at the conference. The acceptance of each paper and each abstract required approval of two experts in the field. The three invited keynote papers also included in the Proceedings are written by well-known engineers and researchers who represent the global earthquake engineering community. They deal with three major areas in the field: structures, geotechnical engineering, and seismology.

We would like to take this opportunity to express our sincere appreciations to the reviewers of the papers who met all the tight deadlines and helped maintain high standards. Thanks are also due to the members of the Organizing Committee and the Program Committee for their efforts. Above all, the authors deserve the appreciations of the Earthquake Engineering Community for their contributions, which present the results from the most current developments in the field.

S.A. Sheikh and S.M. Uzumeri

Préface

La sixième Conférence Canadienne de Génie Sismique entreprend la troisième décennie d'activités depuis la première conférence à l'Université de la Colombie Britannique, en 1971. Depuis, sous les auspices du Comité National Canadien de Génie Sismique (CANCEE), la conférence a fourni un forum unique à des milliers d'experts du génie sismique, incluant ingénieurs, chercheurs, géologues, séismologues et autres professionnels.

Cette conférence, ainsi que d'autres événements similaires à travers le monde, a contribué de façon significative à la compréhension de divers phénomènes reliés au génie sismique, permettant la construction de structures sécuritaires, fiables et économiques. Durant les derniers tremblements de terre importants, le comportement des structures, dimensionnées selon les techniques et codes récents, a témoigné de notre succès à relever les défis que nous impose la nature.

L'intérêt croissant pour cette conférence est mis en évidence par les 179 résumés d'articles reçus pour cette rencontre de trois jours, et provenant de partout au Canada et de 17 autres pays. Parmi ces résumés, 138 furent retenus pour publication possible, sous condition que chaque article sois présenté par un des auteurs lors de la conférence. Au total, 100 articles furent envoyés, et 95 furent choisis pour publication dans les comptes rendus de la conférence. Chaque article a du être consulté et accepté par deux experts dans le domaine. Les trois conférenciers invités sont des ingénieurs connus, représentant la communauté internationale du génie sismique. Leurs présentations, qui figureront dans les comptes rendus de la conférence, concernent trois domaines importants du génie sismique, soient les structures, la géotechnique et la séismologie.

Nous remercions sincèrement les évaluateurs des arcticles, qui ont su respecter les délais exigés et conserver un niveau élevé d'excellence. Nous aimerions également remercier les membres du comité organisateur et du comité de programme pour leurs efforts soutenus. La communauté du génie sismique apprécie grandement la contribution des auteurs qui présentent les résultats des plus récents developpement dans le domaine.

S.A. Sheikh et S.M. Uzumeri

Acknowledgements
Remerciements

The expenses for keynote speakers were met through a grant from the Natural Sciences and Engineering Research Council of Canada (Grant No. CNFO105154). Ontario Hydro and Hatch Associate of Toronto, with their generous financial support, acted as sponsors for the conference. Financial contributions from the Canadian Institute of Steel Construction, Canadian Portland Cement Association, NORR Engineering Limited of Toronto, and Morrison Hershfield Limited of Toronto are also gratefully acknowledged. We would not have been able to organize this conference without financial and personnel support from the Department of Civil Engineering at the University of Toronto. Special thanks are due to Ms. Yasmin Rhemtulla and Ms. Piril Doruk for their efforts in the organization of this conference.

Organizing Committee 6CCEE

Une subvention du Conseil de recherche en sciences naturelles et en génie du Canada (Subvention No CNFO105154) a permis la présentation des discours-thèmes par les conférenciers invités. Le parrainage de la conférence revient à Hydro-Ontario et à Hatch Associate de Toronto, grâce à leur généreux support financier. Les contributions financières de l'Institut canadien de la construction en acier, de l'Association canadienne du ciment Portland, de NORR Engineering Limited, de Toronto, et de Morrison Hershfield Limited, également de Toronto, ont été grandement appréciés. L'organisation de la conférence n'aurait pas été possible sans les supports financier et du personnel du département de génie civil de l'université de Toronto. Des remerciements spéciaux sont adressés à Mesdames Yasmin Rhemtulla et Piril Doruk qui n'ont pas ménagé leurs efforts pour organiser la conférence.

Comité organisateur 6CCEE

Canadian Conferences on Earthquake Engineering

Table of Contents
Tableau des matières

2 Seismicity, mitigation, soil response, and soil-structure interaction / Sismicité, mitigation, réponse des sols, et interaction sol-structure

3 Seismic codes and structures / Codes sismiques et structures

Keynote lectures
Discours d'ouverture

Recent development in design and construction of high-rise reinforced concrete buildings in Japan

Hiroyuki Aoyama[1]

ABSTRACT

Japan experienced a quick development of highrise reinforced concrete frame-type apartment building construction, about 30 stories high, in the last decade. Outline of this development is first introduced in terms of planning of buildings, materials, construction methods, earthquake resistant design and dynamic response analysis. This quick development was made possible by, among others, the available high strength concrete and steel. In an attempt to further promote development of new and advanced reinforced concrete building structures, a five-year national project was started in 1988 in Japan, promoted by the Building Research Institute, Ministry of Construction. Outline of this project is introduced in the second part of this paper. It aims at the development and use of concrete up to 120 MPa, and steel up to 1,200 MPa.

INTRODUCTION

The Building Standard Law in Japan provides design seismic loadings and principal design procedures for buildings up to 60 m in height. Structural design of any building with the height in excess of 60 m is subjected to the review of the Structural Review Committee for Highrise Buildings of the Building Center of Japan, and subsequently a special permit by the Minister of Construction is issued.

As far as reinforced concrete (RC) buildings are concerned, the height had been limited to about 20 m in practice by means of administrative guidance. Any building taller than, say, seven stories had to be constructed by steel structure or composite steel and reinforced concrete (SRC) structure. This administrative guidance was a traditional one, stemming out from public distrust on the seismic resistance of concrete structures ever since 1923 Kanto Earthquake.

In the recent ten years or so, this trend has changed rapidly. There are currently various movements towards the higher RC construction. Among them,

---

I. Professor, Department of Architecture, University of Tokyo.

the most remarkable is the increase of highrise RC frame construction. Kajima Construction Co. broke out the movement. They completed the first highrise RC construction, an 18-story apartment building, in 1974, followed by another 25-story apartment building in 1980. These highrise buildings were realized after a long and extensive effort in research and development of the company. Other construction companies followed, and the number of concrete buildings increased together with the increase of total highrise building construction.

Figure 1 shows the amount of annual highrise construction in Japan, which is the number of annual reception by the Building Center for the review of the Committee for Highrise Buildings. The figure also shows number of SRC and RC buildings in each year. Total number in each year varies from less than 10 to more than 100, reflecting economic fall and rise. Concrete construction takes about 23 percent on the average, however more than half of it is taken up by RC construction in recent years.

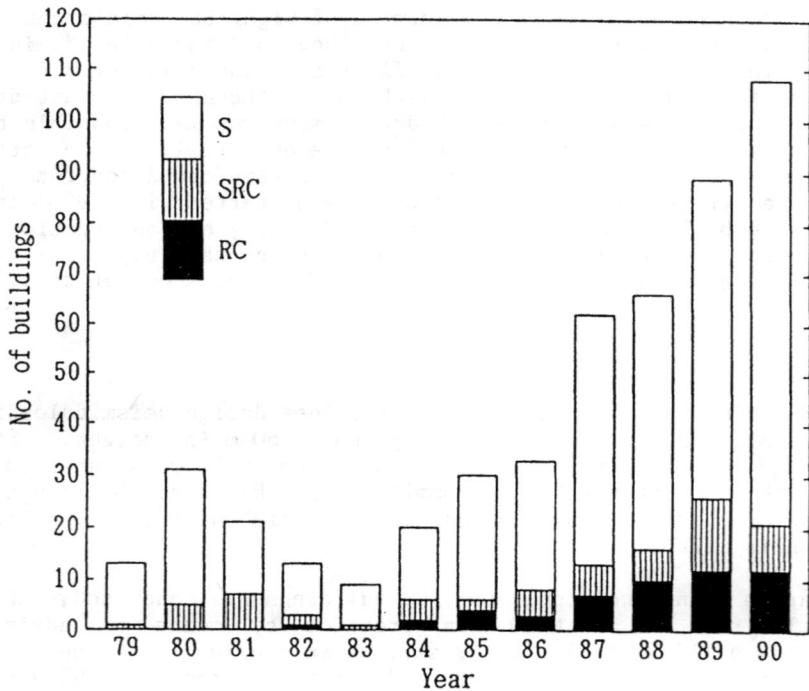

Figure 1. Annual highrise Construction in Japan

The quick development of highrise RC construction owes to many things, but availability of high strength materials was evidently the most fundamental factor. In an attempt to further promote development of new and advanced type of RC construction, the Ministry of Construction started in 1988 a national five year research project entitled "Development of Advanced Reinforced Concrete Buildings using High-strength Concrete and Reinforcement" (usually referred to as "New RC"). This is a very ambitious project, which will probably lead to the realization of highrise RC buildings up to 60 stories, and buildings with wider spans, allowing for use in greater variety.

This paper introduces, in the first half, the current state-of-the-art of structural design and construction of highrise RC buildings, and in the second half, the outline of the New RC national project of the Ministry of Construction.

PLANNING, MATERIALS, AND CONSTRUCTION OF HIGHRISE RC

Floor Plan and Elevation

Highrise RC construction, currently, is used exclusively for apartment houses, because of better habitability provided by concrete. Floor plan of these buildings is generally regular, and symmetric with respect to one or two axes. In almost all cases, all frames in both directions are designed as moment resisting frames. Span length is around 5 m, which is shorter than SRC or steel buildings. The small span is adopted in order to limit the axial load on a column, and thereby reduce the seismic force acting on a column.

The number of stories of highrise RC buildings ranges from 20 to 40 stories. The story height is about 3 m, which is also very small, permissible only for residential buildings. The story height is gradually reduced in upper stories corresponding to the reduction of beam depth, with the minimum clear structural height to the beam soffit of about 2.1 m. The frame elevation is generally quite regular, avoiding sudden change or discontinuity of stiffness in the vertical direction. Most buildings have one-story basement, and the foundation is supported, in most cases, by bearing piles of cast-in-place concrete.

Framing Members

Column section is usually square, with the maximum dimension of about 90 cm at the base of buildings. Axial reinforcement ratio is about 2 to 3 %. To provide effective confinement to the core concrete, columns are provided with one or two of the following types of lateral reinforcement: rectangular or circular spirals, flush butt (FB) rings, closed sub-hoops, high strength deformed PC steel with 1,275 MPa yield stress (Urbon), or welded wire fabric.

To overcome large seismic overturning moment which produces dominating axial forces in exterior columns in lower stories, additional axial bars (core bars) are frequently located in the central portion of these column sections.

Beams are of rectangular sections with relatively large width. Four-leg stirrups are generally used. Urbon stirrups are often used to increase shear resistance.

Materials

All highrise RC buildings use concrete with specified strength much higher than ordinary buildings. In the lowest portion either 36 or 42 MPa concrete is used. Strength is gradually reduced in upper stories. Lightweight concrete is not used.

The use of high strength and large size reinforcing bars is indispensable for highrise RC construction. Longitudinal bars up to 41 mm diameter (D41) with 390 MPa yield stress are used. Lateral reinforcement consists of either D16 bars of 295 MPa steel or high strength deformed PC bars with 1,275 MPa yield stress (Urbon).

Use of Precast Concrete

There are divided opinions as to the advantage of applying precast concrete to highrise RC construction. On one hand, it decreases the amount of on-site labor, thereby decreasing construction time. On the other hand, it increases the number of job types. Since at least a part of concrete must be cast on the site, both labors associated with precast and cast-in-situ concrete must be combined. In fact a wide spectrum in the degree of use of precast concrete is seen in practice, as follows.

(1) All concrete cast on the site. It is an extreme case, but it is probably the most popular method at present for highrise RC construction.

(2) Partial application of precast concrete to floor slabs. When the building has exterior cantilever balconies, they may be most easily constructed as precast elements. For interior floor slabs, composite slabs are often used, consisting of precast panel used also as formwork and cast-in-situ upper course utilized as monolithic diaphragm for earthquake resistance.

(3) Partial application of precast concrete to beams. When adopting precast beams, bar arrangement at the intersection of beams must be organized completely with utmost care in the design stage. Splices of precast beams are located at the beam-column joints, or at the center of spans. Beams are precast only to mid-depth, leaving upper portion for cast-in-situ concrete. This enables the insertion of upper bars after precast members are placed in position, thereby easing the bar arrangement at the beam-column connections. Composite slabs are almost always adopted with precast beams.

(4) Precasting columns. Columns are usually the last member to apply precasting technique, but there are a few applications of precast columns in combination with the cast-in-situ floor system. NMB splice-sleeves or similar splices are used for longitudinal bars. A unique method was also developed to precast columns without longitudinal bars but with sheaths instead. Bare longitudinal bars are installed, concrete unit is lowered, and finally sheaths are grouted for integrity.

Reinforcement Cage Fabrication

For all cast-in-situ columns and beams, reinforcement cages are prefabricated on the ground, for higher construction efficiency and accuracy. Columns are usually prefabricated for each story, and beams in two directions are prefabricated together, in the single or double cross-type shape, with splices at midspans. The anchorage at exterior beam ends are frequently provided by the so-called U-type anchorage, to avoid bar congestion in the beam-column joints.

Re-bar Splices

Lap splices are not used at all in highrise RC construction. Currently following splices are used in practice:
    (1) Gas butt welding with automated welders.
    (2) Enclosed gas-shield arc welding with a parallel root gap.
    (3) Mechanical splice with squeezed steel collars.
    (4) Mechanical splice with infilled steel sleeves.
    (5) Use of screw-type deformed bars and threaded couplers. Either lock nuts or grouting (resin or mortar) is used to tighten couplers. Combination of (4) and (5) is also available.

Formwork and Concrete Casting

For the cast-in-situ concrete, a variety of system formworks are used. Particularly for a floor system, large system formworks are often adopted which combine formworks for two-way beams, floor slabs, and shoring.

Almost all highrise constructions adopt separate concrete casting for columns and floor systems, often called vertical-horizontal (VH) separate casting. Superplasticizer or high performance AE water reducer is used in the concrete mix, to get slump of about 18 cm. Column concrete is usually cast by buckets, while concrete pump is used to cast floor system.

EARTHQUAKE RESISTANT DESIGN AND RESPONSE ANALYSIS

Design Principles and Procedures

As the basic principle of earthquake resistant design, beam hinge mechanism, or strong column-weak beam mechanism, is always assumed. Column hinges are allowed at the bottom of the first story and the top of the uppermost story, and at the exterior columns in the tension side of the lower stories. The beam hinge mechanism is assumed in order to provide large energy dissipating capacity distributed all around the structure.

Earthquake resistant design criteria are summarized in Table 1. These criteria are similar to those for steel or SRC highrise buildings with the height in excess of 60 m. They are not explicitly stipulated in the Building Code. They have been traditionally used in the review by the Building Center of Japan, as a kind of current consensus among structural engineers.

The design procedure consists of two phases, which essentially correspond to the two levels in Table 1. The first phase design is to protect weak links of the structure, that is, yield hinge locations assumed in the mechanism, from forming yield hinges under the action of level 1 earthquake. For this purpose, design seismic loads are determined, usually referring to Building Code and preliminary earthquake response analysis, and members are proportioned to carry forces resulting from the design seismic loads.

The second phase design is to ensure the assumed mechanism to form under

Table 1.   Earthquake resistant design criteria

Seismic hazard level	Level 1	Level 2
Probability of recurrence	Once in lifetime	Possible maximum
Maximum ground velocity (in Tokyo)	25 cm/s	50 cm/s
Member forces	Concrete cracks but no steel yields	Steel yields but no building collapses
Story ductility factor	less than 1	less than 2
Member ductility factor	less than 1	less than 4
Story drift angle	less than 1/200	less than 1/100

the action of level 2 earthquake. Collapse load associated with the mechanism
formation is calculated, which is similar in definition as the ultimate load
carrying capacity in the Building Standard Law. It is generally expected that
this load level exceeds at least one and half times the design seismic loads.
Structural members outside yield hinges are proportioned to forces associated
with the mechanism formation enhanced by appropriate magnification factors.

   A series of nonlinear time history earthquake response analysis are
performed to confirm the design criteria in Table 1.

Design Seismic Loading

   Current Japanese Building Code provides design seismic forces for
buildings up to 60 m in height only. However, considering that the range of
height for recent highrise RC buildings exceeds the limit of 60 m only with a
relatively small margin, it is a common practice for structural engineers to
just extrapolate the provision of Building Code, and modify slightly, as
needed, by a preliminary earthquake response analysis.

   Figure 2(a) shows the design base shear coefficient of highrise RC
buildings against the fundamental natural period from an equation stipulated
in the code, i.e. T_1 = 0.02 h, where h is building height in m. Most design
falls around the code curve for second class (intermediate) soil. Figure 2(b)
shows the design base shear coefficient against calculated elastic fundamental
natural period.   The range shown by dotted curves corresponds to most highrise
construction in Japan, either steel or SRC construction. It seems highrise RC
buildings have slightly lower base shear, as long as they are compared on the
basis of elastic natural period. Probably it would be a more fair comparison
to take natural period based on cracked sections, although it is not a common
practice to do so in Japan.

First Phase Design

   The first phase design consists of structural analysis for design loads
and proportioning of members. Structural analysis is carried out for
permanent loading as well as design earthquake loading. Computer analysis is
normally performed using displacement method, based on the uncracked section,
considering flexural, shear and axial deformation of members, and rigid zones
at member ends. When the structure is susceptible to torsional deformation,

8

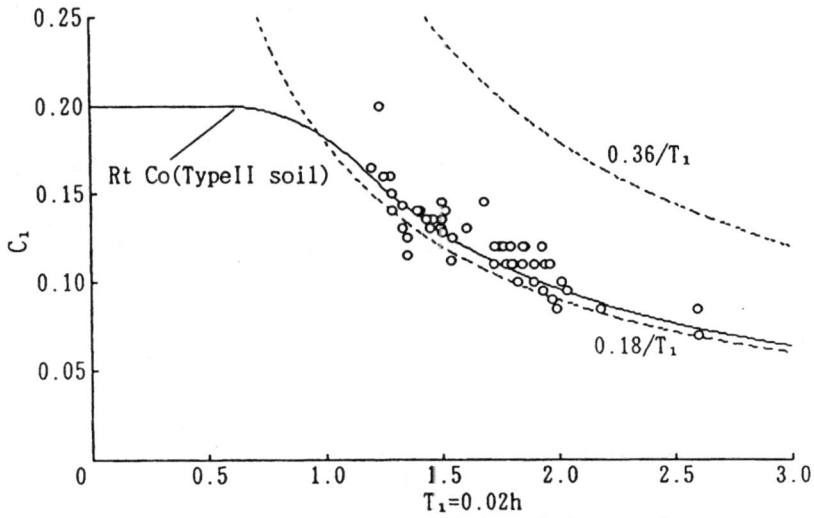

(a) Fundamental period from Code equation

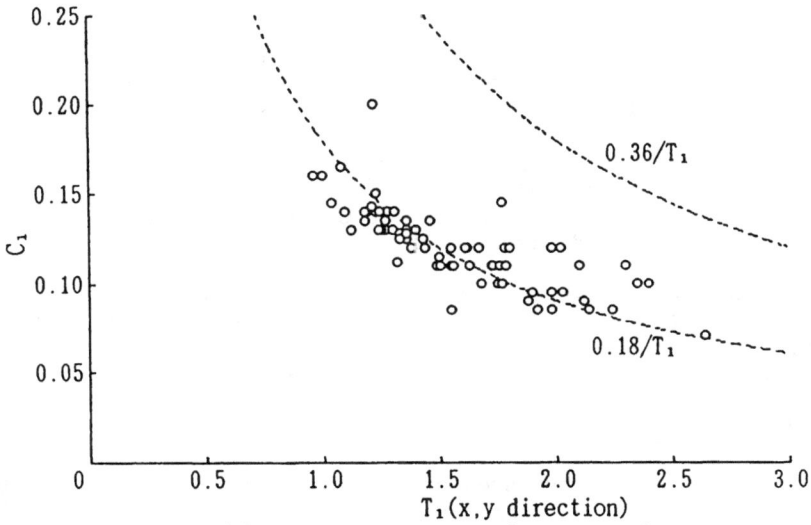

(b) Fundamental period from analysis

Figure 2. Relationship between base shear Coefficient and
fundamental natural Period

9

three-dimensional frame analysis is carried out.

Moment redistribution is applied in some cases, although it is not widely used. The amount of moment redistribution is usually modest, and its appropriateness is demonstrated in the subsequent nonlinear frame analysis, so that no yield hinges would occur under the action of design seismic loads.

Second Phase Design

The second phase design consists of evaluating the ultimate load carrying capacity, and to ensure the formation of assumed mechanism. The ultimate load carrying capacity may be evaluated by limit analysis. However, nonlinear incremental frame analysis is usually performed, which gives not only ultimate capacity but also the primary load-displacement relation for each story for use in the dynamic earthquake response analysis.

For the calculated member forces associated with the mechanism, ultimate strength of each member is investigated whether the assumed mechanism would be actually formed. This consists of the following three points:

(1) Beam ductility. Shear strength of beams must be sufficient to prevent premature shear failure. At the same time, beam end zones must be designed for yield hinges with sufficient rotation capacity. At present, design guidelines in this respect are not sufficiently established in Japan, but the recent publication of Design Guidelines for Earthquake Resistant Reinforced Concrete Buildings Based on Ultimate Strength Concept (AIJ, 1990) is gradually getting popularity among practical design.

(2) Column strength and ductility. Except where yield hinges are expected to occur, columns should be protected against flexure and shear. A practical problem in this respect is how to determine design forces. Forces determined in the inelastic frame analysis correspond to predetermined load profile, but forces during dynamic excitation are subjected to much fluctuation due to ratio of upper and lower story drift, usually referred to as higher mode effect. Furthermore columns must be protected against forces coming from beams in two dimensions. The Guidelines (AIJ, 1990) is serving for practice in this regard also.

(3) Beam-column joints. Prevention of premature joint failure is achieved by restricting shear stress in the connection, and by restricting bond stress along the beam bars passing through the joint. For exterior beam-column joints, beam bar anchorage is checked and is carefully detailed.

Earthquake Response Analysis

Reinforced concrete structures start cracking at a relatively low level of loading. Hence the elastic linear analysis based on the uncracked section serves little in predicting actual behavior. As a simplified analytical model for nonlinear analysis, a lumped mass shear model is almost exclusively used in the time history dynamic response analysis for both level 1 and level 2 earthquake ground motions.

The restoring force characteristics of stories are defined by simplifying the load–displacement relation from incremental frame analysis into an equivalent trilinear relation. Degrading trilinear model or Takeda model is used for hysteresis rules under reversal.

In some cases, so-called flexural shear model is used, in which flexural deformation of overall structure due to overturning moment is separately evaluated and added to the shear deformation which is the frame deformation. The flexural deformation is evaluated on the basis of linear elasticity.

When the building is susceptible to torsional vibration, a dynamic quasi-three-dimensional model is used in the response analysis, which consists of many shear models, or flexural shear models, corresponding to each frame interconnected by rigid diaphragms.

One of the serious drawbacks of shear models is that it cannot predict member ductility factor. Usually it is evaluated indirectly by equating dynamic story drift to the static one in the incremental frame analysis. However, some engineers opt to carry out dynamic frame analysis where inelastic deformation of constituent frame members is directly accounted for in the time history earthquake response analysis.

As for the input earthquake ground motions, the Building Center of Japan recommends the use of waveforms in the following three categories (BCJ, 1986) for any highrise buildings:
  (1) Well known "standard" motions, e.g. El Centro 1940 NS and Taft 1952 EW.
  (2) Records taken at nearby stations, e.g. Tokyo 101 1956 NS for buildings in Tokyo.
  (3) Records containing relatively long period components, e.g. Hachinohe 1968 NS and EW, Sendai TH030 1978 NS and EW.

Earthquake motions are normalized in terms of maximum velocity to the levels as prescribed in Table 1. In many cases, design criteria for story drift angle in Table 1 are found to be the governing criteria.

NATIONAL PROJECT ON ADVANCED RC BUILDINGS

Range of Material Strength

The quick development of highrise RC construction owes to many things, but the development of the use of high strength concrete and high strength, large size reinforcing bars was evidently the most fundamental factor. In an attempt to further promote development of new and advanced RC construction, a national project lasting five years was started by the Ministry of Construction in 1988 (Aoyama, et al., 1990). The project was officially named as "The Project on Development of Advanced Reinforced Concrete Buildings using High-strength Concrete and Reinforcement", but it is usually referred to as "New RC Project".

The range of material strength set out as the target of this project is shown in Fig. 3. The vertical axis shows the yield strength of steel bars,

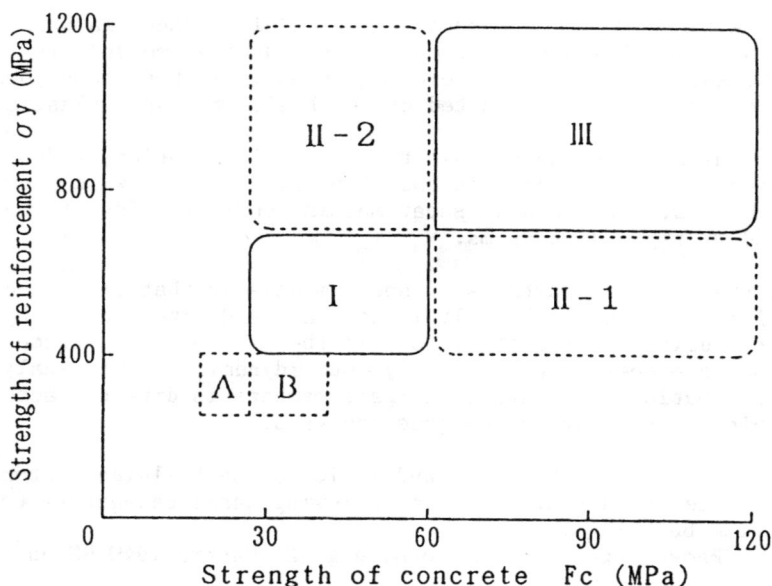

Zone:Type of RC buildings & materials
    A:Low-rise buildings
    B:High-rise buildings
    I:High-strength concrete & reinforcement
    II-1:Ultra high-strength concrete & high-strength reinforcement
    II-2:High-strength concrete & ultra high-strength reinforcement
    III:Ultra high-strength concrete & reinforcement

Figure 3. Strength of Materials and Fields of Research and Development

and the horizontal axis shows the compressive strength of concrete. Small zones A and B in the figure correspond to the ranges for ordinary RC buildings and highrise RC buildings, respectively. As seen in the figure the currently used materials for highrise RC buildings occupy only small zones.

In contrast, the ranges of strength for concrete and steel for this project are much larger. Concrete from 30 to 120 MPa and steel from 400 to 1,200 MPa are included. Comparing the zones for these ranges of materials to zones A and B, it is obviously unrealistic to assume that structural behavior of New RC structures can be understood simply by extrapolating the knowledge of ordinary RC structures. The area in Fig. 3 for the New RC is further divided into four zones, namely zones I, II-1, II-2, and III. Structures in these zones will be studied and developed by somewhat different tactics.

12

Experimental approach is indispensable, but in general, theoretical examination of experimental data will be emphasized in this project. Current technical knowledge on RC structures will also be re-examined.

In some zones in Fig. 3, particularly in zone III, basic problems will have to be re-examined, and hence the project may not yield much practical results. Most practical results are expected in zones I and II-1, because these zones are relatively close to the boundary of the current technology, and simple extrapolation will be effective at least partially.

Objectives of Research and Development and Final Expected Results

The objectives of research and development and the corresponding final results expected in the project are summarized in Table 2. Results will be partly available to refine the current RC technology. In the table under the third objective, the word "guidelines" for structural design and construction do not mean a type of guidelines that will give full details of technology, but it will give only basic principles for design and construction practice. Such a soft type of guidelines is preferred at this stage of the game, as definite and detailed specification-type guidelines often tend to impede development of relevant technology.

Research Organization

The Building Research Institute of the Ministry of Construction is in charge of conducting the project. Research committees were set up in an organization called Japan Institute for Construction Engineering, to organize people from universities, Housing and Urban Development Corporation, makers of cement, admixtures, and steel, and construction companies.

Technical Coordinating Committee (TCC) is the central body for coordinating the entire project. Research Promoting Committee acts as a mediator between the TCC and participating construction companies which are sponsors of the project; i.e., the entire New RC project is financed by the combination of national fund from the Ministry of Construction and contributions from these participating companies. Technical Advisory Board consists of technical authorities of all related engineering fields and acts just as the name implies. The above-mentioned three committees are chaired by the writer.

Under the Technical Coordination Committee, five technical committees were installed. They are in charge of making research programs in detail, implementing research works, and integrating research results in five particular fields. The names and chairmen of these committees are: Concrete Committee, chaired by Prof. F. Tomosawa, University of Tokyo; Reinforcement Committee, chaired by Prof. S. Morita, Kyoto University; Structural Element Committee, chaired by Prof. S. Otani, University of Tokyo; Structural Design Committee, chaired by Prof. T. Okada, University of Tokyo; and Construction and Manufacturing Committee, to start in 1991 and chaired by Prof. K. Kamimura, Utsunomiya University. Technical committees organize working groups as needed.

Table 2. Objectives of research and development and final expected results

| | Final expected results | |
Objectives	For New RC	For current RC
1) Development of high-strength and high-quality materials	Methods for mix proportion method and quality control of concrete (Zone I) Methods for production and arrangement of reinforcements (Zone I) Guidelines for developing materials for Zones II and III	Revision of design method for mix proportion and quality control method for concrete Revision of reinforcement quality standards Revision of upper limits of reinforcement strength
2) Evaluation of properties of members and frames	Methods of analysis	Revision of current methods of analysis
3) Development of design and construction guidelines	Structural design guidelines (Zone I) Construction guidelines (Zone I) Draft guidelines for structural design and construction (Zones II, III)	Revision of structural design methods Revision of construction standards
4) Feasibility study on RC buildings in Zone II-1	New type high-rise buildings	
5) Feasibility study on RC buildings in Zone III	New images of RC buildings	

Items for Research and Development

Following research items were assigned to each technical committees at the onset of the project.

For the Concrete Committee:
(1) Development of materials necessary for making high-strength and super high-strength concrete. Quality standards of the materials.
(2) Physical properties of high-strength and super high-strength concrete.
(3) Mix proportion design, casting and curing works and quality control.

For the Reinforcement Committee:
(1) Development of high-strength and super high-strength steel bars. Mechanical properties of steel bars.
(2) Mechanical properties of confined concrete.
(3) Constitutive equations for RC elements and application of finite element method.

(4) Bond between concrete and steel bars. Anchorage and arrangement of steel bars.

For the Structural Element Committee:
(1) Mechanical properties of beams and columns.
(2) Mechanical properties of shear walls.
(3) Effect of shear force on beams, columns, and shear walls.
(4) Mechanical properties of beam-column joints and frames.
(5) Mechanical properties of foundations.

For the Structural Design Committee:
(1) Methods for modeling and analysis of structural frames in each zone.
(2) Practically feasible types of structures.
(3) Design seismic loads and requirements for structural performance.
(4) Design methodology.

For the Construction and Manufacturing Committee, research items are to be established.

Major Research Topics in the Project

One of the most fundamental and important topics is the ductility of high-strength reinforced concrete. High-strength concrete has been used mainly as countermeasure to high axial stress in the lower story columns of highrise buildings. Maximum number of stories of highrise RC buildings is almost completely determined in practice by the concrete strength, such as 25 stories for 36 MPa, or 30 stories for 42 MPa. From the viewpoint of axial column stress, higher concrete strength is the most vital element in order to realize higher buildings.

However, it has been pointed out that the falling branch of a stress-strain curve of high-strength concrete is more pronounced, and further it is more difficult to improve the falling branch by lateral confinement. It is necessary to develop the most effective method of lateral confinement, and at the same time, to recognize the limitation of improved ductility by means of lateral confinement.

Another problem related to the ductility of high-strength reinforced concrete is the property of high-strength steel. At present super high-strength steel bars of 1,300 MPa specified yield strength are used very frequently in practice for lateral shear reinforcement, but they have never been used as longitudinal reinforcement. In order for high-strength steel bars to be used as longitudinal reinforcement, their quality in terms of stress-strain relationship should be improved. The most important practical problem is how much improvement can be specified and realized by steel manufacturers. In the long run, the ductility of New RC members will have to be lower than ordinary RC members, and hence it will be necessary to develop design philosophy which depends more on strength, but less on ductility, of constituent materials.

Bond between concrete and reinforcing bars is another basic issue in New RC. Demand on bond increases in proportion to the increase of yield strength

15

of steel bars, but the bond capacity does not increase in proportion to the increase of concrete strength. Thus the bond becomes one of the critical problems of New RC structures. In particular, resistance against bond splitting failure is an important subject for bar development.

The use of high strength materials usually results in a reduction of cross section of members, thus reduction of stiffness of members and structures. Among structural design criteria listed in Table 1, that for the drift angle governs in most highrise RC buildings even at present, and it is expected to be more so for New RC buildings if the same design criteria as in Table 1 are maintained. Since the design criterion for drift was established not on any rational basis, it is not easy to challenge for its revision by a rational discussion. But the author believes that we will have to start discussing on it seriously, sooner or later.

CONCLUSION

In this paper the author first reviewed the present state-of-the-art of the design and construction of highrise RC buildings in Japan, a highly seismic country. The quick development of highrise RC construction owes to many factors, but the availability of high strength materials was the most fundamental factor that enabled the development. In the second part of the paper, the author outlined a Japanese national research project entitled "Development of Advanced RC Buildings using High-strength Concrete and Reinforcement (New RC)". It is a project to try to develop high-strength and super high-strength materials for RC structures, and to provide necessary design aids for highrise and other new RC structures. It is expected that by 1993 our knowledge on the reinforced concrete will be substantially enlarged as illustrated in Fig. 3.

REFERENCE

Aoyama, H., Murota, T., Hiraishi, H. and Bessho, S. 1990. Outline of the Japanese national project on advanced reinforced concrete buildings with high-strength and high-quality materials, Proceedings, 2nd International Symposium on Utilization of High-Strength Concrete, Berkeley, CA.
Architectural Institute of Japan (AIJ), 1990. Design guidelines for earthquake resistant reinforced concrete buildings based on ultimate strength concept.
Building Center of Japan (BCJ), 1986. On the earthquake motions for use in dynamic response analysis of highrise buildings, Building Letter, 6, 49-50.

The effects of site conditions on ground motions

W.D. Liam Finn[I]

ABSTRACT

The more important contributions by seismologists and geotechnical engineers over the last 10 years to knowledge of effects of site conditions on ground motions are reviewed and the implications for seismic design are examined. The selection of material for review is based on a judgement of its importance to engineering practice.

INTRODUCTION

The estimation of site specific ground motion parameters for seismic design or microzonation studies is one of the more complex and challenging problems of earthquake engineering.

The effects of local site conditions on the incident wave field have been the focus of major studies by both seismologists and geotechnical engineers. Until the 1980's, these studies proceeded relatively independently with little interaction and with an apparent lack of agreement on some important issues. For example, following the pioneering studies of ground response during the Niigata earthquake of 1964 by Seed and Idriss (1969), geotechnical engineers have been convinced of the importance of nonlinear effects at most soil sites during strong shaking. Since the introduction of the SHAKE program by Schnabel et al. (1972), nonlinear site effects have been taken into account routinely in engineering practice. Yet in a study of the applicability of weak motion amplification factors to the strong motions recorded during the 1989 Loma Prieta earthquake Aki and Ta-Liang Teng (1991) concluded that their study had detected the "pervasive nonlinear effect at sediment sites for the first time seismologically".

Geotechnical engineers have always taken a rather restricted view of site effects, relying almost exclusively on 1-D analysis. They have routinely ignored the effects of surface and buried topography on ground motions which reviews by Aki (1988), Silva (1989) and Faccioli (1991) have shown can be significant from both the seismological and engineering points of view.

I   Department of Civil Engineering, University of British Columbia, Vancouver, B.C., Canada

On important projects geotechnical engineers rely on site specific response analyses to determine site effects. Computed surface motions or response spectra often show significant amplification over the corresponding input quantities. The geotechnical engineer's judgement about site-specific response tends to be based on his experience with a number of individual case histories.

Seismologists, on the other hand, determine the effects of local geology by averaging amplification factors over many soil and rock stations in a network. This process tends to smear out extremes. In addition, seismologists draw their conclusions about geological effects against the background of the standard error in the data. For some motion parameters the averaged effects of local soil conditions may be obscured by the standard error in the data. This may explain why peak ground acceleration has been reported to be independent of site conditions (Aki, 1988), despite obvious contrary findings in individual cases such as accelerations recorded in the lake-bed in Mexico City during the 1985 earthquake (Seed et al., 1988; Finn and Nichols, 1988).

This paper reviews the more important contributions by seismologists and geotechnical engineers over the last 10 years to the understanding of site effects on ground motions and the implications of these for seismic design. The selection of material for discussion is based on a judgement of its potential importance to engineering practice. The paper attempts to integrate both the seismological and geotechnical contributions to the problem of site effects on ground motions and provide a coherent summary of the field for geotechnical earthquake engineers.

The seismic response of ridges and sediment filled valleys will be reviewed first. A knowledge of the complex patterns of motion that can develop in these structures is an essential requirement for interpreting recorded motions and understanding the limitations of 1-D response analysis.

EFFECTS OF TOPOGRAPHY

Aki (1988) used the simple structure of a triangular wedge (Fig. 1a) to illustrate the effects of topography. This structure may be used to model approximately ridge-valley topography as shown in Fig. 1b by Faccioli (1991). An exact solution exists for the wedge for SH waves propagating normal to the ridge and polarized parallel to the ridge axis. Displacement amplification at the vertex is $2/\nu$ where the ridge angle is $\nu\pi$ ($0 < \nu < 2$). In Fig. 1b the amplification of the crest relative to the base is ν_1/ν_2. Thus the simple solution provides a rough estimate of the relative amplification at the crest of the ridge or deamplification in a valley.

Geli et al. (1988) have provided amplification factors in the form of Fourier transfer functions for a smooth-ridge with a shape ratio, defined as height over half-width, $h/L = 0.4$, in terms of the non-dimensional frequency $n = 2L/\lambda$ where λ is the wave length. The factors are shown in Fig. 2 for selected locations along the ridge. Note the broad band amplification at the crest and the increasingly complex nature of the response with

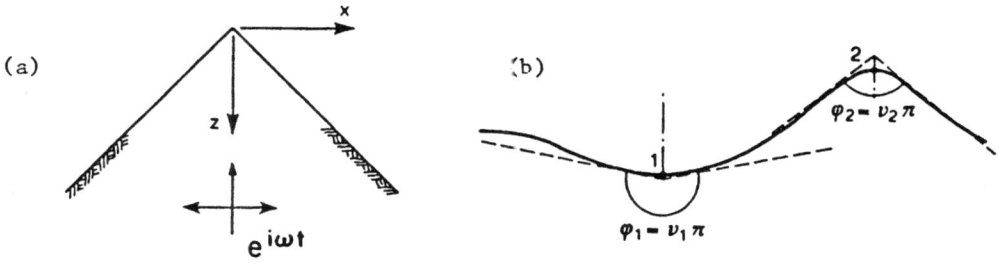

Fig. 1   (a) Approximating a ridge formation by a triangular wedge, (b) Infinite wedge excited by plane SH waves (after Faccioli, 1991).

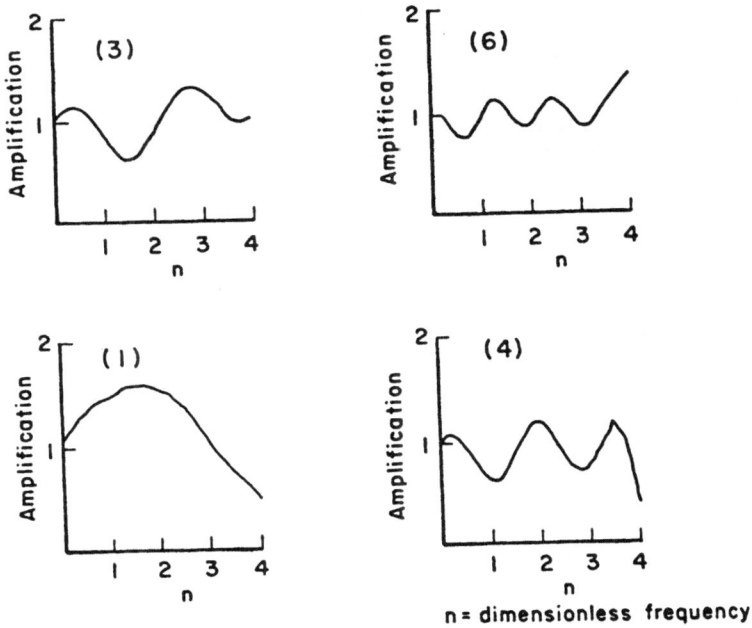

n = dimensionless frequency

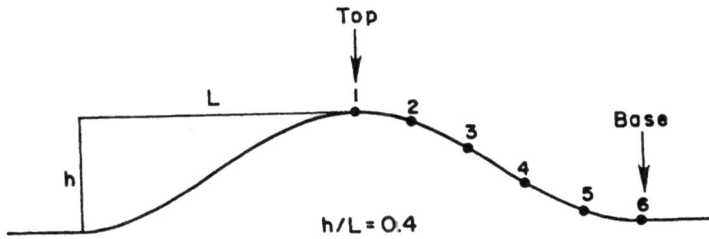

h/L = 0.4

Fig. 2   SH Fourier Transfer Functions to Homogeneous Half-Space Outcrop Motions (after Geli et al., 1988).

distance down slope to the plane below. The interference patterns resulting from the interaction of the incoming waves with the scattered waves in the wedge are reflected in the oscillating pattern of amplification and deamplification at downslope locations.

Observed amplifications range from 2 to 20 in the spectral domain (Bard, 1983) to 5 in the time domain (Griffiths and Bollinger, 1979). Predicted values are generally less, 3 to 4 in the spectral domain to less than 2 in the time domain. The differences have been attributed to the influence of 3-D effects and ridge to ridge interactions (Geli et al., 1988). Faccioli (1991) has suggested that in many cases the topographic site effects are of the same order as the regional variability of motions on hard ground.

A case history illustrating the variation in amplification over a ridge structure is provided by data from the Matsuzaki array in Japan (PWRI, 1986). The mean values and standard error bars of peak accelerations normalized to the crest acceleration for five earthquakes are plotted in Fig. 3 (Jibson, 1987) as a function of elevation. The range in peak accelerations for the five earthquakes is rather limited ranging from a low of a few gals at station 5 to a maximum of about 100 gals at the crest. The amplification of the crest relative to the base is about 2.5. The amplification factor increases rapidly as the crest of the ridge is approached.

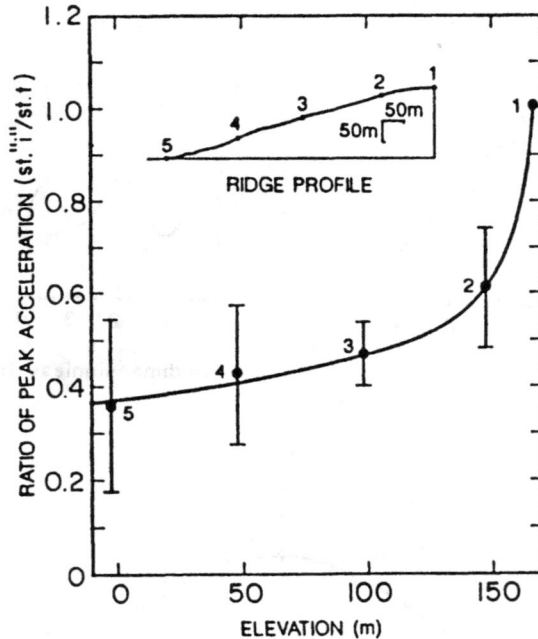

Fig. 3   Relative Distribution of Peak Accelerations Along a Ridge From Matsuzaki Array in Japan (after Jibson, 1987).

20

Amplification of motions at the crest of a ridge relative to the base is also supported by damage patterns during the 1980 Friuli earthquakes in Italy (Brambati et al., 1980) and in the Chilean earthquake of 1985 (Celebi and Hanks, 1986).

The effect of a topographic structure on ground motions depends on the shape ratio of the structure and how the lateral dimensions of the structure are related to wavelengths of the incident motions. Silva and Darragh (1989) show that over the period range of engineering interest, 0.2 Hz to 25 Hz, the range in wavelengths is from 40 m to 5 km assuming a shear wave velocity in rock of 1 km/s. Topographical features with characteristic dimensions in this range have the potential for a significant effect on ground motions depending on the shape ratios.

MOTIONS IN ALLUVIAL VALLEYS

Perhaps more interesting from an engineering point of view is the response of sediment filled valleys, usually the locations of greatest development. Ground motions on these sites generated by shear waves propagating vertically are usually estimated by 1-D shear beam models, using either equivalent linear methods (Schnabel et al., 1972) or nonlinear models (Finn et al., 1978). The sediment-basement rock interface generates surface waves and may trap body waves in the alluvium (Finn and Nichols, 1988; Silva, 1989). These waves amplify the motion and increase the duration over that predicted by 1-D analysis. These effects were very pronounced in the lake-bed motions in Mexico City during the 1985 earthquake.

Bard and Gabriel (1986) calculated the transfer functions for a wide shallow sediment filled valley (h/L < 0.25) shown in Fig. 4. The results are shown for both 2-D and 1-D analyses with a linear gradient in shear wave velocity S with depth and for a 2-D analysis with a constant shear modulus in the sediments. The valley has a shape ratio of 0.1. The frequency n is normalized by the 1-D resonant frequency for the valley centre, S/4h, where h is the depth of the valley at the centre. The 1-D analysis does a very good job of modelling the response from station 5 on, that is just off the sloping edge of the valley but tends to give too sharp a resonance response from the edge of the valley to station 3.

The surface waves generated near the edge are apparently damped significantly by the time station 5 is reached and the remaining effects are swamped by the incoming shear waves. The radically different responses between stations 1 to 5 may result in differential motions, normal to the edge of the valley (Silva, 1989) with implications for the seismic loading of long structures.

Deep narrow valleys with large shape ratios h/L ≥ 0.25 show a different kind of response (Fig. 5). The amplifications for a valley with shape ratio of 0.4 displays several strong maxima instead of the one or two associated with 1-D response. Predictions of motions by 1-D analysis are conservative near the edges but underpredict seriously the response at high frequencies in the middle of the valley.

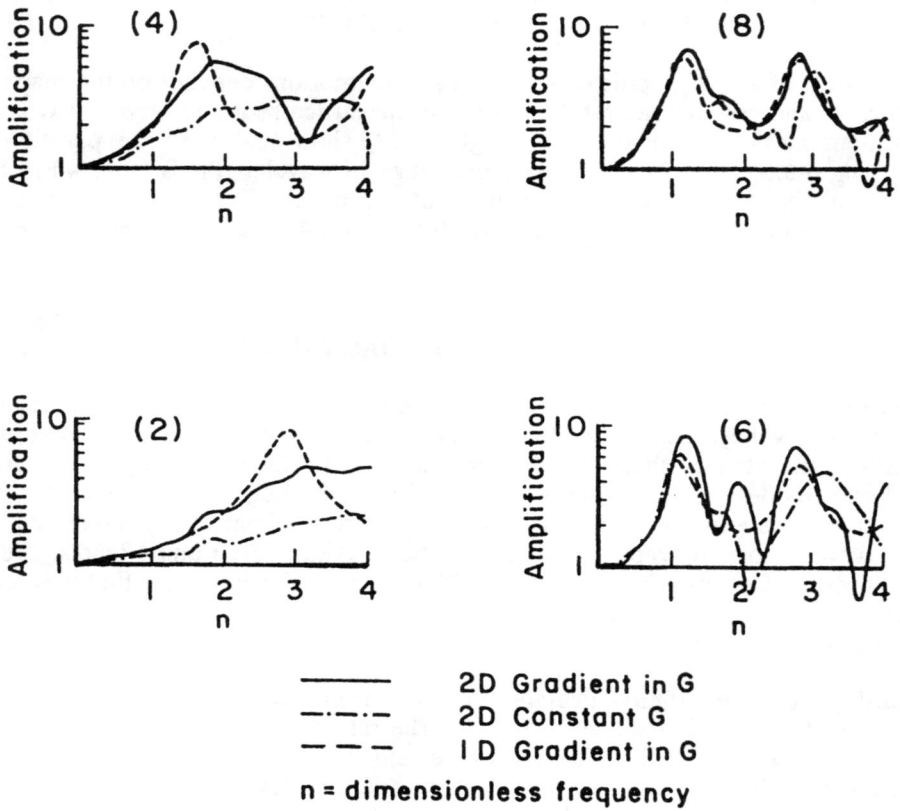

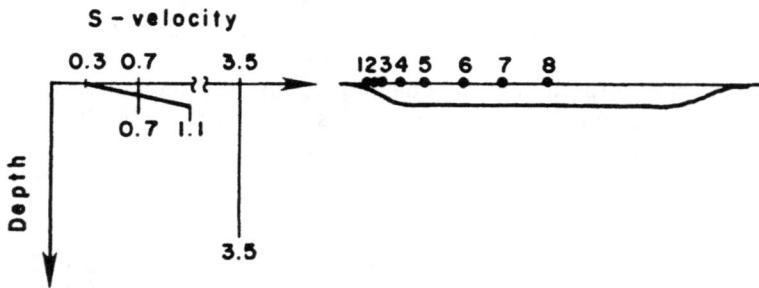

Fig. 4   Smoothed SH Transfer Functions to Homogeneous Half-Space Outcrop Motions for a Wide, Shallow Alluvial Valley with a Shape Ratio of 0.1 (after Bard and Gabriel, 1986).

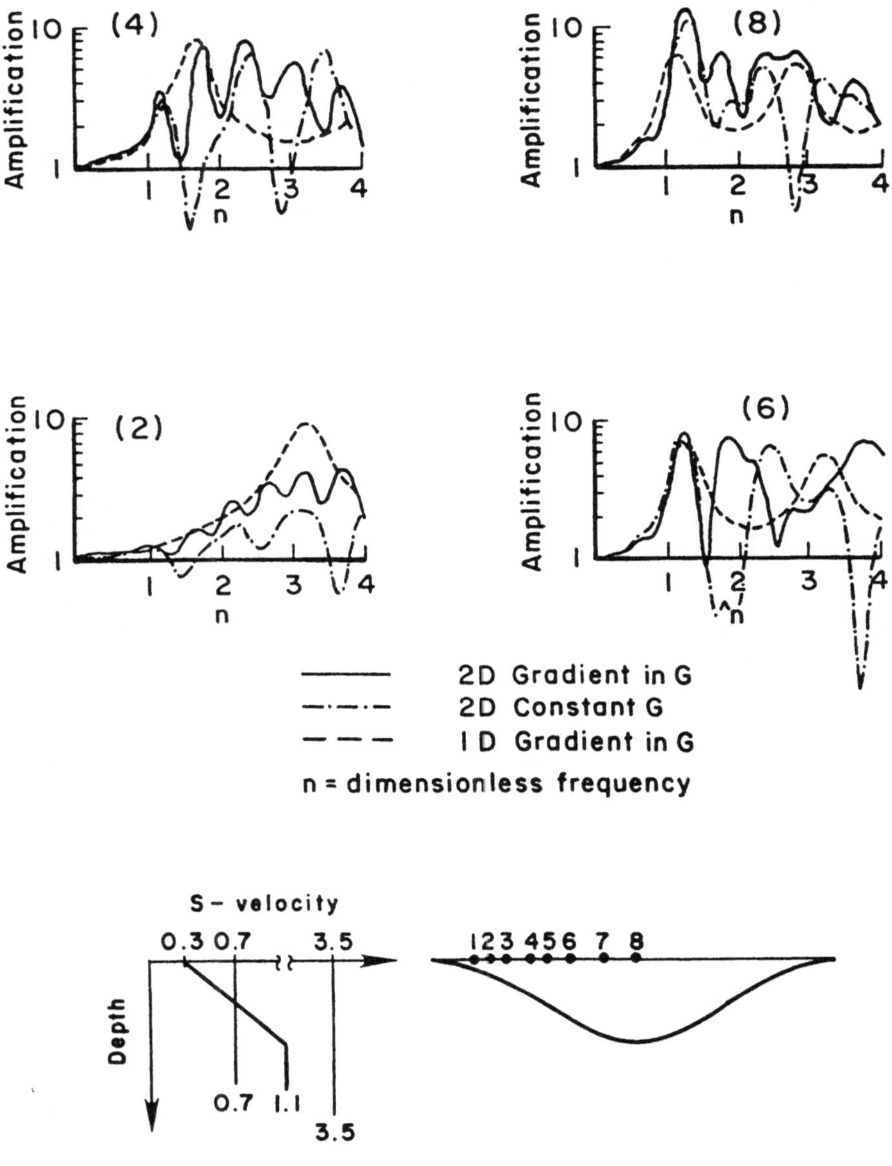

Fig. 5 Smoothed SH Transfer Functions to Homogeneous Half-Space Outcrop Motions for a Valley with a Shape Ratio of 0.4 (after Bard and Gabriel, 1986).

Silva (1989) has summarized the effects of surface topography and sediment-filled valleys on site response in Table 1.

Table 1. 2-Dimensional Geologic Structural Effects (after Silva, 1990)

Structure	Conditions	Type	Size	Quantitative Predictability[a]
Surface Topography	Sensitive to shape ratio, largest for ratio between 0.2 to 0.6. Most pronounced when wavelength $\simeq$ mountain width	Amplification at top of structure amplification and deamplification at base, rapid changes in amplitude phase along slopes	Ranges up to a factor of 30 but generally from about 2 to 10	Poor: generally underpredict size. May be because of ridge interaction and 3-D effects
Sediment-Filled Valleys				
1) Shallow and wide (shape ratio <0.25)	Effects most pronounced near edges. Largely vertically propagating shear wave from edges.	Broad band amplification across valley because of whole valley modes	1-D models may underpredict at higher frequencies by about two near edges	Good: away from edges 1-D works well, near edges extend 1-D amplification frequencies
2) Deep and narrow (shape ratio $\geq$ 0.25)	Effects throughout valley width	Broad band amplification across valley because of whole valley modes	1-D models may underpredict for a wide bandwidth by about 2 to 4 away from edges. Resonant frequencies shifted from 1-D.	Fair: given detailed description of vertical and lateral changes in material properties
3) General	Local changes in shallow sediment thickness	Increased duration	Duration of significant motions can be doubled	Fair
4) General	Generation of long period surface waves from body waves at shallow incidence angles	Increased amplification and duration because of trapped surface waves	Duration and amplification of significant motions may be increased over 1-D predictions	Good at periods exceeding 1 second

[a]Good: generally within a factor of two
Fair: generally within a factor of two to four
Poor: qualitative only, can easily be off by an order of magnitude

An interesting case history which shows the engineering implications of valley effects has been presented by Faccioli (1991). The problem is the estimation of seismic soil displacements to be used as a kinematic loading function for the foundation piles of an expressway bridge near Belluno in North East Italy. The expressway runs parallel to the edge of a broad valley. A cross-section of the edge of the valley is shown in Fig. 6. Also shown are the shear wave velocity profiles close to the axis of the bridge and the design accelerogram. The soil deformations were calculated by 1-D and 2-D methods assuming the design motions were SH waves propagating vertically. The 2-D analysis used an exact solution by Sanchez-Sesma et al. (1989). The soil displacement profiles from the 1-D and 2-D analyses are shown in Fig. 7. The displacements predicted by the 2-D analysis are much larger than those obtained from the 1-D analysis, by a factor of 2 at the ground line. These results suggest that 1-D analysis may seriously underestimate the seismic demand on pile capacity.

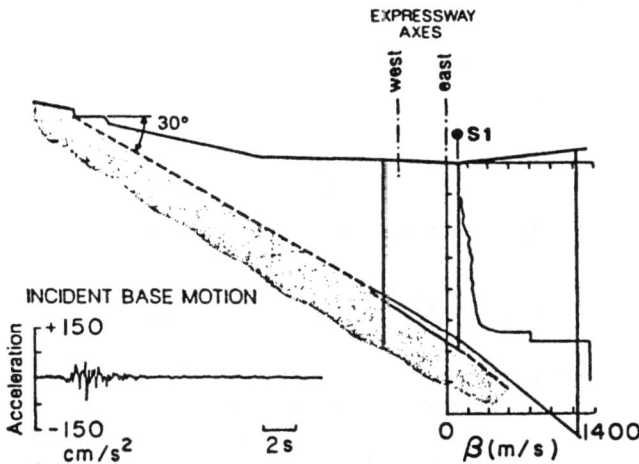

Fig. 6   Cross-Section at Edge of Alluvial Valley Near Belluno, Italy (after Faccioli, 1991).

EFFECTS OF LOCAL SOIL CONDITIONS

The effects of local soil conditions on waves propagating from bedrock to the surface are usually evaluated by 1-D shear beam analysis on the assumption that the site can be modelled as a layered half-space.

The analyses are capable of identifying the more important characteristics of the surface motions; the resonant period of the site, the lengthening of the period with increasing intensity of shaking and the amplification or deamplification of motions at various frequencies.

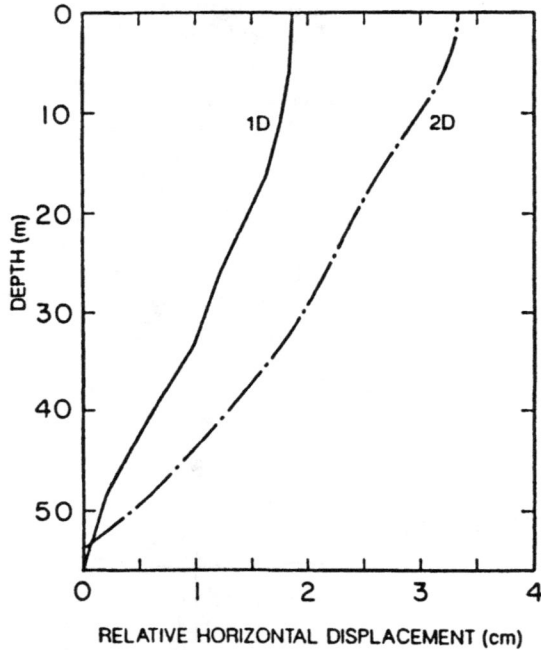

Fig. 7 Comparison of Horizontal Soil Displacements Predicted by 1-D and 2-D Analyses in Soil Profile Near Valley Edge (after Faccioli, 1991).

These effects have been very clearly identified in recent earthquakes. The amplification factors for surface motions recorded at the Treasure Island Site in San Francisco during the Loma Prieta earthquake of 1989 relative to the rock motions at adjacent Yerba Buena Island are shown in Fig. 8. The solid line shows the variation in the

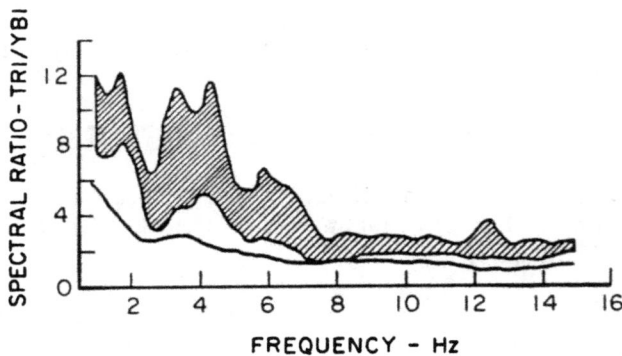

Fig. 8. Amplification Factors for Strong and Weak Motions at Treasure Island Site (after Jarpe et al., 1990).

26

NS spectral ratio for the first 5 seconds of the shear wave in the main shock in the period before liquefaction took place at the site. The shaded area is the 95% confidence region for the NS spectral ratios of 7 aftershocks (Jarpe et al, 1990). The amplification factors are drastically reduced in the strong motion phase although still 2 or greater over a wide frequency band.

Evidence of a significant shift in site period during strong shaking is provided by data from a Japanese site (Tazoh et al., 1988). Ground motions from the Kanagawa-Yamanashi earthquake of August 8, 1988, M_{JMA} = 6.0, were recorded at an epicentral distance of 18 km. The maximum acceleration at the ground surface was 435 gals and 134 gals at the base layer during the main shock. The transfer functions for both weak motions and the main shock between base and surface showed a period shift from 0.33 s for the weak motions to 0.5 s for the main shock. On this evidence, one should be cautious about site periods of peak response deduced from low amplitude events such as microtremors and coda waves. Elton and Martin (1989) took this period shift into account in microzoning Charleston, South Carolina on the basis of site period. They used dynamic site periods deduced from calculated 1-D strong motions.

An alternative to determining ground motion characteristics by dynamic analysis is to establish the parameters empirically from site response to microtremors or frequently occurring low magnitude local earthquakes. Aki and Teng (1991) used coda waves to determine site amplification factors for stations in the central California network operated by the U.S. Geological Survey. The coda waves were associated with local earthquakes between magnitudes 1.8 and 3.5. They found the amplification to be controlled by the geological age of the sediments. Amplification decreased with age from young Quaternary to Tertiary Pliocene sediments. Presumably the stiffness of sediments increased with age.

Aki (1988) found that the site amplification factors for strong motion response spectra showed amplification factors of 2 to 3 relative to rock spectra for periods longer than 0.2 seconds and that the relation was reversed for shorter periods. The site dependent spectra proposed by Seed et al. (1976) showed a similar cross-over near the period of 0.2 seconds.

The amplification factors for weak motions based on coda waves did not show such a cross-over. The data showed consistent amplification at all frequencies up to 12 Hz.

Aki and Teng (1991) applied the weak motion amplification factors derived from coda waves in fundamental studies of strong ground motions recorded during the Loma Prieta earthquake in 1989. For distances less than 50 km, they found a systematic overprediction of peak acceleration. They concluded that, depending on site and level of motion that nonlinear effects became significant in the acceleration range 0.1g - 0.3g. The lower limit agrees with the acceleration level proposed by Seed et al. (1976) as roughly the boundary between amplification and deamplification of rock motions by surface sediments (Fig. 9).

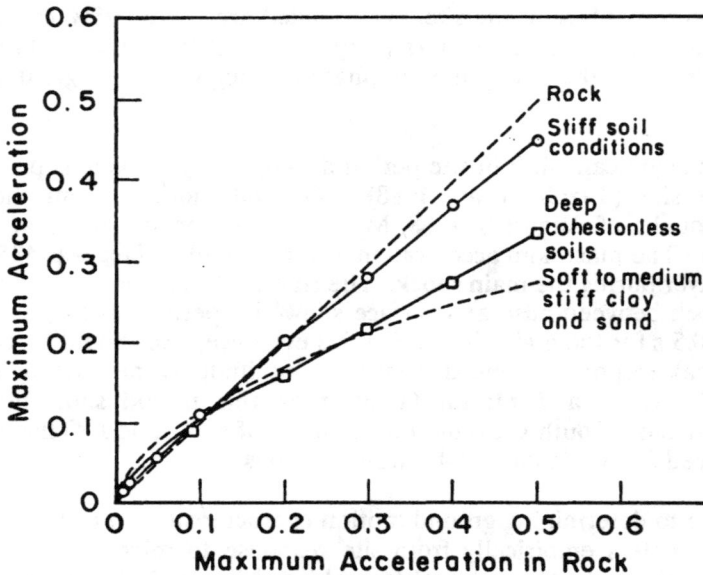

Fig. 9.   The Effects of Site Conditions on Ground Accelerations (after Seed et al., 1976).

Data in Fig. 9 suggest massive deamplification on soft soil sites during strong ground motion.   The response of the soft clays in Mexico City during the 1985 Mexican earthquake and of the soft soil sites in California during the Loma Prieta earthquake in 1989 changed that view dramatically.   Idriss (1990) gives an updated picture of the response of soft soil sites in Fig. 10, based on the Mexico City and Loma Prieta data and on 1-D response analyses using the SHAKE program (Schnabel et al., 1972).   Much greater amplification is now attributed to soft soil sites and the acceleration range over which amplification may occur is raised from 0.1g to 0.4g. Why did the predictions by 1-D dynamic analysis in 1990 change so much from those in 1983 when the same program was used in analysis. The answer lies in a better understanding of the dynamic properties of soft high plasticity clays.

Dynamic response analyses of clay sites in the past were conducted using generalized curves giving the damping ratio and the degradation in shear modulus as functions of shear strain developed by Seed and Idriss (1970).   The degradation in stiffness became significant at very low strains. Consequently, response analyses using strong motion inputs showed high deamplification of soft clay sites. Studies of the Mexico City clay showed that it remained essentially elastic over a much larger range in strain than the generalized curve would suggest. It was thus elastic behaviour that allowed resonance to develop large

28

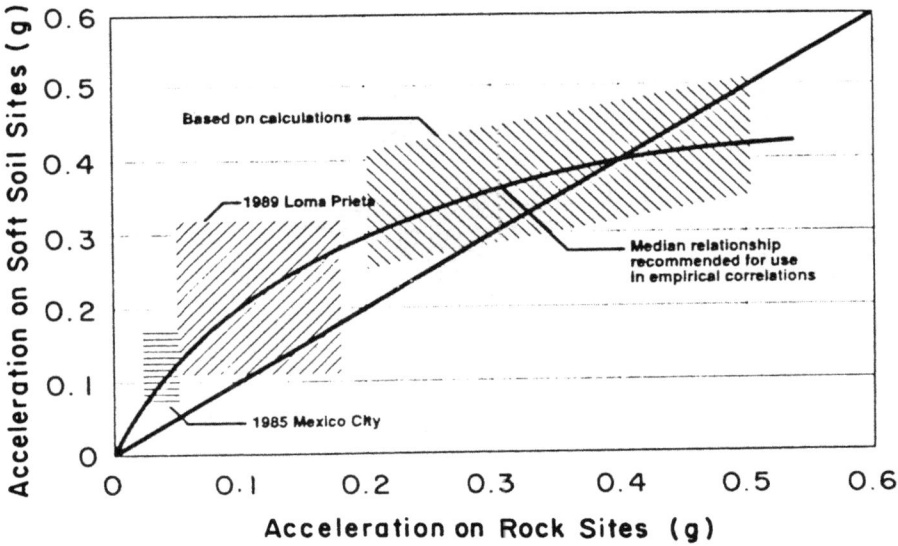

Fig. 10 Comparison of Peak Accelerations in Rock and Soft Clay Sites (after Idriss 1990).

amplifications (Finn and Nichols, 1988). Studies of dynamic soil properties since 1985 shows that the rate of modulus degradation at a given strain level decreases with increasing plasticity. The important lesson to be learned from the ground response in Mexico City is that site specific properties should be used in site response analyses.

But a major problem still remains for dynamic analysis despite all the improvements in the constitutive modelling and methods of analysis, uncertainty in the input motions. This problem cannot be evaded by calculating average motion parameters such as average spectra. The ground motions in Mexico City could not be simulated by following the usual practice of inputting outcrop motions. The rock motions had no preferred direction (Fig. 11), whereas the motions at the SCT site on the lakebed has acquired a strong E-W orientation (Fig. 12). As a result of this, to match the spectra of the E-W motions, the rock input motions had to be increased 2.5 times (Finn and Nichols, 1988).

Idriss (1990) had similar difficulties in simulating the response spectra at the Treasure Island site when using the rock outcrop motions from the nearby Yerba Buena site as input. It would seem that the motions emanating from the rock-sediment interface acquired a directional bias that resulted in substantially underestimating the response spectral ordinates as happened in Mexico City (Finn and Nichols, 1988).

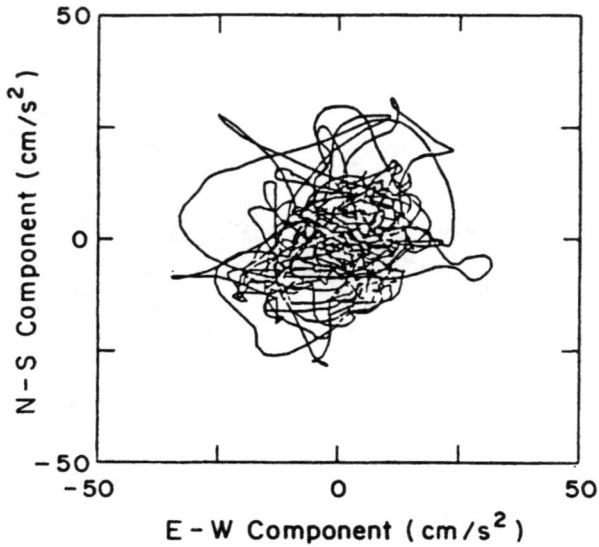

Fig. 11 Accelerations at Rock Site in Mexico City Showing no Directional Bias (after Finn and Nichols, 1988).

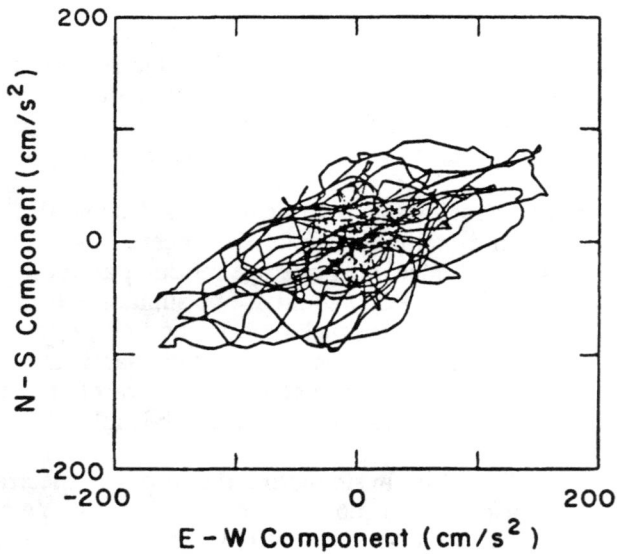

Fig. 12 Accelerations at the SCT Site on the Lakebed in Mexico City Showing Strong Directional Bias (after Finn and Nichols, 1988).

30

ACKNOWLEDGEMENTS

The author is grateful to K. Aki, University of Southern California, Los Angeles, California, and W.J. Silva, Pacific Engineering, El Cerrito, California, for preprints of work in press and very helpful discussions. Thanks are also due to C.B. Crouse, Dames and Moore, Seattle, Washington, for many clarifying discussions on ground motions and response spectra.

REFERENCES

Aki, K. 1988. "Local site effects on strong ground motion." in J. Lawrence Van Thun (Ed.), "Earthquake Engineering and Soil Dynamic (Recent Advances in Ground-Motion Evaluation", ASCE Geotechnical Special Publication No. 20, pp. 103-155.

Aki, K. and Teng, Ta-Liang. 1991. Estimation of local site effects on ground motion. In press, Bulletin Seismological Society of America.

Bard, P.Y. 1983. Les effets de site d'origine structurale en sismologie, modelisation et interpretation, application au resque sismique, These d'Etat, Universite Scientifique et Medicale de Grenoble, France.

Bard, P.Y. and Gabriel, J.C. 1986. "The seismic response of two-dimensional sedimentary deposits with large vertical velocity gradients." Bulletin of the Seismological Society of America, No. 76, pp. 343-346.

Brambati, E. Faccioli, E., Carulli, E., Culchi, F., Onofri, R., Stefanini, R. and Uloigrai, F. 1980. Studio de microzonizzacione sismica dell'are do Tarento (Friuli), Edito da Regione Autonoma Friuli-Venezia, Giulia.

Celebi, M. and Hanks, T. 1986. "Unique site response conditions of two major earthquakes of 1985: Chile and Mexico." Proc. of the Int. Symp. of Eng. Geology Problems in Seismic Areas, IV, Bari, Italy.

Elton, D.J. and Martin, R.J. 1989. Dynamic site periods in Charleston, South Carolina. Earthquake Spectra, Vol. 5, No. 4.

Faccioli, E. 1991. Seismic amplification in the presence of geological and topographic irregularities, Proceedings, 2nd International Conference on Recent Advances in Geotechnical Earthquake Engineering and Soil Dynamics, St. Louis, Missouri, Vol. II, pp. 1779-1797.

Finn, W.D.Liam, Lee, W.K. and Martin, G.R. 1978. An effective stress model for liquefaction. Journal of the Geotechnical Engineering Division, ASCE, Vol. 103, No. GT6, pp. 517-533.

Finn, W.D. Liam and Nichols, A.M. 1988. Seismic response of long period sites: Lessons from the September 19, 1985 Mexican earthquake. Canadian Geotechnical Journal, Feb. 1988, pp. 128-137.

Geli, L., Bard, P.Y. and Jullien, B. 1988. "The effect of topography on earthquake ground motion: a review and new results." Bulletin of the Seismological Society of America, Vol. 78, No. 1, pp. 42-63.

Griffith, D.W. and Bollinger, G.A. 1979. "The effect of Appalachian mountain topography on seismic waves." Bulletin of the Seismological Society of America, No. 69, pp. 1081-1105.

Idriss, I.M. 1990. Response of soft soil sites during earthquakes. Proceedings, H. Bolton Seed Memorial Symposium, Berkeley, California, Vol. II.

Jarpe, S., Hutchings, L., Hauk, T. and Shakal, A. 1989. "Selected strong- and weak-motion data from the Loma Prieta earthquake sequence." Seismol. Res. Letters, 60, pp. 167-176.

Jibson, R. 1987. "Summary of research on the effects of topographic amplification of earthquake shaking on slope stability." U.S. Geological Survey, Open-File Report 87-268, Manlo Park, California, USA.

PWRI (Public Works Research Institute). 1986. "Dense instrument array observation of strong earthquake motion." Ministry of Construction, Tsukubo, Japan.

Sanchez-Sesma, F., Chavez Garcia, F. and Bravo, M. 1988. "Seismic response of a class of alluvial valleys for incident SH waves." B.S.S.A., 78, pp. 83-95.

Schnabel, P.B., Lysmer, J. and Seed, H.B. 1972. "SHAKE: a computer program for earthquake response analysis of horizontally layered sites." Report No. EERC 72-12, University o California, Berkeley, December.

Seed, H.B., Romo, M.P., Sun, J.I., Jaime, A. and Lysmer, J. 1988. "The Mexico earthquake of September 19, 1985 -- relationships between soil conditions and earthquake ground motions." Earthquake Spectra, Vol. 4, No. 4.

Seed, H.B. and Idriss, I.M. 1969. "The influence of soil conditions on ground motions during earthquake." J. Soil Mech. Found. Eng. Div., ASCE, No. 94, pp. 93-137.

Seed, H.B. and Idriss, I.M. 1970. Soil moduli and damping factors for dynamic response analyses. Report No. EERC 70-10, Earthquake Eng. Research Center, University of California, Berkeley.

Seed, H.B., Murarka, R., Lysmer, J, and Idriss, I.M. 1976. "Relationsihps of maximum acceleration, maximum velocity, distance from source, and local site conditions for moderately strong earthquakes". Bulletin of Seismological Society of America, Vol. 66, No. 4, pp. 1323-1342.

Seed, H.B., Ugas, C. and Lysmer, J. 1976. "Site-dependent spectra for earthquake-resistant design." Bulletin of the Seismological Society of America, No. 66, pp. 221-243.

Silva, W.J. and Darragh, R.B. 1989. "Engineering characterization of strong ground motion recorded at rock sites." Report RP2556-48 prepared for the Electric Power Research Institute by Woodward-Clyde Consultants.

Silva, W.J. 1989. Site geometry and global characteristics, state of the art report, Proceedings of Workshop on Dynamic Soil Property and Site Characterization, National Science Foundation and Electric Power Research Institute, Palo Alto, California, November.

Tazoh, T., Sato, K., Shimizu, K. and Hatakeyama, A. 1988. Nonlinear seismic response analysis of soil deposit using strong seismic records. Proceedings, 9th World Conference on Earthquake Engineering, Tokyo, pp. 507-512.

Development of phased strong-motion time-histories for structures with multiple supports

Bruce A. Bolt[1]

ABSTRACT

Complete seismic analysis of critical structures, particularly with long spans such as bridges and elevated viaducts, requires realistic predictions of free-field ground motions at all the interface points on the supporting foundation under design earthquake conditions. Time histories differ slightly in wave form at sequential points because of wave propagation differential velocities, irregularities in rock and soil structure, and wave emission delays at the fault rupture. The effects in the input motions show up as phase shifts in each frequency component of the P, S and surface seismic waves. The result can be measured in terms of coherency factor which is a function of both wave frequency and support separation.

A dynamic response ratio has been defined for a discrete linear structural system in order to allow for the above spatial variations of the ground motions. It has been shown using the SMART 1 array recordings that structural response with differential phasing may be reduced by 25 percent from non-phased input response at 5 Hz for spans of 200 m.

A procedure for the synthesis of sets of phase time-histories is outlined using both time domain and seismic phase response spectral methods. Examples from recent work on phased input for long bridges in the San Francisco Bay are given.

INTRODUCTION

Large structures with a rigid foundation or multiple supports tend to respond

---

I   Professor of Seismology, Department of Geology and Geophysics and Civil Engineering, University of California at Berkeley, CA 94720.

(Luco and Wong, 1986) so as to average the free-field accelerations incident upon the supports (Figure 1). The structural dynamic analysis can be either by suitably phased ("lagged") time histories applied in an appropriate way at each support or by a modal response analysis complete with phase information. The usual response spectrum, as defined by Housner, describes the amplitude of the oscillator motion but does not include information on its phase behavior. However, for large structures, given the observed seismic ambient frequencies, it is necessary to also consider wave phase because wave offsets over inter-support distances produce differential ground acceleration along the base of the structure. The same argument holds for differential rotations.

In all cases, non-vertically propagating seismic waves produce systematic phase shifts between support points, producing significant incoherency which is reinforced by incoherencies produced in other ways.

As data from dense strong motion arrays with absolute times become available, there will be more opportunities to study phase changes and coherency in seismic ground motions over various distances and to estimate this effect on large engineered structures. In the modal analysis option, to facilitate the analysis of the phasing, a definition of a response phase spectrum has been developed (Abrahamson and Bolt, 1985), which is consistent with the response (amplitude) spectrum currently used by engineers. This method allows the routine calculation of the response phase spectrum. The examples given refer to ground acceleration, but the defined response phase is also valid for ground velocity and displacement.

One use of a response phase spectrum is demonstrated by developing a simple method for estimating the dynamic response ratio for large structures using only the response amplitude and phase spectra. The method greatly simplifies the estimation of the response ratio for multiple excitation input points.

The introduction of ground motion phasing in engineering practice is at the present time in its infancy and is probably necessary only in critical cases. At this stage the main approach is largely an empirical one with simple adjustments in onset times to observed strong motion records. The justification of the incorporation of phase shifts, is firstly, that a reduction in response is normally to be expected relative to in-phase motion and secondly, non-linear effects are significant.

WAVE PHASING DEFINED

Time Domain and Spectral Domain

Earthquake ground shaking at a point (or station) on the ground surface consists of a mixture of different elastic wave types with time as the independent variable.

The motion ("time history") is rarely stationary in time but varies in amplitude A, frequency f ($2\pi f = \omega$ radians per sec) and wavelength λ from time window to time window. Nevertheless, the motion can be represented by the superposition of simple harmonic (advancing) plane waves. The wave number k is then a useful concept, where

$$k = 2\pi/\lambda. \tag{1}$$

Linear strong motion array or multiple support analysis can be expressed as filtering followed by a summation. A standard assumption in array analysis is that the record consists of a deterministic signal plus noise. For example,

$$u_j(t) = u(x_j, t) = s(t) + \varepsilon(x_j, t), \tag{2}$$

where $u_j(t)$ is the output (acceleration, velocity, or displacement) of the jth seismometer at position x_j, s(t) is the deterministic signal, and $\varepsilon(x_j, t)$ is the noise. The form of Eq. (2) assumes that the signal arrives simultaneously at each station. This condition can be satisfied by introducing a delay τ_j to the output of the jth seismometer. For a plane wave, the delay is

$$\tau_j = \mathbf{k} \cdot x_j/\omega, \tag{3}$$

where $\mathbf{k}$ is the vector wave number and ω is the frequency of the plane wave.

In some analyses it is sufficient to consider just a single harmonic component of the strong motion record

$$s(x, t) = A \cos(\mathbf{k} \cdot x - \omega t + \delta), \tag{4}$$

where A and δ are constants. The wave velocity is $c = \omega/|\mathbf{k}|$.

In Eq. (4), the angle δ is called the phase angle and, since $S = A \cos \delta$ when $x = 0$, $t = 0$, it clearly represents an advance (or delay) of the whole harmonic with respect to the spatial or temporal origin (see Figure 2). In this discussion we are interested particularly in the phase angle δ.

An alternative representation to that in the time domain given by Eq. (4) above is to transform the motion to an equivalent description in which frequency is the independent variable. This change is usually accomplished mathematically by computing the Fourier transform of u(t) which we could write as $\overline{u}(\omega)$ or simply $u(\omega)$. A similar transformation can be made from location space x to wave number space k. For multiply-supported structures or free-field arrays, the input motions u(x, t) represent a time-dependent, three-dimensional wave field. The wave field may be written as a three-dimensional (generalized) Fourier transform

$$u(x,t) = \int_{-\infty}^{\infty} \int_{-\infty}^{\infty} u(k,\omega) \exp\{-i(k \cdot x - \omega t)\} \, dk \, d\omega,$$

(5)

where k is the wave number vector and ω the frequency. The inverse transform is

$$u(k,\omega) = \frac{1}{(2\pi)^3} \int_{-\infty}^{\infty} \int_{-\infty}^{\infty} u(x,t) \exp\{i(k \cdot x - \omega t)\} \, dx \, dt.$$

(6)

The amplitude-squared ground motion spectrum is given by

$$f_{uu}(k,\omega) = |u(k,\omega)|^2.$$

(7)

To calculate $u(k, \omega)$, the spatial integral in Eq. (6) is replaced by a weighted sum over the sampled station distribution. The weights W_j correspond to the filters adopted in the numerical program. The spectral estimate in Eq. (7) is then

$$f_{uu}(k,\omega) = \left| \sum_{j=1}^{N} W_j \, u(x_j,\omega) \exp\{ik \cdot x\} \right|^2$$

(8)

$$= \sum_{j=1}^{N} \sum_{l=1}^{N} W_j \, W_l \, u(x_j,\omega) \, \bar{u}(x_l,\omega) \exp\{ik \cdot (x_j - x_l)\}$$

$$= W \, \overline{W}^T \, \overline{U}^T(k) \, S(\omega) \, U(k).$$

(9)

In Eq. (9) the factor $S(\omega)$ is called the cross-spectral matrix and as we see plays the most important role in measuring the coherency of the strong ground motion across the site.

It is crucial to note that a Fourier transform is complex ($a + ib$, say). It has both a modulus ($\sqrt{a^2 + b^2}$) and an argument (arctan b/a). Therefore, the complete representation of a time history Eq. (2) is an amplitude spectrum plus a phase spectrum ("a spectrum pair"). In engineering applications with a rigid base or single input point, the response phase spectrum is normally unimportant and is ignored. When the phase of strong-motion time histories becomes a factor, then the complete response spectrum pair must be considered.

It should be noted that the author suggested two decades ago generating new and more suitable time-histories from available accelerograms by cross-over of phase spectra. This substitution procedure preserves the peak motions and spectral power but with a different duration and phasing pattern.

Seismic Wave Considerations

The construction of seismologically realistic time histories (synthetic seismograms) requires that the different types of seismic waves and their interrelationships be correctly incorporated (see Bullen and Bolt, 1985 for a full description). First, seismic waves, like light waves, can be polarized so that, for example, the vertical component of motion may be quite different from either the radial horizontal component (i.e. in the direction from source to station) or the transverse horizontal component.

We need to consider in most applications only three types of seismic waves: compressional (P) waves, shear (S) waves and surface (Love and Rayleigh) waves. The velocities of these waves depend upon the elastic moduli of the rocks and soil through which they propagate. We have

$$v_P = \sqrt{\frac{k + \frac{4}{3}\mu}{\rho}} \, , \qquad\qquad v_s = \sqrt{\frac{\mu}{\rho}}, \qquad\qquad (10)$$

where k is the bulk modulus, μ the rigidity and ρ the density. It follows that $v_P > v_S$ always. The speed of surface waves is always less than or equal to v_S. S waves do not travel in water and all types of waves are damped to various extent in rocks or soil.

It should be mentioned that as P, S or surface waves propagate through complicated rock strata, one type of wave may generate waves of another type. Also, phase shifts (cos δ) may occur as waves are reflected or refracted. In particular a vertically propagating horizontally polarized shear wave (SH) will double its amplitude when reflected from the free ground surface.

THE ELEMENTS OF COHERENCY

Estimation of Coherency

The strong motions (see Figure 2) at the various input points will, for the reasons already explained (such as source emission and path scattering), not be alike in general. A measure of the likeness of two wave trains is called the coherency and quantitative measures can be obtained in the time-domain as follows.

Consider Eq. (9). The factors $U_j(k)$ contain the exponential term which advances the phase of the wave harmonic component at station j with wave number k by the appropriate time delay.

The cross-spectral estimates are often normalized to unit amplitude to reduce site amplification effects. With this normalization, the cross-spectral matrix becomes simply the matrix of exponential phase differences

$$S_{j1}(\omega) = \sum_{m=-M}^{M} a_m \exp\{ i [\varphi(x_j, \frac{\omega+2\pi m}{T}) - \varphi(x_1, \frac{\omega+2\pi m}{T})]\},$$

(11)

where φ is the Fourier phase.

The simplest coherence measures are for two input points in which case the coherency is defined in the frequency domain by

$$\gamma_{12}(\omega) = S_{12}(\omega)/[S_{11}(\omega)S_{22}(\omega)]^{1/2}.$$

(12)

The coherence is defined as the square of the modulus of the coherency and is a normalized measure, $0 \le |\gamma_{12}(\omega)|^2 \le 1$, where a value of 1 indicates complete coherence.

A number of coherency studies have now been published (see Abrahamson and Bolt, 1987) and a few have been incorporated into structural response analyses for critical structures such as the Diablo Canyon Nuclear Power Station as part of soil-structure interaction calculations. One example is given here in Figure 3 from strong motions recorded by the SMART 1 array (Abrahamson and others, 1987). At high frequencies (here above 2 Hz) the recorded motion (mainly the S waves) is dominated by incoherent energy when averaged over a large distance. A practical approach to modelling would be to generate a suite of synthetic ground motions in which the Fourier phase of the incoherent energy is varied in a statistical manner while the Fourier phase of the coherent energy remains deterministic.

CONSTRUCTION OF LAGGED TIME HISTORIES

Source and Site Specification

We are now in a position to outline the basic steps in synthesizing a set of time histories for inputs to a multi-supported structure. It should be stressed that this construction is not unique and that alternative procedures can be adopted at various stages. For example, computations can be performed purely in the time domain or in the spectral domain. At this early stage, each method has its proponents but in the long run the speed of computers will make feasible optimal construction by key cross-over exchanges between the two domains.

The first step is to define from geological and seismological information the appropriate earthquake sources for the site in question. The source specification may be deterministic or probabalistic in concept and may be decided on grounds of acceptable risk (Bolt, 1991) A necessary set of source parameters would be magnitude (moment magnitude), seismic moment, fault location and assumed rupture surface length and depth, fault strike and dip, faulting mechanism (strike-slip, normal, etc.), rupture velocity and rupture rise-time (both usually selected from an empirical regression curve). In fact, much of this latter detail is not essential for reasonably realistic synthesis of phased time histories.

On the other hand, it is essential to specify closely the propagation path distance, the P, S, and surface wave speeds in the vicinity of the site (particularly seismic velocities for the alluvial, deep soil and surficial rock layers). These speeds are needed to calculate the appropriate wave propagation delays between support points (see Eq. (4)) and the angles of approach of the incident waves. Some soil-structure interaction programs already permit calculations of this kind.

Strong Ground Motion Specification

The construction of realistic phased seismograms (Niazi, 1986) can be regarded as a series of iterations from the most appropriate observed strong motion record already available to a set of more specific time-histories which incorporate the physically defined phase patterns. Where feasible, a strong motion accelerogram is chosen which satisfies within allowable limits the seismic source and site specifications described above. Of course, for large near earthquake sources (M > 7.5), there are few or no actual recordings and a synthetic record must first be constructed by scaling and/or numerical modelling.

As the iteration proceeds, certain constraints on each member of the set of phased records must be satisfied. Thus all must have response amplitude spectra that fall within say one standard error of the target spectra. (The response phase spectra (see Figure 5) will, of course, vary within prescribed tolerances also although little research has as yet been done to define such limits.) Similarly, each member of the set must preserve pre-specified peak ground accelerations, velocities and displacements within statistical bounds. The durations of each section (mainly P, S and surface wave portions) of each time-history must also satisfy prescribed source, path, and site conditions.

In order to guide convergence of the iterative process described and to provide a check on the overall result, it is advantageous to have practical seismological

41

advice on the product of the iterations. At the present, more or less automatic convergent algorithms are not available so that experience with observed seismograms and knowledge of the underlying seismic wave theory is especially valuable. In any event, it is important for the model methodology to be fully documented. Plastic overlays showing the phased time-histories and spectra on a standard scale are particularly valuable for checking.

ENGINEERING APPLICATIONS

Seismic Phase Response Spectrum

The effect of spatial variations of the phasing of strong ground motion on large structures where multi-support inputs are appropriate can be demonstrated using strong motion array data. We first need a way to incorporate the structural interaction.

The total response of the structure can be separated into the quasi-static response and the dynamic response. At a given node in the structure, the dynamic response due to the phase shifted inputs is divided by the mean response found using each of the individual support ground motions as rigid base inputs. This ratio, called the "dynamic response ratio", indicates the effect of the spatial variation of the ground motion on the dynamic response of the structure.

Consider a structure with N supports and M structural nodes and assume that the normal modes of the structure are known and that for each structural mode, the participation factor of the kth node to the lth input is also known. These weights, denoted w_{kl}, will depend on the mass and stiffness of the structure as well as the structural mode.

For simplicity, consider the special case of just one structural node with N inputs. The dynamic response of the structure satisfies the differential equation

$$\ddot{r}(t) + 2\xi\omega\dot{r}(t) + \omega^2 r(t) = -\vec{w} \cdot \ddot{u}(t), \qquad (13)$$

where ξ is the damping, ω is the natural frequency, and $\ddot{u}(t)$ is the N length vector of support input accelerations.

Let $R_{\xi,T}(t)$ be the unit impulse response of a single degree of freedom oscillator with period T and damping ξ. The dynamic acceleration response is usually characterized by the maximum of $|\ddot{r}(t)|$. The acceleration response spectrum for Eq. (13) is denoted $SA(\xi, T)$ and is given by

$$SA(\xi,T) = \max_t \left\{ R_{\xi,T}(t) * \vec{w} \cdot \vec{u}(t) \right\},$$

(14)

where * indicates a convolution. This response spectrum is compared to the response spectrum obtained by using rigid base inputs.

Using the lth input at all of the support nodes, the equation of motion becomes

$$\ddot{r}(t) + 2\xi\omega\dot{r}(t) + \omega^2 r(t) = \left(\sum_{k=1}^{N} w_k \right) \ddot{u}_l(t).$$

(15)

Let $SA_l(\xi, T)$ be the acceleration response spectrum for the ground acceleration $\ddot{u}_l(t)$. Then the mean response spectrum of all the inputs is given by

$$\overline{SA}(\xi,T) = \left(\sum_{k=1}^{N} w_k \right) \frac{1}{N} \sum_{l=1}^{N} SA_l(\xi,T).$$

(16)

The dynamic response ratio is given by

$$\Phi^d(\xi,T) = \frac{SA(\xi,T)}{\overline{SA}(\xi,T)}.$$

(17)

Substituting for the response spectra, Eq. (17) becomes

$$\Phi^d(\xi,T) = \frac{\max_t \left| R_{\xi,T}(t) * \vec{w} \cdot \vec{u}(t) \right|}{\left(\sum_{k=1}^{N} w_k \right) \frac{1}{N} \sum_{l=1}^{N} SA_l(\xi,T)}.$$

(18)

This ratio is influenced by both differential amplification of the ground motion due to site effects and by differential phasing of the ground motion. The differential phasing may be due to incoherent waves, nonvertically propagating waves or local site effects. To simplify the interpretation of the dynamic response ratio, site amplification effects are removed by normalizing each support acceleration $\ddot{u}_l(t)$ by the response $SA_l(\xi, T)$. The dynamic response ratio given by Eq. (18) has been evaluated using strong motion array recordings from SMART 1 as ground motion inputs at the supports. Descriptions of the SMART 1 array are given in Bolt et al. (1982). An illustration of the dynamic response ratio and response phase spectra from an earthquake recorded by SMART 1 are shown in Figures 4 and 5 (Abrahamson and Bolt, 1985).

San Francisco Bay Bridges, California

After the 1989 Loma Prieta earthquake, with the fall of the span on the Bay Bridge, much attention has been focussed on the appropriate design criteria for large structures in the vicinity of active faults. The Bay Area Rapid Transit Authority (BART) is in the design stage for extensions of the rail system, some of which pass across or near major active faults. The Golden Gate Bridge District has undertaken a detailed seismic analysis of the Golden Gate Bridge, which might be shaken by an earthquake resembling that in 1906 on the adjacent San Andreas fault. As well, Cal Trans has initiated ground motion studies for its existing and future bridges both in the San Francisco Bay area and in the Los Angeles area.

In all the above cases, analysis must be focussed on relatively long period seismic motions, i.e. from 2 Hz to 5 sec or more. It should be remembered that for wave components (see Eq. (1)) of period 1 sec the S-wave lengths λ incident on the supports are from several hundred to a thousand meters, or of the same order as key bridge dimensions. For this reason, specifications for the studies have called for the consideration of the effects of phased strong-motion inputs and, in some cases, for the incorporation of coherency factors. These studies present a considerable challenge because of the lack of suitable observational material. Few strong motion records are available for large near-earthquakes that measure spatial variation of shaking over distances of hundreds of meters. Also, because of the size of the engineered structures and the relatively large distances between the support points, ground velocity and displacement become of critical importance. The reliability of ground displacement records obtained from accelerograms is a question that needs additional checking.

ACKNOWLEDGEMENTS

A number of examples discussed in the text were worked in collaboration with Dr. N. A. Abrahamson and Dr. J. P. Singh, to whom I extend my thanks. Mr. N. Gregor helped with text preparation. The research was supported by NSF CES88-00457.

REFERENCES

Abrahamson, N. A., B. A. Bolt, R. B. Darragh, J. Penzien and Y. B. Tsai, 1987. The SMART 1 accelerograph array (1980-1987): A review. Earthquake Spectra, 3, 263-284.

Abrahamson, N. A. and B. A. Bolt, 1985. The spatial variation of the phasing of seismic strong ground motion. Bull. Seism. Soc. Am., 75, 1247-1264.

Abrahamson, N. A. and B. A. Bolt, 1987. Array analysis and synthesis mapping of strong seismic motion. In: B. A. Bolt (editor), Strong motion synthetics: Computational techniques series. Academic Press, New York.

Bolt, B. A., 1991. Balance of risks and benefits in preparation for earthquakes. Science, 251, 169-174.

Bolt, B. A., C. H. Loh, J. Penzien, Y. B. Tsai and Y. T. Yeh 1982. Preliminary report on SMART 1 strong motion array in Taiwan. EERC Report No. UCB/EERC-82/13.

Bullen, K. E. and B. A. Bolt, 1985. An introduction to the theory of seismology. Cambridge University Press, New York.

Harichandran, S. and E. H. Vanmarcke, 1986. Stochastic variation of earthquake ground motion in space and time. J. Engin. Mech., 112, 154-174.

Luco, J. E. and H. L. Wong, 1986. Response of a rigid foundation to a spatially random ground motion. Earth. Engin. Struct. Dyn., 14, 891-908.

Niazi, M., 1986. Inferred displacements, velocities and rotations along a long rigid foundation located at El Centro Differential Array site during the 1979 Imperial Valley, California, earthquake. Earth. Engin. Struct. Dyn., 14, 531-542.

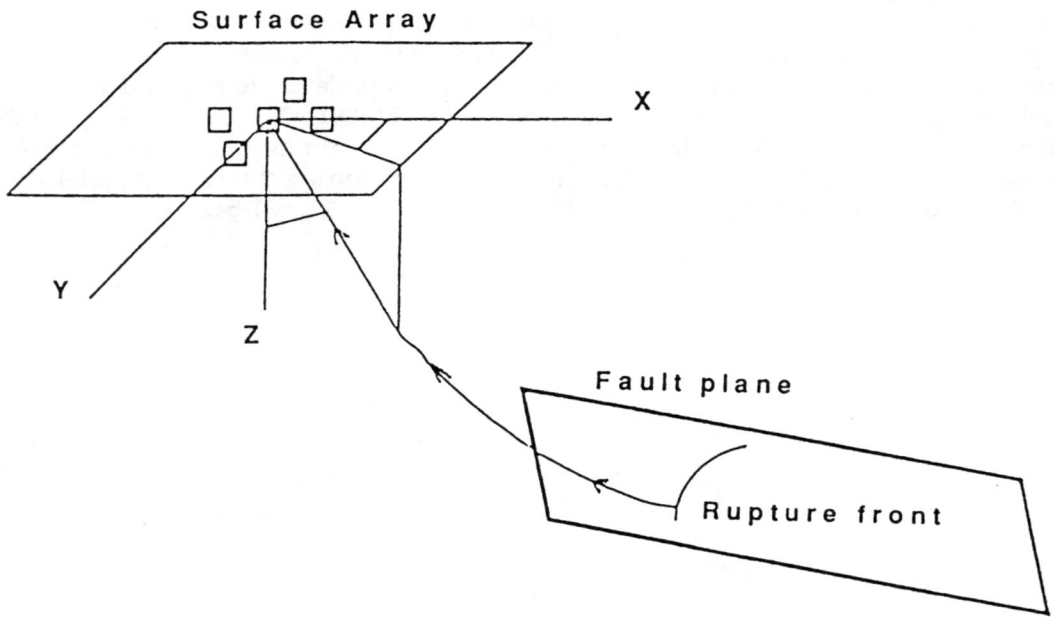

Fig. 1   Passage of a seismic wave from the seismic source to multi-stations in an array or to multi-supports.

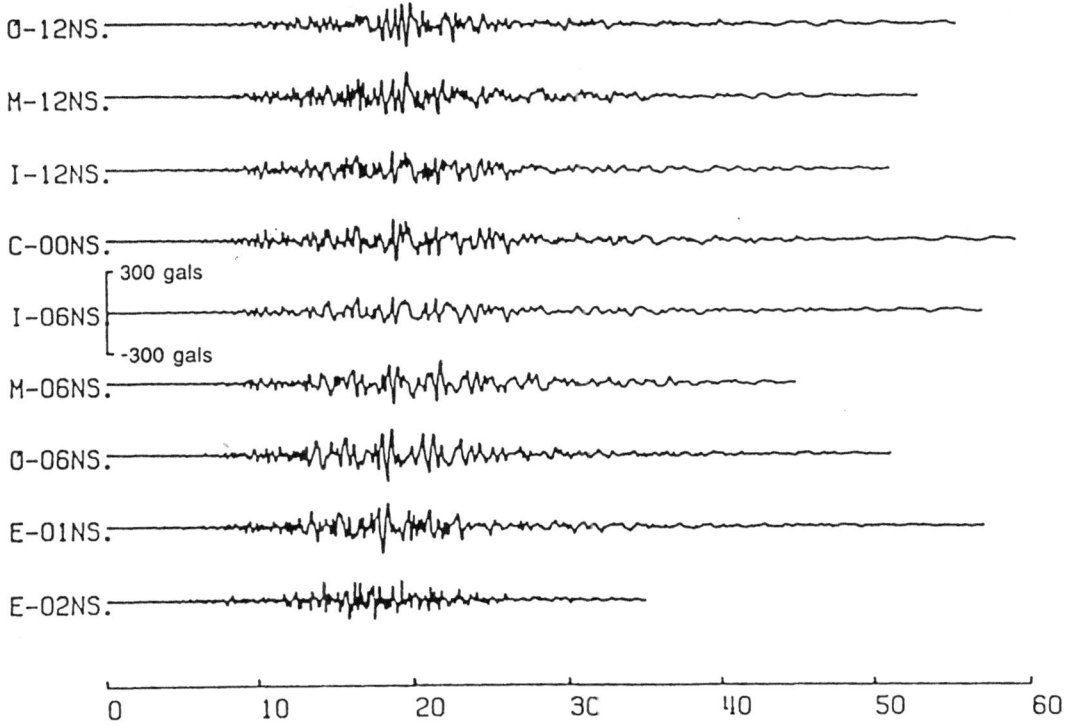

Fig. 2   Differential phasing in the horizontal ground acceleration across the SMART 1 array in Taiwan.   The horizontal scale is in seconds. The distance between the central station C00 and the inner ring (I) is 200 m and the outer ring (O) is 1000 m.

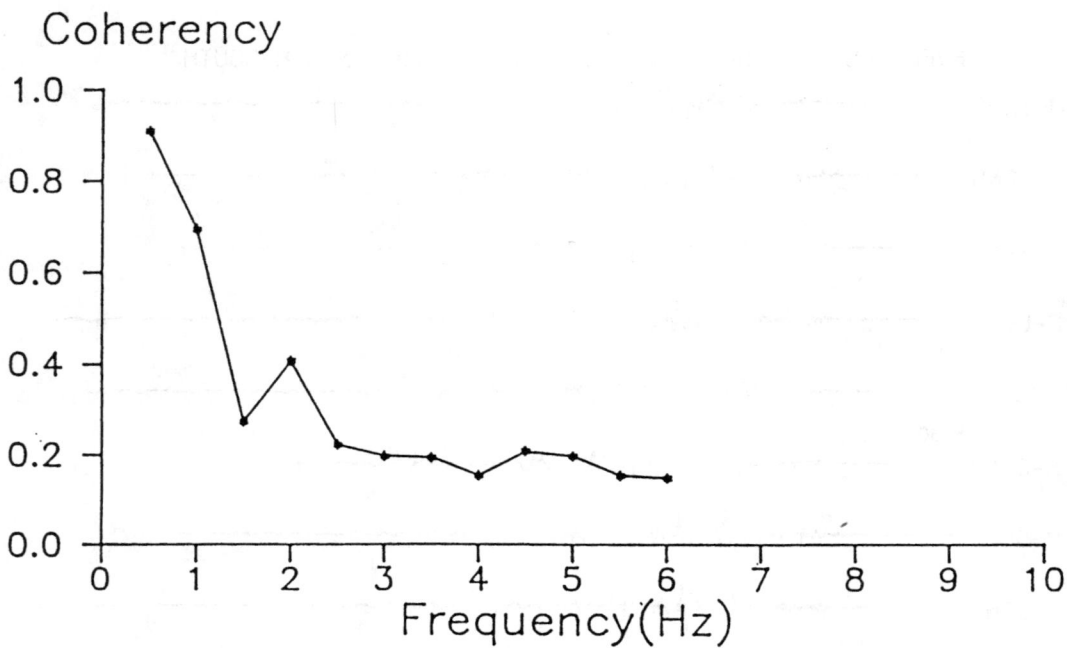

Fig. 3   Calculated coherency values of earthquake number 43, north-south component for all 37 stations of the SMART 1 array. The time window used was from 5 to 10 sec. along the record.

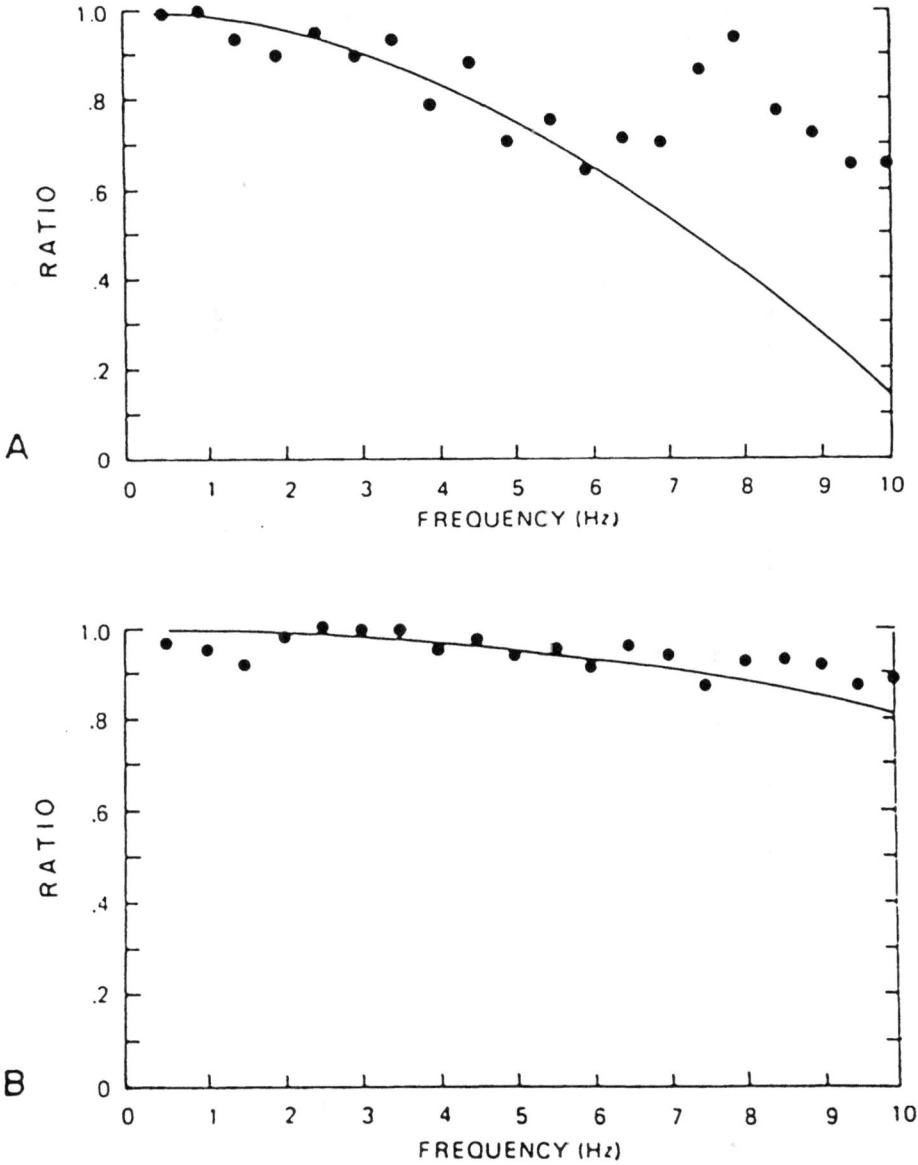

Fig. 4   The fundamental mode (in-phase motion) dynamic response ratio for 200-m support spacing using the tranverse component of acceleration from event 5. The filled circles are the estimated ratios, and the line gives the ration expected for a simple plane wave propagating from the source region. (A) station pair C00-I06. (B) Station pair C00-I03.

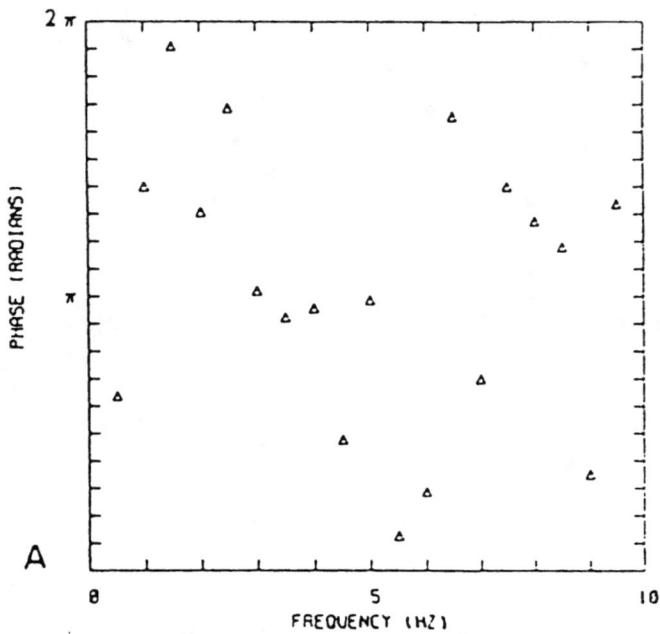

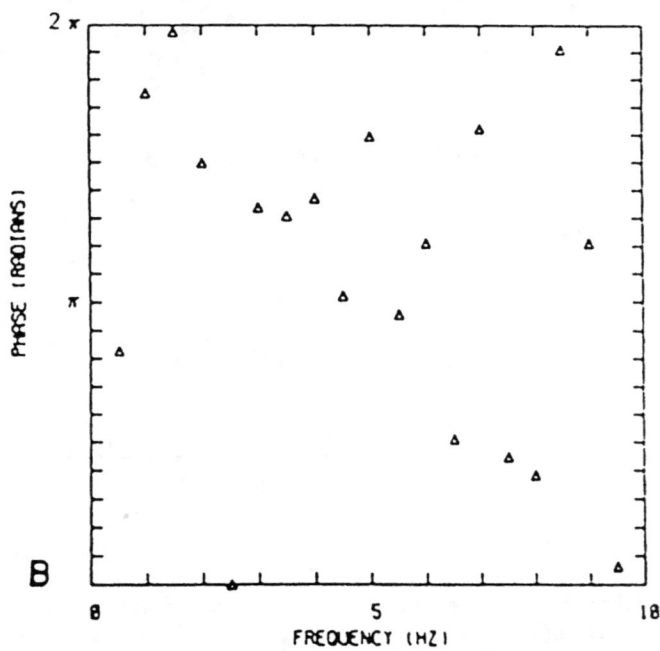

Fig. 5   (A) The response phase spectra $\Psi_\xi^\tau\,(\omega)$ at 5 per cent damping. Station C00, transverse component.   (B) Station I06, transverse component.

1 Response analysis and design of structural components
1 Analyses de réponses et dimensionnement de composantes structurales

Calibrated linear methods for analysis and design of yielding RC structures

J.F. Bonacci[1]

ABSTRACT

This paper explores the development of a method that is useful for design of reinforced concrete (RC) frame structures to resist earthquakes. The method, originally proposed in the 1970s, makes an analogy between viscously damped linear and hysteretic response for the purpose of estimating maximum displacement. Recent dynamic test results are used to extend significantly the calibration of the method.

DRIFT AS A DESIGN CONSIDERATION

Despite the terminology and packaging associated with regular practice in designing for earthquake resistance, what really matters is how a structure performs if it is shaken by a significant earthquake. Figure 1 represents load-displacement responses of two structural systems. Conventional structural design philosophy (Fig. 1a) is understood in terms of demand and supply along the load axis. Some time ago, designers came to a realization that there were several aspects of the earthquake problem (dynamic equilibrium, loading that is unpredictable in terms of amplitude, frequency content, duration, and probability of occurrence) that made it implausible to fit into this conventional outlook. When structural response is idealized as linear (Fig. 1a), the relationship between displacement and load demands are clear, so displacements are of secondary importance and the attention they are given amounts to a verification. When response is nonlinear (Fig. 1b), the displacement axis is the only meaningful indicator of demand. Load demand loses physical meaning once it exceeds the supplied resistance.

All performance-related questions in design for earthquake resistance can be related to displacement demand:

1. Will story drift be too much for attached elements?
2. Will displacement cause collision with adjacent structures?
3. Will drift level demand too much deformability from the elements?
4. Will displacements bring about excessive secondary effects?

If it could be shown that a structure satisfied the above questions, then its supply of resistance along the load axis would be inconsequential.

Many current codes give the impression that design can be managed by control along the load axis in a fashion similar to wind design. Load demand is determined and then reduced by a quantity that is sensitive in an empirical way to the redundancy and deformability inherent to various structural systems. The resistance to be provided is linked to this reduced design load and it is understood that this implies yielding response in the design earthquake. Displacements are calculated for the purpose of checking only after design loads have been set. Displacement response is assumed to be linearly related to the unreduced load demand. Because of the strong parallels to linear design (as say for wind), the framework of this approach conceals the fundamental differences in demand

---

[1]Assistant Professor, Dept. of Civil Engrg., University of Toronto, M5S 1A4, Canada.

for yielding systems and concentrates the designer's efforts to evaluate performance into selection of a few empirical constants. While the parallel with design for other kinds of lateral loads can be convenient, the code approach has the effect of calling attention to the less important axis of response.

Because displacements are directly related to questions about performance, the ability to base earthquake design requirements on realistic displacement demand is essential. This paper reviews a class of methods useful for estimating response of yielding reinforced concrete systems to earthquakes. Results of recent dynamic tests are used to motivate an extensive second look at a comprehensive method for design of RC frame structures that enables the designer to control damage on the basis of a realistic estimate of nonlinear displacement response.

LINEAR EQUIVALENTS OF YIELDING SYSTEMS

Putting aside the uncertainties associated with detailed prediction of earthquake characteristics, the task of computing dynamic response becomes tedious if the subject structure is likely to yield. When this happens, a considerable amount of energy can be dissipated through inelastic action of elements. Algorithms of varied sophistication have been developed to describe hysteresis in reinforced concrete with a certain degree of reliability. But such all-out nonlinear dynamic structural analyses are seldom appropriate for design because they tend to be too specific and out of proportion with the overall design effort.

Several alternative approaches for determining earthquake response have made use of response spectra, which are more intuitive and familiar tools. While it is possible to generate nonlinear response spectra using a selected hysteresis model, these are much more abstract than linear spectra, which form the basis for most design codes. That linear response spectra can be made to furnish values matching the response of a yielding system requires only a superficial argument. Consider a yielding system with initial period T and viscous damping ratio β for which it is known that the maximum response is D_{max}. The hypothetical case depicted in Fig. 2 shows four ways that a linear spectrum can be manipulated to furnish D_{max} by altering the nominal period or damping or both. While the argument is attractive, its feasibility would depend on the development of rules for selecting altered period and damping so that reliable results can be obtained.

The striking analogy that exists between viscous linear and hysteretic systems has been investigated extensively. The object of this section is to trace the development of concepts related to linear analogies of yielding systems. Particular emphasis is given to applications to reinforced concrete as a prelude to the next section, which is an extensive second look at the foundations of the substitute structure method.

Jacobsen (1930) introduced the notion of equivalent viscous damping to approximate the effects of friction proportional to general powers of velocity. The approach was used to estimate response amplitudes for single-degree-of-freedom (SDOF) systems with linear spring stiffness to sinusoidal loads.

Soon after Housner (1956) argued the inevitability and possible benefits of inelastic response of structures in earthquakes, the first suggestions were made to extend the idea of equivalent viscous damping to hysteretic systems (Jacobsen 1960, Ando 1960). Jennings (1964) summarized and compared several ensuing approaches for representing elasto-plastic SDOF oscillators as contrived linear systems. By assigning prescribed values of mass, stiffness and damping ratio, matching of resonant amplitude (and, in some cases, frequency) was achieved for steady-state response to sinusoidal loading. The closed-form nature of the expressions for substitute properties could not be maintained for earthquake problems because the more random exciting force prevented development of steady-state response, which is an essential condition to establishing equivalence in energy dissipation.

Takeda et al (1970) performed static and dynamic tests of SDOF reinforced concrete elements that led to the development of the first hysteresis algorithm tailored specifically to reinforced concrete. The model used a trilinear primary curve and provided expressions for unloading slope as a function of element ductility ratio and previous response history.

By viewing equivalent viscous damping in average fashion rather than on a cycle-by-cycle basis, Gulkan and Sozen (1974) did away with the restriction of steady-state response. Rather than deriving expressions mathematically, values for substitute frequency and damping were deduced from results of

54

a series of dynamic tests of SDOF RC bents. Substitute frequency was taken as the ratio of measured maximum absolute acceleration to measured maximum displacement response, which is related to the apparent stiffness that would be observed in load-displacement relationships. Substitute damping for the yielding RC frames was computed on the assumption that the energy intake of the system over the duration of motion was balanced by a linear dashpot in order for the system to come to rest. This amounted to determination of the appropriate value of β_s in the expression

$$\beta_s \left[2m\omega_s \int_0^{t_f} \dot{x}^2 dt \right] = -m \int_0^{t_f} \ddot{y}\dot{x}\, dt \tag{1}$$

where β_s was the substitute damping ratio, m the mass, ω_s the substitute circular frequency, $\dot{x}$ the relative velocity response, $\ddot{y}$ the base acceleration, and t_f the duration of shaking. It was shown that measured displacement response was suitably approximated from linear spectra of the actual base motion (14 runs sinusoidal, 4 runs simulating El Centro NS 1940, and 10 runs simulating Taft N21E 1952) when the substitute properties (ω_s and β_s) were used. This result was then formulated into a method for determining design base shear based on an admissible ductility ratio, μ. For this purpose, it was suggested to lengthen the period based on cracked sections by $\sqrt{\mu}$ for substitute period (replacing ω_s). Substitute damping was related to μ with an expression patterned after the Takeda (1970) hysteresis model

$$\beta_s = 0.02 + 0.2 \left(1 - 1/\sqrt{\mu} \right) \tag{2}$$

which was shown to provide a reasonable representation of substitute damping ratios deduced from test measurements.

The substitute structure method (Shibata and Sozen 1976) is the extension of Gulkan's SDOF base shear prescription to RC frames with more than one degree of freedom. Substitute period and damping were based on tolerable damage (ductility) ratios, μ_i, for various elements of the frame. In this manner, the designer could establish both the extent and relative distribution of damage to beams and columns. For each element, stiffness was reduced by $1/\mu_i$ times the value for cracked section. Substitute frequency was computed from linear analysis using reduced stiffness values. To compute substitute damping ratios for the full frame for individual modes, substitute damping ratios computed for each element on the basis of μ_i (Eq. 2) were weighted according to relative flexural strain energy associated with the mode shape:

$$\beta_m = \sum_i \left(\frac{P_i}{\sum_i P_i} \right) \beta_{si} \tag{3}$$

where β_m is the smeared modal substitute damping ratio for mode m, P_i is flexural strain energy for an element i in mode m, and β_{si} is the substitute damping ratio for an element i. With substitute properties thus defined, modal responses were read from linear spectra and combined to determine displacements, base shear, and member design forces. The principal benefit of the method was that it based design on issues related to performance (tolerable damage and drift) while requirements could still be stated in terms of design forces, as is customary for most other kinds of loading.

TESTS OF SDOF SYSTEMS

A critical step in applying the concept of equivalent viscous damping to earthquake problems was the shift toward reliance on dynamic test results as a basis for inferred damping. The foundations for the particular method considered in this study are based directly on 28 test runs by Gulkan (1974), for which only half used simulated earthquake motion, and indirectly on 7 test runs by Takeda (1970), for which only 3 used earthquake-like motions.

A recent series of dynamic tests (Bonacci 1989) has provided an extensive set of new data that is ideal for further investigation of linear analogies to yielding RC systems. The study provides 35 new test results for which simulated earthquake motions were used. The results are considered ideal not only because they triple the size of the data set, but also because they reflect the influence of four variables germane to nonlinear dynamic response: initial period and strength of the structure, and frequency content and intensity of the earthquake.

The specimen (Fig. 3) for this parametric study of nonlinear displacement response can be idealized as an inverted pendulum restrained by a flexural spring. The stocky panel, pinned at its base, formed the shaft of the pendulum, and steel plates mounted near the top accounted for most of the mass. The beam extending horizontally from the panel to a roller support (steel pipe column pinned at both ends) functioned as the spring for the pendulum. Detailing of sections was such that beam flexural strength was the weakest link in overall specimen resistance to lateral loads. Specimen initial period was controlled by varying panel height, beam span, and the amount of attached weight. Lateral-load strength was influenced mainly by the ratio of beam reinforcement provided, but also by the spans of the panel and beam. Base motions were patterned after three different earthquake records (Castaic N21E 1971, El Centro NS 1940, and Santa Barbara S48E 1952) to cover a wide range of frequency content. The ordinates of these records were scaled uniformly to vary intensity. A summary of the attributes for each of the test runs is given in Table 1.

Reduction of test results for the purpose of developing a linear response analog followed the procedures used by Gulkan (1974), with a few exceptions. Inspection of Eq. 1 reveals that calculation of substitute damping ratios requires a relative velocity response signal, $\dot{x}(t)$. Because no direct velocity measurements were made and because digital differentiation of recorded displacement response, $x(t)$, would be inherently troublesome, the required signal was computed as the time integral of available acceleration signals

$$\dot{x}(t) = \int \left\{ [\ddot{x} + \ddot{y}](t) - \ddot{y}(t) \right\} \, dt \qquad (4)$$

where $[\ddot{x} + \ddot{y}](t)$ is the absolute acceleration response at the center of mass and $\ddot{y}(t)$ is the base acceleration. The integral was very sensitive to baseline error in the parent signals. Gulkan (1974) applied a parabolic adjustment to the acceleration baseline. In the present study, digital filtering of extreme low-frequency components produced a satisfactory velocity signal. The process, as illustrated in Fig. 4, was to integrate the uncorrected relative acceleration signal ($\ddot{x}$; RHS of Eq. 4); compute the Fourier transform of the uncorrected velocity; filter components with frequency less than 1.25 Hz (the lowest response frequency deduced from zero-crossing rates of all 35 runs); transform back to the time domain. The resulting corrected velocity signals were checked by applying the same process for a second integration in order to compare with measured relative displacement response, $x(t)$. The only noted deviation was the elimination of displacement baseline offset as a result of filtering. Substitute damping ratios computed from 35 test runs (Table 1) ranged from 3 to 20% of critical.

Substitute frequency was computed from apparent stiffness (slope of a line joining unloading points in opposite quadrants) deduced from moment-rotation response of the pendulum in the cycle of peak rotation:

$$\omega_s = \frac{1}{r} \sqrt{k_a/m} \qquad (5)$$

where r is the pendulum radius to the center of mass, m, and k_a is apparent stiffness.

Using these substitute properties (Table 1), an estimate of maximum displacement response (S_d) was made for each test run from a linear spectrum for the recorded base motion. These are compared with measured nonlinear response in Table 1 and Fig. 5, which illustrate that the linear analogy provided accurate estimates of peak response.

Substitute damping ratios are plotted against peak rotation ductility ratios (Table 1), μ, inferred from moment-rotation relationships in Fig. 6 along with the relationship (Eq. 2) proposed by Gulkan (1974). Substitute damping values computed from results of the 35 more recent tests were generally larger than those given by Eq. 2 at all values of ductility ratio. It follows that a revised best-fit equation could be obtained by increasing either or both of the constant (0.02) or linear multiplier (0.20) in the equation, but not by changing the order of the radical ($1/\sqrt{\mu}$). However, no modification to Eq. 2 is proposed for two reasons: (1) if only the current data were available to construct a design equation, a slight underestimate of damping would be judiciously conservative, and (2) many of the results from Gulkan's tests fell below the line given by Eq. 2.

56

CONCLUDING REMARKS

This paper traced the evolution of methods for representing yielding dynamic systems with equivalent linear analogs. Obstacles to applying such methods for earthquake problems in RC were overcome by placing some reliance on experimental results from a relatively small number of dynamic tests. This study considered results from more recent dynamic tests to triple the size of the relevant data set. It was shown that the approach proposed by Gulkan and Sozen (1974) provided an accurate organization of the more recent test results (Bonacci 1989) without need for modification. In itself, this is an obscure result. But its real value is as a "vote of confidence" for the substitute structure method.

Any design routine should be sensitive to the scale that will be used to judge its effectiveness. For building structures in earthquakes, performance questions (such as those listed earlier) are clearly related to displacement response. Most recognized design approaches (building codes, capacity design philosophy) lack sincere consideration of the level of drift response. The method reinvestigated in this paper offers a workable alternative for performance-based design of RC frames in earthquakes.

ACKNOWLEDGMENTS

The experimental work described in this paper was completed at the University of Illinois with the financial assistance of the U. S. National Science Foundation under grant ECE-8418691. The study presented is based on part of the author's Doctoral thesis, which profited from the advice of Prof. M. A. Sozen.

REFERENCES

Ando, N. 1960. Nonlinear vibrations of building structures. Proceedings of the Second World Conference on Earthquake Engineering, Tokyo and Kyoto, Vol. 2.

Bonacci, J. F. 1989. Experiments to study seismic drift of RC structures. Thesis submitted to the Graduate College of the University of Illinois, Urbana, in partial fulfillment of the requirements for the degree of Doctor of Philosophy.

Gulkan, P. and Sozen, M. A. 1974. Inelastic response of RC structures to earthquake motions. ACI Journal, No. 71-41, 604-610.

Housner, G. W. 1956. Limit design of structures to resist earthquakes. Proceedings of the World Conference on Earthquake Engineering, Berkeley, CA, No. 5, 1-13.

Jacobsen, L. S. 1930. Steady forced vibration as influenced by damping. Transactions, ASME, Vol. 52, Part 1, 169-181.

Jacobsen, L. S. 1960. Damping in composite structures. Proceedings of the Second World Conference on Earthquake Engineering, Tokyo and Kyoto, Vol. 2.

Jennings, P. C. 1968. Equivalent viscous damping for yielding structures. J. Engr. Mechs. Div., ASCE, 94(1), 103-116.

Shibata, A. and Sozen, M. A. 1976. Substitute-structure method for seismic design in RC. J. Struct. Div., ASCE, 102(1), 1-18.

Takeda, T., Sozen, M. A. and Nielsen, N. N. 1970. RC response to simulated earthquakes. J. Struct. Div., ASCE, 96(12), 2557-2573.

Table 1. Summary of SDOF experiments (Bonacci 1989).

Test	T_i (1) sec	C (2)	Run	Record (3)	Max ÿ g	μ	ω_s rad/sec	β_s	S_d (4) in	D_{max} in
B-01	0.086	1.34	1	ELC	0.97	1.4	33.5	0.052	0.69	0.58
			2	ELC	1.35	2.5	27.3	0.086	1.26	1.08
			3	ELC	1.54	3.1	24.3	0.106	1.42	1.35
B-02	0.13	0.39	1	ELC	0.78	3.0	15.5	0.127	1.33	1.08
			2	ELC	1.37	5.8	9.9	0.174	2.09	2.12
B-04	0.17	0.33	1	ELC	0.77	3.6	12.9	0.121	1.69	1.40
			2	ELC	1.33	5.4	9.8	0.151	2.30	2.08
B-05	0.14	0.79	1	ELC	0.79	2.0	19.9	0.068	1.59	1.10
			2	ELC	1.34	3.4	14.9	0.133	2.34	1.88
B-06	0.14	0.39	1	CAS	0.49	1.0	21.2	0.062	0.35	0.39
			2	CAS	0.93	2.5	16.0	0.077	0.90	0.90
			3	CAS	1.69	4.9	12.8	0.130	1.38	1.75
B-07	0.17	0.67	1	SAB	0.51	1.2	20.7	0.039	0.90	0.72
			2	SAB	0.70	2.3	16.0	0.105	1.79	1.50
B-08	0.16	0.33	1	SAB	0.30	1.3	17.7	0.052	0.75	0.59
			2	SAB	0.47	3.9	11.9	0.144	1.64	1.73
B-09	0.14	0.39	1	SAB	0.37	1.3	19.3	0.061	0.68	0.49
			2	SAB	0.66	5.5	11.4	0.189	2.12	2.14
B-10	0.17	0.67	1	ELC	0.56	1.6	18.4	0.059	1.18	1.10
			2	ELC	0.95	2.4	14.9	0.104	1.84	1.68
			4	ELC	0.95	2.5	14.9	0.105	1.84	1.74
B-11	0.087	0.67	2	CAS	0.88	2.3	22.7	0.098	0.68	0.71
			3	CAS	1.51	2.9	19.4	0.118	0.99	0.98
			4	ELC	1.51	6.1	14.0	0.199	1.82	2.03
B-12	0.089	1.34	1	CAS	0.66	1.0	34.4	0.037	0.58	0.49
			2	CAS	1.05	1.4	30.8	0.050	0.66	0.63
			3	CAS	1.45	1.8	27.7	0.080	1.08	0.85
B-13	0.091	1.34	1	SAB	0.55	0.8	34.8	0.032	0.37	0.42
			2	SAB	0.88	1.3	31.5	0.044	0.64	0.70
B-14	0.089	0.67	1	SAB	0.40	0.9	32.5	0.044	0.24	0.25
			2	SAB	0.77	2.5	22.7	0.096	0.82	0.85
			3	SAB	0.91	3.5	18.5	0.125	1.30	0.96
B-15	0.14	0.79	1	SAB	0.35	0.8	23.9	0.038	0.60	0.51
			2	SAB	0.78	1.7	19.2	0.067	1.13	1.14
			3	SAB	0.88	2.1	16.7	0.103	1.87	1.46

Notes:
(1) T_i = specimen initial period.
(2) C = specimen lateral strength coefficient = load at yield/weight.
(3) CAS denotes Castaic N21E 1971 motion.
    ELC denotes El Centro NS 1940 motion.
    SAB denotes Santa Barbara S48E 1952 motion.
(4) S_d = displacement from linear spectrum using substitute properties.

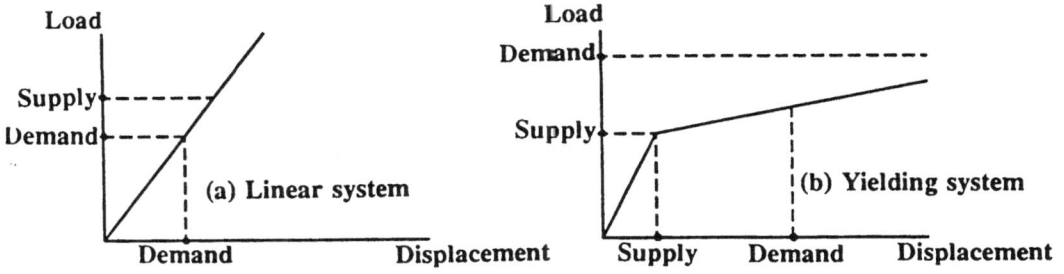

Figure 1. Design philosophy and structural response

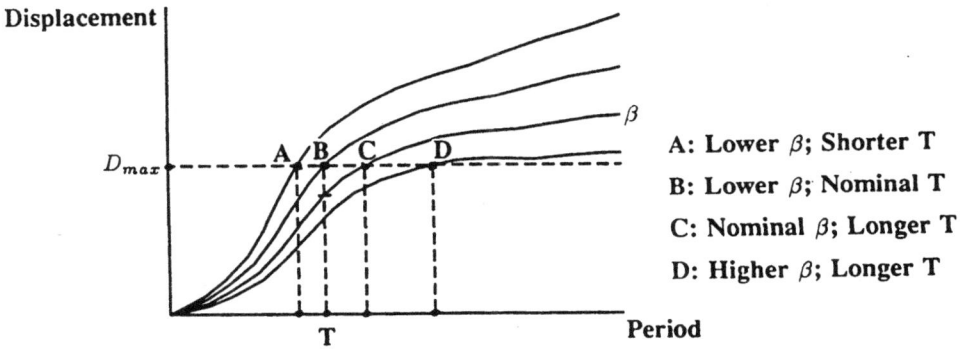

A: Lower β; Shorter T

B: Lower β; Nominal T

C: Nominal β; Longer T

D: Higher β; Longer T

Figure 2. Altered properties for linear response spectrum

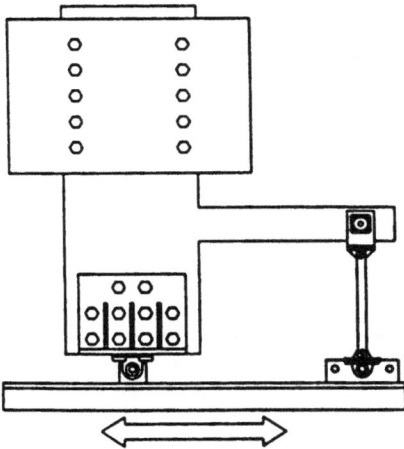

Figure 3. SDOF test specimen

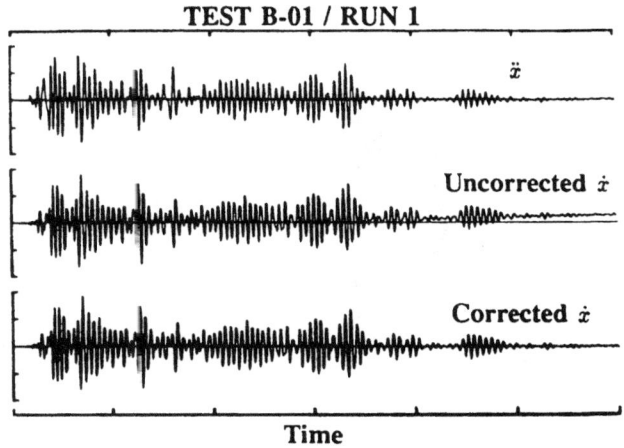

Figure 4. Relative velocity response signal

59

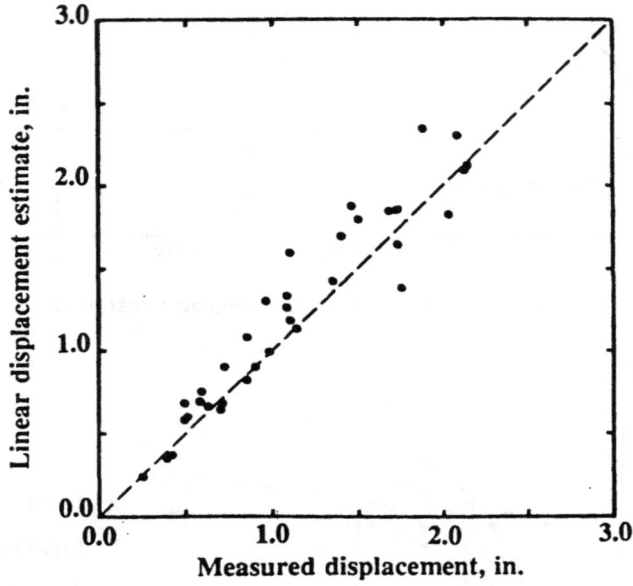

Figure 5. Evaluation of estimated displacements

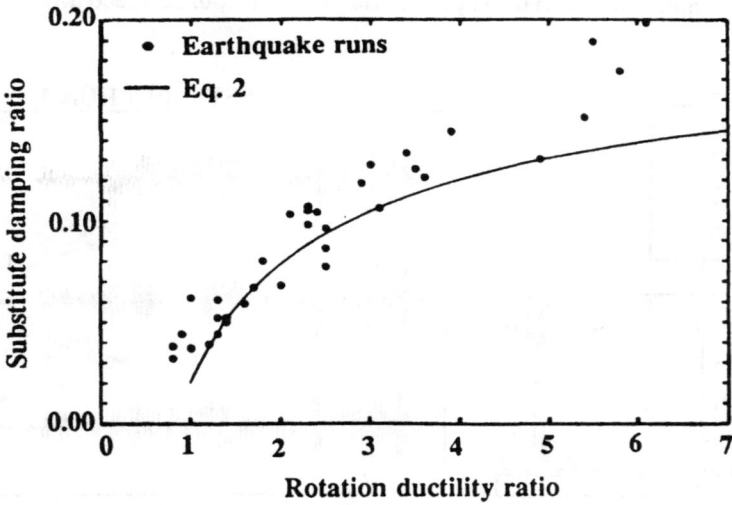

Figure 6. Substitute damping ratio relationship

Strength and ductility demand in relation to velocity spectrum

Steven Y.W. Kuan[I] and Noel D. Nathan[II]

ABSTRACT

Recent analytical studies on filtered earthquakes such as the Mexico City 1985 event have shown that inelastic displacements can be significantly greater than the elastic ones. It is found that the response of a yielded structure is highly dependent on its strength, the initial structural period, and the characteristics of the velocity spectrum of the ground motion. An equation relating inelastic response to the spectral velocity spectrum is made based on energy principles to explain both the equal- and unequal-displacement phenomena.

INTRODUCTION

Current practice in seismic-resistant design allows structures to yield during severe earthquakes provided that sufficient ductility is available for the large deformations. By assuming pure plastic behaviour in the inelastic range, the response of the structure can be idealized as a bilinear force-displacement plot as shown in Fig. 1.

It has been shown that, for a structure of any strength, the maximum inelastic displacement during an earthquake is very close to the displacement that would have been attained if the structure had remained elastic (Blume, Newmark, Corning, 1961). This idea is generally accepted and is incorporated in most design codes including the National Building Code of Canada 1990.

The "equal-displacement" phenomenon, however, is based on response to typical earthquakes that have very low predominant periods of vibration. Recent earthquakes in Mexico City and Loma Prieta have demonstrated the damaging effects of filtered earthquakes. This type of earthquake has its predominant period shifted to higher values by the local soil conditions. The acceleration response spectra for Taft S69E 1952, representing a typical earthquake, and for Mexico City SCT EW 1985 are shown together in Fig. 2. The effects of these different characteristics on the inelastic

---

[I]Graduate Student, and [II]Professor, Civil Engineering Dept., University of British Columbia, 2324 Main Mall, Vancouver, B.C., Canada

response of a structure will be examined, with the hope of developing some general concepts of inelastic response to earthquake excitations.

INELASTIC RESPONSE ENVELOPES

In the lateral load-displacement plot for any given structure, the line joining the maximum displacements at different yield levels is the inelastic response envelope (see Fig. 3). It indicates the inelastic behaviour of the structure having that particular initial period during that particular earthquake. Therefore, the envelope is a function of the initial structural period and of the characteristics of the earthquake. As mentioned earlier, for typical "normal" earthquakes, the envelope is defined by an almost vertical line. This is demonstrated by the envelopes in Fig. 4 obtained for the Taft ground motion using the dynamic program DRAIN-2D (Kanaan and Powell, 1973) on elasto-plastic single-degree-of-freedom systems. In this diagram, the results for each structural period are normalized with respect to the corresponding elastic values of load and displacement. Note that all the envelopes are very close to each other and stay reasonably vertical.

When the same dynamic analysis is carried out using the Mexico City ground motion, a different picture emerges. Figure 5 shows the resulting normalized inelastic response envelopes. It can be seen that not all the envelopes are vertical; in fact, they can be vertical, sloping outward, or sloping inward towards the elastic response line.

In Fig. 4, the line OA represents elastic behaviour. A line OB at a slope equal to that of OA divided by μ defines the maximum displacement reached by structures of ductility μ. In the idealized case, when the response envelope is a straight vertical line, it is evident that a structure of any initial period must be designed to yield at C in order to achieve a particular ductility of μ. It is also evident from the geometry that the "Force Reduction Factor", R, by which the elastic load at A must be reduced to give the design load at C is equal to μ.

The situation is clearly very different in Fig. 5, representing the Mexico City earthquake. The Force Reduction Factor is no longer equal to the ductility; indeed for the point A on that figure, one would have to use an R of about 2 to achieve a ductility of 4.

An explanation for the differences in the inelastic response envelopes can be found in the behaviour of an inelastic structure in relation to the characteristics of the earthquake. As the structure yields, it becomes less stiff, so that its fundamental period of vibration increases. Depending on the nature of the earthquake, then, it imparts more or less energy to this changed structure; if more, the ductility demand will be high; if less, the ductility demand will be lower. This phenomenon was observed in the Mexico City earthquake, and the increase in energy imparted to structures was related by Mitchell (1987) to the rising acceleration response spectrum, reproduced here as Fig. 6. It will be shown in the next section, however, that the velocity response spectrum is a more direct indicator of the ductility demand.

INELASTIC RESPONSE ENVELOPES FROM VELOCITY SPECTRA

The relationship between the Force Reduction Factor and the desired ductility depends upon the inelastic response envelope, as shown by Fig. 5. Therefore a method of determining the position of the latter would be useful.

As shown in Fig. 7, the response of an inelastic structure is represented as a bilinear curve. An "equivalent elastic" structure can be defined by a straight line of slope less than the actual stiffness, reaching the same maximum deflection as the inelastic structure. The slope of this line, or the "effective stiffness" of the equivalent elastic structure, is obtained by setting the area under the curve equal to that under the inelastic curve. The period of this equivalent elastic structure then suggests the period shift, as the real structure becomes inelastic in the earthquake.

The areas referred to above represent the maximum energy of deformation stored in the structure during the earthquake; by conservation of energy, this is essentially equal to the maximum kinetic energy achieved by the system, which is proportional to the square of the velocity. Thus, the load-displacement diagrams are indicative of the maximum velocities reached by the systems.

Applying these relationships to the original elastic structure and the equivalent elastic structure, and noting the relationship of the latter to the inelastic structure, one is able to express the Force Reduction Factor in terms of the ductility and the spectral velocities before and after the period shift:

$$R = \frac{V_1}{V_2} \sqrt{2\mu-1} \tag{1}$$

This equation shows the influence of the velocity response spectrum on the inelastic response envelope. If the velocity response spectrum is horizontal, so that V_1 equals V_2, the inelastic response envelope is given by curve H on Fig. 8, and the Force Reduction Factor is given by

$$R = \sqrt{2\mu-1} \tag{2}$$

This is the well-known equal-energy criterion; it implies that the energy under the inelastic force-displacement curve is equal to that under the original elastic structure curve. If the velocity response spectrum slopes downwards, as it does for "typical" earthquakes, then $V_2 < V_1$, the Force Reduction Factor is increased, and the inelastic response envelope is seen to lie closer to the initial stiffness line. The equal-displacement criterion applies to a line of this type. If the spectrum slopes upwards, as it frequently does when the ground motion has been filtered by local site conditions, then $V_2 > V_1$, R is reduced, and the inelastic response envelope moves further from the initial stiffness line, outside that for the equal energy case.

In order to apply these concepts, it is necessary to determine the period shift, so that V_1 and V_2 can be found. Turning again to Fig. 7, the stiffness of the equivalent elastic structure can be related to the initial stiffness and the ductility demand on the inelastic structure; this enables one to determine the period of this structure in terms of the initial period:

$$T_2 = \frac{\mu}{\sqrt{2\mu-1}} T_1 \qquad (3)$$

The period shift is given here in terms of the ductility demand; a similar relationship was previously presented by Iwan (1980), and a reasonable agreement between the two is shown on Fig. 9.

The validity of the equation for the determination of the inelastic response envelopes has been tested on three earthquake records: Taft S69E 1952, Mexico City SCT EW 1985, and an artificial strong ground shaking record for Richmond, British Columbia. Ductility values of 2 and 4 were used in the tests. The calculated Force Reduction Factors, Rc, were compared with the actual factors Ra, obtained by inelastic time-step analyses of elasto-plastic single-degree-of-freedom systems. The results were good when the velocity spectrum sloped downward, but when it sloped upwards the values were in error by a factor as large as 5.

An examination of Eq. 1 reveals that the effects of the change in structural damping during yielding, and of the actual shape of the velocity spectrum over the range of the period shift (instead of merely the two end values) have not been included. To account for these two effects, a modification factor is included:

$$R = f_N \frac{V_1}{V_2} \sqrt{2\mu-1} \qquad (4)$$

From a plot of the ratio of the calculated factors from Eq. 1 and the actual reduction factors against the secant slope of the velocity spectrum, the modification factor, f_N, is found to be a function of this slope and the ductility. Empirical values of f_N for ductilities of 2 and 4 were determined and are shown in Fig. 10. The limiting value of 1 for zero and negative slopes is imposed since no modification is required for horizontal and downward sloping spectra.

CONCLUSIONS

Studies of filtered earthquakes have revealed the possibility of non-vertical inelastic response envelopes occurring for structures with low to medium periods. The most important type of envelopes are those that slope outwards as they demand a lower force reduction factor to achieve a certain ductility in the structure. This force reduction factor can be estimated by a simple formula once the initial period, the ductility desired, and the characteristics of the velocity response spectrum are known.

REFERENCES

Blume, J.A., Newmark, N.M., and Corning, L.H. 1960. Design of Multistory Reinforced Concrete Buildings for Earthquake Motions. Portland Cement Association, Skokie, Illinois.

Iwan W.D. 1980. "Estimating Inelastic Response Spectra from Elastic Spectra", Earthquake Engineering and Structural Dynamics, Vol. 8, 375-388.

Kanaan, A.E. and Powell, G.H. 1973. "DRAIN-2D: A General Purpose Computer Program for Dynamic Analysis of Inelastic Plane Structures", Report No. UCB/EERC/73-6, Earthquake Engineering Research Center, University of California, Berkeley.

Mitchell, D. 1987. "Structural Damage due to the 1985 Mexican Earthquake", 5th Canadian Conference in Earthquake Engineering, Ottawa, 87-111.

National Building Code of Canada 1990. National Research Council of Canada, Ottawa.

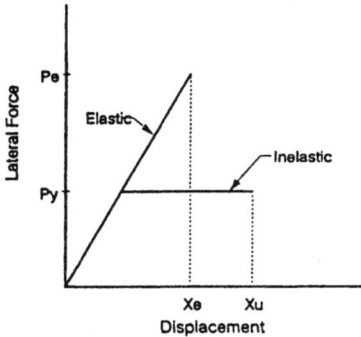

Figure 1.   Seismic response of elasto-plastic structures.

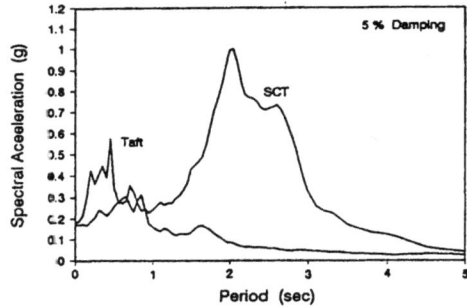

Figure 2.   Sample acceleration response spectra.

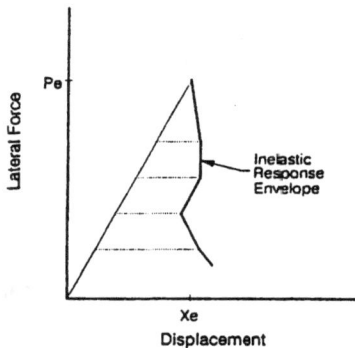

Figure 3.   Inelastic response envelope.

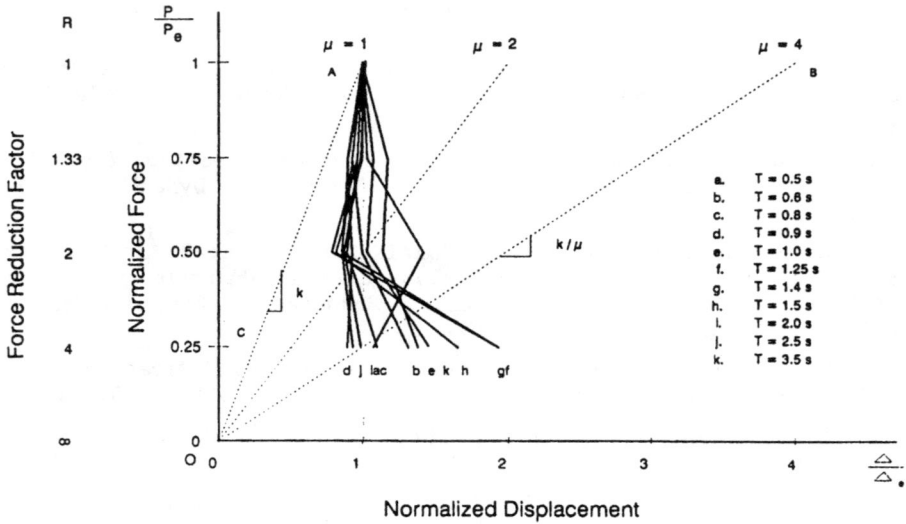

Figure 4. Normalized inelastic response envelopes for Taft S69E 1952.

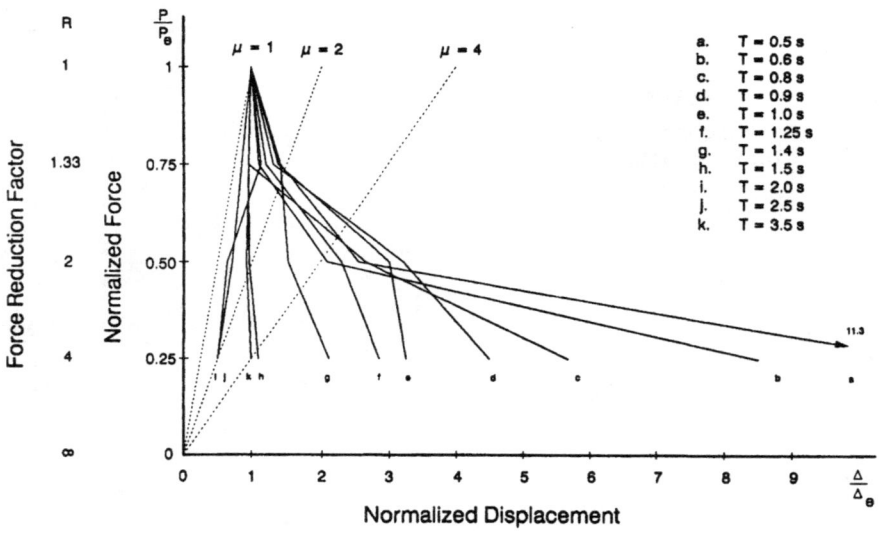

Figure 5. Normalized inelastic response envelopes for Mexico City SCT EW 1985.

66

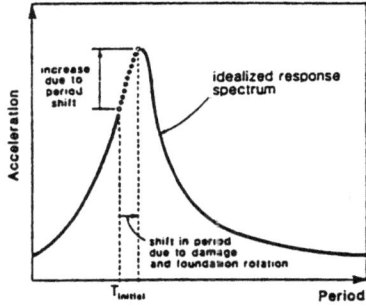

Figure 6. Energy increase from shift in period.

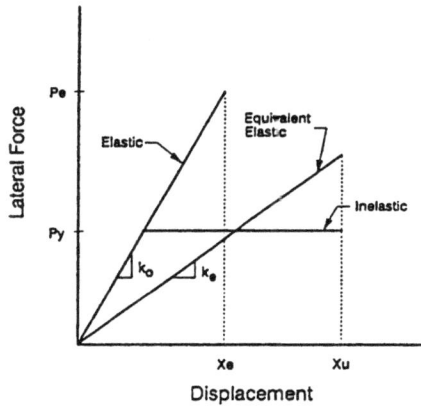

Figure 7. Equivalent elastic response.

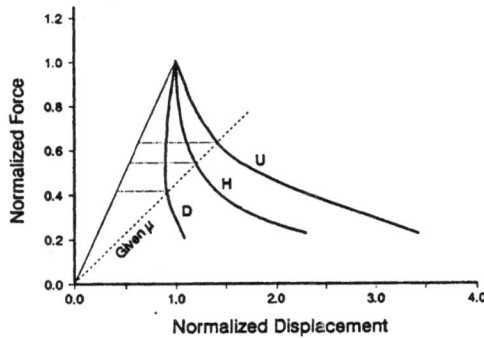

Figure 8. Types of inelastic response envelope.

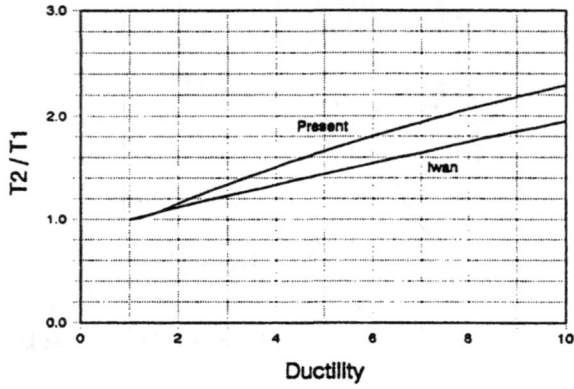

Figure 9. Ratio of equivalent to original periods of vibration.

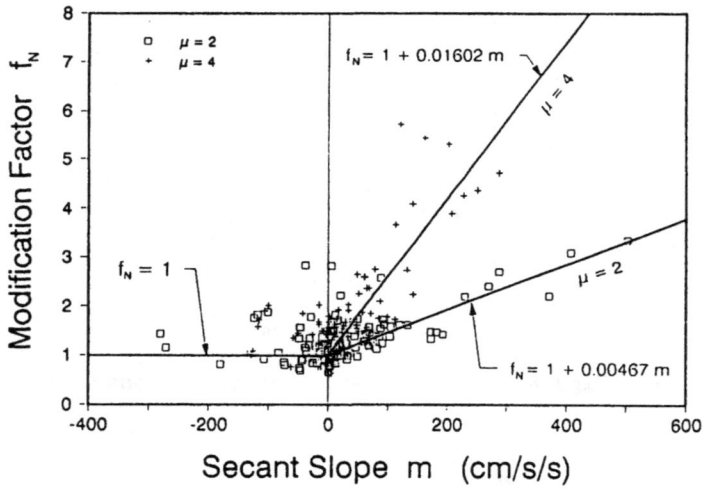

Figure 10. Modification factors for calculated force reduction factors.

On the hysteretic to input energy ratio

Tomaž Vidic[I], Peter Fajfar[II], and Matej Fischinger[III]

ABSTRACT

The hysteretic to input energy ratio E_H/E_I has been studied in a parametric study of SDOF systems. The most important structural parameters were varied and ground motions of very different characteristics were taken into account. When using viscous damping proportional to the instantaneous stiffness, the maximum hysteretic to input energy ratio amounted to 0.9 and 0.8 in the case of 2 and 5 per cent damping, respectively. These values correspond to systems with the natural period shorter than the predominant period of ground motion. They are practically independent of the hysteretic behaviour, the strength (with the exception of very high strength) and the input motion. In the case of a system with the natural period longer than the predominant period of the input motion, E_H/E_I can be approximately determined as a function of the strength of a system and of the characteristics of the input motion using simple formulae proposed in the paper.

INTRODUCTION

The aseismic design philosophy for usual building structures relies strongly on energy dissipation. The energy input to a structure subjected to strong ground motion is dissipated in part by inelastic deformations (hysteretic energy E_H) and, in part, by viscous damping which represents miscellaneous damping effects other than inelastic deformation (damping energy). Only E_H is assumed to contribute to structural damage.

Different investigators have studied input energy (E_I) and proposed procedures or formulae for determining it quantitatively. A very brief review is given in the next chapter. This paper deals predominantly with the ratio of the hysteretic energy to the input energy. This ratio represents a convenient parameter for determination of hysteretic energy, provided that input energy is known.

A few investigators have studied the E_H/E_I ratio recently. Zahrah and Hall (1982) have found out that the proportion of input energy dissipated by yielding increases as viscous damping decreases and as the displacement ductility of a system increases. Similar observations have been made by Akiyama (1985). Based on these observations he proposed a formula for determining E_H/E_I as a function of viscous damping and strength rather than ductility. As a rough approximation, Akiyama proposed a simplified equation, neglecting the influence of strength. Another expression for E_H/E_I as a function of viscous damping and cumulative ductility was proposed by Kuwamura and Galambos (1989). Nariyuki et al (1989) have observed that the relation E_H/E_I versus period is not influenced by the difference in the earthquake ground motion provided that the periods are properly scaled. In this paper results of a parametric study of single-degree-of-freedom (SDOF) systems are presented. Based on these results simple formulae for determining the E_H/E_I ratio are proposed.

I   Graduate Student, II Professor   University of Ljubljana, Department of Civil Engineering,
III Assistant Professor   Jamova 2, Ljubljana, Slovenia, Yugoslavia

INPUT ENERGY

From research carried out by different investigators (Akiyama 1985, Zahrah and Hall 1982, Fajfar et al 1989) is known that maximum input energy, which is imparted to systems with fundamental periods in the vicinity of the predominant period of the ground motion, is a stable parameter. It is only scarcely dependent on the strength of the structure, on hysteresis and on damping. A shift of the input energy curves to the shorter periods occurs as the strength and/or the "fatness" of the hysteresis loops decrease. The shift can be physically explained by a change in the effective period due to inelastic behaviour. It is larger in the case of strongly non-linear behaviour, typical for a system with lower strength and hysteresis with low "fatness". When a proper scaling of the accelerograms is used, the maximum input energy is not very dependent on ground motion, either (Fajfar et al 1989, 1990).

Formulae for the numerical calculation of input energy as a function of ground motion parameters have been proposed by Kuwamura and Galambos (1989), Fajfar et al (1989), and Uang and Bertero (1990). A preliminary comparative statistical study performed by the authors has indicated that the best prediction of the input energy can be generally obtained by a corrected version of the formula proposed by Kuwamura and Galambos. This formula, however, defines input energy as a function of the integral of the squared ground acceleration which may be difficult to predict in a design procedure. According to Fajfar et al, input energy per unit mass can be estimated as a function of the peak ground velocity v_g and the duration of strong ground motion t_D (defined according to Trifunac and Brady 1975, in seconds)

$$E_I / m = 2.2 \, t_D^{0.5} \, v_g^{\,2} \tag{1}$$

Eq. 1 yields acceptable results for the majority of different types of ground motion. Only two of the basic ground motion parameters which can be routinely predicted in the design procedure are included in the formula. The input energy according to Eq. 1 represents the maximum input energy corresponding to the region in the vicinity of the predominant period of the ground motion. The value of this period can be estimated by a formula proposed by Heidebrecht (1987)

$$T_1 = 4.3 \, v_g / a_g \tag{2}$$

where a_g is the peak ground acceleration. Usually it is conservatively assumed (Akiyama 1985, Kuwamura and Galambos 1989) that input energy is constant in the whole medium- and long-period ranges (i.e. for all periods longer than T_1). Uang and Bertero (1990) called attention to the difference between "absolute" and "relative" energy formulations, which is important in the very short and very long period ranges. In this study, the "relative" energy formulation has been used. The relative input energy is defined as the work done by the equivalent force (mass multiplied by ground acceleration) on the equivalent fixed-base system. The effect of the rigid body translation of the structure is neglected.

PARAMETRIC STUDY

To find general purpose expressions for hysteretic to input energy ratio, E_H/E_I, it is necessary to identify the influence of the most important strong motion and structural parameters on the response of a structure. In this study, input ground motion, as well as initial stiffness (period), strength, hysteretic behaviour and damping of SDOF systems were varied. All values of energies were determined at the time equal to the time at the end of the ground motion plus two initial periods of the system. The study represents a continuation of a study on seismic demand performed by the same authors. A detailed description of the parameters can be found in (Fajfar et al 1989). Here, only a very brief overview will be given.

The influence of input motion has been studied using five different groups of records in order to take into account ground motions of basically different types. Standard records from California and records from Montenegro, Yugoslavia, 1979, are representative for "standard" ground motion. The main characteristics of the Friuli, Italy, 1976, and Banja Luka, Yugoslavia, 1981, records is the short duration of the strong ground motion.

The predominant periods of these records are short and fairly narrow-banded, and the peak ground velocity to peak ground acceleration ratios are small. The 1985 Mexico City records represent ground motions of very long duration, with long predominant periods and high peak ground velocity to peak ground acceleration ratios. Totally, 40 horizontal components of records obtained at 20 different stations have been used.

The period range from 0.1 to 2.5 s was considered. The value of the strength parameter η, which is defined as the yield resistance F_y divided by the mass of the system and by the peak ground acceleration a_g

$$\eta = F_y / (m\, a_g) \tag{3}$$

was varied from 1.5 to 0.2. Both mass-proportional viscous damping and viscous damping proportional to instantaneous stiffness (2 and 5 per cent) were used. Altogether, eight different hysteresis models were investigated. Six of them simulate predominantly flexural behaviour: elastic-plastic, bilinear, the Q-model and three variants of Takeda's model with trilinear envelope (they differ according to unloading stiffness and envelope shape). The shear behaviour is simulated by two variants of the shear-slip model.

The main results of the study are presented in Figs. 1 to 4. The hysteretic to input energy ratio is plotted as a function of the initial (elastic) period of the system. In different figures the influence of damping, strength, hysteresis model and input motion are shown. The results of the parametric study of different structural parameters are presented as mean values from 20 "standard" records (the U.S.A. and Montenegro groups). The medium-period range for these records is roughly between periods 0.5 s and 2.0 s. The inelastic system with $\eta = 0.6$, Q-hysteresis and 5 per cent of damping proportional to instantaneous stiffness was chosen as the basic "average inelastic system" which is supposed to represent an average reinforced concrete structure designed according to the codes. In this system, strength, hysteresis and damping were varied one by one.

As shown in the figures, the E_H/E_I ratio has generally its peak values in the short-period range, where the periods are shorter than the predominant period of ground motion T_1. In the medium- and long-period range, where the periods are longer than T_1, the E_H/E_I ratio decreases as the period increases. The zero value of E_H/E_I, indicating elastic behaviour, is reached at the period which will be denoted by T_0.

One of the most important parameters influencing the E_H/E_I ratio is damping. The result of a decrease in viscous damping is an increase of the E_H/E_I ratio in the whole period range (Fig. 1). It is important to realize that the E_H/E_I ratio is strongly influenced by the mathematical modeling of viscous damping. Damping may be assumed to be proportional either to mass or to stiffness. In linear analysis both approaches yield the same results. In nonlinear analysis, however, the stiffness degrades with damage. Consequently, damping related to the instantaneous stiffness tends to decrease, and damping related to the mass tends to increase with degrading of stiffness. In the case of inelastic analysis mass-proportional damping dissipates more energy and is thus more effective in reducing hysteretic energy. This tendency is larger for a system in the short-period range. Systems with longer period experience less stiffness degradation and the influence of the damping model is less important. Similar conclusions have been obtained by Otani (1981). He stated that it was not probable to expect mass-proportional damping in a real reinforced

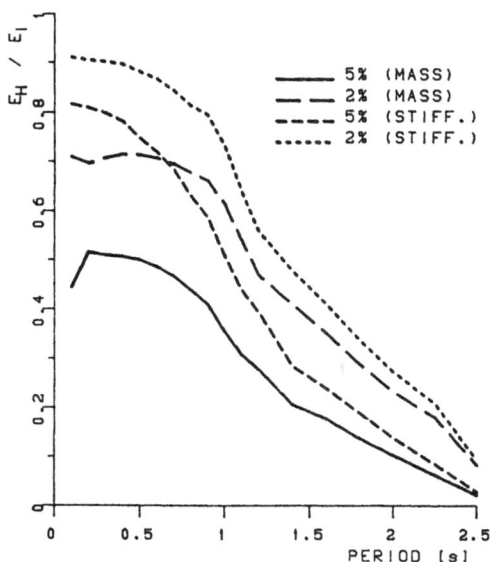

Figure 1. E_H/E_I versus T. Influence of damping

71

concrete structure. Consequently, in our study mainly the stiffness-proportional damping was used. This decision was supported by the observation that the stiffness-proportional damping eliminated the influence of the hysteretic behaviour to a great extent (Fig. 2).

The authors believe that the concept of viscous damping in nonlinear analysis is questionable. In the nonlinear range, it seems to be reasonable to define the damping, which includes the effect of energy dissipation other than hysteretic energy, as a portion of the input energy. More experimental and analytical work on this subject is needed.

The influence of different hysteretic behaviour can be observed in Fig. 2. It can be seen that the type of the hysteretic behaviour is very important in the case of the mass-proportional damping (Fig. 2b) and has a surprisingly small influence in the case of the instantaneous-stiffness-proportional damping (Fig. 2a). An important difference can be observed only in the case of Takeda's hysteretic rules where trilinear envelopes were used. The period T in the figures corresponds to the initial stiffness, i.e. the stiffness before cracking in the case of a trilinear envelope. If an equivalent stiffness or the yield stiffness was considered at the horizontal axis, the difference between the two sets of curves, corresponding to bilinear and trilinear envelopes, would disappear.

The influence of the strength of a system can be seen in Fig. 3. Usually it has been believed that E_H/E_I decreases with an increase in strength (or with a decrease in ductility). It was surprising to find out that this is not always the case. The maximum E_H/E_I ratio is, with exception of a system with very high strength ($\eta > 1$), a reasonably stable quantity practically independent of the strength of a system. The decrease of the E_H/E_I ratio in the medium- and long-period range, however, strongly depends on the strength. A system with low strength experiences large inelastic excursions and dissipates more hysteretic energy than a system with high strength and the same stiffness.

Mean values of E_H/E_I for different groups of records are shown in Fig. 4. It can be seen that the maximum E_H/E_I values are more or less the same for all groups of records. The values of the predominant period T_1 and of the period T_0, however, strongly depend on the type of ground motion.

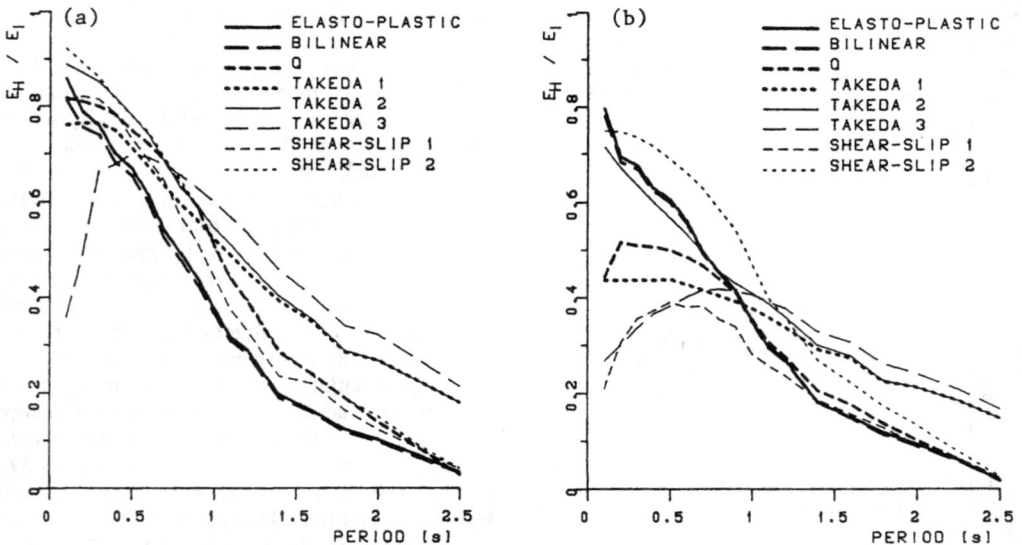

Figure 2. E_H/E_I versus T. Influence of hysteretic behaviour.
(a) Damping proportional to instantaneous stiffness. (b) Damping proportional to mass

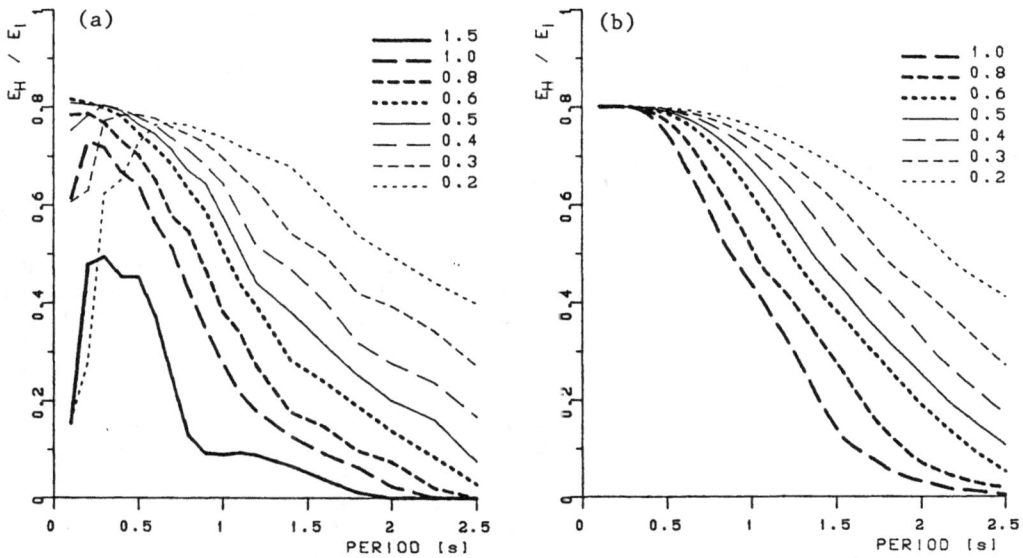

Figure 3. E_H/E_I versus T. Influence of the strength of a system.
(a) Mean curves for 20 "standard" records
(b) Mean of curves determined according to Eq. 4 for "standard" records

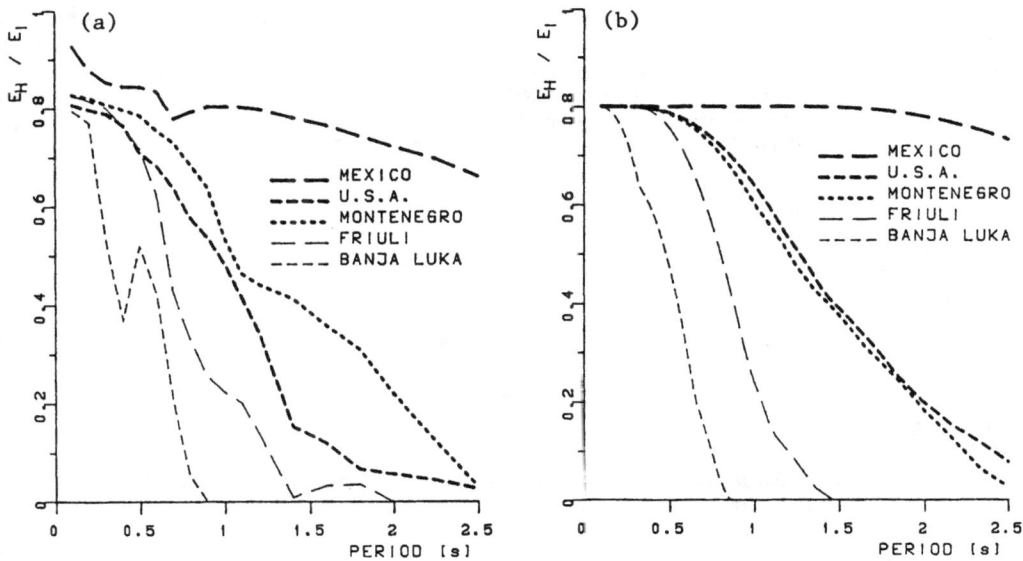

Figure 4. E_H/E_I versus T. Influence of input motion.
(a) Mean curves for different group of records
(b) Mean curves determined according to Eq. 4 for different groups of records

73

APPROXIMATE PROCEDURE FOR DETERMINATION OF THE E_H/E_I RATIO

Based on observations obtained in the parametric study a simple approximate relation between E_H/E_I and period T is proposed (Fig. 5)

$$\frac{E_H}{E_I} = \begin{cases} C & T \le T_1 & (4a) \\ C\,(T_o^2 - T^2 - 2T_oT_1 + 2T_1T)/(T_o - T_1)^2 & T_1 \le T \le T_o & (4b) \\ 0 & T \ge T_o & (4c) \end{cases}$$

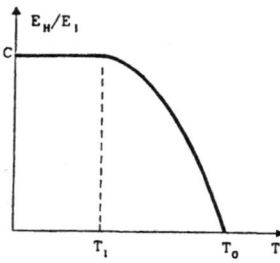

Figure 5. E_H/E_I versus T. Proposed relation

T_1 is the transition period between the short- and medium-period region (Eq. 8) which is usually considered as the predominant period of the ground motion (Eq. 2). T_o is the period at the transition between inelastic ($T < T_o$) and elastic behaviour ($T > T_o$). It depends on the strength of a system and on the characteristics of ground motion, as discussed in detail in the next subchapter. C is a constant depending on damping and, in the case of high strength, on strength. In the case of $\eta \le 1.0$ the values $C = 0.8$ and $C = 0.9$ are proposed for 5 and 2 per cent of instantaneous-stiffness-proportional damping, respectively. The choice of the quadratic function in the region between T_1 and T_o is based on empirical observations.

Determination of T_o

The zero value of the E_H/E_I ratio, indicating elastic behaviour, is reached at the period which was denoted by T_o. The elastic spectra of the Newmark-Hall type can be used in the process of determination of T_o. In the Newmark-Hall spectra (Newmark and Hall, 1982), the whole frequency range is divided into three ranges: the short-period, the medium-period and the long-period range. The limits between the three regions are defined by the transition periods T_1 and T_2. In these regions the pseudo-accelerations A_p, the pseudo-velocities V_p and the relative displacements D are determined as the products of the ground motion parameters (maximum ground acceleration a_g, velocity v_g, and displacement d_g) and the corresponding elastic amplification factors c_a, c_v and c_d (T is the period of the system):

$$A_p = c_a a_g \qquad T \le T_1 \tag{5}$$
$$V_p = c_v v_g \qquad T_1 \le T \le T_2 \tag{6}$$
$$D = c_d d_g \qquad T \ge T_2 \tag{7}$$

The transition periods are expressed by

$$T_1 = 2\pi\,(c_v v_g)/(c_a a_g), \qquad T_2 = 2\pi\,(c_d d_g)/(c_v v_g) \tag{8,9}$$

The following relations between A_p, V_p and D are known

$$V_p = D\,(2\pi/T), \qquad A_p = V_p\,(2\pi/T) = D\,(2\pi/T)^2 \tag{10,11}$$

The behaviour of a bilinear SDOF oscillator will be elastic, if the yield strength F_y is at least equal to the product $m A_p$ (m is the mass of the system)

$$F_y = m A_p \tag{12}$$

or larger. Using Eq. 3, Eq. 12 may be written as

$$\eta a_g = A_p \tag{13}$$

74

In the derivation of the formula for T_o two possibilities will be considered. It will be assumed that T_o is in the medium- and long-period range, respectively. First, general expressions will be derived. Then, constants proposed by the authors will be employed.

a) Medium-period region ($T_1 \leq T_o \leq T_2$)

From Eqs. 6, 11 and 13 it follows

$$T_o = (2 \pi c_v / \eta) (v_g / a_g) \qquad (14)$$

Using Eq. 8, Eq. 14 can be rewritten

$$T_o = (c_a / \eta) T_1 \qquad (15)$$

Eqs. 14 and 15 represent two forms of a general formula for T_o, provided that T_o is located in the medium-period region. According to Newmark and Hall (1982) the mean amplification factor c_a amounts to 2.12 in the case of 5 % damping. In the study reported in this paper, the amplification factor

$$c_a = 1.56 \, t_D^{0.25} \qquad (16)$$

where t_D is the duration of strong ground motion in seconds, was used. Eq. 16 follows indirectly from the results of the parametric study performed by the authors (Fajfar et al 1989) and from Eq. 2.

b) Long-period region ($T_o \geq T_2$)

Formulae derived here will be used, if T_o determined according to Eqs. 14 or 15 is greater than T_2. From Eqs. 7, 11 and 13 it follows

$$T_o^2 = (4 \pi^2 c_d / \eta) (d_g / a_g) \qquad (17)$$

Using Eqs. 8 and 9, Eq. 17 becomes

$$T_o^2 = (c_a / \eta) T_1 T_2 \qquad (18)$$

In our study, c_a according to Eq. 16, T_1 according to Eq. 2, and T_2 according to Eq. 19 (Fajfar et al 1989) were used.

$$T_2 = (13 / t_D^{0.25}) (d_g / v_g) \qquad (19)$$

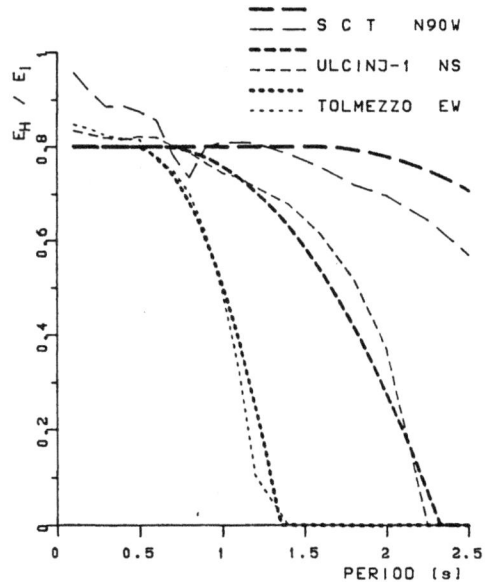

Figure 6. E_H/E_I versus T. Comparison of "exact" relations (thin lines) and curves obtained according to Eq. 4 (thick lines) for three records (SCT, Mexico City 1985; Ulcinj, Montenegro 1979; Tolmezzo, Friuli 1976)

VERIFICATION OF THE APPROXIMATE PROCEDURE

The proposed procedure for determining the E_H/E_I ratio has been verified by a comparison of approximate results with the E_H/E_I values computed for a number of strong ground motions. In Fig. 6 the results for the "average inelastic system" ($\eta = 0.6$, Q - hysteresis, 5 % damping) subjected to three very different ground motions are shown. A comparison of mean curves for systems with different strength subjected to 20 "standard" records is shown in Fig. 3. A similar comparison is shown in Fig. 4, where the influence of different groups of records can be examined. A good correlation can be observed in all figures. An exception represents the group of Californian records. The less favourable correlation in this case might be attributed to the choice of the high-pass filter in the correction procedure for accelerograms, which has an important influence on the maximum ground displacement d_g (Fajfar et al 1989) and therefore also on T_o, if located in the long-period region (Eq. 17).

CONCLUSIONS

Based on results of a parametric study of inelastic SDOF systems, the following main conclusions and results have been obtained.

1. The hysteretic to input energy ratio E_H/E_I depends on the amount of viscous damping and also on the modeling of damping. The damping proportional to instantaneous stiffness seems to simulate the real structural behaviour more realistically than the mass-proportional damping. In addition, it practically eliminates the influence of the hysteretic behaviour on E_H/E_I.

2. The maximum values of E_H/E_I are observed in the short-period systems with the natural period shorter than the predominant period of the ground motion T_1. They practically do not depend on the input motion, on the strength of the system (with the exception of a system with high strength $\eta > 1$), and in the case of the instantaneous-stiffness-proportional damping, on hysteretic behaviour. Values of E_H/E_I for usual inelastic systems ($\eta \le 1$) amount to 0.8 and 0.9 in the case of 5 and 2 per cent damping proportional to instantaneous stiffness, respectively. It is reasonable to assume a constant value of E_H/E_I in the short-period region.

3. In systems with the natural period longer than T_1 (medium- and long-period systems) the value of E_H/E_I decreases as T increases. The zero value is reached at the period T_o. The ratio T_o/T_1 depends mainly on the strength of a system. The influence of the characteristics of ground motion has been also observed. Simple formulae for T_o have been derived. The E_H/E_I ratio between the periods T_1 and T_o can be approximated by a quadratic function.

AKNOWLEDGEMENTS

The results presented in this paper are based on work supported by the Ministry for Research of Slovenia and by the U.S.-Yugoslav Joint Fund for Scientific and Technological Cooperation, in cooperation with the NSF and University of California, Berkeley (co-principal investigator S. Mahin, Grant JF 797 NSF), as well as Stanford University (co-principal investigator H. Krawinkler, Grant JF 794 NSF).

REFERENCES

Akiyama, H. 1985. Earthquake-resistant limit-state design for buildings. University of Tokyo Press.

Fajfar, P., Vidic, T. and Fischinger, M. 1989. Seismic demand in medium- and long-period structures, Earthquake Eng. Struct. Dyn., 18, 1133-1144.

Fajfar, P., Vidic, T. and Fischinger, M. 1990. A measure of earthquake motion capacity to damage medium-period structures. Soil Dyn. Earthquake Eng., 9, 236-242.

Heidebrecht, A.C. 1987. Private communication.

Kuwamura, H. and Galambos, T.V. 1989. Earthquake load for structural reliability. J. Struct. Engrg., ASCE, 115, 1446-1462.

Nariyuki, Y., Hirao, K. and Ohgishi, K. 1989. Study on relation between Fourier spectra of earthquake motion & energy response spectra of SDOF systems. Proc. 2nd East Asia-Pacific Conf. on Structural Engineering & Constructions, Chiang Mai, 1503-1509.

Newmark, N.M. and Hall, W.J. 1982. Earthquake spectra and design. Earthquake Engineering Research Center, Berkeley, CA.

Otani, S. 1981. Hysteresis models of reinforced concrete for earthquake response analysis. Journal of the Faculty of Engineering, The University of Tokyo, 36, 125-159.

Trifunac, M.D. and Brady, A.G. 1975. A study of the duration of strong earthquake ground motion. Bull. of the Seism. Soc. of Am., 65, 581-626.

Uang, C.-M. and Bertero, V.V. 1990. Evaluation of seismic energy in structures. Earthquake Eng. Struct. Dyn., 19, 77-90.

Zahrah, T.F. and Hall, W.J. 1982. Seismic energy absorption in simple structures. Structural Research Series No. 501, Civil Engineering Studies, University of Illinois, Urbana-Champaign, Illinois.

Application of perturbation techniques in the analysis of braced frames

Motohide Tada[I], Kazuo Inoue[II], and Susumu Kuwahara[III]

ABSTRACT

The efficiency of applying the incremental perturbation method to combined nonlinear problems was particularly reported by Yokoo, Nakamura and Uetani. (1976) And this successful approach has been extended by Ishida and Morisako (1987) to one dimensional finite element method.

For the sake of studying a macro behaviour of steel frames, this paper deals with the formulation of an ordered set of perturbation equation for a combined nonlinear constitutive equation using strain hardening general yield hinge model to lead to more efficiency on the capacity of computers, a higher speed of calculation and simpler procedures in modeling.

Some examples of braced frames are analysed by the present method, and good correspondence between the results of test and those of analysis on post buckling load - displacement relationship is successfully achieved.

INTRODUCTION

While material and geometrical nonlinear analysis for steel skeleton structures has been done by several investigators, using a series of linear approximations within small increments, the efficiency of applying the incremental perturbation method to combined nonlinear problems was particularly reported by Yokoo, Nakamura and Uetani.(1976) This successful approach has been extended by Ishida and Morisako (1987) to one dimensional finite element method.

However, strain hardening general yield hinge method (Inoue and Ogawa 1978) proves to be more useful than one dimensional finite element method in studying a macro behaviour of frames in terms of the capacity of computers, the speed of calculation and the simple procedures in modeling.

Thus this paper deals with the formulation of an ordered set of perturbation equation for a combined nonlinear constitutive equation using strain hardening general yield hinge model, and some examples of analysis are shown.

I   Assistant Professor , Osaka University , Japan

II Associate Professor , Osaka University , Japan

III Graduate Student , Osaka University , Japan

FORMULATION OF PERTURBATION EQUATIONS

Primary Conditions

The primary conditions for the present analysis are as follows.
1) As a deflection function for each member, a linear expression for the longitudinal direction and the third order polynomial expression for transverse direction are assumed.
2) Yielding can occur only at the ends of the member, and the behaviour during plastic deformation will be characterized by the kinematic hardening rule by Prager. (1955)

In this paper, the m-th coefficient of Taylor series of any variable x will be denoted by $\overset{(m)}{x}$. So x can be expanded about a parameter t as follows.

$$x = \overset{(0)}{x} + \overset{(1)}{x} t + \overset{(2)}{x} t^2 + \cdots\cdots\cdots \tag{1}$$

Hence the following relation is given by differentiation.

$$\left.\frac{d^m x}{d t^m}\right|_{t=0} = m!\, \overset{(m)}{x} \tag{2}$$

Elastic Constitutive Equation

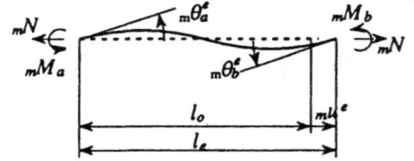

Figure 1

The elastic relation of member-end forces and member-end elastic deformations are provided by following equations, considering the effect of shortening by bending. (Jennings 1963)

$$_mN = EA\left[\frac{_m u^e}{l_e} + \frac{1}{30}\left\{2\left(_m\theta_a^e\right)^2 - {_m\theta_a^e}\,{_m\theta_b^e} + 2\left(_m\theta_b^e\right)^2\right\}\right] \tag{3a}$$

$$_mM_a = \left[\frac{4EI}{l_e} + \frac{2EA}{15}\,_m u^e + \frac{EA\,l_e}{225}\left\{2\left(_m\theta_a^e\right)^2 - {_m\theta_a^e}\,{_m\theta_b^e} + 2\left(_m\theta_b^e\right)^2\right\}\right]_m\theta_a^e$$
$$+ \left[\frac{2EI}{l_e} - \frac{EA}{30}\,_m u^e - \frac{EA\,l_e}{900}\left\{2\left(_m\theta_a^e\right)^2 - {_m\theta_a^e}\,{_m\theta_b^e} + 2\left(_m\theta_b^e\right)^2\right\}\right]_m\theta_b^e \tag{3b}$$

$$_mM_b = \left[\frac{2EI}{l_e} - \frac{EA}{30}\,_m u^e - \frac{EA\,l_e}{900}\left\{2\left(_m\theta_a^e\right)^2 - {_m\theta_a^e}\,{_m\theta_b^e} + 2\left(_m\theta_b^e\right)^2\right\}\right]_m\theta_a^e$$
$$+ \left[\frac{4EI}{l_e} + \frac{2EA}{15}\,_m u^e + \frac{EA\,l_e}{225}\left\{2\left(_m\theta_a^e\right)^2 - {_m\theta_a^e}\,{_m\theta_b^e} + 2\left(_m\theta_b^e\right)^2\right\}\right]_m\theta_b^e \tag{3c}$$

Where superscript e denotes the elastic component of the deformation. Let $_1d^e$ and $_1p$ define by the following equations.

$$_1d^e = \left\{_m u^e,\, {_m\theta_a^e},\, {_m\theta_b^e}\right\}^T \qquad\qquad _1p = \left\{_mN,\, {_mM_a},\, {_mM_b}\right\}^T \tag{4, 5}$$

Let the right term of Eqs.3 be function f_i, rewriting them as follows.

$$_1p_i = f_i\left(_m u^e,\, {_m\theta_a^e},\, {_m\theta_b^e}\right) \quad,\quad (i = 1, 2, 3) \tag{6}$$

With Taylor expanding of Eq.6 about a parameter t which denotes the progress of time, and with comparison among the same ordered terms up to the third order, the following equations are obtained.

$$_1\overset{(0)}{p}_i = \overset{(0)}{f}_i \qquad\qquad _1\overset{(1)}{p}_i = \sum_{j=1}^{3} \frac{\partial f_i}{\partial\,_1d_j^e}\,_1\overset{(1)}{d}_j^e \tag{7a, 7b}$$

78

$$\overset{(2)}{_1p_i} = \sum_{j=1}^{3} \frac{\partial f_i}{\partial _1d_j^e}\,\overset{(2)}{_1d_j^e} + \frac{1}{2!}\sum_{j=1}^{3}\sum_{k=1}^{3} \frac{\partial^2 f_i}{\partial _1d_j^e\,\partial _1d_k^e}\,\overset{(1)}{_1d_j^e}\,\overset{(1)}{_1d_k^e} \tag{7c}$$

$$\overset{(3)}{_1p_i} = \sum_{j=1}^{3} \frac{\partial f_i}{\partial _1d_j^e}\,\overset{(3)}{_1d_j^e} + \frac{1}{2!}\sum_{j=1}^{3}\sum_{k=1}^{3} \frac{\partial^2 f_i}{\partial _1d_j^e\,\partial _1d_k^e}\left(\overset{(1)}{_1d_j^e}\,\overset{(2)}{_1d_k^e} + \overset{(2)}{_1d_j^e}\,\overset{(1)}{_1d_k^e}\right)$$
$$+ \frac{1}{3!}\sum_{j=1}^{3}\sum_{k=1}^{3}\sum_{l=1}^{3} \frac{\partial^3 f_i}{\partial _1d_j^e\,\partial _1d_k^e\,\partial _1d_l^e}\,\overset{(1)}{_1d_j^e}\,\overset{(1)}{_1d_k^e}\,\overset{(1)}{_1d_l^e} \tag{7d}$$

It is noted that Eq.7a expresses the equilibrium condition at $t = 0$. Matrix expression of Eqs.7 provides the following ordered set of perturbation equation about the elastic constitutive equation;

$$\overset{(m)}{_1p} = K^e\,\overset{(m)}{_1d^e} + \overset{(m)*1}{_1p} \tag{8}$$

where

$$\overset{(m)*1}{_1p} = \sum_{n=1}^{m-1} \frac{m-n}{(n+1)!}\,\overset{n}{_2S}\,\overset{(m-n)}{_1d^e} \quad (m \ge 2) \tag{9}$$

and K^e, $_2^1S$, $_2^2S$ are the matrices whose ij component is defined as follows.

$$K_{ij}^e = \frac{\partial f_i}{\partial _1d_j^e} \qquad \overset{1}{_2S}_{ij} = \left\{ \frac{\partial^2 f_i}{\partial _1d_j^e\,\partial _1d_1^e} \quad \frac{\partial^2 f_i}{\partial _1d_j^e\,\partial _1d_2^e} \quad \frac{\partial^2 f_i}{\partial _1d_j^e\,\partial _1d_3^e} \right\}\overset{(1)}{_1d^e} \tag{10a, 10b}$$

$$\overset{2}{_2S}_{ij} = \left\{ \overset{2}{_2S}_{ij1} \quad \overset{2}{_2S}_{ij2} \quad \overset{2}{_2S}_{ij3} \right\}\overset{(1)}{_1d^e}$$

$$\overset{2}{_2S}_{ijk} = \left\{ \frac{\partial^3 f_i}{\partial _1d_j^e\,\partial _1d_k^e\,\partial _1d_1^e} \quad \cdots \quad \frac{\partial^3 f_i}{\partial _1d_j^e\,\partial _1d_k^e\,\partial _1d_3^e} \right\}\overset{(1)}{_1d^e} \tag{10c}$$

Plastic Constitutive Equation

Now we consider the case when yield hinges form at either end of the member. Let the member-end deformation, $_1d = (_mu,\ _m\theta_a,\ _m\theta_b)^T$, and member-end force be divided into components at each end, a and b. And we define as follows;

$$_1d_{(a)} = \left\{\ _mu_a \quad _m\theta_a\ \right\}^T \qquad _1d_{(b)} = \left\{\ _mu_b \quad _m\theta_b\ \right\}^T \tag{11a, 11b}$$

$$_1p_{(a)} = \left\{\ _mN \quad _mM_a\ \right\}^T \qquad _1p_{(b)} = \left\{\ _mN \quad _mM_b\ \right\}^T \tag{12a, 12b}$$

where

$$_mu = _mu_a + _mu_b \tag{13}$$

According to to Prager (1955), if the yield surface is drawn with the axes of the stresses divided by yield stresses and those of plastic strains multiplied by yield stresses, the subsequent yield surface will move parallel in the direction of plastic strain without changing its shape and scale as Fig.2. Yield surfaces are defined at both a and b ends of the member, and following equation is provided when yielding occurs at both ends.

$$F\left(_1p_{(a,b)} - \alpha_{(a,b)}\right) = 0 \tag{14}$$

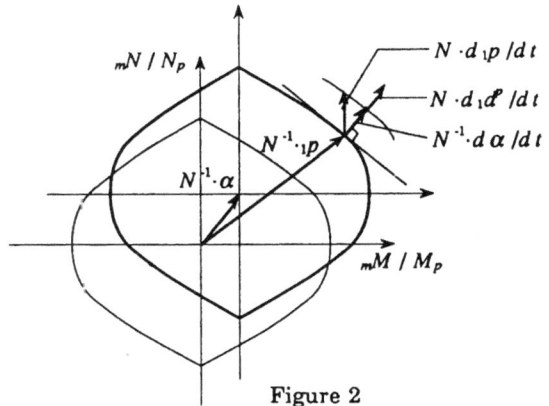

Figure 2

The sign (a,b) in Eq.14 expresses that similar equations are provided for both a and b ends. $\alpha_{(a)}$ and $\alpha_{(b)}$ express the movement vector of the origin of the yield surface respectively at a end and b end and according to Prager's kinematic hardening rule they are related to plastic deformation velocity $_1 d^p/dt$ with scalar amount μ, that is;

$$\frac{d\,\alpha_{(a,b)}}{dt} = \mu_{(a,b)}\,N\,N\,\frac{d_1 d^p_{(a,b)}}{dt} \tag{15}$$

where N is the diagonal matrix with the components of yield stresses N_p and M_p. The yield condition that the stress point remains on the subsequent yield surface provides the following equation.

$$\left|\frac{d_1 d^p_{(a,b)}}{dt}\right|^T \left(\frac{d\,\psi_{(a,b)}}{dt} - \frac{d\,\alpha_{(a,b)}}{dt}\right) = 0 \tag{16}$$

Taylor expansion of Eq.15 provides an ordered set of perturbation equation as follows;

$$\overset{(m)}{\alpha}_{(a,b)} = \frac{1}{m}\,N\,N\,\overset{(m-1)}{\mu}_{(a,b)}\,\,{}_1\overset{(1)}{d}^p_{(a,b)} + \overset{(m)*1}{\alpha}_{(a,b)} \tag{17}$$

where

$$\overset{(m)*1}{\alpha}_{(a,b)} = N\,N\sum_{n=2}^{m}\frac{n}{m}\,\overset{(m-n)}{\mu}_{(a,b)}\,\,{}_1\overset{(n)}{d}^p_{(a,b)} \quad (m \geq 2) \tag{18}$$

Similarly, Eq.16 may be expanded to Taylor series and the substitution of Eq.17 provides the following ordered set of equation about μ ;

$$\overset{(m-1)}{\mu}_{(a,b)} = \left({}_1\overset{(1)}{d}^p_{(a,b)}{}^T\,N\,N\,\,{}_1\overset{(1)}{d}^p_{(a,b)}\right)^{-1}\,m\,\,{}_1\overset{(1)}{d}^p_{(a,b)}{}^T\,{}_1\overset{(m)}{p}_{(a,b)} + \overset{(m-1)*1}{\mu}_{(a,b)} \tag{19}$$

where

$$\overset{(m-1)*1}{\mu}_{(a,b)} = \left({}_1\overset{(1)}{d}^p_{(a,b)}{}^T\,N\,N\,\,{}_1\overset{(1)}{d}^p_{(a,b)}\right)^{-1}\left\{-m\,\,{}_1\overset{(1)}{d}^p_{(a,b)}{}^T\,\overset{(m)*1}{\alpha}_{(a,b)}\right.$$
$$\left. + \sum_{n=2}^{m} n\,(m-n+1)\,\,{}_1\overset{(n)}{d}^p_{(a,b)}{}^T\left({}_1\overset{(m-n+1)}{p}_{(a,b)} - \overset{(m-n+1)}{\alpha}_{(a,b)}\right)\right\} \quad (m \geq 2) \tag{20}$$

The velocity of the member-end deformation $d_1 d/dt$ is given as a summation of the elastic component $d_1 d^e/dt$ and the plastic component $d_1 d^p/dt$, that is

$$\frac{d_1 d}{dt} = \frac{d_1 d^e}{dt} + \frac{d_1 d^p}{dt} \tag{21}$$

If the stress points at both a and b ends are on the singular point of the yield surface (for example, on the intersection of $F_1 = 0$ and $F_2 = 0$), the velocity of plastic deformation is provided as follows according to the generalized plastic flow rule;

$$\frac{d_1 d^p}{dt} = \phi\,\frac{d\,\zeta}{dt} \tag{22a}$$

that is

$$\begin{Bmatrix} \dfrac{d_m\mu^p}{dt} \\[2mm] \dfrac{d_m\theta_a^p}{dt} \\[2mm] \dfrac{d_m\theta_b^p}{dt} \end{Bmatrix} = \begin{bmatrix} \dfrac{\partial F_{1(a)}}{\partial_m N} & \dfrac{\partial F_{2(a)}}{\partial_m N} & \dfrac{\partial F_{1(b)}}{\partial_m N} & \dfrac{\partial F_{2(b)}}{\partial_m N} \\[2mm] \dfrac{\partial F_{1(a)}}{\partial_m M_a} & \dfrac{\partial F_{2(a)}}{\partial_m M_a} & 0 & 0 \\[2mm] 0 & 0 & \dfrac{\partial F_{1(b)}}{\partial_m M_b} & \dfrac{\partial F_{2(b)}}{\partial_m M_b} \end{bmatrix} \begin{Bmatrix} \dfrac{d\,\zeta_{1(a)}}{dt} \\[2mm] \dfrac{d\,\zeta_{2(a)}}{dt} \\[2mm] \dfrac{d\,\zeta_{1(b)}}{dt} \\[2mm] \dfrac{d\,\zeta_{2(b)}}{dt} \end{Bmatrix} \tag{22b}$$

When the yield hinge formation is different from that in the above case, $d_1 d^P/dt$ may be provided by adequate squeezing of ϕ and $d\zeta/dt$.

Let $d\alpha_o/dt$ be defined by the following equation as a product of the elastic stiffness matrix K^e, the deformation velocity $d_1 d/dt$ and strain hardening coefficient τ.

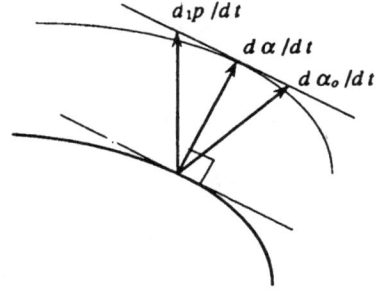

Figure 3

$$\frac{d\alpha_o}{dt} = \tau K^e \frac{d_1 d}{dt} \tag{23}$$

Although $d\alpha_o/dt$ is generally unequal to $d\alpha/dt$ except in the case of uniaxial loading, let $d\alpha/dt$ approximate to the normal component of $d\alpha_o/dt$ (normal to the yield surface). In this case, the yield condition that the stress point remains on the subsequent yield surface may be written as follows. (refer to Fig.3)

$$\phi^T \left(\frac{d_1 p}{dt} - \frac{d\alpha_o}{dt} \right) = 0 \tag{24}$$

Taylor expansion of Eq.22 provides the following ordered set of equation;

$$_1 \overset{(m)}{d}{}^P = \overset{(0)}{\phi}\ \overset{(m)}{\zeta} +_1 \overset{(m)^{\bullet}1}{d}{}^P \tag{25}$$

where

$$_1 \overset{(m)^{\bullet}1}{d}{}^P = \sum_{n=1}^{m-1} \frac{n}{m}\ \overset{(m-n)}{\phi}\ \overset{(n)}{\zeta} \tag{26}$$

Similarly, from Eq.21 the following equation is provided.

$$_1 \overset{(m)}{d}{}^e =_1 \overset{(m)}{d} -_1 \overset{(m)}{d}{}^P \tag{27}$$

By substituting Eqs.25 and 27 for Eq.8, $_1 \overset{(m)}{p}$ is expressed as follows.

$$_1 \overset{(m)}{p} = K^e \left(_1 \overset{(m)}{d} - \overset{(0)}{\phi}\ \overset{(m)}{\zeta} \right) - K^e {}_1 \overset{(m)^{\bullet}1}{d}{}^P +_1 \overset{(m)^{\bullet}1}{p} \tag{28}$$

From Eqs.23 and 24, the following equations are respectively provided.

$$\overset{(m)}{\alpha}_o = \tau K^e {}_1 \overset{(m)}{d} \tag{29}$$

$$m \overset{(0)}{\phi}{}^T {}_1 \overset{(m)}{p} + \sum_{n=1}^{m-1} n \overset{(m-n)}{\phi}{}^T {}_1 \overset{(n)}{p} = m \overset{(0)}{\phi}{}^T \overset{(m)}{\alpha}_o + \sum_{n=1}^{m-1} n \overset{(m-n)}{\phi}{}^T \overset{(n)}{\alpha}_o \tag{30}$$

Substituting Eqs.28 and 29 for Eq.30 provides the following equation;

$$\overset{(m)}{\zeta} = (1 - \tau) C^{-1} \overset{(0)}{\phi}{}^T K^e {}_1 \overset{(m)}{d} + \overset{(m)^{\bullet}1}{\zeta} \tag{31}$$

where

$$\overset{(m)^{\bullet}1}{\zeta} = C^{-1} \left\{ - \overset{(0)}{\phi}{}^T K^e {}_1 \overset{(m)^{\bullet}1}{d}{}^P + \overset{(0)}{\phi}{}^T {}_1 \overset{(m)^{\bullet}1}{p} + \sum_{n=1}^{m-1} \frac{n}{m} \overset{(m-n)}{\phi}{}^T \left({}_1 \overset{(n)}{p} - \overset{(n)}{\alpha}_o \right) \right\} \qquad (m > 2) \tag{32}$$

$$C = \overset{(0)}{\phi}{}^T K^e \overset{(0)}{\phi} \tag{33}$$

By substituting Eq.31 for Eq.28, we can obtain the ordered set of the constitutive equation under loading as follows;

$$\overset{(m)}{_1p} = K^p \overset{(m)}{_1d} + \overset{(m)\cdot2}{_1p} \tag{34}$$

where

$$K^p = K^e \left\{ I - (1-\tau) \overset{(0)}{\phi} C^{-1} \overset{(0)}{\phi}{}^T K^e \right\} \tag{35}$$

$$\overset{(m)\cdot2}{_1p} = -K^e \left\{ \overset{(0)}{\phi} \overset{(m)\cdot1}{\zeta} + \overset{(m)\cdot1}{_1d}{}^p \right\} + \overset{(m)\cdot1}{_1p} \tag{36}$$

I : unit matrix

EXAMPLES OF ANALYSIS

The frames considered as examples are the two-story braced frames as shown in Fig.4 which are the specimens for a static alternate loading test we conducted. (Igarashi, 1986) A node is set at the mid-point of each bracing member which is rigidly fixed at ends to the frame. Initial deflection of each bracing member e is set to have the amount of $l/500$ in the lateral direction at the mid-point, where l is the length of the bracing member. The characteristics of the member, either at the panel zone or at the region where the gusset plate is connected, are considered as the stiffness being double and the strength being infinite. The

Table 1

Member	Section	A (cm²)	I (cm⁴)
		Np (ton)	Mp (t cm)
Beam	H - 125 x 125 x 6.5 x 9	27.2	774
		85.7	442
Column	H - 150 x 150 x 7 x 10	36.6	1510
		108	668
Bracing	WH - 80 x 80 x 6 x 6	13.2	49.3
		52.5	58.3

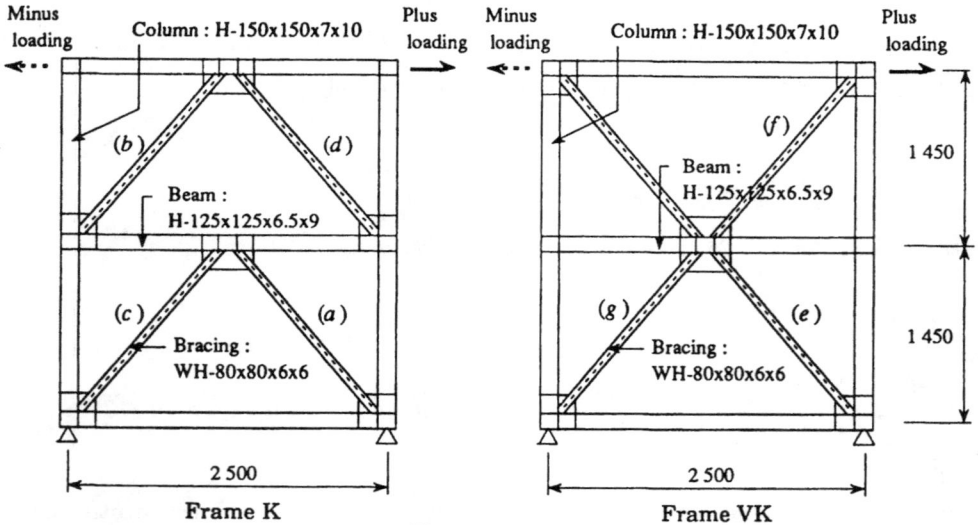

Figure 4

parabolic yield functions are assumed for all members.

Ordered set of perturbation equations up to the third order are considered in the present analysis. The length of increment for step-by-step analysis is determined so that the tolerance limit (Yokoo, 1976) be 0.01 for the horizontal and vertical displacements at the joints, among the column, the beam and the bracing members, and at the mid-point of bracing members. Yielding and unloading of the member and the stress point reaching to the singular point during loading also determine the length of increment.

General stiffness equation is solved by controlling the displacement largest in the previous increment. The direction of controlled displacement is determined so that the yield hinge occurring at the last do not unload. The resulting post buckling behaviour that both load and displacement decrease is successfully achieved and expressed in the analysis.

The results of test and present analysis on the relationship between load and the average relative story rotation are shown in Fig.5. Although the analytical result, compared with the test result, has sharp peaks around the buckling points owing to the hinge method, the relative story rotation and the load level when buckling occurs correspond well with the test result. (a) - (g) in Fig.5 shows the occurrence of buckling. Each buckling member corresponds to the bracing member denoted by (a) - (g) in Fig.4. All buckled members in the analysis correspond with those buckled in the test for the frame K. But, for the frame VK, bracing member (g) has buckled at the first minus-loading in the analysis, while the bracing member (f) has buckled in the test. It is considered that because the stress levels of both (f) and (g) members are nearly equal, the slightest effect on the stress level caused by some deviations from the test specimens and the mathematical models can easily change the collapse mechanism. In fact, the stress level of the member (f), when the member (g) was buckled in the analysis, was $(N/Np)^2 + (M/Mp) = 0.987$, where 1.0 is the yielding stress level. Although the collapse mechanisms are different, the analytical result corresponds well with the test result on the the load - average relative story rotation relationship afterwards in this example.

The load - average relative story rotation relationship for the frame K, analysed under the condition that the initial deflection e is equal to zero, is shown in Fig.6. The initial buckling load rises about 13% more compared to the case of $e = l$ /500. The bracing members (c) and (d) have not buckled. The reason for the member (c) is that the right end of the member (c) yielded at the point of sign $\triangle$ in the figure because of the rising of the initial buckling load and it experienced the plastic deformation in the direction opposite to the buckling mode.

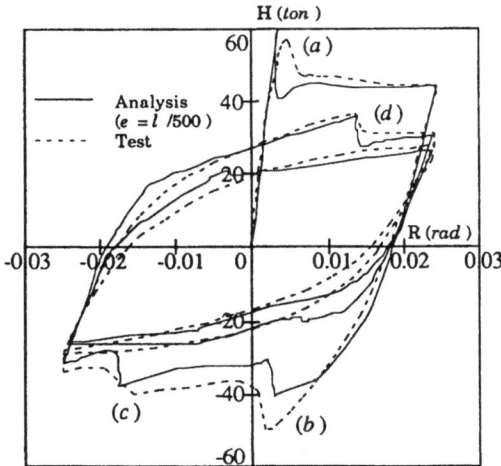

Figure 5 (a) Frame K

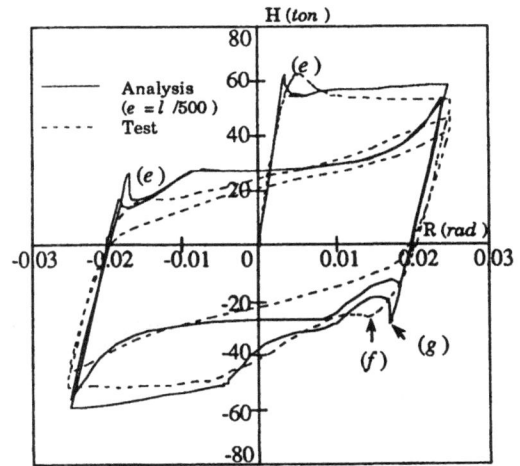

Figure 5 (b) Frame VK

The reason for the member (d) is considered as follows. The upper end of the right hand column at the second story yielded at the point of sign ▲ and experienced great plastic deformation when $e = 0$ because of the member (c) not buckling, while the section was immediately unloaded when $e = l/500$. This caused the change of the moment distribution at the following plus-loading phase and obstructed the buckling of the member (d). Thus, the difference in the initial deflection amounts of the member causes not only the load level of the initial buckling but also the ensuing load - displacement relationship.

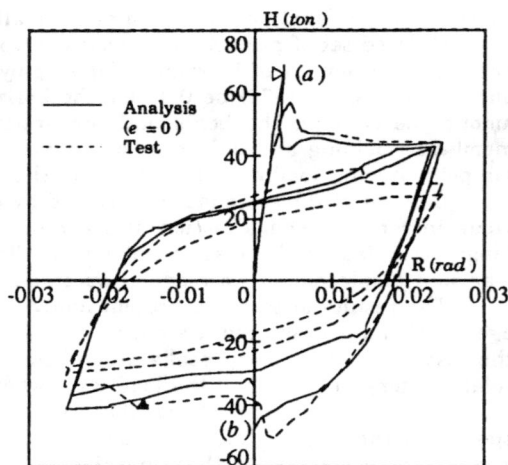

Figure 6

CONCLUSION

Conclusions obtained are summarized as follows.
1. The ordered set of perturbation equation of combined nonlinear constitutive equation can be obtained for the strain hardening general yield hinge model.
2. The present method of analysis gives good estimation for the load - displacement relationship of the braced skeleton when it is accompanied by buckling.
3. The load - displacement relationship analysed by general yield hinge method has sharp peaks around the buckling points because the model cannot express the gradual expansion of the plastic region.
4. The initial deflection of the member effects the initial buckling load level and can also effect the load - displacement relationship afterwards.

ACKNOWLEDGMENTS

The authors are grateful to associate professor Dr. Kouji Uetani of Kyoto university and assistant professor Kiyotaka Morisako of Kyoto Institute of Technology for their help in introducing the incremental perturbation method.

REFERENCES

Igarashi, S., Inoue, K., Shimizu, N., Katayama, T., Watanabe, T., Segawa, T. and Hisatoku, T. 1986. Experimental study on plastic strength of K and VK-type braced steel frames. Summaries of Technical Papers of Annual meeting, A.I.J., 1003-1006.

Inoue, K. and Ogawa, K. 1978. A study on the plastic design of braced multi-story steel frames. Trans. of A.I.J. No.268, 87-98.

Ishida, S. and Morisako, K. 1987. Application of incremental perturbation method to one dimensional combined materially and geometrically nonlinear finite element method. J. Struct. and Const. Eng. A.I.J. No.397, 73-82

Jennings, A. 1963. Frame analysis including change of geometry. Proc. of ASCE, No.ST3, 627-644

Prager, W. 1955. The theory of plasticity; A survey of recent achievement. Proc. of Inst. Mech. Eng. Vol.199, 41-57

Yokoo, Y. Nakamura, T. and Uetani, K. 1976. The incremental perturbation method for large displacement analysis of elastic-plastic structures. Int. J. Num. Meth. Eng. Vol.10, 503-525

Dynamic instability in buildings

Dionisio Bernal[1]

ABSTRACT

This paper presents the development of a Single Degree of Freedom (SDOF) model for the calculation of instability thresholds in multistory buildings. The model is restricted to buildings that can be idealized as two dimensional structures but allows consideration of arbitrary mechanism failure modes. Validation of the model is done by computing ground motion scale factors leading to instability and comparing them with results derived from a general multidegree of freedom formulation. The comparisons indicate that the SDOF model provides good predictions of instability, provided the mechanism that controls dynamically is not too disparate from the one used in deriving the model parameters.

INTRODUCTION

The action of vertical forces acting through lateral deformations is known as the P-delta effect. Under static conditions the P-delta effect can be interpreted in terms of equivalent story shears which add to those resulting from the lateral forces (MacGregor and Hage 1977). During seismic response, however, the lateral forces are not prescribed but dependent on the properties of the structure and the P-delta effect is more appropriately interpreted as a reduction in the lateral stiffness.

Although second order effects invariably lead to increases in deformations for static loading, the dynamic response with P-delta effects included may or may not be larger than the first order solution. In particular, studies by Jennings and Husid (1970), Takizawa and Jennings (1975), and Bernal (1990), have shown that gravity typically has a small effect on inelastic response, except when the strength of the structure is near a certain critical value below which the response grows unbounded, indicating failure from instability. Fig.1 illustrates the maximum response versus yield strength for an elastoplastic SDOF system with a first order elastic period of 1.0 second and 5 % of critical damping. Part (a) shows results computed with and without gravity for El Centro and part (b) illustrates the same for Pacoima Dam.

I Assistant Professor, Northeastern University, Boston MA 02115

The small influence of gravity in the inelastic response, except when the strength is close to the instability threshold can be readily observed in both cases.

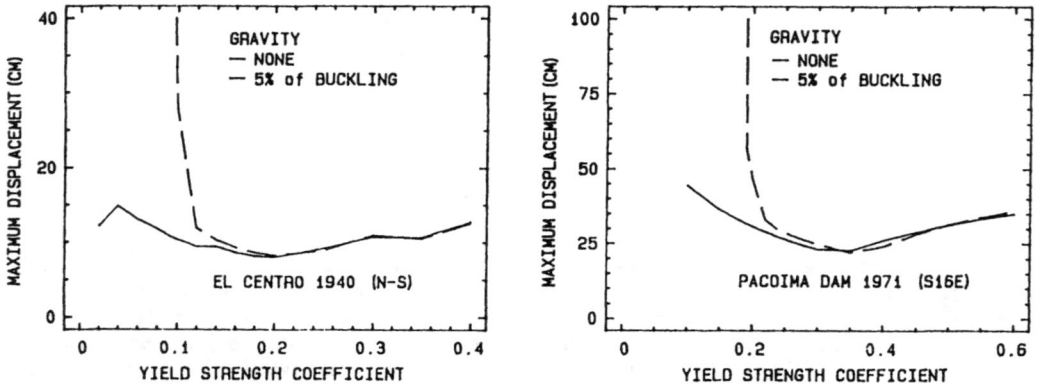

Figure 1. Maximum response of SDOF system (period = 1 sec, 5 % damping)

A conclusion in the preceding discussion is that the importance of the P-delta effect for seismic loading does not stem from amplifications in the inelastic response but derives from the potential for dynamic instability during significant inelastic excursions. The derivation of a SDOF model to asses the strength level associated with the threshold of instability is presented in this paper. Information on instability limit spectral ordinates, as well as additional discussion on the SDOF model, may be found in Bernal (1990).

EQUIVALENT SINGLE DEGREE OF FREEDOM MODEL FOR STABILITY ANALYSIS

We restrict the treatment to buildings that can be modelled as having one horizontal degree of freedom per story. For these class of structures the equations of motion can be written in incremental form as:

$$M\Delta\ddot{u} + C\Delta\dot{u} + (K_t - K_g)\Delta u = -M1\Delta\ddot{x}_h \qquad (1)$$

where M and C are the mass and damping matrices, K_t and K_g are the tangent and geometric stiffness matrices, u is the horizontal displacement vector, $\ddot{x}_h$ is the horizontal ground acceleration, and the symbol Δ is used to indicate increment.

The standard approach to obtain a reduction of Eq.1 to one with a SDOF is to assume that the displacement vector is given by the product of a constant shape times an amplitude. It is possible, however, to use a variable shape function, provided the variations are made dependent on the amplitude of the generalized displacement (Biggs 1964). The use of a variable shape conduces to a model that can accommodate arbitrary failure mechanisms and is thus selected here. In particular, the derivation is based on the shapes resulting from a nonlinear static analysis for a preselected lateral load pattern. A brief discussion on the selection of the load pattern is presented later.

Derivation

Consider a structure subjected to a statically applied lateral load of increasing amplitude. Although the load may have an amplitude dependent distribution, we require the shape of the load increments to be constant between yielding events. It follows, accepting the typical idealization of concentrated plastic hinges, that the incremental displacement shape between events is also constant. For any interval of linear behavior between events one can write;

$$K_t \Delta u = p_i L_i \tag{2}$$

and

$$\Delta u = \Delta Y \phi_i \Gamma_i \tag{3}$$

where L_i is the incremental load pattern after the formation of i hinges and p_i is its amplitude; Y is the generalized displacement, ϕ_i is the shape after the formation of i hinges, and Γ_i is a normalizing parameter (chosen for convenience in the derivation as);

$$\Gamma_i = \frac{\phi_i{}^T M 1}{\phi_i{}^T M \phi_i} \tag{4}$$

By substituting Eqs.2 to 4 into Eq.1 the equations of motion can be reduced to a SDOF. Performing these substitutions and premultiplying by $\phi_i{}^T$ one can write, after some manipulation;

$$\Delta \ddot{Y} + 2\omega_0 \xi_i \ \Delta \dot{Y} + \Delta S_a - \omega_0^2 \theta_i \ \Delta Y = -\Delta \ddot{x}_h \tag{5}$$

where

$$\xi_i = \frac{c_i}{2 m_i \omega_0} \tag{6}$$

$$\theta_i = \frac{k_{gi}}{m_i \omega_0^2} \tag{7}$$

$$\Delta S_a = \frac{p_i \phi_i{}^T L_i}{\Gamma_i \ m_i} \tag{8}$$

and

$$m_i = \phi_i{}^T M \phi_i \tag{9}$$

$$c_i = \phi_i{}^T C \phi_i \tag{10}$$

$$k_{gi} = \phi_i{}^T K_g \phi_i \tag{11}$$

and ω_0 is the elastic (first order) natural circular frequency, given by;

$$\omega_0 = \frac{\phi_0{}^T K_0 \phi_0}{m_0} \tag{12}$$

87

In the preceding expressions ξ_i is the damping ratio, θ_i the stability coefficient, S_a the generalized resistance per unit mass and, in Eq.12, K_0 is the elastic stiffness matrix (the subscript 0 is used for elastic conditions).

The computation of the skeleton for the damping, resistance and geometric terms as a function of the generalized displacement Y can be carried out from the results of the monotonically increasing lateral load analysis. These calculations lead to a piece-wise constant damping and stability terms and to a piece-wise linear resistance per unit mass. Although the cyclic behavior of these terms can in principle be derived by tracing the incremental shapes under reversed static loading, this is impractical and unwarranted. Instead, it is appropriate to idealize the skeletons using a small number of parameters and then associate them with a hysteretic rule.

Generalized Resistance per Unit Mass

A bilinear idealization of the generalized resistance can be defined by the initial slope and the maximum value S_{au}; the initial slope is equal to the square of the elastic natural circular frequency. To derive an expression for S_{au} consider the relationship between S_a in the SDOF system and the static base shear in the structure.
The incremental base shear, ΔV, is given by;

$$\Delta V = p_i 1^T L_i \tag{13}$$

dividing Eq.13 by Eq.8 one gets;

$$\Delta S_a = \frac{\Delta V}{M_{ei}} \tag{14}$$

where

$$M_{ei} = \frac{1^T L_i \Gamma_i m_i}{\phi_i^T L_i} \tag{15}$$

Since M_{ei} is not constant, the ultimate resistance per unit mass can only be computed "exactly" by adding increments. It can be shown, however, that little error results if S_{au} is computed as the ratio of the base shear capacity V_u to the elastic value of the effective mass, (Bernal 1990). It is opportune to note that, for buildings with uniform mass distribution, M_{e0} can be taken as 90 % of the total mass without introducing undue error.

Gravity Effect (Stability Coefficient)

The stability coefficient (Eq.7) reflects the influence of gravity in the equation of motion. In the study reported by Bernal (1990), it is shown that this parameter is rather insensitive to the deformation pattern and that only large changes in shape associated with the formation of partial mechanisms lead to significant departures from the elastic value. On this basis it is reasonable to idealize the fluctuations by assuming that θ equals the elastic value for the linear part of the idealized resistance and that it takes the value associated with the mechanism shape in the plastic range.

For buildings having a reasonably uniform distribution of the mass along the height the following formulas apply (Bernal 1990);

$$\theta_0 = \frac{3N\ g\tau}{(2N+1)\omega_0^2\ h} \tag{16}$$

and

$$\theta_m = \frac{\Omega\ g\tau}{\omega_0^2\ h} \tag{17}$$

where τ is the ratio of total weight to inertial weight (typically calculated at the first story level), g is the acceleration of gravity, h is the total height of the structure, N is the number of stories and Ω is given by;

$$\Omega = \frac{(1 - G/2h - E/h)}{G/h\ (1 - 2G/3h - E/h)} \tag{18}$$

where the parameters E and G depend on the mechanism shape and are defined in Fig.2. In deriving Eq.16 a straight line was used to approximate the elastic shape and Eq.18 was derived by replacing the discrete mass distribution with a continuous one. As expected, the error introduced by the continuum formulation decreases with the number of stories. For one and two story buildings θ_m should be evaluated using Eq.7.

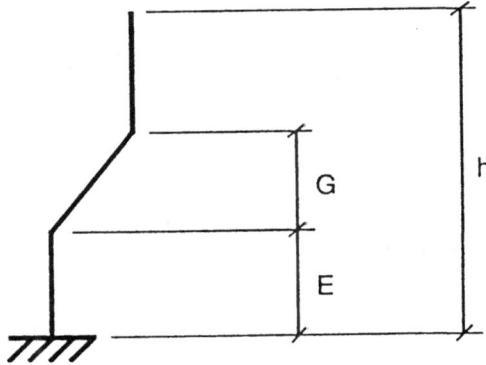

Figure 2. Parameters that define the mechanism shape

It is possible to combine the two stability coefficients into a single equivalent value. Fig.3(a) illustrates a plot of the skeleton of the idealized bilinear effective restoring force with the effect of gravity included in accordance with the two θ idealization previously discussed. In part (b) of the figure, another curve, based on a single stability coefficient, $\bar{\theta}$, is shown. It is apparent by comparing these two plots that they can be made identical by defining equivalent parameters as;

$$\bar{\theta} = \frac{\theta_m}{Q} \tag{19a}$$

$$\bar{\omega}_0^2 = \omega_0^2 \, Q \qquad (19b)$$

and

$$\bar{S}_{au} = S_{au} \, Q \qquad (19c)$$

where

$$Q = 1 - \theta_0 + \theta_m \qquad (19d)$$

by inspection of Eq.5 the equivalent damping is;

$$\bar{\xi} = \frac{\xi}{\sqrt{Q}} \qquad (19e)$$

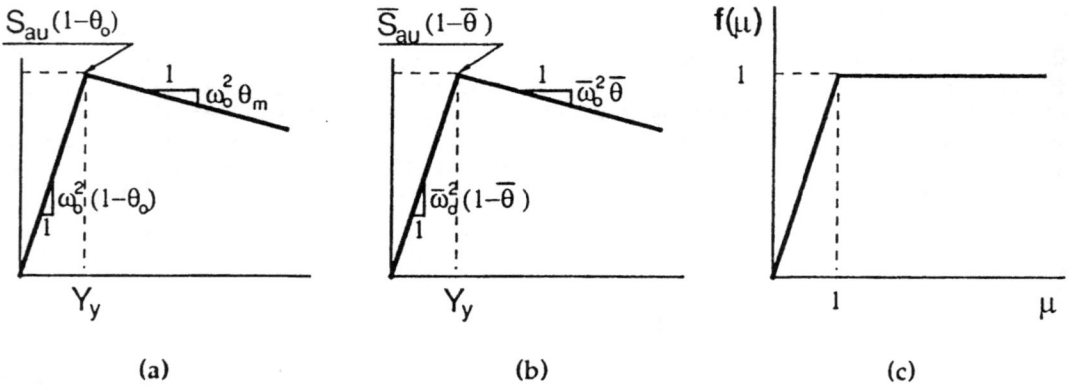

Figure 3. Effective restoring force per unit mass: a) for a two θ model b) equivalent single θ system; c) Normalized first order resistance function.

Using the equivalent parameters and normalizing the equation of motion with respect to the yield displacement Y_y (Fig.3) one can write;

$$\Delta\ddot{\mu} + 2\bar{\omega}_0\bar{\xi}\,\Delta\dot{\mu} + \bar{\omega}_0{}^2(\Delta f(\mu) - \bar{\theta}\Delta\mu) = -\frac{\bar{\omega}_0^2 \Delta\ddot{x}_h}{\bar{S}_{au}} \qquad (20)$$

where

$$\mu = \frac{Y}{Y_y} \qquad (21)$$

and $f(\mu)$ is as shown in Fig.3(c)

Eq.20 can be solved numerically to compute the minimum value of $\bar{S}_{au}$ at which the response becomes unstable. It is perhaps appropriate to note that dividing $\bar{S}_{au}$ by a factor is identical to scaling the ground motion and thus the results from Eq.20 can be interpreted in terms of minimum strength for a given record or as a scale factor leading to instability for a fixed value of the available strength $\bar{S}_{au}$.

90

Selection of the Lateral Load Pattern

It can be shown numerically that, provided there is no change in the form of the controlling kinematic mechanism, the shape of the lateral load has very little influence in the parameters of the equivalent SDOF model. Since most buildings have mechanism shapes that do not vary for a rather wide range of lateral load distributions, the selection of the load pattern is typically not critical. Notwithstanding, research on an approach to define the lateral load pattern, applicable to buildings with sensitive mechanisms, seems warranted.

CORRELATION BETWEEN INSTABILITY PREDICTIONS DERIVED WITH THE EQUIVALENT SDOF SYSTEM AND THOSE OBTAINED USING MDOF MODELS

Comparison of results derived using the equivalent SDOF system and those obtained from MDOF structural representations can be used to asses the accuracy of the simplified model. The earthquake scale factor corresponding to an instability failure is used here for the comparisons. A ten story three bay frame (Anderson and Bertero 1989), a ten story single bay frame (Anderson and Bertero 1969) and a 4 story 3 bay frame (Pique 1976) are chosen for the calculations. Each frame is analyzed using four ground motions. The load pattern chosen to calculate the SDOF parameters is triangular up to the fourth story and uniform for higher levels. This pattern is tentatively recommended for buildings with reasonably uniform mass distribution, on the basis of results from an ongoing study.

Table 1 summarizes the results. The table lists the smallest value of the scale factors at which the structures failed from instability for each one of the ground motions. The factors for the MDOF formulation are shown in the upper part of each box and were obtained using DRAIN-2D (Kannan and Powell 1973). The predicted instability scale factors from the SDOF model (in parenthesis) were computed from Eq.20 using the program INEL1 (Bernal 1990).

Table 1.   Comparisons between SDOF and MDOF predictions of scale factors leading to instability

	El Centro (1940) S00E Comp.	I.V.C. (1979) S50W Comp.	Pacoima (1971) S16E Comp.	Taft (1952) S69E Comp.
10 Story 3 bay frame	4.75 (5.12)	1.63 (1.60)	2.25 (2.15)	8.10 (8.10)
10 Story 1 bay frame	4.20 (4.54)	1.50 (1.43)	1.90 (1.88)	8.00 (7.88)
4 Story 3 bay frame	3.10 (3.10)	1.40 (1.55)	1.40 (1.60)	5.60 (6.28)

Values in parenthesis are from the equivalent SDOF model

As can be seen the SDOF predictions are generally in good agreement with the values obtained from the MDOF formulation.

CONCLUSIONS

An equivalent SDOF model to investigate dynamic instability in multistory buildings is presented. The model allows for fluctuations in the dominant shape of the structure and is thus applicable to buildings having arbitrary kinematic mechanisms. Although derived using the sequence of shapes that result from the static application of an increasing lateral load, it is found that the elastic and the mechanism shapes dominate the behavior of the model and, as such, can be used to arrive at a simplified formulation. It is essential, however, that the load pattern chosen lead to a mechanism that approximates the one that actually controls dynamically. Aside from this requirement, the results from the equivalent SDOF model are insensitive to the shape of the loading used. Given a building, definition of the equivalent SDOF requires only the natural period and the shape of the expected failure mode.

Comparisons of SDOF model predictions of instability with results derived using complete MDOF formulations showed good correlation. It should be noted, however, that in the cases tried, the static mechanism provided a reasonable approximation to the manner in which the buildings failed dynamically. Research to establish a general practical approach to predict dynamic failure mechanism modes is needed.

The practicality of the SDOF model presented is fully realized when it is used in conjunction with collapse response spectra to calculate minimum strengths to prevent instability in terms of key ground motion parameters.

ACKNOWLEDGMENT

Support for this research was provided by the National Science Foundation under Grant CES-8708707. This support is gratefully acknowledged.

REFERENCES

Anderson, J. C. and Bertero, V. V. 1987. Uncertainties in establishing design earthquakes. J. Struct. Eng., ASCE, 113 (8), 1709-1724.

Bernal, D. 1990. P-delta effects and instability in the seismic response of buildings. Report No CE-90-14, Department of Civil Engineering, Northeastern University, Boston.

Biggs, J.M. 1964. Introduction to structural dynamics. McGraw-Hill.

Jennings, P.C. and Husid, R. 1968. Collapse of yielding structures under earthquakes. J. Eng. Mech., ASCE, 94 (EM5), 1045-1065.

Kannan A.E., and Powell, G.H. 1975. DRAIN-2D A general purpose computer program for dynamic analysis of inelastic plane structures. Report No EERC 73-6 and EERC 73-22, Earthquake Engineering Research Center, University of California, Berkeley.

MacGregor, J.G. and Hage, S.E. 1977. Stability analysis and design of concrete frames. J. Struct. Eng., ASCE, 103 (ST10), 1953 -1970.

Pique, J. 1976. On the use of simple models in nonlinear dynamic analysis. Report R76-43, Dept. of Civil Engineering, Massachusetts Institute of Technology, Cambridge.

Takizawa, H. and Jennings, P. 1980. Collapse of a model for ductile reinforced concrete frames under extreme earthquake motions. Earth. Eng. & Struct. Dyn., (8), 117-144.

A study on earthquake pounding between adjacent structures

K. Kasai[I], B.F. Maison[II], V. Jeng[III], D.J. Patel[IV], and P.C. Patel[IV]

ABSTRACT

Analytical research on pounding is presented. It includes development of pounding dynamic analysis programs, parameter studies on building pounding response as well as appurtenance response, a spectrum method to obtain peak pounding responses, actual case studies, and a spectrum method to determine required building separations to preclude pounding.

INTRODUCTION

Pounding Incidents. - Investigations into past earthquake damage have shown that collisions between adjacent buildings during earthquakes have been one of the causes of severe structural damage. This collision, commonly called 'structural pounding' occurs during an earthquake when, due to their different dynamic characteristics, adjacent buildings vibrate out of phase and there is insufficient separation distance between them.

Many incidents of seismic pounding have been reported to date. Pounding of adjacent buildings has made damage worse, and/or caused total collapse of the buildings. The earthquake that struck Mexico City in 1985 has revealed the fact that pounding was present in over 40% of 330 collapsed or severely damaged buildings surveyed, and in 15% of all cases it led to collapse (Rosenblueth and Meli 1986). This earthquake illustrated the significant seismic hazard of pounding by having the largest number of buildings damaged by its effect during a single earthquake (Bertero 1986). The writers have surveyed the damage due to pounding in the San Francisco Bay area during the recent 1989 Loma Prieta Earthquake (Kasai and Maison 1991). Significant pounding was observed at sites over 90 km from the epicenter thus indicating the possible catastrophic damage that may occur during future earthquake having closer epicenters.

Code Provisions Regarding Pounding. - Past aseismic codes did not give definite guidelines to preclude pounding. Because of this and due to economic considerations including maximum land usage requirements, there are many buildings world-wide which are already built in contact or extremely close to one another that will suffer pounding damage in future earthquakes. The 1990 Uniform Building Code (UBC) based on the 1988 provisions by the Structural Engineers

---

[I] Assist. Prof., Dept. Civil. Eng., Illin. Instit. of Tech., Chicago, IL 60616.
[II] Structural Engineer, SSD Engineering Consultants Inc., Berkeley, CA 94704.
[III] Adjct. Assist. Prof., Dept. Civil Eng., Illin. Instit. of Tech., Chicago, IL 60616.
[IV] Grad. Assist., Dept. Civil Eng., Illin. Instit. of Tech., Chicago, IL 60616.

Association of California (SEAOC) gives more detailed requirements for building separations, yet the SEAOC commentary (Section 1H.2.1) indicates the need for research to verify the recommendations. The resulting required separations are "large" and controversial from both technical (difficulty in using expansion joint) and economical (loss of land usage) views.

Research Needs. - Based on above, research into the pounding problem is necessary in order to determine proper seismic hazard mitigation practice for already existing buildings as well as new buildings. The previous research is somewhat limited in scope and it is generally not in a readily usable form for practicing engineers (Maison and Kasai 1990). Continued research is urgently needed in order to provide the engineering design profession with practical means to evaluate and mitigate the effects of pounding. The following describes the writers' research.

POUNDING TIME HISTORY ANALYSIS PROGRAMS

Idealization. - The writers have developed two micro-computer pounding analysis programs SLAM and SLAM-2, which are made publicly available (Maison and Kasai 1988, 1990). The programs idealize buildings as three-dimensional (3-D) multi-degree-of-freedom (MDOF) systems. The SLAM program assumes that a building laterally collides with a rigid adjacent building and the SLAM-2 (Fig. 1) program considers that both buildings are flexible. Pounding is assumed to occur at a single floor level having a rigid diaphragm. The pounding problem is idealized as having two linear states: <u>in State 1</u>, the buildings vibrate without contact, and <u>in State 2</u> the buildings are in contact. A nonlinear problem results as the response oscillates from one linear state to the other. These idealizations were made as a starting point in the pounding investigation in order to make the problem manageable, while retaining important pounding dynamic characteristics.

Advantages. - Unlike most of the past research based on the single-degree-of-freedom (SDOF) modeling, the 3-D MDOF approach used by SLAM and SLAM-2 give information on the distributions of important response quantities such as story displacements, drifts, shear, overturning moments (OTM's), torques, and accelerations through the height of the multi-story building modeled. The programs employ a theoretically exact solution scheme (Maison and Kasai 1990), and they are also computationally efficient.

BUILDING RESPONSE UNDER POUNDING

SLAM-2 Analyses Results. - An existing 15-story steel moment resistant frame building (Building A, period = 1.13 sec) is assumed to collide with an adjacent flexible 8-story building (Building B, period = 0.8 sec.). Three cases are considered for the floor mass of Building B: one-third; one; and three times the floor mass of Building A. SLAM-2 dynamic analyses were conducted using 0.4g artificial earthquakes (Kasai et al. 1990). Throughout this paper, local contact stiffness (Fig. 1) is set to 50000 k/in considering the past studies (Maison and Kasai 1990). Fig. 2 shows the case where buildings A and B have the same floor mass. Pounding is a severe load condition. Sudden stopping of displacement (Fig. 2(a)) at the pounding level results in large and quick acceleration pulses in the opposite direction, and the peak floor accelerations can be more than 10 times those from the no-pounding case (Fig. 2(b)).

In both buildings, pounding produces peak drifts, shears, OTM's, torques, and accelerations at various story levels that are greater than those from the no pounding case (e.g., Fig. 3). Building midheight pounding (Building A) increases shears above pounding level as well as accelerations in the vicinity of the impact (Figs. 3(b) and (c)). Building top level pounding (Building B) decreases the peak shears over the entire building height with the exception of the stories in the vicinity of the impacts (Figs. 3(b) and (c)). As the difference in the relative mass increases, the adverse effects of pounding increase in the building having the lessor mass. These locations of pounding amplified responses correspond to the observed damage locations in the recent earthquake (Kasai and Maison 1991).

The flexible adjacent building cases studied have many trends that are similar to those from a rigid adjacent building case. The rigid adjacent building case, therefore, was the first subject of the writers' study, results of which are discussed below.

SIMPLIFIED THEORY BASED ON SPECTRAL ENERGY

Background. - Through numerous SLAM analytical studies, the writers found that the non-linear pounding peak response of SDOF as well as MDOF systems is not sensitive to the details of the particular earthquake history as long as the earthquakes have a common spectrum characteristics (Kasai et al. 1990). Based on this, the following method to predict the peak pounding responses were developed.

No Pounding and Fixed Spring Systems. - The technique is based on response spectrum analyses of two basic linear systems (Fig. 4): <u>(1) no pounding system (the building vibrating without contact)</u>, and <u>(2) fixed spring system (the building vibrating in continuous contact with the adjacent structure)</u>. The peak response of the pounding system is predicted by considering the distribution of earthquake energy in both systems in the form of kinetic energy and strain energy in each linear state.

Method. - The peak pounding responses of MDOF system are calculated as follows (Kasai et al. 1990):

$$\{u^-\} = \alpha\{u_{np}\} \qquad , \text{ and } \qquad \{u^+\} = \beta\{u_{np}\} + \gamma\{u_{fs}\} \qquad (1)$$

in which $\{u^-\}$ and $\{u^+\}$ = the peak negative and positive displacement vectors, respectively; $\{u_{np}\}$ and $\{u_{fs}\}$ = the peak displacement vectors obtained from commonly used multimode response spectrum analysis of the no pounding system and the fixed spring system, respectively. The separation ratio β is defined as the ratio of the at-rest separation distance divided by the peak displacement of the no-pounding system at the corresponding story level. The α and γ are obtained from simple equations consisting of the kinetic energies as approximately computed using the first modal participation factor and earthquake pseudo-velocity spectra (Kasai et al. 1990). Estimations of the other peak pounding responses such as drifts, shears, and OTM's can be made in a similar manner.

Theoretical Predictions and SLAM Analysis Results. - The theory was verified by more than 500 case studies (Kasai and Patel 1990) comparing the theoretical results to those from SLAM analyses. Figs. 5(a) and (b) illustrate the good accuracy of the theory for predicting MDOF pounding system peak response for various separations in a mid-height pounding case and a top pounding case, respectively. Note that vertical location of pounding significantly influences the distribution of story peak responses through the height of the building, and that the shears remain almost the same with the separation ratio from 0 to 2/3.

BUILDING APPURTENANCES UNDER POUNDING

Building Appurtenance Damage. - The writers observed damage to building appurtenances such as electrical and mechanical equipments, building parapets, and curtain walls which was caused by pounding of buildings during Loma Prieta earthquake (Kasai and Maison 1991). As discussed earlier (Fig. 2(b)), the peak floor accelerations can be more than 10 times those from the no-pounding case. It was also found that a rigid adjacent building case gives the results similar to those from a relatively heavy flexible adjacent building case (e.g., 1-mass and 3-mass building cases in Fig. 3(c)). The following studies consider the rigid adjacent building case (Kasai et al. 1990).

Floor Acceleration Response Spectra for Pounding Case. - The floor acceleration response spectra (FARS) at the top pounding level of the 15-story building are shown in Fig. 6. They indicate that pounding is especially harmful for equipment or secondary systems having short periods (≤1.0 sec). This effect is not covered by existing industrial design spectra. For example, see the Network Equipment-Building System (NEBS) design spectrum given by Bell Communication Research (BELLCORE 1988), which is very close to the FARS of no-pounding case. The FARS in the pounding case can be as much as 30 times higher than those in no-pounding case. Based on these, the commonly considered method of designing the secondary systems to have shorter periods to reduce the system response may be effective only when no pounding occurs, but would be significantly unconservative in a pounding condition.

Neglecting the effect of damping, the acceleration $\ddot{u}_i$ of i-th pounding level during pounding (State 2) is approximately expressed from equilibrium as:

$$\ddot{u}_i = [\ V_{i+1} - V_i - k_s(u_i - s)\ ]\ /\ m_i \qquad (u_i > s) \qquad (2)$$

where V_i, u_i, and m_i = story shear, displacement, and mass of the i-th floor level, respectively, k_s = local contact stiffness (Fig. 1), and s = at-rest separation distance. The writers have found that the peak $\ddot{u}_i$ at State 2 is approximately obtained by substituting into Eq. 2 the peak V_i, peak V_{i+1}, and peak u_i that are estimated using the simplified method explained earlier (Eq. 1).

The writers have also found that the ratio between pounding FARS and no-pounding FARS, hereby defined as a spectrum amplification, remains very stable regardless of different separation ratios (0 to about 2/3) and earthquakes types (Fig. 7) (Kasai et al.). Because of this effect and considering Eq. 2, a simplified method of obtaining pounding FARS seems possible.

CORRELATIVE STUDY ON EXISTING BUILDINGS DAMAGED FROM POUNDING

The writers are conducting correlative pounding analyses of actual buildings damaged during Loma Prieta earthquake. The following describes sample analytical study conducted for two buildings.

Mission Street, San Francisco. - The 10 story building is constructed of thick masonry walls (13 inch thickness) combined with 9 steel plane frames. It was built in 1904. This building experienced severe pounding with an adjacent massive 5 story building which occupies most of the city block. Pounding was located at the 7th level in the 10 story building and at the roof level in the 5 story building (Fig. 8(a)). Only 1 to 1.5 inches building separation is present. The 10 story building suffered structural damage above the pounding elevation as evidenced by the large diagonal shear cracks in the masonry piers (Kasai and Maison 1991).

A 3D-dynamic analysis SLAM model for the building consists of combined steel frames and masonry walls. The building pounded near the corner of the building (Fig. 8(a)). Fig. 8(b) shows an analysis result using a 0.4g artificial earthquake. Note the large shear above pounding level and large torsion developed due to pounding. The pounding analysis results appear to explain the observed damage.

15th Street, Oakland City Center. - The building is a 7-story residential apartment building. It was built in 1913, and consists of 15 reinforced concrete primary plane frames as well as concrete in-fill shear walls that are typically of 6 inches thick at exterior and interior locations. The building has zero separation with the adjacent building, and pounding occurred at its 3rd level (Fig. 9(a)). The building has a rectangular base plan for the ground, mezzanine, and the 2nd level. Above the 2nd floor it takes the form of a "T" in plan with the stem of the "T" pointing north (Fig. 9(a)). A 3D-dynamic analysis SLAM model of the building includes 41 different types of columns, 22 different types of beams and 83 different types of shear panels. Fig. 9(b) shows example analysis results. Note again, the increase of shear and torsion due to pounding.

SEPARATION DISTANCE TO PRECLUDE POUNDING

Spectrum Difference Method. - Based on random vibration theory and considering a first mode approximation for displacements of elastic multi-story buildings, the writers have found a simplified method to obtain an accurate estimate of the required building separation, s, to preclude pounding. In contrast to the commonly known spectrum modal combination method, it is called a <u>spectrum difference method</u>. The method considers the difference of vibration phase between the adjacent buildings. i.e.,

$$s = \sqrt{u_A^2 + u_B^2 - 2\,\rho_{AB}|u_A||u_B|} \qquad (3)$$

In which u_A and u_B are the peak lateral displacements at the possible pounding location under the no pounding condition in Building A and B, respectively, the magnitude of which are simply obtained by the commonly used spectrum approach. The ρ_{AB} is a cross-correlation coefficient. The Eq. 3 is analogous to the double sum combination (DSC) rule commonly used in response spectrum analysis, except that a "minus" ρ_{AB} instead of a "plus" ρ_{AB} is used. Therefore, Eq. 3 is called the <u>double difference combination (DDC) rule</u>. The ρ_{AB} is obtained by substituting the fundamental periods and damping ratios for Buildings A and B into the expressions such as given by Der Kiureghian (1980). Other combination rules <u>commonly</u> known may be the square-root-of-sum-of-squares (SRSS) rule (i.e., $s = \sqrt{u_A^2 + u_B^2}$) or the absolute sum (ABS) rule (i.e., $s = |u_A| + |u_B|$). The use of ABS rule is implied by SEAOC/UBC.

Accuracy of Proposed Rule. - Fig. 10 illustrates the performance of the three rules as compared with time history analysis results using 1940 El Centro earthquake. When the adjacent buildings have the same period, they vibrate in-phase and the absolute relative displacement between them becomes minimum, thus, s becomes minimum and equal to $|u_A - u_B|$, indicating the small separation required to avoid pounding. If additionally the buildings have the same height, then s = 0 (Figs. 10(b) and (c)). These are accurately predicted by the DDC rule, whereas the SRSS and ABS rules are erroneous.

If the adjacent buildings have large damping, s becomes significantly smaller (Figs. 10(c) and (f)), which suggests the feasibility of using interior damper to reduce relative building motions. This is due to the fact that larger damping does not only result in smaller $|u_A|$ and $|u_B|$, but also promotes in-phase vibration of the two adjacent buildings subjected to the same earthquake excitation. Again, the proposed DDC rule captures this effect through the cross-correlation coefficient ρ_{AB}, whereas the SRSS and ABS rules are erroneous. Accuracy of the proposed method has been verified by using 15 different earthquakes as well as varying the period, damping, and height of the buildings.

CONCLUSION

Pounding is a more severe load condition than the case where it is ignored. Continued research is urgently needed in order to provide the engineering design profession with practical means to evaluate and mitigate the effects of pounding. Pursuant to this need, the writers are conducting further research on pounding.

ACKNOWLEDGEMENT

The study presented here is a part of the United States National Science Foundation (NSF) sponsored research project (Grant No. BCS-9003579). The cognizant NSF program official for this support is Dr. Shih-Chi Liu. The support is gratefully acknowledged. The writers also thank A. Jagiasi and S. Alyasin for their assistance in this study.

REFERENCES

Bell Communication Research 1988. "Network Equipment-Building System (NEBS) Generic Equipment Requirements", Issue 3, Piscataway, NJ 08854-4196.
Bertero, V.V. 1986. "Observation of Structural Pounding," Proc. Internat. Conf. : The Mexico Earthquake - 1985, ASCE.
Der Kiureghian, A. 1980. "A Response Spectrum Method for Random Vibrations", EERC

Report, 80/15, Earthq. Eng. Res. Center, U. of Calif. Berkeley.

Kasai, K., Maison, B.F., and Patel, D.J. 1990. "An Earthquake Analysis For Buildings Subjected to a Type of Pounding," Proc. 4th U.S. Nat. Conf. Earthq. Eng.

Kasai, K., Jeng, V., and Maison, B.F. 1990. "The Significant Effects of Pounding Induced Accelerations on Building Appurtenances," Proc. Seism. Design and Perform. of Equipmt. and Nonstruct. Compts. in Bldgs. and Indust. Structs. Appl. Tech. Council Seminar ATC-29.

Kasai, K., and Patel, D. 1990. "A Proposed Method of Evaluating Response For a Type of Collision Between Buildings", IIT-CE-91-02, Illin. Inst. of Tech.

Kasai, K., and Maison, B.F. 1991. "Structural Pounding Damage," Loma Prieta Earthquake Reconnaissance Report, Chapter 6, Struct. Engnrs. Assoc. of Calif.

Kasai, K. and Maison, B.F. 1991. "Observation of Structural Pounding Damage from 1989 Loma Prieta Earthquake", Proc. 6th Canad. Conf. on Earthq. Eng.

Maison, B.F., and Kasai, K. 1990. "Analysis for Type of Structural Pounding," J. of Struct. Eng., ASCE, 116(4).

Maison, B.F., and Kasai, K. 1988. "SLAM: A Computer Program for the Analysis of Structural Pounding". Nat. Inf. Serv. for Earthq. Eng., U. of Calif.

Maison, B.F., and Kasai, K. 1990. "SLAM-2: A Computer Program for the Analysis of Structural Pounding". Nat. Inf. Serv. for Earthq. Eng., U. of Calif.

Rosenblueth, E., and Meli, R. 1986. "The 1985 Earthquake: Causes and Effects in Mexico City," Concrete International. 8(5).

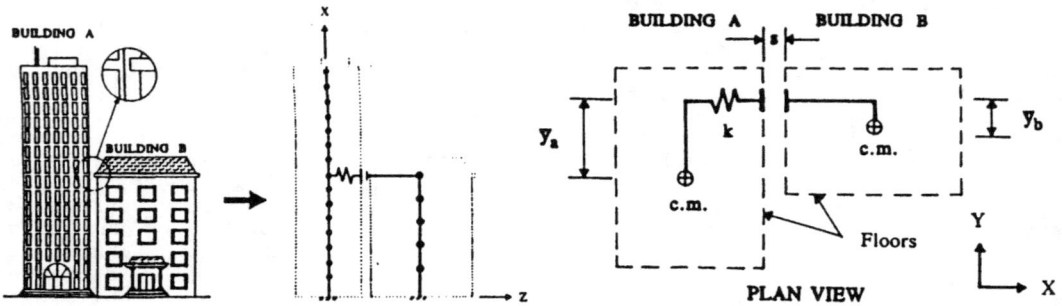

Fig. 1   The pounding problem and SLAM-2 idealization.

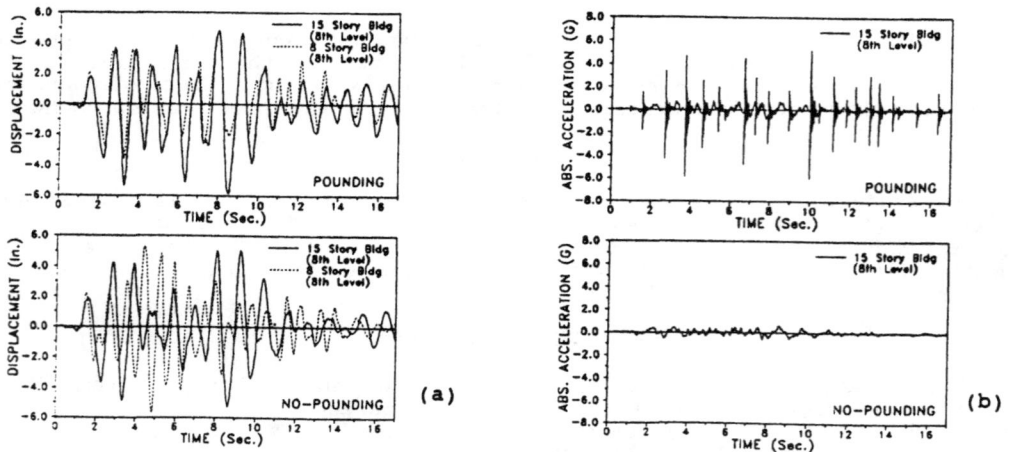

Fig. 2   Pounding case (15-story against 8-story building) vs. no-pounding case.

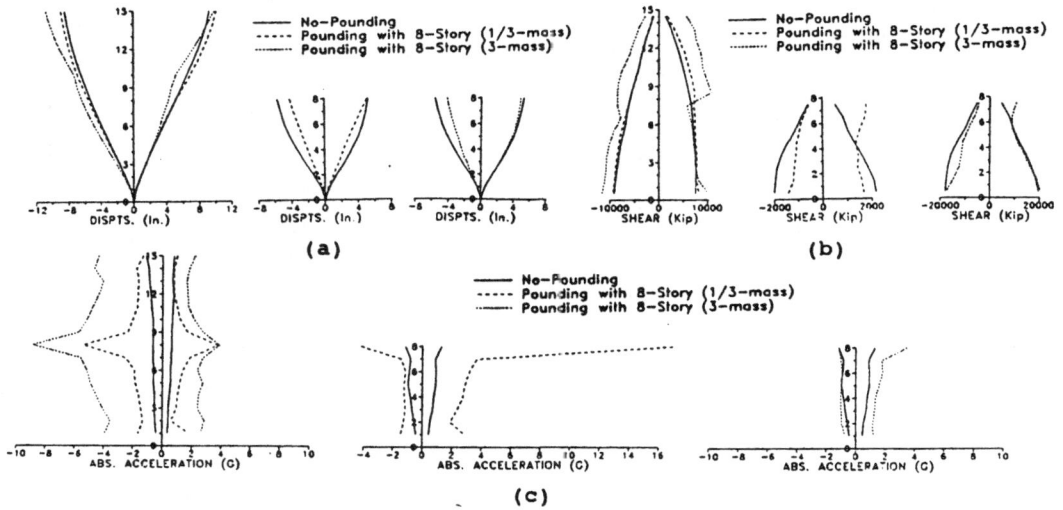

Fig. 3 Response envelops (pounding case vs. no-pounding case).

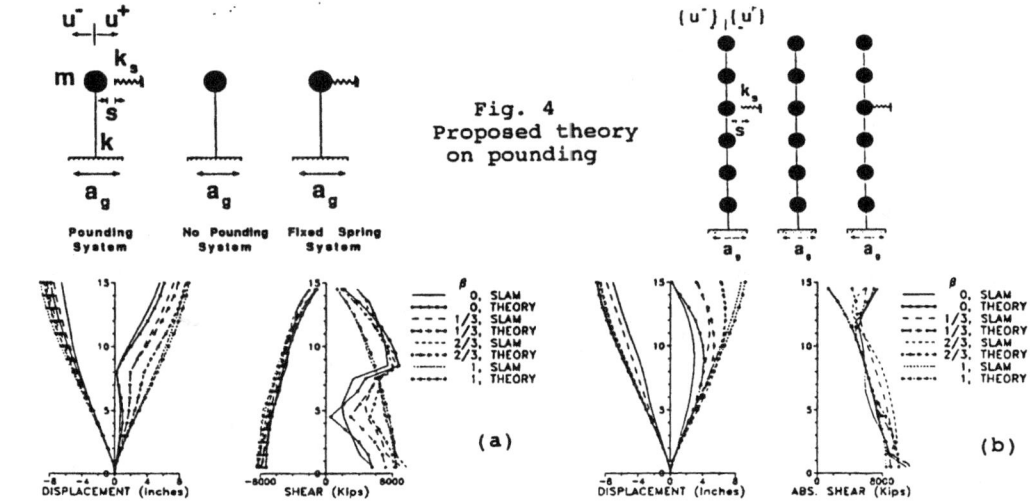

Fig. 4
Proposed theory
on pounding

Fig. 5 Theory vs. average of SLAM-analysis results (six art. earthqs., 0.4g)

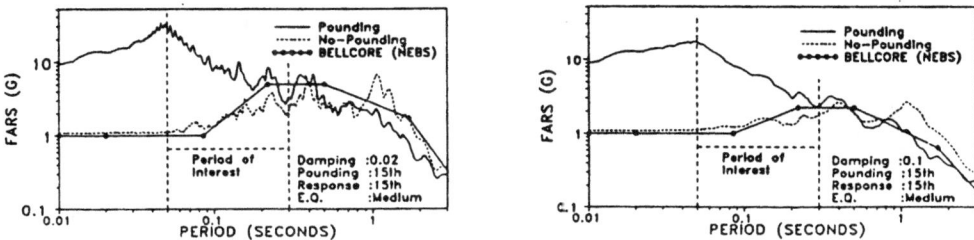

Fig. 6 FARS : floor acceleration response spectra (15-story building)

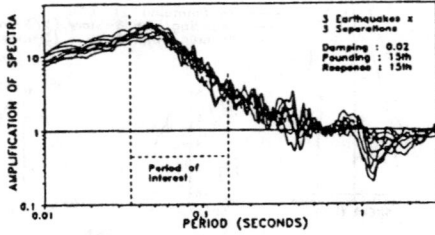 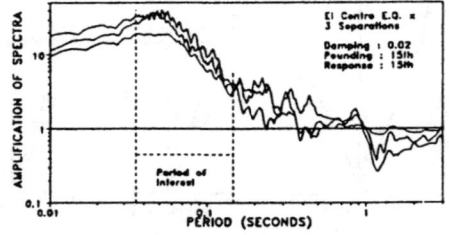

Fig. 7 Spectrum amplification (15-story building).

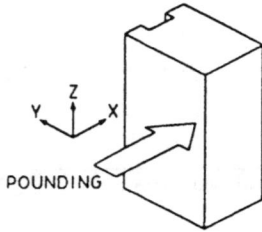 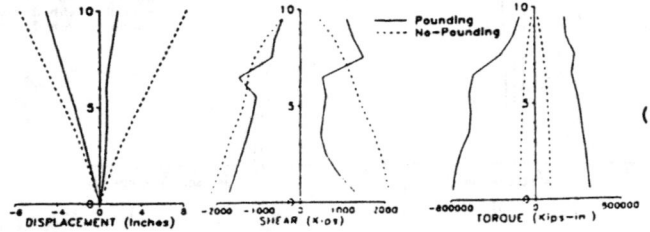

Fig. 8 Example building, Mission Street, San Francisco.

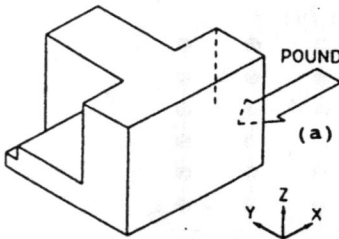 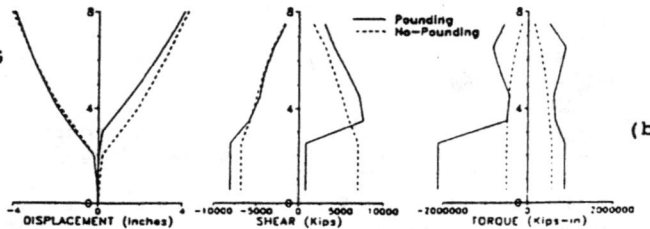

Fig. 9 Example building, 15th Street, Oakland City Center.

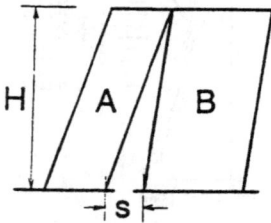 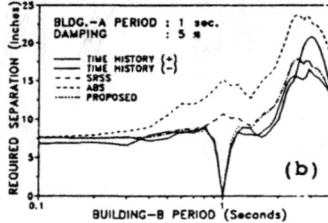 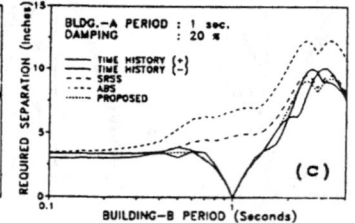

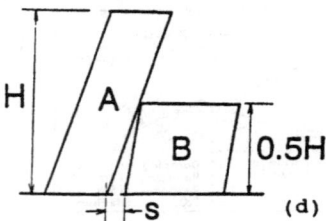 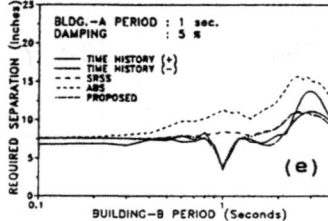 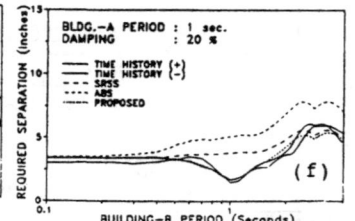

Fig. 10 Building separations required to preclude pounding.

Seismic response of non-symmetric structures using the 1990 NBCC

René Tinawi[1] and Yves Cadotte[II]

ABSTRACT

The seismic response of three non-symmetric structures is evaluated using a 3-D dynamic spectral analysis. The analyses are calibrated to the static base shear obtained using the 1990 NBCC. For the spectral analysis, three dynamic degrees of freedom are used at the centre of mass of each floor. In applying the NBCC, four techniques are proposed for evaluation of the centre of rigidity at each floor and corresponding torsional moments. Results show, for a given global base shear, different distributions between the various lateral load-resisting elements for the various techniques.

INTRODUCTION

When the centre of rigidity (CR) of a structure is not coincident with its centre of mass (CM), torsional effects are introduced. If a 3-D dynamic analysis is performed, with 3 degrees of freedom at the CM of each floor level, there is no need to evaluate the structure's CR. However, the obtained global base shear must be calibrated to the 1990 NBCC static base shear. When the 1990 NBCC is used for non-symmetric structures, the torsional moments at the CR are obtained by multiplying the forces at each floor level by the design eccentricity e_x where

$$e_x = 1.5e + .1D \tag{1}$$

$$e_x = .5e + .1D \tag{2}$$

The value of e is simply the distance between the CM and CR and the second term accounts for the accidental eccentricity which represents 10% of the width of the structure perpendicular to the earthquake direction. Four methods are used to evaluate the position of the CR.

Relative rigidities (Method A)

The most common method, proposed by Blume et al (1961), defines the X and Y coordinates of the CR, at level r, by

$$X_{CR} = \sum_{i=1}^{n} X_i\, k_{yi} \Big/ \sum_{i=1}^{n} k_{yi} \tag{3}$$

[1] Professor, Dept. of Civil Engineering, Ecole Polytechnique, Montreal.
[II] Presently with the Structural Engineering Dept., Lavalin, Montreal.

$$Y_{CR} = \sum_{i=1}^{n} Y_i \, k_{xi} / \sum_{i=1}^{n} k_{xi} \qquad\qquad (4)$$

k_{xi} and k_{yi} is the stiffness of the lateral load resisting element i at X_i and Y_i and n is the total number of such elements.

Force and moment applied at each level (Method B)

It uses the same principle as Method A and was proposed by Cadotte (1990). It consists of applying systematically on a 3-D model, arbitrary forces F_x and F_y in both X and Y directions as well as a moment M_z at the CM of each floor level. The eccentricity is then given from the corresponding diaphragm rotations obtained at each level by:

$$e_x = \Theta_{Fx} \, M_z / \Theta_{Mz} \, F_x \qquad\qquad (5)$$

$$e_y = \Theta_{Fy} \, M_z / \Theta_{Mz} \, F_y \qquad\qquad (6)$$

Positions for overall zero rotations (Method C)

This method, proposed by Stafford-Smith and Vezina (1985) and by Cheung and Tso (1986), requires that for all positions of the CR at all levels of the buildings, a pure translation occurs with no torsional rotation. This method can be applied using 2D models for the lateral load-resisting elements connected by rigid links. From the applied lateral loads P_x and P_y, the shears Q_x or Q_y are evaluated, for element i, in both X and Y directions and at a given level r. The CR for given lateral loads is simply:

$$X_{CR,r} = \sum_{i=1}^{n} (Q_{yi,r} - Q_{yi,r+1}) \, X_i / P_{y,r} \qquad\qquad (7)$$

$$Y_{CR,r} = \sum_{i=1}^{n} (Q_{xi,r} - Q_{xi,r+1}) \, Y_i / P_{x,r} \qquad\qquad (8)$$

Positions for overall zero rotations in 3-D (Method D)

This method is identical to the previous one except that the modelling of the lateral load-resisting elements is performed is 3-D. Therefore the displacements perpendicular to the direction of load application and the rotations at floor levels are prevented.

NUMERICAL EVALUATION OF CENTRE OF RIGIDITIES

The four methods described above are evaluated numerically for three different structures that have a torsional eccentricity in one of the two directions.

Building 1 is a 25-storey structure with an important vertical set-back at about one-third of its height (88,8m). Four lateral load-resisting elements are present in the Y-direction while three elements are placed symmetrically in the X-direction as shown in Fig. 1a. In order to simplify the model, floors have been lumped together. Fig. 1b shows the position of the CM vertically as well the CR using the four methods described earlier. It should be noted that methods C and D yield identical results and the value of CR can undergo wild excursions outside the building enveloppe.

Building 2 is a 27-storey structure shown in Fig. 2a with a height of 97,2m. The two cores provide the major lateral load-resisting elements. Core #1 is continuous throughout the height of the building while core #2 stops half-way. Fig. 2b shows the position of the CR. Once again, Method D shows erratic values but nonetheless consistent with the definition for the CR. Method C has been omitted as this structure cannot be appropriately analyzed in 2D due to the core/frame interactions in 3D.

Building 3 is a parking structure with eleven stories and a height of 27,5m. The lateral load-resisting elements consists of 3 cores and bracings have been added at the perimeter of the structure as shown in Fig. 3a. This structure has a double eccentricity but only the earthquake in the Y direction has been considered here. The other direction X has been analysed by Cadotte (1990). Due to the difficulty in evaluating the stiffness of the peripheral frames which have a flexural as well as a shear contribution, Method A has been discarded for evaluation on the CR. Fig. 3b shows the CR for the three other techniques. Even though the CR for methods C and D are within the structure their vertical positions vary quite significantly.

EVALUATION OF TORSIONAL RESPONSE

The NBCC requires the evaluation of the torsional moments by multiplying the storey force by the design eccentricities evaluated using equations (1) and (2). Considering the very wild differences in the position of the CR using different techniques, it is essential for a designer to appreciate the impact of the torsional moments and the corresponding internal distributions on the lateral load resisting elements. Furthermore, some techniques concerning the evaluation of the CR do create excessively large torsional moments of opposite signs (fortunately) but leaves the designer wondering about their validity. Furthermore, the NBCC (Clause 4.1.9.1(24)) states explicitly that when the CR and CM of the different floors do not lie approximately on a vertical line, a dynamic analysis shall be carried out to determine the torsional response. However, the base shear from a dynamic analysis must be calibrated to the static base shear. When a 3-D multi-modal spectral analysis is performed where modes are combined, such a calibration is not quite obvious.

Calibration of a dynamic analysis

The procedure is presented in point-form and is summarized below:

1) A 3-D model using three dynamic degrees of freedom at the CM of each floor is created. No allowance is made for the accidental eccentricity of the mass but it could easily be accounted for.

2) The periods, mode shapes, for at least the first ten modes and the corresponding generalized masses M^* and participation factors Γ_x, Γ_y and Γ_θ for each mode are obtained.

3) For a given response spectrum, using an annual probability of 0,0021, and earthquake direction for the site considered, the spectral modal accelerations S_a are evaluated. No allowance is made for the ductility by introducing reduced spectra.

4) For an earthquake in X direction, the base shears and torsions are evaluated for each mode i.

$$V_{xi} = M_i^* \, \Gamma_{xi}^2 \, S_{ai} \qquad (9)$$

$$V_{yi} = M_i^* \, \Gamma_{xi} \, \Gamma_{yi} \, S_{ai} \qquad (10)$$

$$T_{\theta i} = M_i^* \, \Gamma_{xi} \, \Gamma_{\theta i} \, S_{ai} \qquad (11)$$

Similar expressions are obtained for an earthquake in a different direction. The torsional spectrum has not been considered in this study but could be introduced (Awad and Humar, 1984).

5)   For each earthquake direction, the modal base shears are combined using either RSS or CQC techniques. This base shear corresponds to the translational excitation. For example, for an earthquake in X

$$V_{DYN} = (\sum_{i=1}^{n} V_{xi}^2)^{1/2} = (\sum_{i=1}^{n} M_i^* \Gamma_{xi}^2 S_{ai})^{1/2} \qquad (12)$$

6)   Using the 1990 NBCC, the base shear formula is given by:

$$V_{90} = .6(vSIFW)/R \qquad (13)$$

a calibration factor α is obtained where

$$\alpha = V_{90} / V_{DYN} \qquad (14)$$

7)   Finally for each lateral load resisting element, using the combined modal spectral response, the results of the dynamic analysis are multiplied by α.

It is obvious that the dynamic base shear, for a given direction, must be evaluated using Eq. (12). If a software does not provide such values or the required information to calculate V_{DYN} manually, the combined modal results for the individual elements cannot be used to evaluate the dynamic base shear due to loss of signs in the modal combinations.

Table 1 shows the results of the base shears using the above procedure assuming the buildings exist in Montreal and the force reduction factor R=1. Similarly I and F were set to unity. For the spectral analysis using 5% damping, the ground motion was assumed to have a PHA=.18g and a PHV=.097m/s. The values for α vary from .90 to 1.31 depending on the building configuration, torsional coupling as well as the contribution of higher modes.

DISCUSSION OF RESULTS

Table 2 shows the resultant torsional moment M_1 and M_2, obtained using Eqs. (1) and (2), for the various methods of evaluating the eccentricity. Only the effect of the governing torsional moment and the shear, at each floor level, is shown in the figs. 1c, 1d, 1e, 2c, 2d, 3c and 3d. Other details are given by Cadotte (1990).

For Building 1, the static and dynamic analysis shows large differences in the torsional response. In Figs. 1c and 1d, the shears in elements 1 and 2 are overestimated by a factor of nearly 2 while element 3, shown in Fig. 1e and is furthest from the CM, has its base shear underestimated by about the same factor.

For Building 2, the shears in cores 1 and 2 are close at the base as shown in Figs. 2c and 2d. Where the largest difference occurs is at level 15 where core No. 2 is discontinued. The dynamic analysis shows a smoother transition while all the static methods yield a much larger discontinuity.

For Building 3, the shears in elements 9 and 11 are shown in Figs. 3c and 3d. The results are surprisingly similar at the base since little difference is shown in the torsional moments using methods B and C. Also, as shown in Fig. 3b the CM and CR are within the bounds of the elements considered.

CONCLUSIONS

For non-symmetric structures where the centre of rigidity and centre of mass do not lie approximately on a vertical line, a 3D dynamic analysis procedure is presented. This method has the advantage of avoiding the complex evaluation of CR of the structure and yields, for a calibrated base shear to the static base shear, a better internal distribution of the shears arising from the torsional response. The higher flexural or flexural-torsional modes are accounted for. The only missing item in this procedure is the evaluation of a torsional response spectrum.

REFERENCES

Awad, A.M. and Humar, J.L., 1984. Dynamic response of buildings to ground rotational motion. Canadian Journal of Civil Engineering, 11, 48-56.

Blume, J., Newmark, N.M., Corning, L.H., 1961. Design of multi-storey reinforced concrete buildings for earthquake motions. Portland Cement Association, Chicago, Illinois.

Cadotte, Y., 1990. Réponse séismique des bâtiments asymétriques. M.Sc. thesis, Ecole Polytechnique, University of Montreal Campus.

Cheung, V.W.-T. and Tso, W.K., 1986. Eccentricity in irregular multi-storey buildings. Canadian Journal of Civil Engineering, 13, 46-52.

National Building Code of Canada, 1990, Associate Committee on the National Building Code, National Research Council of Canada, Ottawa.

Stafford-Smith, B. and Vézina, S., 1985. Evaluation of centres of resistance in multi-storey building structures. Proc. Inst. Civ. Engrs., Part 2, 1979, 623-635.

TABLE 1. Calibration factors for a 3D dynamic analysis

Building	Quake Direction	1990 NBCC		Dynamic		α
		T(s)	V_{90} (kN)	T(s)	V_{DYN} (KN)	
1	Y	2,50	25 300	3,06	19 300	1,31
2	X	2,53	15 900	2,38	17 740	0,90
3	X	0,95	21 900	1,25	22 960	0,95

TABLE 2. Resultant torsional moments at base using 1990 NBCC

Building	Method	M_1 (kN.m) eq.1	M_2 (kN.m) eq.2
1	A	-1 260 000	- 45 000
	B	- 711 000	94 000
	C	171 000	197 000
2	A	- 134 400	44 500
	B	- 79 100	63 000
	D	- 57 100	63 900
3	B	256 200	- 73 400
	C	227 700	- 37 500

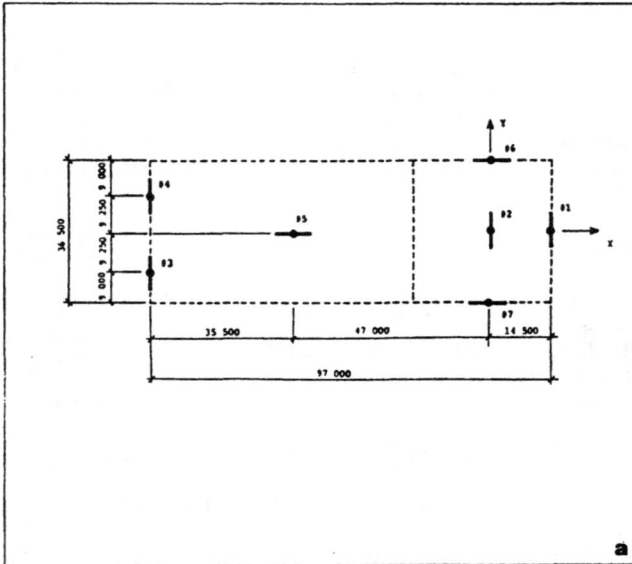

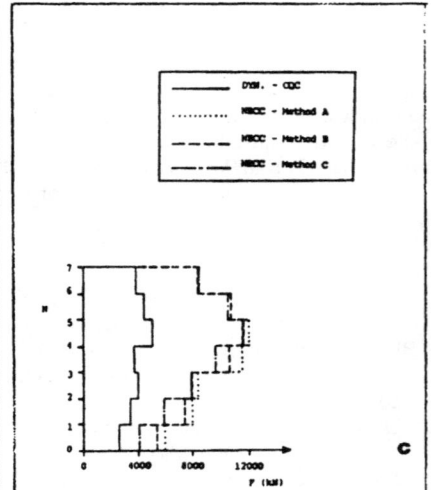

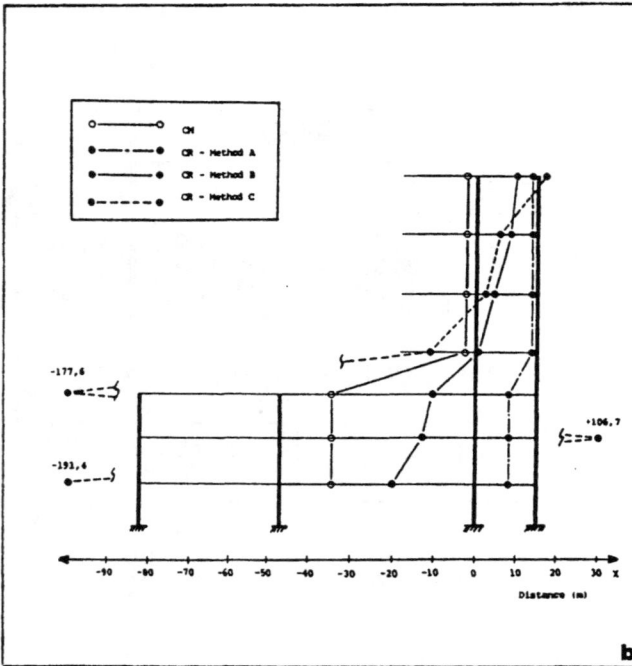

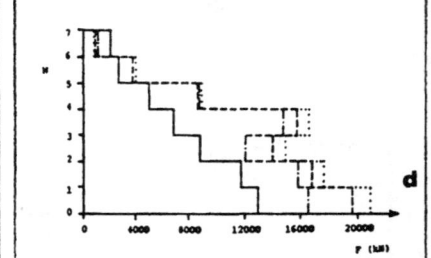

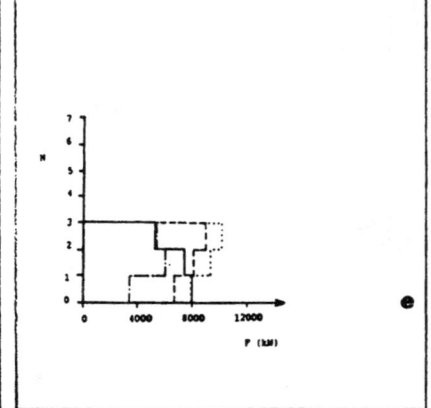

Figure 1. Building 1 - 25 stories. (a) plan; (b) position of CM and CR; (c) shears in element 1;
(d) shears in element 2; (e) shears in element 3.

106

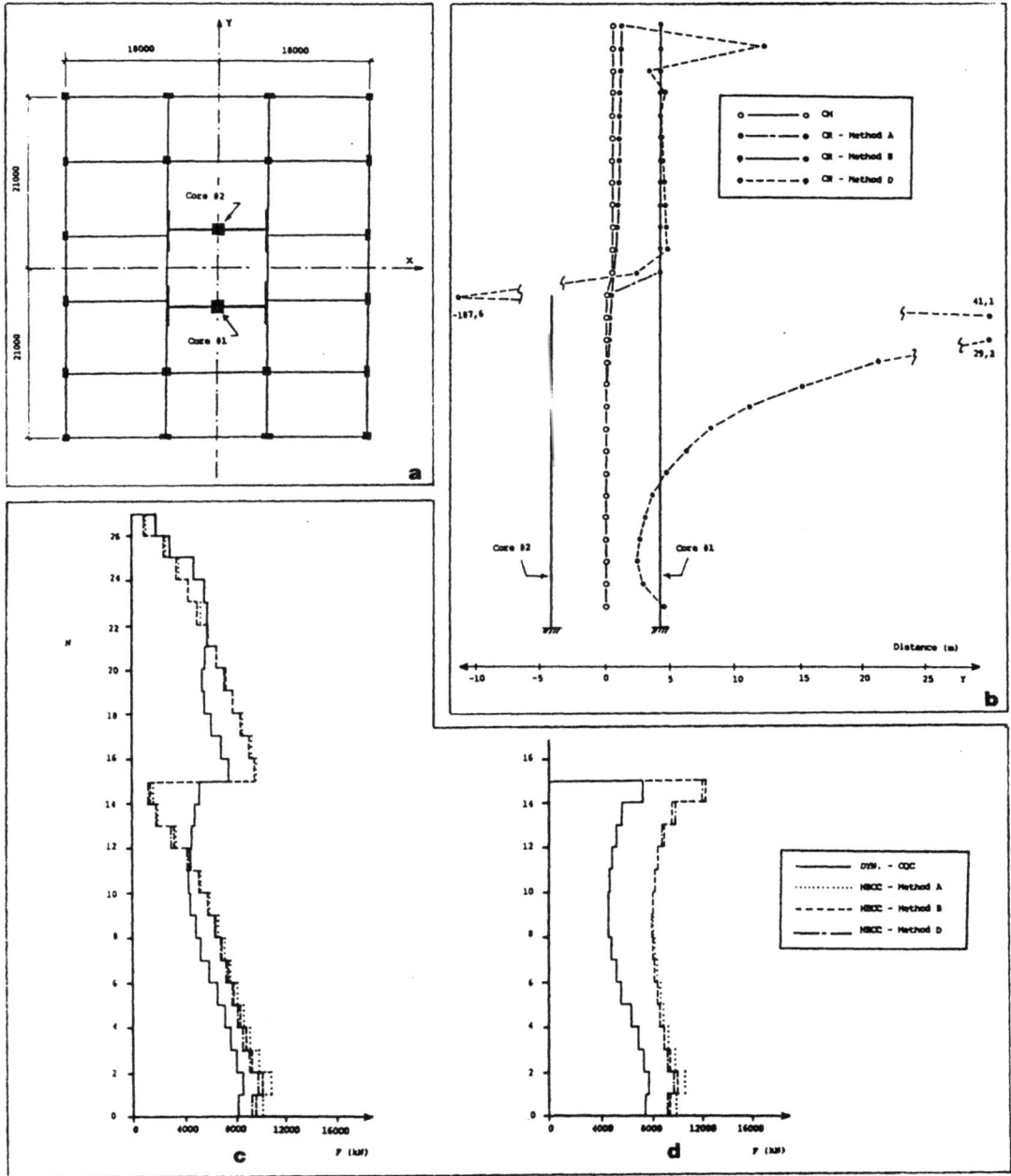

Figure 2. Building 2 - 27 stories. (a) plan; (b) position of CM and CR; (c) shears in core #1; (d) shears in core #2.

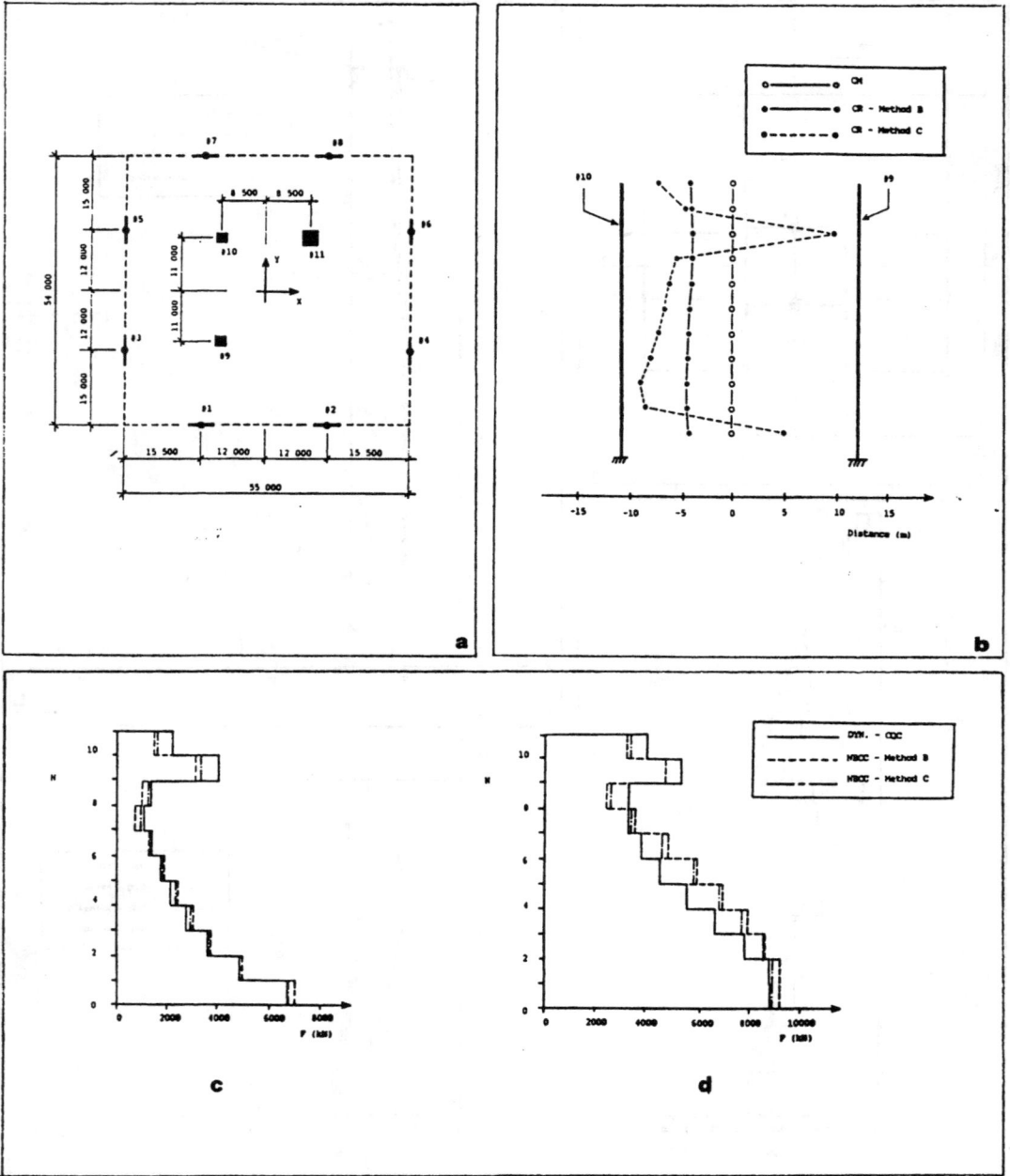

Figure 3. Building 3 - 11 stories. (a) plan; (b) position of CM and CR in Y-Z plane; (c) shears in element #9; (d) shears in element #11.

Seismic response of symmetric structures having unbalanced yield strengths in plan

Michel Bruneau[I] and Stephen A. Mahin[II]

ABSTRACT

Symmetric structures, with coincident centers of mass and stiffnesses in plan, can be excited torsionally in the inelastic domain due to unbalanced distribution of their yield strength. Under some circumstances, this can produce a magnification of the otherwise expected ductility demand of the lateral load resisting structural element having the lesser strength. A parametric study has been conducted to investigate the circumstances where this amplification becomes significant, and results of this research are presented herein.

INTRODUCTION

Current seismic retrofitting strategies emphasize the need to eliminate the eccentricities in plan of existing structures by adding new lateral load resisting structural elements (LLRSEs), and by calibrating the stiffnesses of the new elements such as to minimize these eccentricities. While the resulting retrofitted structures become symmetric in the elastic sense, the new structural elements used are often of a different type than the existing ones. This results in dissimilar yield strengths between the various LLRSEs, and a transient state of torsional response is consequently expected to develop when the structure will be excited in the inelastic range. It is noteworthy that such dissimilarities can also be present in many types of new or existing structures, simply as a consequence of other engineering or architectural decisions; thus the particular structural behavior described above can be equally attributable to various other causes.

While other researchers have investigated the effect of torsional instability of symmetric systems *[Tso and Asmis 1971, Tso 1975, Antonelli et.al. 1981, Pekau and Syamal 1984]* and the effect of seismic wave motions characteristics in exciting torsional modes in otherwise symmetric structures *[Newmark 1969, Awad and Humar 1984]*, little research has focused

I Assistant Professor, Civil Engineering Department, University of Ottawa, Ottawa, Ontario, Canada, K1N 6N5, Tel.: (613) 564-3432

II Professor of Civil Engineering, University of California, Berkeley, California, 94720, USA

on elastically symmetric structures having considerable dissimilarities in yield strength and force-displacement relationships. The results reported herein are from research conducted at the University of California, Berkeley, from 1985 to 1987 [Bruneau and Mahin 1987]. In the time elapsed since, results from similar research has been published by Pekau and Guimond [1988]. Both pieces of work are complementary as they address the same problem in a different perspective. As the problem of inelastic seismic response of torsionally coupled structures is currently receiving a renewed and considerable attention by the research community, it is felt that the present reporting of the original work by the authors is worthwhile, and long overdue.

In a first approach to this broad subject, the present study was limited to the consideration of idealized LLRSEs of identical force-displacement relationship having dissimilar yield strengths. The results of a parametric study of different simple monosymmetric initially symmetric structures having two LLRSEs are presented following.

EQUATIONS OF MOTION AROUND THE CENTER OF MASS

The general equations of motion around the center of mass for single-story torsionally coupled structures are well known and have been derived by others, [Awad and Humar 1984, among many]. From these equations are obtained the following parameters generally describing torsionally coupled structures, which are also adopted for this study.

$$\Omega \;=\; \omega_\theta / \omega_x \;=\; T_x / T_\theta \tag{1}$$

$$\omega_x^2 \;=\; K_x / m \tag{2}$$

$$\omega_\theta^2 \;=\; K_\theta / mr^2 \tag{3}$$

K_X and K_θ are the structure's translational (along X) and rotational (around θ) stiffnesses. The mass of the floor is m, and its radius of gyration r. ω_x and ω_θ are the translational and torsional uncoupled frequencies, T_x and T_θ the corresponding uncoupled periods, and Ω the ratio of those uncoupled frequencies. The reader not familiar with those equations should refer to Bruneau and Mahin 1987. For linear elastic perfectly symmetric structures, an uncoupling of the torsional and translational response is possible. This uncoupling would persist until dissymmetric yielding of the LLRSEs occurs, at which point a transient state of torsional coupling is established. The corresponding instantaneous equations of motion could be established [Bruneau and Mahin 1987].

METHODOLOGY OF THE PARAMETRIC STUDY

This parametric study was performed in order to assess the significance of the torsional coupling developing in the inelastic range when symmetric structures consist of LLRSE's of identical force displacement relationship, but dissimilar yield strengths; in particular, the effect of various parameters on the element ductility demand of the initially symmetric structure were investigated when equivalent SDOF systems achieved preselected target ductility values, μ_T.

In this study, simple structures having two LLRSEs equidistant from the center of mass are used. The torsional stiffness of individual LLRSEs is neglected. All floor diaphragms are assumed to be infinitely rigid in their own plane. Elements in the orthogonal direction are ignored for the sake of simplicity. The structure studied is illustrated in Fig. 1. Simple bi-linear inelastic element model was chosen. The introduction of more sophisticated modelling was not warranted at this stage, but due consideration has been given to other non-linear element models elsewhere *[Bruneau and Mahin 1987]*. Providing both LLRSEs are of the same force-displacement relationship type, the element model has been found to have little influence on the conclusions of this study. Strain-hardening was set to 0.5% of the initial elastic stiffness of the elements, making the element model almost elasto-perfectly plastic. The damping was chosen to be of the Rayleigh type, arbitrarily set at 2% of the critical damping for each of the true elastic frequencies of the structures analyzed. For the initially symmetric structures used in this study, the LLRSE's yield level combinations are expressed as "Ry and x(Ry)" (with corresponding yield displacements Δ_Y and $x(\Delta_Y)$, x being a fraction or multiplier of the reference yield strength Ry. For $x \neq 1.0$, the resulting mismatch between the yield strengths of the LLRSEs produces the inelastic torsional response of interest in this study. The strong and weak elements are obviously defined as those having the largest and smaller yield strengths respectively. For $x = 1.0$, the resulting structures constitute equivalent SDOF systems whose inelastic response provides a basis for comparing element ductility demands.

The study was performed for ten values of uncoupled period T_x (0.1, 0.2, 0.3, 0.4, 0.6, 0.8, 1.0, 1.2, 1.6, and 2.0 seconds), six values of the ratio of uncoupled frequencies Ω (0.4, 0.8, 1.0, 1.2, 1.6, and 2.0), two SDOF target ductility levels μ_T (4 and 8), and four element yield combinations ("0.8 and 1.0 Ry", "1.0 and 1.2 Ry", "1.0 and 1.5 Ry", "1.0 and 2.0 Ry"). The equivalent SDOFs used for comparison were all selected to yield at Ry.

The four chosen yield level differences bracket many possible situations. Ultimately, further increasing the difference between the strong and weak elements could lead to permanent elastic response of the strong element with no further changes in element response. Note that although the difference in yield levels are herein assumed to result from the difficulty, or impossibility, in achieving similar stiffnesses and yield levels in different LLRSEs, this difference also implicitly considers the difficulty in accurately predicting the yield strength of some types of structural systems. Further, the intent is to assess the significance of over or under-estimating the yield strength of one element, and consequently, systems of different ultimate translational strengths will be compared in this process.

The following methodology was adopted for the parametric study.

1. Equivalent SDOF systems were selected to have a period equal to the uncoupled translational period T_x, the only period of initially symmetric structures excited before

initiation of yielding. These SDOF were designed such that they shared the same hysteretic characteristics and same yield displacement Δ_Y.

2. Normalized strength factors necessary for each SDOF system to attain target ductilities μ_T of 4 and 8 were calculated for each earthquake record considered. Constant ductility inelastic response spectra were constructed, from which the required strength factors, as a function of SDOF period, were read. Normalized strength factors are defined as:

$$\eta = R_Y / m\, a_{MAX} = K_X\, \Delta_Y / m\, a_{MAX} = \omega_X^2 \Delta_Y / a_{MAX} \qquad [4]$$

where a_{MAX} is the peak ground acceleration of a particular earthquake record, R_Y is the yield strength of the SDOF, and m is the mass of the equivalent SDOF system. For this study, ductility demand μ is defined as the maximum displacement, in absolute value, divided by the yield displacement. For simplicity in this study, peak ground accelerations were scaled as necessary, for fixed values of element model properties, to satisfy the imposed target ductility condition. These steps were taken to ensure that the SDOF structures were insensitive to variations in ground motion intensity. While this departs from a design approach, it ensures that any period-dependency observed in the calculated ductility amplification ratios (see item 4 following) is only attributable to the inelastic torsional coupling phenomena, and not to the seismic input spectral characteristics.

3. For the earthquake excitation levels calculated in the previous step, the same structures were reanalysed considering the unequal element yield strengths. The maximum inelastic element displacements were then calculated, as well as the corresponding element ductility demands. It is noteworthy that, as yield displacement is proportional to yield strength, equal maximum displacements will result in larger ductility for the weak element. Ductility demand of LLRSEs were selected as the response value of interest in this study.

4. The ductility demands calculated for each individual initially symmetric case analyzed above were then divided by the ductility demands obtained from their respective equivalent inelastic SDOF system, to obtain a ratio of the ductilities [indicated "Ductility Amplification Ratios" on all figures herein]. This amplification ratio provides a normalization over the selected target ductilities. It is believed that the ductility amplification ratios for each element of the two-element structures provide the best quantitative measure of the damage sensitivity of the structures.

To provide results mostly independent of the particular characteristics of a single earthquake, five different earthquake records (El Centro 1940 N-S, Olympia 1949 N-S, Parkfield 1966 S16E, Paicoma Dam 1971 N65E, and Taft 1952 N21E) were considered, and the mean, and mean-plus-one-standard-deviation, of response values were calculated.

RESULTS OF THE PARAMETRIC STUDY AND OBSERVATIONS

Figs. 2 and 3 present the results from step 4 above. These plots show the mean ductility amplification ratios of the weak [Fig. 2] and strong [Fig. 3] elements for a target ductility of 4, for the five earthquake records described. Results pertaining to the mean-plus-one-standard-deviation of the ductility amplification ratios, as well as results for a selected target ductility of 8, are presented elsewhere *[Bruneau and Mahin 1987]*.

Considering the nature of ductility measurements in earthquake engineering, and the accuracy expected in ductility prediction of this kind, it might be said that element ductility amplification ratios of 1.25 or less are not considered significant, ductility amplification ratios from 1.25 to 1.5 are considered of moderate importance, and ratios above 1.5 are judged to be of major importance. Following this arbitrary convention, the following can be observed.

1. The weak element ductility amplification ratios for the case "0.8 & 1.0 Ry" are always at least of moderate importance, and often of major importance. This amplification is most severe for cases with small periods or large Ω values (and most significantly a combination of both), with ductility amplification ratios ranging from 2 to 4 for the mean response [Fig. 2]. Amplifications were somewhat expected since the ultimate translational strength of the "0.8 Ry and Ry" structures are less than that of their reference systems; nevertheless, the rather large magnification of weak element ductilities obtained remain impressive.

2. When the yield level of one element is superior to that of the reference SDOF, the weak element ductility amplification ratios are mostly non-affected until Ω becomes larger than 1.6 for the mean response. In that case, the response is also seen to slightly increase along with the yield stresses differentials. The increase in weak element ductility amplification ratios, despite the increased ultimate translational strength of the structures, is surprising. It implies that the added torsional behavior induced by the increase in yield level differential more than overcomes the benefit one might associate with the increase in strength (or balances it in the best case). Increases of 100% are seen for large Ω and large yield level differences, and much larger ductility amplification ratios, often up to 2.5, were observed for single earthquake excitation results. Thus, there is no guarantee that an unbalanced increased strength in a symmetric structure decreases ductility. It should be noted that at some point, further increase in yield level differential would produce no additional change in response for either elements, as the strong element would reach permanently elastic behavior.

3. The strong element ductility amplification ratios are all less than 1.0, except in the "0.8 & 1.0 Ry" case where the lower structures ultimate translational strengths make larger inelastic deformations also possible in the stronger element. Ductility amplification ratios of moderate importance can be noticed in the case of low periods (T ≤ 0.2) and low Ω values (Ω ≤ 0.8) [Fig. 3]. It is noteworthy that the decrease in strong element ductility amplification ratios occurring with the increase in the ultimate translational strength of

the structures is partly a consequence of the increase of the yield level of the strong element; i.e. an increase in yield level (corresponding to an equivalent increase in the yield displacement), will produce an effective reduction in the ductility demand for a given magnitude of displacement. A value of the strong element ductility amplification ratio below 1.0 reflects that situation; it does not imply that the strong element remains elastic, but simply that ductility demand is less than that of its corresponding SDOF system.

4. The observed ductility amplification ratios are generally independent of target ductility levels. Demonstration of this is presented elsewhere *[Bruneau and Mahin 1987]*.

5. The structural period is seen to have no significant influence on the ductility amplification ratios of initially symmetric structures, except for the "0.8 and 1.0 Ry" case, where weak element ductility amplification ratios are generally larger for structures with small periods [Fig. 2]. This is expected as equivalent SDOF systems have been calibrated to target ductilities with yielding set at Ry. The case "0.8 and 1.0 Ry" having smaller ultimate translational strength than its equivalent SDOF system, the natural tendency of short period structures to have larger response than more flexible structures, typical of earthquakes for the West Coast of the United States, resurfaces.

6. For the methodology followed herein, the element yielding at Ry (i.e. the strong one in the case "0.8 Ry and Ry," and the weak one in the other cases) will always have the same inelastic response as the SDOF system yielding at Ry when $\Omega = 1.0$, and therefore the element ductility amplification ratios will always be 1.0 in that particular case. This rather interesting phenomenon can be accurately predicted by theory, and is explained in great detail elsewhere *[Bruneau and Mahin 1987]*.

7. As seen from Figs. 2 and 3, weak element ductility amplification ratios tend to increase with larger Ω, while strong element ductility amplification ratios tend to decrease accordingly. This can be explained by the lower resistance to angular motion provided by structures with larger Ω values, as explained in more details elsewhere *[Bruneau and Mahin, 1987]*. Obviously, this increase in weak element ductility amplification ratios with Ω would not be observed as consistently when looking at the response under a given earthquake excitation, on account of the particular characteristics proper to any single earthquake record, but it is a clear trend that can be observed from the presented results for the mean responses to the five earthquake excitations used in this study. Although there is a few instances in Fig. 2 where the weak element ductility amplification ratios decreases for step increases in Ω, most of these decreases are of negligible magnitudes, and principally occur for low Ω values and large dissimilitudes in element yield strengths.

Based on previous observations, the following design recommendation can be formulated: For structural systems which can be idealized within the restrictions of this study, assuming the yield strength of the LLRSEs are dissimilar and can be estimated, the ductility demand of the weaker element is expected to exceed by approximately 50% the ductility demand of a SDOF of similar yield strength, if Ω is larger than 1.2. The designer expecting to limit the ductility demand on structural members in those cases should reduce its target ductility demand by 30% (1/1.5 = 0.67).

CONCLUSIONS

For the simple initially symmetric structures studied having unbalanced yield strengths in plan, a transient torsional response is created by the desynchronizing in inelastic element response, despite the existence of symmetry in the elastic domain. Resulting element ductility amplification ratios will remain low provided the ratio of uncoupled frequencies Ω is not excessively large (preferably 1.2 and lower) and the yield strength of the weaker element in the initially symmetric structure is not less than the yield strength of the equivalent SDOF. This conclusion is seen to remain valid for all translational periods and level of seismic excitation.

REFERENCES

1. Antonelli, R. G., Meyer, K. J., Oppenheim, I. J., 1981. Torsional Instability in Structures, International Journal of Earthquake Eng. and Soil Dyn., Vol. 9, pp. 221-237.
2. Awad, A. M., Humar, J. L., 1984. Dynamic Response of Buildings to Ground Rotational Motion, Canadian Journal of Civil Engineering, Vol. 11, No. 1, pp. 48-56.
3. Bruneau, M., Mahin, S. A., 1987. Inelastic Seismic Response of Structures with Mass or Stiffness Eccentricities in Plan, Report No. UCB/EERC-87/12, Earthquake Engineering Research Center, University of California, Berkeley.
4. Newmark, N. M., 1969. Torsion in Symmetrical Buildings, 4th World Conference on Earthquake Engineering, Vol. 2, Santiago, Chile, pp. A3-19 to A3-32.
5. Pekau, O. A., Guimond, R., 1988. Accidental Torsion in Yielding Symmetric Structures, Proc. of the Ninth World Conference on Earthquake Eng., Vol. V, pp. 85-90.
6. Pekau, O. A., Syamal, P. K., 1984. Non-Linear Torsional Coupling in Symmetric Structures, Journal of Sound and Vibrations, Vol. 94, No. 1, pp. 1-18.
7. Tso, W. K., 1975. Induced Torsional Oscillations in Symmetrical Structures, International Journal of Earthquake Eng. and Soil Dyn., Vol. 3, pp. 337-346.
8. Tso, W. K., Asmis, K. G., 1971. Torsional Vibration of Symmetrical Structures, Proc. of the First Canadian Conference on Earthquake Engineering, Vancouver, pp. 178-186.

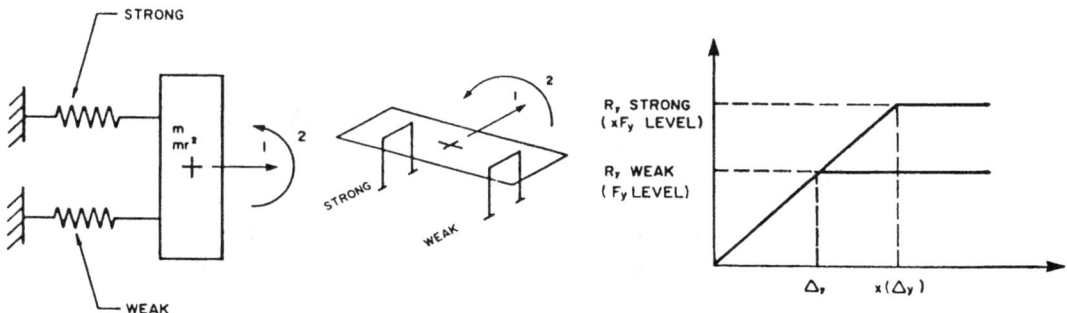

Figure 1: Two LLRSE system - Computer model, physical model and constitutive model.

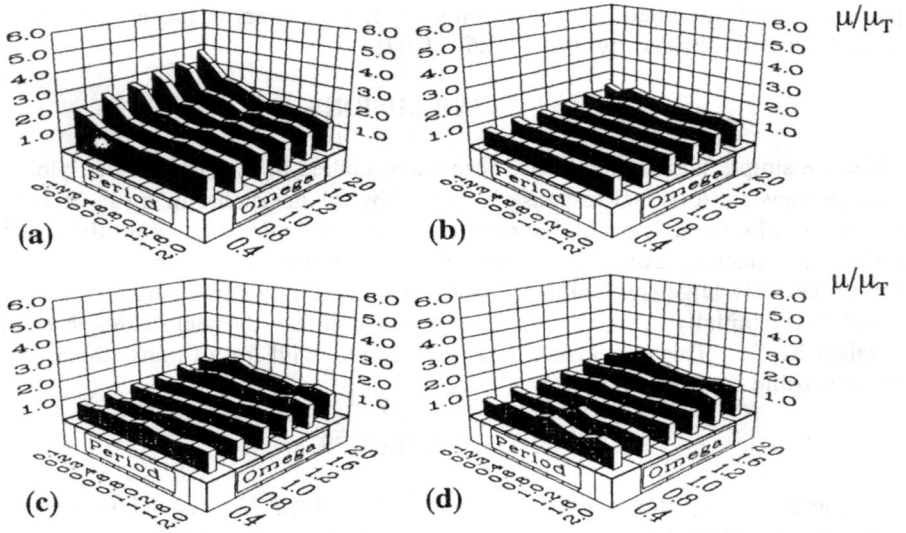

Figure 2: Mean weak LLRSE ductility amplification ratios for target ductility (μ_T) of 4 and yield strengths combinations (a) 0.8 and 1.0 R_Y, (b) 1.0 and 1.2 R_Y, (c) 1.0 and 1.5 R_Y, (d) 1.0 and 2.0 R_Y.

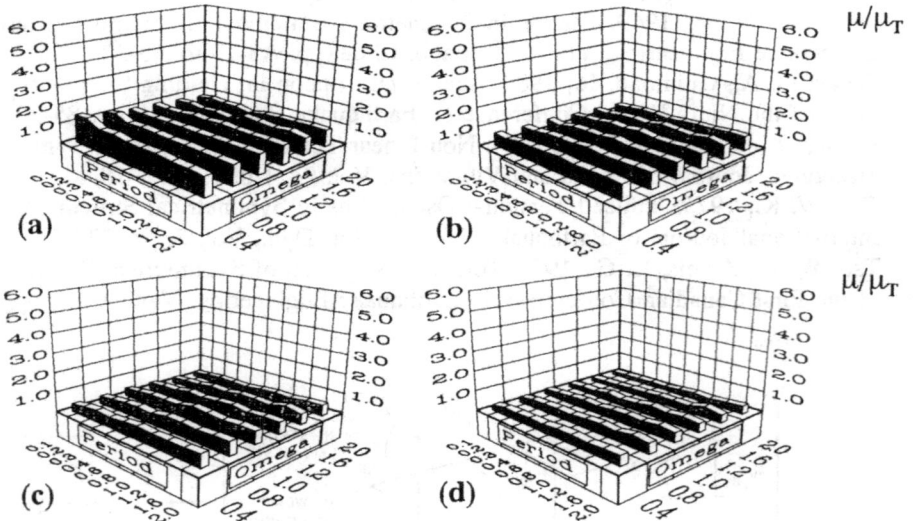

Figure 3: Mean strong LLRSE ductility amplification ratios for target ductility (μ_T) of 4 and yield strengths combinations (a) 0.8 and 1.0 R_Y, (b) 1.0 and 1.2 R_Y, (c) 1.0 and 1.5 R_Y, (d) 1.0 and 2.0 R_Y.

Seismic damage concentration in steel building structures with weak beam-to-column joint panels

Takashi Hasegawa[I], Hiroyuki Yamanouchi[II], and Hiroshi Akiyama[III]

ABSTRACT

It is, in general, recommended that beam-to-column joint panels of moment-resisting steel building structures should remain elastic under seismic design forces representing medium earthquake effects. To keep this requirement, particularly in using H-shaped beams and columns, strengthening of beam-to-column joint panels with cover-plates or stiffeners is often needed. On the other hand, when no strengthening of beam-to-column joint panels is given, the seismic failure mode of this type of structures may be specified by the yielding of the weak joint panels prior to that of the adjacent beams or columns. Here, there arises a significant question whether weak joint panels are preferable or not under severe earthquake ground motions. This paper gives an answer to the above question through the dynamic time-history response analysis with an appropriate structural modeling, and by using a concept of seismic energy input.

INTRODUCTION

In a moment-resisting steel building structure, the strength of beam-to-column joint panels is a key factor that would determine the failure mode of the structure under severe earthquake ground motions. In high seismic zones such as almost all areas of Japan, it is recommended that the beam-to-column joint panels should be strengthened to remain elastic under seismic design forces representing the load effect of medium earthquakes. However, such strengthening as with cover plates or stiffeners usually requires difficult and tedious design work, and further the effectiveness of the strengthening is sometimes doubtful.

On the other hand, when no strengthening of joint panels is given, the panel yield strength is, probably in almost all cases, less than the larger yield strength of beams and columns. Thus, the seismic failure mode in those cases may be determined by the joint panel yielding prior to the yielding of the adjacent beams and columns; this type of failure is hereinafter called "joint panel failure mode". According to a lot of test data, however, beam-to-column joint panels have generally a large capacity of energy absorption owing to stable shear deformation. This capacity may be used effectively to preclude a moment-resisting steel structure from collapse under severe earthquakes.

I   Research Engineer, Structural Dynamics Division, Building Research
    Institute (B.R.I.), Tsukuba, Ibaraki, Japan
II Head of Structural Dynamics Division, B.R.I.
III Associate Professor, University of Tokyo, Japan

At the present time, the "joint panel failure mode" is not yet recognized to be recommendable since the effect of joint panel strength on the overall seismic behavior has not yet fully understood. Recently a few studies have been carried out regarding the effect of strength of joint panels on the strength and ductility of steel frames (Nakao 1990; Tanaka 1990). In these studies, ratios of panel yield strength to the lesser yield strength of adjacent beams and columns are presented as a recommendable lower limit. However, the ratios should be determined through a parametric study on the effect of panel strength on the characteristics of seismic damage concentration in structures.

The primary object of this paper is then to investigate the influence of strength of beam-to-column joint panels on seismic behavior of steel structures and to propose a minimum strength of joint panels, which is required to keep the structures sound under both medium and severe earthquake conditions.

ANALYSIS

Analytical models

The basic dynamic system used for the analysis is a single column model having masses concentrated into the center of beam-to-column joint panels as shown in Fig. 1(a). This model is assumed to be extracted from a horizontally infinite and equally spanned moment-resisting plane frame as shown in Fig. 1(b). Further, it is assumed that each particle of mass has the quantity of m. As characteristics of the members used in the model, only flexural deformation is considered for beams and columns, and only shear deformation for joint panels.

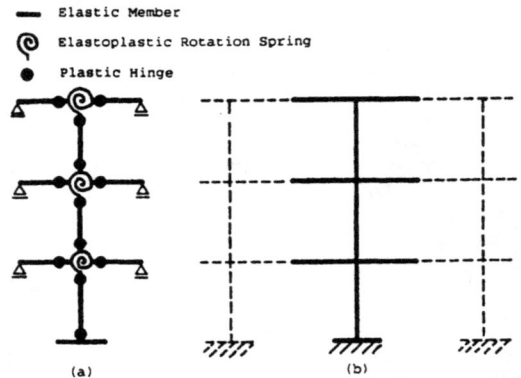

— Elastic Member
ⓞ Elastoplastic Rotation Spring
● Plastic Hinge

Fig. 1 Vibration Model for Analyses

In addition, axial forces in the columns are neglected. As inelastic properties of the members, plastic hinges are assumed to be formed at the ends of beams and columns. Furthermore, at the joint panels elastoplastic rotation springs representing shear deformation of the panels are considered. The restoring force characteristics of these inelastic elements are assumed to have elastoplastic bi-linear relationships.

Three groups of analysis

The analysis of this study is divided into three groups; Analyses A, B and C. The purposes and major parameters of each analysis are briefly summarized in Table 1.

The Analysis A was carried out to investigate the effect of failure mechanisms on total input energy into the model. Three different failure modes, that is, the column failure mode, beam failure mode and joint panel failure mode were independently implemented in the analysis. Namely, to make one particular failure mode predominant, plasticization of inelastic elements concerned with other failure modes was not allowed in setting the model.

The purpose of the Analysis B was to make clear the relation between the joint panel strength and the damage concentration into the structure in which the yielding of joint panels is ahead of the yielding of beams and columns.

Table 1 Purposes and Parameters for Analyses

	Analysis(A)	Analysis(B)	Analysis(C)
Story N	5,10	5	5
Base shear coefficient;α_1	0.1 0.2 0.3	0.2	0.25
Rpy	---	0.3 - 1.3	0.3 - 1.2
Beam/column strength ratio	---	1/1.4 - 1.4	1/1.2 1.2
Fundamental natural period	1.0(s)	1.0(s)	0.57 - 0.77(s)
Second slop of bi-linear	column,beam; 0% panel; 0%	column,beam; 0% panel; 3%	column,beam; 0.5% panel; 2%
Input earthquake ground motion unit(m/sec^2)	Hachinohe EW (Max1.83)	Hachinohe EW (Max1.83)	Hachinohe EW 0.5m/sec(Max.3.02) 0.25m/sec(Max.1.54) El Centro NS 0.5m/sec(Max.4.06) 0.25m/sec(Max.2.03)
Purpose of analysis	Effect of failure mechanism on input energy	Relation between Rpy and damage concentration	Seismic performance in terms of η and interstory drift

Here, the panel strength was represented by a new important parameter, Rpy, which is defined as follows (also see Fig.2);

$$R_{py} = {}_pM_{yi} \ / \ min \left[\left({}_{Lb}M_{yi} + {}_{Rb}M_{yi} \right), \left({}_{Uc}M_{yi} + {}_{Lc}M_{yi+1} \right) \right] \qquad (1)$$

where min[A,B] means the lesser of A and B, and where LbMyi and RbMyi are yield moments of left and right hand side beams adjacent to the joint panel, and UcMyi and LcMyi+1 are yield moment of the upper and lower columns adjacent to the joint panel.

In this analysis, the values of Rpy were varied from 0.3 to 1.3. Also in this analysis, the ratios of beam strength to column strength were varied from 1.0/1.4 to 1.4 considering realistic combinations of beams and columns.

In the Analysis C, the model structures in which the joint panel yielding was ahead of the yielding of beams and columns were dealt with to obtain the seismic performance in terms of the maximum interstory drift and accumulated plastic deformations of the inelastic hinge elements. Further, to investigate the seismic performance under both medium and severe earthquakes, the maximum input velocities of 0.25m/sec and 0.5m/sec were chosen. As input waves, the two recorded ground motions, Hachinohe Earthquake 1968(EW) and El Centro Earthquake 1940(NS) were linearly scaled to have the two maximum velocities.

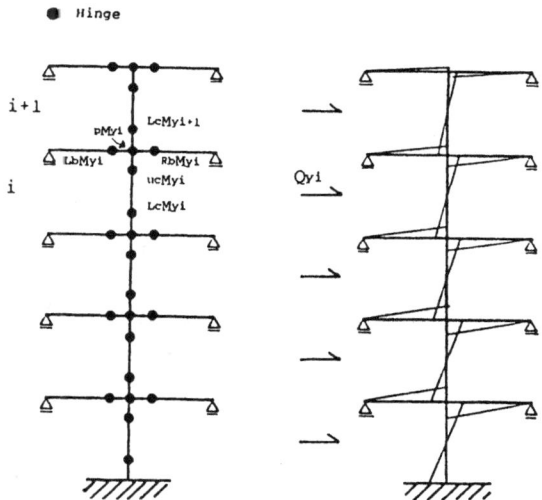

Fig. 2 Prototype Model Structure

Total input energy

Fig. 3 shows the most significant result of the Analysis A; this figure deals with the model structures having the base shear coefficient $\alpha 1 = 0.2$ and the number of stories N = 5. Here, the total input energy into the structures was normalized as the form of velocity VE which is defined as

$$V_E = \sqrt{\frac{2E}{M}} \tag{2}$$

Fig. 3   Total Input Energy

where E is the total input energy and M is the total mass of the system. From this figure, it can be seen that the total input energy represented by VE depends on only the fundamental natural period T of the system and does not on the predominant failure modes. Also in this figure, the dashed line designates the VE values obtained from a damped one-mass system with h = 0.1. It is then found that the plotted values on the five-mass system are close to the dashed line.

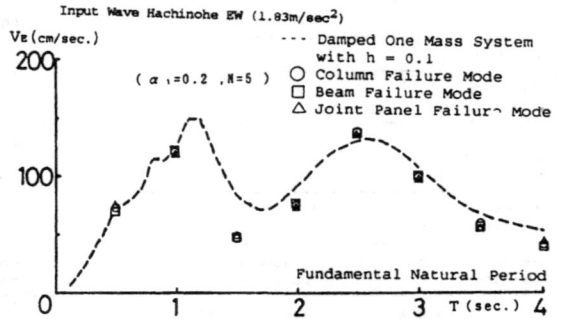

Damage concentration

Before discussing the results of the Analyses B and C, we should describe the structural model in more detail. Fig. 2 shows the prototype model structure in which the yield moments of beams and columns determined by the moment distribution corresponding to the optimum yield-shear distribution Qyi along the height of the structure (Akiyama 1985). Then, the ratios among the yield strengths of beams, columns and joint panels were varied from those of the prototype to investigate the damage distribution.

Fig. 4   Damage Distribution along the Height of Model Structures

Here, the above ratio for each story was set to be identical for one analyzed structure.

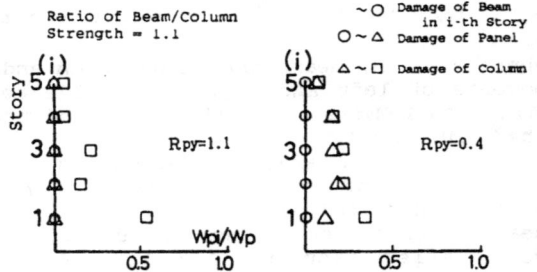

Fig. 4 shows an example of the results of the Analysis B. The two analyzed structures of the figure had 1.1 times stronger beam than those of the prototype. Furthermore, the panel yield ratios, Rpy, were set to be 1.1 and 0.4. The ordinate of the figure designates the story number, and the abscissa indicates the degree of the damage concentration into the i-th story, Wpi/Wp, where these symbols are defined as follows; Wpi is the absorbed energy by inelastic hysteresis into the i-th story, which is calculated by

$$W_{pi} = {}_cW_{pi} + \frac{K_i}{K_i+K_{i+1}}\left({}_pW_{pi}+{}_bW_{pi}\right) + \frac{K_i}{K_i+K_{i-1}}\left({}_pW_{pi-1}+{}_bW_{pi-1}\right) \tag{3}$$

where Ki is the story stiffness of the i-th story, bWpi is the total energy absorbed into the right- and left-hand side beams of the i-th story, pWpi is the absorbed energy of the panel of the i-th story and cWpi is the sum of the absorbed energy at the both ends of the i-th story column, $\text{UcWpi} + \text{LcWpi}$. Furthermore, the total input plastic energy into a structure, Wp, is given by

$$W_p = \sum_{i=1}^{N} W_{pi} \tag{4}$$

From Fig. 4, it can be seen that the damage into the panels having the strength ratio of $\text{Rpy} = 0.4$ is obviously increased in each story, compared with the case of $\text{Rpy} = 1.1$ where there is no damage in the panels. Moreover, from this figure, we can find that in the case of $\text{Rpy} = 0.4$ the extent of damage concentration into the column at the first story becomes considerably less than that in the case of $\text{Rpy} = 1.1$; this means that the damage distribution along the height of the structure becomes uniform by decreasing the panel strength, Rpy.

Damage concentration factor

To study more clearly the peculiarity of damage concentration resulted from the change of Rpy, a new factor, damage concentration factor, n, should be introduced. This factor is included in the equation that defines the damage distribution over all stories as follows (Akiyama 1985);

$$\frac{W_{pk}}{W_p} = \frac{S_k \cdot \bar{P}_k^n}{\sum_{j=1}^{N} S_j \cdot \bar{P}_j^n} \tag{5}$$

where
n = damage concentration factor, $P_j = (\alpha_j/\alpha_1)/\bar{\alpha}_j$,
$\bar{\alpha}_j$ = optimum yield-shear force coefficient distribution, and

$$S_j = \left(\sum_{k=j}^{N} m_k/M \right)^2 \bar{\alpha}_j^2 \left(K_1/K_j \right).$$

The damage concentration factor, n, is then obtained from Eq. 5 by using the results of response analysis where only the story shear coefficient of the k-th story, α_k, is changed from the optimum one, $\bar{\alpha}_k$. That is, the damage concentration factor, n, is expressed as

$$n = - \ln \left\{ \frac{b(1-a)}{a(1-b)} \right\} / \ln P_d \tag{6}$$

where a is given by W_{pk}/W_p with the optimum shear coefficient $\bar{\alpha}_k$ and b is W_{pk}/W_p with the changed shear coefficient, $\alpha_k = P_d\bar{\alpha}_k$. The values of a and b can be concretely attained by response analyses. The factor, n, is then considered to be an index designating the damage concentration into the k-th story having a certain weakness in story shear strength. In a shear-type multi-story structure, the value of n is found to be 12, and as the flexural component is gradually incorporated into the overall structural behavior, the n-value becomes smaller than 12 (Akiyama 1985).

Now, the factor n was evaluated on the structure models having different Rpy-values from 0.3 to 1.3. Herein, the coefficient Pd and the story number k were chosen to be 0.8 and 3, respectively. Needless to say, the factor n would depend on Pd and k. Thus, the story number, k = 3 is determined by the result of the past study in which the number, 3, gives the most severe damage concentration in five-story models as a safety side evaluation (Akiyama 1985), and also the selection of Pd = 0.8 gives an appropriate value of n which can designate reasonably damage concentration (Akiyama 1985).

121

Fig. 5 depicts the obtained values of n as a function of the panel strength ratio, Rpy, associated with the ratios of the beam yield strength to column yield strength. From this figure, it can be seen that in the region of Rpy<1.0, there is a tendency for damage concentration into the third story to lessen as Rpy becomes smaller.

When Rpy>1.0, on the other hand, the damage concentration does not depend on Rpy, and does depend on only the strength ratios between beams and columns; this is because the stronger joint panels do not exhibit plasticization. Thus, in this range of Rpy, the damage concentration becomes smaller as the ratio of beam strength to column strength becomes smaller.

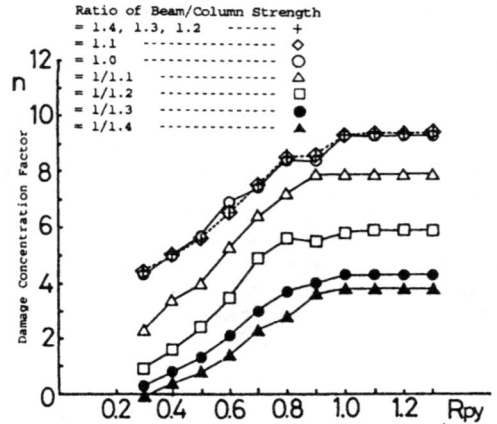

Fig. 5 Relationship between Rpy and n

Seismic performance

The analysis C was carried out to examine the seismic performance of structures having weak joint panels in terms of plastic deformations of hinge- and spring-elements and interstory drifts. Here, to indicate the amount of plastic deformations of hinge and spring-elements, we used the " cumulative inelastic deformation ratio, η " which is defined as follows;

$$\eta = \frac{_jW_{pi}}{_jM_{yi} \cdot _j\theta_{yi}}$$

(7)

where j expresses each member of the column, beam and joint panel, and where $_jM_{yi}$ and $_j\theta_{yi}$ is the yield moment and yield rotational angle of the i-th story's j member.

Now, Figs. 6 demonstrates some examples of the analytical results in terms of obtained η regarding inelastic elements in beams, columns and joint panels; this figure shows just the results obtained against the input of El Centro record having the scaled maximum velocity of 0.5m/sec. In this analysis, the panel yield ratios, Rpy, were set to be 1.2, 1.0, 0.7 and 0.3, and the strength ratio of beams to columns was chosen to be 1.2.

Looking at Fig. 6, we can say that η of the panels becomes larger as Rpy becomes smaller, and corresponding to this inclination, η of beams and columns decreases. Furthermore, it can be found that in the case of Rpy >1.0, the values of η of the columns become larger than those in the case where Rpy<1.0. These values exceed the value of η=12 which is generally accepted as the upper limit of H-shaped columns (S.C. 1990).

On the other hand, when Rpy is 0.7, the values of η of the columns do not reach the upper limit except for the bottom of the first story column, since the weak panels absorb a considerable amount of energy. Therefore, it can be said that weak panels is effective to reduce the damage of beams and columns, in particular to prevent the failure of columns.

Fig. 7 shows the relationships between Rpy and the average of η of all panels in each model. In addition, recent data obtained from monotonic loading tests on cruciform beam-to-column joint specimens (Matsumoto 1990) are

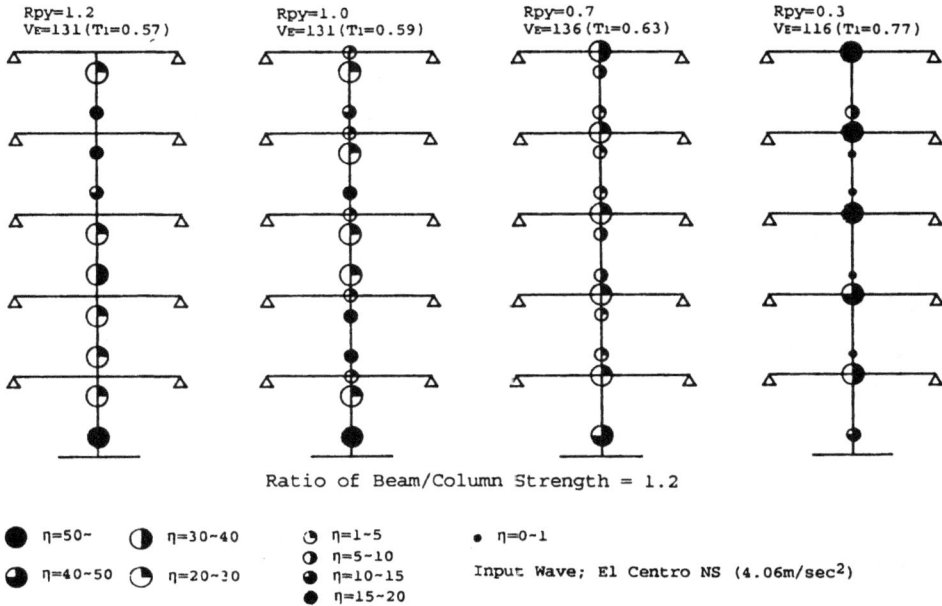

Ratio of Beam/Column Strength = 1.2

η=50~   η=30~40   η=1~5   η=0-1

η=40-50   η=20-30   η=5-10

η=10-15   Input Wave; El Centro NS (4.06m/sec²)

η=15-20

Fig. 6   η of Each Member against Severe Earthquake

also plotted in this figure. By these experimental data, the upper limit of η to lose load-carrying capacity is around 75, whereas the analytical η reaches about 80 at Rpy of 0.3. Thus, by this comparison, the joint panel with Rpy=0.3 seems to be critical against such severe earthquake as the scaled Hachinohe. However, the experimental η of 75 was attained under monotonic loading so that attainable η under reversed and alternative cyclic loading simulating seismic responses would be larger than η=75, because the monotonic deformation capacity of panels can be demonstrated in both the plus and minus loading directions. By this consideration, even when the panel yield ratio, Rpy, is 0.3, a dangerous failure of the panel would not occur in the structure having the base shear coefficient of 0.25.

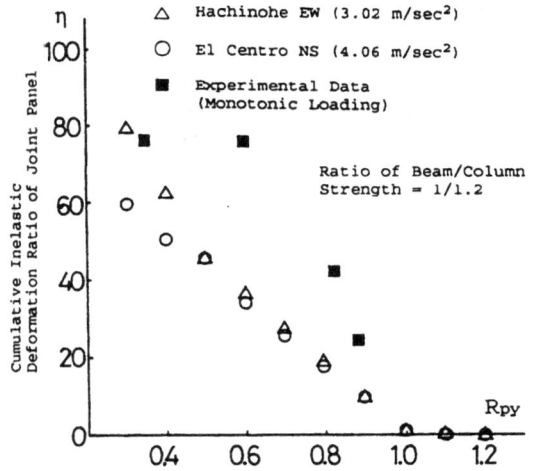

Fig. 7   Relationships between Rpy and η of Joint Panels

As the other result of the Analysis C, Fig. 8 shows the relationships between Rpy and interstory deflection against medium earthquakes. In this part of the analysis, the panel yield ratio, Rpy, were 0.3, 0.4, 0.5, and 0.8. The used earthquake records were the El Centro and Hachinohe whose maximum velocities were scaled to 0.25 m/sec to represent a medium level of earthquake ground motions.

From this figure, it would be concluded that Rpy should be larger than 0.5 if the upper limit of interstory drift angle is required to be 1/200 against medium earthquake as a serviceablity limit.

CONCLUSIONS

The influence of strength of beam-to-column joint panel on seismic damage concentration in a steel building structure with shear deformation of joint panel was investigated through the three analyses, and the following conclusions were obtained.

1) The total input energy represented by VE depends on only the fundamental natural period T of the system and does not on the predominant failure modes. And the VE values obtained from the response analysis on a damped one mass system with h=0.1 can represent these values.

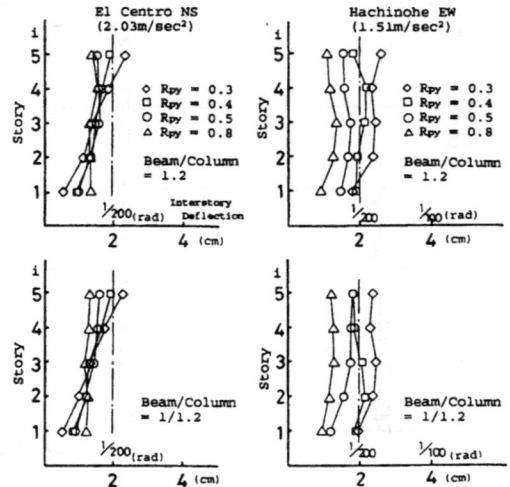

Fig. 8 Relationships between Rpy and Interstory Deflections and Drift Angles

2) Regarding the relation between Rpy and seismic damage concentration, in the region of Rpy<1.0, there is a tendency for damage concentration into the particular story to lessen as Rpy becomes smaller. When Rpy>1.0, on the other hand, the damage concentration does not depend on Rpy, and does depend on only the strength ratios between beams and columns. The damage concentration becomes smaller as the ratio of beam strength to column strength becomes smaller.

3) From the analysis to examine the seismic performance of structure having weak joint panels, it was cleared that η of the joint panels becomes larger as Rpy becomes smaller, and that corresponding to this inclination, η of beam and column decreases. On the other hand, Rpy should be larger than 0.5 against a medium earthquake as a serviceablity limit.

REFERENCES

Akiyama, H. 1985. Earthquake-resistant limit-state design for buildings. University of Tokyo Press.

Matsumoto, S., Nakao, M. et al. 1990. Experiment on energy absorbing of weak panel type frame sub-assemblages composed of H-shapes. Proceedings of Annual Conference of Architectural Institute of Japan (AIJ), 1609-1610.(in Japanese)

Nakao, M. 1990. Structural characteristics of H-shaped beam-to-column joint. Kenchiku gijutsu, No.464, 124-130.(in Japanese)

Structural Committee. 1990. Ultimate strength and deformation capacity of building in seismic design (1990). AIJ.(in Japanese)

Tanaka, A. 1990. Seismic design of beam to column joints in steel structures. AIJ, symposium, 1-11.(in Japanese)

Tanaka, A. 1990. New tendency of structural design and design of beam-to-column joints in steel structures. Kenchiku gijutsu, No.464, 106-114.(in Japanese)

Seismic analysis of structurally interconnected steel frames

A. Filiatrault[I], B. Folz[II], and H.G.L. Prion[III]

ABSTRACT

This paper evaluates the seismic performance of a pair of closely spaced plane steel frames interconnected by a horizontal structural link to prevent pounding during earthquake excitations. Friction damping capability is incorporated into the modelling of this link in order to also determine its potential for dissipating the seismic energy. The frames, one three-storey, the other eight-storey, are individually designed according to the 1990 edition of the National Building Code of Canada. A parametric study, utilizing nonlinear time-history dynamic analysis, is performed to determine the influence of the slip load for the structural link when the frames are excited by artificial accelerograms representative of an upper bound of the potential ground motion in Vancouver. It is shown that a structural link with friction damping capabilities reduces the ductility demands of both frames compared to a purely elastic link.

INTRODUCTION

Structural damage caused by two buildings, or different parts of the same building, impacting one another during an earthquake has been observed on numerous occasions over the past several decades. For example, during the 1972 Managua earthquake the third floor of the Grant Hotel in downtown Managua, completely collapsed when hit by the roof level of an adjacent two-storey building (Berg and Degenkolb, 1973). Also, the fourteen-storey Westward Hotel suffered damage when it pounded against its low-rise six-storey wing during the 1964 Alaska earthquake (National Academy of Sciences, 1973). More dramatic pounding failures were observed recently in Mexico City in 1985 (EERI, 1985), and in Santa Cruz during the 1989 Loma Prieta earthquake (Asteneh et al., 1989). The problem of pounding is particularly acute in many large cities located in seismically active regions where, due to land usage requirements, buildings are constructed in very close proximity to each other.

Although pounding may constitute one of the primary sources of structural damage during an earthquake, limited research has been conducted on the subject. A literature review reveals that the effect of pounding has been considered in investigations on the response and/or collapse of particular buildings (Mahin et al., 1976; Wada et al., 1984; Wolf and Shrikerud, 1980). Also, some valuable insights on the problem of pounding have been obtained recently from an analytical study (Anagnostopoulos, 1988). In particular, it was shown that for structures aligned in series the response of the end structures can be magnified while the response of the interior structures can be reduced. This result agrees well with observed damage patterns.

[I]Assist. Prof., Dept. Civil Eng., Ecole Polytechnique, Montréal, QC;
[II]Research Eng., Dept. Civil Eng. Univ. of B.C., Vancouver, B.C.;
[III]Assist. Prof., Dept. Civil Eng., Univ. of B.C., Vancouver, B.C.

Most modern building codes address the problem of pounding by requiring that adjacent structures be either separated by the sum of their anticipated individual deflections or be connected to each other. Although it seems clear that connecting buildings together will force them to vibrate as a unit and thereby reduce the pounding potential, very limited information is available on the actual seismic response of interconnected structures. The elastic vibrational response of coupled plane frames was recently investigated (Westermo, 1989). However, the effect of the coupling on the inelastic structural response was not included in this study and therefore the results may not be realistic since most structures will undergo inelastic deformations under severe earthquakes.

This paper attempts to shed some further light on the problem of interconnected buildings by considering a case study. The inelastic earthquake response of a pair (three and eight storey) of closely spaced, code designed, plane steel frames interconnected by a horizontal link is computed using nonlinear time-history dynamic analysis. In addition, the energy dissipating potential of the structural connection is investigated by considering a linkage system that incorporates friction damping capabilities.

STRUCTURAL MODELS AND ASSUMPTIONS

Building Models

The pair of plane steel framed building models considered in this case study consisted of an eight-storey frame connected to the roof of a three-storey frame. The frames were designed individually as ductile moment resisting frames (R_v=4) according to the static method of the 1990 edition of the National Building Code of Canada (NBCC, 1990). The new seismic detailing requirements of the CAN/CSA-S16.1-M89 (CSA, 1989) Canadian steel code was incorporated into the design. The "weak-beam strong-column philosophy" was utilised by specifying the beams as the critical elements. The resulting structural properties along with some of the design assumptions are detailed in Fig. 1. Note for simplification that the two structures considered have identical floor elevations. For the case where floors of adjacent buildings are not in horizontal alignment, a special vertical beam can be used to span two adjacent floors on one of the buildings (Westermo, 1989). The horizontal connecting link could then be attached to this beam. Horizontal linkage is obviously preferable so as not to introduce any supplementary uplift on the buildings.

Structural Interconnection

For the purpose of this case study the axial stiffness of the linkage system was set equal to the axial stiffness of the more flexible connecting beams (W310x60). In order to investigate the potential of dissipating seismic energy through the linkage system, friction (Coulomb) damping devices were incorporated into their design as illustrated in Fig. 2. In practice, the idealized hysteretic behaviour shown in Fig. 2. could be achieved by inserting heavy duty brake lining pads between clamped steel surfaces. An earlier experimental investigation by the first author has shown that this type of friction damping device produces very stable non-deteriorating hysteresis loops (Filiatrault and Cherry, 1987).

Earthquake Ground Motions

The structural models were subjected to a set of three earthquake records. which represents an upper bound of the potential ground motion in Vancouver originating from a subduction earthquake off the west coast of Vancouver Island. These accelerograms were generated from random vibration theory (Filiatrault and Cherry, 1988) based on a magnitude M_L of 8.5 and an epicentral distance of 150 km. The peak acceleration of each record was 0.35g with a duration of 25 seconds.

126

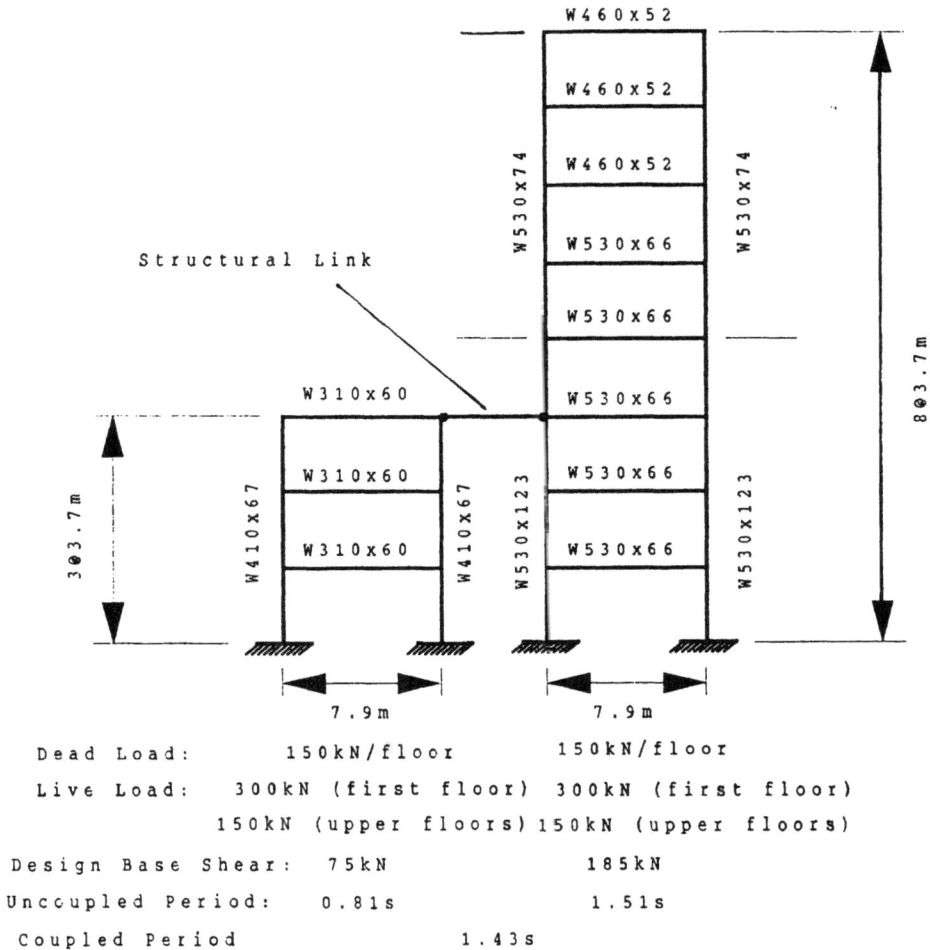

Figure 1 - Structural Models for Case Study.

Dead Load:	150kN/floor	150kN/floor
Live Load:	300kN (first floor)	300kN (first floor)
	150kN (upper floors)	150kN (upper floors)
Design Base Shear:	75kN	185kN
Uncoupled Period:	0.81s	1.51s
Coupled Period	1.43s	

The absolute acceleration response spectra corresponding to the seismic events are presented in Fig. 3.

Structural Analysis

All structural responses were obtained from inelastic time-history dynamic analyses using a microcomputer version of the well-known general purpose program DRAIN-2D (Kannan and Powell, 1973). Flexural and axial deformations were monitored in the structural members and the interaction between axial forces and moments at yield were taken into account by means of standardized yield interaction surfaces for steel members. No viscous damping was considered in the

127

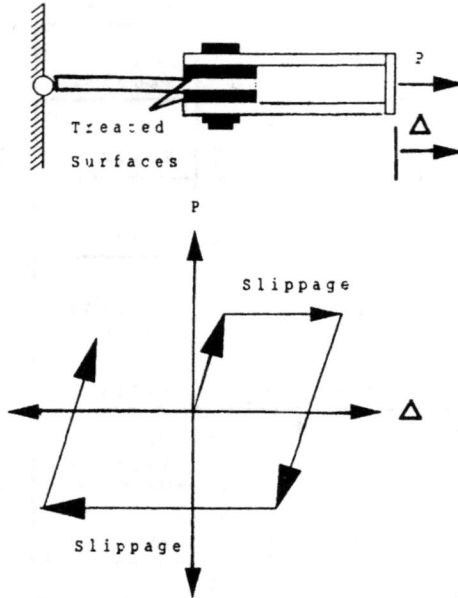

Figure 2 - Friction Damped Structural Link.

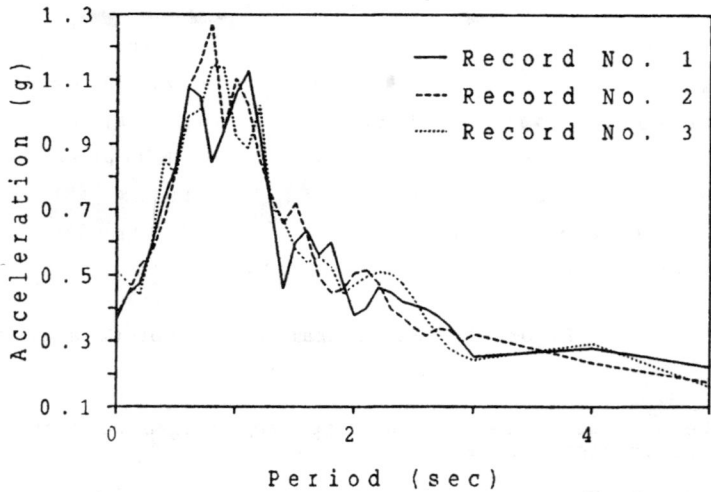

Figure 3 - Acceleration Response Spectra for Seismic Events (5% Damping).

structures so that the proportion of energy dissipated by friction through the linkage system could be more easily identified. Rigid foundations were assumed and soil-structure interaction was neglected. The full dead load was applied to the members prior to the shaking.

128

RESPONSE OF UNCOUPLED SYSTEM

The first step in the analysis was to evaluate the response of the uncoupled structures to the seismic events considered. From these results, minimum required separation distances to avoid pounding were obtained. These distances (S_r) are presented in Table 1. For comparison, the separations required for Vancouver by the National Building Code of Canada (NBCC, 1990) are also presented. Correlation of the code results with the full dynamic inelastic response results is surprisingly good for this case. This is mainly due to the conservative code equations for evaluating the fundamental period of the structures. The code formula yields a value of 0.3 sec and 0.8 sec for the three and eight-storey frames respectively, while the computed periods are much longer as noted in Fig. 1. These shorter periods required by the code yield larger design forces and therefore the drifts are overestimated.

Table 1 - Minimum Required Separation Distances to Avoid Pounding of the Uncoupled Structures.

Seismic Event	S_r (mm)
Record #1	235
Record #2	290
Record #3	240
Average	255
National Building Code of Canada	290

INFLUENCE OF SLIP LOAD OF LINKAGE SYSTEM

The energy dissipated by the friction damped interconnection is simply equal to the product of the slip load and the total slip travel. For very high slip loads the energy dissipated by friction will be zero, as there will be no slippage. In this situation the buildings will respond as an elastically coupled system. If the slip load is very low, large slip travels will occur but the amount of energy dissipated by the linkage system will again be negligible. In this case, the structures will approach the behaviour of the uncoupled system. Between these extremes, there may be an intermediate value of the slip load which would result in optimum energy dissipation and thereby minimize structural response. This intermediate value is defined as the "Optimum Slip Load". An optimum slip load study was carried out for the building pair excited by the three earthquake records considered in this investigation. A series of inelastic time-history dynamic analyses was performed for different values of slip loads in the elastic range of the linkage system. Figure 4 presents the results of the optimum slip load study. The results are given in terms of the base shear coefficient C and the slip load ratio S, which are defined as

$$C = \frac{\textit{Maximum Base Shear}}{\textit{Weight of Building}}$$

$$S = \frac{\textit{Slip Load of Linkage System}}{\textit{Yield Load of Linkage System}}$$

[1]

where the yield load of the linkage system is simply the product of the cross-sectional area of the link with its yield stress.

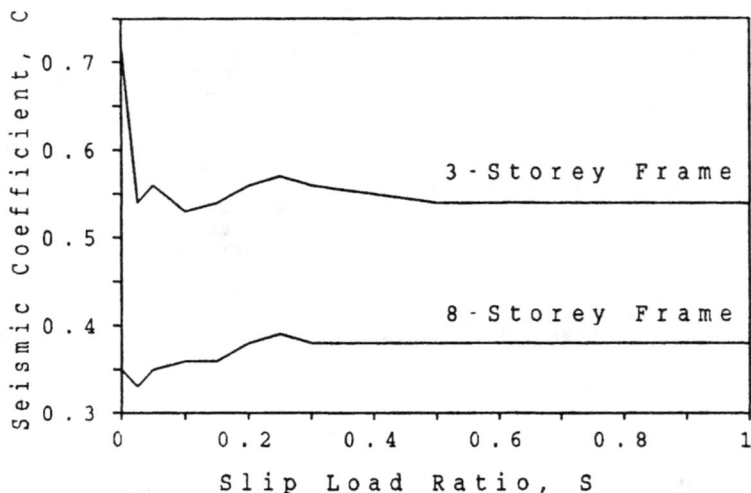

Figure 4 - Optimum Slip Load Study (Average of 3 Records).

For the cases considered the axial loads induced in the linkage system were small and therefore small slip loads were required to cause slippage of the friction devices. Quantitatively for slip loads higher than 50% of the yield load of the link, no slippage was observed. From Fig. 4 it can be seen that the slip load does not significantly influence the base shear developed by the structures. For the three-storey frame, a reduction of base shear of 26% is observed for a slip load ratio S=0.10, compared to 25% for an elastic link. For the eight-storey frame the elastic linkage increases the base shear by 9% while for S=0.025, the base shear is reduced by 6%. While it is not obvious that the use of a friction link is beneficial, a value of S=0.025, which induced a reduction of base shear in both frames, will be used for comparison in the subsequent detailed analyses.

COMPARATIVE DETAILED SEISMIC RESPONSES

In this section the distribution of structural damage in terms of inelastic response in the various members of the two structures after the end of each earthquake is investigated. Three different linkage configurations are compared: 1) the uncoupled system (S=0.0); 2) the elastically coupled system (S>0.50) and 3) the friction damped coupled system (S=0.025).

The plastic hinge distributions are illustrated in Fig. 5 along with the ductility demands. The response quantities presented represent averages over the three earthquake records. A plastic hinge is recorded in Fig. 5 when the section yielded in at least one of the earthquakes.The ductility ratio is based on an assumed plastic hinge length. When an element end moment is in the plastic regime, the ductility ratio is defined as one plus the ratio of the plastic curvature developed during the largest excursion into the plastic range to the yield curvature for the section. The plastic curvature is obtained by assuming that the plastic hinge rotation takes place on a length equal to the depth of the member. It can be seen from Fig. 5 that the uncoupled frames behave according to the code design philosophy, i.e. the plastic hinges are forming in the beams first. However significant ductility demands (>5) occur in the lower beams and base columns of the three-storey frame and in the upper beams of the eight-storey

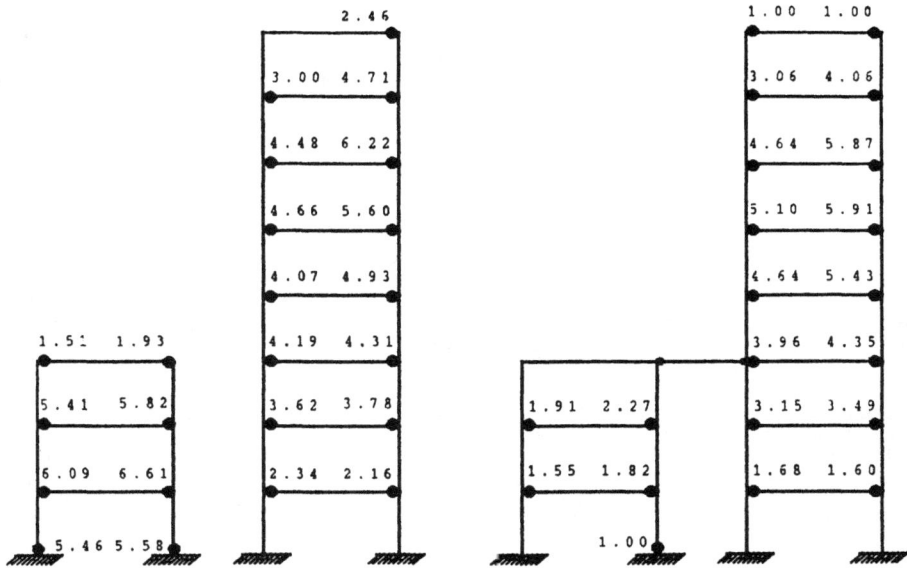

a) Uncoupled   b) Friction Damped Coupled

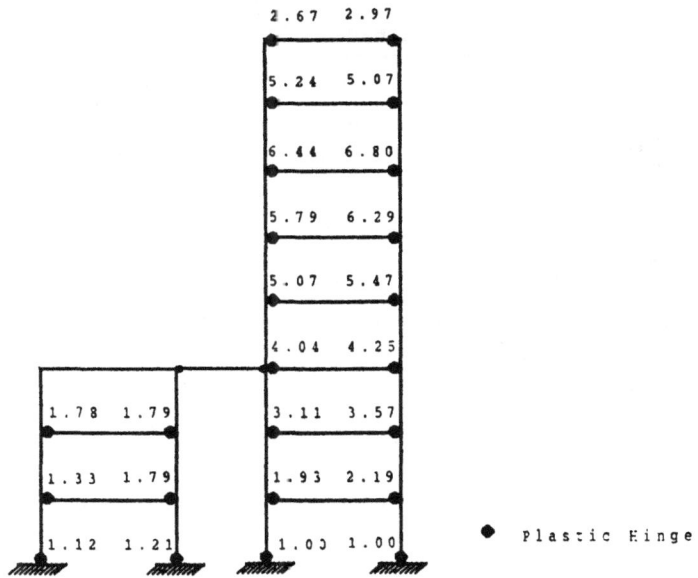

c) Elastically Coupled

Figure 5 - Plastic Hinge Distributions and Ductility Demands.

131

structure. Both the friction damped and the elastically coupled systems reduce significantly the ductility demands in the members of the three-storey frame. Also a significant reduction in ductility is observed in the upper beam of the eight-storey friction damped coupled frame compared to it elastically coupled counterpart. The elastic link introduces an abrupt change in the lateral stiffness of the coupled system above the connection level. The friction damped connection dissipates some of the seismic energy and also limits the force transfer between the two frames. These effects contribute to the protection of the upper floors.

CONCLUSION

This paper presents, to the authors' knowledge, the first investigation into the inelastic seismic response of structurally interconnected buildings. Based on the limited case study presented herein, the use of a friction damped connection offers promises in preventing pounding between adjacent buildings and in reducing the ductility demands of the structures as a result of a strong earthquake. Particularly the levels above the connection point in the taller structure benefit the most by the introduction of the friction damped link compared to a purely elastic link. This paper represents only a preliminary study and further research investigations are required to more fully understand the dynamic behaviour of structurally interconnected buildings and to develop appropriate design guidelines.

REFERENCES

_____ 1990. National building code of Canada 1990. National Research Council of Canada, Ottawa, Ontario, Canada.
_____ 1973. The great Alaska earthquake of 1964 - engineering. National Academy of Sciences, Washington, D.C., USA.
_____ 1985. Impressions of the Guerrero-Michoacan Mexico earthquake of 19 September 1985. EERI Preliminary Reconnaissance Report.
Anagnostopoulos, S. 1988. Pounding of buildings in series during earthquakes. Earthq. Eng. Struct. Dyn., 16, 443-456.
Astaneh, A., Bertero, V.V., Bolt, B.A., Mahin, S.A., Moehle, J.P. and Seed, R.B. 1989. Preliminary report on the seismological and engineering aspects of the October 17, 1989 Santa Cruz (Loma Prieta) earthquake. Report No. UCB/EERC-89/14, EERC, University of California, Berkeley, CA, USA.
Berg, G.V. and Degenkolb, J.H. 1973. Engineering lessons from the Managua earthquake, American Iron and Steel Institute Report.
Canadian Standards Association (CSA) 1989. Steel structures for buildings - limit states design. CAN3S16.1-M89, Rexdale, Ontario, Canada.
Filiatrault, A. and Cherry, S. 1987. Performance evaluation of friction damped braced steel frames under simulated earthquake loads. Earth. Spect., 3, 57-78.
Filiatrault, A. and Cherry, S. 1988. Design of friction damped steel plane frames by energy methods. Reports No. EERL-88-01, Dept. of Civil Eng. University of British Columbia, Vancouver, B.C.
Kannan, A.E. and Powell, G.M. 1973. DRAIN-2D, A general purpose computer program for dynamic analysis of inelastic plane structures. Report No. EERC 73-6, EERC, University of California, Berkeley, CA, USA.
Mahin, S.A., Bertero, V.V., Chopra, A.K. and Rollins, R.G. 1976. Response of Olive View hospital during the San Fernando earthquake. Report No. EERC-76/22, EERC, University of California, Berkeley, CA, USA.
Wada, A., Shinoyaki, Y. and Nakamura, N. 1984. Collapse of building with expansion joints through collision caused by earthquake motion. Proc. 8th World Conf. Earthq. Eng., San Francisco, IV,855-862.
Westermo, B.D. 1989. The dynamics of interstructural connection to prevent pounding. Earthq. Eng. Struc. Dyn., 18, 687-699.
Wolf, J.P. and Shrikerud, P.E. 1980. Mutual pounding of adjacent structures during earthquakes. J. Nucl. Eng. Des., ASCE, 57, 253-275.

Seismic design of concentrically braced frames – code comparisons

R.G. Redwood[I] and A.K. Jain[II]

ABSTRACT

Several codes and standards provide seismic design provisions for concentrically braced steel frames. Some of these are recent, and are based on research results produced over the last decade. The requirements for Canada, Japan, New Zealand, Europe and the USA are compared. The basic design philosophy, loading and member detailing requirements are considered. Similarities and differences are summarized, and areas needing further study are identified.

INTRODUCTION

Research and experience in earthquakes has shown that ductile response of concentrically braced steel frames is feasible in spite of two characteristics that militate against their ability to absorb energy under severe seismic loading. The low redundancy compared with that of other possible structural systems limits the potential for redistribution of load following inelastic deformation, and the fact that load is resisted principally by members subject to axial force, rather than shear or bending, increases the possibility of member instability controlling the response under severe load. In recent years rational approaches to design for ductile behaviour have been incorporated in several codes and standards. In view of the appearance of these provisions for the first time, it is the objective of this paper to summarize and compare the existing requirements for concentrically braced frames and to determine the extent of uniformity and to comment on significant differences. The codes and standards considered are those of Canada (NBCC 1990 and CSA 1989), the USA (UBC 1988), New Zealand (NZS 1984, 1989), Europe (Eurocode 8, 1988 (draft)) and Japan (BSL, 1981).

Prior to this new generation of codes, the design of concentrically braced frames (CBF) for seismic load in Europe and North America was not subject to many restrictive requirements. This was particularly true of detailed member design, there being for example no limitations on slenderness ratio. The principal

---

[I] Professor, McGill University, Montreal, Canada.

[II] Professor, University of Roorkee, India.

effect of the code provisions was to control the member design loads in a variety of ways. The low fundamental period of the stiff CBF system naturally leads to higher response loads, and high values of the structural system coefficients were assigned due to the limited ductility of CBF. These coefficients usually considered some structural system effects, such as the difference between tension-only and tension-compression bracing. In addition, higher loads for brace members were recommended in some codes to enhance the elastic response and reduce the ductility demand (SEAOC 1976).

Research on the physical behaviour of bracing members (for example the work of Jain, Goel and Hanson (1980), Popov and Black (1981) and Astaneh, Goel and Hanson (1986)), has provided a greatly improved understanding of detailing necessary to achieve high ductility, and non-linear modelling of the brace behaviour has provided information on the response of different structural systems. This information provides the basis for the significant changes incorporated in some recent editions of codes and standards.

Codes differ in the extent of detail specified. NZS (1989) for example provides, in Part 1, the design philosophy and general requirements, and in Part 2 gives very detailed means of compliance with Part 1. On the other hand, Eurocode 8 gives loading and performance requirements, and does little to specify the means of achieving these requirements. Others, such as UBC (1988) and CSA (1989) give specifications for some of the major ductile details. These different approaches leave some comparisons made below incomplete. Many of the seismic provisions described are not applicable in regions of low seismic risk, and the discussion relates only to the most severe seismic regions.

In the following comparisons, two aspects are of primary interest: first the basic philosophy used in the design approach, and the resulting specification of loading, and secondly, the detailed design requirements for members and connections considered necessary to achieve the implied level of available ductility.

CLASSIFICATION OF CBF AND PRINCIPLES OF DESIGN

Canadian design (NBCC 1990) classifies CBF as 'ductile', 'nominally ductile' or in a third category for which no special provisions are made for ductile behaviour, which can be called 'nominally elastic'. Detailed provisions for the first two of these are given in CSA (1989) and impose limitations on overall brace slenderness ratio, brace cross-section width-to-thickness ratios, stitch fasteners for built-up members, framing configurations, and loading on beams and columns. The expected behaviour is that ductility is provided by braces yielding in tension or in flexure under compression; other members of the structure remain essentially elastic, except for beam flexure under certain circumstances. The braced frame is considered in isolation, that is, there are no differences between a CBF alone and one participating in a dual framing system. The base shear magnitude for the nominally elastic CBF is double that of the ductile CBF.

BSL (1981) of Japan classifies brace members in four categories, those with excellent, good, fair and poor ductility. The highest base shear coefficient is 1.43 times the lowest. In providing interpretative material, the commentary on the BSL (Ishiyama, 1985) classifies braced frames according to three parameters: the brace overall slenderness ratio as in the four categories of the BSL, the extent to which the CBF participates in a dual system with a moment resisting frame (MRF), and the ductile classification of the MRF in the dual system. Many different combinations of the three parameters are possible leading to a total of 28 different loading categories.

The New Zealand standard NZS (1989) requires that CBF be designed so that "energy is dissipated through compression and/or flexural yielding of the braces, whereas the beams and columns shall remain elastic." In addition the Standard provides for energy dissipation in tension-only braces in tension-only systems. Four categories of ductilty demand are defined: 'fully ductile CBF'; 'CBF with limited ductility'; 'nominally elastic' and 'fully elastic CBF'. Members are categorized into those subject to high, low, very low and no ductility demand, and in general, as a minimum, the member categories must be used in the corresponding structure category. Seismic load is related to brace slenderness, frame configuration and number of storeys. For structures over one storey, nine different loading categories are specified and the ratio of structural coefficient for the least ductile to the most ductile CBF is 2.82, but for similar brace slenderness ratios the value is 1.27.

Eurocode 8 was available only in early draft form and consequently it is possible that significant changes from those discussed herein will be made. The draft contains provisions for three categories of frame, 'diagonal bracings' (defined as tension-only), 'V-bracing', which includes Chevron braces, and 'non-dissipative' systems such as K-bracing, where braces intersect within the column height. Diagonal tension-compression bracing is not mentioned. Energy dissipation is to occur in braces only. Design load depends upon the above configurations and on the degree of structural regularity, which is defined both in plan and elevation. The ratio of proposed design load for the non-dissipative system and for the most highly ductile CBF system is 4, and between the highest and lowest dissipative systems is 2.27.

UBC (1988) gives requirements that are almost identical with those of SEAOC (1988). Except for light framed bearing wall systems with tension-only bracing, the UBC provisions apply to all CBF in US seismic zones 2, 3 and 4, and thus frames with lesser ductility are not recognized as a separate category. However the braces in Chevron systems are subject to special treatment as outlined below, and K-braces are forbidden in severe seismic zones. When combined with a MRF incorporating fully ductile details, the dual system structural coefficient is reduced to 80% of the CBF alone.

Most of the codes reviewed incorporate special provisions for Chevron braced frames in view of the problems anticipated in their response, as outlined for example by Nordenson (1984) and Khatib et al. (1988). Thus Eurocode (1988), CSA (1989) and NZS (1989) assign them to a different category from diagonal braced frames, requiring respectively 2, 1.5 and 1.33 times the base shear of the latter, as well as imposing other detailing requirements. UBC (1988) assigns the same base shear to all CBF but requires Chevron brace members to be designed for 1.5 times the calculated member load. Since the maximum force that can be delivered by the braces to the columns must be considered in column design this is almost the same as imposing a higher load on the complete structure. Only the earlier BSL (1981) appears not to treat Chevron bracing differently from diagonal bracing.

All the above design methods are based on the brace acting as the principal, or often the only, energy dissipating element, unless the CBF is part of a dual system. The classification of CBF (and thus the design base shear) varies from a single category to many categories related principally to brace slenderness. The latter allows a close gradation in design load with variation in the main parameters. Dual action of a CBF with a ductile MRF results in a reduced design load (compared with the CBF alone) in Japan, Europe and the USA. In Canada the design load in a dual system is equal to that for the least ductile component of the system. In New Zealand, while NZS (1989) comments on the improved inelastic performance of typical dual systems, NZS (1984) suggests a procedure that uses the structural load coefficients corresponding to the systems acting separately.

Some restrictions are imposed on the bracing systems by requirements in UBC (1988) that neither tension nor compression braces in any planar braced frame should carry more than 70% of the total shear in any storey. A similar restriction exists in CSA (1989) for ductile CBF, and in NZS (1989) the difference between the seismic shear carried by tension and compression braces must not exceed 10%. Eurocode (1988) imposes a limit on tension-only bracing so that the difference in resistance in opposite directions of loading does not exceed 10% of the average of these resistances.

Among the codes considered there is substantial uniformity in the understanding of the behaviour of ductile MRF, and in the codification of requirements for ductile behaviour of MRF. A comparison of the structural system loading coefficient for the CBF with that for the ductile MRF is therefore of interest, and is shown in Table 1. Chevron bracing is excluded, since this system is not considered ductile by some codes, or is assigned a significantly higher design load in others.

In nearly all cases some restrictions related to height (or number of storeys) are placed on CBF in the more severe seismic regions. These may prohibit use of low ductility CBF (Canada, NZ) or prohibit the system altogether (NZ); increase the design load (Canada); require a dual structural system (USA), or require special analysis (NZ) or special analysis and governmental approval (Japan). These criteria are too numerous to detail here, but in the comparisons made, it is assumed that the height is less than the value that triggers these special provisions.

Table 1.   Comparison of CBF and MRF Design Loads

	Ductile CBF/ Ductile MRF
Canada	1.33
Japan	1.25-1.40
New Zealand (>1 storey)	1.38-1.88
Europe	1.25-1.50
USA	1.50

136

MEMBER DESIGN - BRACES

Slenderness ratio

Compression braces in the most ductile category of CBF are subject to the slenderness ratio limits given in Table 2 where the slenderness parameter $\lambda=(KL/r)\sqrt{(F_y/\pi^2 E)}$ is shown. The very low slenderness specified in New Zealand and Japan, where three ranges are utilized, influences other aspects of design and should be born in mind when comparing these most ductile systems with those of the other codes.

Table 2. Limitations on brace slenderness (λ)

	Most ductile	Other categories
Canada	1.35	1.35-2.47
Japan	0.35	0.35-0.63 & 0.63-1.41
New Zealand	0.45	0.45-0.90 & 0.90-1.35
Europe	1.50	1.50
USA	1.35	1.35

Width-thickness ratios

Limiting values of these ratios for the most ductile braces are given in Table 3. Wide differences are apparent, reflecting the fact that this is an active subject of research and while there is awareness of the severe curvatures that occur at hinges in buckling braces, there is not yet sufficient information or a concensus on appropriate width-thickness limitations.

Compression strength reduction factor

Due to residual curvatures and the Bauschinger effect, a buckled strut has a reduced compressive resistance on second and subsequent loadings to its compressive ultimate strength, even after the yield load in tension has been applied. For this reason SEAOC (1988) recommends a reduction factor, equivalent to $1/(1+0.35\lambda)$, be applied to the brace compressive resistance. This factor is used in UBC (1988), for nearly all braces, in CSA (1989) for the most ductile category, and in NZS (1989) for Chevron braces. For braces and columns the latter also limits the ratio of factored design force to yield load to 0.5 and 0.7 for the two most ductile of the member categories.

Built-up members

The individual components of built-up members are susceptible to local buckling following overall buckling under cyclic load (Goel and Aslani 1989), and

Table 3. Width-thickness limits for ductile braces ($b/\sqrt{F_y}/t$)

Origin[1]	Flanges	Webs	RHS	Angle
Canada	116	670	336	115
New Zealand	136	512	350	.136
Europe	169	598	567	N.A.
USA[2]	249	664	624	200

[1] Japanese values not available.
[2] These are specified in UBC (1988); more stringent requirements are under consideration.

so some limitations have been placed on component slenderness ratios. UBC (1988) limits the component KL/r of a brace to 0.75 of that of the overall member, and CSA (1989) uses 0.5. These codes, as well as NZS (1989), specify design loads for intermediate connections in these members.

BRACE CONNECTIONS

In all codes considered, the basic design load for brace connections in ductile CBF is the brace tensile yield load. This is modified in several ways. Overstrength factors are incorporated by Eurocode (1988) and NZS (1989). For the former this is 1.20 and for the latter range from 1.35 to 1.70, depending on steel grade and ductile category of the brace. CSA (1989) requires the factored connection resistance to exceed the (unfactored) brace yield load, which is equivalent to imposition of an overstrength factor of between 1.11 and 1.33.

BSL (1981) requires the connection "not to fail before braces yield," and UBC (1988) requires connection design for the brace tensile strength. The latter restricts the net to gross area ratio in bolted connections, and CSA (1989) includes an impact factor for tension-only bracing. Three codes, CSA, NZS and UBC, give an upper limit on the connection design load, applicable when braces are over-designed. These are respectively the nominally elastic response load, the fully elastic value, and for UBC an estimate of the maximum response load in the brace (which is related to the elastic response value), but limited also to the "maximum force that can be transferred to the brace by the system."

COLUMNS AND BEAMS

CSA (1989), NZS (1989) and UBC (1988) require columns to be designed for forces corresponding to brace yield loads, with NZS including the brace overstrength factor only at the level of the column considered. In each case the load need not exceed the maximum code specified response load, this being defined as for the brace, above, and each being related to the elastic response value of the column load. Eurocode (1988) defines the column (and beam) design load as the product of 1.20, the calculated load, and an additional factor equal to the

138

ratio of brace resistance to brace calculated load. The minimum value of this factor in all floors is to be used. The column is therefore designed for a load 20% greater than that at first brace yield. BSL (1981) makes no specific mention of column or beam design loads.

NZS (1989) requires beams that "transfer axial load developed in the braces" to be designed to develop brace overstrength. CSA (1989) specifies that beams shall be designed to resist the forces due to brace yielding, and "redistributed loads due to brace buckling or yielding shall be considered." Again, such loads need not exceed the nominally elastic load if the beam participates in the lateral load resisting system.

CONCLUSIONS

The most recent of the codes considered have many similar features reflecting recent research results, and commentary material indicates the influence of the SEAOC Lateral Force Recommendations in some of these standards. The New Zealand standard is the most comprehensive of those considered, and provides a complete, rational approach using capacity design. The other codes are more succinct and make use of many capacity design principles, although not as uniformly as NZS (1989). Areas of similarity are many, whereas significant differences also exist, largely reflecting subjects of current research interest. The Japanese regulations date from 1981, and since considerable study of braced frames has taken place since then, much of it in Japan, it is probable that BSL (1981) no longer reflects all aspects of Japanese CBF design.

The concensus of the recent codes can be summarized as follows:
- CBF design treats the brace as the energy dissipating element; its slenderness ratio is restricted, and seismic design load increases with this ratio; the design load may be from 1.25 to 1.88 of that for a DMRF; when part of a dual structural system, the design load can be reduced
- seismic shears should be resisted partly by members in tension and partly by those in compression, and the proportion carried by each should be balanced
- due to the anticipated poor response of Chevron braced frames, special design requirements are required
- brace connections must be designed to carry the tensile yield load of the brace, and the forces induced in other members by this load must be considered in their design
- brace cross-section elements must conform to severely restricted width-thickness ratios to permit high inelastic strains.

Not all the codes subscribe to all these features. The main differences and other considerations not embraced in the above are:
- the widely different local buckling requirements (Table 3)
- categories of brace slenderness: is one or several appropriate?
- overstrength factors are not used in all codes: should they be and if so what magnitude?
- the upper limit on brace connection design load based on the elastic response value could be exceeded during inelastic response
- approaches to the design of tension-only bracing differ significantly
- beams not active in the lateral load resisting system during elastic response may be called upon to carry substantial axial loads during

139

inelastic response: should they be considered explicitly? (see CSA 1989)
- the accumulation of column loads due to yielding braces: there is a very
  small probability of simultaneous yielding; a load combination rule is needed
- the benefits assigned to a CBF acting in a dual system differ considerably.

ACKNOWLEDGEMENT

The authors acknowledge with appreciation the contribution of Vadiraj
Channagiri in accumulating data for this paper.

REFERENCES

CSA 1989. Limit States Design of Steel Structures. CAN/CSA-S16.1-M89. Canadian
Standards Association, Rexdale, Ontario.

BSL 1981. Japan Building Standard Law. See IAEE: Earthquake resistant
regulations: a world list - 1988. IAEE Tokyo, 1026 pp.

Eurocode 1988. Eurocode 8 (draft). Structures in Seismic Regions - Design. Part
1. General and Buildings. Commisson of the European Communities.

Goel, S.C, and Aslani, F. 1989. Seismic behavior of built-up bracing members.
Proc. ASCE Structures Congress, Steel Structures, San Francisco. pp. 335-344.

International Association for Earthquake Engineering. 1988. Earthquake resistant
regulations: a world list - 1988. IAEE, Tokyo. 1026 pp.

Ishiyama, Y. 1985. Comparison of NBC (1985) seismic provisions with Japanese
seismic regulations (1981), DBR Paper No. 1297, National Research Council of
Canada, Ottawa, Ontario. 37 pp.

Jain, A., Goel, S. and Hanson, R.D. 1980. Hysteretic cycles of axially loaded
members, Jour. of the Struct. Div, Proc. ASCE Vol. 106, No ST8, pp. 1777-1795.

Khatib, I. F., Mahin, S. A. and Pister, K.S. 1988. Seismic behaviour of
concentrically braced steel frames. Report No. UCB/EERC-88/01. Earthquake
Engineering Research Center, Berkeley, Calif. 222 pp.

NBCC 1990. National Building Code of Canada, 1990. Associate Committee on the
National Building Code, National Research Council of Canada, Ottawa, Ontario.

Nordenson, G. J. P. 1984. Notes on the seismic design of steel concentrically
braced frames. Proc. 8th World Conference on Earthquake Engineering, Vol. 5, San
Francisco. pp 395-402.

NZS 1984. Code of Practice for General Structural Design and Design Loadings for
Buildings. NZS 4203:1984. Standards Assoc. of New Zealand, Wellington.

NZS 1989. Steel Structures Code: Parts 1 and 2. NZS 3404:1989. Standards
Association of New Zealand, Wellington.

SEAOC 1976. Recommended Lateral Force Requirements and Commentary, Seismology
Committee, Structural Engineers Association of California, San Francisco.

_____ 1988. Recommended Lateral Force Requirements and Tentative Commentary,
Seismology Committee, Structural Engineers Association of California.

UBC 1988. Uniform Building Code, International Conference of Building Officials,
Whittier, California.

Behavior of open-web steel joists in moment-resisting frames

Terrence F. Paret[I] and Brian E. Kehoe[II]

ABSTRACT

Many lightweight steel structures are constructed using open-web steel truss joists as part of the gravity load carrying system. Recently, these joists have also been used as part of moment-resisting frames. These frames are relatively new adaptations of older gravity framing systems and are unusual in that they forego the use of standard rolled steel shapes in favor of proprietary structural elements. The behavior of the typical moment-resisting connections between vertical and horizontal framing members of these systems are complex and bear little resemblance to the connections in more standard systems. Since these systems have little or no history of seismic resistance, little is known about their actual behavior during earthquakes. Frame drifts and ductility are affected by many construction features not found in more typical systems. This paper presents analytical studies of open web moment-resisting frames that explore features of these systems related to vertical and lateral loading. Experimental data on the behavior of components is also presented. Recommendations are made regarding detailing of the connections. The analysis presented in this paper is based on reviews of several structures which have specified steel truss joists in moment frames in both single-story and multi-story buildings.

SYSTEM DESCRIPTION

The moment frames which are the subject of this study have several typical features: open-web steel truss joists, square tubular columns and welded connections. Open-web steel joists are generally proprietary items which are ordered from a manufacturer. Framing plans usually specify truss joists at regular spacing in one direction and truss joists as girders in the other direction. Truss joists used as girders are larger versions of the same system

[I]Senior Engineer, Wiss, Janney, Elstner Associates, Inc., 2200 Powell Street, Suite 925, Emeryville, CA 94608.

[II]Senior Engineer, Wiss, Janney, Elstner Associates, Inc., 2200 Powell Street, Suite 925, Emeryville, CA 94608.

and will be described as truss girders for this study. The details characterized below apply to both truss joists and truss girders.

The sizes of the components and the details of the fabrication are not available to the engineer. The designer usually supplies the loads, spans and other design criteria to the manufacturer and the sizes and details of the members are then determined by the manufacturer. Many of these joists are composed of steel angles for the top and bottom chord member and steel angles or round bars as the diagonal web members as shown in Fig. 1. To achieve the

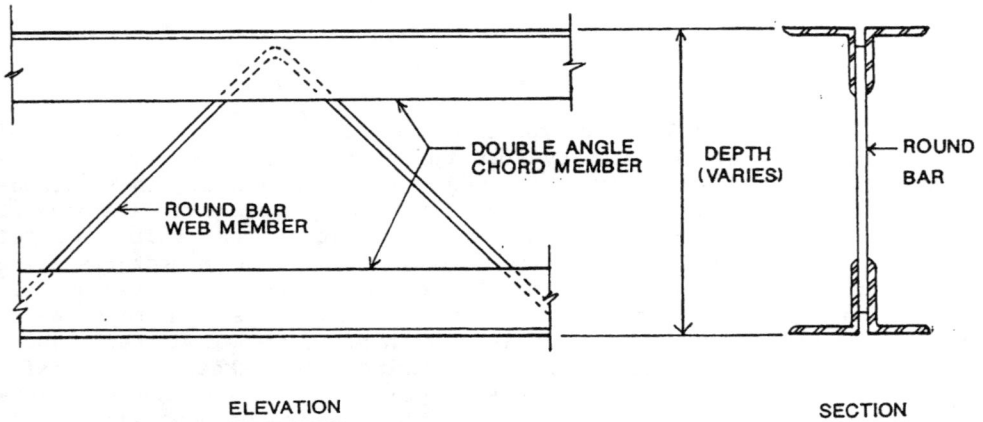

Figure 1. Detail of Open-Web Joist

greatest economy, the chord members are usually very light sections, with thicknesses as thin as 1/8 inch.

Square tubular columns are typically used for the vertical elements of these structural systems. This allows the column to be part of the lateral force resisting system in both directions and also provides efficiency in axial load carrying capacity. The tubes are supported on a steel base plate which can be set below the level of the floor slab. Where this occurs, the base plate may act with the floor slab to provide fixity at the base of the column. If crack control joints are provided in the slab around the column, the slab may not provide supplemental fixity.

CONNECTIONS

The connection details are critical elements of the design of truss joists within moment frame structures. In contrast to wide flange frames, separate connections are provided for the top and bottom chords of the truss joist. A force couple develops between the top and bottom chord connections. The design of the connections can be provided by either the engineer or the fabricator. When designing a building using this system, the responsibility for the design of the connections should be clearly specified in order to insure that the connections will be designed for all anticipated loads.

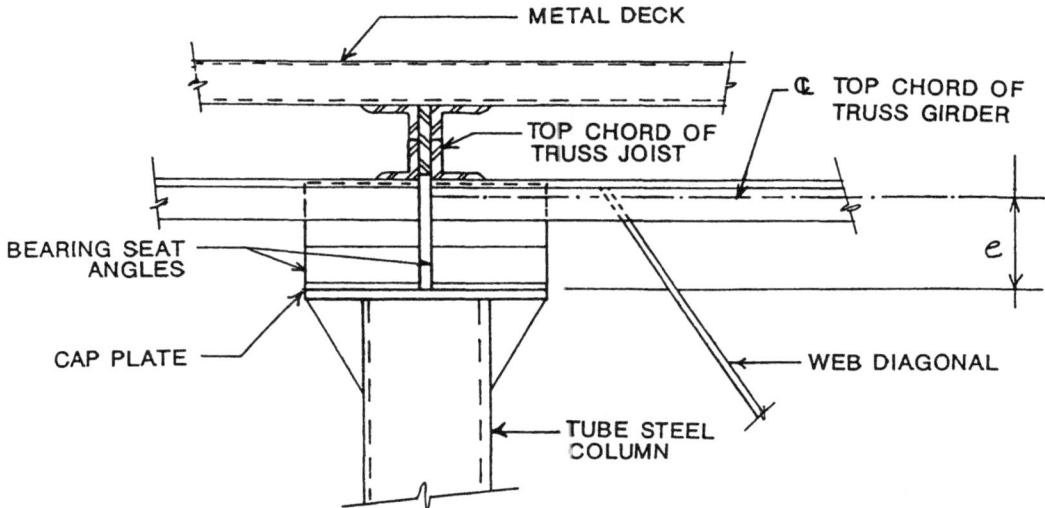

Figure 2. Top Chord Connection

In single story structures, or the top floor of multi-story structures, the top of the columns are typically fitted with a cap plate. The end of the truss-joist is supported vertically by inverted bearing seat angles which rest either on truss girders or on top of this cap plate. An example of this connection is shown in Fig. 2. In this connection, there is an eccentricity which is developed between the force in the top chord of the joist and the weld to the bearing seat angle and the column cap plate. It is important that the design of the welds of the joist bearing angles and the girder bearing angles account for this eccentricity.

In addition, the eccentricity of the top chord connection also causes bending in the top chord member of the truss joist. The bending moment is greatest in the length of the top chord between the end of the joist chord and the first panel point of the diagonal web member. Although the truss joists are proprietary and are not designed by the engineer, the engineer should confirm that the fabricator understands the behavior of the truss resulting from this eccentricity.

A typical connection of the bottom chord to the column consists of a flat stabilizer plate, oriented vertically and welded to the column face. The double angle bottom chord is field welded to this plate, usually after the application of all dead loads, as shown in Fig. 3. This allows the joist to act as a simply supported beam for resisting dead loads, and is intended to provide a fixed end condition for resisting live and lateral loads. Under the influence of lateral load, this connection develops a concentrated force from the chord on the middle of the wall of the tube where the stabilizer plate is attached. Where this concentrated force would cause excessive bending stresses in the wall of the tube, a doubler plate may be added to the face of the tube.

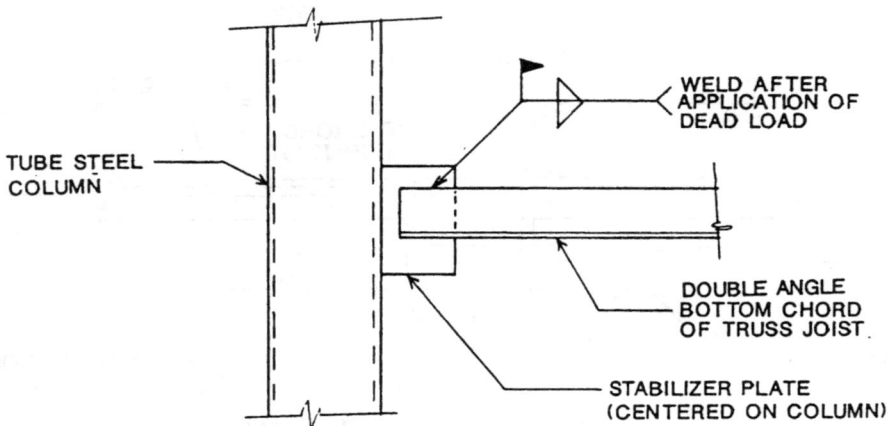

Figure 3.   Bottom Chord Connection

The allowable and ultimate loads for this type of bottom connection cannot be easily determined by conventional analysis procedures. Various approximations of the effective width of the wall of the tube suggest substantially different allowable loads for the tube. Finite element analysis of this connection has been used to approximate the load at which first yielding could occur and also to indicate the initial stiffness of the connection. This analysis, however, does not indicate how much reserve strength is in the connection nor the stiffness of the connection after initial yield. Yield line methods provide indications of total strength but cannot identify first yield or the stiffness characteristics of the connection.

To verify the actual strength and stiffness of this connection, two test specimens were fabricated and tested. Two 40 inch long tube steel columns, each 12 inch square with 1/2 inch thick walls were fitted with stabilizer plates, as shown in Fig. 4. One the specimens was fitted with a 1/2 inch thick doubler plate to strengthen the tube wall to which the stabilizer plate was welded, as shown in Fig. 5. The testing procedure called for the application of monotonically increasing load to each specimen until first local yield was indicated by strain gages. The load was then removed. Each specimen was subsequently cycled 6 times to the load corresponding to first local yield to assess the response of the welds and the tube face to pseudo-seismic loading. Finally, the specimens were loaded to failure. A plot of the load-defection curve for the specimen with the doubler plate is shown in Fig. 6 and the plot for the specimen without the doubler plate is shown in Fig. 7. The load at which first yield was measured is 26.5 kips and 17.8 kips for the two specimens, as shown on the load-deflection plots.

The load-displacement curves for the test specimens indicate that these connections have a considerable ultimate capacity. However, the connections are very soft and substantial deflections are required in order to engage the

144

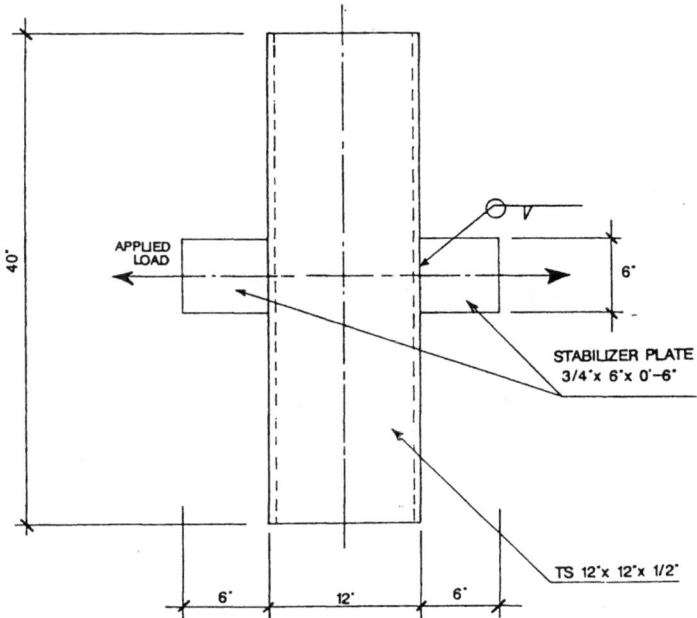

Figure 4. Test Specimen

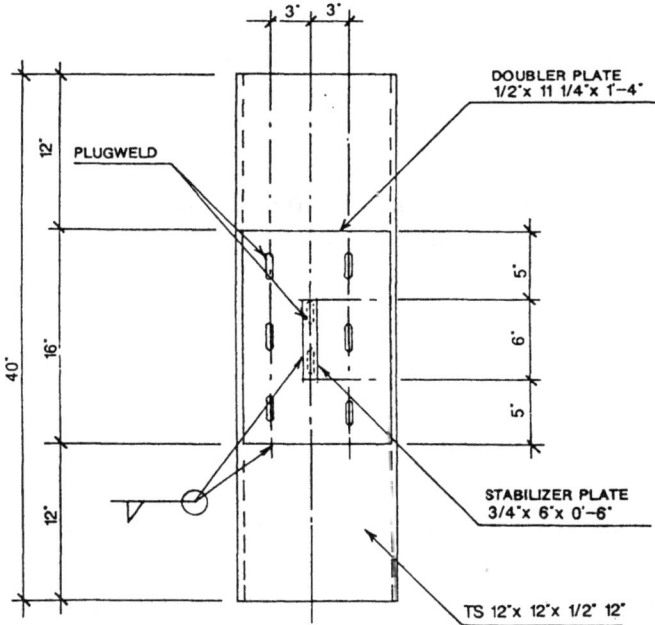

Figure 5. Test Specimen with Doubler Plate

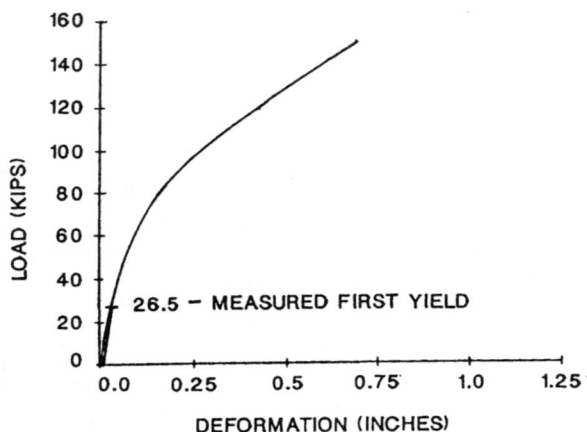

Figure 6. Load-Deflection Curve for Test Specimen with Doubler Plate

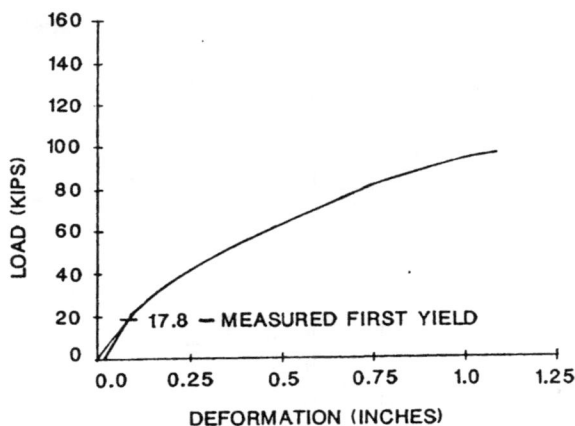

Figure 7. Load-Deflection Curve for Test Specimen without Doubler Plate

full strength. The results also indicate that the addition of doubler plates not only increases the ultimate capacity, but also the overall stiffness. It is important to recognize that these deflection characteristics influence the behavior of the truss under vertical loads as well as the behavior of the frame under lateral loads.

Other connection configurations for the bottom chord can be used which will increase the stiffness and strength of the connection. One method of reducing the force to the bottom chord connection is to use a diagonal "kicker"

146

from the bottom chord of the truss joint to the column. The moment in the joist, however, drops off away from the column too quickly for the kicker to be of any benefit for the forces in the joist. Additional problems can arise if the bottom chord of the truss joist is not braced out-of-plane where the kicker is attached. A seat angle can be provided for the bottom chord. The chord is then field welded to the seat angle after application of the dead load. The horizontal leg of the angle acts as a stiffener for the wall of the tube. This detail can pose a fit-up problem if the seat angle, which is shop welded, does not align vertically with the truss joist.

SEISMIC FORCE PERFORMANCE

The current seismic design guidelines in the Uniform Building Code (1988 Edition), provide for two types of moment-resisting frames: Special moment-resisting space frames and ordinary moment-resisting space frames. The use of truss joists as part of the moment resisting space frame typically do not meet the ductility requirements to be considered as a special moment-resisting space frame. Very little information is available on the performance of the lateral load resisting open-web joist systems under seismic loads. Analysis of moment-resisting frames using truss joists indicates a number of critical elements affecting the performance.

The lateral load transfer in this system varies from typical moment-resisting frame systems. In most cases, truss joists span in one direction and are supported at the ends by bearing on truss girders spanning in the perpendicular direction. The metal deck rests on top of the truss joists and is not be directly connected to the truss girders (see Fig. 2). For lateral loading aligned in the direction of the girder frames, load originating in the deck must be transferred through out-of-plane bending of the joist ends. The top chord of the joist ends should be designed with this in mind. Similarly for lateral load in the direction of the joist frames, load from the deck must travel through out-of-plane bending of the girder chords and bearing angles to reach the columns.

Analysis of the lateral deflections of the building should account for the deformations due to any flexibility of the connections. Since the distance between the top and bottom connections can be as large as 36 inches, the response of the system is significantly affected by deformations in this area. This is similar to the inclusion of panel zone deformations when analyzing typical moment framed structures.

An eigenvalue analysis, taking into account the stiffness and mass properties of the building, can be performed to determine the dynamic characteristics of the structure. The flexibility of the frame gives the structure a longer fundamental period of vibration compared to a similar building using wide flange members as bending elements. Building code formulas for determining the seismic design base shear will produce a lower design base shear, and thus lower forces throughout the structure. Some building codes, such as the Uniform Building Code, specify limits on the computed period to avoid underestimating the actual seismic forces. The additional flexibility of the system that results from deformation within the joint is difficult to incorporate into a global building analysis.

Because of the excessive flexibility of the structure, drift can also be a problem in this system. Approximate analysis methods may not yield accurate estimates of lateral deflection. Excessive drift can cause extensive nonstructural damage during an earthquake. Nonstructural elements, which may not have been considered as part of the lateral force resisting system, can become engaged because of the flexibility, and act to resist forces. This can result in a change in the anticipated force distribution.

CONCLUDING REMARKS

Open-web steel truss joists are being used as bending elements in moment resisting frames. Little information is presently available on the performance of this type of system for resisting lateral loads due to earthquakes. Until such information has been gathered, the designers who specify this system should be aware of the potential problems associated with this system.

The critical feature of the truss joist system is the connections of the truss joists to the columns. At the top connection, an eccentricity between the chord member and the attachment to the column can occur when a bearing seat is used at the end of the joist. The eccentricity can also cause bending in the chord member of the joist or girder. Stabilizer plates used for the bottom chord connection result in concentrated forces being developed at the wall of the tube. These details must be analyzed to determine the stiffness and capacity of the connections. The stiffness of the bottom connections can also have an effect on the vertical capacity of the joists. Flexibility of the bottom connection may not provide the anticipated fixity at the ends of the joists or girders, thereby allowing them to behave as simply supported beams.

Structures which utilize truss joists as part of moment resisting frames are more flexible than typical systems. Excessive drifts can result from the flexibility of the moment frames. The flexibility of the connections must be included in an lateral analysis of the building. The designer should evaluate the effect of the increased drift on the building to verify that the nonstructural elements are capable of withstanding the anticipated lateral deflections.

Finally, the responsibility for the performance of the overall structural system rests with the design engineer. Although the design of the truss joists is provided by the fabricator, the engineer should be aware of the significance of the features used in the system. The engineer must also clearly define the responsibilities for the design of connections and other details. The design engineer should work closely with the fabricator to ensure that all of the anticipated forces are considered.

The concepts presented herein were generated during structural reviews of existing structures. Details of construction identified as typical may not be for structures other than those reviewed by the authors. Other details may result in seismic resisting systems of greater or lesser adequacy.

Local buckling and early fracture of cold-formed steel members

Hiroyuki Yamanouchi[I]

ABSTRACT

Although recent investigations are well proceeding on basic properties of cold-formed members with square or rectangular tubing sections. Namely, early fracture of those members is much anticipated to take place under severe seismic loadings. This significant problem is directly related to the seismic safety of steel structures using the members with cold-formed tubing sections. This paper deals with, firsts a review of recent studies related to the problem. Then, an example of early fracture of a cold-formed bracing member is shown associated with an analysis that indicates the effect of the fracture on the overall structural behavior. Finally, possible measures and research needs against the danger are proposed based on related studies.

INTRODUCTION

Early fracture of cold-formed steel members is anticipated to occur by future major earthquakes. This is because: 1) recently cold-formed square or circular tubing sections have come to be often used as columns and bracing members; and 2) severe stress and strain in the critical sections of such members, induced by severe earthquakes, particularly associated with local buckling. Needless to say, investigations are proceeding well on basic properties of cold-formed steel sections and on the behavior of cold-formed members (Karren, 1967; Kato, 1978; Kato, 1986). However, less attention has been given to the crisis of cold-formed members with square or rectangular tubing sections.

In this paper, first of all, the past studies on this critical problem will be reviewed. Then, some remarkable findings from the US/Japan full-scale seismic test (Yamanouchi, 1989) and analysis will be briefly discussed; early fracture of one bracing member took place actually in the test. In addition, the current design requirements for such bracing members will be critically discussed. Finally, several practically possible measures against such premature fracture will be discussed on the basis of related studies. Also in this discussion, the technical problems to be solved as soon as possible will be identified considering the prevailing use of cold-formed sections as bracing members or other major structural members.

I Head of Structural Dynamics Division, Building Research Institute, Tsukuba, 305 Japan

REVIEW OF RELATED STUDIES

It is well known by recent studies that square or rectangular tubular members are susceptible to local buckling. Thus, limitations of width-to-thickness ratio are stipulated in design codes of several countries for these sections. However, there have been few excellent experimental studies focusing on the local buckling and fracture of square tubular members under cyclic loading. In particular, the local buckling and following cracking is very sensitive to the so-called scale-effect. Thus, to attain experimental knowledge on these problems, it is significant to use full-scale test specimens that generally need large-scale testing facilities and high experimental costs.

Before the US/Japan full-scale test, Gugerli and Goel (Gugerli, 1982) carefully tested full-size braces with cold-formed rectangular tubing sections. According to this study, local buckling occurred in the initial stage of severe inelastic cycling. After this, complete rupture took place in a few inelastic cycles. Consequently, the study found that: 1) the width-thickness ratio was the most significant key factor to the influence and severity of local buckling; 2) severe local buckling adversely affected the fracture life of the cold-formed tubular members; and 3) as the slenderness ratios of the bracing members became smaller, the fracture of the members occurred earlier with strains much concentrated in locally buckled sections. Further, Tang and Goel (Tang, 1987) developed a criterion to predict fracture lives of rectangular tubular braces in terms of normalized cyclic displacements, on the basis of bracing member tests conducted by Liu (Liu, 1987). According to the study, the empirical fracture life is inversely proportional to the square of width-thickness ratio and is proportional to the slenderness ratio and the width-depth ratio of the section. Recently, Lee and Goel (Lee, 1987) have proposed a refined fracture criterion for square or rectangular tubular bracing members. In this formulation, the following assumptions based on the experimental results have been added to the Tang's formulation: 1) the fracture life increases as yield strength decreases; 2) the fracture life is independent of the slenderness ratio, KL/r; 3) the loading history until the first overall buckling has no effect on the fracture life; and 4) large tension forces in a member have more dominant effect on the fracture life than small tension forces. The above studies identified pivotal factors influencing the fracture life of square or rectangular tubular members.

EARLY FRACTURE OF COLD-FORMED BRACING MEMBERS IN FULL-SCALE TEST STRUCTURE

Findings from Full-Scale Test

The bracing members of the US/Japan full-scale six-story test structure (Yamanouchi, 1989) were made of the cold-formed square hollow sections having the steel of ASTM A500 Grade B. The principal dimensions of the bracing members are listed in Table 1.

By the test results from the Moderate test (2.5m/sec^2 peak input), several bracing members exhibited slight softening in their cyclic behavior as shown in Table 2. Further, these braces had overall buckling in the Final test (5.0m/sec^2 peak input). Immediately after these buckling, local buckling occurred at critical sections. At last, at 11.135sec in the input seismic record, the north brace at the third story level completely ruptured (Yamanouchi, 1989).

150

To discuss the test results on the damaged braces in more detail, the strain energy absorbed in the braces during the Final test is listed in Table 2. This amount of energy has been obtained from the total area of the axial force versus displacement hysteresis loops. The second column of this table shows the values of the energy normalized by the volume per unit length for each brace. These values indicate approximately the energy dissipated in plastic hinges in each braces, since inelastic deformations concentrate in the hinges located at the mid-span and both the ends of the brace. Looking at these normalized energy, the values on the north brace in the third story and both the braces in the second story are much more dominant than those on the other braces. On the other hand, although the north brace in the first story marks a considerably larger amount of energy (Table 2), no local buckling was observed. This is chiefly because the width-thickness ratio of the section of this brace is quite smaller than that of the other brace sections as listed in Table 1.

Another peculiarity of the brace rupture is the fast pace of crack propagation. Fig. 1 shows the time history of the axial displacement of the ruptured brace, associated with the crack growth recorded by the observation. By this figure, the brace underwent only three inelastic displacement cycles with the amplitude of 6 to 8 times the yield displacement, Δy. In addition, it was observed that the crack appeared initially at the corner of the concave side of the locally buckled section; the crack then propagated rapidly toward the compressed flange and both the webs as well. Then, the cracks came rapidly into the flange plate, leading to the complete rupture of the brace.

By this premature fracture of the brace in the full-size test that simulated seismic responses, it is authentic to anticipate similar failure in real structures by severe earthquake ground motions.

Analytical Response of Test Structure after Brace Rupture

The Final test was terminated just after the rupture of the north brace at the third story, at 11.135 sec in the seismic record. Thus, it is not only interesting but also practically significant to know or presume how the response of the test structure would have proceeded after the brace rupture if the test was continued. One way to attain this knowledge is to perform a dynamic response analysis that can simulate pertinently the response of the structure after the brace fracture. In the analysis of this paper, thus, the north brace of the third story was assumed to lose completely its axial resistibility at the same time as the experimental rupture time (11.135sec). A variation of the DRAIN-2D computer program was used for this analysis.

As a result, Fig. 2 shows the time history of interstory drift of the third story. Before the brace rupture, the drift varied chiefly in the negative region; this means that the north brace received cyclically larger compressive deformations than the south brace that was mainly under tension. By the analysis, around 15 sec, the maximum interstory drift angle of 1/52.5 would then occur in the positive side. Up to this moment, probably other braces would have ruptured one after another as noted by Tang and Goel (Tang, 1987). By their analysis, the maximum story drift angle of 1/19 would occur after 16 sec in the fifth story, associated with a large shift of the drift to the negative direction. The above maximum story drift angle at the fifth story level implies the collapse or very dangerous state of the structure. From these analyses it can be recognized that the rupture of a bracing member makes the interstory drift of the story extremely large. Of course, this peculiarity of response depends on the dominant period or spectral content of input earthquake ground motions. Certainly the used earthquake record (the 1978 Miyagi-Ken-Oki Earthquake) has the predominant period of about 1.0 sec so that the response would become large since the major periods are elongated by the brace rupture. As a conclusion, the rupture of bracing members will put a braced steel structure in jeopardy with high possibility.

CURRENT DESIGN REQUIREMENTS

Table 3 shows the international comparison about the width-thickness ratio limitations for box-section columns together with Kato's proposal for cold-formed box-section columns (Kato, 1987). In this table, the ductility required for members is classified into three grades in the Japanese codes. That is, the classes I, II, and III correspond to the ductility of η = 6, 1.5 and 0, respectively, where η is defined as η = $(\Delta u - \Delta y)/\Delta y$, and where Δy is the maximum flexural deformation limited by local buckling and Δu is the yield flexural deformation of members. Obviously, η = 0 means that local buckling occurs almost simultaneously with yielding. In addition the ductility in the classification adopted by the codes of ECCS (ECCS, 1984), New Zealand (NZ, 1985), AISC (AISC, 1978) and CSA (CSA, 1989) is not clear in the above definition. Therefore, the correspondence to the Japanese classes is only referential. Now, from Table 3 it can be seen that only the code of New Zealand satisfies Kato's proposal. Further, the values regulated in each code in this table are exactly the same as those on hot-formed or welded box sections (not indicated in Table 3). This implies that the peculiar proper-ties of cold-formed members, which is discussed in the following section, are not reflected by the regulations cited in Table 3.

Consequently, the present code conditions concerning the cold-formed square or rectangular tubular members are inconsistent with the crucial results by recent investigations.

POSSIBLE MEASURES AND FUTURE RESEARCH NEEDS

In view of the circumstances mentioned above, structural designers should examine the following measures immediately:

1) Avoid using cold-formed square or rectangular tubing sections as bracing members, if other sections such as H-shaped sections are usable; or otherwise hot-formed or built up square (or rectangular) tubular sections may be adopted;
2) Use cold-formed sections with small width-thickness ratios (compact section);
3) Fill cold-formed tubing members with concrete to delay local buckling or to lessen its severity;
4) Anneal cold-formed members, before installing, to change somewhat mechanical properties of the members.

The item 1) is the easiest way, if it is allowed. By the past studies (6), H-shaped sections have even longer fracture lives than cold-formed square hollow sections.

Regarding the item 2) Tang and Goel (Tang, 1987) proposed a 50 percent reduction of the width-thickness ratio specified by the AISC Specifications (Part 2: Plastic Design) for compression members in order to minimize the

152

effect of local buckling and delay the fracture. That is, the proposed value is (B-2t)/t=14. Also, Uang and Bertero (Uang. 1986) suggested a reduction in the B/t ratio to the value of 18, through their scale model test. Further, Kato commented in his paper (Kato, 1987) that the B/t ratio of cold-formed square tubular bracing members should be smaller than his proposal, B/t=20 for columns. Consequently, considering the above proposals and the past experimental results (Gugerli, 1982; Liu, 1987; Lee, 1987), the width-thickness ratio for cold-formed square or rectangular tubular members can be understood to be less than 16 for braces and 20 for columns in terms of B/t to preclude early fracture under severe inelastic cyclic displacements.

Relating to B/t ratios, recently in Japan, cold-formed square sections having thick wall-thickness up to 40mm by pressing have widely been used for columns in medium- and high-rise buildings. Although sufficient experimental data have not yet been attained, local buckling was considerably delayed in the case of B/t ≦ 20 in a few tests which simulated realistic columns (Nakamura, 1990).

The effectiveness of the item 3) has been experimentally confirmed by Liu (Liu, 1987) and Lee (Lee, 1987). By their studies, the presence of concrete was not able to delay the occurrence of local buckling itself. However, it changed the local buckling mode and reduced the severity of local buckling. Then, the fracture life was much elongated owing to the delay of the crack initiation.

In a cold-formed square or rectangular tubing section, yield strength generally has a higher value than that of the pre-formed virgin steel sheet (Kato, 1988), as shown in Fig. 3. This is due to severe plastic cold-work in forming the section. In particular, the yield strength at the corner of the section is extremely heightened, whereas the tensile strength at the corner is less increased (Fig. 3). Thus, the yield ratio, defined as the ratio of yield strength to tensile strength, is very large at the corner resulting in poor strain capacity that leads to early cracking.

Now, under the above discussion, the measure of the item 4) can be conceived as an effective way to delay cracking. Theoretically, by annealing cold-formed members the high yield strength, high yield ratio and lack of deformability at corner portions may be lessen, since residual stresses or strains confined in sections can be released by annealing (Aoki, 1985). However, the effect of annealing on delaying the local buckling initiation cannot be expected (Aoki, 1985; Kato, 1980).

REFERENCES

AISC, 1978, "Specification for Design, Fabrication and Erection of Structural Steel for Buildings," American Institute of Steel Construction, Chicago, Illinois, November.

Aoki, H. Narihara, H., Nakamura, K., and Kurosawa, T., 1985, "Stub-Column Test of Cole-Forming Steel Tubes with Square Hollow Section," Proceedings of Annual Conference of AIJ, Tokai, Japan, October. (in Japanese)

BSL, 1981, "Guidelines of Structural Calculation Based on the Revised Enforcement Order under Building Standard Law," Housing Bureau, Ministry of Construction, the Building Center of Japan, February.

CSA, 1989, "Limit State Design of Steel Structures," CSA-S16·1, Canadian Standard Association.

ECCS, 1984, "Recommendation for Steel Structures in Seismic Zones," ECCS.

Gugerli, H. and Goel, S.C., 1982, "Inelastic Cyclic Behavior of Steel Bracing Members." Report No. UMEE 82R1, University of Michigan, Ann Arbor, Michigan, January.

Karren, K.W., 1967, "Corner Properties of Cold-Formed Steel Shapes," J. of Struct. Div., ASCE, ST1, Vol. 93, February.

Kato, B., and Aoki, H., 1978, "Residual Stresses in Cold-Formed Tubes," Journal of Strain Analysis, Vol 13, No. 4, England.

Kato, B., Aoki, H. and Narihara, H., 1986 "Residual Stresses in Square Steel Tubes Introduced by Cold-Forming and the influence on Mechanical Properties," IIW/AIJ Joint International Meeting on Safety Criteria in Design of Tubular Structures, Tokyo, July.

Kato, B., 1987, "Deformation Capacities of Tubular Steel Members Governed by Local Buckling," Journal of Structural Engrg., AIJ, No. 378, August. (in Japanese)

Kato, B., Aoki, H., and Kurosawa, T., 1988, "Plastic Strain History and Residual Stresses Locked in Cold-Formed Square Steel Tubes," Journal of Structural Engrg., AIJ, No.385, March. (in Japanese)

Kato, B., and Nishiyama, I., 1980, "Local Buckling Strength and Deformation Capacity of Cold-Formed Steel Rectangular Hollow Section." Journal of Structural Engineering, Architectural Institute of Japan, No. 294, August. (in Japanese)

Liu, Z., and Goel. S.C., 1987, "Investigation of Concrete-Filled Steel Tubes under Cyclic Bending and Buckling," Report No. UMCE 87-3, University of Michigan, Ann Arbor, Michigan, April.

Lee, S., and Goel. S.C., 1987, "Seismic Behavior of Hollow and Concrete-Filled Square Tubular Bracing Members," Report No. UMCE 87-11, University of Michigan, Ann Arbor, Michigan, December.

Nakamura, H., 1990, "Structural Performance of Square Hollow Sections by Cold-Forming Press Work as Columns," The Structural Technology, No. 11 and 12. (in Japanese)

NZ, 1985, "Papers Resulting from Deliberations of the Society's Discussion Group for the Seismic Design of Steel Structures, Bulletin of the New Zealand National Society for Earthquake Engineering, Vol 18, No. 4.

Tang, X., and Goel, S.C., 1987, "Seismic Analysis and Design Considerations of Braced Steel Structures," Research Report UMCE 87-4, University of Michigan, Ann Arbor, Michigan, April.

Uang, C.M., and Bertero, V.V., 1986, "Earthquake Simulation Tests and Associated Studies of a 0.3-Scale Model of a Six-Story Concentrically Braces Steel Structure," Report No. UCB/EERC-86/10, University of California, Berkeley. California, December.

Yamanouchi, H., Midorikawa, M., Nishiyama, I., Watabe. M., 1989, "Seismic Behavior of Full-Scale Concentrically Braced Steel Building Structure," ASCE, Journal of Structural Engineering, Vol. 115, No. 8, August.

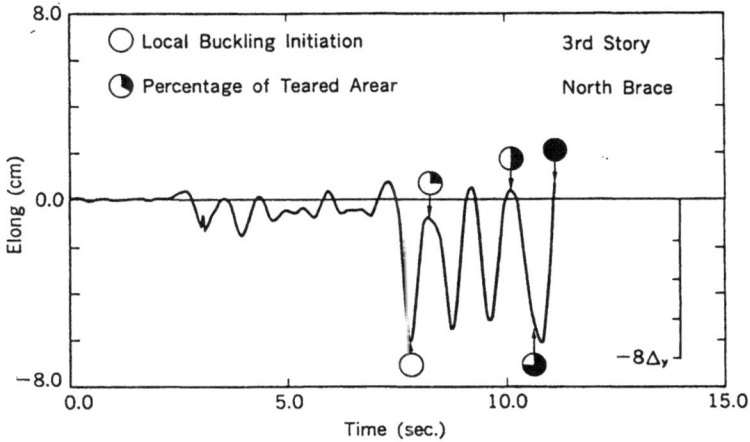

Fig. 1 Rapid Propagation of Crack after Local Buckling

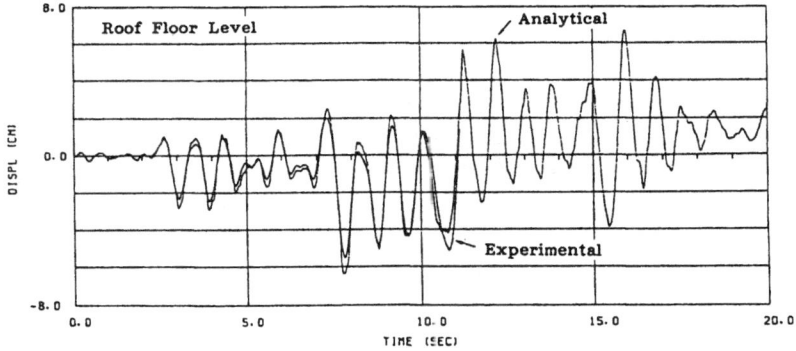

Fig. 2 Analytical Time History of Interstory Drift after Brace Rupture

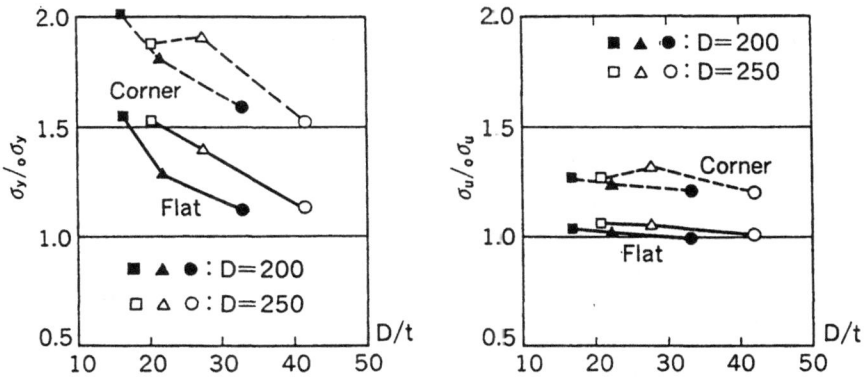

Fig. 3 Increase in Yield Stress and Maximum Stress of Cold-Formed
Square Tube (Ratio to Virgin Steel Sheet)

Table 1 Geometry of Square Tubular Bracing Members Used for Phase I Test

Story & Side	Sizes (inch)	Sectional Area (cm^2)	B/t	(B/2t)/t	Length (cm)	Radius of Gyration(cm)
6 S/N	4x4x1/5.56	17.21	22.2	20.2	442.3	3.94
5 S/N	5x5x1/5.56	21.85	27.8	25.8	441.0	4.98
4 S/N	5x5x1/4	29.61	20.0	18.0	435.7	4.88
3 S/N	6x6x1/4	36.06	24.0	22.0	434.0	5.92
2 S/N	6x6x1/4	36.06	24.0	22.0	432.9	5.92
1 S / N	6x6x1/2	66.84	12.0	10.0	513.4 / 512.4	5.61

Note: (1) B=width of section, and t=wall thickness
(2) Brace length designates face-to-face length, not based on center-line dimension

Table 2 Absorbed Energy in Bracing Members

Story & Side	Absorbed Energy (tonf cm)	Normalized Energy(tonf/cm^2)
6 S	—	—
N	—	—
5 S	70	3.20
N	180	8.24
5 S	143	4.83
N	500	16.9
3 S	393	10.9
N	1321	36.6
2 S	1875	52.0
N	1750	48.5
1 S	214	3.20
N	1790	26.8

Table 3 Comparison of Width-Thickness (B/t) Limitations

	Class	AIJ	BSL	ECCS	NZ	AISC	CSA	Kato's Study
Column of Cold Formed Box Section (SHS)	I	27.0	27	28	19.5	30.0	27.4	20.0
	II	31.0	32	—	23.4	—	34.2	29.6
	III	41.0	41	—	31.1	37.1	43.7	36.0

Effect of seismic input on hydrodynamic forces acting on gravity dams

A.M. Jablonski[1] and J.H. Rainer[1]

ABSTRACT

A 2D constant boundary element reservoir model is used to obtain hydrodynamic forces due to harmonic horizontal and vertical excitations. Energy dissipation due to both infinite radiation and damping along the bottom of the reservoir has been incorporated. The Fast Fourier transformation (FFT) technique is then used to find a response to a given seismic input through the frequency domain analysis. This gives the time history of the resultant hydrodynamic force acting on the dam face. The results indicate that the total response in hydrodynamic forces depends not only on the geometry of the reservoir but also on the nature of the seismic input (frequency content, ground acceleration and duration of strong motion). Critical depths, at which maximum reservoir response occurs, are identified for various seismic excitations.

INTRODUCTION

A number of studies have been carried out on the seismic response of 2D gravity dam-reservoir-foundation deterministic models using both finite element and boundary element methods (Hall and Chopra, 1980; Hanna and Humar, 1982; Liu, 1984; Humar and Jablonski, 1988). Some attempts have been made to introduce non-deterministic models (Cheng, Yang and Niu, 1990). However, little attention has been given to the effect of different seismic inputs on hydrodynamic forces acting on gravity dams.

The analytical 2D model of the system, based on the model first introduced with the FEM (Hall and Chopra, 1980) , is used in this study. The energy loss in the outgoing waves has been modelled by assuming that beyond a certain distance upstream of the dam the reservoir has a uniform rectangular shape. Boundary element discretization is limited to the irregular region of the reservoir in the vicinity of the dam face. For the regular but infinite region a one-dimensional finite element solution is employed. The effect of the reservoir foundation damping is also included based on a simplified boundary condition which models the absorption of the reservoir bottom using a one-dimensional compressive wave propagation model (Hall and Chopra, 1980). This model has been incorporated into a boundary element solution for small vibrations of the reservoir (Humar and Jablonski, 1988). This paper presents the results of more recent study on the application of BEM

---

[1] Structures Section, Institute for Research in Construction, National Research Council of Canada, Ottawa, Ontario K1A 0R6

and the Fast Fourier transformation technique to the seismic analysis of a 2D infinite reservoir bounded by a gravity dam. At first the given system is solved for horizontal and vertical harmonic excitations with the use of the previously developed and recently modified computer program BEMC2D (Humar and Jablonski, 1988). Then, the Fast Fourier transformation technique is used to obtain the combined response of the system for horizontal and vertical components of two different earthquake acceleration records: the Saguenay earthquake of 1988 and the El Centro earthquake of 1940. The objective of this study is to show the effect of seismic input on hydrodynamic response of the reservoir. The critical depth for obtaining maximum reservoir response is identified for the above mentioned earthquake records.

BOUNDARY ELEMENT FORMULATION FOR
HARMONIC RESPONSE

The details of the development of the boundary element model of the reservoir can be found in the earlier publications (Liu, 1984; Humar and Jablonski, 1988). Only a few of the important assumptions and boundary conditions are summarized here.

The concrete gravity dam and reservoir are modelled as a two-dimensional system in which only planar vibrations are considered. The base of the dam and the bottom of the reservoir may undergo the prescribed acceleration history due to horizontal and vertical components of the ground motion. Dam-foundation interaction effects are not included in this model, but the bottom of the reservoir may be flexible or almost rigid. Small vibration theory is used to solve the problem. The governing equations are the linearized Navier-Stokes equations, which are valid for non-viscous and compressible water. For harmonic motion of the dam and foundation, uniformly distributed along the dam and the reservoir bottom, the Navier-Stokes equations are reduced to the Helmholtz equation:

$$\frac{\partial^2 p}{\partial x^2} + \frac{\partial^2 p}{\partial y^2} + k^2\, p = 0 \tag{1}$$

where $k = \omega/c$ the reservoir wave number, $p =$ the water pressure, $\omega =$ the excitation frequency, $c =$ velocity of sound in water.

Along the reservoir bottom, a special boundary condition originally developed by Hall and Chopra (1980) is used,

$$\frac{\partial p}{\partial n}(s',\omega) = -\frac{w}{g}\, a_n(s') - i\,\omega\,\gamma\, p(s',\omega) \tag{2}$$

in which $s' =$ the coordinate along the reservoir bottom, $\gamma = w/(w_r c_r) =$ the damping coefficient of the reservoir bottom , $w_r =$ specific weight of the bottom material, $w =$ specific weight of water, $c_r =$ compression wave velocity, travelling in the direction normal to the reservoir bottom interface; and $a_n(s') =$ the harmonic excitation amplitude along the outward normal.

Damping of the reservoir bottom can be expressed with help of a wave reflection coefficient α_R by

158

$$\gamma \cdot \frac{\begin{matrix} 1 & 1 & \alpha_R \\ c & 1 - \alpha_R \end{matrix}}{} \tag{3}$$

For a rigid reservoir bottom, $\alpha_R = 1.0$, and $\gamma - 0$. i.e. damping is absent. For $\alpha_R = 0$, and $\gamma = 1/c$, this represents full damping. The wave reflection coefficient may be expressed in the following form:

$$\alpha_R = \frac{1 - w\ c/(w_r\ c_r)}{1 + w\ c_i/(w_r\ c_r)} \tag{4}$$

Along the transmitting boundary, the boundary condition has been derived based on the one-dimensional finite element model of the infinite region (Hall and Chopra, 1980; Humar and Jablonski, 1988). It has been successfully used in both the FEM and BEM formulations. For a horizontal and/or vertical harmonic ground motion this leads to a solution of an eigenvalue problem. This solution is then coupled with the boundary element solution for the finite irregular region of the reservoir adjacent to the dam. Compatibility of the pressure and pressure gradients is then imposed at the interface of the regular and irregular regions. The reservoir damping in the infinite region is also included.

RESPONSE TO AN ARBITRARY GROUND MOTION

When the excitation history is given in terms of a function of the acceleration $a(t)$ and $\bar{F}(\omega)$ represents the harmonic response of the system to a unit acceleration, $e^{i\omega t}$, the so-called harmonic synthesis (superposition of all responses) may be expressed in the following form:

$$F(t) = \frac{1}{2\pi} \int_{-\infty}^{\infty} A(\omega)\ \bar{F}(\omega)\ e^{i\omega t} d\omega \tag{5}$$

where

$$A(\omega) = \int_{-\infty}^{\infty} a(t)\ e^{-i\omega t} dt \tag{6}$$

Equation 6 is the Fourier Transform of the acceleration function $a(t)$, while Eq. 5 is the Inverse Fourier Transform of the product of the frequency functions $A(\omega)$ and $\bar{F}(\omega)$. The continuous expressions given by Eqs. 5 and 6 can be converted into finite sums in the following form:

$$\{F(m\Delta t)\} = \frac{\Delta\omega}{2\pi} \sum_{n=0}^{N-1} \{A(n\Delta\omega)\ \bar{F}(n\Delta\omega)\} e^{i\frac{2\pi n m}{N}} \tag{7}$$

$$\{A(n\Delta\omega)\} = \Delta t \sum_{m=0}^{N-1} \{a(m\Delta t)\} e^{-i\frac{2\pi n m}{N}} \tag{8}$$

Then, the Fast Fourier Transformation (FFT) technique is used to evaluate Eqs. 7 and 8. A computer program has been developed to calculate the response of the linear system subjected to a given excitation (Jablonski, 1990).

PARAMETRIC STUDY OF RESERVOIR RESPONSE

Two sets of parametric studies were carried out: 1. variation of reservoir boundary conditions; 2. variation of reservoir depth. For both sets of calculations 16 seconds of ground motions were used from the Saguenay earthquake of November 25, 1988, and the 1940 El Centro earthquake in California. Further details of the motions are given in Table 1.

Variation of reservoir damping and shape

Various reservoir bottom absorption conditions were chosen, as represented by the reflection coefficients $\alpha_R = 0.975$, 0.75, and 0.50. The reservoir has a water depth H of 100 m at the dam. One reservoir is of rectangular shape of infinite length. For the other two, the reservoir bottom slopes from a depth of of $H_D = 100$ m at the dam to $H_1 = 50$ m over a distance L of 200 m from the dam; thereafter the water depth is constant at 50 m. The rectangular reservoir has a fundamental resonance frequency of 3.6 Hz, the sloping reservoir, 4.4 Hz. The horizontal and vertical components of seismic ground motions were applied both individually and simultaneously in order to observe the influence of excitation direction. The results of the numerical study are presented in Table 2, giving the ratio of the peak resultant hydrodynamic force to the hydrostatic force (maximum positive and negative values) on the vertical dam face.

Variation of reservoir depth

Rectangular reservoirs of depths $H = 36$ m, 77.6 m, and 113.8 m were selected so that the resonance frequency of the reservoir coincides with specific peaks or valleys in the vertical response spectra of the two seismic motions shown in Figs. 1 and 2. No horizontal motion was applied for this set of calculations. The resonance frequencies are 10 Hz, 4.6 Hz, and 3.2 Hz corresponding to above water depths. The results of resultant peak hydrodynamic force ratios on the dam for two reservoir bottom reflection coefficients, $\alpha_R = 0.975$ and 0.75, are shown in Table 3. Representative time variations of the resultant hydrodynamic force ratios are shown in Figs. 3 and 4.

DISCUSSION OF RESULTS

From Table 2, the peak force ratios for the rectangular reservoir are seen to be primarily influenced by excitation in the vertical direction. For the sloping reservoir bottom, however, excitation in the horizontal direction becomes more significant, generally exceeding the results from the vertical component for the El Centro motion. This effect depends also on the spectral content of the vertical and horizontal components of ground motion as can be seen by comparing the results for the same geometry, Nos. 2 and 3 in Table 2. The

Saguenay results, No. 2, are always higher for the vertical excitation than the horizontal or the combined ones, whereas for the El Centro excitation, No. 3, the horizontal one predominates. In all cases, the maximum peak from the combined horizontal and vertical excitation is close to the maximum of either of the two individual excitations.

The reflection coefficient for the reservoir bottom is seen to have a significant effect on the peak pressure on the dam. For all results in Table 2, reductions in peak force ratios by factors of 3 to 4 occur for $\alpha_R = 0.75$ as compared to $\alpha_R = 0.975$. Only small further reductions occur for $\alpha_R = 0.5$. The absorption properties of the reservoir bottom are therefore of major importance in relation to the peak dynamic pressures on the dam.

The peak resultant pressures as a function of the variation of the reservoir depth are presented in Table 3. For the shallowest reservoir, Case 1 and for $\alpha_R = 0.975$, the peak hydrodynamic force from the El Centro motion exceeds substantially the hydrostatic force. This case coincides with a peak in the response spectrum curve shown in Fig. 2. For the lower reflection coefficient $\alpha_R = 0.75$, the peaks of hydrodynamic forces then become less than the hydrostatic ones, but are still of substantial amplitude. The results for the Saguenay earthquake are quite moderate in comparison. For Case 2 in Table 3, the results for Saguenay earthquake exceed those from the El Centro earthquake for the higher α_R. For Case 3, the deepest reservoir, the El Centro excitation produces larger pressures than Saguenay. As was the case for the results presented in Table 2, substantial reductions in peak hydrodynamic forces occur, by factors of 2 to 3, when the reflection coefficient changes from $\alpha_R = 0.975$ to 0.75.

CONCLUSIONS

1. The damping characteristics of the reservoir bottom have a significant effect on the peak hydrodynamic forces on the dam caused by seismic excitations.
2. The geometry of the reservoir affects the relative contributions to the total hydrodynamic pressure arising from horizontal and vertical excitations.
3. For obtaining maximum pressures on the dam, the critical depth of the reservoir varies for different seismic records and may be derived with the help of response spectra. For the parameters chosen in this study, the seismically induced hydrodynamic pressures acting on gravity dams can be a considerable fraction of static water pressure, and in certain cases, can exceed it.

REFERENCES

Cheng, A.H.-D., Yang, C.Y., Niu, T.P. 1990. Earthquake Reliability Analysis of Dam-Reservoir-Foundation System Using Boundary Element Method , Proceedings of Fourth U.S. National Conference on Earthquake Engineering, May 20-24, 1990, Palm Springs, California, Vol. 3, pp. 85-94 .

Hall, J.F., Chopra, A.K. 1980. Dynamic Response of Embankment Concrete-Gravity and Arch Dams Including Hydrodynamic Interaction. Report No. UCB/EERC-89/39, Earthquake Engineering Research Center, University of California, Berkeley, California.

Hanna, Y.G., Humar, J.L. 1982. Boundary Element Analysis of Fluid Domain, J. Eng. Mechanics. ASCE, Vol. 108, 1982, pp. 436-450.

Humar, J.L., Jablonski, A.M. 1988. Boundary Element Reservoir Model for Seismic Analysis of Gravity Dams, Earthquake Engineering and Structural Dynamics, Vol. 16, 1988, pp. 1129-1156.

Jablonski, A.M. 1990. Seismic Hydrodynamic Forces Acting on 2D Models of Gravity Dams - User's Guide to Computer Programs. Internal Report No. 602, Institute for Research in Construction, National Research Council of Canada, November 1990.

Liu, P.L.-F. 1984. Hydrodynamic Pressures on Rigid Dams During Earthquakes, J. Hydraulic Eng., ASCE, Vol. 110, 1984, pp. 51-64.

Table 1. Characteristics of earthquake records

Earthquake Record	Component	Peak Acceleration	Notes
M5.7, Saguenay, November 11, 1988 Site 16, Chicoutimi, Quebec	Vertical	$A_V = -1.005$	High frequency content
	Horizontal, 214°	$A_H = 1.045$	
M6.6, Imperial Valley, May 18, 1940 El Centro, California	Vertical	$A_V = 2.063$	Strong shaking
	S-E Horizontal	$A_H = 3.417$	

Table 2. Maximum amplitudes of hydrodynamic forces for various reservoirs

No.	Reservoir Model	Reservoir Fundamental Frequency [Hz]	Earthquake Motion	$\alpha_R = 0.50$			$\alpha_R = 0.75$			$\alpha_R = 0.975$		
				Horiz.	Vert.	Comb.	Horiz.	Vert.	Comb.	Horiz.	Vert.	Comb.
1	Rectangular Infinite H = 100 m	3.6	Saguenay, 1988	0.031	0.080	0.079	0.038	0.115	0.119	0.049	0.164	0.180
				-0.039	-0.063	-0.065	-0.039	-0.109	-0.114	-0.044	-0.179	-0.196
2	Inclined Infinite $H_D = 100$ m $H_I = 50$ m	4.4	Saguenay, 1988	0.016	0.049	0.049	0.021	0.095	0.091	0.045	0.245	0.231
				-0.019	-0.042	-0.041	-0.023	-0.091	-0.092	-0.042	-0.253	-0.233
3	Inclined Infinite $H_D = 100$ m $H_I = 50$ m	4.4	El Centro, 1940	0.144	-0.056	0.132	0.174	0.121	0.165	0,343	0.304	0.363
				-0.164	-0.-089	-0.143	-0.172	-0.121	-0.180	-0.300	-0.300	-0.396

Table 3. Maximum amplitudes of hydrodynamic forces on dam
from vertical excitation in rectangular reservoirs

Case	Reservoir Depth H [m]	Reservoir Fundamental Frequency [Hz]	Earthquake Motion	Ratio Hydrodynamic to Static Force	
				$\alpha_R = 0.975$	$\alpha_R = 0.75$
1	36	10	Saguenay	0.269	0.128
				-0.261	-0.124
			El Centro	1.814	0.689
				-1.610	-0.588
2	77.56	4.6	Saguenay	0.460	0.155
				-0.463	-0.148
			El Centro	0.403	0.179
				-0.420	-0.224
3	113.83	3.2	Saguenay	0.120	0.076
				-0.114	-0.066
			El Centro	0.348	0.203
				-0.333	-0.165

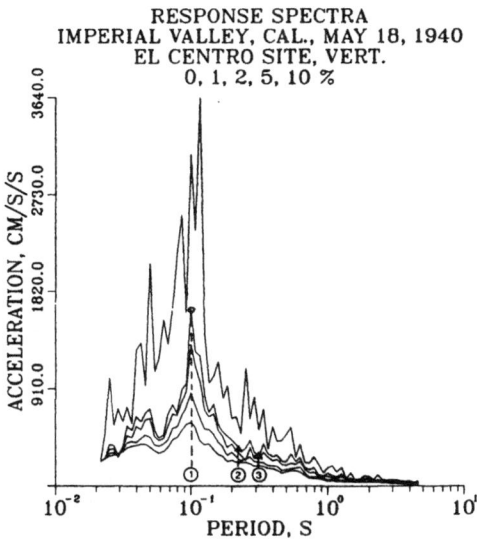

RESPONSE SPECTRA
IMPERIAL VALLEY, CAL., MAY 18, 1940
EL CENTRO SITE, VERT.
0, 1, 2, 5, 10 %

RESPONSE SPECTRA
SAGUENAY, QUEBEC, NOV. 11, 1988
SITE 16: CHICOUTIMI, VERT.
0, 1, 2, 5, 10 %

Figure 1. The El Centro Earthquake response
spectra, vertical component, showing
Cases 1, 2 and 3 of Table 3

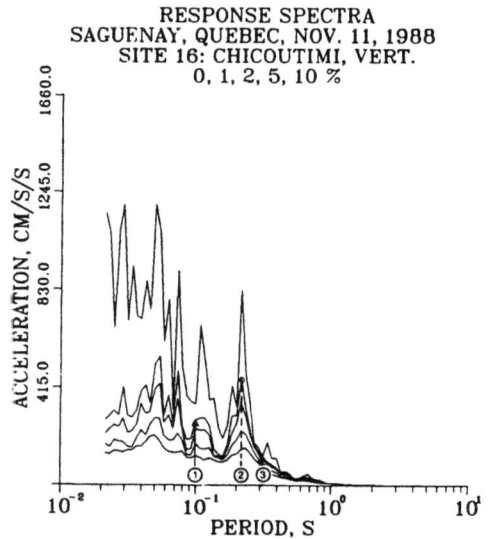

Figure 2. The Saguenay Earthquake response
spectra, vertical component, showing
Cases 1, 2 and 3 from Table 3

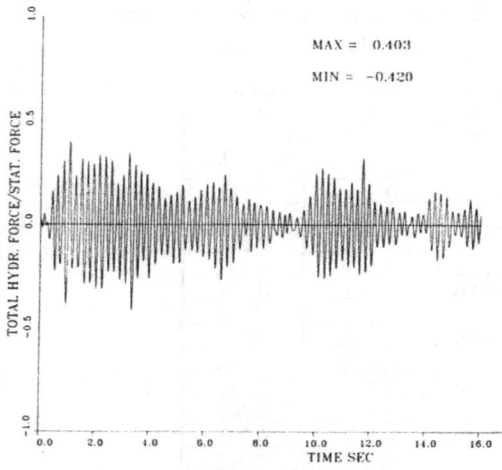

Figure 3. Time history of hydrodynamic force
for Case 2, $H = 77.56\ m$ due to the El Centro
Earthquake, vertical component

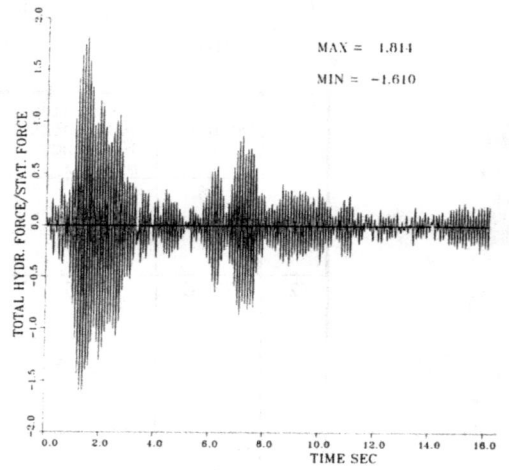

Figure 4. Time history of hydrodynamic force
for Case 1, $H = 36\ m$ due to the El Centro
Earthquake, vertical component

Boundary element analysis of the seismic response of gravity dams

R. Chandrashaker[I] and J.L. Humar[II]

ABSTRACT

A boundary element technique for the seismic response analysis of gravity dam-reservoir-foundation system is presented. Analysis is carried out in the frequency domain and a sub-structure approach is used. The reservoir substructure is divided into two parts: a near field of specified irregular geometry modeled by boundary elements and an infinite far field of uniform cross section modeled by 1D finite elements. The energy absorption along the reservoir bed is modeled by an approximate wave propogation model. The dam is discretized by finite elements. The foundation soil, idealized as an elastic half plane, is modeled by boundary elements. To enhance the computational efficiency of the procedure, a frequency independent fundamental solution is adopted for the reservoir, and the equations of the motion are transformed to the first few modal coordinates of the dam-foundation system. The complex frequency response functions of Pine Flat dam are obtained for various conditions of reservoir and foundation. A response time history analysis is carried out for the Taft ground motion.

INTRODUCTION

Analytical procedures for determining the seismic response of gravity dams are now fairly well developed. Since the response of the dam is affected both by the reservoir induced hydrodynamic forces and interaction with the flexible foundation, a substructuring technique is most effective in the analysis. The dam and the reservoir substructures are usually represented by finite elements, while the foundation is treated as elastic or viscoelastic halfspace. A finite element representation of the foundation is also possible provided a rigid boundary exists or an artificial boundary that prevents the reflection of outgoing waves can be introduced.

The reservoir model must account for the energy lost in waves radiating towards infinity and through the absorption effects along the reservoir bottom. To model the infinite radiation, the reservoir is divided into two parts: a near field of specified irregular geometry and a far field of uniform cross-section. The latter is usually modeled by 1D finite elements (Hall and Chopra 1980). The energy dissipation through partial wave absorption in a flexible reservoir bottom can be accounted for by using a 1D wave propogation model proposed by Hall and Chopra (1980).

In recent years boundary element method (BEM) has emerged as an effective alternative to

---

[I] Graduate Student

[II] Professor and Chairman, Dept. of Civil Engineering, Carleton University, Ottawa, Canada K1S 5B6

the finite element method (FEM) for the seismic analysis of gravity dams. Analytical procedures based on BEM have been developed for reservoir vibration by Hanna and Humar (1982), and Humar and Jablonski (1988). In a recent study, Medina et al. (1990) have used the BEM for the response analysis of the complete dam-reservoir-foundation system. However, their study extends only to the computation of frequency response functions; and analytical solutions for real earthquake motions have not been obtained.

The present study extends the BEM procedures to the response analysis for a realistic earthquake motion using a Fourier synthesis approach. In the analysis for a real ground motion of random nature, it is important to pay attention to the computational efficiency of the procedure. The traditional BEM uses a frequency dependent fundamental solution, implying that the BEM matrices must be evaluated for each excitation frequency, a computational task of considerable magnitude. The present study uses a frequency independent fundamental solution (Nardini and Brebbia 1984; Tsai 1987) for the near field of the reservoir, the far field being modeled by 1D finite elements. To minimize the computations, the reservoir bottom absorption is represented by 1D wave propagation model and the far field of the reservoir is assumed to possess rigid bottom. The dam is discretized by finite elements. The foundation soil is treated as a half plane, modeled by boundary elements along the foundation surface (Abdalla 1984). The effect of cross coupling between the foundation portion below the reservoir and that below the dam has been found to be small by other researchers (Fenves and Chopra 1984) and is therefore ignored. The analysis is carried out in the frequency domain using Fast Fourier Transform (FFT).

Brief formulations of the substructure models are provided in the following section. The boundary element procedure is then applied to the analysis of Pine Flat dam, for its response to Taft ground motion.

ANALYTICAL FORMULATIONS

Reservoir Hydrodynamic Force

The 2D wave equation governing the reservoir fluid vibration (Fig. 1) is given by

$$\nabla^2 p + k^2 p = 0 \tag{1}$$

where p is the hydrodynamic pressure, k the wave number $= \Omega/c$, Ω the excitation frequency, and c the wave velocity in water. By using the weighted residual technique, with the weighting function p_i^* chosen to be equal to the static solution for a unit source at point i, Eq. 1 can be transformed into an integral equation involving integrals on the boundary and the domain integral $k^2 \int \nabla^2 p.p_i^* dA$. This domain integral can also be transformed to boundary integrals by assuming the following particular solution for p

$$p(x,y,\Omega) = \sum_{m=1}^{M} \alpha_m(\Omega) f_m(x,y) \tag{2}$$

in which $\alpha_m(\Omega)$ is an unknown coefficient, and $f_m(x,y) = \nabla^2 \psi^m(x,y)$ is a suitably chosen harmonic function. By discretizing the near field boundary into a series of segments termed boundary elements, with the pressure and pressure gradients as well as ψ and its derivative $\eta = \partial\psi/\partial n$ assumed to vary in a prescribed manner over each element, the boundary integral equations, can be represented in a discretized form as

$$\mathbf{H}\mathbf{p} - \mathbf{G}\mathbf{q} = k^2 \left(-\mathbf{H}\psi + \mathbf{G}\eta \right) \alpha \tag{3}$$

166

where $\mathbf{H}$ and $\mathbf{G}$ are matrices of the boundary integrals involving $q_i^* = \partial p_i^*/\partial n$ and p_i^* respectively, $\mathbf{p}$ is the vector of pressures and $\mathbf{q}$ the vector of pressure gradients.

In the present formulation, p_i^* is given by

$$p_i^* = -(1/2\pi)\ln\left(r_{ij}/r_{ij}'\right) \tag{4}$$

and function f is chosen as

$$f_j(x_i, y_i) = r_{ij}' - r_{ij} \tag{5}$$

where r_{ij} and r_{ij}' are the distances to the field point j respectively from the source point and the mirror image of the source point. The solution p_i^* automatically satisfies the zero pressure condition on the surface of the reservoir.

The vector α can be expressed in terms of $\mathbf{p}$ and a matrix $\mathbf{F}$ containing the values of f_m at the field points. Substitution in Eq. 3 gives

$$\left(\mathbf{H} - k^2\tilde{\mathbf{M}}\right)\mathbf{p} = \mathbf{G}\mathbf{q} \tag{6}$$

where $\tilde{\mathbf{M}} = (\mathbf{G}\eta - \mathbf{H}\psi)\mathbf{F}^{-1}$

The respective boundary conditions along the dam face and reservoir bottom are

$$q(s, \Omega) = -\rho_w a_n(s) \tag{7a}$$
$$q(s', \Omega) = -\rho_w\, b_n(s') - \iota\Omega\beta p(s', \Omega) \tag{7b}$$

where ρ_w is the mass density of water; a_n the acceleration of the dam face along the normal; β the wave absorption parameter $= \rho_w/\rho_r c_r$; ρ_r and c_r the mass density and the compression wave velocity of the reservoir bottom; b_n the free field acceleration of reservoir bottom along the normal; and s' the coordinate along reservoir bottom. It is perhaps more meaningful to represent the wave absorption along the reservoir bottom through a wave reflection parameter $\tilde{\alpha}$ given by $\tilde{\alpha} = (1 - \beta c)/(1 + \beta c)$. A value of $\tilde{\alpha} = 1.0$ represents a nonabsorptive reservoir bottom.

Along the interface with far field, vectors $\mathbf{p}$ and $\mathbf{q}$ are related by (Hall and Chopra 1980)

$$\mathbf{p} = \mathbf{\Gamma}\, v + \rho_w\, \mathbf{\Gamma}\, \mathbf{K}^{-2}\, \mathbf{\Gamma}^T\, \mathbf{d} \tag{8a}$$
$$\mathbf{q} = -\mathbf{\Gamma}\, \mathbf{K}\, v \tag{8b}$$

where γ_n and λ_n are respectively the eigenvectors and eigenvalues of the far field; $\mathbf{\Gamma}$ the matrix of eigenvectors; v is a vector of modal coordinates; $\mathbf{K}$ a diagonal matrix with elements $\kappa_1, \kappa_2, \cdots, \kappa_L$; L the number of finite element nodes on the interface boundary; $\kappa_n = \sqrt{\lambda_n - k^2}$ and $\mathbf{d}$ a vector with only one non-zero entry equal to an input acceleration a_y corresponding to the base node.

Foundation Impedance Matrix

The boundary integral equations for an homogeneous, isotropic and linear elastic soil region is given by

$$\beta^P u_j(P) + \int_S T_{ji}^*(P, Q)u_i(Q)\, ds - \int_S t_i(Q)\, U_{ji}^*(P, Q)ds = 0 \tag{9}$$

167

where P is the source point and Q the field point; $\beta^P = 0.5$ for P on a smooth boundary; u_i and t_i denote displacement and traction vectors; U_{ji}^* and T_{ji}^* are the fundamental solutions (Abdalla 1984); S is the boundary of the soil domain and indicial notations are used. Since the kernels in the integrals decay rapidly with distance, only the dam-foundation interface along with a finite portion of the soil surface needs to be discretized. The infinite radiation is automatically accounted for by the fundamental solution.

The complex valued impedance matrix of the foundation, required in the substructure analysis, is obtained by first evaluating the flexibility matrix corresponding to the degrees of freedom of the dam-foundation interface nodes and then taking its inverse. For a horizontal surface, with equally spaced interface nodes, the complete flexibility matrix can be constructed from 2 sets of analyses, one for a unit harmonic horizontal traction and other for a unit harmonic vertical traction applied at any one node.

Coupled Dam-Reservoir-Foundation System

The equation of motion of the dam in frequency domain is given by

$$\left\{ -\Omega^2 \mathbf{M} + (1 + \iota\eta_d)\mathbf{K} \right\} \mathbf{U}^l(\Omega) = -\mathbf{E}^l + \mathbf{R}^l(\Omega) \tag{10}$$

where $\mathbf{M}, \mathbf{K}$ represent respectively the mass and the stiffness matrices of the dam substructure including the dam-foundation interface nodes; η_d is the hysteretic damping factor; $\{\mathbf{U}^l(\Omega)\}$ is the nodal displacement vector relative to the free field; $l = x, y$ is the direction of excitation; $\{\mathbf{E}^x\}^T = \{m_1\ 0\ m_2\ 0\ .....\ m_N\ 0\}$ and $\{\mathbf{E}^y\}^T = \{0\ m_1\ 0\ m_2\ .....\ 0\ m_N\}$; $\{\mathbf{R}^l(\Omega)\}$ is the nodal load vector, having non-zero quantities corresponding to the upstream face and the base DOF of the dam. The nodal load vector is composed of two vectors: $\mathbf{R}_h^l$ denoting the hydrodynamic forces along the upstream face of the dam, and $\mathbf{R}_b^l$ the forces along the dam-soil interface nodes. The latter are given by

$$\mathbf{R}_b^l(\Omega) = -\mathbf{S}_f(\Omega)\mathbf{U}_b^l(\Omega) \tag{11}$$

where $\mathbf{S}_f(\Omega)$ is complex valued soil impedance matrix.

The size of the problem represented by Eq. 10 can be reduced by using the Rayleigh Ritz method, in which the first Nm undamped mode shapes of the dam-foundation system are used as Ritz vectors. The transformed equations become

$$\left\{ -\Omega^2 \mathbf{I} + (1 + \iota\eta_d)\boldsymbol{\Lambda} + \boldsymbol{\Phi}^T \tilde{\mathbf{S}}_{0f}(\Omega)\boldsymbol{\Phi} \right\} \mathbf{Z}^l(\Omega) = -\boldsymbol{\Phi}^T \mathbf{E}^l + \boldsymbol{\Phi}^T \mathbf{R}_h^l(\Omega) \tag{12}$$

where

$$\tilde{\mathbf{S}}_{0f}(\Omega) = \begin{pmatrix} 0 & 0 \\ 0 & \mathbf{S}_f(\Omega) - (1 + \iota\eta_d)\mathbf{S}_f^r(\Omega_0) \end{pmatrix}$$

$\boldsymbol{\Lambda}$ is the diagonal matrix of the squared frequencies ω_j^2, $Z_j^l(\Omega)$ is the jth modal coordinate; ω_j and ϕ_j are the natural frequencies and mode shapes; $\mathbf{S}_f^r(\Omega_0)$ is the real part of the soil impedance matrix corresponding to a small frequency value Ω_0.

The hydrodynamic force vector $\mathbf{R}_h^l$ can be expressed as

$$\mathbf{R}_h^l(\Omega) = \mathbf{R}_o^l(\Omega) - \Omega^2 \sum_{j=1}^{Nm} Z_j^l(\Omega)\mathbf{R}_j^f(\Omega) \tag{12}$$

168

where $R_o^l(\Omega)$, $R_j^f(\Omega)$ are the nodal forces along dam's upstream face equivalent to the hydrodynamic pressures $p_o^l(s,\Omega)$ and $p_j^f(s,\Omega)$ respectively. Pressures $p_o^l(s,\Omega)$ are obtained from boundary conditions corresponding to a rigid dam and free field earthquake motion, while the pressures $p_j^f(s,\Omega)$ are obtained with the motion of the upstream face of the dam being equal to that in the jth mode.

EARTHQUAKE ANALYSIS OF PINE FLAT DAM

To illustrate the analytical procedure developed here, the seismic response of a non-overflow section of Pine Flat Dam, California is analysed for Taft ground motion. The cross section dimensions and the finite element mesh for the analysis of this dam are identical to those given in Fenves and Chopra (1984). For the dam material, the Young's modulus of elasticity $E_d = 3.25 \times 10^6$ psi, unit weight=155 lb/ft^3, Poisson's ratio = 0.2, and the hysteretic damping factor $\eta_d = 0.1$. The foundation is idealized as a homogeneous, isotropic, elastic half-plane. The properties of the foundation material are $E_f = 3.25 \times 10^6$ psi, so that the elastic modulii ratio $E_f/E_d = 1$, unit weight=165 lb/ft^3, and Poisson's ratio = 1/3. The full reservoir has a constant depth of 381 ft. The pressure wave velocity is 4720 ft/sec, and the unit weight of water is 62.4 lb/ft^3. Two values of wave reflection coefficient $\tilde{\alpha} = 1.0$ and $\tilde{\alpha} = 0.75$ are considered. The length of the near field L_f is taken as 0.1H, for the case of fully rigid reservoir bottom and as 2.5H for the case of absorptive bottom. The dam-foundation interface along with a length of 400 ft on either side of the dam base is discretized for the BEM analysis of the foundation.

Taft ground motion of July 1952 earthquake is chosen as free-field input, its S69E and vertical components acting along the transverse and vertical direction of the dam respectively. Using the analytical procedure discussed, the seismic response analysis of the dam section to the Taft motion is carried out considering all the significant interaction components. The analysis is performed by adopting the first 5 mode shapes of the dam, for a rigid foundation case, and 10 mode shapes of the associated dam-foundation system, when the foundation is flexible.

The complex frequency response functions for the absolute horizontal crest acceleration and the time history of the horizontal crest displacements for different conditions of foundation, reservoir water, bottom absorption and excitation direction are presented. For the frequency domain analysis, 2048 time steps of 0.02 s are used. The first 20 s of the excitation is taken as equal to the ground acceleration values for the Taft motion, the remaining period is made up of a grace band of zeros to avoid the aliasing errors inherent in discrete Fourier transforms.

The complex frequency response functions for the crest acceleration due to horizontal excitation are shown in Fig. 2. Reservoir bottom absorption reduces the peak response near the fundamental frequency for both rigid and flexible foundation, although the effect is not as pronounced in the latter case. The absorption effects are less significant at higher frequencies. Flexibility of foundation causes a reduction in the crest acceleration. The frequency response functions due to vertical excitation are shown in Fig. 3. The response exhibits infinite peaks for a non-absorptive bottom, whether or not the foundation is flexibile. Wave absorption at reservoir bottom results in a drastic reduction in the peak response. As for a horizontal excitation, foundation flexibility causes a reduction in the overall crest acceleration.

Figure 4 presents the first 15 s of time history of the horizontal crest displacement of Pine Flat dam subjected to Taft ground motions. It is apparent from these results that the responses to the S69E (transverse) component of Taft motion are quite similar for $\tilde{\alpha} = 1.0$ (non-absorptive) and 0.75 (absorptive reservoir bottom), irrespective of the foundation flexibility. This implies that the response due to the transverse ground motion is less sensitive to the reservoir bottom absorption. For the case of vertical excitation of the system with non-absorptive bottom, the peaks in the complex frequency response function render the results of an FFT analysis invalid because of aliasing. Hence, only results corresponding to absorptive bottom ($\tilde{\alpha} = 0.75$) are presented. The overall effect of foundation flexibility is an increase in crest displacement, both for horizontal and a vertical excitation.

For vertical excitation, the approximate 1D wave absorption model (Hall and Chopra 1980) has been found to compare well with a more rigorous model studied by Medina et al. (1990). For horizontal excitation, Medina has cited that the approximate model underestimates the complex frequency response. However, since the reservoir bottom absorption plays only a minor role in the response of a dam to realistic horizontal excitation, the approximation involved in the 1D model has very little effect on the final response.

CONCLUSIONS

The boundary element procedure developed in the present study is applicable to the seismic response analysis of a general gravity dam-reservoir-foundation system of arbitrary geometry. The computational efficiency of the procedure is enhanced by the use of a frequency independent model of the reservoir, and the tranformation of the equations of motion to modal coordinate of a dam-foundation system. The analytical results show that the effect of reservoir bottom absorption is not very significant for the response of the dam to a horizontal excitation, but is important in the response to a vertical excitation. In either case, the approximate 1D wave propogation model yields reasonably accurate results and a more rigorous approach for modeling the reservoir bottom may not be necessary. On the other hand, use of approximate model results in considerable improvement in the computational efficiency.

References

1. Abdalla, M.H.M. 1984. Dynamic Soil-Structure Analysis by Boundary Element Method. Ph.D. Thesis, Carleton Univ., Ottawa, Canada.

2. Fenves G. and Chopra, A.K. 1984. Earthquake Analysis and Response of Concrete Gravity Dams. Report No. UCB/EERC-84/10, Univ. of California, Berkeley.

3. Hall, J.F. and Chopra, A.K. 1980. Dynamic Response of Embankment Concrete-Gravity and Arch Dams Including Hydrodynamic Interaction. Report No. UCB/EERC-80/39, Univ. of California, Berkeley.

4. Hanna, Y.G. and Humar, J.L. 1982. Boundary Element Analysis of Fluid Domain. J. Eng. Mech., ASCE, 108(EM2), 436-450.

5. Humar, J.L. and Jablonski, A.M. 1988. Boundary Element Reservoir Model for Seismic Analysis of Gravity Dams. Inter. J. Earthq. Eng. Struct. Dynamics, Vol. 16, 1129-1156.

6. Medina, F., Dominguez, J. and Tassoulas, J. L. 1990. Response of Dams to Earthquakes Including Effects of Sediments. J. Struct. Eng., ASCE, 116(11), 3108-3121.

7. Nardini, D. and Brebbia, C.A. 1984. Boundary Integral Formulation of Mass Matrices for Dynamic Analysis. B.E. Research, Vol. 16, Ed. Brebbia, C.A., Spr.-Verlag, NY, 191-208.

8. Tsai, C.S. 1987. An Improved Solution Procedure to the Fluid Structure Interaction Problem as Applied to the Dam-Reservoir System, Ph.D. Thesis, State Univ. of NY, Buffalo.

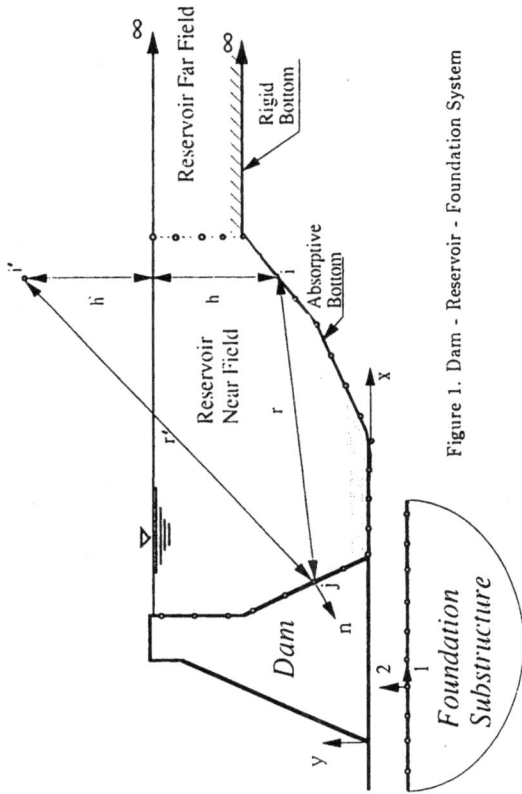

Figure 1. Dam - Reservoir - Foundation System

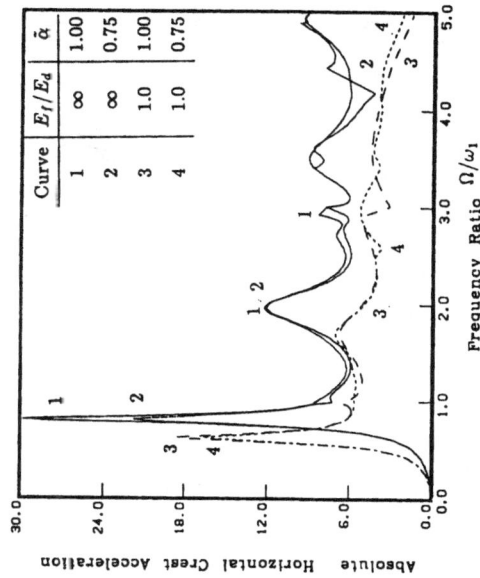

Figure 3 Response Function to Vertical Excitation

Curve	E_f/E_d	$\tilde{\alpha}$
1	∞	1.00
2	∞	0.75
3	1.0	1.00
4	1.0	0.75

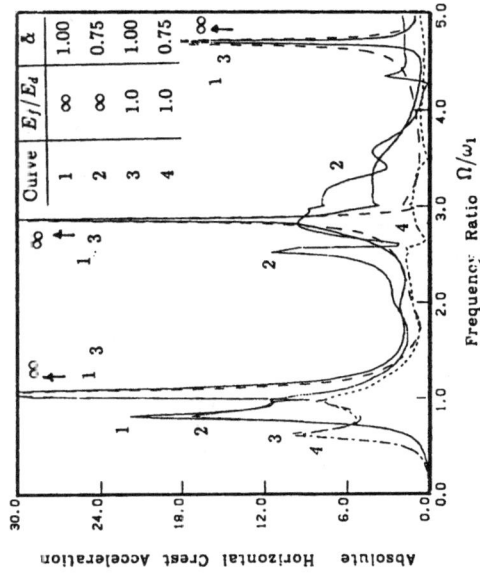

Figure 2 Response Function to Horizontal Excitation

Curve	E_f/E_d	$\tilde{\alpha}$
1	∞	1.00
2	∞	0.75
3	1.0	1.00
4	1.0	0.75

171

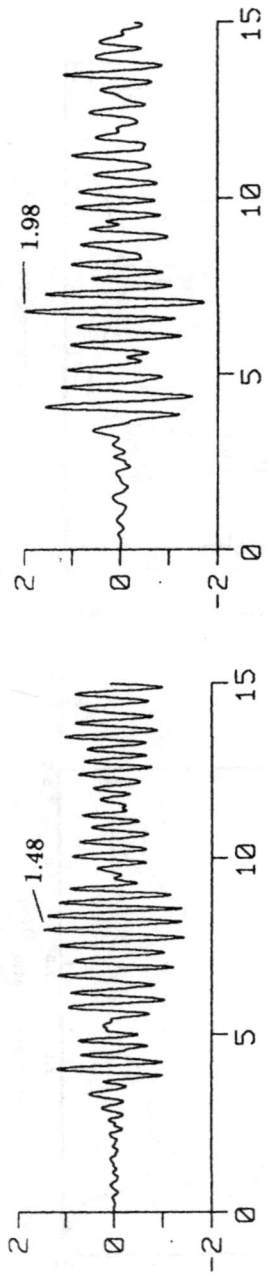

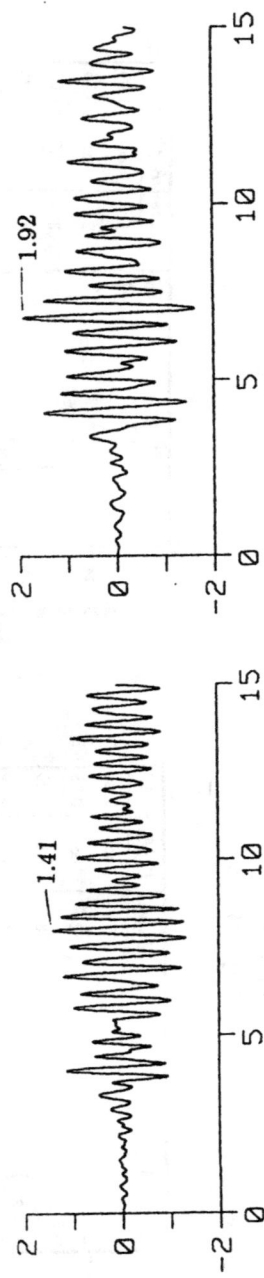

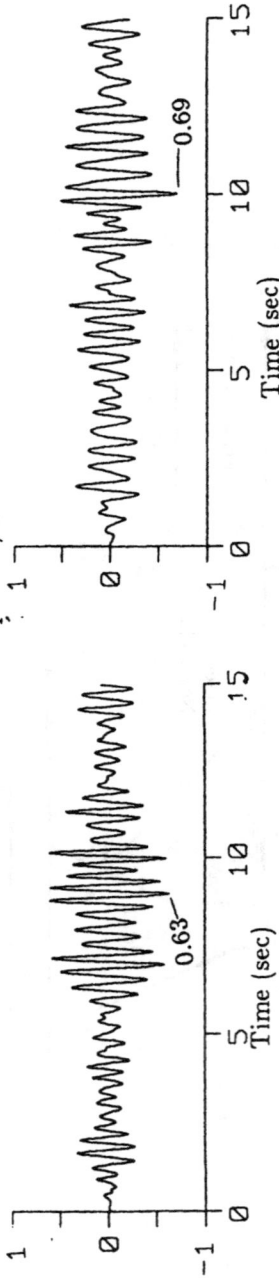

Figure 4 Displacement Time History of Pine Flat Dam to Taft Ground Motion

172

A nonlinear formulation for the earthquake responses of dam-reservoir systems

C.S. Tsai[I] and G.C. Lee[II]

ABSTRACT

A nonlinear formulation for fluid-structure interactions based on the velocity potential is developed. In this formulation the convective accelerations, nonlinear surface waves and exact transmitting boundary condition are included. Nonlinear behaviors of dam-reservoir systems subjected to constant accelerations, harmonic and actual earthquake ground motions are also investigated.

INTRODUCTION

The linear analysis of hydrodynamic pressures acting on rigid dams was first reported by Westergaard (1933). By neglecting the water compressibility, Chwang (1983) addressed the nonlinear hydrodynamic pressures on a rigid plate when the system was subjected to a short period of constant accelerations. Hung and Wang (1987), using the finite difference method and primitive variables for the governing equations, obtained the nonlinear hydrodynamic pressure on a rigid dam subjected to ground motions.

In this paper a nonlinear formulation, based on the velocity potential, for fluid-structure interactions is proposed. In the formulation the convective terms, nonlinear surface waves and exact transmitting boundary conditions, developed by Tsai et al. (1990a, 1990b, 1990c), are included. The reservoir is divided into two fields, the near and far fields (see Fig. 1). The near field is considered as a nonlinear area, and the behavior in the far field is linear. The final discretized matrices are symmetrical, even the nonlinearity of the near field are involved. The nonlinear responses of the system subjected to constant accelerations, harmonic and actual earthquake ground motions are also presented.

GOVERNING EQUATION FOR THE NEAR FIELD

Assuming that water is inviscid, the equations of motion, in terms of primitive variables, for the reservoir can be written as

$$\frac{\partial u}{\partial t} + u\frac{\partial u}{\partial x} + w\frac{\partial u}{\partial z} = -\frac{1}{\rho}\frac{\partial P}{\partial x} \tag{1}$$

$$\frac{\partial w}{\partial t} + u\frac{\partial w}{\partial x} + w\frac{\partial w}{\partial z} = -\frac{1}{\rho}\frac{\partial P}{\partial z} - g \tag{2}$$

I. Research Assistant Professor, Department of Civil Engineering, State University of New York at Buffalo, NY, 14260
II. Professor and Dean of Engineering and Applied Science, 412 Bonner Hall, State University of New York at Buffalo, NY, 14260

The Bernoulli equation can be used to replace the momentum equations, if irrotational wave is assumed; that is,

$$\frac{\partial \phi}{\partial t} + \frac{P}{\rho} + \frac{q^2}{2} + gz = 0 \tag{3}$$

The continuity equation is

$$\frac{\partial u}{\partial x} + \frac{\partial w}{\partial z} = -\frac{1}{\rho C^2} \frac{\partial P}{\partial t} \tag{4}$$

where $q^2 = (\frac{\partial \phi}{\partial x})^2 + (\frac{\partial \phi}{\partial z})^2$, $u = \frac{\partial \phi}{\partial x}$, $w = \frac{\partial \phi}{\partial z}$, $\rho =$ the mass density, $\phi =$ the velocity potential, $g =$ the gravity acceleration, $C =$ the velocity of sound in water, u and $v =$ the velocities of the fluid in a system of coordinates x and z, and $P =$ the pressure.
Differentiation of Eq. 3 with respect to t results in

$$-\frac{1}{\rho}\frac{\partial P}{\partial t} = \frac{\partial^2 \phi}{\partial t^2} + \frac{\partial}{\partial t}(\frac{q^2}{2}) \tag{5}$$

Substitution of Eq. 4 into Eq. 5 yields the nonlinear governing equation given by

$$\frac{\partial u}{\partial x} + \frac{\partial w}{\partial z} = \frac{1}{C^2}\frac{\partial^2 \phi}{\partial t^2} + \frac{1}{C^2}\frac{\partial}{\partial t}(\frac{q^2}{2}) \tag{6}$$

Eq. 6 can be rewritten, in terms of the potential velocity ϕ, as

$$\frac{\partial^2 \phi}{\partial x^2} + \frac{\partial^2 \phi}{\partial z^2} = \frac{1}{C^2}\frac{\partial^2 \phi}{\partial t^2} + \frac{1}{C^2}\frac{\partial}{\partial t}(\frac{q^2}{2}) \tag{7}$$

BOUNDARY CONDITIONS OF THE NEAR FIELD

Two boundary conditions for the free surface are given as follows.
(i) the kinematic boundary condition

$$\frac{\partial \phi}{\partial n} = \frac{\partial \eta}{\partial t}n_z \tag{8}$$

(ii) the dynamic boundary condition

$$\frac{\partial \phi}{\partial t} + \frac{q^2}{2} + g\eta = 0 \tag{9}$$

The other boundary conditions are

$$\frac{\partial \phi}{\partial n} = \bar{u}_n \tag{10}$$

and

$$\phi = \bar{\phi} \tag{11}$$

where η is the displacement of the free surface in the z direction; n_z is the z-component of the unit normal direction, n, on the free surface.
If we set

$$\eta^* = \eta n_z \tag{12}$$

then its derivative with respect to time t is

$$\frac{\partial \eta^{\cdot}}{\partial t} = \frac{\partial \eta}{\partial t} n_z + \eta \frac{\partial n_z}{\partial t} \tag{13}$$

Substitution of Eq. 13 into Eq. 8 results in the kinematic boundary condition on the free surface; that is,

$$\frac{\partial \phi}{\partial n} = \frac{\partial \eta^{\cdot}}{\partial t} - \eta \frac{\partial n_z}{\partial t} \tag{14}$$

Substitution of Eq. 12 into the dynamic boundary condition, Eq. 9, yields

$$\frac{\partial \phi}{\partial t} + \frac{q^2}{2} + \frac{g}{n_z} \eta^{\cdot} = 0 \tag{15}$$

Applying the Galerkin's method to Eq. 7, one obtains

$$\int_A \mathbf{N}^T \left(\frac{\partial^2 \phi}{\partial x^2} + \frac{\partial^2 \phi}{\partial z^2} \right) dA = \frac{1}{C^2} \int_A \mathbf{N}^T \frac{\partial^2 \phi}{\partial t^2} dA + \frac{1}{C^2} \int_A \mathbf{N}^T \left[\frac{\partial}{\partial t} \left(\frac{q^2}{2} \right) \right] dA \tag{16}$$

Integration of the term on the left hand side by parts yields

$$\int_S \mathbf{N}^T \frac{\partial \phi}{\partial n} dS - \int_A \left(\frac{\partial \mathbf{N}^T}{\partial x} \frac{\partial \phi}{\partial x} + \frac{\partial \mathbf{N}^T}{\partial z} \frac{\partial \phi}{\partial z} \right) dA$$
$$= \frac{1}{C^2} \int_A \mathbf{N}^T \frac{\partial^2 \phi}{\partial t^2} dA + \frac{1}{C^2} \int_A \mathbf{N}^T \left[\left(\frac{\partial \phi}{\partial x} \frac{\partial}{\partial x} + \frac{\partial \phi}{\partial z} \frac{\partial}{\partial z} \right) \frac{\partial \phi}{\partial t} \right] dA \tag{17}$$

Introduction of the shape function N into Eq. 17 leads to

$$\int_A \left(\frac{\partial \mathbf{N}^T}{\partial x} \frac{\partial \mathbf{N}}{\partial x} + \frac{\partial \mathbf{N}^T}{\partial z} \frac{\partial \mathbf{N}}{\partial z} \right) \Phi \, dA + \frac{1}{C^2} \int_A \mathbf{N}^T \mathbf{N} \, dA \, \ddot{\Phi}$$
$$= \int_S \mathbf{N}^T \frac{\partial \phi}{\partial n} dS - \frac{1}{C^2} \int_A \mathbf{N}^T \left[\left(\frac{\partial \phi}{\partial x} \frac{\partial}{\partial x} + \frac{\partial \phi}{\partial z} \frac{\partial}{\partial z} \right) \frac{\partial \phi}{\partial t} \right] dA \tag{18}$$

Rearranging Eq. 18, the following equation can be obtained

$$\mathbf{M} \ddot{\Phi} + \mathbf{K} \, \Phi = \mathbf{B} - \mathbf{E} \tag{19}$$

The matrices $\mathbf{M}, \mathbf{K}, \mathbf{B}$ and $\mathbf{E}$ in Eq. 19 are defined as follows

$$\mathbf{M} = \frac{1}{C^2} \int_A \mathbf{N}^T \mathbf{N} \, dA \tag{20}$$

$$\mathbf{K} = \int_A \left(\frac{\partial \mathbf{N}^T}{\partial x} \frac{\partial \mathbf{N}}{\partial x} + \frac{\delta \mathbf{N}^T}{\partial z} \frac{\partial \mathbf{N}}{\partial z} \right) dA \tag{21}$$

$$\mathbf{B} = \int_S \mathbf{N}^T \frac{\partial \phi}{\partial n} \, dS \tag{22}$$

and

$$\mathbf{E} = \frac{1}{C^2} \int_A \mathbf{N}^T \left[\left(\frac{\partial \phi}{\partial x} \frac{\partial}{\partial x} + \frac{\partial \phi}{\partial z} \frac{\partial}{\partial z} \right) \frac{\partial \phi}{\partial t} \right] dA \tag{23}$$

Applying the Galerkin's method to Eq. 15, one obtains

$$\int_{S2} \mathbf{N}^T \frac{\partial \phi}{\partial t} \, dS + \int_{S2} \mathbf{N}^T \left(\frac{q^2}{2} \right) dS + g \int_{S2} \mathbf{N}^T \frac{1}{n_z} \eta^{\cdot} \, dS = 0 \tag{24}$$

Introduction of the shape function into Eq. 24 yields

$$\int_{S2} \mathbf{N}^T \mathbf{N}\, dS\, \dot{\mathbf{\Phi}} + \int_{S2} \mathbf{N}^T (\frac{q^2}{2})\, dS + g \int_{S2} \frac{1}{n_z} \mathbf{N}^T \mathbf{N} \underline{\dot{\eta}}\, dS = 0 \tag{25}$$

For convenience, Eq. 25 can be rewritten in the following matrix form

$$\mathbf{H}\dot{\mathbf{\Phi}} + \mathbf{G}\, \underline{\dot{\eta}} = -\mathbf{L} \tag{26}$$

where

$$\mathbf{H} = \int_{S2} \mathbf{N}^T \mathbf{N}\, dS \tag{27}$$

$$\mathbf{G} = g \int_{S2} \frac{1}{n_z} \mathbf{N}^T \mathbf{N}\, dS \tag{28}$$

$$\mathbf{L} = \frac{1}{2}\int_{S2} \mathbf{N}^T (q^2)\, dS = \frac{1}{2}\int_{S2} \mathbf{N}^T[(\frac{\partial\phi}{\partial x})^2 + (\frac{\partial\phi}{\partial z})^2]\, dS = \frac{1}{2}\int_{S2} \mathbf{N}^T[(\frac{\partial\eta}{\partial t}n_z)^2 + (\frac{\partial\phi}{\partial s})^2]\, dS \tag{29}$$

Substituting Eq. 14 into Eq. 22, the boundary condition on the free surface is given by

$$\mathbf{B}_2 = \int_{S2} \mathbf{N}^T \frac{\partial\phi}{\partial n}\, dS = \int_{S2} \mathbf{N}^T \frac{\partial\eta^\cdot}{\partial t}\, dS - \int_{S2} \mathbf{N}^T \eta \frac{\partial n_z}{\partial t}\, dS$$

$$= \int_{S2} \mathbf{N}^T \mathbf{N}\, dS\, \underline{\dot{\eta}} - \int_{S2} \mathbf{N}^T \eta \frac{\partial n_z}{\partial t}\, dS = \mathbf{H}\, \underline{\dot{\eta}} - \mathbf{V} \tag{30}$$

where

$$\mathbf{V} = \int_{S2} \mathbf{N}^T \eta \frac{\partial n_z}{\partial t}\, dS \tag{31}$$

Combining 19 and 26, the following matrix can be obtained

$$\begin{bmatrix} M_{11} & M_{12} & M_{13} & 0 \\ M_{21} & M_{22} & M_{23} & 0 \\ M_{31} & M_{32} & M_{33} & 0 \\ 0 & 0 & 0 & 0 \end{bmatrix} \begin{Bmatrix} \ddot{\Phi}_1 \\ \ddot{\Phi}_2 \\ \ddot{\Phi}_3 \\ \bar{\eta}^\cdot \end{Bmatrix} + \begin{bmatrix} 0 & 0 & 0 & 0 \\ 0 & 0 & 0 & 0 \\ 0 & 0 & 0 & 0 \\ 0 & -H & 0 & 0 \end{bmatrix} \begin{Bmatrix} \dot{\Phi}_1 \\ \dot{\Phi}_2 \\ \dot{\Phi}_3 \\ \dot{\eta}^\cdot \end{Bmatrix}$$

$$+ \begin{bmatrix} K_{11} & K_{12} & K_{13} & 0 \\ K_{21} & K_{22} & K_{23} & 0 \\ K_{31} & K_{32} & K_{33} & 0 \\ 0 & 0 & 0 & -G \end{bmatrix} \begin{Bmatrix} \Phi_1 \\ \Phi_2 \\ \Phi_3 \\ \eta^\cdot \end{Bmatrix} = \begin{Bmatrix} B_1 - E_1 \\ B_2 - E_2 \\ B_3 - E_3 \\ L \end{Bmatrix} \tag{32}$$

Substitution of Eq. 30 into Eq. 32 leads to

$$\begin{bmatrix} M_{11} & M_{12} & M_{13} & 0 \\ M_{21} & M_{22} & M_{23} & 0 \\ M_{31} & M_{32} & M_{33} & 0 \\ 0 & 0 & 0 & 0 \end{bmatrix} \begin{Bmatrix} \ddot{\Phi}_1 \\ \ddot{\Phi}_2 \\ \ddot{\Phi}_3 \\ \bar{\eta}^\cdot \end{Bmatrix} + \begin{bmatrix} 0 & 0 & 0 & 0 \\ 0 & 0 & 0 & -H \\ 0 & 0 & 0 & 0 \\ 0 & -H & 0 & 0 \end{bmatrix} \begin{Bmatrix} \dot{\Phi}_1 \\ \dot{\Phi}_2 \\ \dot{\Phi}_3 \\ \dot{\eta}^\cdot \end{Bmatrix}$$

$$+ \begin{bmatrix} K_{11} & K_{12} & K_{13} & 0 \\ K_{21} & K_{22} & K_{23} & 0 \\ K_{31} & K_{32} & K_{33} & 0 \\ 0 & 0 & 0 & -G \end{bmatrix} \begin{Bmatrix} \Phi_1 \\ \Phi_2 \\ \Phi_3 \\ \eta^\cdot \end{Bmatrix} = \begin{Bmatrix} B_1 - E_1 \\ -V - E_2 \\ B_3 - E_3 \\ L \end{Bmatrix} \tag{33}$$

It should be noted that nodes on the free surface are denoted by subscript 2, at the interface of the near and far fields by 3. The remaining nodes are denoted by subscript 1.

EXACT TRANSMITTING BOUNDARY CONDITION

The effect of radiation damping in the time-domain analyses is treated following the development by Tsai et al. (1990a, 1990b, 1990c). The procedure will be described briefly in the following :

The governing equation is

$$\nabla^2 \phi = \frac{1}{C^2} \ddot{\phi} \tag{34}$$

The velocity potential should satisfy : (1) the compatability equations at the interface of the near and far fields, (2) the rigid boundary condition at the floor of the reservoir, (3) the radiation condition for the infinite fluid domain, (4)the boundary condition on the free surface, $\frac{\partial \phi}{\partial t} = 0$, if the surface wave is neglected, and (5) zero initial conditions for the velocity potential.

The solution of the velocity potential, $\Phi_3(t)$, at the interface of the near and far fields, is therefore given by

$$\Phi_3(t) = -C \int_0^t \Psi \Xi(\tau) \Psi^T \mathbf{G}_3 \frac{\partial \Phi_3(\tau)}{\partial n} \, d\tau \tag{35}$$

where $\Xi(\tau)$ is an $M \times M$ diagonal matrix with mth diagonal term $= J_0(\lambda_m C(t - \tau))$, $t > 0$. Ψ is the normalized modal matrix with respect to $\mathbf{G}_3$ at the interface of the near and far fields. λ_m is the mth eigenvalue associated with mth eigenvalue. $\mathbf{G}_3 = \int_{S3} \mathbf{N}_3^T \mathbf{N}_3 \, dS$.
The time axis is divided into N equal intervals Δt, then Eq. 35 can be rewritten in the following form

$$\Phi_3(N\Delta t) = \mathbf{F}(N\Delta t) - \mathbf{R}\mathbf{G}_3 \frac{\partial \Phi_3(N\Delta t)}{\partial n} \tag{36}$$

For the special case of constant temporal variation and zero initial conditions, the matrices $\mathbf{F}$ and $\mathbf{R}$ are defined as

$$\mathbf{F}(N\Delta t) = \sum_{n=1}^{N-1} \Psi \mathbf{Q} \Psi^T \mathbf{G}_3 \frac{\partial \Phi_3(n\Delta t)}{\partial n} \tag{37}$$

and

$$\mathbf{R} = \Psi \Upsilon \Psi^T \tag{38}$$

where $\mathbf{Q}$ is an $M \times M$ diagonal matrix with mth diagonal term Q_{mm}

$$Q_{mm} = \frac{-1}{2\lambda_m} \int_{\lambda_m C(N-n-1)\Delta t}^{\lambda_m C(N-n+1)\Delta t} J_0(\tau) \, d\tau = \frac{-1}{2\lambda_m} \left[\frac{\pi \tau}{2} \{ J_0(\tau) \mathbf{H}_{-1}(\tau) + \mathbf{H}_0(\tau) J_1(\tau) \} \right]_{\tau=\lambda_m C(N-n-1)\Delta t}^{\tau=\lambda_m C(N-n+1)\Delta t} \tag{39}$$

and Υ is also an $M \times M$ diagonal matrix with mth diagonal term Υ_{mm} given by

$$\Upsilon_{mm} = \frac{1}{2\lambda_m} \int_0^{\lambda_m C\Delta t} J_0(\tau) \, d\tau = \frac{1}{2\lambda_m} \left[\frac{\pi \tau}{2} \{ J_0(\tau) \mathbf{H}_{-1}(\tau) + \mathbf{H}_0(\tau) J_1(\tau) \} \right]_{\tau=0}^{\tau=\lambda_m C\Delta t} \tag{40}$$

where $\mathbf{H}_\nu(\tau)$ is the Struve's function of order ν.

It is noted that coefficients in the matrix $\mathbf{R}$ are constant if the time step Δt is constant. Substitution of Eqs. 22 and 36 into Eq. 33 yields

177

$$
\begin{bmatrix} M_{11} & M_{12} & M_{13} & 0 \\ M_{21} & M_{22} & M_{23} & 0 \\ M_{31} & M_{32} & M_{33} & 0 \\ 0 & 0 & 0 & 0 \end{bmatrix} \begin{Bmatrix} \ddot{\Phi}_1 \\ \ddot{\Phi}_2 \\ \ddot{\Phi}_3 \\ \ddot{\eta} \end{Bmatrix} + \begin{bmatrix} 0 & 0 & 0 & 0 \\ 0 & 0 & 0 & -H \\ 0 & 0 & 0 & 0 \\ 0 & -H & 0 & 0 \end{bmatrix} \begin{Bmatrix} \dot{\Phi}_1 \\ \dot{\Phi}_2 \\ \dot{\Phi}_3 \\ \dot{\eta} \end{Bmatrix}
$$
$$
+ \begin{bmatrix} K_{11} & K_{12} & K_{13} & 0 \\ K_{21} & K_{22} & K_{23} & 0 \\ K_{31} & K_{32} & (K_{33} + R^{-1}) & 0 \\ 0 & 0 & 0 & -G \end{bmatrix} \begin{Bmatrix} \Phi_1 \\ \Phi_2 \\ \Phi_3 \\ \eta \end{Bmatrix} = \begin{Bmatrix} B_1 - E_1 \\ -V - E_2 \\ R^{-1}F - E_3 \\ L \end{Bmatrix} \tag{41}
$$

It should be noted that the matrices of the system in Eq. 41 are symmetrical although the convective terms, nonlinear surface waves and exact transmitting boundary condition are included. The incremental form of the formulation can be obtained from Eq. 41. Iteration procedures are adopted for each time step in this study to obtain numerical solutions.

NUMERICAL EXAMPLES

Hydrodynamic pressures, in excess of the hydrostatic pressure, acting on a vertical rigid dam, having a reservoir of $180m$ in height with flat floor extending to infinity, subjected three different types of loadings are given to illustrate the nonlinear behaviors of the dam-reservoir system. The extending length of the near field in the upstream direction and the sound velocity in water are taken as $360m$ and 1438.656 m/sec respectively.

1. Constant Accelerations

When the system is excited by a constant acceleration, $a = 0.5g$, very significant nonlinear effects, shown in Fig. 2, are observed. The hydrodynamic pressure for the nonlinear case is much higher than that for the linear case. This is because the nonlinear effect, shown in Eq. 3, is proportional to the square of the velocity which is monotonously increasing with time for this special case of constant accelerations.

2. Harmonic Ground Motions

If the system is subjected to a harmonic ground motion, $a \sin 12t = 0.5g \sin 12t$, the results in Fig. 3 show insignificant nonlinear effects when compared to linear responses. In this case the velocities of the fluid particles along the dam face are also a harmonic function. Therefore, the nonlinear effect is insignificant, even though the excitation frequency, $12 rad/sec$, is near the first natural frequency of the reservoir, 12.555 rad/sec.

3. Earthquake Ground Motions

An actual earthquake ground motion, 1940 El Centro, is applied to the dam-reservoir system. The response of the system, shown in Fig. 4, also shows negligible nonlinear effect.

CONCLUSIONS

A new nonlinear formulation for the fluid-structure interactions is presented. The proposed formulations based on the velocity potential yield symmetrical matrices, although convective accelerations, nonlinear surface waves and the exact transmitting boundary condition are involved. The nonlinear behaviors of the dam-reservoir system subjected to different type of ground motions in the upstream-downstream direction are also examined. The nonlinear effects for the two dimensional case of the rigid dam-reservoir system are very much dependent on the loading types.

REFERENCES

Chwang, A. T. 1983. Nonlinear hydrodynamic pressure on an accelerating plate. J. of

Physics Fluids, 26(2), 383-387.

Hung, T. K. and Wang, M. H. 1987. Nonlinear hydrodynamic pressure on rigid dam motion. J. of Engineering Mechanics, ASCE, 113(4), 482-499.

Tsai, C. S., Lee, George C. and Ketter, Robert L. 1990a. A semi-analytical method for time-domain analyses of dam-reservoir interactions. International Journal for Numerical Methods in Engineering, 29, 913-933.

Tsai, C. S. and Lee, George C. 1990b. "Method for the transient analysis of three-dimensional dam-reservoir interactions." J. of Engineering Mechanics, ASCE, 116 (10), 2151-2172.

Tsai, C. S., Lee, George C. and Yeh, C. S. 1990c. Transient responses of 3-D fluid structure interactions. " Advances in Boundary Elements Methods in Japan and USA, Editors : M. Tanaka, C. A. Brebbia and R. Shaw, Computational Mechanics Publications, Southampton UK and Boston USA, Vol. 7, 183-199.

Westergaard, H. M. (1933). Water pressures on dams during earthquakes. Transactions, ASCE, 98, 418-433.

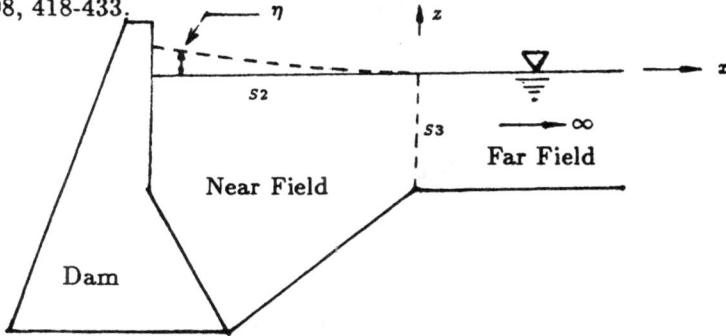

Fig. 1 Dam-Reservoir System

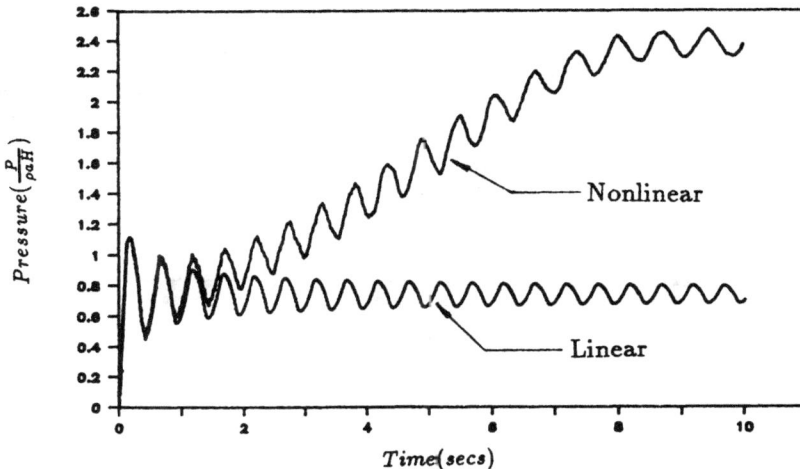

Fig. 2 Comparison of hydrodynamic pressure at bottom when system is subjected to constant acceleration

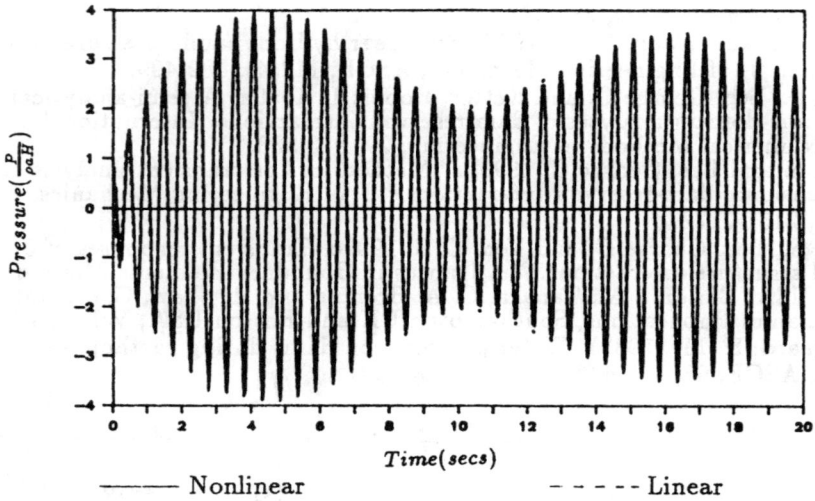

Fig. 3 Comparison of hydrodynamic pressure at bottom when system
is subjected to harmonic ground motion

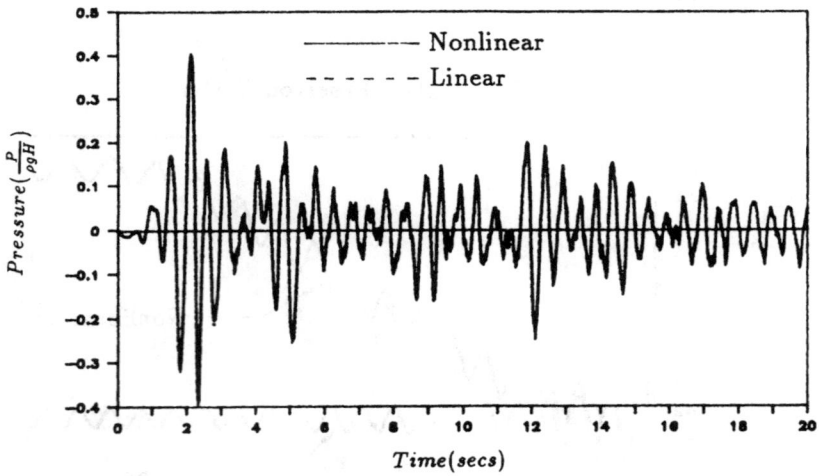

Fig. 4 Comparison of hydrodynamic pressure at bottom when system
is subjected to earthquake ground motion

Dam-reservoir interaction during earthquake

S.K. Sharan[I]

ABSTRACT

A novel technique developed for the finite element modelling of reservoir vibration is implemented in the earthquake response analysis of gravity dams to include the effects of energy dissipations due to the radiation and absorption of pressure waves in the reservoir. When the proposed radiation condition is used to model an infinitely long reservoir as a finite one of relatively very short length, some minor discrepancies are observed in the results for the complex frequency response of hydrodynamic pressures at frequencies near the second cut-off frequency of the reservoir vibration. The effect of such discrepancies on the earthquake response of dams is found to be negligible. Example analyses are conducted by using a past earthquake data to demonstrate the effectiveness of the proposed technique and to study the effects of hydrodynamic interaction on the earthquake response of dams.

INTRODUCTION

The response of a dam to an earthquake depends very significantly on the effects of dam-reservoir interaction, compressibility of water in the reservoir and the absorption of pressure waves at the bottom of the reservoir (Chopra 1970; Chakrabarti and Chopra 1973; Fenves and Chopra 1983; Chopra 1987). In order to include these effects in the analysis of a dam having an arbitrary geometry, the complete dam-reservoir system must be discretized (Sharan and Gladwell 1985a). A computational difficulty arises in such a discretization because in most of the practical situations, reservoirs are infinitely long. Even if a relatively very large length of the reservoir is discretized, use of the conventional Sommerfeld radiation condition (Zienkiewicz and Newton 1969) does not produce satisfactory results (Humar and Roufaiel 1983). This difficulty may be circumvented by coupling the finite element model with continuum solutions (Hall and Chopra 1982) or boundary elements (Humar and Jablonski

I Professor, School of Engineering, Laurentian University, Sudbury, Ontario, Canada.

1989). Recently, a highly effective, very efficient and simple radiation condition was developed (Sharan 1985b, 1987) for the finite modelling of infinite reservoirs having a rigid bottom. The technique was then extended to include the effects of the deformability of reservoir bottom (Sharan, to be published). In these studies, the dam was idealized as being rigid or deformable plates and the analyses were limited to the complex frequency response of hydrodynamic pressures. The objective of this paper is to implement the technique in the earthquake response analysis of two-dimensional concrete gravity dams impounding reservoirs having a deformable bottom. The analysis is conducted in the frequency domain and the response of dams to an earthquake is obtained by using the fast Fourier transform.

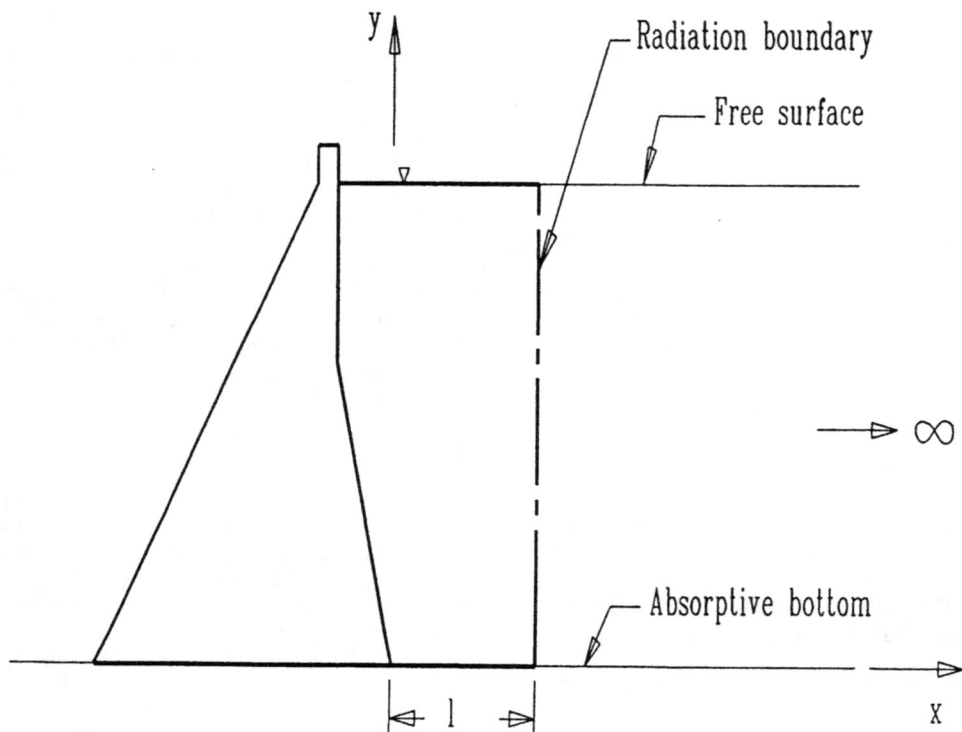

Figure 1. A dam-reservoir system

FINITE ELEMENT ANALYSIS

By considering displacements and pressures to be the basic nodal unknowns for the dam and the reservoir, respectively, the finite element discretization (Zienkiewicz and Newton 1969) of a dam-reservoir system (Fig. 1) leads to the following coupled equations for the complex frequency response (Sharan 1987):

$$[A_{dr}]\{\bar{\delta}\} = -[M_{dr}]\{\bar{a}_g\}$$ (1)

$$[B_r]\{\bar{p}\} = -\rho[S_{dr}]^T(\{\bar{a}_g\} - \omega^2\{\bar{\delta}\})$$ (2)

$$[A_{dr}] = [K_d] - \omega^2[M_{dr}] + i\omega[C_d]$$ (3)

$$[B_r] = [K_r] - \omega^2[M_r] + [C_r]$$ (4)

$$[M_{dr}] = [M_d] + \rho[S_{dr}][B_r]^{-1}[S_{dr}]^T$$ (5)

$$\{\bar{a}_g, \bar{\delta}, \bar{p}\} = \{a_g, \delta, p\}e^{i\omega t}$$ (6)

In the above equations, $[K_d]$, $[C_d]$ and $[M_d]$ are the stiffness, damping and mass matrices for the dam and $[K_r]$, $[C_r]$ and $[M_r]$ are the corresponding matrices for the reservoir; $[S_{dr}]$ is the coupling matrix; $\{a_g, \delta, p\}$ are the vectors of nodal ground accelerations, displacements and hydrodynamic pressures, respectively; ρ is the density of water in the reservoir; ω is the circular frequency of vibration, t is the time variable and $i = \sqrt{-1}$.

Damping matrix for the reservoir

Energy dissipation in the reservoir is caused by (i) dam-reservoir intraction, (ii) radiation and (iii) absorption of pressure waves at the bottom of the reservoir. The first type of damping is modelled by dam-reservoir coupling and the remaining two types are modelled through the damping matrix $[C_r]$ in Eq. 4. Implementation of the absorbing boundary condition (Fenves and Chopra 1983) and the proposed radiation condition (Sharan, to be published) for a horizontal ground motion leads to the following form of the damping matrix:

$$[C_r] = i\omega q[C_{ra}] + \zeta[C_{rt}]$$ (7)

$$q = \frac{(1-\alpha_r)}{c(1+\alpha_r)}$$ (8)

$$\zeta = \frac{1}{d}\sqrt{\beta - (\frac{\omega d}{c})^2}$$ (9)

183

where $[C_{ra}]$ and $[C_{rt}]$ are damping matrices corresponding, respectively to the absorbing and truncation boundary, α_r is the reflection coefficient of pressure waves, c is the velocity of acoustic waves in water, d is the depth of water at the truncation surface and β is the first root of the equation

$$e^{2i\beta} = \frac{\omega q d - \beta}{\omega q d + \beta}$$
(10)

PARAMETRIC STUDY

The effectiveness and efficiency of the proposed method was tested by analyzing two different dams having the following dimensions (Fig. 2): (a) the Pine Flt dam (h = 121.9 m, d = 116.1 m, b = 95.8 and t = 9.8 m) to represent a high dam and (b) the Angostura dam (h = 48.5 m, d = 44.9 m, b = 42.2 m and t = 3.0 m) to represent a low dam having an inclined upstream face. Three different values of α_r (= 1.0, 0.75, and 0.5) and three different locations of the truncation boundary given by ℓ/d (= 0.5, 1.0 and 2.0) were considered. A detailed description of material and geometrical properties of the systems may be found elsewhere (Sharan, 1985a). Three-noded triangular finite elements were used for the discretization and Fig. 2 shows a typical finite element model.

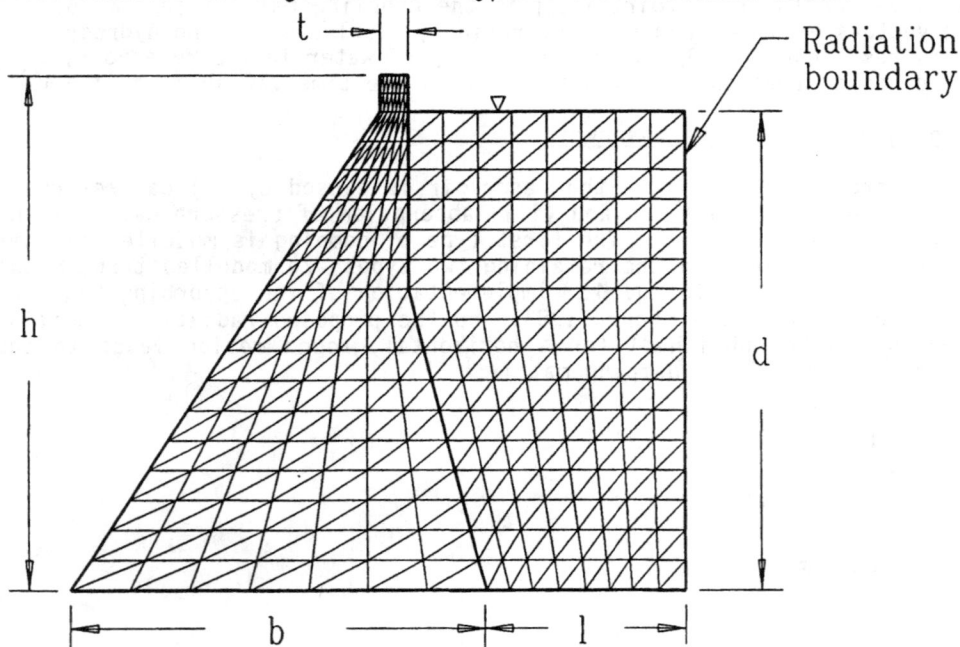

Figure 2. A typical finite element model of Angostura dam-reservoir system (ℓ/d = 0.5)

Table 1. Complex frequency response of hydrodynamic forces on (a) Pine
flat dam and (b) Angostura dam

Dam	ω/ω_1	α_r	$|\bar{F}_{dyn}|/F_{stat}$			
			Sommerfeld radiation		Proposed radiation	
			$\ell/d=0.5$	2.0	0.5	2.0
(a)	0	1.00	1.52	1.05	1.05	1.05
		0.75	1.52	1.05	1.05	1.05
		0.50	1.52	1.05	1.05	1.05
	1	1.00	2.88	11.95	5.86	5.90
		0.75	2.64	6.24	5.63	5.63
		0.50	2.42	4.26	4.10	4.10
	2	1.00	0.58	0.53	0.52	0.52
		0.75	0.55	0.54	0.51	0.51
		0.50	0.53	0.54	0.51	0.52
	3	1.00	0.43	0.51	0.43	0.51
		0.75	0.47	0.53	0.46	0.52
		0.50	0.50	0.52	0.49	0.51
	4	1.00	0.18	0.20	0.18	0.20
		0.75	0.18	0.19	0.18	0.19
		0.50	0.18	0.18	0.17	0.18
(b)	0	1.00	1.27	0.98	0.98	0.98
		0.75	1.27	0.98	0.98	0.98
		0.50	1.27	0.98	0.98	0.98
	1	1.00	2.37	6.44	29.60	30.42
		0.75	2.17	4.16	4.27	4.27
		0.50	1.97	3.00	2.95	2.96
	2	1.00	0.55	0.49	0.49	0.49
		0.75	0.54	0.51	0.49	0.49
		0.50	0.54	0.52	0.50	0.50
	3	1.00	0.72	0.80	0.73	0.81
		0.75	0.74	0.75	0.74	0.74
		0.50	0.75	0.73	0.74	0.72
	4	1.00	0.09	0.10	0.10	0.10
		0.75	0.10	0.12	0.10	0.12
		0.50	0.10	0.14	0.11	0.14

Table 2. Complex frequency response of horizontal displacements at the
crest of (a) Pine Flat dame and (b) Angostura dam

Dam	ω/ω_1	α_r	$\|\bar{\delta}_{dyn}\|/\delta_{stat}$			
			Sommerfeld radiation		Proposed radiation	
			$\ell/d=0.5$	2.0	0.5	2.0
(a)	0	1.00	9.6	8.5	8.5	8.5
		0.75	9.6	8.5	8.5	8.5
		0.50	9.6	8.5	8.5	8.5
	1	1.00	33.8	94.7	28.4	28.7
		0.75	33.3	57.0	49.0	49.0
		0.50	33.0	46.0	44.3	44.3
	2	1.00	7.1	6.7	7.0	7.0
		0.75	7.4	7.0	7.2	7.2
		0.50	7.6	7.4	7.4	7.4
	3	1.00	5.0	4.9	5.0	4.9
		0.75	4.9	4.6	4.9	4.6
		0.50	4.7	4.7	4.8	4.8
	4	1.00	1.2	1.2	1.2	1.2
		0.75	1.3	1.4	1.3	1.4
		0.50	1.4	1.5	1.4	1.5
(b)	0	1.00	11.1	10.4	10.4	10.4
		0.75	11.1	10.4	10.4	10.4
		0.50	11.1	10.4	10.4	10.4
	1	1.00	32.3	60.5	173.0	178.8
		0.75	31.5	44.4	43.9	44.0
		0.50	30.8	37.7	37.0	37.2
	2	1.00	11.3	11.0	11.3	11.3
		0.75	11.6	11.2	11.4	11.4
		0.50	11.9	11.5	11.6	11.6
	3	1.00	7.0	5.9	6.9	5.8
		0.75	6.8	6.1	6.8	6.0
		0.50	6.7	6.6	6.8	6.6
	4	1.00	2.1	2.1	2.1	2.1
		0.75	2.2	2.2	2.2	2.2
		0.50	2.2	2.3	2.2	2.3

Complex frequency response

Tables 1 and 2 show absolute values of the horizontal components of the total hydrodynamic force F_{dyn} (in excess of the hydrostatic force F_{stat}) and the dynamic displacement δ_{dyn} (in excess of the static displacement δ_{stat}) at the crest of dams subjected to a horizontal ground acceleration $a_g = ge^{i\omega t}$, g being the acceleration of gravity. Results are presented for a few typical values of the frequency of excittion normalized with respect to the first cut-off frequency ω_1 (= 0.5 π c/d) of the reservoir. The value of $\omega = 3\omega_1$ corresponds to the second cut-off frequency of the reservoir. In these tables, a comparison is made between results obtained by using the proposed and the conventional Sommerfeld radiation conditions for two different locations of the truncation boundary. By imposing the proposed radiation condition, almost identical results were produced for ℓ/d = 0.5, 1.0 and 2.0. However, some minor discrepancies were observed for frequencies near the second cut-off frequency of the reservoir. With the use of the conventional Sommerfeld radiation, no convergence could be achieved even for larger values of ℓ/d, particularly for the case of α_r = 1 and ω/ω_1 = 1 which is of greater importance in the earthquake response analysis.

Earthquake response analysis

The south-component of El Centro, 1940 earthquake (Hudson and Brady 1971) was used for the analysis. The duration of earthquake considered was 10.24 sec and 8192 time increments were used for the fast Fourier transform. Results for the maximum crest displacement δ_{max} (in excess of δ_{stat}) and its time of occurence t_m were obtained by using the proposed technique and as shown in Table 3, results for ℓ/d = 0.5 and 2.0 were found to be almost identical for all the cases analyzed. The effect of α_r was found to be very significant for the Pine Flat dam and almost insignificant for the Angostura dam.

Table 3. Response of (a) Pine Flat and (b) Angostura dams to El Centro earthquake obtained by using the proposed method

Dam	α_r	ℓ/d	$\delta_{max}/\delta_{stat}$ (at the crest)	t_m (sec)
(a)	1.00	0.5	6.7	2.32
		2.0	6.7	2.32
	0.75	0.5	6.4	2.32
		2.0	6.4	2.32
(b)	1.00	0.5	6.1	2.64
		2.0	6.2	2.64
	0.75	0.5	6.0	2.48
		2.0	6.0	2.48

CONCLUSIONS

A novel radiation condition is implemented in the finite element analysis of an infinite reservoir-dam system. The technique is found to be highly effective and very efficient. Although the effectiveness of the proposed radiation condition is slightly reduced for fequencies near the second cut-off frequency of the reservoir, its effect on the earthquake response of dams is found to be insignificant. Based on the limited analysis, it is concluded that the effect of absorption of pressure waves at the bottom of a reservoir may be very significant.

ACKNOWLEDGEMENT

The financial support provided by the Natural Sciences and Engineering Research Council of Canada is gratefully acknowledged.

REFERENCES

Chakrabarti, P. and Chopra, A.K. 1973. Earthquake analysis of gravity dams including hydrodynamic interaction. Earthq. Eng. Struct. Dyn., 2, 143-160.

Chopra, A.K. 1970. Earthquake response of concrete gravity dams. J. Eng. Mech. Div. ASCE, 96(EM4), 443-454.

Chopra, A.K. 1987. Earthquake analysis, design, and safety evaluation of concrete dams. Proc. Fifth Can. Conf. Earthq. Eng., Ottawa, 39-62.

Fenves, G. and Chopra, A.K. 1983. Effects of reservoir bottom absorption on earthquake response of concrete gravity dams. Earthq. Eng. Struct. Dyn., 11 (6), 809-829.

Hall, J.F. and Chopra, A.K. 1982. Two dimensional dynamic analysis of concrete gravity and embankment dams including hydrodynamic effects. Earthq. Eng. Struct. Dyn., 10, 305-332.

Hudson, D.E. and Brady, A.G. 1971. Strong motion earthquake accelerograms. Report No. EERL 71-50, Vol. II, Part A. California Institute of Technology, Pasadena.

Humar, J.L. and Roufaiel, M. 1983. Finite element analysis of reservoir vibration. J. Eng. Mech., ASCE, 109, 215-230.

Humar, J.L. and Jablonski, A.M. 1988. Boundary element reservoir model for seismic analysis of gravity dams. Earthq. Eng. Struct. Dyn., 16, 1129-1156.

Sharan, S.K. and Gladwell, G.M.L. 1985a. A general method for the dynamic response analysis of fluid-structure systems. Comput. Struct., 21(5), 937-943.

Sharan, S.K. 1985b. Finite element modeling of infinite reservoirs. J. Eng. Mech., ASCE, 111, 1457-1469.

Sharan, S.K. 1987. A non-reflecting boundary in fluid-structure interaction. Comput. Struct., 26, 841-846.

Sharan, S.K. (to be published). Efficient finite element analysis of hydrodynamic pressure on dams. Comput. Struct.

Zienkiewicz, O.C. and Newton, R.E. 1969. Coupled vibrations of a structure submerged in a compressible fluid. Int. Symp. Finite Element Tech., Stuttgart, 359-379.

Simplified three-dimensional analysis of concrete gravity dams

A. El-Nady[I], A. Ghobarah[II], and T.S. Aziz[II]

ABSTRACT

A new procedure is developed for the analysis of concrete gravity dams. This procedure has the capability of including the three dimensional effect with the simplicity of a typical two dimensional analysis. The effect of the adjacent monoliths is represented by interaction shear forces. The results of this new approach were compared to that of a typical three dimensional analysis and high accuracy was obtained. The main advantage of this procedure is the computer cost saving and the simplicity of dealing with beam elements as well as including the interaction of the adjacent monoliths during the ground motion.

INTRODUCTION

The possiblity of catastrofic failure of concrete gravity dams due to earthquakes presents a hazard for life as well as substantial economic losses in seismically active regions. Existing dams were designed using the equivalent static method which neglects the dynamic properties of the dam and the exciting ground motion. Most of the previous studies on gravity dams used a two dimensional planar model to represent the structure (Chopra 1980 and 1984). Such a two dimensional analysis of the dam is based on the assumption that during an earthquake the expansion joints fail and the monoliths vibrate independently. However, this is not the case in most of the gravity dams where keyed expansion joints between the monoliths are used. A three dimensional analysis of the dam neglecting the expansion joints and simplifying the dam geometry as a rectangular cross- section was performed by Rashed (1985). He indicates that a reduction up to 50% of the response over the two dimensional analysis could be obtained depending on the length/height ratio of the dam. Although many approximations have been introduced in the study, it shows the importance of the third dimension of the dam structure. As the traditional three dimensional analysis of dams, based on the theory of elasticity, is expensive and practically impossible to manage, a new approximate model needs to be developed. This model should save time and effort and at the same time gives reliable results.

Concrete gravity dams in general consist of a number of monoliths (about 15 m wide each) built across the river and separated by expansion joints. These monoliths are fixed at the ground and are subjected to vertical and lateral in plane loads. Depending on the type of the expansion joint used

I   Research Assistant, Civil Eng. Dept., McMaster University, Canada
II   Professor, Civil Eng. Dept., McMaster University, Canada

in the dam, shear forces may develop among the monoliths. however, as the expansion joints are designed to allow for the axial displacement of the monoliths, generally no axial forces are expected. The idea of the new model is to represent the effect of adjacent monoliths, on the one under consideration, by shear forces transmitted through the expansion joints. Therefore, each monolith is considered to be fixed to the ground and connected to other monoliths by links. These links can only transmit shear forces between monoliths and their properties depend on the type and behaviour of the expansion joint used in the dam.

In fact, the relative lateral displacement of the monoliths is the main factor affecting the shear forces developed across the expansion joint. this concept is implemented in the model by using links to connect the monoliths. The forces developed in these links depend on the relative displacement of monoliths.

A plane stress analysis is the most suitable model for each monolith. However, as the number of monoliths in each dam is large, using a finite element model for each monolith will need a large number of degrees of freedom. As a result and as a further simplification, it is suggested that a beam model is used to represent the monoliths for the purpose of calculating the relative displacements among the monoliths. This model provide a great saving in the number of degrees of freedom without loss of accuracy. It provides the capability of including any effects which may change along the longitudinal axis of the dam as long as it can be incorporated in the beam model.

ANALYSIS PROCEDURE

A concrete gravity dam is normally constructed with joints in order to prevent tensile longitudinal cracks caused by chemical or thermal contraction of concrete and to facilitate the construction work. As each monolith has its own vibration characteristics, they tend to vibrate in different manners. As a result, the motion of the dam may become discontinuous at every monolith causing the water stop plate to break. In order to prevent discontinuous movement between adjacent monoliths, keys are introduced at the joint between the monoliths. Keys also serve the purpose of distributing the stresses in the concrete three dimensionally.

The simplified procedure introduced in this study considers all the monoliths along the dam axis. Each monolith is represented by an equivalent cantilever beam fixed at the ground level and connected with other monoliths by horizontal beams. As the depth / span ratio of the equivalent cantilever beams is high, the effect of shear deformations is included in the procedure. The horizontal beams represent the expansion joints among the monoliths and its properties should be evaluated to give the same behaviour as that of the joints used in the actual dam construction. Although, the procedure is developed for the analysis of concrete gravity dams on rigid foundations and subjected to horizontal ground motion, it could be extended to include the flexibility of the soil and other components of the ground motion.

It is assumed that the expansion joints in gravity dams are capable of transmitting shear forces among the monoliths. Only the horizontal shear developed in the beams connecting the monoliths is considered. The properties of these beams are important and should represent the behaviour of the expansion joints. Although in this study, rigid joints have been assumed, it is possible to incorporate other types of joints and to represent any nonlinear behaviour which may occur. To evaluate the equivalent moment of inertia of these beams the following formula is used:

$$I = \frac{L^3}{12E} \left[\frac{1}{\dfrac{L_1^3}{3EI_1} + \dfrac{L_1}{GA_r}} \right] \tag{1}$$

where:

I is the equivalent moment of inertia for the connecting beams
L is the length of the connecting beams
E is the modulus of elasticity of concrete
G is the shear modulus of concrete
L_1 is the length of monolith
I_1 is the moment of inertia of the monolith
A_r is the shear area of the monolith

This formula is developed based on equating the lateral deformations within the monolith to that of a cantilever beam with concentrated load at the edge. The purpose of allowing for such deformations is to simulate the longitudinal deflection behaviour of the dam. In fact, as shear forces develop along the edges of each monolith, it is expected that some longitudinal deformations will take place even within each monolith.

Several methods are available for including the hydrodynamic effects in the dynamic response analysis of concrete gravity dams. The equivalent static method, developed by Westergaad's (1933) is the simplest method available. However, this method underestimates the hydrodynamic forces on the dam as it neglects the water compressibility and dam flexibility. Another method is to model the fluid using a continuum approach or a finite element approach. For the purpose of this study, an approximate technique is used to simulate the hydrodynamic effects considering the horizontal component of an earthquake [Chopra, (1978)]. In this method the effect of the fluid interaction is assumed to be the same as added masses on the upstream face of the dam. The main advantages of this method are the saving in calculation time and effort and the fact that no special programming is needed. The main assumption in this approach is that the response of the dam will be mainly due to the fundamental mode of vibration. This is a realistic assumption as the concrete gravity dams are short vibration period structures.

The dynamic analysis procedure uses a displacement finite element formulation for the dam with added masses to approximate the hydrodynamic effects. The equations of motion of the system are:

$$[M^*]\ddot{\underline{v}} + [C]\dot{\underline{v}} + [K]\underline{v} = [M][R]\ddot{\underline{u}}_g + \underline{R}_m \qquad (2)$$

$$[M^*] = [M] + [M_a] \qquad (3)$$

Where,
[M], [C] and [K] are the mass, damping and stiffness matrices of the dam
$\underline{v}$ is the vector of nodal displacements
[R] is the influence matrix for the ground motion components.
$\ddot{\underline{u}}_g$ is the vector of ground acceleration
$[M_a]$ is the added mass matrix
ω_s fundamental natural circular frequency of dam without water
R_m is the vector of shear forces transmitted through expansion joints

The added mass representing the hydrodynamic effects are estimated by the formula:

$$m_a(y) = \frac{P(y, \omega_s)}{\psi(y)} \qquad (4)$$

191

$$P(y, \omega_s) = \frac{2w}{gH} \sum_{n-1}^{\infty} \frac{I_n}{\sqrt{\lambda_n^2 - \dfrac{\omega_s^2}{c^2}}} \qquad (5)$$

$$I_n = \int_0^H \psi(y) \cos \lambda_n \, y \, dy \qquad (6)$$

$$\lambda_n = [(2n - 1)\pi]/2H \qquad (7)$$

$\psi(y)$ = The shape of the fundamental mode of vibration of the dam without water.
w   = Unit weight of water
c   = The velocity of sound in water
H   = Water depth
g   = Gravity acceleration
x,y   represent the longitudinal and the vertical spatial coordinates, respectively

The damping ratio for the equivalent system should be modified according to the following formula:

$$\xi_s = \frac{\omega_s}{\omega} \xi \qquad (8)$$

Because the dam - reservoir interaction lowers the fundamental natural vibration frequency of the dam, ω_s, the damping ratio for the equivalent system ξ_s is less than the damping ratio ξ for the dam alone (Chopra, 1978).

DISCUSSION OF RESULTS

The results of the simplified procedure were compared to those obtained by a typical three dimensional analysis for an idealized dam structure shown in figure (1). The concrete in the dam is assumed to be homogenous, isotropic and linear elastic, with the following properties: unit weight = 24.3 kN/m^3, shear modulus = 14.74×10^6 kPa, which corresponds to a modulus of elasticity = 34.45×10^6 kPa and Poisson's ratio = 0.17. The earthquake ground motion record used in this study is the S69E component of the (1952) Taft earthquake recorded at the Lincoln School Tunnel. This record is considered to be an intermediate frequency earthquake with an estimated dominant circular frequency content of ω = 12.5 rad/s (Naumoski 1988). The linear analyses of gravity dam response to seismic ground motion were performed using the finite element code SAP IV. Different cases of the ratio H_1/H_2 were considered to study the effect of sudden change in span on the accuracy of the model. Figure (2) shows the simplified three dimensional model of this structure.

The natural frequencies of the structure with different H_1/H_2 ratios evaluated by the new model as well as the traditional three dimensional finite element analysis are summarized in table (1). It is noted that the accuracy of the first five modes evaluated by the simplified procedure, S.P., is high. The case $H_1/H_2 = 0.0$ represents a uniform soil profile under the dam, i.e. no shear forces transmitted through the expansion joints as the response of each monolith to the ground motion is the same. However, increasing the ratio H_1/H_2 leads to an increase in the shear forces among the monoliths. As a part of testing the new procedure the displacements at different points of the crest are compared. The horizontal component of the Taft ground motion is used in this study. Figures (3),(4) and (5) show the time history of the maximum displacements evaluated by the new procedure and the error relative to the traditional three dimensional analysis for the cases H_1/H_2 = 0.0, 0.25 and 0.5 respectively. A reasonable agreement between the two approaches is observed in the aforementioned figures.

Figure (6) shows the maximum displacement profile for the dam crest evaluated by the two procedures for the case H_1/H_2 = 1.0. Good accuracy is achieved using the new procedure compared to the three dimensional finite element. Figure (7) shows the same relation for the case H_1/H_2 = 0.5. It is noted that the results start diverging somewhat as the ratio H_1/H_2 increases. For practical purpose the observed accuracy is considered acceptable as the ratio H_1/H_2 normally does not exceed 0.5.

CONCLUSIONS

A new cost-effective procedure for the analysis of concrete gravity dams including the interaction effects of monoliths, is introduced. The procedure is simple, powerful and is demonstrated to give high accuracy compared to typical three dimensional analysis. As a result of this simplification, other parameters which have not been considered before, such as soil profile under the dam, the change of ground motion along the dam axis and the type of the expansion joint used in the analysis can be studied.

REFERENCES

Chopra, A., 1978.'Earthquake Resistant Design of Concrete Gravity Dams', Journal of the Structural Division, ASCE, Vol. 104, No. ST6, pp. 953-971.

Chopra, A.K., 1987.'Earthquake Analysis, Design and Safety Evaluation of Concrete Dams', 5th Canadian Conference of Earthquake Engineering, Ottawa, Ontario.

Chopra, A.K., Chakrabati, P. and Gupta, S., 1980. 'Earthquake Response of Concrete Gravity Dams Including Hydrodynamic and Foundation Interaction' Report No. UCB/EERC-80/01, Earthquake Engineering Research Center, University of California, Berkely, California.

El-Nady, A., Ghobarah, A. and Aziz, T., 1990.'Effect of High Frequency Earthquakes on The Response of Concrete Gravity Dams', CSCE annual conference, Hamilton, Ontario, Vol. IV-1, pp. 59-71.

Fenves, G. and Chopra, A.K., 1984.' Earthquake Analysis and Response of Concrete Gravity Dams', Report No. UCB/EERC-84/10, Earthquake Engineering Research Center, University of California, Berkeley , California.

Mlakar, P.F., 1987.'Nonlinear Analysis of Concrete Gravity Dam   to Strong Earthquake-Induced Ground Motion', Computers and Structures, Vol. 26, No. 1/2, pp. 165-173.

Naumoski, N., Tso, W.K. and Heidebrech, A., 1988.'A Selection of Represntative Strong Motion Earthquake Records Having Different A/V Ratios', EERG Report 88-01, Earthquake Engineering Research Group, McMaster University.

Rashed, A. and Iwan, W., 1984. 'Earthquake Response of Short length Gravity Dams', Proceeding of Seventh World Conference on Earthquake Engineering, California, pp. 355-362.

Table 1. Natural frequencies of the test structure

Mode No.	$H_1/H_2 = 0.0$		$H_1/H_2 = 0.25$		$H_1/H_2 = 0.5$	
	S.P.	3-D	S.P.	3-D	S.P.	3-D
1	120.9	121.1	105.8	102.8	95.2	92.5
2	169.6	171.0	156.8	161.5	150.7	157.2
3	268.8	264.5	256.0	255.4	244.0	242.3
4	367.3	304.9	297.5	298.8	249.1	280.9
5	369.2	374.9	315.2	319.6	269.5	295.3

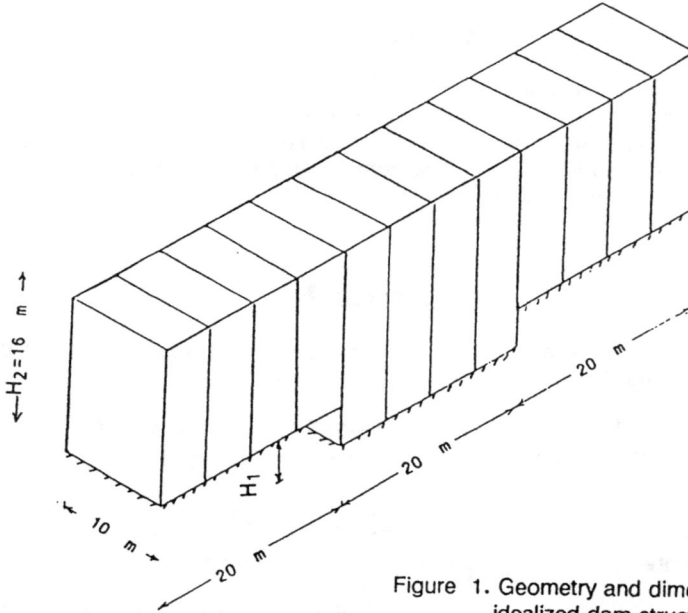

Figure 1. Geometry and dimensions of an idealized dam structure

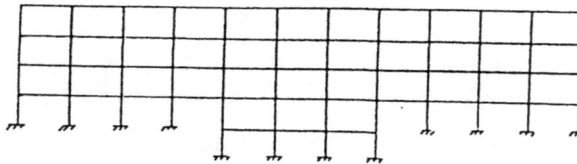

Figure 2. Modelling of the dam structure

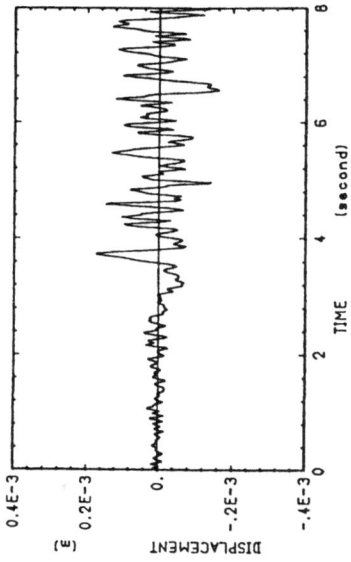

a) Simplified analysis

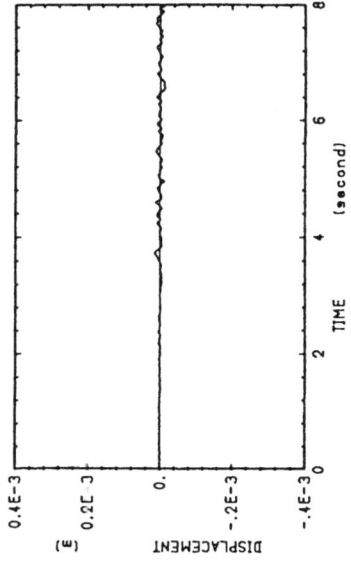

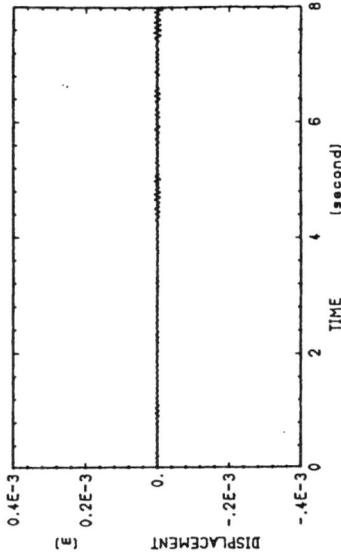

b) Variation from traditional analysis

Figure 3. Time history of the crest displacement
$H_1/H_2 = 0.0$

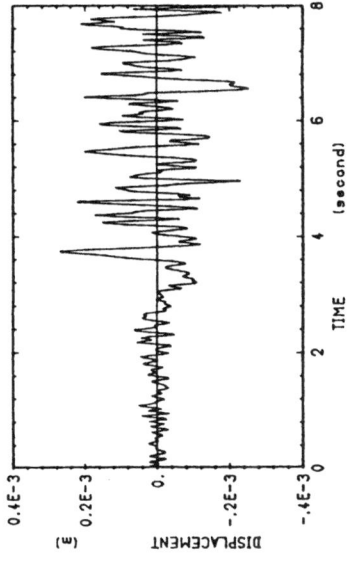

a) Simplified analysis

b) Variation from traditional analysis

Figure 4. Time history of the crest displacement
$H_1/H_2 = 0.25$

195

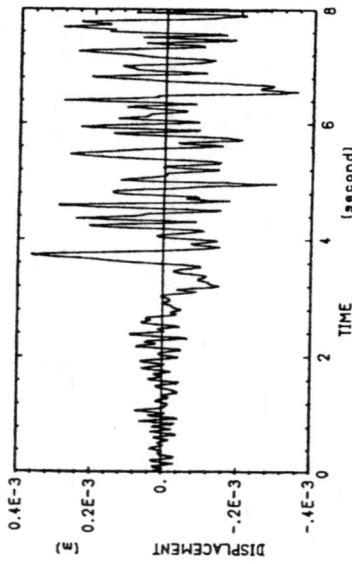

Figure 6. Crest displacement profile of
the test structure
$H_1/H_2 = 0.5$

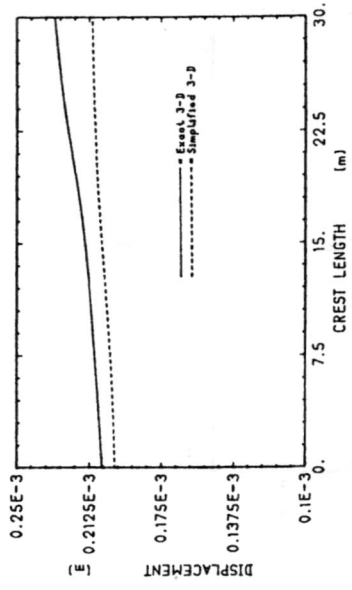

Figure 7. Crest displacement profile of
the test structure
$H_1/H_2 = 1.0$

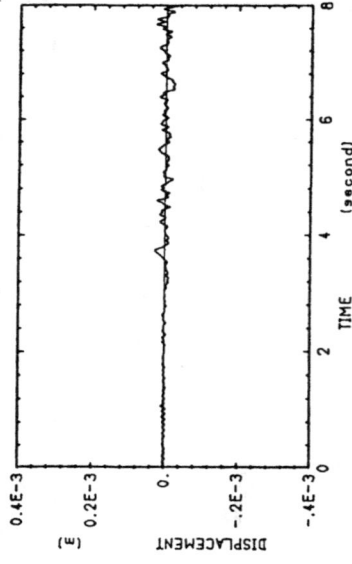

a) Simplified analysis

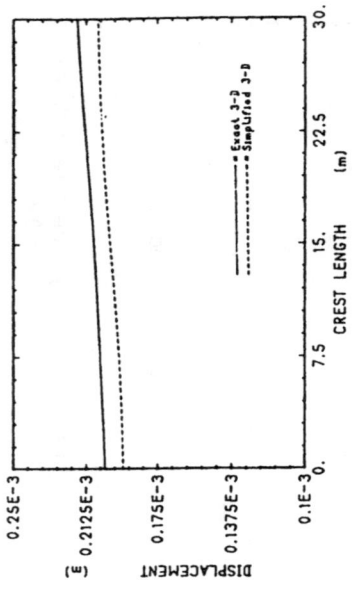

b) Variation from traditional analysis

Figure 5. Time history of the crest displacement
$H_1/H_2 = 0.5$

196

Seismic crack propagation in concrete gravity dams

O.A. Pekau[I] and V. Batta[II]

ABSTRACT

Crack propagation in a concrete gravity dam under seismic loading is analyzed. The boundary element method together with the principles of linear elastic fracture mechanics is employed to analyze the dynamic behaviour of a discrete crack. Propagation of the crack is monitored in a time step-by-step procedure. The formulation is employed in a study of the cracking of the Koyna dam under the 1967 Koyna earthquake, in which the effects of various parameters on the propagation process and the final crack profile are examined. The analytical results obtained appear to be in agreement with the general field observations made following the Koyna earthquake.

INTRODUCTION

Concrete gravity dams can be significantly affected by the loading produced by earthquake-induced ground shaking which may result in tensile cracking of the dam. This has been amply demonstrated by past experiences, including those of the Koyna and Hsingfengkiang dams, and also predicted by dynamic response analyses (Chopra 1987). Although these analyses identified the regions of the dam where large tensile stresses would occur, their formulations could not predict the extent of the subsequent cracking which the dam may suffer during a particular ground motion.

Prediction of the extent of a crack and its significance for the future safety of a dam poses formidable analytical difficulties. Nevertheless, the practical importance of obtaining such data to prevent catastrophic failure is becoming increasingly evident, particularly with scaling-up of the MCE for most of the existing dams. The required analytical procedure should therefore not only be accurate in its numerical representation of the crack problem but should also be efficient in studying dynamic crack propagation.

[I] Professor and Chairman, Dept. of Civil Engineering, Concordia University, Montreal, Quebec, Canada.

[II] Ph.D. student, Dept. of Civil Engineering, Concordia University, Montreal; also Structural Engineer, Shawinigan Lavalin Inc., Montreal, Quebec, Canada.

COMPUTATIONAL PROCEDURES

Dynamic fracture analyses generally employ either the finite element or the boundary element method to discretize the structure, together with a discrete or smeared approach to model the crack itself. However, most of the available studies have used the finite element method and, to date, the boundary element method has found only a limited application in this area. The available finite element procedures are those by Graves and Derucher(1987) and Mlakar(1987), who used a smeared crack approach to study the developement of tensile cracks in concrete gravity dams subjected to ground motion. While the model used by Graves and Derucher was limited to small cracks only, Mlakar used three different examples of dams in the United States to show that, under a strong earthquake, cracks may penetrate the entire cross-section of a dam. However, the geometry (length and width) as well as the exact location of these cracks could not be obtained from the analysis because of the use of the smeared crack approach in the formulation.

The discrete crack approach, on the other hand, not only has the potential of determining the above characteristics but is also more suited to model the physical behaviour of unreinforced concrete during and after cracking. This is because the concrete would develop a distributed (smeared) cracking pattern only if reinforcing steel is present. Employing discrete cracks, Skrikerud and Bachmann(1986) and Saouma et al(1990) have recently presented finite element analyses of concrete dam cracking. However, in spite of a correct physical representation of the crack these studies still require the redefinition of the finite elements in the vicinity of the crack tip. Moreover, as shown by Skrikerud and Bachmann(1986), the predicted crack pattern is extremely sensitive to the nodal density of the finite element mesh. Recently, for a propagating discrete crack in concrete dams, Pekau et al(1991) have presented a methodology employing the boundary element technique together with the discrete crack approach. In this study the dynamic response was obtained by the mode superposition technique and the nonlinear crack-closing behaviour was modelled using load pulses.

PRESENT FORMULATION

In the present study the frequency independent dynamic boundary integral formulation, presented by Nardini and Brebbia(1982), is employed. Isoparametric boundary elements are used for the discretization of the structure which is divided into subdomains in such a way that the two crack flanks belong to adjacent domains (Blandford et al 1981). This procedure, known as the multidomain discretization scheme, provides a mathematically correct crack representation and allows for the subsequent stress intensity factor computations using the displacement correlation technique. Analytical singularities exhibited by tractions in front of the crack tip are accounted for by using traction singular quarter-point elements. The latter are obtained by a suitable transformation of the isoparametric boundary elements.

Using the above discretization scheme, boundary integrals for each subdomain are evaluated and the governing equations of dynamic equilibrium are assembled. A time step-by-step solution of these equations is obtained by direct integration. Time integration using the Houbolt method is employed in the present study because the method is unconditionally stable and provides numerical damping for higher modes.

Dynamic fracture analysis

Dynamic analysis of a structure with discrete crack and subjected to an oscillatory time-dependent loading, as imposed by an earthquake, also requires additional analytical modelling of the crack closing behaviour. This is because, when the crack closes, the crack flank nodes in the analytical model tend to overlap, which is inconsitent with the true physical behaviour. Thus, while an open crack is accurately represented in the model, the crack closure is modelled by introducing dimensionless springs between the nodes on the crack surfaces. During the subsequent opening phase these springs are removed and the structural stiffness is appropriately modified to once again obtain traction free crack surfaces. The stress intensity factors are obtained from the crack flank displacement data and the maximum tensile strain theory (Pekau et al 1991) is used to determine the direction and times of propagation of the discrete crack.

PARAMETRIC INVESTIGATION

Various parameters which generally influence the seismic behaviour of concrete gravity dams comprise the presence of gravity loads, reservoir interaction, foundation interaction, reservoir bottom absorption and the direction in which the earthquake loading is applied. In addition, the value of the dynamic fracture toughness for concrete and the length by which the discrete crack is extended at each stage of crack propagation also have a very significant influence on the resulting crack profile. It should be noted that the selection of the length of crack extension is a limitation of all discrete crack formulations. While a large value of the extension may result in forcing the crack profile in a certain direction, too small an extension would, on the other hand, be numerically inefficient.

Although a number of the above effects were considered in this study, results are presented here for only a few of these parameters. For the example application which follows, two different parametric cases were defined. In Case 1, the gravity loads due to the weight of the concrete as well as the effect of the reservoir were not considered. The earthquake loading was applied in both the horizontal and vertical directions and the influence of the length of the piecewise crack extension on the computed crack profile was studied. In Case 2, the gravity loads were included and the analysis was conducted for a full reservoir. Thus, both the hydrostatic as well as the hydrodynamic interaction of the impounded water were incorporated in the dynamic analysis. The latter was modelled as an added mass considering the water as incompressible. In this case also, the ground motion was applied in both directions. The dam was considered fixed at the base in both cases and the effect of reservoir bottom absorption was not included.

DISCUSSION OF RESULTS

Using the procedure presented in this paper, a parametric dynamic fracture analysis of the Koyna dam subjected to the Koyna earthquke of Dec. 11, 1967 was conducted. Since the ground motion of the Koyna earthquake was

recorded by a series of strong motion accelerographs located in a gallery
close to the foundation of the dam, relevant comparison between analytical
results and the observed behaviour can also be drawn.

The Koyna dam is 854 m long and 103 m high at the tallest non-overflow
monolith and has a cross-section which is not typical of a gravity dam (see
Fig. 1). As shown in Fig. 2, cracking of the dam during the Koyna earthquake
resulted in roughly horizontal cracks on both the upstream and downstream
faces near elevation 66.5 m and there was inconclusive evidence that some of
these cracks might have penetrated through the entire width of the dam. Based
on this observation, an initial crack 1.0 m in length was assumed in the
present analysis at elevation 66.5 m. on the downstream face of the dam.
Occurrence of crack initiation on the downstream face was also predicted by
Chopra(1987). The boundary element discretization employed for the cracked
dam is shown in Fig. 1.

Linear elastic material properties, valid for the entire cross-section,
were used for the concrete: modulus of elasticity = 31×10^3 MPa, Poisson's
ratio = 0.2 and mass density = 2400 kg/m^3. Damping equal to 5% of critical
was assumed in the first two modes of vibration. The dynamic fracture
toughness for concrete was assumed to be 1.96 MPa.m$^{1/2}$.

Figures 3(a) and 3(b) show the time histories of horizontal displacement
and acceleration at the dam crest for Case 1. Corresponding plots of stress
intensity factors are shown in Figs. 4(a) and 4(b). Fig. 5 shows the computed
crack profile for this case. The crack profile is shown for two different
values of the assumed crack extension lengths, namely 0.5 m and 1.0 m. As
evident in Fig. 5, almost identical profiles are predicted for the two
extension lengths. A piecewise extension of 1.0 m was thus used in the
subsequent analyses.

The response results for Case 2 are presented in Figs. 6(a) and 6(b),
while Figs. 7(a) and 7(b) depict the computed stress intensity factors for
this case. From Fig. 7(a), it can be seen that the gravity loads of the
present case oppose opening of the crack, which remains closed until
approximately 1.95 sec, at which time the first major peak of ground
acceleration occurs. This is unlike Case 1, where the crack opens and closes
a number of times before finally breaking through to the opposite face of the
dam at 1.95 sec. The predicted crack profile for Case 2 is shown in Fig. 8.
It can be seen that the crack penetrates through the entire width of the dam
and reaches the upstream face near elevation 65.0 m which is consistent with
the observed behaviour. Furthermore, compared to Case 1 (no static loads),
the profile of Case 2 shows that inclusion of these forces does not have a
profound influence on the final pattern of cracking for the Koyna dam.
Rather, static loads affect primarily the times at which the crack is induced
to propagate during seismic excitation. This was evident from the previous
comparison of Figs. 4 and 7.

FINAL REMARKS

Based on the results of this study, it appears that the present boundary
element formulation is well suited for the dynamic crack propagation analysis
of mass concrete structures such as gravity dams. Moreover, various

analytical and physical parameters affect the pattern of the predicted cracking. These involve primarily the fracture toughness of concrete, the assumed length of the incremental crack extensions and the loading conditions of the dam.

REFERENCES

Blandford, G.E., Ingraffea, A.R. and Ligget, J.A. 1981. Two-dimensional stress intensity factor computations using the boundary element method, International Journal for Numerical Methods in Engineering, 17, 387-404.

Chopra, A.K. 1987. Earthquake analysis, design and safety evaluation of concrete dams, 5th Canadian Conference on Earthquake Engineering, 39-62.

Graves, R.H. and Derucher, K.N. 1987. Interface smeared crack model analysis of concrete dams in earthquakes, Journal of Engineering Mechanics Division, ASCE, 113(11), 1678-1693.

Mlakar, P.F. 1987. Nonlinear response of concrete gravity dams to strong earthquake-induced ground motion, International Journal of Computers and Structures, 26(1/2), 165-173.

Nardini, D. and Brebbia, C.A. 1982. A new approach to free vibration analysis using boundary elements, 4th International Conference on Boundary Element Methods, Springer Verlag, Berlin, 312-326.

Pekau, O.A., Zhang, C. and Feng, L. Seismic fracture analysis of concrete gravity dams, Accepted for publication in Earthquake Engineering and Structural Dynamics

Saouma, V.E., Bruhwiler, E. and Boggs, H.L. 1990. A review of fracture mechanics applied to concrete dams, Dam Engineering, 1, 41-57.

Skrikerud, P.E. and Bachmann, H. 1986. Discrete crack modelling for dynamically loaded, unreinforced concrete structures, Earthquake Engineering and Structural Dynamics, 14, 297-315.

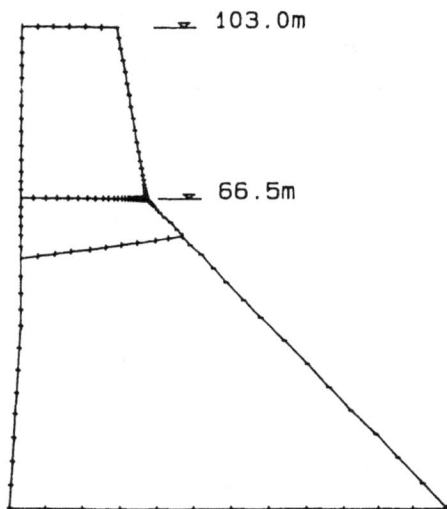

Figure 1. Koyna dam and BE mesh

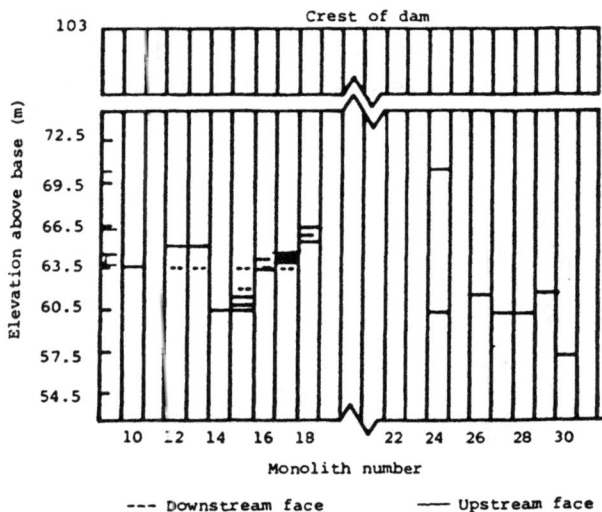

Figure 2. Observed pattern of cracking in Koyna dam

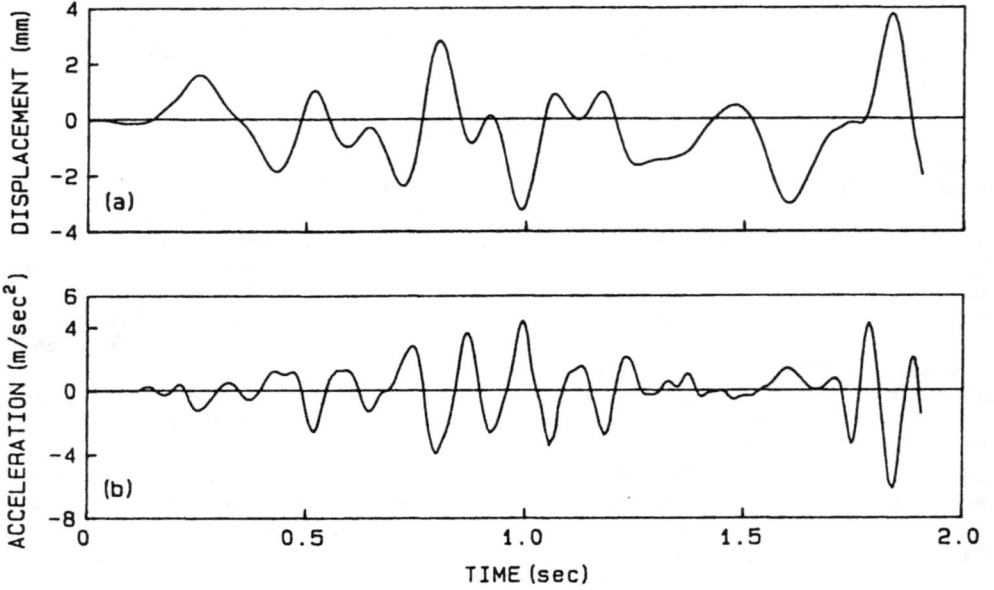

Figure 3. Case 1 time histories of response: (a) crest displacement;
(b) crest acceleration

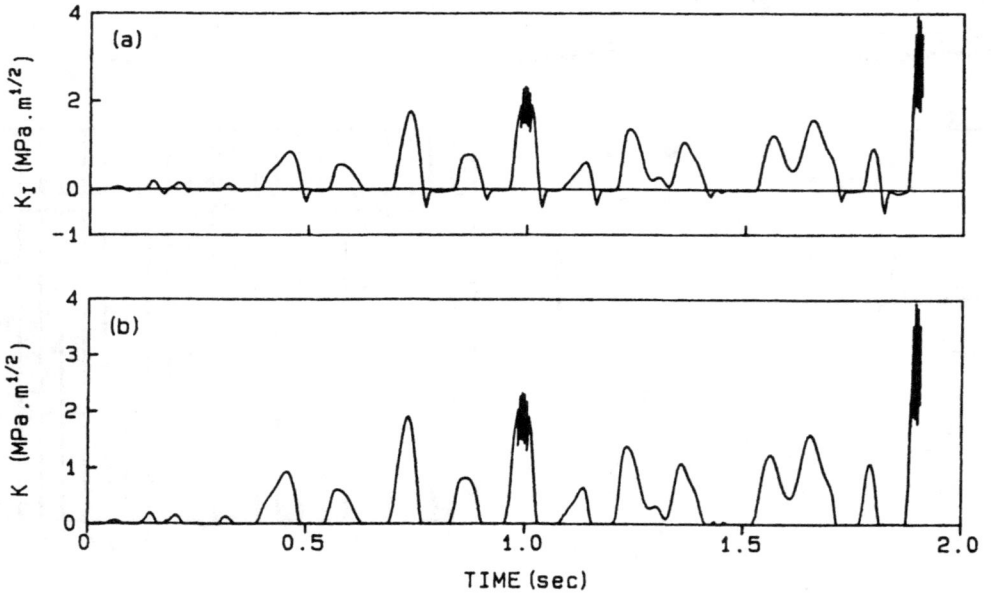

Figure 4. Case 1 time histories of stress intensity factors:
(a) crack opening mechanism; (b) combined opening
and sliding mechanism

202

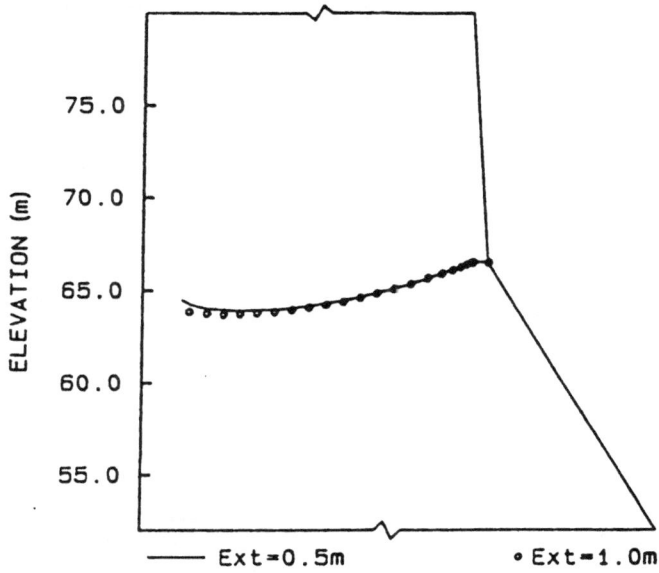

Figure 5.   Effect of crack extension length on crack
profile

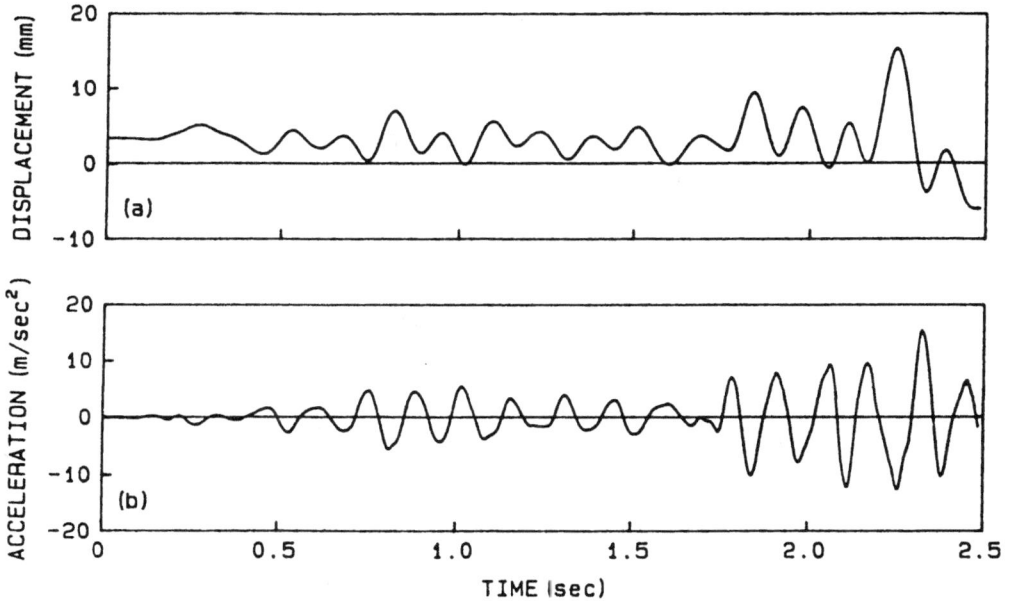

Figure 6.   Case 2 time histories of response: (a) crest displacements;
(b) crest acceleration

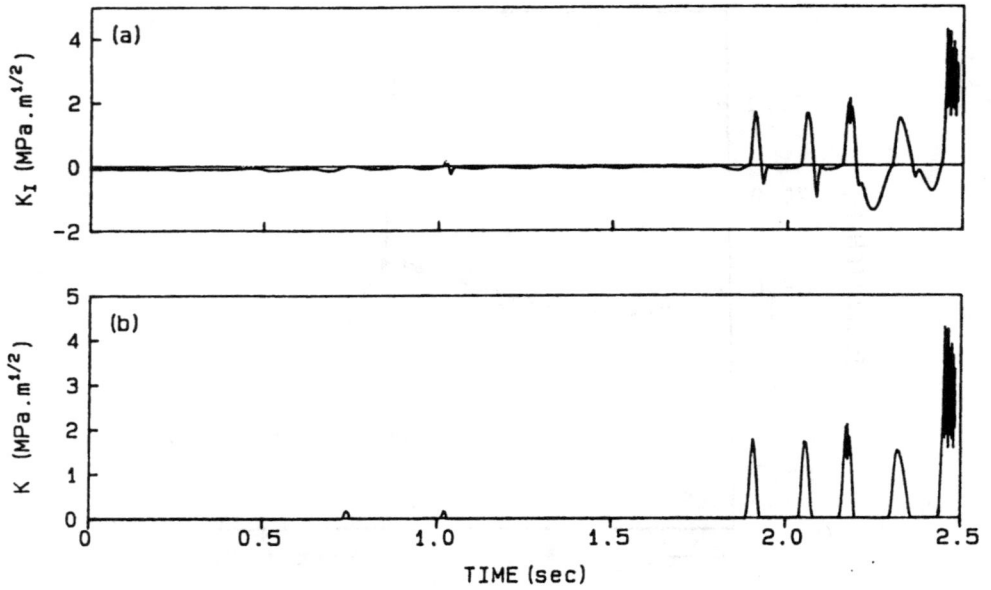

Figure 7. Case 2 time histories of stress intensity factors:
(a) crack opening mechanism; (b) combined opening
and sliding mechanism

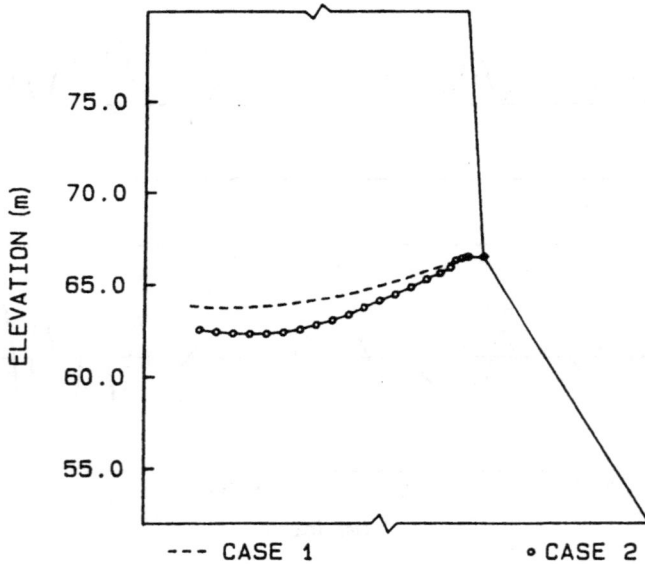

Figure 8. Comparison of crack profiles for: Case 1 - no
static loads; Case 2 - static loads included

Classical formulation of the impact between bridge deck and abutments during strong earthquakes

Emmanuel Maragakis[I], Bruce Douglas[II], and Spiros Vrontinos[III]

ABSTRACT

The impact between bridge decks and abutments has been the source of extensive damage to highway bridges during the 1971 San Fernando and other recent earthquakes. In this paper, a preliminary study related to the effects of impact energy losses on the dynamic response of bridges is presented. The focus of this study is .the development of analytical computer models for the formulation of the problem based on the classical impact theory, the performance of some parametric studies to identify the most important parameters, and the exploration of modeling the impact effects by using simpler techniques such as equivalent hysteretic dampers.

INTRODUCTION

The impact between bridge decks and the abutments during strong earthquake shaking, is a phenomenon that has attracted research interest during recent years. This impact affects primarily bridges with seat type abutments, and has been the source of serious damage to both decks and abutments in recent earthquakes. The 1971 San Fernando, California earthquake provides a particularly relevant example.

Many aspects of this interesting phenomenon have been investigated in recent years by several researchers. A short description of these studies and an appropriate reference list are provided by Maragakis et al. (1990). In these studies, the kinematic mechanism of the phenomenon has been analyzed and explained; and several parametric studies have been performed in order to identify the role of the most important parameters associated with this impact.

One aspect that has been neglected in all the previous studies is the

I Associate Professor, Civil Engineering Dept., University of Nevada, Reno

II Professor, Civil Engineering Dept., University of Nevada, Reno

III Graduate Student, Civil Engineering Dept., University of Nevada, Reno

energy loss associated with the impact between the deck and the abutments. In all the studies related to this phenomenon that have been performed so far, as well as in almost all bridge dynamic analysis programs used by bridge researchers and engineers, the abutments were represented with linear or nonlinear springs, and in some cases with a combination of springs and hysteretic dampers. Although this type of representation for the abutments allows for the modeling of the foundation stiffness as well as for energy losses due to material or radiation soil damping, it does not take into account the energy losses due to the severe collision between the bridge deck and the abutments. The objective of this paper is the presentation of the initial part of a study that has the following major goals: (i) To investigate the effects that energy losses directly related to the collision between a bridge deck and its abutments have on the response of the bridges, and (ii) To develop modeling techniques for these losses that can be incorporated in already existing bridge analysis programs and be used easily by bridge researchers and engineers.

FORMULATION OF THE PROBLEM

In order to investigate the importance that the impact energy losses have on the response of the bridge, two simplified models were developed.

In the first model, hereafter called "model 1", the bridge structure is represented with the system shown in Fig. 1. Based on this figure, one can see that the bridge deck is represented by a rigid mass supported by a translational spring, which accounts for the resistance of the bridge pier. The abutments are represented with translational springs and gaps. An abutment spring is activated after the closure of the corresponding abutment gap, and it is de-activated after the opening of the gap. The stiffness of the abutment springs were evaluated based on a method developed recently for the estimation of the stiffness of longitudinal abutment springs (Maragakis et al. 1990). One should note that this model does not take into account impact energy losses. Its response will be compared with the response of the impact model that will be described next, so that the importance of impact can be investigated. It should also be noted that all the springs are assumed to be linear and that no soil-structure interaction effects are considered. The model is excited at the bottom of the central mass. For the evaluation of its response to a dynamic loading, a computer program was written for the solution of the incremental equations of motion using the Wilson's θ method.

In the second model, hereafter called "model 2", the bridge structure is represented with the system shown in Fig. 2. The major difference between the two models is that in model 2 the abutment masses are included in the representation of the abutment system. After the closure of either one of the abutment gaps, an impact between the bridge deck and the corresponding abutment takes place. Therefore, in this model the effects of energy losses due to impact can be considered. It should be mentioned here, that the abutment springs represent the resistance in the longitudinal direction of the soil masses behind and underneath the abutments. The stiffnesses of these springs should be much higher in compression than in tension. However, for purposes of simplicity in the initial stage of this study, the values of

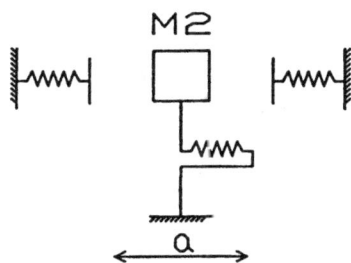

Figure 1. Model 1

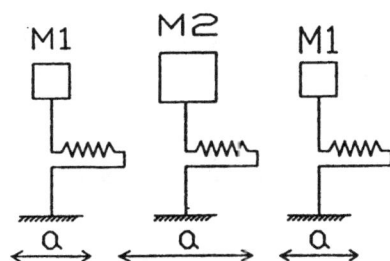

Figure 2. Model 2

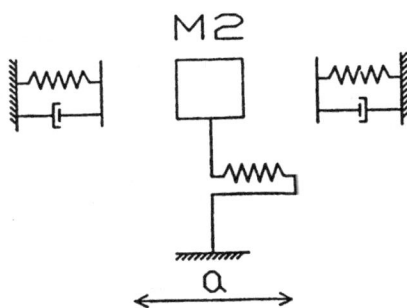

Figure 3. Model 3

the compressive and tensile stiffnesses of these springs were assumed to be equal. The masses of the model are excited at their base. For the evaluation of the response of the system to a dynamic loading a computer program was developed. In this program, the Wilson's θ method was used for the solution of the incremental equations of motion of the three masses when no impact between them is taking place. However, when impacts between the bridge deck mass and either one of the abutment masses occur, the equations from the classical theory of impact between moving masses were used (Maragakis et al. 1990).

A SIMPLER MODEL FOR THE EFFECTS OF THE IMPACT

Using the classical impact theory to evaluate the dynamic response of the bridge when the impact effects between the bridge deck and the abutments are taken into account, required two major modifications: (i) The vibrations of the abutments had to be considered since the abutment mass is an important parameter when the effects of the impact are taken into consideration, and (ii) The solution algorithm had to be modified to evaluate the response of the bridge-abutment system when impacts between the deck and the abutments occur. In the case of dynamic analysis computer programs for large bridges, these modifications will require substantial changes to the whole program. In the majority of the available bridge analysis programs, the bridge abutments are represented by an equivalent spring-damper system. For this reason, the classical formulation of the impact problem cannot be used in these programs.

To simplify the solution of this problem, a third model ("model 3"), was developed. This model is shown in Fig. 3. From this figure, it can be seen that the major difference between this model and model 1 is the representation of the abutment system. In model 3 a viscous damper has been added in the system representing each abutment. The role of this damper is to provide an equivalent representation of the impact between the bridge deck and the corresponding abutment. The value of the damper will be evaluated based on the equality of energy losses of the bridge deck mass between models 2 and 3. For a given base excitation, a value of the damper will be found such that the total energy loss of the bridge deck mass in model 2 due to the impact, will be equal to the same total energy loss in model 3 due to the viscous damper when subjected to the same duration of excitation. To accomplish this, a computer program was written which changes the value of damping in model 3, until equality of energy losses between models 2 and 3 for a certain ground excitation has been reached. After such a value is found, the response of the deck in model 3 is compared to the response of the deck in model 2, in order to find out if the criterion used for evaluating the damping coefficient is adequate to produce a reasonable correlation between the two responses.

RESULTS OF ANALYTICAL STUDIES

To check the validity of the computer programs developed herein, several special cases were analyzed using all three models. Details about these cases are provided by Maragakis et al. (1990).

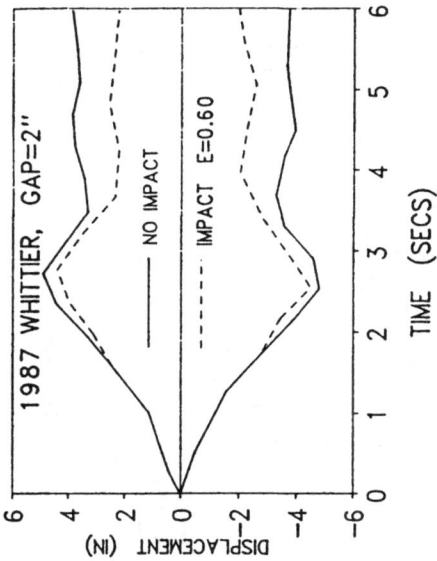

Figure 5. Deck displacement envelopes of models 1 & 2 for stiff soil

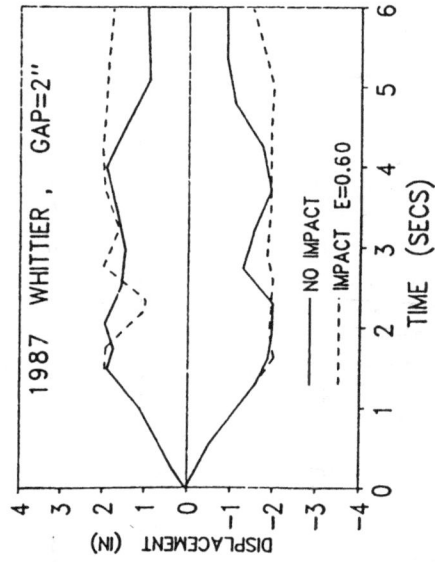

Figure 7. Variation of deck displacement with abutment stiffness for the El Centro earthquake

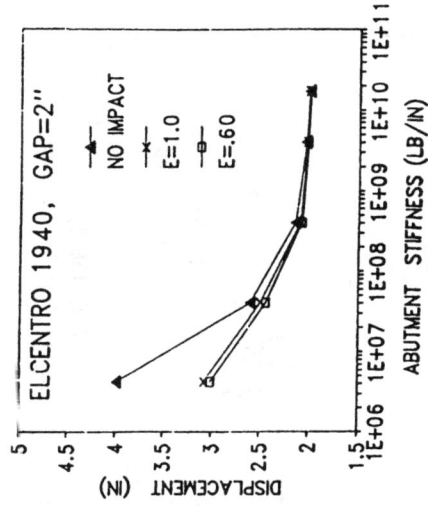

Figure 4. Deck displacement envelopes of models 1 & 2 for soft soil

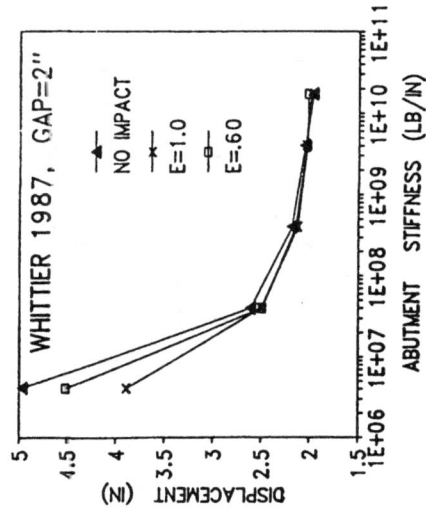

Figure 6. Variation of deck displacement with abutment stiffness for the Whittier earthquake

209

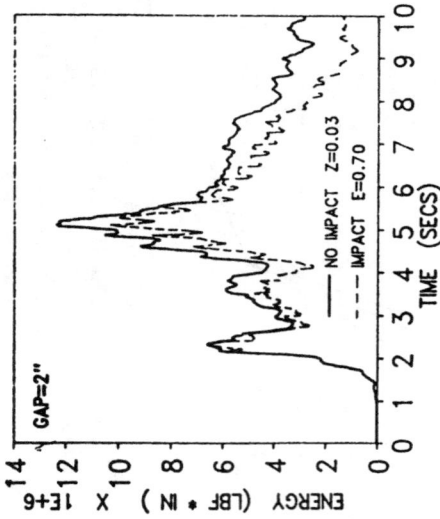

Figure 8. Energy levels of models 2 & 3

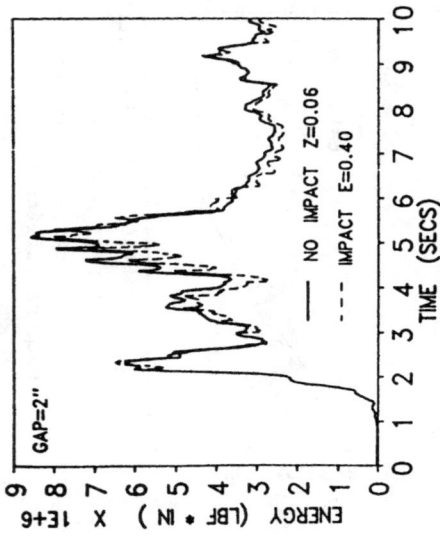

Figure 9. Energy levels of models 2 & 3

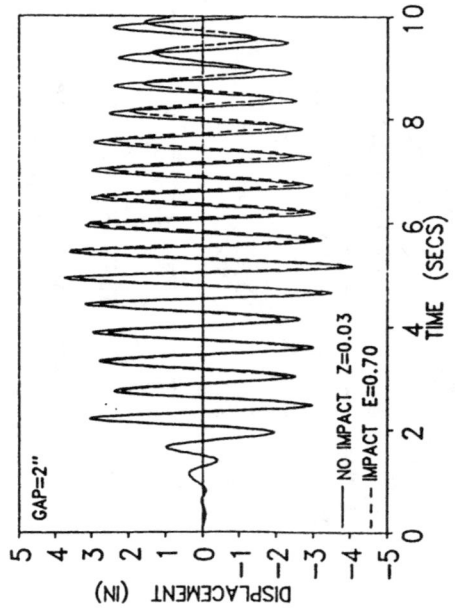

Figure 10. Displacement time histories of models 2 & 3

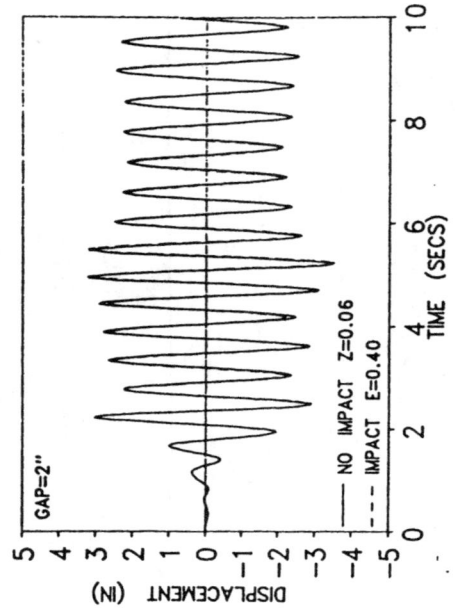

Figure 11. Displacement time histories of models 2 & 3

210

Figures 4-5 show a comparison of the displacement envelopes of the middle mass for a coefficient of restitution equal to 0.6. In both cases, the models considered have been excited by the 1987 Whittier earthquake. It can be seen that the degree of the impact on the response depends on the stiffness of the abutment soil spring. This is more obvious in Figs. 6-7 that show the maximum displacement of the middle mass for several values of the abutment springs for these different earthquake excitations. One can see that, for softer springs, the effect of the impact on the maximum response is more significant.

Figures 8-11 show the results of a preliminary study related to using model 3 for a simpler representation of the effects of the impact. Figure 8 shows the variation of the energy of the middle mass in the time domain in models 2 and 3. In model 2 the coefficient of restitution, e, was equal to 0.7, while in model 3 the damping, Z, ratio was equal to 3%. The damping coefficient in model 3 was evaluated from the assumed damping ratio based on the mass and stiffness properties of the middle mass. Figure 9 shows the comparison between the displacement time history responses of the middle mass corresponding to the cases described for Fig. 5. Figures 10-11 are similar to Figs. 8-9 with the only exceptions being that the values of the coefficient of restitution and the damping ratio were equal to 0.4 and 6% respectively. From these figures, it is evident that a better correlation between the energy time histories (Fig. 9) results in a better correlation between the displacement time histories (Fig. 11). This is consistent with the criterion used for estimating the equivalent damping ratio, which was described earlier. In all the cases discussed in Figs. 8-11, the models were excited by the first ten seconds of the 1940 El Centro earthquake while the mass properties of the bridge deck and the abutments were evaluated based on the properties of Nichols Road Overcrossing, a two span reinforced concrete bridge located in Riverside, California (Maragakis et al. 1990).

CONCLUSIONS - RECOMMENDATIONS FOR FUTURE RESEARCH

Based on the results of this initial study described above, the following conclusions can be drawn:

i)   The energy losses due to impact influences the dynamic response of the bridge deck mass (middle mass), and they should be considered when the dynamic response of bridges with seat type abutments is evaluated.

ii) The degree to which impact influences the response of the bridge depends on the size of the abutment gap, the mass ratio between the bridge deck and the abutments, the stiffness of the abutment springs, and the coefficient of restitution used for calculating the impact energy losses. Extensive parametric studies will be required to accurately determine the sensitivity of the response to these parameters.

iii) The procedures for finding an equivalent viscous damper to allow simple modeling of the effects of impact, based on the criterion of the equality of energy losses between models 2 and 3, produced very promising results. Study of more cases is required in order to perfect the method, identify any limitations or parameter sensitivities that it might have, and improve the efficiency of the computer algorithm.

ACKNOWLEDGEMENTS

The project leading to this study was funded by grant No. BCS-8718009 from the National Science Foundation. The financial support provided by the Civil Engineering Department of the University of Nevada, Reno is also gratefully appreciated.

REFERENCES

Maragakis, E., Douglas, B., Vrontinos, S., "Simple Modeling of the Impact Energy Losses Occurring Between the Bridge Deck and Abutments," Proceedings of the 6th U.S.-Japan Bridge Engineering Workshop, Lake Tahoe, Nevada, May 1990.

Design of bridges to satisfy ductility requirements

Ergin Atimtay[I]

ABSTRACT

Design codes permit plastic hinging of bridge piers under the design earthquake, provided that collapse does not occur; that is enough ductility must be available. However, ductility is not quantified and "How much ductility is enough ductility?" remains as a basic question for the designer to answer.

The ductility of the cross-section of a reinforced concrete member is obtained by developing the thrust-moment-curvature relationship. Then, a plastic analysis is performed to determine the sway of the supurstructure at the initiation of yielding. Thus, curvature and sway are related to each other. Now, the designer can decide on the desired sway ductility and calculate the corresponding curvature ductility. Then, the dynamic analysis of the bridge is performed in two steps, first without and then with abutment-superstructure interaction. The ductility demand thus obtained is compared with the available ductility. Obviously, available ductility must be greater than the demand. If not, design parameters must be changed to satisfy this constraint.

INTRODUCTION

Satisfying ductility requirements has been an essential goal of design against earthquakes. According to AASHTO (1983), bridges must resist small to moderate intensity ground shaking in the elastic range. For design earthquakes, the design philosophy accepts plastic hinging in the piers, provided that collapse does not occur. AASHTO further requires that such plastic hinging in the piers must be repairable. Damage to the foundations and joints is not acceptable.

However, design codes do not really define how the essantial ductility requirements can be met. The amount of confinement, splice requirements, etc. are carefully stated, but the following questions still ask for answers in the designer's mind: How much ductility is provided by a certain amount of confinement in a cross-section with a certain geometry? How can the elusive concept of ductility be quantified? Given a structural system, how much ductility is to be provided?

(I) Prof.Dr.Building Science, Middle East Technical University, Ankara, TÜRKİYE.

CROSS-SECTIONAL DUCTILITY

The cross-sectional ductility of a reinforced concrete member is commonly expressed by the thrust-moment-curvature (N-M-K) relationship. N-M-K relationships are obtained by solving a set of equations which are dictated by equilibrium, the proper stress-strain relationship, and compatibility conditions. The degree of confinement of the cross-section can be accounted for by modifications in the stress-strain relationship of concrete.

A computer program to give the N-M-K relationship has been developed (Atimtay, Cengizkan, 1975), which is another addition to many available. It's validity has been checked against test data and found to be very acceptable.

Being underreinforced, reinforced concrete beams exhibit a definite yield point. However, with the application of the axial load, the N-M-K- relationship begins to get non-linear and the definite yield moment gradually disappears. For bridges in earthquake zones, the level of axial load is kept small, usually under 0.2. At such axial load levels, the yield point of the member can still be defined easily on the N-M-K relationship. A typical N-M-K relationship for a bridge pier subject to low axial load levels is shown in Fig.1.

DUCTILITY DEMAND

It is common to design bridges with a gap between the superstructure and the abutment. This enables the superstructure to displace longitudinally under temperature effects and small to moderate intensity ground shaking. Additionally the abutment can be designed as free-standing where only active soil pressure behing the abutment can be considered under service conditions.

If designed in conformance with the AASHTO requirements, the bridge piers will hinge under the design earthquake, thus magnifying the superstructure sway. As a result, the gap between the superstructure and the abutment will close. The superstructure will bang against the abutment wall and the passive soil resistance behind the wall will be mobilized. Consequently, this mobilized passive soil pressure will restrain the superstructure sway. For the above described behavior to take place, the bridge pier will have to hinge at the pier-footing connection putting a substantial demand on ductility. To be able to meet this demand in the structural design context, ductility must be quantified. This requirement leads to the following question: Given a certain longitudinal sway, how much ductility must be provide so that the AASHTO requirements of earthquake damage and repairability can be satisfied?

The elevation of a typical bridge is shown in Fig.2. Seat type abutments have been employed to allow free movement of the superstructure under temperature effects and small to moderate intensity ground shaking. The superstructure may be continuous or simply supported over the piers. The piers act as supports for the superstructure and no moments are transferred between the superstructure and the piers. This bridge type is very common and popular, where the contractor slip-forms the piers and the superstructure is prefabricated. Once the piers are completed, the superstructure is dropped in place (or pushed-out) and seismic keys are cast-in-situ to prevent lateral movement of the superstructure.

The piers may be multi-column bents or wall-type piers (blade piers).The contractors prefer wall-type piers because they can be slip-formed much more easily. Preferable they may be from a contractor's point of view, pier walls present special problems for the designer.

Under the design earthquake, the only location in the pier wall where hinging can occur is the pier-footing connection. Obviously, with hinges at the pier-footing connections, the structure will be transformed into a mechanism. Consequently, the superstructure will sway until it bangs against the abutment back wall.

DESIGN PROCEDURE

The above described behavior will necessitate a two-phase analysis procedure. Firstly, the bridge must be considered as a structure which can sway freely without any interaction with the abutment. At this stage, the amount of sway and the gap between the superstructure and the abutments must be carefully monitored. Under moderate intensity ground shaking, the piers must not reach yielding. The author prefers to use a ratio of "service load moment" to yield moment, $M_s/M_y=0.7$. In other words, the maximum moment at the pier-footing connection, which is developed when the bridge makes longitudinal sways under a moderata intensity earthquake, should not exceed 70% of the yield moment of the pier.

In the second phase, the bridge must be considered as a system which interacts with the abutment back wall. The passive soil pressure behind the abutment wall may be considered as elastic springs which stabilize the super-structure sway. Here, the designer is faced with a tough question to answer: What is the "spring constant equivalent" of the select fill material behind the back wall which is in passive pressure mode?

Caltrans (1987) makes suggestions about the stiffness coefficients for average abutment backfill conditions:

    ks = Soil stiffness per linear length of wall
       = 36 kN/m (200 k/in) based on material with
         Vs = 0.3 m/sec.
         (effective height of wall ≠ 2500 mm)

It is clear to the engineer that the soil behind the back wall cannot sustain sresses after certain limits are exceeded. Here again, Caltrans recommendation will be followed, that maximum effective soil stress of 375 kN/m^2, should not be exceeded. The preliminary abutment stiffness should be adjusted after dynamic analysis, if the resulting forces on soil or displacement of the abutment are found to be excessive.

The structural model after the superstructure makes contact with the abutment back wall is shown in Fig.3. When the structure sways under the design earthquake, it hits the left abutment back wall mobilizing the full passive pressure. When it reverses the sway and moves away from the abutment, no soil pressure acts on the structure. Then, it hits the right abutment back wall and the full passive pressure is again mobilized. To account for this behavior, one-half of the total soil resistance is considered at each abutment. This is necessary to compute the correct dynamic properties of the bridge system. However, attention should be paid to the point that the resulting forces from the analysis should be doubled when designing the abutments.

QUANTIFYING DUCTILITY

Consider a pier as shown in Fig.4. with the corresponding moment and curvature distributions (Park and Paulay, 1975). The displacement of the pier at its top can now be calculated.

$$\Delta_u = (\frac{Ky}{2} \cdot \frac{2L^3}{3}) \cdot (Ku - Ky) \, Lp \, (L - 0.5 \, Lp) \tag{1}$$

$$\Delta_y = \frac{Ky}{2} \cdot \frac{2L^3}{3} \tag{2}$$

Ky = curvature corresponding to yielding
Ku = curvature at the end of the post-yielding range
Δu = displacement at the end of the post-yielding range
Δy = displacement corresponding to yielding
L = length of the pier
Lp = equivalent length of the plastic hinge

The displacement ductility ratio μ and curvature ductility ratio Ku/Ky can be calculated as follows :

$$\mu = \frac{\Delta u}{\Delta y} = 1 + \frac{(Ku - Ky)}{Ky} \cdot \frac{Lp \, (L - 0.5 \, Lp)}{L^2/3} \tag{3}$$

$$\frac{Ku}{Ky} = \frac{L^2 \, (\mu - 1)}{3Lp \, (L - 0.5 \, Lp)} \tag{4}$$

Earthquake resistant structures should posses a displacement ductility ratio of at least 4. So, assuming $\mu = 4$, Table 1 can be prepared to give the curvature ductility demands. It should be noticed that Lp/L is a variable. In other words, the length of the plastic hinging influences the ductility demand greatly.

Since the equivalent length of the plastic hinge Lp is typically in the range of 0.5 – 1.0 times the member depth, the length of the pier closely controls the curvature ductility demand. For very high piers, Lp/L will be small, thus creating the greatest demand on ductility. This demand should be met by the available ductility as given by the N-M-K relationship.

CONCLUSION

A design method has been developed which considers the interaction of the superstructure with the abutments. The total longitudinal sway of the structure puts a demand on ductility of the pier at the pier-footing connection. This ductility demand is quantified. The design method also shows how this quantfied ductility demand can be met by making use of the thrust-moment-curvature relationship of the pier. Cross-sectional properties can be changed as necessary to provide the required ductility.

REFERENCES

1. "Guide Specifications for Seismic Design of Highway Bridges (1983)", American Association of State Highway and Transportation Officials, 1983, 11pp.

2. "Bridge Memo to Designers Manual, 1987", State of California, Department of Transportation, 1987, 10-15 pp.

3. Atimtay, Ergin and Cengizkan, Kemal, "A Numerical Approach to Creep of Reinforced Concrete", METU Journal of Pure and Applied Sciences, Vol.8, No.3, 1975, 365 pp.

4. Park, Robert and Paulay, Thomas, Reinforced Concrete Structures, John Wiley and Sons, New York, 1975, 569 pp.

APPENDIX-NUMERICAL EXAMPLE

The typical underpass as shown in Fig.5 will be designed.
Width of superstructure = 11500 mm
Width of abutment   = 11000 mm
Weight of superstructure = 237 kN/m
Select pier cross-section as shown in Fig.5.
Normal force on pier   = 8053 kN
Develop N-M-K relationship of the pier for materials C20 and S420, as shown in Fig.6. This will provide the cracked EI, the moment capacity of the pier and the corresponding Ky.
Perform intermediate level eartquake analysis by the Response Spectrum Method.
Use effective EI (AASHTO Eq. 4-1) in dynamic analysis.
Superstructure sway = 63 mm
Max. pier moment   = 17000 kN-m
M/Mu = 0.67 (about 70%, say OK)
Change pier geometry if unacceptable.

$$\Delta_y = \frac{Ky}{2} \cdot \frac{2L^3}{3} = \frac{5.02 \times 10^{-3}}{2} \cdot \frac{2 (7)^2}{3} = 0.0849 \text{ m} = 84.9 \text{ mm}$$

Choose μ = 4
Total Δ u = 84.9 x 4 = 339.6 mm
Check if this total sway can be sustained by the pier

Maximum displacement of the pier must be 75 mm (CALTRANS).
Perform design eartquake analysis considering the passive pressure of the soil behind the abutment back wall as equivalent elastic springs (Fig.3).
Soil stiffness = 1264000 kN/m (CALTRANS)
Effective soil strength = 9525 kN (CALTRANS).
Abutment movement = 75.8 mm ~ 75 mm OK.
Total EQ force = 7582 kN
Total EQ force on pier = 3714 kN (AASHTO 4.8.2)
Do not use overstrength factor to calculate pier moment capacity when checking if soil strength is exceeded or not.
Total force on abutment wall = 7582 - 3714 = 3868 kN < 9525 kN
The constraints related with the abutment movement and soil strength are satisfied.
Check is enough ductility is provided in the pier to sustain a total longitudinal displacement of 339.6 mm.
Take length of plastic hinge L_p = 0.65(b) = 0.65 x 1.0 = 0.65 m.
L_p/L = 0.65/(7-1) = 0.108 (Considering hinge lengths)
K_u/K_y = 11.0 (by interpolation, Table 1)
Calculate curvature ductility from N-M-K of pier (Fig.8).
K_u/K_y = 10.67 < 11.0 but close. Say OK.

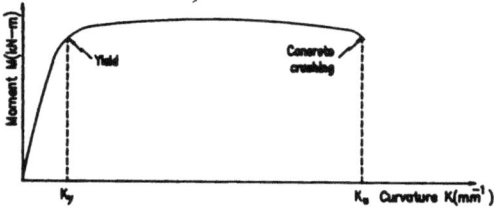

Figure 1. Typical N-M-K relationship for low axial load levels.

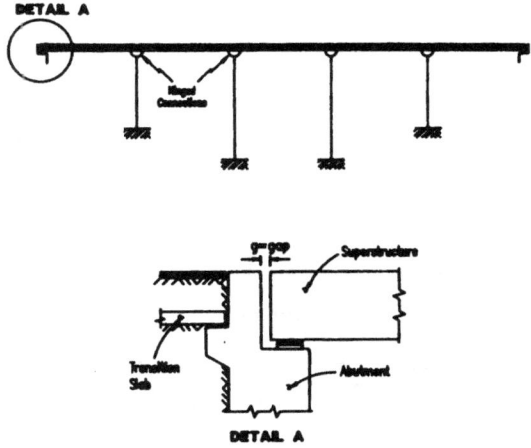

Figure 2. The elevation of a typical
bridge.

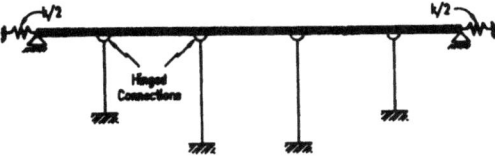

Figure 3. The structural model of
bridge-abutment interaction

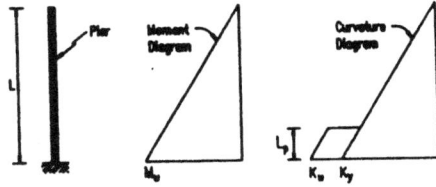

Fifure 4. The pier at ultimate
condition (Ref.3)

Table 1. Ductility demands for μ = 4

L_p/L	0.05	0.1	0.15	0.20	0.25	0.30	0.35
K_u/K_y	21.5	11.5	8.2	6.6	5.6	4.9	4.5

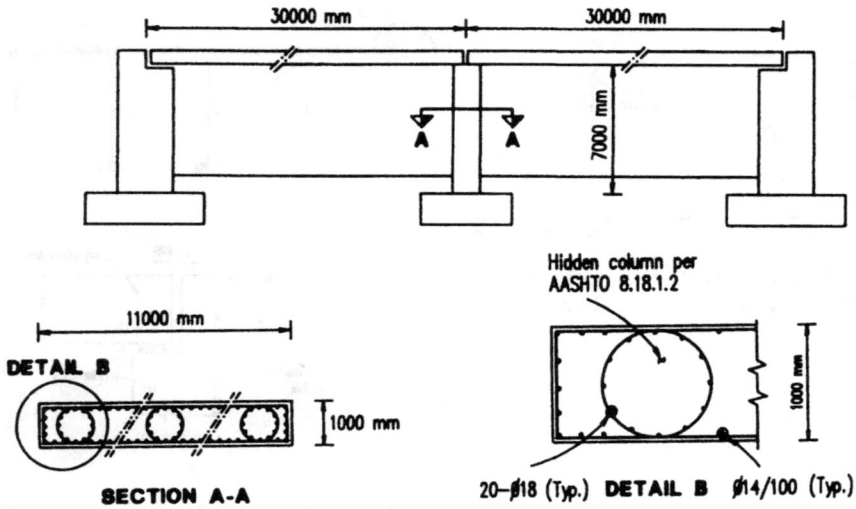

Figure 5. Typical underpass of the design example.

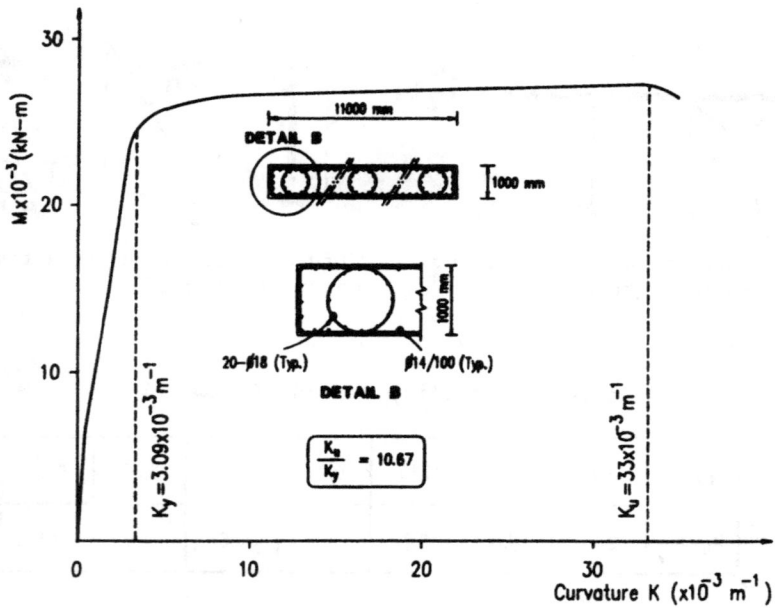

Figure 6. N-M-K relationship of pier used in the example.

Modified response spectrum approach for multiply-supported secondary systems

A. Saudy[I], A. Ghobarah[II], and T.S. Aziz[II]

ABSTRACT

An alternative technique has been developed to evaluate the ordinates of the Cross Cross Floor Spectra (CCFS). The technique properly accounts for the dynamic interaction, tuning and non-classical damping characters of the combined Primary-Secondary (P-S) systems. The approach can estimate the peak response of the tuned, non-classically damped P-S systems more accurately. In the analysis, two fictitious oscillators are attached to the primary system in the course of evaluating the ordinates rather than attaching only one oscillator to the primary system as was previously suggested. A model for the combined P-S systems is analyzed by the original and the proposed techniques. The results are compared with the response values obtained using coupled dynamic analysis. The proposed technique proved to be more accurate in estimating the peak response of the secondary system, specially in cases of tuned, non-classically damped P-S systems.

INTRODUCTION

In industrial facilities and nuclear power plants, relatively light structures are normally attached to heavier ones. Normally, the lighter structures are considered as secondary systems to the supporting structures which are the primary systems. The secondary systems are generally attached to the primary ones at several attachment points. The seismic behaviour of multiply supported Multi-Degree-Of-Freedom (MDOF) secondary systems has received considerable attention due to the vital role such systems play in regard to safety. To investigate the seismic behaviour of multiply supported secondary systems two aspects need to be addressed. Firstly, the dynamic characteristics of the combined primary-secondary (P-S) system have to be accounted for. These characteristics include : dynamic interaction, tuning, non-classical damping and spatial coupling. Secondly, in addition to the different support accelerations, the attachment points normally undergo differential support motions. This, in turn, will lead to increased stresses in the secondary system.

In general, the different approaches adopted in the seismic analysis of MDOF multiply supported secondary systems can be classified as coupled and uncoupled analyses. In theory, the exact response of a general secondary system can be obtained by using standard methods of coupled dynamic analysis for the combined P-S system. Due to many practical difficulties in carrying out a coupled dynamic analysis,

---

[I]Graduate student, Department of Civil Engineering and Engineering Mechanics, McMaster University, Hamilton, Ontario.

[II]Professor, Department of Civil Engineering and Engineering Mechanics, McMaster University, Hamilton, Ontario.

an uncoupled approach is traditionally adopted, (Amin et al, 1971; Lee et al, 1983; Shaw, 1975; Vashi, 1975). In the uncoupled analysis, complex structures are commonly subdivided into two subsystems; primary and secondary; then, analyzed separately. Such analysis completely ignores the dynamic characteristics of the combined P-S system. Moreover, the seismic responses of the secondary system are partitioned into two components; dynamic and pseudo-static.

Recently, the Cross Cross Floor Spectrum (CCFS) technique has been developed based on the principles of random vibration and stochastic analysis, (Asfura et al, 1986). Although the technique is based on an uncoupled analysis approach, it attempts to account for some of the dynamic characteristics of the combined P-S system. This approach proved to be inaccurate in predicting the response of tuned, non-classically damped combined P-S systems.

The objectives of this study are to manifest the sources of error in the original CCFS approach and to propose a modified CCFS technique that properly accounts for the dynamic interaction and tuning.

STATEMENT OF THE PROBLEM

Consider the model for the combined P-S system subjected to base excitation $u_g(t)$ as shown in Fig. 1. It is assumed that both the primary and secondary systems are linear elastic, viscously and classically damped. The secondary system is attached to the primary system at various points. The primary system would respond to the base excitation differently at the attachment points. Accordingly, the secondary system will be subjected to different or multiple accelerations. These accelerations are normally different in both phase and amplitude. In addition, the attachment points will exhibit differential movements which would cause stresses in the secondary systems.

Based on the principles of random vibration and stochastic analysis, the CCFS approach has been developed by Asfura and Der Kiuregian (1986). In their work, they employed the idea of attaching two fictitious oscillators that have two frequencies of those of the secondary system; i.e. ω_i and ω_j, at two support points (floors) of the primary system; i.e. K and L, Fig. 2. Accordingly, modal combination rules were suggested to predict responses at the rth degree of freedom of the secondary system independently on the primary system. These rules lead to the mean of the peak acceleration at the rth degree of freedom of the secondary system as;

$$E\ [\ddot{U}_{r,\max}] - [\sum_{i-1}^{n} \sum_{j-1}^{n} a_{ri}\ a_{rj} \sum_{K-1}^{n_s} \sum_{L-1}^{n_s} b_{iK}\ b_{iL}\ S_{KL}^{a}(\omega_i,\ \xi_i;\ \omega_j,\ \xi_j)]^{1/2} \tag{1}$$

In a similar manner, the mean of the peak relative displacement of the rth degree of freedom is given by

$$E\ [v_{r,\max}] - [\sum_{i-1}^{n} \sum_{j-1}^{n} a_{ri}\ a_{rj} \sum_{K-1}^{n_s} \sum_{L-1}^{n_s} b_{iK}\ b_{iL}\ S_{KL}^{v}(\omega_i,\ \xi_i;\ \omega_j,\ \xi_j)]^{1/2} \tag{2}$$

where

$$a_{ri} - \frac{\phi_{ri}}{m_i\ \omega_i^2} \tag{3}$$

and

$$b_{iK} - \sum_{m-1}^{n} \phi_{mi}\ k_{c_{mK}} \tag{4}$$

$S_{KL}(\omega_i, \xi_i; \omega_j, \xi_j)$ is the ordinate of a "cross-oscillator cross-floor" response spectrum associated with the motions of the Kth and L^{th} floors of the primary system, Fig. 3. In other words, it is the mean of the peak response associated with the covariance of the responses of the two oscillators. ϕ_i, ω_i, ξ_i and m_i are the mode shape, modal circular frequency, modal damping factor and modal mass associated with the ith mode of vibration of the secondary system. K_c is the conventional coupling stiffness matrix between the two subsystems.

Theoretically, the ordinates of the CCFS are evaluated directly in terms of

- the input ground response spectrum, and
- the modal properties of the primary system.

The problem could be stated here as : how to evaluate such ordinates so that the dynamic characteristics of the combined P-S system could be properly accounted for?

In the original approach, a technique was suggested to evaluate the ordinates of the CCFS which account for the interaction and tuning effects. A mass value has been assigned to each oscillator. The mass value is calculated to bring about a shift in the nearest frequency of the primary system similar to that which actually takes place in the combined P-S system. Thus, according to a formula which is based on a tuning criterion, (Igusa et al, 1983), the mass values for the different oscillators are determined depending on the attachment point of each oscillator. Suppose that "N" is the number of degrees of freedom of the primary system, "n" is the number of degrees of freedom of the secondary system and "n_a" is the number of the attachment points supporting the secondary system. Thus, in order to evaluate a CCFS ordinate, an (N+2) DOF system, as that shown in Fig. 2, is studied. The (N+2) DOF system was replaced in the original approach by two (N+1) DOF systems, Fig. 4. Accordingly, ($n \times n_a$) different systems are analyzed. Each of these systems consists of the original primary system and an oscillator representing one of the secondary system modes. Thus, each system is an (N+1) DOF system. The effect of the non-classical damping character has been approximately accounted for based on another tuning formula, (Igusa et al, 1983). Finally, a modal combination rule for evaluating the CCFS ordinates $S_{KL}(\omega_i, \xi_i; \omega_j, \xi_j)$ is developed. In this rule, a correlation coefficient, (Der Kiureghian, 1980), that accounts for the cross modal correlation between the two (N+1) DOF systems is employed.

It was observed that the CCFS approach gives accurate results in case of detuned secondary systems whether the damping is classical or non-classical. In case of tuned secondary systems, the CCFS approach overestimates the response. The error is greatly increased in the tuned, non-classical cases, (Asfura et al, 1986). A more accurate technique has to be developed in order to account properly for the compound effect of tuning and non-classical damping.

ALTERNATIVE TECHNIQUE FOR EVALUATING CCFS ORDINATES

It is believed that the major sources of error arising in case of adopting the original technique to analyze tuned P-S systems could be attributed to the negligence of the great dynamic interaction in case of tuning. Although, the interaction effect is approximated by assigning mass values to the oscillator in each (N+1) DOF system, it is believed that, still in cases of tuned P-S systems, that effect is not considered properly. The alternative technique presented here is based on the fact that the (N+2) DOF system that has two oscillators with equal frequency can not be replaced with two similar (N+1) DOF systems. It is clear that the dynamic interaction between the two oscillators themselves is completely neglected if the technique of the (N+1) DOF system is followed. Moreover, the multiple tuning situation which arises due to coincidence of frequencies of the two oscillators and one (or more) of the frequencies of the primary system has also been ignored. To account for those neglected effects, the original (N+2) DOF system rather than the two (N+1) DOF systems has to be adopted in evaluating the CCFS ordinates.

To account for both the interaction and tuning effects, the idea of assigning equivalent mass values to the oscillators is again adopted. For the case of two oscillator with detuned frequencies, a mass value is assigned to each oscillator. Each mass value is equal to the that of a corresponding oscillator in an (N+1) DOF system. It has to be mentioned that this (N+1) DOF system is composed of the primary system to which is attached the oscillator at the same floor level as that in the analyzed (N+2) DOF system. For the case of two oscillators with tuned frequencies, each mass value is related to that of the corresponding (N+1) DOF system with a reduction factor (α). The factor (α) is introduced to account for the tuning effect between the two oscillators and the multiple tuning between the two oscillators and the primary system in case of tuned P-S systems. Several cases were analyzed to quantify the reduction factor (α). It was found that a value that ranges between 0.9 and 1.0 is suitable for the cases of tuned P-S combined systems. For the cases of detuned P-S systems, a value of 0.75 could be used.

Accordingly, the proposed technique for evaluating the CCFS ordinates can be summarized in the following two steps:

1 - Determination of the dynamic modal properties of the (N+2) DOF systems. The modal frequencies and mode shapes are determined through direct analysis rather than employing perturbation techniques in order to achieve better accuracy. A reduction factor (α) is applied to the equivalent masses assigned to the two oscillators when they have equal frequencies.

2 - Determination of the cross cross floor spectrum (CCFS) ordinates utilizing a modal combination rule to combine the modal responses of the $(N+1)^{th}$ and $(N+2)^{th}$ degrees of freedom in each (N+2) DOF system.

NUMERICAL EXAMPLE

A model for the combined P-S systems is selected for analysis in order to examine the validity of the proposed (N+2) technique for evaluating the ordinates of the CCFS. The same model was analyzed before by Asfura and Der Kiureghian, (1986) subjected to an idealized ground response spectrum defined as

$$S_a(\omega, \xi) = g \; [\frac{\pi \omega}{2000\xi}]^{1/2} \tag{5}$$

The idealized acceleration ground response spectrum and the N-S component of ElCentro earthquake are employed as seismic inputs.

For comparison purposes, the theoretically "exact" responses have been obtained as well as the responses determined following the original (N+1) technique. The "exact" responses are determined by a coupled analysis of the combined P-S system.

A schematic representation of the model is shown in Fig. 5. The properties of the primary system are given in the same figure. Four cases are considered to account for both effects of tuning and non-classical damping. These cases are detuned, classically damped; detuned, non-classically damped; tuned, classically damped and tuned non-classically damped. Two different sets of mass and stiffness ratios were adopted to achieve the tuned and detuned cases. The sets of mass and stiffness ratios, the frequencies of the primary system and the modal damping factors for the two subsystems are tabulated in Table 1 for the all four cases.

Tables 2 and 3 summarize the estimated peak acceleration and displacement responses at the nodes of the secondary system subjected to the idelized ground response spectrum and to ElCentro earthquake. The coupled analysis as well as the two CCFS techniques are employed to determine the peak response of the model. The percentages of error in estimating the peak responses by the (N+1) and (N+2) technique are also tabulated. For the detuned cases, the reduction factor (α) is assumed 0.75 while for the tuned cases, (α) is assigned to 0.9 and 1.0 for classical and non-classical damping respectively.

It can be observed that the percentages of error in estimating the peak responses of the secondary system following the (N+1) technique is large, specially for the tuned, non-classically damped cases. The error percentage reaches 12% in the analyzed model when subjected to the idealized ground response spectrum. The error percentage exceeds 50% when subjected to ElCentro earthquake. It is also noticed that a more accurate response could be achieved by following the proposed (N+2) technique. The error percentages drops to less than 5% when the model is analyzed under the idealized ground response spectrum. When the model is analyzed under the ground acceleration history of ElCentro earthquake, the error percentages drops to about 20%. Employing the reduction factor (α) in the tuned, non-classically damped cases greatly improves the predicted peak responses. In other words, using the (N+2) technique implies that more refined ordinates of the CCFS could be developed. In the mean time, the detuned cases indicate that a reduction factor (α) of approximately 0.75 is essential to get accurate predictions of the peak response by the (N+2) technique. The percentages of error in those cases are comparable to those in the corresponding cases analyzed by the (N+1) technique.

CONCLUSIONS

A modified CCFS approach that accounts for the dynamic interaction, tuning and non-classical damping was presented. An alternative technique for evaluating the ordinates of the cross cross floor spectra has been developed. While the original technique is based on analyzing a number of (N+1) DOF systems, the proposed technique is based on the analysis of a number of (N+2) DOF systems. Neglecting the tuning between the two oscillators themselves and the multiple tuning situation between the two oscillators and the primary system were found to be the major sources of error in estimating the response of tuned, non-classically damped combined P-S systems. The estimated peak responses of the tuned, non-classically P-S systems have been greatly improved by using a reduction factor (α) applied to the equivalent masses assigned to the two oscillators when they have equal frequencies. The reduction factor (α) ranges between 0.9 and 1.0 for the tuned P-S systems. For the detuned systems, the reduction factor (α) may be assigned a 0.75 value. The percentages of error in case of tuned P-S systems significantly drop as a result of using this modified CCFS. This behavior is true whether the seismic input is in the form of a ground response spectrum or a ground time history.

REFERENCES

Amin M., Hall W. J., Newmark N. M. and Kassawara R. P. 1971. Earthquake response of multiply connected light secondary sysems by spectrum methods. ASME 1st Nat. Congress on Pressure Vessels and Piping, San Francisco.

Asfura A. and Der Kiureghian A. 1986. Floor Response Spectrum Method for Seismic Analysis of Multiply Supported Secondary Systems. Earthq. Eng. & Struc. Dyn. (14), 245-265.

Der Kiureghian A. 1980. Structural Response to Stationary Excitations. J. Eng. Mech., ASCE, 106(EM6), 1195-1213.

Igusa T. and Der Kiureghian A. 1983. Dynamic Analysis of Multiply Tuned and Arbitrarily Supported Secondary Systems. Report No. UCB/EERC-83/07, University of California, Berkeley.

Lee M. C. and Penzein J. 1983. Stochastic Analysis of Structures and Piping Systems Subjected to Stationary Multiple Support Excitations. Earthq. Eng. & Struc. Dyn. (2),99-110.

Shaw D. E. 1975. Seismic Structural Response Analysis of Multiple Support Excitation. Proc. 3rd Int. Conf. on Structural Mechanics in Reactor Technology (SMiRT), K7/3, London.

Vashi K. M. 1975. Seismic Spectral Analysis of Structural Systems Subjected to Nonuniform Excitations at Supports. Proc. 2nd ASCE Specialty Conference on Structural Design of Nuclear Power Plants Facilities, (1-A), 188-211, New Orleans, Louisiana.

Table 1. Properties of the studied model

	Primary System Properties	Secondary System Properties			
		Detuned Classical Damped	Detuned Non-Classical Damped	Tuned Classical Damped	Tuned Non-Classical Damped
Frequencies in (rad/s)	4.025	16.106	16.106	4.025	4.025
	11.750	22.361	22.361	5.588	5.588
	18.522	33.993	33.993	8.494	8.494
	23.794	38.730	38.730	9.678	9.678
	27.139	45.662	45.662	11.410	11.410
Modal Damping Factor	0.05	0.05	0.02	0.05	0.02
m/M	-----	0.02	0.02	0.03203	0.03203
k/K	-----	0.05	0.05	0.005	0.005

Table 2. Estimated peak response
(subjected to the idealized ground response spectrum)

	DOF	Coupled	CCFS				Coupled	CCFS			
			(N+1)		(N+2)			(N+1)		(N+2)	
		Acc. (g)	Acc. (g)	%Error	Acc. (g)	%Error	Dis. (cm)	Dis. (cm)	%Error	Dis. (cm)	%Error
Detuned Classical Damped	1	0.4784	0.467	-2.38	0.475	-0.71	25.60	25.60	0.0	25.70	-0.39
	2	0.4572	0.442	-3.32	0.456	-0.26	23.45	23.45	0.0	23.54	-0.39
	3	0.3956	0.398	0.61	0.398	0.61	20.50	20.50	0.0	20.50	0.00
	4	0.4065	0.401	-1.35	0.406	-0.12	16.87	16.78	-0.58	16.87	0.00
	5	0.3606	0.354	-1.83	0.359	-0.44	12.65	12.46	-1.55	12.56	-0.72
Detuned Non-Classical Damped	1	0.5224	0.505	-3.33	0.519	-0.65	25.70	25.60	-0.38	25.70	0.00
	2	0.5061	0.485	-4.17	0.509	0.57	23.54	23.45	-0.42	23.54	0.00
	3	0.4079	0.409	0.27	0.411	0.76	20.70	20.50	-0.95	20.50	-0.97
	4	0.4588	0.448	-2.35	0.463	0.92	16.97	16.78	-1.16	16.68	-1.71
	5	0.4141	0.400	-3.40	0.410	-0.99	12.75	12.56	-1.54	12.46	-2.28
Tuned Classical Damped	1	1.0093	1.070	6.01	1.010	0.07	67.98	69.85	2.74	67.79	-0.28
	2	1.3863	1.460	5.32	1.360	-1.90	91.63	95.16	3.85	89.96	-1.82
	3	1.1043	1.150	4.14	1.080	-2.20	73.48	75.73	3.07	71.81	-2.27
	4	1.3550	1.410	4.06	1.320	-2.58	86.62	90.45	4.42	84.37	-2.60
	5	0.9365	0.967	3.26	0.908	-3.04	58.96	61.12	3.66	57.29	-2.83
Tuned Non-Classical Damped	1	1.2844	1.420	10.56	1.290	0.44	84.07	90.35	7.47	85.45	1.64
	2	1.7790	1.990	11.86	1.760	-1.07	114.68	126.55	10.35	115.76	0.94
	3	1.4181	1.570	10.71	1.400	-1.28	92.12	101.04	9.69	92.31	0.21
	4	1.7688	1.950	10.24	1.740	-1.63	110.17	122.63	11.31	109.87	-0.27
	5	1.2325	1.340	8.72	1.210	-1.83	75.24	83.29	10.69	74.56	-0.90

Table 3. Estimated peak response
(subjected to ElCentro earthquake)

	DOF	Coupled Acc. (g)	CCFS (N+1) Acc. (g)	%Error	(N+2) Acc. (g)	%Error	Coupled Dis. (cm)	CCFS (N+1) Dis. (cm)	%Error	(N+2) Dis. (cm)	%Error
Detuned Classical Damped	1	0.2592	0.246	-5.09	0.253	-2.39	11.8716	11.8701	-0.01	11.8701	-0.01
	2	0.2217	0.211	-4.83	0.218	-1.67	10.8482	10.8891	0.38	10.8891	0.38
	3	0.2149	0.221	2.84	0.218	1.14	9.5422	9.5648	0.24	9.5648	0.24
	4	0.2675	0.267	-0.19	0.271	1.31	7.9186	7.9460	0.35	7.9559	0.47
	5	0.2685	0.265	-1.30	0.270	0.56	6.0404	6.0430	0.04	6.0528	0.21
Detuned Non-Classical Damped	1	0.2821	0.261	-7.48	0.273	-3.23	11.8742	11.8701	-0.03	11.8701	-0.03
	2	0.2473	0.229	-7.40	0.245	-0.93	10.8509	10.8891	0.35	10.8891	0.35
	3	0.2184	0.224	2.56	0.224	2.56	9.5425	9.5648	0.23	9.5648	0.23
	4	0.2874	0.280	-2.57	0.291	1.25	7.9210	7.9461	0.32	7.9657	0.56
	5	0.2873	0.276	-3.93	0.285	-0.80	6.0428	6.0430	0.00	6.0528	0.17
Tuned Classical Damped	1	0.3825	0.509	33.07	0.420	9.80	20.8457	31.0977	49.18	26.5851	27.53
	2	0.4854	0.693	42.77	0.560	15.37	26.1432	42.8697	63.98	35.2179	34.71
	3	0.3782	0.548	44.90	0.446	17.93	20.4017	34.0407	66.85	28.1547	38.00
	4	0.4901	0.680	38.75	0.565	15.28	25.0189	41.3001	65.08	33.8445	35.28
	5	0.3529	0.475	34.60	0.402	13.91	17.1287	27.9585	63.23	23.0535	34.59
Tuned Non-Classical Damped	1	0.4841	0.598	23.53	0.473	-2.29	25.2169	35.0217	38.88	28.8414	14.37
	2	0.6870	0.833	21.25	0.665	-3.20	35.6884	49.6386	39.09	40.1229	12.43
	3	0.5648	0.670	18.63	0.546	-3.33	28.8765	39.5343	36.91	32.3730	12.11
	4	0.7464	0.836	12.00	0.707	-5.28	37.8870	48.9519	29.21	40.3191	6.42
	5	0.5313	0.586	10.30	0.506	-4.76	26.0791	33.2559	27.52	27.6642	6.08

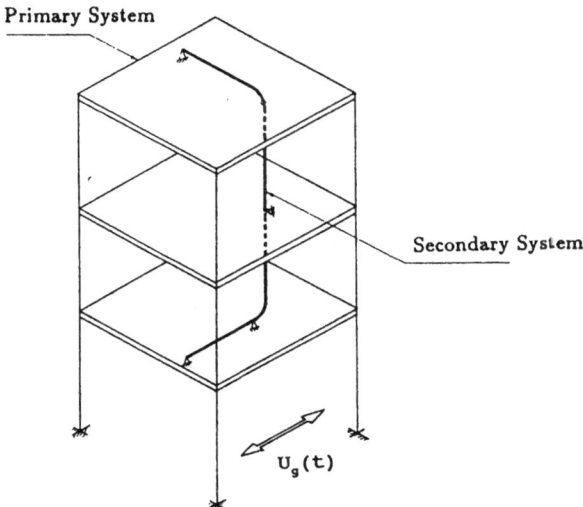

Figure 1.   Combined Primary-Secondary (P-S) Systems

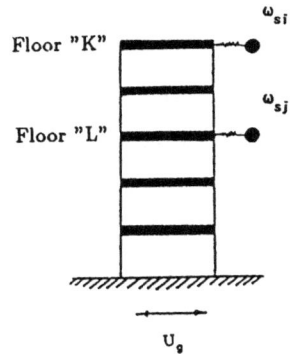

Figure 2.   Primary-Double Oscillator System

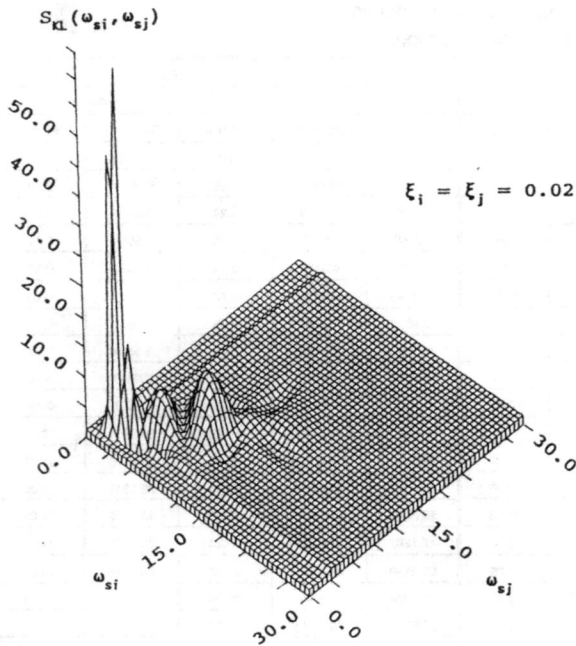

Figure 3.   Cross Cross Floor Response Spectrum (CCFS)

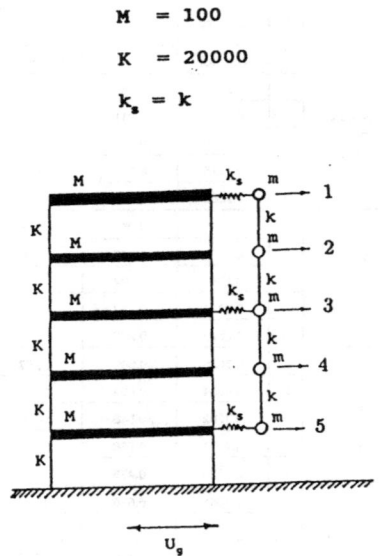

$\xi_i = \xi_j = 0.02$

$M = 100$

$K = 20000$

$k_s = k$

Figure 5.   Model Considered in The Analysis

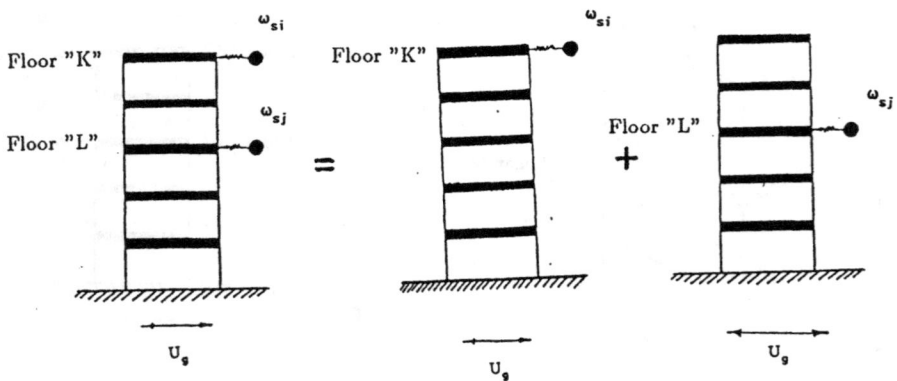

Figure 4.   Replacement of The (N+2) DOF System with Two (N+1) DOF Systems

Parameter studies of statically tilted unanchored cylindrical tanks

David T. Lau[I] and Ray W. Clough[II]

ABSTRACT

In recent years many thin-shell cylindrical tanks have been damaged during earthquakes. The seismic uplift response problem of unanchored tanks has been studied under the simpler static tilt condition. This paper quantifies the sensitivity of the uplift behaviour to some of the important paramters which characterize the tank system, namely, the tilt angle, the height-to-radius ratio, the tank shell and bottom plate thickness, the stiffening effect of the top rim wind girder and bottom toe ring, and the roof type. The discussion on the uplift resistance mechanism is based on the response quantities related to observed seismic tank damage, such as the vertical uplift displacement, the extent of uplift in the bottom plate, and the tank shell membrane and bending stresses near the base.

INTRODUCTION

Thin shell cylindrical tanks are very efficient structures, and have applications in almost every major industry. They are simple in form and relative easy to construct. However, in recent years, many cylindrical tanks have been damaged during earthquake, sometimes with serious environmental ramifications, thus raising concerns about the seismic safety of these structures.

In the design standards AWWA (1984) and API (1988), which cover the design and fabrication of welded cylindrical steel tanks, the static strength design is based on the allowable hoop stress from the membrane theory. However, the seismic behaviour of cylindrical tanks is very different from the static behaviour, in particular for the unanchored ones. The seismic resistance mechanisms involved were not until recently fully understood. Based on field studies of seismic damages (Benuska ed. 1990), and observations from shaking table tests (Manos and Clough 1982), it has been determined that an unanchored tank uplifts significantly at the edge when subjected to even moderate horizontal base motion. When the tank liquid content is subjected to horizontal strong ground motion, hydrodynamic forces are generated acting on the tank wall and bottom plate, in addition to the hydrostatic pressure. A lateral resultant force and overturning moment develops at the tank base due to the liquid inertia effect and sloshing, and the tank uplifts on one side in response to this overturning moment.

[I]Assistant Professor, Dept. of Civil Engineering, Carleton University, Ottawa, Ont. K1S 5B6
[II]Nishkian Professor of Structural Engineering, Emeritus, Univ. of Calif., Berkeley, CA 94720.

STATIC TILT UPLIFT

In general, the dynamic uplift mechanism of an unanchored tank is a very complex nonlinear problem. Recognizing the many uncertainties in the seismic behaviour of cylindrical tanks, the uplift problem has been studied here under the simpler static tilt condition (Clough and Niwa 1979, Manos and Clough 1982, Lau and Clough 1989, Peek 1986, Haroun et al. 1990) produced by tilting the support platform from its initial horizontal position as shown in Fig.1. The partially filled tank thus is subjected to an effective static lateral acceleration component which simulates the effect of the horizontal acceleration component of an earthquake. The resultant static lateral force and overturning moment caused uplifting of the unanchored tank on one side, as shown in Fig.2. This deformation has all the important characteristics of the dynamic uplift response, and makes it possible to develop understanding of the uplift mechanism involved. As noted in the figure, there is separation of the bottom plate from the supporting foundation even on the contact side, because of the base joint rotation caused by the hydrostatic pressure pushing out on the tank wall. A similar condition applies axisymmetrically around the tank bottom in the initial horizontal position. Details of the static tilt uplift behaviour have been discussed in an earlier paper (Lau and Clough 1989).

This paper quantifies the sensitivity of the uplift behaviour to various parameters characterizing the tank system based on the study of the static tilt uplift model.

METHOD OF ANALYSIS

The tank system considered in the present study consists of a cylindrical shell of radius a and height t_l, partially filled to a depth of d_w. The thickness of the cylindrical shell and the bottom plate are h_t and h_b respectively. As specified in the design standards, the bottom plate protrudes 2 in. (50.8 mm) outside the shell wall. The cylindrical tank also has a top rim stiffening wind girder. The support platform is statically tilted to an angle ϕ with the horizontal plane.

An effective analysis procedure based on the substructure concept has been developed to study the static tilt behaviour. Putting the method in the finite element context, the uplifting cylindrical tank is divided analytically into substructure super elements, namely the cylindrical shell, the bottom plate, the top wind girder and the bottom toe ring. Each element is formulated by Ritz discretization using derived displacement shape functions. Large displacement theory is employed in the modeling of the bottom plate, of which both the membrane and flexural mechanisms are considered. A nonlinear equilibrium equation is obtained by applying the principle of virtual work, which is then solved by iteration using current updated tangent stiffness. The bottom plate contact pattern is also established by iteration. The details of the formulations can be found in the references (Lau and Clough 1989; 1990).

PARAMETER STUDIES

In general, the geometry of a cylindrical tank is characterized by its height-to-radius (aspect) ratio. The top may be open with a flexible floating roof, or it may have a fixed welded roof typically of conical shape. The flexibility of the tank system is related to, in addition to the tank geometry, the plate thickness and the stiffening effect of the wind girder.

The critical parameter variations considered in the present study are the height-to-radius ratio, the plate thickness of both the tank shell and the bottom plate, the roof type and the stiffening effect of the wind girder and bottom toe ring. The influences of these parameters on the uplift behaviour of unanchored tanks are studied.

Tilt Angle

The static uplift loading, induced by tilting the tank rigid support platform, is directly related to the tilt angle, which may be looked upon as related to the intensity of the ground motion in the dynamic case. Therefore by increasing the tilt angle in static tilt analysis, the results obtained should give an indication about the sensitivity of the unanchored tank uplift mechanism to the ground motion intensity.

Static tilt uplift analysis results are presented for a broad 1/3 scale welded aluminum model tank, tested in a previous investigation (Manos and Clough 1982). The model tank, denoted here as the standard case, is 0.08 in. (2.03 mm) thick, 12 ft. (3.66 m) in diameter and 6 ft. (1.83 m) high, with a height-to-radius ratio of one. In the present study, the static tilt angle varies from 0 to 16 degrees with 5 ft. depth of water in the tank.

The variations with tilt angle of the maximum base uplift (Pt.A in Fig.2), the extent of the bottom plate uplift (e in Fig.2), the tank shell membrane stresses at the contact pivot point (Pt.B in Fig.2), and the bending stresses on the uplifted side (Pt.A in Fig.2), are presented in Figs.3(a)-(d) respectively.

It can be deduced from Figs.3(a) and (b) that the overturning moment, induced by the unbalanced liquid pressure distribution and directly related to the tilt angle, must reach a minimum level before any base rim uplift can occur. This minimum level of overturning moment is necessary to overcome the liquid pressure acting on the tank bottom, which resists the tendency to uplift. Fig.3(b) also clearly indicates that the maximum allowable uplift extent of 7% of the radius assumed in the design standards does not seem to correlate with the analysis results.

Studies of seismic tank damage show that many tanks failed because of buckling of the tank shell near the base rim, due to the high concentration of compressive axial membrane stress along the contact portion of the base rim. In Fig.3(c), the compressive membrane stresses are normalized by the tensile membrane hoop stress of a similar but free cylindrical shell subjected to the liquid pressure loading in the initial horizontal position. The shell design provision specified in the design standards is largely based on this reference hoop membrane stress. As shown in the figure, the hoop membrane stress increases more rapidly than the axial stress at the contact point, suggesting that perhaps at large tilt angle, the hoop stress may become an important factor in determining the buckling strength of the tank shell.

Another common seismic failure mode is the tearing or rupture at the bottom plate–tank shell connection. This kind of failure may be initiated by the high bending moment and curvature near the base of the tank shell caused by the restraining action of the bottom plate (Fig.3(d)). It is noted here that the magnitude of the bending stresses at the joint are at least one order of magnitude higher than that of the membrane stresses. Consequently, the effect on the critical buckling load of the tank shell curvature resulting from flexural deformation should also be considered.

Height-to-Radius (Aspect) Ratio

Perhaps the single most important factor affecting the seismic behaviour of an unanchored tank system is its geometry, described by the height-to-radius ratio parameter. From the experience of seismic damage study and shaking table tests, tall tanks typically have a higher tendency to uplift as compare to broad tanks, and thus are more susceptible to seismic damage. The following parameter study seems to agree with this observation. In the present study, the height-to-radius ratio varies from 0.5 to 5, keeping the capacity of the tank and the liquid volume constant. The variations of the same response parameters with the aspect ratio are presented in Figs.4(a)-(d).

As noted in the figures, there is a sudden increase in the uplift response of tanks with height-to-radius ratio between 2 and 3. However, this sudden change in the behaviour does not seem to reflect as significant in the relation between the extent of bottom plate uplift and the aspect ratio shown in Fig.4(b). It is noted in Fig.4(c) there is a corresponding drop in the tank shell membrane stresses on the contact side, possibly due to the combined effects of the lateral displacement and the out-of-round deformation in the tall tank. In comparison, the bending stresses presented in Fig.4(d) increase monotonically with the aspect ratio. In summary, it seems that the height-to-radius ratio of 2 may be conveniently adopted to classify tanks into the two categories of broad and tall tanks, of which tall tanks are more sensitive to lateral acceleration with larger uplift deformations.

Bottom Plate and Tank Shell Thickness

In this parameter study, the tank shell thickness is first kept unchanged, while the bottom plate thickness is modified; then both the tank shell and the bottom plate thickness are varied simultaneously. Because there is no design strength requirement for the bottom plate under the horizontal static condition, but the design standards specify that the bottom plate thickness shall not exceed the bottom shell course thickness, hence the bottom plate thickness is reduced from the standard case of 0.08 in. (2.03 mm) to a smaller value of 0.03 in. (0.76 mm) in order to study the relative contribution to the uplift resistance mechanism derived from the bottom plate. The results are presented in Figs.5(a)-(d).

In both cases, uplift increases as the structure becomes more flexible when the plate thickness is reduced. However, the extent of the bottom plate that uplifted does not change significantly with the reduction in the plate thickness. Similar behaviour is observed for the tank shell axial membrane stress shown in Fig.5(c). On the contrary, the membrane hoop stress increases substantially when a thinner bottom plate is chosen. The results presented in Fig.5(d) seems to indicate that reducing the bottom plate thickness alone tends to decrease the flexural deformation of the tank shell, whereas reducing also the tank shell thickness will have the opposite effect. In summary, a thicker and stiffer bottom plate will reduce the amount of vertical uplift around the base rim, but at the same time will also increase the bending stresses at the bottom plate–tank shell connection, due to the often neglected interaction between the two components.

Stiffening Wind Girder, Bottom Toe Ring and Roof Type

The stiffening effect of the wind girder at the top rim and toe ring around the tank base is studied by varying the stiffness of the two elements in the analysis model.

In summary, the wind girder has the effect of reducing the top rim out-of-round deformation, which is closely related to the warping distortion at the tank base. It should be noted here that the vertical component of the base uplift deformation is essentially similar to the warping distortion of the tank shell. Therefore, a stiffer wind girder increases the overall uplift resistance of the tank system and reduces the amount of uplift. Analytically, a tank roof with rigid diaphragm can be simulated by very stiff wind girder, essentially eliminating the top rim out-of-round deformations.

The only significant effect of increasing the bottom toe ring stiffness is the reduction of the tank shell membrane hoop stress through the reduction of the tank shell out-of-round deformations at the base. As discussed before, this reduction in hoop stress may affect the buckling strength of the tank shell, even though the buckling criterion is often thought to be associated largely with the magnitude of the compressive axial stress.

CONCLUSIONS

The uplift behaviour of unanchored tanks has been investigated in previous studies using the static tilt model. This paper quantifies the sensitivity of the uplift behaviour to some of the important parameters which characterize the tank system. It is found that important aspects of the uplift mechanism have not been adequately considered in the design standards. For instance, the extent of uplift in the bottom plate clearly exceeds the maximum allowable assumed in the design standards even for a moderate tilt angle. It is also found that there is a substantial increase in stresses and displacements associated with uplift when the tank's height-to-radius ratio is increased from 2 to 3. It is also shown the relative thickness of the bottom plate and the tank shell base course has a significant effect on the uplift behaviour. Furthermore, tanks with a rigid roof have greater uplift resistance than open-top tanks with flexible floating roof due to limiting the interaction between the bottom plate and tank shell. Finally, through this parameter study, a better understanding of the uplift behaviour has been achieved, which provides a basis for solving the dynamic uplift problem. A more detailed paper on the parameter study of unanchored tank uplift is currently in preparation.

REFERENCES

AWWA D100-84. 1984. AWWA Standard for Welded Steel Tanks for Water Storage. American Water Works Association, Denver, Colorado.

API Standard 650. 1988. Welded Steel Tanks for Oil Storage. American Petroleum Institute, Washington, D.C.

Benuska,L. ed. 1990. Loma Prieta Earthquake Reconnaissance Report. Earthquake Spectra, EERI, supplement to vol.6.

Haroun,M.A., Hossam,S.B., Bains,G.P. 1990. Static Tilt Analyses of Unanchored Tanks. Proceedings Fourth U.S. National Conference on Earthquake Engineering, Palm Springs, California, EERI, vol.3, 157-166.

Lau,D.T., Clough,R.W. 1989. Static Tilt Behavior of Unanchored Cylindrical Tanks. EERC-89/11, Earthquake Engineering Research Center, Univ. of Calif., Berkeley, CA.

Lau,D.T., Clough,R.W. 1990. Use of Ritz Shape Functions in Analysis of Uplifting Cylindrical Tanks. Proceedings Fourth U.S. National Conference on Earthquake Engineering, Palm Springs, California, EERI, vol.3, 167-176.

Manos,G.C., Clough,R.W. 1982. Further Study of the Earthquake Response of a Broad Cylindrical Liquid-Storage Tank Model. Report No. UCB/EERC-82/07,Earthquake Engineering Research Center, Univ. of Calif., Berkeley, CA.

Clough,R.W., Niwa,A. 1979. Static Tilt Tests of a Tall Cylindrical Liquid Storage Tank. Report No. UCB/EERC-79/06,Earthquake Engineering Research Center, Univ. of Calif., Berkeley, CA.

Peek,R. 1986. Analysis of Unanchored Liquid Storage Tanks under Seismic Loads. EERL Report No.86-01, Calif. Inst. of Tech., Pasadena, CA.

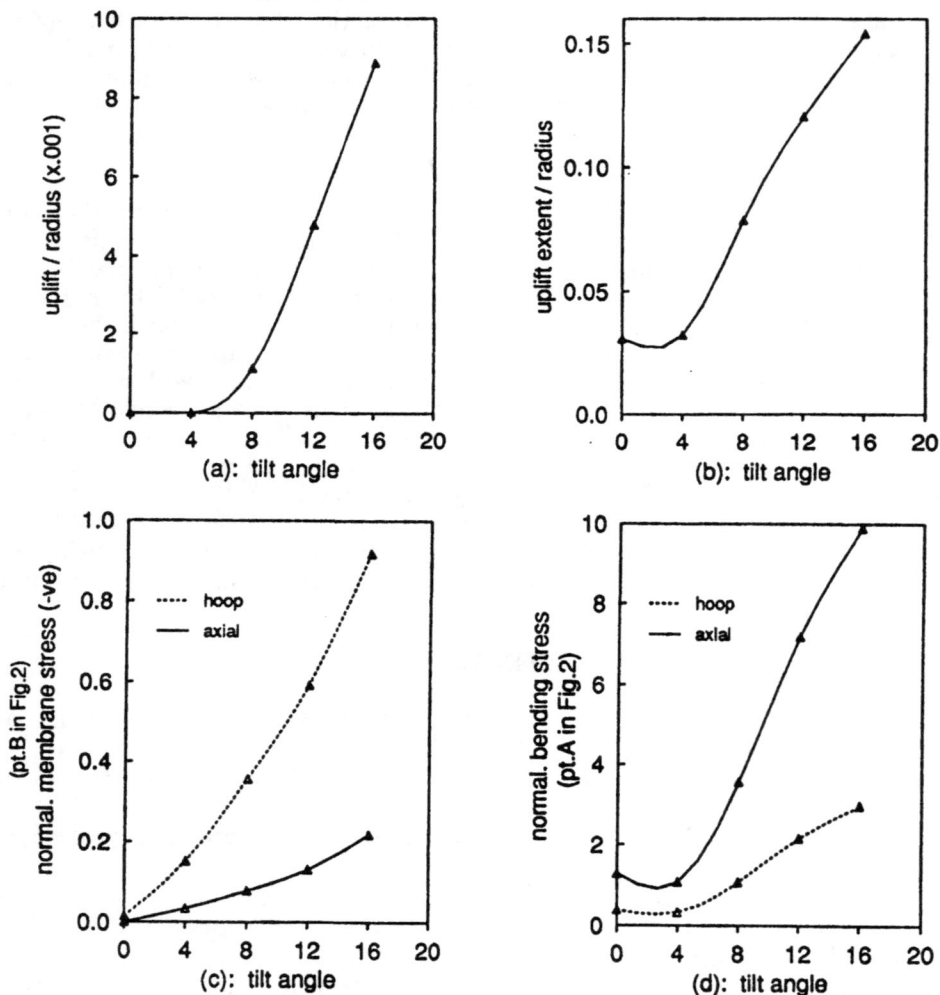

Fig.3. Parameter study of tilt angle

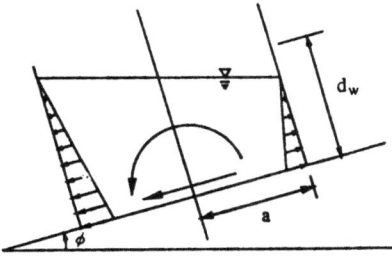

Fig.1. Static tilt tank

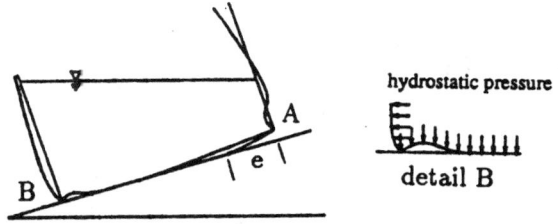

Fig.2. Uplift deformation

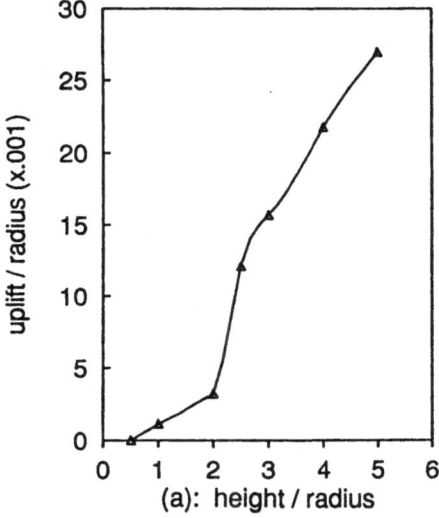

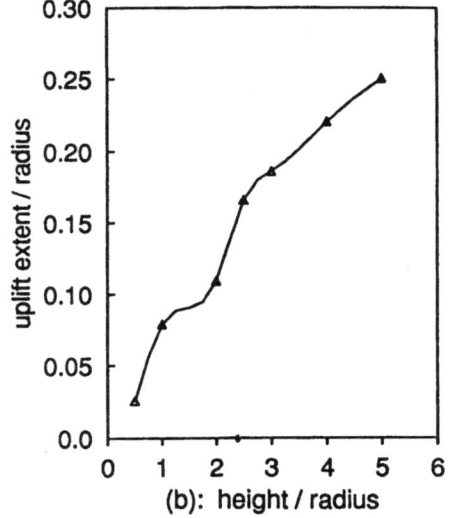

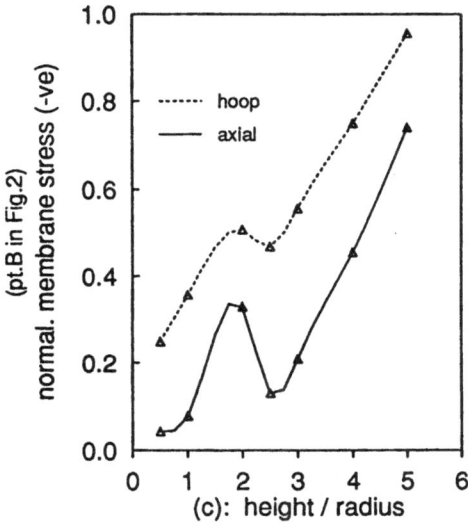

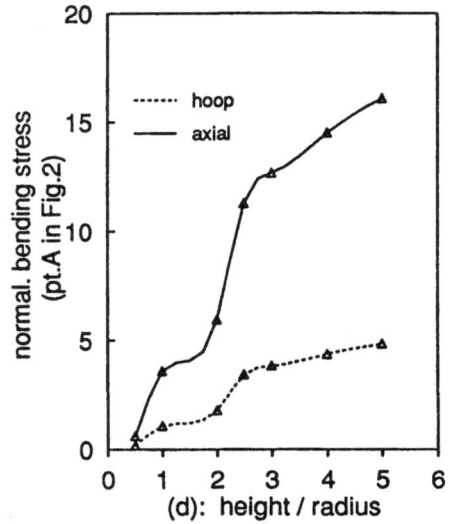

detail B

Fig.4. Parameter study of height-to-radius ratio (8 deg. tilt)

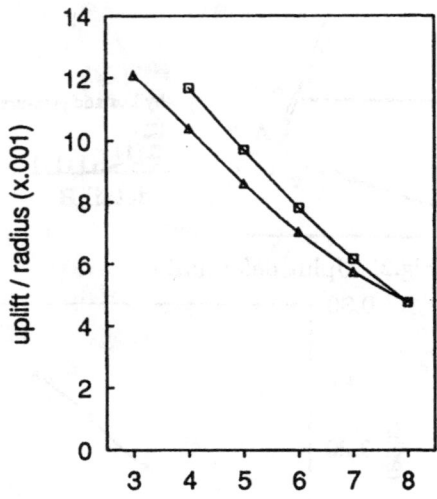

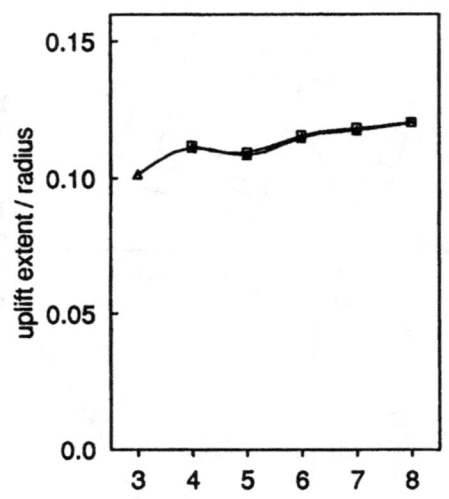

(a): plate thickness (x.01 in)

(b): plate thickness (x.01 in)

△ bottom plate thickness only

□ tank shell & bottom plate thickness

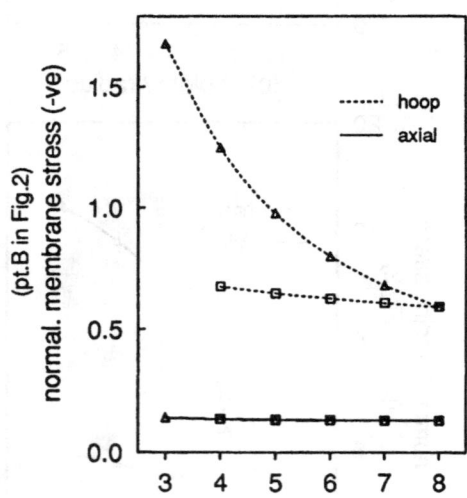

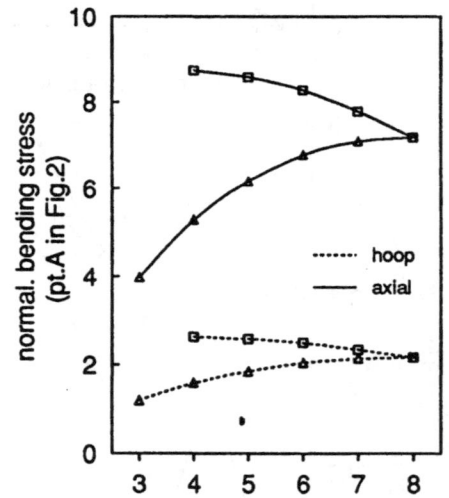

(c): plate thickness (x.01 in)

(d): plate thickness (x.01 in)

Fig.5. Parameter study of plate thickness (12 deg. tilt)

Experimental study on the seismic response of telecommunications equipment supported on access floor

C.M. Wong[1] and W.K. Tso[2]

ABSTRACT

An experimental study was carried out to investigate the dynamic properties and the seismic response of telecommunications equipment supported on access floors. A total of three access floor systems from three manufacturers were used in the test program. A commercially available telecommunications cabinet was used as a representative equipment. The test program consisted of three phases. First, static tests were performed to determine the lateral load-deflection characteristics of the access floors by themselves. Second, shake table tests were performed to determine the frequency and damping characteristics of the equipment cabinet and the combined equipment-access floor systems. Lastly, the equipment and the combined equipment-access floor systems were tested on a shake table under different level of excitation compatible with the building floor response spectra specified in the NEBS criteria of the telecommunications industry. The experimental results showed that the response motion of the equipment cabinet was amplified significantly by the access floor. Equipment units supported on access floors without stringers are more prone to be damaged due to the possibilities of collapse of the access floor in a strong earthquake.

INTRODUCTION

Access floor systems have been widely used to support equipment in telecommunications central offices. There are two major issues associate with the seismic safety of telecommunications equipment on access floors. First, most access floor systems are designed to support mainly gravity load; hence, their seismic performance which is dependent of their lateral strength and stiffness is unknown. Second, an access floor system tends to amplify the earthquake-induced building floor motion; as a result, the equipment supported on it would be subjected to a more severe shaking than if it were supported directly on a building floor. To address to the first issue, static tests were carried out to investigate the lateral strength and stiffness properties of three commercially available access floor systems. The second issue was addressed by carrying out shake table testing using a commercially available telecommunications equipment cabinet on different access floor systems.

In order to perform shake table testing to study the seismic response of the equipment cabinet on different access floors, it is necessary to use an input excitation which represents the upper-bound building floor motions induced by earthquakes. In North America, such input excitation can be derived from the floor

---

[1] Post Doctoral Fellow, Dept. of Civil Engineering and Engineering Mechanics, McMaster University, Hamilton, Ontario.

[2] Professor, Dept. of Civil Engineering and Engineering Mechanics, McMaster University, Hamilton, Ontario.

response spectra shown in Fig. 1a. These spectra are given in the NEBS (Network Equipment Building System) criteria (Bell Communications Research, 1988) which is the most commonly employed industry standard in the telecommunications industry. A typical artificially generated acceleration time history compatible with the NEBS 2% damped spectrum is shown in Fig. 1b. The time history has a duration of 31 seconds and a maximum acceleration of 1 g. This time history is a representative of the potential earthquake-induced floor motion at the top stories of buildings situated in zone 3 of Uniform Building Code. The associate displacement time history obtained from numerical integration of this acceleration time history is shown in Fig. 1c. This displacement time history was used as the input for the shake table in the experiment.

DESCRIPTION OF ACCESS FLOOR SYSTEMS AND EQUIPMENT CABINET

Access floor systems manufactured by three different suppliers (referred as supplier A, B, and C in the subsequent discussions) were used in this study. They all have a finished floor height of 20" (50.8 cm). Access floor systems of suppliers A and B are similar. They consist of an assembly of three components made of galvanized steel, pedestals, stringers and panels. Each pedestal has a square base plate to which a tubular stem is connected by spot-welding. The stringers are bolted onto the pedestals by means of screws. The stringers and the pedestals form a 2' (61 cm) square-grid understructure on which the floor panels are rested. The floor panels can either rest on the understructure or be bolted at its four corners on the understructure by means of screws. The access floor system of suppliers C is a stringerless system. It also has 2' (61 cm) square floor panels which rest on pedestals laid out in a 2' (61 cm) square pattern. The pedestals and panels are made of cast aluminum.

The equipment cabinet used in the study is a commercially available switching equipment used commonly in telecommunications central offices. The electronic gears in the cabinet were replaced by an equivalent amount of mass. The dimensions of the cabinet are 42x26" (107x66 cm) and 6' (1.83 m) in height. The entire cabinet weighs 1660lb (7.42 KN).

TESTING PROGRAM AND SET-UPS

To determine the lateral strength and stiffness properties of the access floors, static tests were done on a single-panel module of each of the access floor systems. An overview of the test set-up and the instrumentation involved is shown in Fig. 2. The lateral load was applied by means of a hydraulic jack. A load cell and two LVDT's (linear variable displacement transducer) were used to monitor the applied load and the lateral displacement of the access floor respectively. The loading history comprised of several cycles of loading and unloading followed by a monotonically increasing loading until the specimen had been failed.

To study the dynamic amplification effect of the equipment cabinet on access floors, the cabinet by itself and the combined cabinet-access floor systems were tested dynamically on a shake table using random excitation and artificially generated earthquake excitation compatible with the 2% damped NEBS spectrum. For the random excitation test, a low level (maximum displacement of 0.05") random shaking was applied to the specimen. The transfer function for the acceleration signal measured by the accelerometers at the top of the cabinet and that at the shake table were monitored by a dual channel spectrum analyzer. The frequencies and damping ratios of the specimen were determined from the transfer function. For the NEBS excitation test, several levels of the NEBS excitation given by the time history shown in Fig.1b were used. The accelerations and displacements at various locations on the specimen and at the shake table were recorded by a microcomputer based data acquisition system.

For the shake table test carried out for the equipment cabinet alone, the equipment was anchored by four expansion fasteners on a concrete slab which was in turns mounted on the shake table. The set-up and instrumentation involved in the shake table tests for the combined cabinet-access floor systems are shown in Fig. 3. A 3 panel by 3 panel access floor was used to support the equipment cabinet. The cabinet's mounting on access floor was achieved by a through-bolting scheme which is shown also in Fig. 3. In this scheme, four 1/2" (12.7 mm) steel rods were passed through predrilled holes on the floor panels and anchored the cabinet directly to the shake table.

STATIC TEST ON ACCESS FLOOR SYSTEMS

The static test results for the three access floors are presented in terms of load-deflection curves shown in Fig. 4. All three systems exhibit linear behaviour at small loading. No well-defined yield point can be found on the three load deflection curves. The overall lateral behaviour of the access floors provided by suppliers A and B are very similar. Both systems failed in a very brittle manner from the tearing of welds which connect the pedestal stems and the base plates. After the initial tearing of the welds, no additional loading could be sustained. The static lateral behaviour of the access floor provided by supplier C was different from the other two. Under lateral load the system exhibits a ductile behaviour with large amount of displacement after yielding. The failure of the system was due to the yielding of the pedestal stems. Fig 5. shows the large distortion of the access floor of supplier C after the ultimate load had been reached.

The lateral stiffness and ultimate strength values for the three access floor systems tested are tabulated in Table 1. The stiffness values are similar for the two access floors with stringers (i.e., system A and B). The values of stiffness and strength of system C are much smaller than the other two systems. The reason for this is twofold. First, the access floor components of system C are made of a weaker material; namely, cast aluminum. Second, no stringer was used in the system to provide the frame action which can enhance the stiffness of the overall system; as a result, the lateral stiffness is only due to the individual stiffness of the pedestals acting as a cantilever.

RANDOM EXCITATION TESTS

Random excitation tests were performed to determine the natural frequencies and damping ratios of the equipment cabinet mounted on a concrete slab. The cabinet was tested in both the side-to-side and the back-to-front directions. The cabinet responded in a single mode in both directions and the frequencies and damping ratios for the two directions are tabulated in Table 2. The cabinet has a higher frequency in the back-to-front direction.

Random excitation tests were also performed for the various combination of the equipment cabinet and the three access floor systems. Two lateral response modes were found for each of the equipment-access floor configurations. Since the second mode frequency is in general higher than 15 Hz which is beyond the frequency range of the upper-bound floor motion described by NEBS, only the fundamental mode results will be discussed here. The fundamental frequencies and the damping ratios for each of the test configurations are tabulated in Table 2.

By comparing the frequency of the equipment cabinet on a concrete slab, it is apparent that the access floors cause the frequencies of the equipment cabinet to be reduced drastically. Additional experimental results for 12" FFH access floors presented in Ref. 2 shows that in general the reduction is larger for a more laterally flexible access floor system (Wong, 1990). The implication of such reduction is that even though the cabinet by itself is well designed such that its fundamental frequency is higher than the frequency range contained in the NEBS specified motion, when it is put on an access floor, its frequencies may be reduced such that it is vulnerable to the NEBS shaking. The cabinet's frequencies on the 20" FFH access floor are indifferent to the orientation of the cabinet although it has a much higher frequency in the back-to-front than the side-to-side orientation when it is supported on a concrete slab. The reason for this is that the equipment cabinet is much stiffer than the 20" access floors; as a result, the equipment cabinet behaves essentially as a rigid mass on the access floors.

NEBS TIME HISTORY TESTS

NEBS time history tests were performed for the equipment cabinet supported on a concrete slab and on the three access floor systems. Shown in Fig. 6 are two acceleration traces for the response at the top of the equipment cabinet. These traces were obtained from the tests done for the equipment cabinet mounted in the side-to-side orientation on a concrete slab and on access floor A. Both acceleration traces were obtained using an input excitation of 25% of the full NEBS time history given in Fig. 1. It can be seen from Fig. 6 that the equipment cabinet's response on the access floors is much larger than that on the concrete slab. The level of amplification of the response depends largely on the fundamental frequencies of the systems considered. The fundamental frequency for the equipment on the concrete slab alone is 8.5 Hz and on the access floor is 4.6 Hz. As indicated by the NEBS spectra in Fig. 1a, a decrease in fundamental frequencies from 8.5 Hz to 4.6 Hz is associated with an increase of spectral acceleration values. As a result, the equipment response is higher when supported on the access floor than on the concrete slab.

The equipment cabinet on the various access floors was tested until the access floors were damaged by gradually increasing the intensity of the NEBS shaking. None of the access floor systems could sustain more than a 50% intensity of NEBS shaking without damage while supporting the equipment cabinet. The access floors of suppliers A and B failed under the NEBS shaking in a similar manner as the failure depicted in the static tests. The failure was due to the tearing of the spot-welds at the base plate of the pedestals. A typical damaged pedestal is shown in Fig. 7. System A can sustain a slightly higher NEBS shaking than system B due to stronger spot-welded based plate connections. Although in some tests more than half the pedestal stems were completely severed from the base plates, no collapse of the access floor occurred because the stringers were holding all the pedestal stems together. Access floor system C failed under the NEBS shaking in a totally different way from the failure observed in the static tests. The failure was due to the sharing off of the pedestal heads as shown in Fig. 8. Since the pedestal heads were the only support for the floor panels, the shearing off of several pedestal heads directly underneath the equipment cabinet led to a collapse of the access floor under the weight of the equipment cabinet. The collapsed access floor can be seen in Fig. 9. Such catastrophic failure was primary due to the lack of a stringer system which stabilizes the pedestal and provides a more even distribution of the imposed seismic force to the pedestals. It was observed in the tests for one of the access floors with a stringer system that in spite of 90% of the pedestal stems were severed from their base plates no collapse occurred.

In order to relate the results of the NEBS time history test to that of the static test, the maximum floor displacements under different level of NEBS shaking are indicated on the static lateral-load displacement curves. Fig. 10 shows the load-deflection curve for the access floor system A. The maximum floor displacement denoted in the figure were obtained from the NEBS test performed for the equipment cabinet in the back-to-front direction on the access floor system A. For the 50% NEBS excitation test where the access floor was damaged, the load induced by the maximum displacement was very close to the ultimate load. The correlation with the dynamic test results implies that it is viable to use the static test results of a simple one-panel access floor module to evaluate the seismic safety margin of an access floor system given that the upper-bound access floor displacement can be estimated with reasonable accuracy.

CONCLUSIONS

In this paper the results of an experimental study undertaken to investigate the seismic response of telecommunications equipment supported on access floors were presented. Several conclusions can be drawn from the results.
1. Access floors utilize spot-welded base plate connections are in general lack of the ductility and strength to support heavy telecommunications equipment in region with high seismic risk.
2. Stringerless access floors systems should not be employed for seismic applications because they are prone to catastrophic collapse. A stringer system helps to prevent the collapse of an access floor even when

majority of the pedestals were severed from their base plates.

3. It is viable to use the static test results of a simple one-panel access floor module, to evaluate the seismic safety margin of an access floor system given that the upper-bound access floor displacement can be estimated with reasonable accuracy.

4. Only the fundamental mode of response of the combined equipment-access floor system is important in earthquake consideration.

5. The dynamic response of an equipment cabinet can be amplified significantly when it is mounted on an access floor. This implies that equipment which has been seismically qualified to a certain level of NEBS shaking on a concrete slab may not be qualified for the same level of shaking on access floors.

REFERENCES

1. Bell Communications Research, "Network Equipment Building System (NEBS) - General Equipment Requirements", Technical Reference TR-EOP-000063, Issue 3, Morristown, New Jersey, March 1988.

2. C.M. Wong, Seismic Response of Telecommunications Equipment Supported on Access Floors, Ph.D. Thesis, McMaster University, Hamilton, Ontario, Canada, 1990.

Table 1. Lateral stiffness and strength of different access floor systems

System	A20	B20	C20
Stiffness (KN/mm)	0.16	0.15	0.093
Ultimate strength (KN)	2.8	2.1	2.0

Table 2. Fundamental frequencies and damping ratios

System	Side-to-side				Back-to-front			
	Cabinet alone	A20	B20	C20	Cabinet alone	A20	B20	C20
Freq. (Hz)	8.91	4.7	4.5	4.09	13.2	4.6	4.53	4.03
Damp. (%)	0.70	1.6	1.7	3.0	1.23	1.7	1.9	3.4

Fig. 1 NEBS spectra criteria and time histories compatible with the 2% damped NEBS spectrum

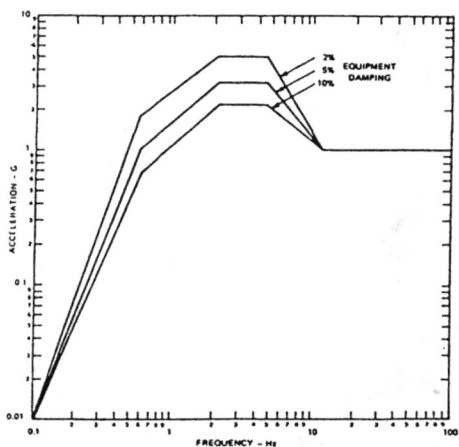

a. NEBS floor response spectra

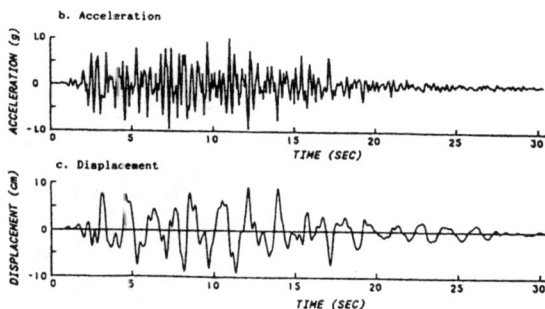

b. Acceleration

c. Displacement

Fig. 2 Static test set-up

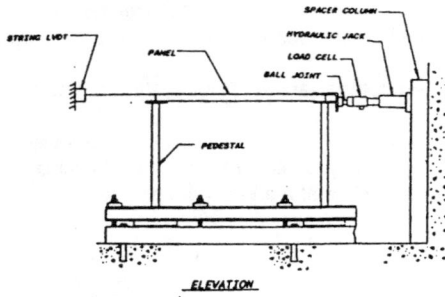

Fig. 3 Shake table test set-up for
        equipment-access floor system

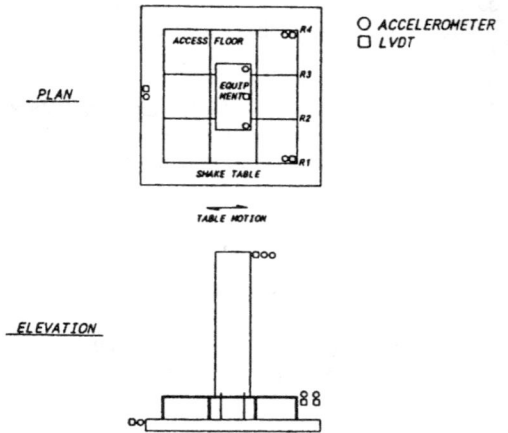

Fig. 4 Load-deflection curves for different access floors

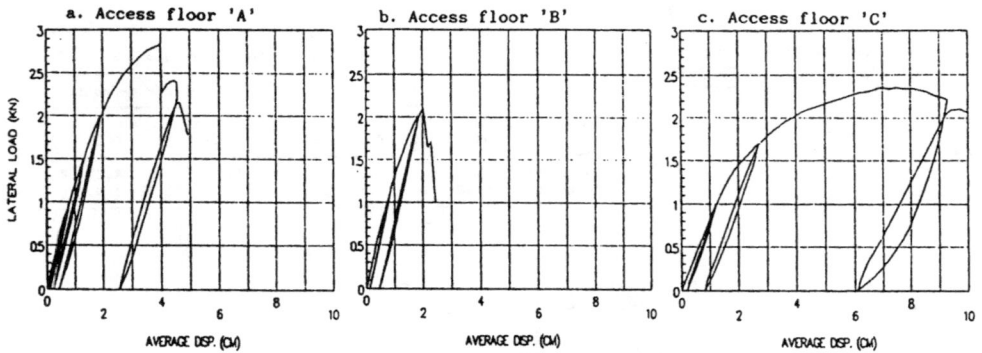

Fig. 5 Damaged access floor 'C'
        after static test

242

Fig. 7 Damaged pedestal of access floor 'B' after NEBS time history test

Fig. 6 Acceleration response at the top of the equipment-cabinet under 25% NEBS excitation

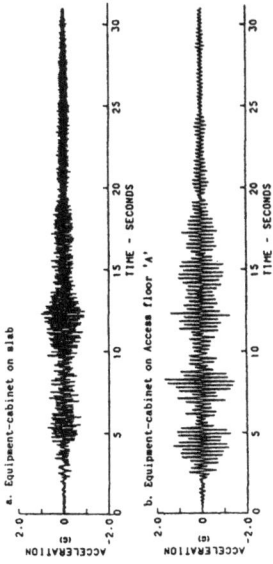

a. Equipment-cabinet on slab

b. Equipment-cabinet on Access floor 'A'

Fig. 10 Correlation of results between static test and dynamic test for access floor 'A'

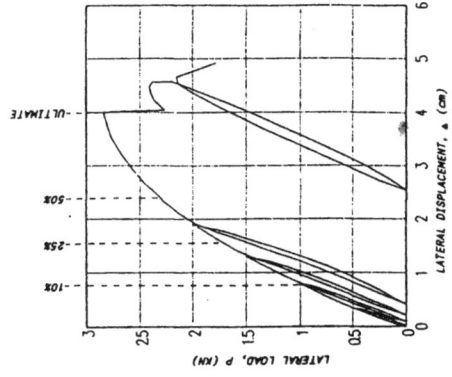

Fig. 9 Collapsed access floor 'C'

Fig. 8 Sheared off pedestal head

243

Mathematical modeling of seismic isolators

L.J. Billings* and R. Shepherd**

ABSTRACT

Base isolation is a relatively new design strategy used to protect constructed facilities from seismic shaking. It has yet to be validated by satisfactory response being achieved in strong earthquakes. Also adequate methods of predicting, at the design stage, the degrading properties of isolators particularly under biaxial horizontal motion are still under development. This paper summarizes current design practices in which an isolator is treated as an equivalent elastic column at the base of a building and, for the analysis of concurrent horizontal orthogonal motions as two columns sharing the same space concentrically, one bending in one horizontal direction and the other in the perpendicular horizontal direction. Various sophisticated analysis techniques including finite element representations of individual isolators are reviewed and the need for an accurate model incorporating damping and stiffness degradation is emphasized. Progress in the development of such a model is reported.

INTRODUCTION

The principle of reducing the dynamic effects of seismic ground motions by providing a building with a flexible lower storey was popular at one stage of the evolution of earthquake resistant design. However the poor behavior of many flexible first-storey buildings emphasized the difficulties in avoiding catastrophic failure at zones of sudden change of stiffners and strength in the load path. More recently the concept of genuine seismic isolation using a damped flexible mechanism between the ground and the bottom of a structure has received considerable attention (Kelly 1986). The main concepts of base isolation are to lower the fundamental frequency of the structure below the predominant frequencies of the seismic excitation and to provide a mechanism for energy dissipation within the isolator.

In simplistic terms seismic isolation can be thought to be a trade-off of acceleration for displacement. The inertia loads on an isolated structure will be lowered, as also will be the interstorey drifts, at the expense of increased movement at the interface where the isolators are installed. Clearly seismic isolation is not applicable to all situations. Whereas it may very well

---

*   Graduate Student, Dept. of Civil Eng., UCI. California   92717
** Professor, Dept. Civil Eng., University of California, Irvine.

be advantageous in the case of a constructed facility on a firm site, where the excitation from earthquakes will be of relatively high frequency, it is likely to be unsuited for application to soft sites with characteristic long period seismic responses.

Significant development of seismic isolation has occurred in the last decade as the result of advances in several areas. An improved understanding of the effects of site foundation conditions on incoming seismic waves, together with a much enlarged data base of earthquake records, has prompted more reliable predictions of site specific seismic movements. Advances in computer based mathematical modelling have permitted more comprehensive analyses to be undertaken at the design stage and, probably of greatest importance, the development of a variety of isolator units including laminated bearings using high-damping elastomer layered between metal plates has resulted in components of practical applicability being available.

New buildings incorporating base isolation have been completed in Japan, New Zealand and the U. S. A.(Anderson 1990). Bridges have been isolated, notably in New Zealand, as have nuclear power stations in France and South Africa. Some existing bridges and buildings have been retrofitted with base isolators as a means of reducing the earthquake forces applied to understrength structures. Clearly an increasing awareness of the potential benefits of seismic isolation is reflected in this activity. Future applications will be determined largely by the degree of field success enjoyed by existing installations in mitigating the effects of strong motion earthquakes and by the availability of analysis and design procedures which will ensure economical and reliable configurations.

CURRENT DESIGN CONSIDERATIONS

Elementary design considerations of base isolators can commence with an awareness that elastomeric bridge bearings may be sized (AASHTO 1983) using an allowable compressive stress of 1,000 to 2,000 p.s.i. whereas failure is expected at an order of magnitude greater pressure. Once an approximate plan area, A, is selected the unidirectional horizontal stiffness, K, of a simple bearing can be predicted, when G is the shear modulus and H is the height of the rubber in the bearing as:

$$K = \frac{GA}{H} \qquad (1)$$

Practical considerations of overall stability and the need to provide relatively high vertical stiffness require that most base isolators comprise alternate layers of steel shims and elastomeric material. When the elastomer layers are thin relative to the bearing diameter, the bearings are referred to as having high shape factor (HSF). The shape factor S is defined as the compressive area divided by the circumferential area free to bulge in a single layer. For circular bearings, when D is the diameter of the steel shim and t is the thickness of one layer of elastomer,

$$S = \frac{D}{4t} \tag{2}$$

Since the stiffness of the metal shims is so much greater than that of the layer of elastomers, the horizontal stiffness equation (1) may be modified for a bearing having n layers of elastomer each of thickness t to be

$$K = \frac{GA}{nt} \tag{3}$$

Derham (1982) has proposed that the vertical stiffners K_v of bearings having shape factors of about 10 (i.e. medium shape factors) can be determined when E_c, the effective compression modulus equals $5.6GS^2$, to be

$$K_v = \frac{E_c A}{nt} \tag{4}$$

Kelly et al. (1990) have suggested modifications of this expression for high shape factor bearings.

Several variations on standard cylindrical or square laminated isolators have been developed including those with lead plugs through the center, to increase energy dissipation, and initially slack vertical chains connecting top and bottom plates, to limit the sideways distortion. This last development appears to have been prompted, in some degree at least, by awareness of the potential for roll-out and instability of bearings under the combined action of significant vertical loads and very large (greater than 100%) horizontal shear strains. Such high strains will result in a reduction of the effective area resisting vertical loads and contribute to the measured non-linear behavior at high strains of tested bearings. Additional contributions to observed non-linear characteristics are provided by the inelastic properties of the elastomer and, in the case of the isolators with lead plug inserts, by the yielding of this component.

Initial designs of seismic isolation systems may be undertaken on the basis of the assumptions of extended elastic behavior summarized above. Refinements may involve ETABS (Wilson 1975) type analyses and even non-linear investigations using coding such as ANSR (Oughourlian 1982). However current limitations on the available data base of non-linear material characteristics, including the properties of the yield surfaces of a three dimension model of an isolator, seriously restrict viable inelastic analyses. Consequently current design techniques are conservative to ensure an adequate factor of safety against collapse or instability of elastomeric bearings. This situation is unlikely to change until results are available of definitive experimental and theoretical studies of the mechanics of deformation of such components.

ADVANCES IN MODELLING

Several researchers have attempted to extend the simplistic models of base isolators to more valid representations. One such effort (Yasaka 1989) depicts the behavior of an isolator subject to biaxial shearing forces with either a Kinematic Hardening model or a Multiple Shear Spring model. The first of these is limited to bi-linear hysteretic characteristics whereas the actual properties of an isolator are more complicated. This restriction prompted the development of the second model in which several inelastic spring elements are equally spaced radially. The restoring force in each spring element q may be expressed

$$q = q(E, \dot{E}) \tag{5}$$

Where E and E of an element are respectively the deflection and rate of deflection.

The deflection E_i of an element i is given by

$$E_i = u_x \cos\theta; u_y \sin\theta_i \tag{6}$$

where u_x and u_y are the deflection in each of the x and y directions and θ_i is the direction angle of the i th element. Also

$$\theta_i = (\frac{\pi}{N})_i \tag{7}$$

where N is the number of spring elements. The greater is N, the more accurate is the model.

The total restoring forces F_x on F_y along the x and y axes are also functions of the deflections and deflection rates viz:

$$F_x = \sum_{i=1}^{N} q(E_i, \dot{E}_i) \cos\theta_i \tag{8}$$

and

$$F_y = \sum_{i=1}^{N} q(E_i, \dot{E}_i) \sin\theta_i \tag{9}$$

Substitution of equation (6) into equations (8) and (9) allows expression of the restoring forces in terms of the orthogonal deflections u_x and u_y, i.e.

$$F_x = \sum_{i=1}^{N} q(u_x \cos\theta_i, \dot{u} \cos\theta_i) \cos\theta_i \tag{10}$$

The function of q which satisfies equations (5) and (10) can be found if the uniaxial restoring force F_x is known. The Kinematic hardening and Multiple Sheer spring models have been compared using for different types of loading namely uniaxial, circular, oval and eight shaped. Good correlation was obtained providing N was larger than six.

Despite the fact that modelling is complicated by inherent material and geometric properties at the large deformations experienced, coupled with the difficulty of evaluating the constitutive property of a material such as rubber which varies from sample to sample (Mark 1982), attempts have been made to undertake finite element analyses. The composite layered system can be treated as a series of discrete units, with each element containing only one type of material, or composite with the system represented as an equivalent homogeneous orthotropic continuum. The discrete analysis necessitates detailed geometry but allows direct calculation of the local stresses and strains at the layer interfaces and the edges of the bearings. The composite approach is computationally more economic at the expense of providing less detailed results.

Seki (1987) has reported analyses undertaken at the Technical Research Laboratory of the Bridgestone Corporation. Strip biaxial testing provided details of the strain energy density function used to set up a two dimensional finite element mesh. It is claimed that the analyses agree well with experimental results.

In recognition that the constitutive laws for natural and synthetic rubber are not established reliably, earlier work (Lim 1987) at the University of California, Davis, modelled elastomer as nonlinear elastic material which is adequate except in cases where loads are cyclic and damping is important. More recent research (Herrmann 1988) has resulted in the development of a composite three dimensional finite element procedure based on an equivalent homogeneous continuum. Both geometric and material non-linearities are allowed for and a unit cell concept is utilized. Verification has been undertaken by comparing results of the finite element predictions with those of discrete analyses of simple configurations and of experimental measurements of actual bearing behavior. However Herrmann (1988) stresses that residual uncertainty exists regarding the adequacy of the non-linear elastic characterization used for the elastomer. He also emphasizes that, in the case of cyclic loading of bearings, analyses need to incorporate considerations of material inelasticity.

Several finite element methods for modelling layered elastomeric isolator bearings have been developed at the Argonne National Laboratory (Kulak 1989). The analysis of low to medium shape factor bearings is tackled by a discrete approach in which there is only one material per element and each steel and elastomer layer is represented by several elements through their thickness. It is envisaged that future development will involve a composite element such as Herrmann's and will be better suited to the analysis of high shape factor components. Kulak and Wang (1989) note the present limitation of only a bi-linear elasto-plastic constitutive equation being available in the non-linear spring element for simulating the horizontal response of the isolator and indicate that efforts are being made to develop a constitutive relation appropriate to the response of a laminated elastomer bearing.

OBJECTIVES OF PRESENT WORK

In order that the state-of-understanding of the mechanics of the deformation of elastomeric bearings at high shear strain can be advanced several important aspects need to be addressed.

Improved understanding is necessary of the distribution of stress and strain under both tension and compression particularly at the rubber-steel interface of each laminate and around any holes in the laminates. Techniques are needed to predict the axial stiffness, in both tension and compression, for bearings with high shape factors and any variation of this stiffness with increasing shear strain. The change in overall height with increasing shear strain has to be established as has the effect of creep on both axial deformation and stress or strain relaxation within the elastomer. Methods of establishing both the critical displacement at which rollover is imminent and the critical buckling load at high shear strain are needed. The use of such mechanical coupling of the bearing to the rest of the structure as is provided by dowels or bolts and the effects of these boundary conditions on other characteristics requires further investigation.

Physical testing of prototype bearings is clearly desirable however such programs tend to be expensive and the current requirement of verification testing of samples of production components could be reduced significantly if better substantiated mathematical models were available.

Some closed form mathematical solutions which are available are limited in scope and are not directly applicable to the large shape factors and high aspect ratios that are typical of many base isolators. Non-linear finite element models appear to offer considerable promise. Once a verified computer code is available a series of parameter studies is likely to provide improved understanding of the mechanics of elastomeric base isolators and thereby lead to improved design methodologies for these components.

ONGOING RESEARCH

A finite element model of an elastomeric bearing is being studied using PATRAN. This is a general purpose, 3-D Mechanical Computer Aided Design (MCAE) tool developed by PDA Engineering that allows finite element representation of an object before construction. It uses extensive graphics capabilities to aid in interpreting input and output data. PATRAN can link with virtually every major finite element analysis program, along with its own P/FEA finite element code.

Initially, version 2.3 of this coding was available on an Apollo Domain 3500. A low shape factor bearing was chosen for analysis using PATRAN and P/FEA since the discretization would not be excessive. This bearing, shown in Figure 1, was tested at U.C. Berkeley. The results were published by Tajirian, et al.(1990).

A 3-D, F.E.M. of the bearing was discretized with a total of more than 600 elements. Vertically, one element was used to represent the steel shims and end plates, while the rubber was represented using four elements. The rubber was assumed to be isotropic, linear elastic, and incompressible. Therefore, Poisson's ratio, ν, was assumed to equal 0.5. The published experimental results were used to evaluate the rubber's shear modulus G. G was set equal to 120 psi, and with ν = 0.5, the elastic modulus, E, was calculated to be 360 psi. It was established that if ν was adjusted to 0.49983, close to the assumed 0.5, PATRAN's results for a linear, static analysis matched the experimental results very closely.

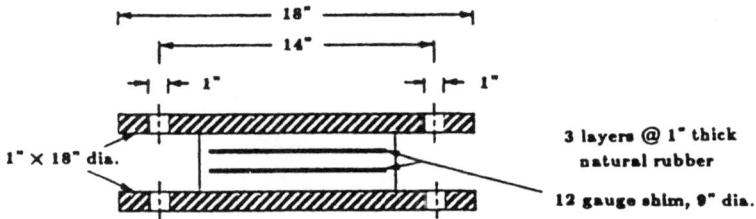

Quarter Scale Bolted Bearing (Tajirian, F., et al. 1990)

The availability of a DEC 5000/200 system together with the release of PATRAN's latest version 2.4 has allowed larger element meshes to be undertaken. Currently, the mesh includes 9 elements in each rubber layer, and Poisson's ratio was changed to 0.49998 yielding results that correspond to the experimental results. However, the finer mesh does affect the internal stress and strain values and distribution. This was expected since the P/FEA coding does not allow for large strain theory. The finer mesh also reduced the vertical stiffness of the bearing, hence Poisson's ratio was increased until the experimentally derived stiffness was obtained.

It is concluded that the P/FEA coding is valuable for obtaining an approximate solution to the internal stress and strain distribution. However, more accurate solutions are possible with other analysis codings. It is planned to substitute P/FEA with MARC analysis which is specifically tailored toward elastomeric materials. It is expected that analyzing the above bearing will result in some realistic stress strain distributions. If the analysis is successful, it is intended to model larger bearings with higher shape factors.

ACKNOWLEDGEMENTS

Financial support from the National Earthquake Engineering Research Center, State University of New York at Buffalo, and the continuing advice and encouragement of Dr. I.G. Buckle, Deputy Director of the Center, is acknowledged with gratitude.

REFERENCES

AASHO, 1983. Standard Specifications for Highway Bridges. AASHTO, Washington D.C.

Anderson, T.L. (Editor) 1990. Earthquake Spectra. Special Volume, Vol. 6, No. 2, Earthquake Engineering Research Institute, El Cerrito, California.

Derham, C.J. 1982. The Design of Laminated Bearings. Proceedings of the International Conference on Natural Rubber for Earthquake Protection of Buildings and Vibration Isolation, Malaysia.

Herrmann, L. R. et al. 1988. Non-Linear Behavior of Elastomeric Bearings - FE Analysis and Verification. J. Engrg. Mech., ASCE 114, 11, 1831-1847.

Kelly, J.M. 1986. Aseismic Base Isolation: Review and Bibliography. Soil Dynamics and Earthquake Engineering, Vol. 5, No. 3.

Kelly, J.M., I.D. Aiken, and F.F. Tajirian 1990. Mechanics of high Shape Factor Elastomeric Seismic Isolation Bearings. Report UCB/EERC 90-01, University of California, Berkeley.

Kulak, R. F. and C.V. Wang 1989. Design and Analysis of Seismically Isolated Structures. Proceedings of the First International Seminar on Seismic Base Isolation of Nuclear Power Facilities, San Francisco, California.

Lim, C.K. and L.R. Herrmann 1987. Equivalent Homogeneous FE Model for Elastomeric Bearings. J. Engrg. Mech., ASCE, 113, 1, 106-125.

Mark, J. and L. Joginder, (Editors) 1982. Elastomers and Rubber Elasticity. American Chemical Society, Washington, D.C.

Oughourlian, C. V. and G.H. Powell 1982. ANSR-II, General Computer Program for Nonlinear Structural Analysis. Report UCB/EERC, 82-21, University of California, Berkeley.

Seki, W. et al. 1987. A Large-Deformation Finite-Element Analysis for Multilayers Elastomeric Bearings. Rubber Chemistry and Technology, 60, 5, 856-869.

Tajirian, F.F., I.D. Aiken and J.M. Kelly, 1990. Earthquake Spectra. Special Volume, Vol. 6, No. 2, Earthquake Engineering Research Institute, El Cerrito, California.

Wilson, E.L. and H. Dovey 1975. "ETABS", Three Dimensional Analysis of Building Systems. Report UCB/EERC 75-13, University of California, Berkeley.

Yasaka, A., et al., 1989. Biaxial Hysteresic Model for Base Isolation Devices. Kajima Corporation, Tokyo.

Response of buried pipelines located through liquefied and non-liquefied ground

Masakatsu Miyajima[I] and Masaru Kitaura[II]

ABSTRACT

The earthquake damage data in the 1964 Niigata Earthquake and the 1983 Nihonkai-Chubu Earthquake show that the pipeline damage were found near the boundary between the liquefied and non-liquefied areas. In this paper, main factors affecting the pipe failure located through both liquefied and non-liquefied ground are considered in connection with subsidence, buoyancy and seismic motion which is different from the motion of neighboring sites. Formulae obtained by using beam theory are presented and the response characteristics of pipelines are discussed. The results suggest that serious attention should be paid to vibration-induced strains during liquefaction.

INTRODUCTION

The liquefaction of sandy soil caused much damage to buried pipeline systems during the 1989 Loma Prieta Earthquake in the U.S.A. Pipeline damage induced by liquefaction has repeatedly occurred in several earthquakes, for example, in the 1964 Niigata Earthquake, the 1983 Nihonkai-Chubu Earthquake and so on. Effects of liquefaction are classified into three types as follows: Loss of bearing capacity and generation of buoyancy, large ground displacement, and large dynamic response of ground. These effects seem to be great at the boundary between liquefied and non-liquefied sites because of sharp change of the ground characteristics. It was very interesting to note that all of the damage to cast iron pipe occurred at liquefied sites and most of those occurred near the boundary between the liquefied and non-liquefied sites during the 1983 Nihonkai-Chubu Earthquake (Kitaura and Miyajima 1986).

The purposes of the present paper are to clarify pipe behavior through a boundary between liquefied and non-liquefied sites and to discuss characteristics of the failures of the pipelines buried in such areas.

I Assistant Professor, Kanazawa University, 2-40-20 Kodatsuno Kanazawa 920 JAPAN.

II Professor, Kanazawa University, ditto.

MATHEMATICAL TREATMENT

Fig. 1 shows analytical models for pipelines buried through a boundary between liquefied and non-liquefied ground. In case 1, the pipeline is assumed to be subjected to subsidence of the liquefied ground, while in case 2 buoyancy effect is a governing factor. In case 3, the pipeline is subjected to a seismic motion which is different from the motion of the neighboring zones. In this last case, the superficial layer of the ground above the liquefied layer can be regarded as a horizontally vibrating elastic plate which is subjected to a periodical driving force at both ends. In these three analytical models, the perfectly elastic behavior is assumed for the pipe material and the pipe motion is analyzed in the two-dimensional plane. Characteristics of the pipe strains are investigated using these simplified mathematical models. In this investigation, the responses of the pipelines can be estimated by solving the following differential equations with suitable boundary conditions.

(1) Case 1

According to the schematic representation of case 1, as shown in Fig. 1 and upon the above several assumptions, the differential equations for case 1 may be expressed as follows;

$$EI \frac{d^4 v_1}{dx^4} + K_{v1} \, v_1 = K_{v1} \, V_1 \qquad\qquad (0 < x < l) \qquad (1)$$

$$EI \frac{d^4 v_2}{dx^4} + K_{v2} \, v_2 = 0 \qquad\qquad (x > l) \qquad (2)$$

The boundary conditions are

$$\frac{dv_1}{dx} = 0, \quad \frac{d^3 v_1}{dx^3} = 0 \qquad\qquad (x = 0) \qquad (3)$$

$$v_1 = v_2, \quad \frac{dv_1}{dx} = \frac{dv_2}{dx}, \quad \frac{d^2 v_1}{dx^2} = \frac{d^2 v_2}{dx^2}, \quad \frac{d^3 v_1}{dx^3} = \frac{d^3 v_2}{dx^3} \qquad (x = l) \qquad (4)$$

$$v_2, = 0, \quad \frac{dv_2}{dx} = 0 \qquad\qquad (x = \infty) \qquad (5)$$

respectively.

(2) Case 2

Supposing that the transverse displacement in the pipeline depends on buoyancy effect, the governing differential equations for case 2 may be given by

$$EI \frac{d^4 v_1}{dx^4} + K_{v1} \, v_1 = F \qquad\qquad (0 < x < l) \qquad (6)$$

$$EI \frac{d^4 v_2}{dx^4} + K_{v2} \, v_2 = 0 \qquad\qquad (x > l) \qquad (7)$$

The boundary conditions Eqs. (3), (4) and (5) are also available for the problem of case 2.

(3) Case 3

Since the schematic model of case 3 is taking account of the liquefaction in the superficial layer of the ground, including the original model presented by Nishio et al. (1987), the differential equation can be written in a form of

$$EA\frac{d^2u_1}{dx^2} - K_{u_1} u_1 = -K_{u_1} U_1 \qquad\qquad (0 < x < l) \qquad (8)$$

$$EA\frac{d^2u_2}{dx^2} - K_{u_2} u_2 = -K_{u_2} U_2 \qquad\qquad (x > l) \qquad (9)$$

The boundary conditions are

$$\frac{du_1}{dx} = 0 \qquad\qquad (x = 0) \qquad (10)$$

$$u_1 = u_2, \quad \frac{du_1}{dx} = \frac{du_2}{dx} \qquad\qquad (x = l) \qquad (11)$$

$$\frac{du_2}{dx} = 0 \qquad\qquad (x = L) \qquad (12)$$

where U_1 in Eq. (8) and U_2 in Eq. (9) are expressed as follows;

$$U_1 = \frac{\cos(\frac{2\pi x}{cT})}{\cos(\frac{2\pi l}{cT})} U_{s0} \qquad\qquad (13)$$

$$U_2 = U_{s0} \qquad\qquad (14)$$

where u, v = longitudinal and transverse displacements in the pipeline, U, V = displacements in the ground, E = young's modulus of the pipe material, A = cross-sectional area of the pipe, I = area moment of inertia of the pipe, K_u, K_v = equivalent spring constants of the longitudinal and transverse motions, F = a force caused by the buoyancy effect, c = longitudinal wave velocity which is given as $c^2 = E_s/\rho$ in which E_s is Young's modulus of soil and ρ is soil density, U_{s0} = displacement amplitude in non-liquefied superficial layer, T = period of shaking, $2l$ = width of the liquefied zone and L = pipe length. Subscripts 1 and 2 respectively correspond to the two sections.

APPLICATION OF ANALYTICAL SOLUTIONS TO THE PRACTICAL CASES

Soil spring constant shows non-linear characteristics with increasing soil displacement. However, the soil spring constant is assumed to be unchangeable irrespective of the relative displacement for simplicity in this paper. This unchangeable value is called as equivalent soil spring constant. The equivalent soil spring constant for liquefiable ground depends on the degree of liquefaction of the soil. Initially, the bending pipe stresses due to subsidence of the ground induced by liquefaction are investigated in relation to the ratio of the equivalent soil spring constant K_1/K_2 for case 1. Fig. 2 shows the relationship between the maximum bending pipe stresses due to subsidence of the liquefied ground, and the width of the liquefaction zone. The pipelines used in this analysis are steel pipelines whose physical properties are listed in Table 1. The magnitude of subsidence of the liquefied ground is assumed to be 20 cm, and it is equivalent to a 2% subsidence of 10m liquefied layer. Yoshimi et al. concluded that the average vertical strain as a result of consolidation under the weight of the soil was 1% to 3% when a horizontal layer of loose saturated sand was liquefied to a depth of approximately 5m to 20m (Yoshimi et al. 1975). The experimental results presented by Lee and Albaisa (1974) also agree with the results indicated by Yoshimi et al. (1975). It can be seen from Fig. 4 that the higher the ratio of the equivalent soil spring constant is, which means smaller degree of liquefaction, the greater the maximum bending pipe stresses are. It is interesting to note that the maximum bending pipe stress for each ratio of the equivalent soil spring constant does not increase in areas of liquefied ground where the width exceeds 10m. The maximum bending stress exceeds the allowable bending stress of steel pipe $(4,200 kgf/cm^2)$ when the ratio of the equivalent soil spring constant is greater than 0.1, particularly at a width of liquefied zone less than 5 m except for the region near 0m. Fig. 3 shows the relationship between the displacement of the pipe and the width of the liquefied zone. It can be seen from this figure that the relative displacement between the ground and pipe is observed for relatively narrow width of the liquefied ground or low ratio of the equivalent soil spring constant. In case 1, the location where the maximum bending stress occurs, is nearer the boundary between the liquefied and non-liquefied ground, for greater width of the liquefied ground or higher ratio of the equivalent soil spring constant.

In case 2, the buoyancy effects of the pipelines during liquefaction are estimated by using the method described by Kitaura and Miyajima (1985). Fig. 4 shows the relationship between the maximum bending pipe stress due to buoyancy effects induced by liquefaction and the excess pore water pressure. Fig. 5 illustrates the relationship between the displacement of the pipe and the width of the liquefied ground. In this study, the equivalent soil spring constant of the liquefied soil is estimated by the empirical equation proposed by Yoshida and Uematsu (1978). It was noted that the pipe reached the ground surface in the area where the width of liquefied ground exceeded 10 m as shown in Fig. 5. Therefore, the maximum bending pipe stresses disappear for great width and high excess pore water pressure in Fig. 4. It is evident from Fig. 4 that the greater the width of the liquefied ground is, the higher the maximum bending stress are, that is, the greater the probability of failure of the pipe due to buoyancy effects. It is also interesting to note that the maximum bending pipe stresses occurred in the non-liquefied ground in case of great width of the liquefied ground. In these analyses, it is assumed that the duration of the liquefaction process is lengthy enough to allow the occurrence of pipe deformation. However, since the duration of liquefaction induced by an actual earthquake depends on the local soil conditions, the above may actually not be the cases in reality. In order to evaluate the pipeline response due to buoyancy effects more precisely, the dynamic analysis indicated by Kitaura et al. (1987) is preferable.

In case 3, the distribution of the axial pipe strain is shown in Fig. 6, in relation to the ratio of the equivalent soil spring constant K_1/K_2. Fig. 7 illustrates the relationship between the maximum axial pipe strain and the ratio of the equivalent soil spring constant. The conditions for the ground and the magnitude of an earthquake used in these analyses are summarized in Table 2. It is evident from these figures that the maximum axial strains decrease and the location where the maximum strains occurs approaches the boundary of the ground with a decrease in the ratio. During liquefaction processes, not only the equivalent soil spring constant varies but also longitudinal wave velocity c varies. Therefore, the longitudinal wave velocity is assumed to be proportional to the fourth root of $(1 - r)$, where r is the excess pore water pressure ratio because the longitudinal wave velocity is proportional to the square root of Young's modulus of soil, and the shear modulus of the soil is proportional to the square root of $(1 - r)$. Moreover, Young's modulus is proportional to the shear modulus of the soil. Fig. 8 shows the distribution of the axial pipe strain in relation to the excess pore water pressure ratio. Fig. 9 illustrates the relationship between the maximum axial pipe strain and the excess pore water pressure ratio. Excess pore water pressure higher than 0.9 is not included in this analysis because the longitudinal wave cannot be transmitted in such soft ground. It is evident from these figures that the maximum axial pipe strain occurs at the liquefied ground near the boundary between the liquefied and non-liquefied ground. The maximum axial pipe strain corresponding to the value of 0.2 of excess pore water pressure ratio is markedly great. This can be explained as follows: Fig 10 shows the magnification ratio of response displacement in a superficial layer. In this analysis, $l/(cT/2)$ is less than 1.0; therefore, the magnification ratio increases sharply with an increase in $l/(cT/2)$, i.e. with a decrease in c. However, in this case the magnification ratio decreases because $l/(cT/2)$ is greater than 1.0. Effects of the resonance of the liquefied ground are great when excess pore water pressure ratio is equal to 0.2. Furthermore, Fig. 10 suggests that care should also be taken in evaluating the resonance of the liquefied ground for greater width of liquefied ground than that in this analysis. The above results obtained by mathematical analyses suggest that the probability of failure is high at the boundary between the liquefied and non-liquefied ground for each cause of pipe failure. These findings agree with the experimental results presented by Kitaura and Miyajima (1983).

CONCLUSIONS

Response characteristics of pipelines located through both liquefied and non-liquefied ground were clarified based on mathematical analysis. The following can be concluded:
(1) One of the response characteristics of pipelines subjected to subsidence of ground is that the smaller the degree of liquefaction in the superficial ground, the greater the maximum bending pipe stresses. Moreover, the higher bending pipe stress occurs in areas of smaller width of the liquefied ground.
(2) The effects of buoyancy on pipe response are great in areas of great width of the liquefied ground and a high degree of liquefaction in the superficial ground.
(3) Resonance of the superficial layer of ground has great influence on the axial pipe strains during liquefaction processes. This suggests that great consideration should be given to vibration-induced strains during liquefaction.

ACKNOWLEDGMENTS

This study is supported in part by the Ogawa Foundation's Research Grant and the Grant-in-Aid for Scientific Research from Ministry of Education, Science and Culture of Japan.

REFERENCES

Kitaura, M. and Miyajima, M. 1983. Dynamic Behaviour of a Model Pipe Fixed at One End During Liquefaction, Proceedings of JSCE, No. 336, pp. 31-38 (in Japanese).

Kitaura, M. and Miyajima, M. 1985. Dynamic Behavior of Buried Model Pipe During Incomplete Liquefaction, Journal of Structural Engineering, JSCE, Vol. 31A, pp. 421-426 (in Japanese).

Kitaura, M. and Miyajima, M. 1986. Assessment of Safety of Pipelines Subjected to Soil Liquefaction, Preliminary Report of IABSE Symposium TOKYO 1986, Vol. 51, pp. 133-140.

Kitaura, M., Miyajima, M. and Suzuki H. 1987. Response Analysis of Buried Pipelines Considering Rise of Ground Water Table in Liquefaction Processes, Proceedings of JSCE, No. 380/I-7, pp. 173-180.

Lee, K. L. and Albaisa, A. 1974. Earthquake Induced Settlements in Saturated Sands, Proceedings of ASCE, Vol. 100, No. GT4, pp. 387-406.

Nishio, N., Tsukamoto, K. and Hamura, A. 1987. Model Experiment on the Seismic Behavior of Buried Pipeline in Partially Liquefied Ground, Proceedings of JSCE, No. 380, pp. 449-458 (in Japanese).

Yoshida T.and Uematsu, M. 1978. Dynamic Behavior of a Pile in Liquefaction Sand, Proceedings. of the 5th Japan Earthquake Engineering Symposium-1978, pp. 657-663 (in Japanese).

Yoshimi, Y., Kuwabara, F. and Tokimatsu, K. 1975. One-dimensional Volume Change Characteristics of Sands under Very Low Confining Stresses, Soils and Foundations, JSSMFE, Vol. 15, No. 3, pp. 51-60.

Table 1. Physical properties of steel pipe.

Outer diameter	(mm)	406.4
Thickness	(mm)	6.0
Young's modulus	(kgf/cm²)	2.1×10^6
Specific gravity		7.85

($1 kgf/cm^2 = 98 kPa$)

Table 2. Conditions of ground and magnitude of earthquake.

Longitudinal wave velocity c	170(m/s)
Period of shaking T	0.5(s)
Acceleration in superficial layer α	200(gal)
Displacement amplitude of non-liquefied superficial layer	$U_s = (T/2\pi)^2\alpha$

258

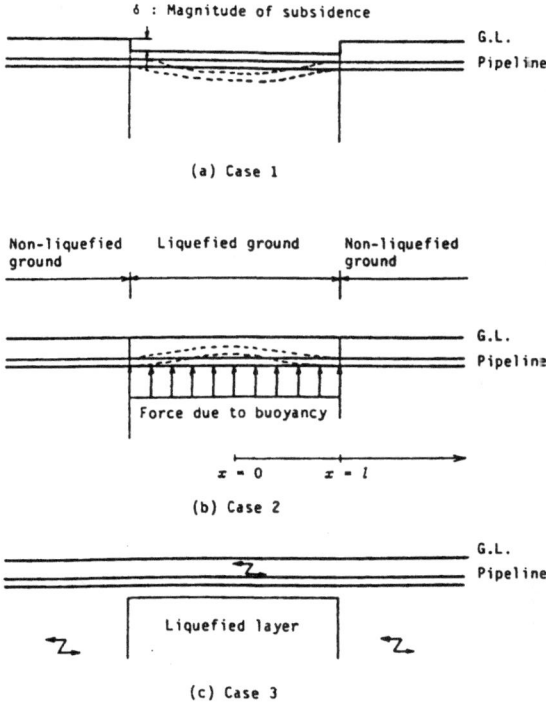

(a) Case 1

Non-liquefied ground | Liquefied ground | Non-liquefied ground

(b) Case 2

(c) Case 3

Figure 1. Analytical models for buried
pipelines located between
liquefied and non-liquefied
ground.

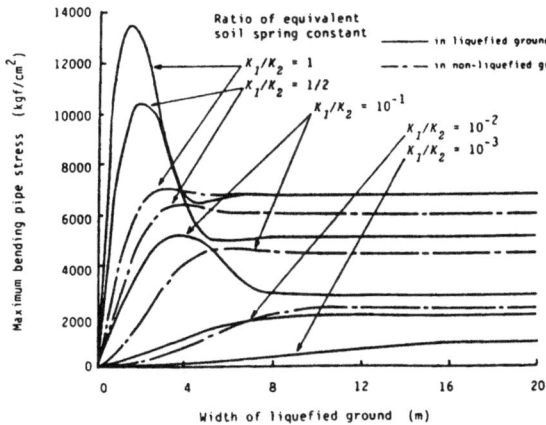

Figure 3. Distribution of pipe displacement
due to subsidence.

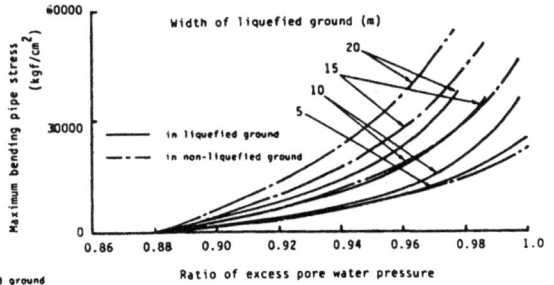

Figure 4. Relationship between maximum
bending pipe stress due to
buoyancy effects induced by
liquefaction and excess pore water
pressure ratio.

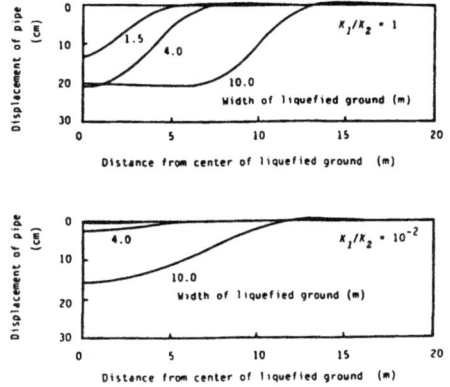

Figure 2. Relationship between maximum
bending pipe stress due to
subsidence of liquefied ground
and width of liquefied zone.

259

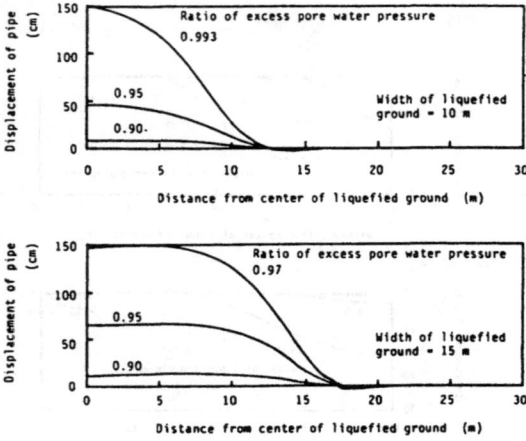

Figure 5. Distribution of pipe displacement due to buoyancy effects.

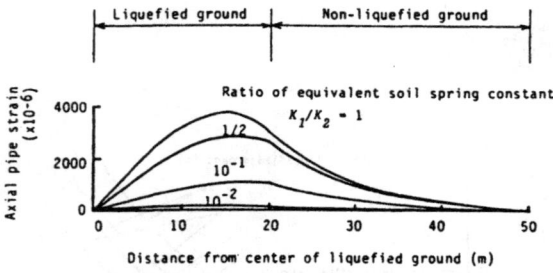

Figure 6. Distribution of axial pipe strain in relation to the ratio of equivalent soil spring constant.

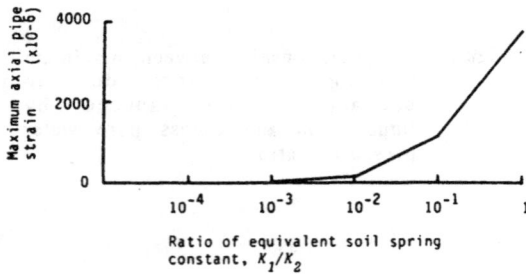

Figure 7. Relationship between maximum axial pipe strain and the ratio of equivalent soil spring constant.

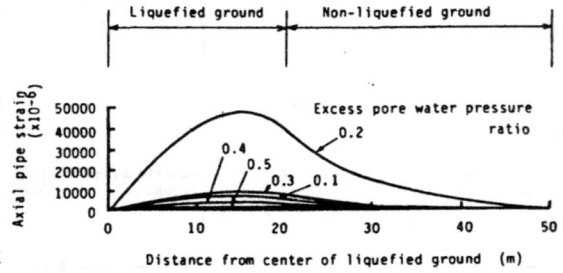

Figure 8. Distribution of axial pipe strain in relation to excess pore water pressure.

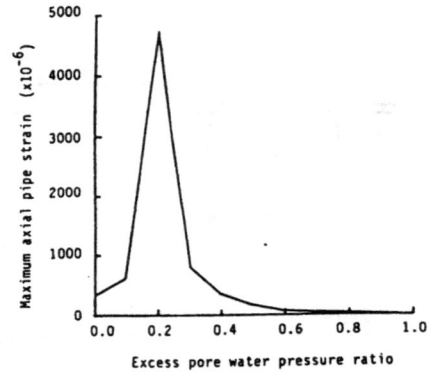

Figure 9. Relationship between maximum axial pipe strain and excess pore water pressure ratio.

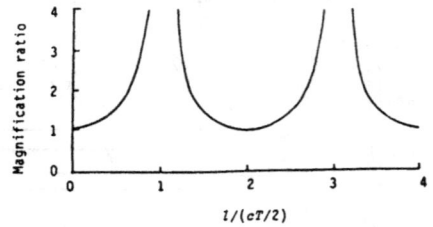

Figure 10. Magnification ratio of response displacement in superficial layer.

260

Seismic damage potential to ductile and nominally ductile concrete frames

K.A. Hamdy[I], W.K. Tso[II], and A. Ghobarah[II]

ABSTRACT

Two reinforced concrete ductile moment resisting frames and two nominally ductile frames are designed according to the current Canadian concrete design code. For each type of frame design, one frame is 4 storeys high while the other is ten storeys high. The frames are analyzed dynamically using the computer program DRAIN-2D. The response parameters investigated are the storey drifts, the beam rotational ductility demands and the beam shear stresses. The above mentioned parameters are used to compare the equivalence of the seismic behaviour of nominally ductile frames and that of ductile frames. It was found that the behaviour of nominally ductile frames is less favourable than that of ductile frames mainly because the response shear stresses in the beams of nominally ductile frames are much larger than the design values.

INTRODUCTION

Moment resisting frames (MRF) are the most commonly used framing system for reinforced concrete structures. According to the current Canadian practice, designers are given two options for the seismic design of reinforced concrete frames (National Building Code of Canada, NBCC 1990). The first option is to design a ductile frame, which involves special design and detailing provisions to insure ductile behaviour. The second option is to design a nominally ductile frame. This option involves designing for twice the seismic lateral load as that for ductile frames, but without taking all the special provisions for good detailing in the design of the frame members. By allowing such a choice, the code implies that either type of frames will provide equivalent seismic performance under the design level earthquake disturbance.

The objective of this investigation is to compare the equivalence of the seismic behaviour of ductile and nominally ductile frames. To achieve this aim, two four storey and two ten storey frames are designed according to NBCC 1990 and CAN3-A23.3-M84. Each of the frames is subjected to earthquake records normalized to the design velocity. To avoid dependence on the characteristics of a single record, 15 records are used and the results are discussed based on a statistical analysis of the individual responses. The response parameters investigated are the total and interstorey drifts, the beam rotational ductility demands and the beam shear stresses.

[I] Research assistant, Civil Eng. Dept., McMaster University, Canada.

[II] Professor

STRUCTURAL CONFIGURATION

Each of the two buildings considered, consists of 3 bays in the E-W direction and 7 bays in the N-S direction. The bay widths are 8.0 m in each direction. One building is four storeys high and the other is ten storeys high. The storey height is 3.5 m. The concrete compressive strength is 30 MPa and the steel yield strength is 400 MPa. The seismic loading is assumed to be acting in the E-W direction. The typical interior E-W frame of each building is designed once as a ductile frame and once as a nominally ductile frame. This has resulted in four different frames.

DESIGN LOADINGS

The frames are designed for the critical combinations of gravity and seismic loads as per NBCC 1990. For seismic base shear calculations, the frames are assumed to be located in Quebec City. The force modification factor, R, is chosen according to the ductility level for which the frames are designed. For ductile frames R = 4.0 and for nominally ductile frames R = 2.0 .

DESIGN OF FRAME MEMBERS

For comparison purposes, the member dimensions are taken to be the same for all frames. The reinforcement ratio differs due to the variation of the seismic lateral load and the design approach. Figs. 1 and 2 show the member dimensions and the reinforcement ratios of the frames.The different approaches used in the design of ductile and nominally ductile frames are briefly described below.

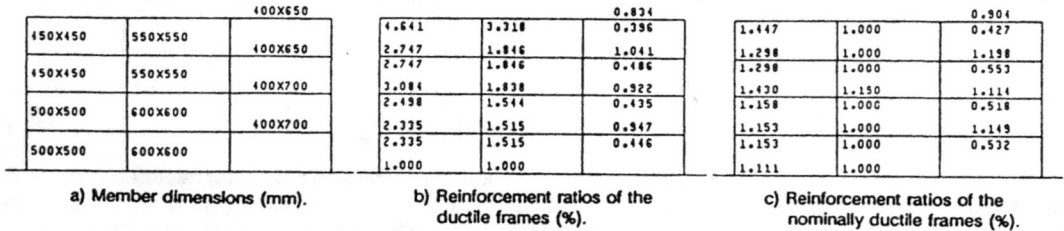

a) Member dimensions (mm).

		beam
450X450	550X550	400X650
450X450	550X550	400X650
500X500	600X600	400X700
500X500	600X600	400X700

b) Reinforcement ratios of the ductile frames (%).

		0.834
4.641	3.318	0.396
2.747	1.846	1.041
2.747	1.846	0.486
3.084	1.838	0.922
2.498	1.544	0.435
2.335	1.515	0.947
2.335	1.515	0.446
1.000	1.000	

c) Reinforcement ratios of the nominally ductile frames (%).

		0.904
1.447	1.000	0.427
1.298	1.000	1.198
1.298	1.000	0.553
1.430	1.150	1.114
1.158	1.000	0.518
1.153	1.000	1.149
1.153	1.000	0.532
1.111	1.000	

Figure 1. Member dimensions and reinforcement ratios of the four storey frames.

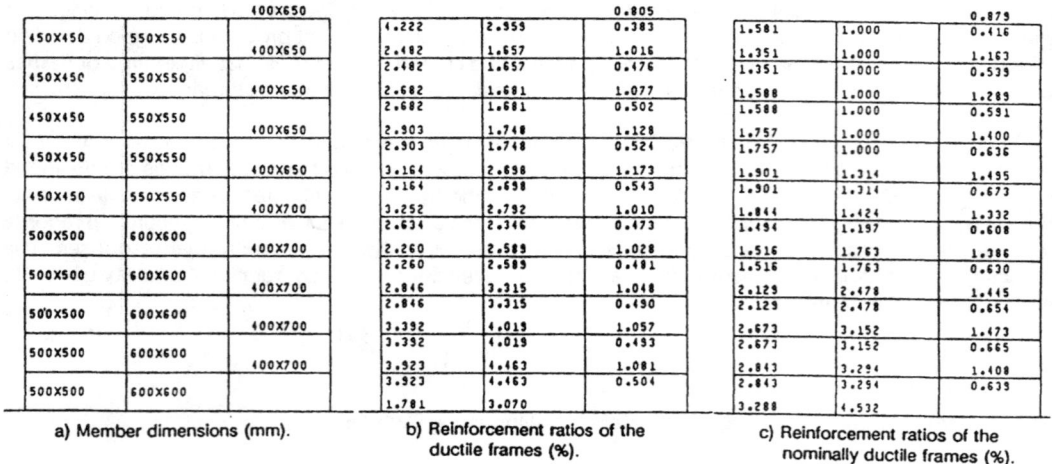

a) Member dimensions (mm).

		beam
450X450	550X550	400X650
450X450	550X550	400X650
450X450	550X550	400X650
450X450	550X550	400X650
450X450	550X550	400X650
500X500	600X600	400X700
500X500	600X600	400X700
500X500	600X600	400X700
500X500	600X600	400X700
500X500	600X600	

b) Reinforcement ratios of the ductile frames (%).

		0.805
4.222	2.955	0.383
2.482	1.657	1.016
2.482	1.657	0.476
2.682	1.681	1.077
2.682	1.681	0.502
2.903	1.748	1.128
2.903	1.748	0.524
3.164	2.698	1.173
3.164	2.698	0.543
3.252	2.792	1.010
2.634	2.346	0.473
2.260	2.589	1.028
2.260	2.589	0.481
2.846	3.315	1.048
2.846	3.315	0.490
3.392	4.019	1.057
3.392	4.019	0.493
3.923	4.463	1.081
3.923	4.463	0.504
1.781	3.070	

c) Reinforcement ratios of the nominally ductile frames (%).

		0.879
1.581	1.000	0.416
1.351	1.000	1.163
1.351	1.000	0.533
1.588	1.000	1.289
1.588	1.000	0.591
1.757	1.000	1.400
1.757	1.000	0.636
1.901	1.314	1.455
1.901	1.314	0.673
1.844	1.424	1.332
1.494	1.197	0.608
1.516	1.763	1.386
1.516	1.763	0.630
2.129	2.478	1.445
2.129	2.478	0.654
2.673	3.152	1.473
2.673	3.152	0.665
2.843	3.294	1.408
2.843	3.294	0.635
3.288	4.532	

Figure 2. Member dimensions and reinforcement ratios of the ten storey frames.

Ductile frames

The main aim of designing ductile frames is to avoid brittle failure and storey side-sway mechanisms. The seismic design provisions specified in chapter 21 of CAN3-A23.3-M84 (1984) are to be followed. The main features of the design methodology are i) strong columns-weak beams, ii) design shear forces based on the probable strength of probable plastic hinges and iii) good detailing.

Nominally ductile frames

No special seismic design provisions are considered in the design of these frames. All the design actions are directly obtained from the results of the elastic static analysis. Detailing requirements are far less stringent than those of ductile frames.

DISCUSSION OF THE DESIGN RESULTS

The reinforcement ratios in the beams of nominally ductile frames are larger than those of the beams of ductile frames. Nevertheless, the columns of ductile frames usually contain more steel than those of nominally ductile frames This is the result of the strong column-weak beam requirement for ductile frames design. Thus, the overstrength possessed by the frames may vary significantly due to the variation in the design methodology. According to NBCC 1990, the overstrength factor (1/U) is assumed to be 1.67 for all structures regardless of type or design methodology. In order to determine their overstrengths, the designed frames were analyzed under a monotonically increasing static lateral loading. The lateral loads are distributed over the height according to NBCC 1990. The overall base shear-top displacement relationships of the frames are shown in figs. 3 and 4. For the ductile frames, the overstrength factor was 4.5 for the four storey frame and 3.0 for the ten storey frame. For the nominally ductile frames, the overstrength factor was 1.8 for the four storey frame and 1.7 for the ten storey frame. While the lateral overstrength factor of the ductile frames is significantly higher than that assumed in NBCC 1990, the corresponding factor of the nominally ductile frames is of the same order as that assumed in NBCC 1990. It should be noted that the design base shear for nominally ductile frames is twice that of ductile frames, therefore the actual lateral strengths of the frames designed based on either approach are comparable.

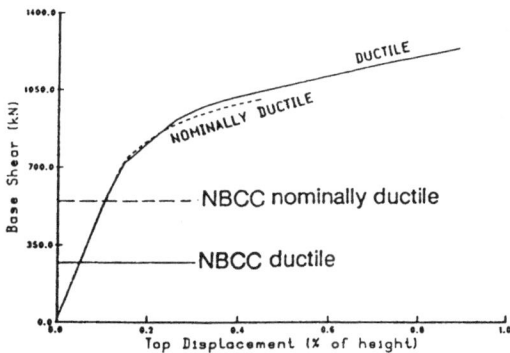

Figure 3. Base shear-top displacement curves of the four storey frames.

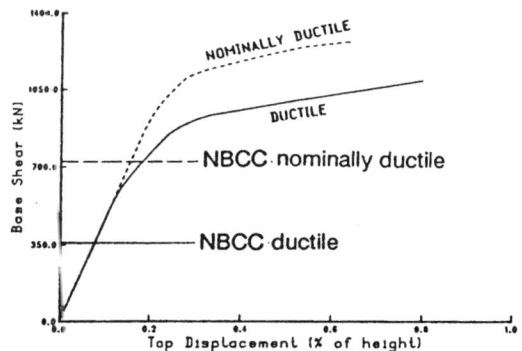

Figure 4. Base shear-top displacement curves of the ten storey frames.

263

DYNAMIC ANALYSIS PROCEDURE

The DRAIN-2D computer program (Kanaan and Powell 1973) in the dynamic analysis of the frames. The members of the ductile frames are modelled by the dual-component bilinear model developed by Clough, Benushad and Wilson (1965). The use of stable loops may be justified by the good detailing of the members of ductile frames. The members of the nominally ductile frames are modelled by elastic elements with nonlinear rotational springs at their ends. Strength deterioration and pinching are introduced in the hysteresis behaviour of the springs. The used hysteresis model was developed by Chung, Meyer and Shinozuka (1987). A typical hysteresis loop is shown in fig. 5. Fifteen records in the high acceleration-to-velocity (a/v) range (Naumoski et al. 1988) were selected as the ground motion input for this study. In a seismic region where $Z_a > Z_v$ according to the Canadian seismic zoning, the ground motions are expected to have high frequency contents and the records with high a/v ratio are considered representative of the expected ground motions.

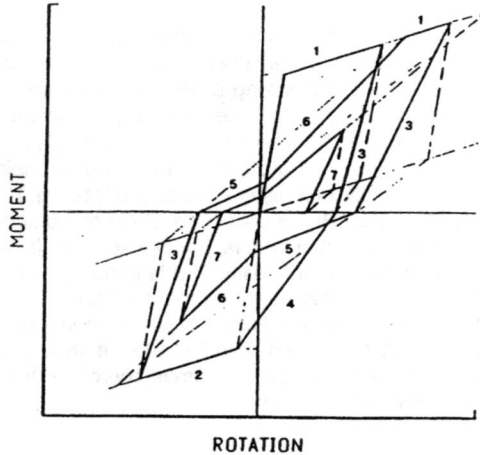

ROTATION

Figure 5. Typical moment-rotation relationship for the members of nominally ductile frames. (Chung, Meyer and Shinozuka, 1987)

DISCUSSION OF THE ANALYSIS RESULTS

The results discussed below are the average results of the dynamic response of the frames under the fifteen ground motion records used in this study.

Total and interstorey drifts

As shown in figs. 6 and 7 there is very small difference between the drifts of ductile and nominally ductile frames. This mainly stems from the fact that the same sections were used for both design options.

Rotational ductility demand for beams

The maximum rotational ductility demands at the beam ends of are shown in fig. 8. It can be seen that the ductility demand for ductile frames is larger than that for nominally ductile frames. Nevertheless, the difference is smaller that what would be implied by the difference in the design base shear of the two frames. (base shear of a ductile frame is half that of a nominally ductile frame).

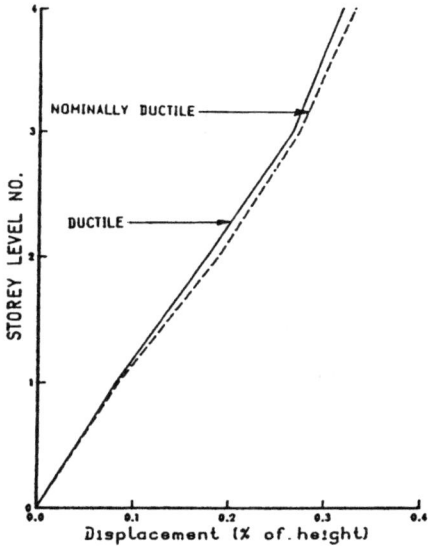

a) Four storey frames.

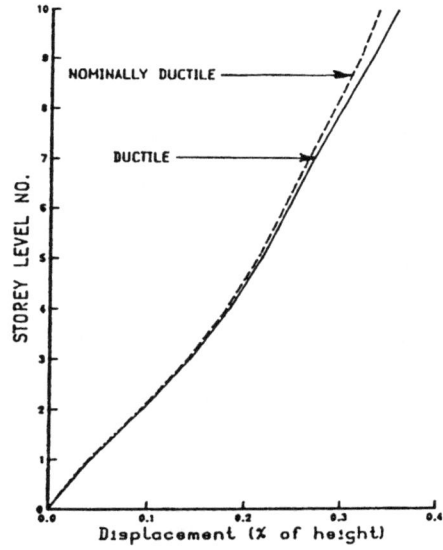

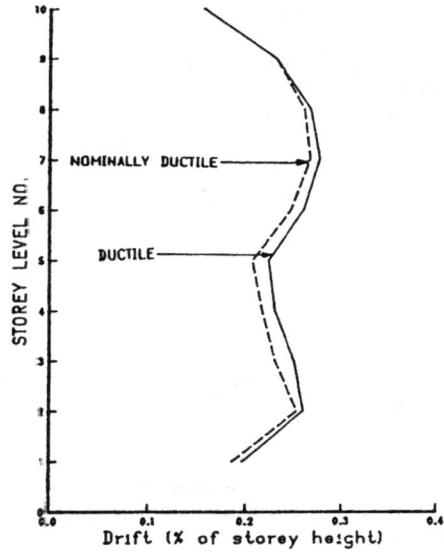

b) Ten storey frames.

Figure 6. Lateral displacements of the frames (% of height).

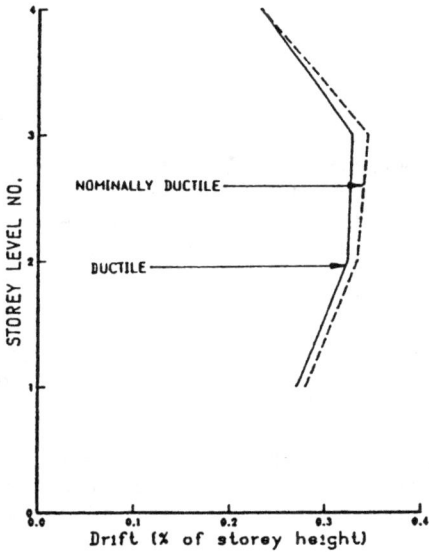

a) Four storey frames.

b) Ten storey frames.

Figure 7. Interstorey drifts of the frames (% of storey height).

265

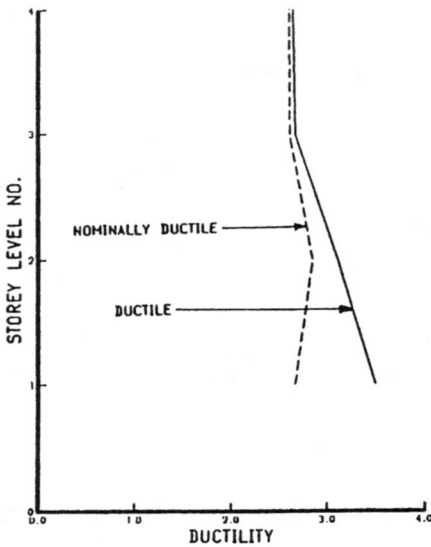

a) Four storey frames.   b) Ten storey frames.

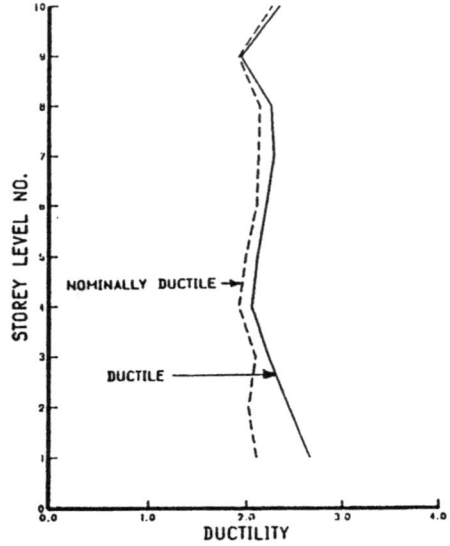

Figure 8. Beam rotational ductility demands.

Maximum shear stresses in the beams

Figs. 9 and 10 show a comparison between the beam design shear stresses and the maximum shear stresses attained during the response for the four and ten storey frames respectively. It can be seen that in ductile frames the response shear stresses exceed the design values by only 6 percent. In nominally ductile frames the response shear stresses exceed the design values by 30 percent. It should be mentioned here that in ductile frames, the provided transverse reinforcement is usually larger than that required for shear, because of the other detailing requirements. But in nominally ductile frames, the provided transverse reinforcement in the beams is usually just sufficient to resist the shear stresses. Therefore, the response shear stress being larger than the design value for nominally ductile frames should be of concern.

CONCLUSIONS

The seismic behaviour of ductile and nominally ductile reinforced concrete frames is compared. The response parameters investigated were the total and interstorey drifts, the beam ductility demands and the beam shear stresses. From the results of this investigations, the following conclusions could be drawn;

1) The behaviour of ductile and nominally ductile frames is similar in terms of deflections and ductility demands. Also, the difference between the ductility demands on the beams of ductile frames and those of nominally ductile frames is smaller than what would be implied by the difference in the design seismic base shear.

2) The beams of nominally ductile frames undergo shear stresses which are much larger than their design values. Therefore, the design shear forces used for nominally ductile frames should be increased from those obtained directly from the elastic static analysis. This approach would be consistent with the

266

Figure 9. Comparison between the design and response shear stresses in the beams of the four storey frames.

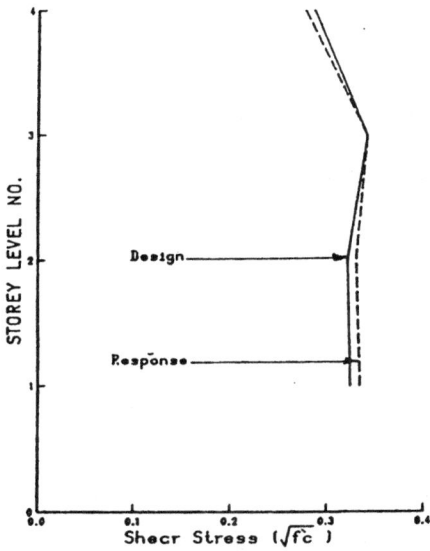

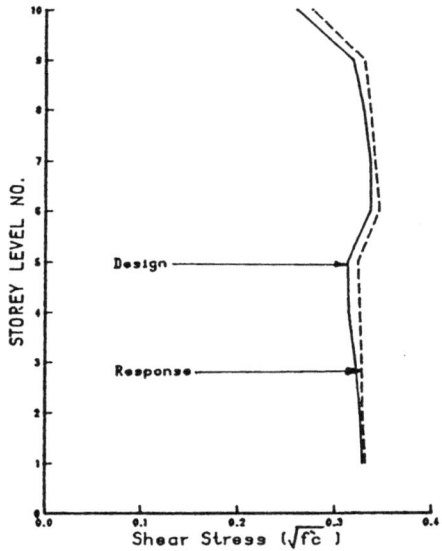

a) Ductile frame.

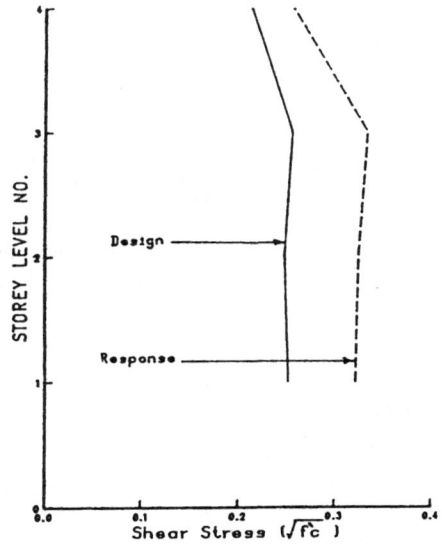

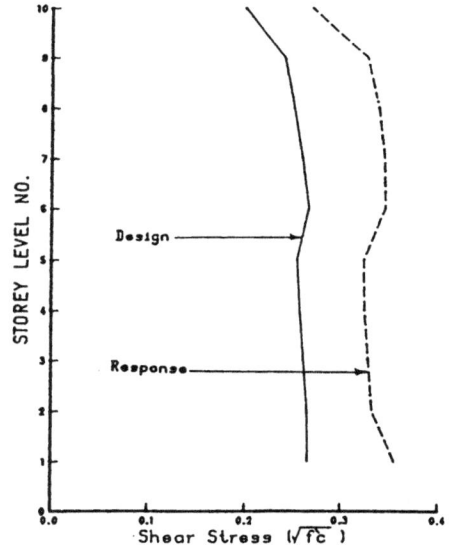

b) Nominally ductile frame.

Figure 10. Comparison between the design and response shear stresses in the beams of the ten storey frames.

a) Ductile frame.

b) Nominally ductile frame.

procedure employed by the New Zealand code (Standards association of New Zealand 1982) for the design of concrete structures with limited ductility. This will reduce the possibility of a non-desirable shear failure resulting in large cracks and irreparable damage.

REFERENCES

Associate committee on national building code 1990. National Building Code of Canada. Ottawa, Canada.

Canadian standards association 1984. Design of concrete structures for buildings. CAN3-A23.3-M84. Rexdale, Ontario, Canada.

Chung, Y.S., Meyer, C. and Shinozuka, M. 1987. Seismic damage assessment of reinforced concrete members. Report No. NCEER-87-0022, National center for earthquake engineering research, Buffalo, N.Y., U.S.A.

Clough, R.A., Benushad, K.L. and Wilson, E.L. 1965. Inelastic earthquake response of tall buildings. Proceedings, third world conference on earthquake engineering, New Zealand, II, 68-84.

Kanaan, A. and Powell, G.H. 1973. General purpose computer program for inelastic dynamic response of plane structures. Report No. EERC 73-6, Earthquake engineering research center, University of California, Berkeley, U.S.A.

Naumoski, N., Tso, W.K. and Heidebrecht, A.C. 1988. A selection of strong motion earthquake records having different a/v ratios. Report No. EERG 88-01, Earthquake Engineering Research Group, McMaster University, Hamilton, Ontario, Canada.

Standards Association of New Zealand 1982. Code of practice for the design of concrete structures. Wellington, New Zealand.

Seismic performance of R.C. structures in the Eastern U.S.A.

Bahram M. Shahrooz[1] and Rafael Muvdi[2]

ABSTRACT

Seismic performance of a R.C. frame designed according to current design provisions for structures located in moderate seismic risk was investigated analytically. A number of actual ground motions recorded in the northeast U.S. and southeast Canada, and artificially-generated earthquakes were utilized for this purpose. The dominant frequencies of the actual earthquakes were found to be several times larger than lower-mode frequencies of the structure. The level of excitation was not significant to cause structural or nonstructural damage. When the structure was subjected to a "rare" earthquake, which was obtained by amplifying the time scale of one of the recorded motions, the structure sustained inelastic action in the beams while the columns remained mostly elastic. The computed inter-story values indicate damage to nonstructural elements. In general, the structure appears to have adequate strength, but the overall stiffness is somewhat low.

INTRODUCTION

The state of art and practice in earthquake resistant design of structures has seen significant developments in the past two decades. Most of the changes can be traced to experimental and analytical studies conducted for better understanding of the behavior of the components that provide significant lateral resistance. The emphasis of previous investigations has been mostly on mitigating seismic hazard in regions with high seismicity, with little attention to other regions. Nevertheless, historical records and recent seismic events (e.g., the earthquake of 25 November 1988 in the Saguenay, Quebec with a felt area as far as Boston) indicate the possibilities of ground motions with appreciable magnitudes in regions currently designated having low to moderate seismic risk. The seismic performance of structures in the Eastern and Central U.S. has become an important issue. Considering that only a limited number of significant earthquakes have occurred since the installation of strong motion instrumentation in these regions, a comprehensive database is still not available.

As a result, most of the available studies have focused on the response of structures under artificially-generated ground motions which are anticipated in the Eastern U.S. (Sidel et al. 1989). Furthermore, the performance of beam-column joints in buildings designed primarily for gravity loads have been investigated experimentally (Pessiki et al. 1990). Dynamic response of a 1/6 model of a lightly

---

[1]Assistant Professor of Civil Engineering, University of Cincinnati, OH

[2]Research Assistant, Dept. of Civil and Environmental Engineering , University of Cincinnati

reinforced concrete frame under Taft S69E 1952 record was investigated through shake table tests (El-Attar et al. 1990). Considering the paucity of sufficient data on the seismic response of reinforced concrete structures in zones with moderate seismic risk, the research reported herein was undertaken. Using actual and artificially-generated ground motions, the behavior of a six-story reinforced concrete frame with a setback was investigated analytically.

TEST STRUCTURE

The structure was a six-story moment resisting reinforced concrete frame with a fifty percent setback at the midheight, as shown in Fig. 1. The structure was assumed to be an imaginary office building located in UBC Zone 2, and it was designed following the ACI 318-89 (ACI 1989) and UBC-88 (UBC 1988) provisions for frames located in moderate seismic risk regions. The design forces due to wind loading were found to be larger than those computed under earthquake loads specified by building codes. The members were detailed according to common practice for such structures. Inter-story drifts, estimated crack widths, and member deflections were within limits required by building codes (ACI 318-89 and UBC-88). The structure appears to be somewhat flexible; the computed first-mode vibration period was found to range between 0.7 to 1.0 sec. (obtained by using uncracked or cracked gross section properties, respectively) which is larger than the values expected for reinforced concrete frames similar to the test structure.

GROUND MOTIONS

The geological conditions of the northeast U.S. are similar to those in the adjoining regions in Canada, and hence seismic motions in these two areas are expected to have close attributes. In this study, records from Canada and the northeast U.S. with magnitudes of 5 or larger were used. These ground motions include 25 November 1988 Saguenay, Quebec; 31 March 1982 Miramichi, New Brunswick; 6 May 1982 Miramichi, New Brunswick; and 19 January 1982 Franklin Falls, New Hampshire with peak accelerations of 0.125g, 0.34g, 0.11g, and 0.031g, respectively. All of these records were measured within 100 km from the epicenter. As seen from the pseudo acceleration response spectra and Fourier amplitude spectra (Fig. 2), the earthquakes have more energy at higher frequencies, or for a narrow range of frequencies. A similar observation has been made for other Eastern North America earthquakes (Atkinson 1989), and has been attributed to site effects, crustal conditions, or source effects. For example, although the 31 March 1982 Miramichi record has a sizable peak acceleration (0.31g), its dominating frequency is approximately 25 Hz. At such high frequency, only very stiff building structures or bridges seem to be prone to damage. By comparing the first-mode frequency of the test structure (ranging between 1.0 to 1.4 Hz) and the dominant frequency of the ground motions, the level of excitation for the structure would be limited. As a result, artificially-generated ground motions with more energy closer to the frequencies of the test structure were also considered.

Based on the earthquake source model proposed by Boore and Atkinson (Boore and Atkinson 1987), three records with epicentral distances of 80 km, 160 km, and 640 km were derived using data from the East Coast. The records were generated for hard rock conditions, i.e., shear wave velocity was taken as 3.5 km/sec., and the records were assumed to have a moment magnitude of 7. The peak acceleration drops as the distance from the epicenter is increased, but the records farther from the epicenter indicate more energy at smaller frequencies (Fig. 3). Considering the peak accelerations and frequency contents, the record with epicentral distance of 80 km was selected for the analytical studies.

ANALYTICAL MODEL

Using the actual and artificially-derived ground motions, the response of the test structure parallel to the setback was investigated. The structure was modeled as two plane frames representing the interior frame and the two exterior frames. The floor slabs were assumed to act as rigid floor diaphragms. The beam-column joints were assumed to be rigid, i.e., rigid end zones equal to the column width and beam depth were used for the beams and columns, respectively. The additional rotational flexibilities due to possible reinforcement slip were not considered.

The analyses were conducted by using the computer program DRAIN-2D (Kannan and Powell 1973) which was modified to incorporate a trilinear, unsymmetrical version of Takeda's model commonly used to simulate stiffness degradation of reinforced concrete members. Hence, it was possible to define different moment strengths and stiffnesses under negative and positive bending. The contribution of the floor slab to the strength and stiffness of the supporting beams (particularly under negative bending moment) was modeled using techniques described elsewhere (Chern 1990). Moment-axial load interaction was ignored for the columns, and both the columns and beams were modeled by a yield interaction surface which neglects axial load. Member stiffnesses were computed based on flexural deformation only. Effects of gravity loads on element strength were considered by initializing the member end forces equal to those under gravity loads. Viscous damping was assumed to be proportional to the mass and original stiffness, and a damping ratio equal to 0.05 was used for the first two modes.

RESPONSE OF THE TEST STRUCTURE

The response of the test structure was gauged in reference to the value of inter-story drift, magnitude of roof lateral displacement, peak base shear, and damage pattern which may be inferred by formation of plastic hinges. The inter-story drift profiles over the height of the structure are plotted in Fig. 4. The largest value of drift occurred in the fifth floor when the structure was subjected to the 25 November 1988 Saguenay record. The maximum computed drift (0.14% of floor height) is approximately ten times smaller than the limit of 1.5 percent of inter-story height which is normally considered acceptable (Algan 1983). For the maximum inter-story drift sustained by the test structure, no damage to the structural or nonstructural elements is expected. The observed good performance is despite small stiffness of the structure (first-mode period ranging between 0.7 to 1.0 sec). The ground motions could not apparently excite the structure significantly. The peak values of base shear and roof displacement also indicate a similar observation (Table 1). For example, the largest base shear sustained by the structure was 0.027W (W = total weight) comparing to 0.04W under design wind loads. The analyses indicate that none of the elements experienced inelastic action.

It should be noted that the analyses did not account for brittle failure modes such as pullout of discontinuous bottom beam bars in the beam-column joints or column splice failure. Experimental tests on beam-column connections found in reinforced concrete structures similar to the one studied herein indicate that although failure would be eventually initiated by pullout of discontinuous bars, such connections have rather stable hysteresis loops for drifts up to 1.5 to 2 percent story height (Pessiki et al. 1990). For this range of drifts, the response of connections with continuous and discontinuous bottom beam bars were observed to be similar. Furthermore, connections with discontinuous bottom bars could sustain joint shear stresses as large as 80 percent of the value resisted by those with continuous bars through the connection. Lightly-confined column splices were also found to perform adequately. Such local failure modes would not apparently occur for the range of drifts considered in the analyses. Hence, even though the effects of poor detailing could not be simulated analytically, the good performance of the test structure concluded based on the dynamic analyses would likely remain valid.

To examine the behavior of the structure under "rare" ground motions, the 25 November 1988 Saguenay record was changed such that the frequency of the structure would be closer to the dominant frequencies of this record. For this purpose, the time scale of the actual acceleration history was arbitrarily amplified by a factor of 5. Under the altered Saguenay record, the structure experienced a maximum base shear equal to 0.16W, and the largest inter-story drift occurring at the fifth floor indicates damage to nonstructural elements. Some plastic hinges (Fig. 5) formed in the structure, but the level of inelastic action (approximately quantified by rotational ductility demands) was not significant. Despite the level of excitation (Table 1), the structure exhibited adequate strength. Nevertheless, the structure appears to be somewhat

271

flexible with potential damage to partitions and other nonstructural elements over a number of floors.

SUMMARY AND CONCLUSIONS

The behavior of an imaginary reinforced concrete frame located in regions with moderate seismicity was investigated analytically. Actual recorded motions from earthquakes in the northeast U.S. and southeast Canada, and artificially-generated earthquakes were used for this purpose. The actual records exhibit more energy at high frequencies, considerably larger than the natural frequencies of the structure. Within the limitations of the modeling techniques, the following conclusions may be drawn.

(1) When subjected to the actual recorded motions, the structure was not excited significantly. The peak base shear was about half of the base shear under the design wind loads, and the inter-story drifts were much smaller than values considered to cause damage. Damage to structural and nonstructural elements would be unlikely. The structure exhibited adequate strength and stiffness.

(2) For the level of drifts experienced by the structure, the effects of poor details such as discontinuous beam bottom bars through beam-column connections or lightly-confined column splices appear to be minimal.

(3) Under a "rare ground motion", plastic hinges were formed mainly in the beams. The level of inelastic action was rather limited indicating that the structure has sufficient strength. Considering the likelihood of damage to nonstructural elements at several floors, the overall stiffness of the structure appears to be somewhat low.

ACKNOWLEDGEMENTS

The research reported in this paper was supported partially by the University of Cincinnati Research Council. Professor Paul A. Friberg at NCEER ground motion facility at Lamont-Doherty geological observatory provided the recorded ground motions, and generated the artificial records. His assistance is gratefully appreciated. Professor Matej Fischinger at University EK in Ljubljana, Yugoslavia is thanked for providing the trilinear Takeda model used in this study.

REFERENCES

Algan, B.B., 1983, "Drift and Damage Considerations in Earthquake-Resistant Design of Reinforced Concrete Buildings," Ph.D. dissertation submitted to the Graduate College, University of Illinois, Urbana.

Atkinson, G.M. 1989, "Implications of Eastern Ground-Motion Characteristics for Seismic Hazard Assessment in Eastern North America," *Earthquake Hazards and the Design of Constructed Facilities in the Eastern United States*, The New York Academy of Sciences, Vol. 558, pp.128-135.

Boore, D. and Atkinson, G.M. 1987, "Stochastic Prediction of Ground Motion and Spectral Response Parameters at Hard-Rock Sites in Eastern North America," *Bulletin of the Seismological Society of America*, Vol. 77, pp. 440-467.

"Building Code Requirements for Reinforced Concrete (ACI 318-89)", ACI Committee 318, American Concrete Institute, Detroit.

El-Attar, A. et al. 1990, "Shake Table Test of a 1/6 Scale 2-Story Lightly Reinforced Concrete Building", *Proceedings*, Fourth U.S. National Conference on Earthquake Engineering, Vol. 2, Palm Springs, California, pp. 767-776.

Kannan, A.E. and Powell, G.H. 1973, "DRAIN-2D: A General Purpose Computer Program for Dynamic Analysis of Inelastic Plane Structures," *Report No. UCB/EERC-73/6*, Earthquake Engineering Research Center, University of California at Berkeley, Berkeley.

Pessiki, S.P. et al. 1990, "Seismic Resistance of the Beam-Column Connection Region in Lightly-Reinforced Concrete Frame Structures," *Proceedings*, Fourth U.S. National Conference on Earthquake Engineering, Vol. 2, Palm Springs, California, pp. 707-716.

Sidel, M.J., Reinhorn, A.M., and Park, Y.J. 1989, "Seismic Damageability Assessment of R/C Buildings in Eastern U.S.," *Journal of the Structural Division*, ASCE, Vol. 115, No. 9, pp. 2184-2203.

Chern, S.P. 1990,"An Analytical Model for Simulating the Contribution of Floor Slab in Slab-Beam-Column Systems," Thesis submitted to the Division of Graduate Studies and Research of the University of Cincinnati.

Uniform Building Code 1988, International Conference of Building Officials, Whittier, California.

Table 1. Extreme values

Ground motion	Roof displacement (in.)	Base shear (kips)
25 Nov. 88 Saguenay	0.5 (0.058)[*]	19 (2.4)[+]
31 March 82 Miramichi	0.3 (0.035)	6.9 (0.86)
6 May 82 Miramichi	0.1 (0.012)	6.2 (0.77)
19 Jan. 82 Franklin Falls	0.005 (0.0006)	2.5 (0.31)
Artificial record	0.58 (0.067)	21.6 (2.7)
Modified Saguenay	7.6 (0.88)	126 (16)

[*] Roof displacement as a percentage of total height.
[+] Base shear as a percentage of total weight.

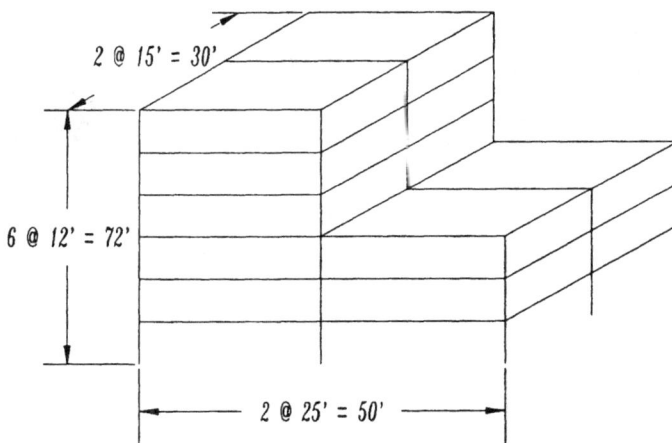

2 @ 15' = 30'

6 @ 12' = 72'

2 @ 25' = 50'

Fig. 1 Overall geometry of test structure.

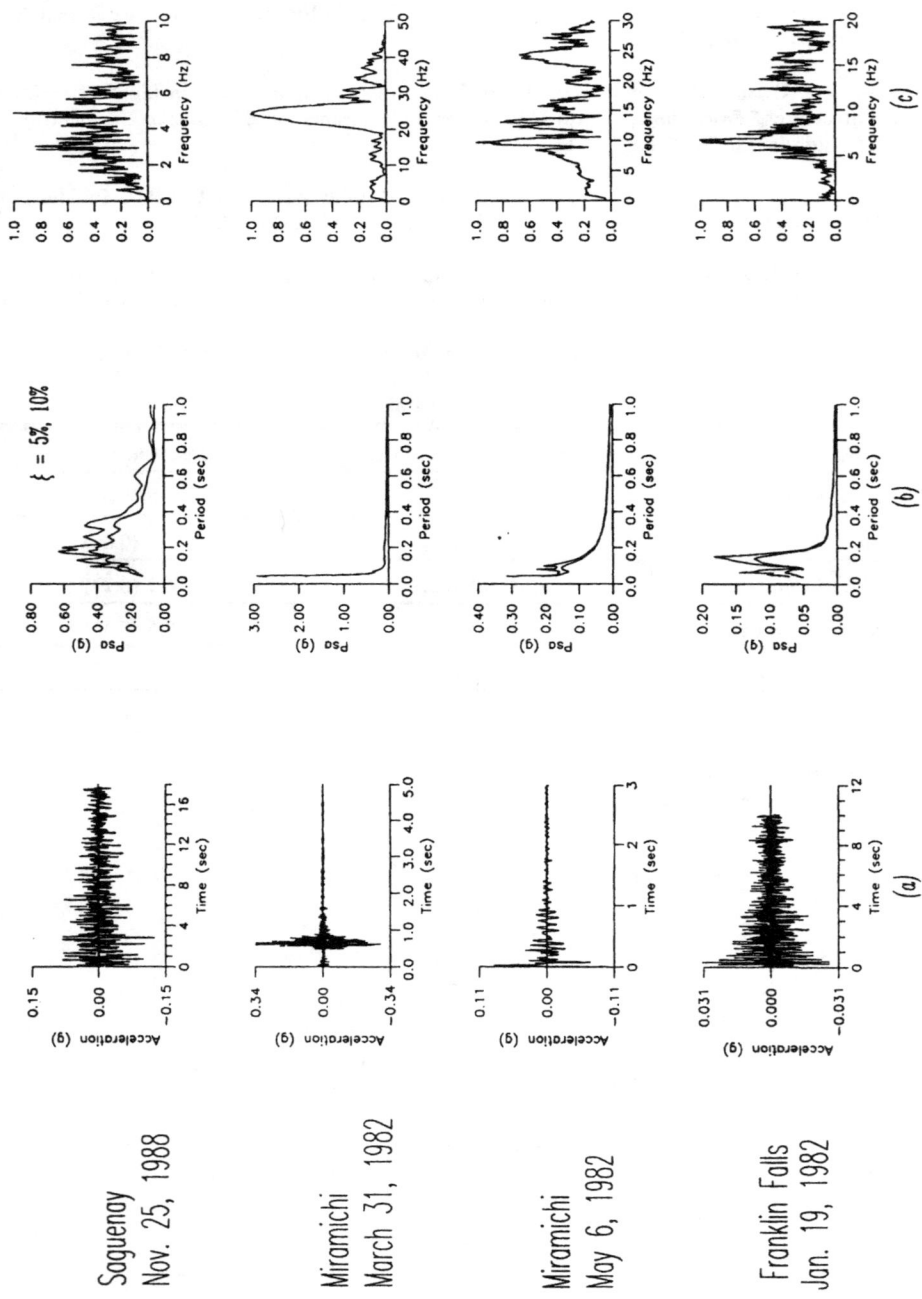

Fig. 2 Characteristics of recorded motions.

274

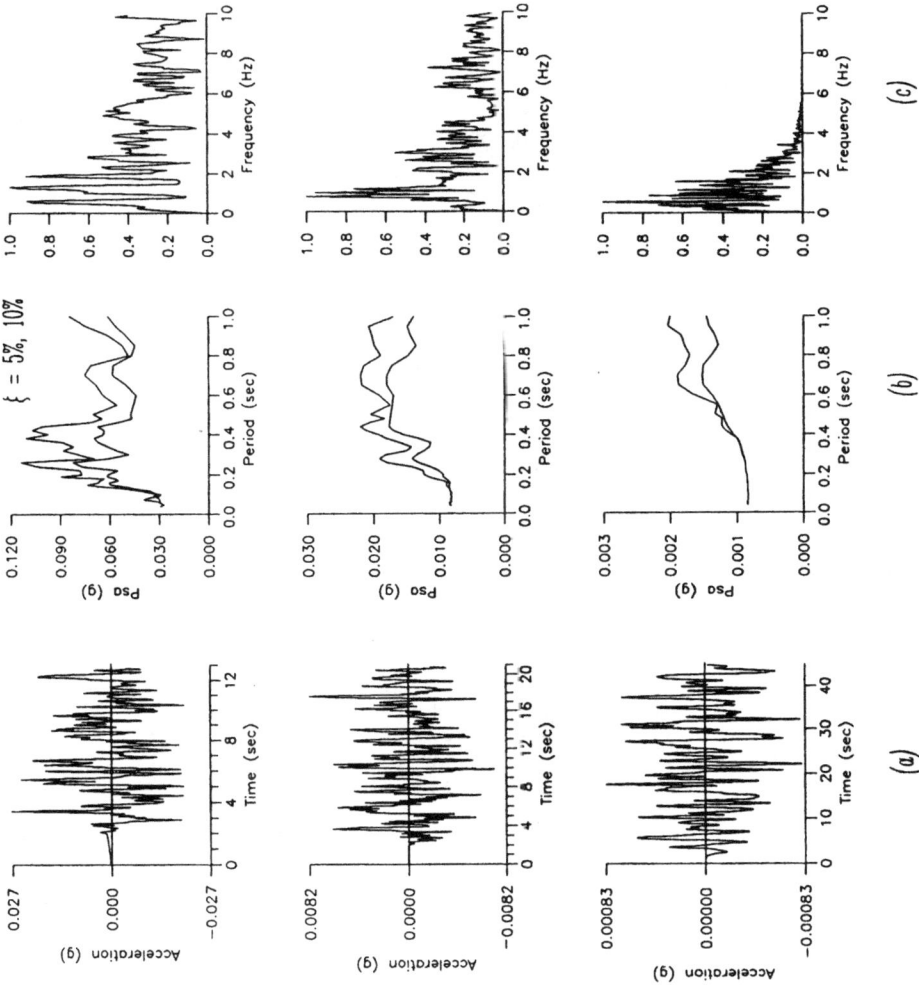

Fig. 3 Characteristics of the artificial records.

275

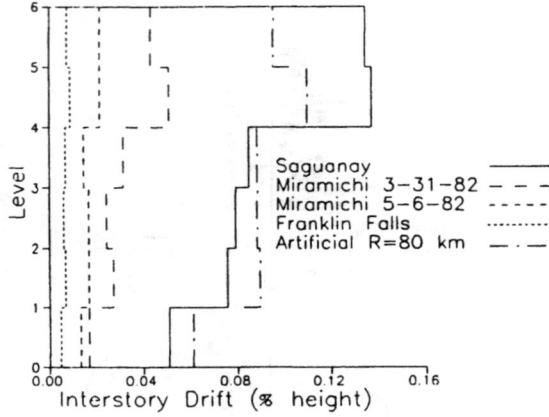

Fig. 4 Inter-story drift profiles.

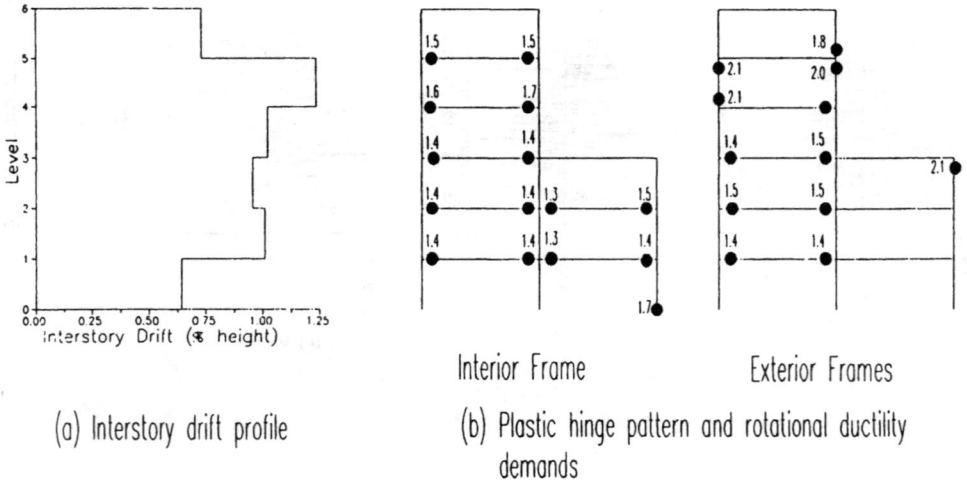

(a) Interstory drift profile

Interior Frame

Exterior Frames

(b) Plastic hinge pattern and rotational ductility demands

Fig. 5 Structural response to modified Saguenay.

276

Retrofit techniques for masonry chimneys in seismic zones

T. Baumber[I] and A. Ghobarah[II]

ABSTRACT

Unreinforced masonry chimneys sustained severe damage during recent earthquakes. The failure or collapse of such structures during a seismic event may result in the loss of industrial production as well as the loss of human life. It is now necessary to retrofit existing masonry chimneys as many were built either according to older codes or before current seismic codes were in place. It was found that the use of a reinforcing steel cage is the most effective retrofit system. This system can be designed and adapted to resist high levels of lateral inertia forces. The post-tensioning retrofit system was found not to be a reliable technique. This system worked adequately for short, slightly tapered chimneys but not for tall, highly tapered chimneys.

INTRODUCTION

Unreinforced masonry chimneys sustained severe damage during recent earthquakes. The susceptibility of masonry chimneys to severe damage was dramatically observed during the 1976 Tangshan (China) earthquake event (China Academy of Building Research, 1986). The failure or collapse of such structures during earthquakes may result in the loss of industrial production as well as the loss of human life. Many of the existing masonry chimneys were built according to older codes and long before the current seismic code provisions were available. Damage to chimneys was observed both at the base and at the top third of the height. The causes of these two types of failures were investigated in a recent study by Baumber and Ghobarah (1990). Now it is necessary to retrofit existing masonry chimneys to improve their seismic performance. The objective of this study is to evaluate retrofitting systems and investigate their effect on the seismic response of masonry chimneys.

A retrofit technique that was used in China involved the application of steel angles to the exterior of the chimney. Niu (1981) developed analytical expressions for the strength of the retrofitted structure. In his analysis, Niu smeared the reinforcing steel into a shell around the exterior of the chimney. Shu-quan (1981) also used this technique but did not present any details concerning the theoretical capacity.

[I]Graduate Student, Department of Civil Engineering and Engineering Mechanics, McMaster University, Hamilton, Ontario, Canada L8S 4L7

[II]Professor, Department of Civil Engineering and Engineering Mechanics, McMaster University, Hamilton, Ontario, Canada L8S 4L7

The two most common retrofit methods were investigated in this study. The first method involves the addition of a reinforcing steel cage to the chimney's exterior. The reinforcing steel will resist the tensile loads once the applied stress exceeds the tensile strength of the unreinforced masonry. The second retrofit technique involves the addition of a post-tensioning stress into the structure. The post-tensioning will induce initial compression into the masonry. This will allow the chimney to resist higher moments and thus the lateral load carrying capacity of the chimney will be effectively increased. Sketches of the two retrofit systems are presented in Fig. 1.

REINFORCING STEEL CAGE RETROFIT SYSTEM

The retrofit system involves the attachment of equally spaced bundles of reinforcing steel (e.g. 32 bundles). The large number of bar bundles is intended to spread the reinforcement as uniformly as possible around the circumference of the chimney. These bundles are fastened to the chimney by welding them to steel hoops which are attached to the chimney. The area of each bundle is dependent on the bending moment that must be resisted by the structure. In this system, the steel cage and chimney work together to resist the earthquake induced loads.

A dynamic analysis was performed utilizing a lumped mass approach. The masses of the masonry and steel were lumped at the mid-point of the mass segment. The analysis is carried out as outlined by Baumber and Ghobarah (1990).

Stress Distribution

Unreinforced masonry behaves in a brittle manner when subjected to earthquake loads. To avoid failure, unreinforced masonry must remain elastic. Any excursion beyond the elastic range will result in cracking of the masonry assemblage and thus causes failure. The reinforcing steel cage resists the tensile loads. Once the applied loads reach levels that exceed the tensile strength of the masonry, the masonry cracks and the steel then begins to resist all the tensile loads.

The stress distribution upon cracking of the masonry is assumed to be linear across the section. The maximum allowable design stress on the compression side is 0.32 f'm, where f'm is the ultimate compressive stress of the masonry assemblage. This design level is given by the Canadian Masonry Code, CAN3-S304-M84, 1984. The masonry is assumed to carry no tensile load. The stress in the steel is converted to an equivalent stress in the masonry by dividing by the ratio of the elastic moduli of steel to masonry.

Theoretical Strength

The theoretical strength can be determined from the known section and material properties. The steel is assumed to yield in tension just as the masonry reaches its' compressive capacity. The location of the neutral axis and the forces in both the steel and the masonry can then be determined for this balanced condition. For equilibrium, the sum of these forces must equal the applied axial load. In the case of chimneys, the applied axial load arises from both the self weight of the chimney and from forces induced by the vertical component of the earthquake ground motions. Equilibrium conditions are determined by a trial and error procedure. The force balance is adjusted by altering the area of steel required.

Periods of Vibration

Several chimneys of various practical dimensions were analyzed. The chimney configurations are listed in Table 1. The vibrational characteristics of the structure are altered when the mass and stiffness of the reinforcing steel cage is included in the dynamic analysis. Typically the retrofit increases the moment of inertia at the base of the structure by approximately ten percent. The mass of the base element is typically increased by approximately five percent. When the section is heavily reinforced, the increase in the moment of inertia of the base and in the mass of the base element can be as large as twenty five percent.

For small amounts of reinforcing steel, the effect of the reinforcing cage can be neglected in the analysis of the structure. As the size of the reinforcing cage increases, the effect on the vibrational periods increases. The periods of three chimney configurations with and without a reinforcing cage are presented in Table 2. The large change in the magnitude of the periods of some of the taller chimneys warrants an analysis which takes into account the effect of the stiffness and mass of the reinforcing cage.

Effectiveness of Technique

The effectiveness of this technique is best illustrated through the use of an example. A chimney with a height ratio of 20 and a diameter ratio of 3.0 was selected (configuration 13). This chimney's fundamental mode was the strongest participant in the response when subjected to intermediate a/v ratio ground motion records. When the chimney was subjected to high a/v ratio ground motion records, the second mode was predominant.

The critical section in the design of the reinforcing steel is that at the base of the chimney. This is the location of the maximum induced bending moment. This section is critical when the structure is subjected to both intermediate and high a/v ratio events. Previous work (Baumber and Ghobarah, 1990) showed that for high a/v ratio events the maximum bending stress occurred at a section at the top third of the chimney's height. The bending stress at this section is at a maximum even though the bending moment is not. This is a result of the taper of the chimney.

The reinforcing steel designed to resist the base moment is continued along the entire height of the chimney for convenience in the analysis. Ideally, the reinforcement should be decreased with height as the earthquake loads imposed on the structure decrease with height. This will provide a design that is economical as well as capable of resisting the forces induced by the earthquake ground motions.

When the resistance offered by the retrofit system is compared to the bending moments experienced by the original geometry, the system was found to perform adequately. The induced bending moments are always less than the resistance offered by the retrofitted section. This comparison is presented in Fig. 2.

This retrofit system is an adequate means of strengthening masonry chimneys. The system is effective for chimneys of small height ratios. These chimneys experience relatively low earthquake induced bending moments. A small amount of steel is required for the chimney to be able to resist the applied loads. The moment capacity and maximum applied bending moments for four chimney configurations are presented in Table 3.

POST TENSIONING RETROFIT SYSTEM

This retrofit technique involves the application of a compressive force onto the existing structure. This force is applied to the chimney by post-tensioning tendons placed parallel to the height. These tendons are anchored at both the foundation and the top of the chimney to a steel hoop. The tendons are not bonded to the exterior of the chimney and thus only provide a means to apply a compressive force.

Post-Tension Force

The magnitude of the post-tensioning force is limited by the compressive capacity of the smallest cross section at the top of the chimney. This section has no earthquake applied moment and therefore all the compression on the section is due to the post-tensioning force. Other cross sections will have earthquake induced bending stresses. The post-tensioning stress at these locations will be lower than at the top of the structure. The post-tensioning force is constant, however, for tapered chimneys, the cross-sectional area increases thus decreasing the applied stress.

In the design of the post tension force, the compressive stress at any section should not exceed the allowable given by the Canadian Masonry Standard. The limiting allowable post tension force for the majority of the chimney configurations used was approximately 8000 kN. For some of the configurations, this allowable force is slightly smaller.

Periods of Vibration

When the chimney is post tensioned, there is an axial force acting on the structure. This axial force will result in a reduction in the structure's stiffness. When the chimneys were analyzed taking the effect of the axial force into consideration, an increase in the magnitude of the periods of vibration did occur. The periods for three cases are presented in Table 4. The largest increase was for the fundamental period of the chimney configuration 16. The increase when compared to the period of the same configuration without the effects of the axial force considered was 1.68 percent. The higher modes were effected to an even lesser degree. This increase was deemed to be insignificant.

Effectiveness of the Technique

This retrofit technique was found to be adequate for structures that have a fundamental mode which is the strongest participant in the response. Short chimneys with a slight taper benefit the most from this retrofit system. The area of the base is small enough that the limiting allowable post tension force can still create a large compressive stress relative to the magnitude of the induced bending stress. This compressive stress will be able to reduce the created tension to a level which is below the masonry's allowable limit. The stress distribution for chimney configuration 13 subjected to the ground motions of the Imperial Valley record is presented in Fig. 3. The bending stress shown is the combination of the compressive stress which arises from the post-tensioning force and the stresses induced by the earthquake ground motions. As the height ratio increases, the effectiveness of this technique diminishes.

For chimneys with a second mode which participates strongly in the response, this retrofit system was also found to be effective. The stress distribution of chimney configuration 13 when subjected to the Tangshan ground motion is also presented in Fig. 3. The post-tensioning force is large enough that it was able to create significant compression stresses in the cross section. This compression again allowed the tension stress experienced by the cross section to be below the allowable limit.

SUMMARY

Observed failures of unreinforced masonry chimneys during earthquakes clearly show the need for retrofitting existing chimneys. The attachment of a reinforcing steel cage to the exterior of the chimney is an effective retrofit technique. This technique is able to resist any induced forces from any type of earthquake ground motion when an adequate amount of steel is utilized. Dynamic analysis of the structure including the steel cage is usually not required. The stiffness increase due to the presence of the cage does not significantly effect the periods of vibration. Also, this technique can be implemented both readily and at a reasonable cost.

The post tensioning system is not as an effective retrofit technique. The application of the technique is limited by the fact that the force induced by the post tensioning is dependant on the area of the smallest cross section of the chimney and the ultimate compressive strength of the material. This technique is adequate for short, slightly tapered chimneys. As the taper increases, the post tensioning stress at the base is small enough to significantly reduce the tensile stress.

REFERENCES

Baumber, T.A. and Ghobarah, A. 1990. Seismic Response of Tall Masonry Chimneys. The Annual Conference of the Canadian Society of Civil Engineers CSCE, Vol. IV, pp. 72-84, Hamilton, Canada.

CAN3-S304-M84. 1984. Masonry Design for Buildings. A National Standard of Canada. Canadian Standard Association.

China Academy of Building Research. 1986. The Mammoth Tangshan Earthquake of 1976 Building Damage Photo Album. China Academic Publishers.

Niu, Zezhen. 1981. Calculating Methods of Strengthened and Repaired Brick Masonry Structures for Earthquake Resistance. Final Proceedings: FRC-US Joint Workshop on Earthquake Disaster Mitigation through Architecture, Urban Planning and Engineering; Beijing, November 2-6, pp. 380-395.

Shu-quan, Zhang. 1981. Repair of Damaged Structures in Tangshan City. Final Proceedings: FRC-US Joint Workshop on Earthquake Disaster Mitigation through Architecture, Urban Planning and Engineering; Beijing, November 2-6, pp. 396-409.

Table 1. Chimney configurations

CONFIGURATION DESIGNATION	DIAMETER RATIO BOTTOM OD* / TOP OD	HEIGHT RATIO HEIGHT / TOP OD
1	1.5	10
13	3.0	20
16	3.0	25
17	3.5	25

* O.D. refers to outside diameter.

Table 2. Periods of vibration

CASE	MODE	RETROFIT (SEC)	NO RETROFIT (SEC)	% CHANGE
01	FIRST	0.18713	0.19413	3.61
	SECOND	0.04014	0.04284	6.30
	THIRD	0.01562	0.01685	7.27
16	FIRST	0.51351	0.52201	1.63
	SECOND	0.13365	0.14928	10.47
	THIRD	0.05530	0.06401	13.61
17	FIRST	0.43440	0.43593	0.35
	SECOND	0.12421	0.13056	4.86
	THIRD	0.05313	0.05697	6.72

Table 3. Resistance provided by reinforcing cage

CASE	STEEL AREA * (mm sq.)	RESISTANCE (kN m)	INDUCED MOMENT (kN m)
01	1400	8555	7544
13	1500	32638	29426
16	6000	63167	60627
17	2500	55395	52903

* Total steel area of one bundle, chimney reinforced by 32 such bundles

Table 4. Post tension vibrational period

CASE	MODE	RETROFIT (SEC)	NO RETROFIT (SEC)	% CHANGE
01	FIRST	0.19679	0.19413	1.37
	SECOND	0.04300	0.04284	0.39
	THIRD	0.01687	0.01685	0.14
16	FIRST	0.53080	0.52201	1.68
	SECOND	0.15046	0.14928	0.80
	THIRD	0.06421	0.06401	0.32
17	FIRST	0.44096	0.43593	1.15
	SECOND	0.13136	0.13056	0.62
	THIRD	0.05712	0.05697	0.27

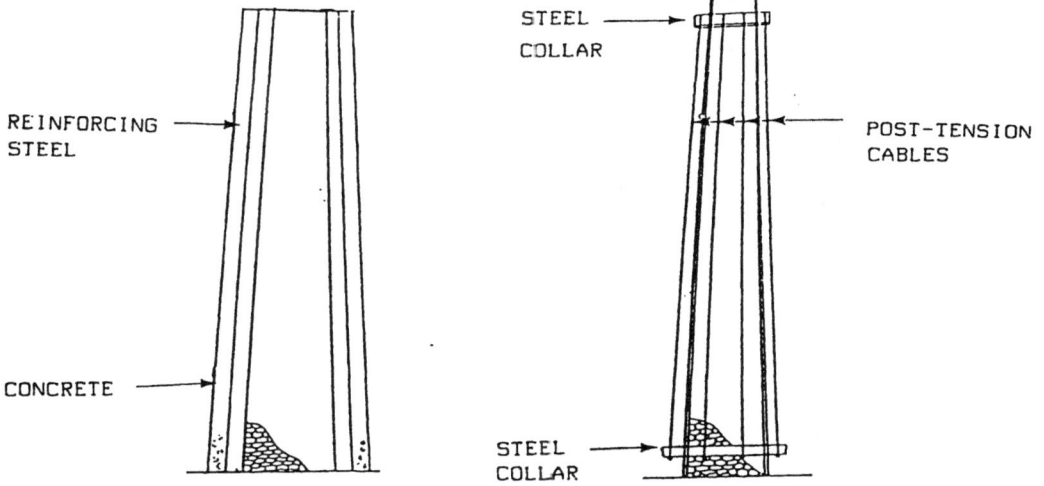

(a) Reinforcing Steel Cage   (b) Post-Tensioning

Fig. 1 - Sketches of Retrofit Techniques

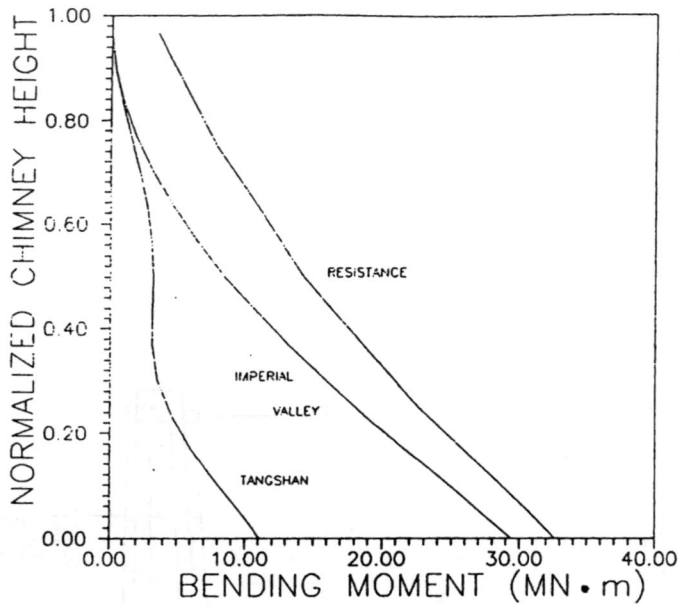

Fig. 2 - Resistance Provided using Reinforcing Steel Cage

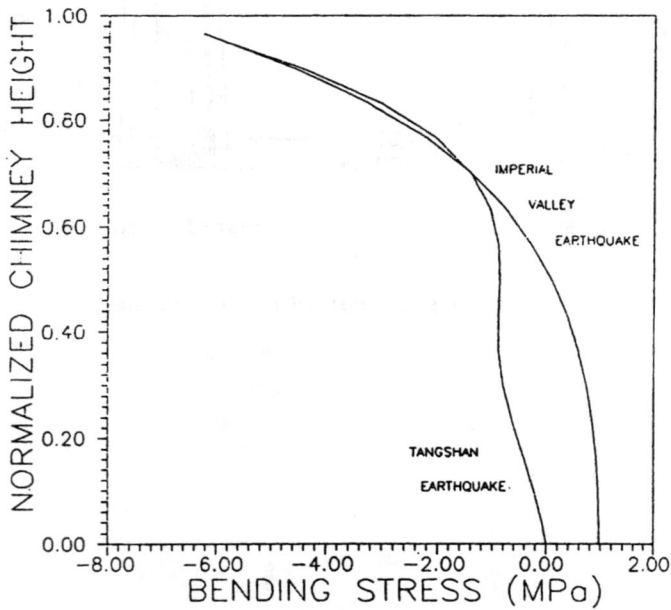

Fig. 3 - Bending Stress Distribution on Tension Side,

Post-Tension Technique

Elastic analysis of infilled frames using substructures

Janez Reflak[I] and Peter Fajfar[II]

ABSTRACT

The behaviour of reinforced concrete and steel frames subjected to horizontal loading may be strongly influenced by the presence of masonry infill. In the paper, a mathematical model based on the finite element idealisation is proposed for the linear analysis. Each infill is treated as a substructure and all degrees of freedom corresponding to the infill, with the exception of those at the contact with the frame, are eliminated using the static condensation procedure. The lengths of contact correspond to the condition where separation between frame and infill occured in the infill frame. The procedure is computationally effective for the linear static and free-vibration analysis and preserves the versatility of a finite element approach. It can be easily applied in the practice by using a computer program which includes substructuring option, e.g. SAP84.

INTRODUCTION

In many countries, steel and/or reinforced concrete structures are filled by brick or concrete block masonry. A continuous contact is usually provided on all sides between frame and infill wall. The experience both from real earthquakes and experiments has proved that the behaviour of frames subjected to horizontal loading may be strongly influenced by infill. Consequently, for a realistic simulation of the actual behaviour infill should be included in mathematical models.

Ordinary building structures subjected to strong ground motion will deform into inelastic range. The inelastic dynamic analysis, however, is too demanding for practical design procedures. The design of earthquake resistant structures is based on linear methods which can, in many cases, not only closely simulate the structural behaviour under minor to moderate loading, but also be used for an approximate simulation of the nonlinear behaviour.

Two types of mathematical models have been widely applied for the linear analysis of infilled frames. An equivalent diagonal strut represents the simplest model which can be used to simulate overall effects of infill after partial separation of frame and infill. The basic idea of the model was proposed by Polyakov (1957), and further developed by other researchers, (e.g. Stafford Smith 1966, Stafford Smith and Carter 1969, Mainstone 1971). Another possibility is to model infill walls with finite elements. Such a model is much more versatile, provides results also on local levels and can be easily used for an infill wall with arbitrary openings and for parapet walls. A finite element model was first used by Mallick and Severn (1967) and later applied in different variants by other researchers (e.g. Moss and Carr 1971, Mallick and Garg 1971, Riddington and Stafford Smith 1977, King and Pandey 1978).

I   Assistant Professor   University of Ljubljana, Department of Civil Engineering,
II Professor   Jamova 2, Ljubljana, Slovenia, Yugoslavia

Some attempts have been made to develop different models, e.g. a multidiagonal model (Thiruvengadam 1985), a model with an eccentric diagonal (Žarnić 1990), a model with two diagonals (Schmidt 1990), and a panel model (Axley and Bertero 1979). In the latter approach the infill was first modeled by finite elements. Then, all degrees of freedom, with the exception of the displacements and rotations of the corner nodes, were eliminated by the static condensation procedure. By using different assumptions, four different stiffness matrices, representing different types of infill, were developed in closed forms. The procedure is computationally effective but restricted to the predefined types of infill frames.

The procedure described in this paper represents a combination of computing efficiency (not much lower as in the case of a diagonal strut model) and versatility of a finite element approach. Infill walls are modelled with finite elements and treated as substructures which are connected to frame only in a few points. Very large building structures can be efficiently analysed by using a general purpose finite element program with substructuring option (e.g. SAP84). In the paper the mathematical model and the method of analysis are described and some numerical examples are presented.

MATHEMATICAL MODEL

The mathematical model and the method of analysis described in this paper can be in principle applied for any type of infill frames at any load level. We will study, however, only infill frames where no special connectors are provided to ensure contact between frame and infill. In such structures, separation between frame and infill occurs due to differences between flexural deformations of the frame and shear deformations of the infill panel (Fig. 1) which produce a tension failure of the connection. This separation may occur at a load level of approximately half of the ultimate capacity of the infill frame. If the linear analysis is used, mathematical models typically simulate the situation after bond separation (e.g. model with equivalent diagonal strut). This situation can be easily reproduced in a finite element analysis, if no tension strength is assumed at the connection between frame and infill.

In the mathematical model, applied in our study for static and free-vibration analysis, beam elements with 3 degrees of freedom per node will be used to model the frame and rectangular plane stress elements with 4 nodes and 2 degrees of freedom per node (panel elements) will be used to model the infill. Both types of elements will be rigidly connected in the common nodes (the same displacements will be assumed). No connection between beam and panel elements will be provided in the regions where separation between frame and infill occurs.

The zones and the lengths of contact between frame and infill can be determined by iteration. First, a rigid contact is assumed in all common nodes. Then, the rigid connection between beam and panel elements is removed in all regions where tension occurs at the connection. The procedure rapidly converges towards the final stage. It should be noted, however, that the real length of contact depends not only on structural parameters which are included in the analysis, but also on the variations in the quality of material and of workmanship which cannot be taken into account in the computation. Fortunately, the results of parametric studies have demonstrated that the main results (with the exception of the stresses in the corners of infill) are only scarcely dependent on small variations of the lengths of contact. For these reasons, it is not reasonable to attempt to calculate "exact" lengths of contact. In usual cases, the length of contact α of a column can be determined using the formula proposed by Stafford Smith (1966) (Fig. 1)

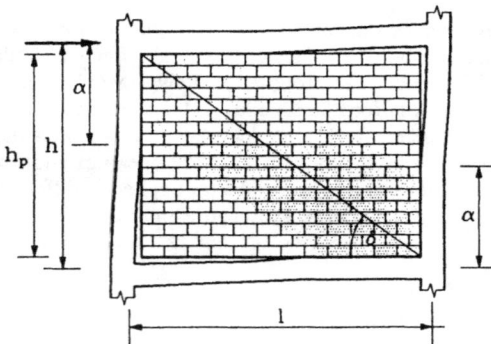

Figure 1. Infilled frame (after separation between frame and infill)

$$\frac{\alpha}{h} = \frac{\pi}{2 \lambda h} \tag{1}$$

where λh is a nondimensional parameter dependent on the ratio of the stiffness of the infill to the stiffness of the frame

$$\lambda h = h \sqrt[4]{\frac{E_p t_p \sin 2\delta}{4 E I h_p}} \tag{2}$$

E_p and E are the modulus of elasticity of infill and frame, respectively, I is the moment of inertia of the column, t_p is the thickness of the infill. The meaning of other parameters can be seen in Fig. 1.

The length of contact of a beam is less important and can be either assumed to be approximately equal to one half of the length of the beam or determined according to the formulae analogous to Eqs. 1 and 2. In the case of a partial infill (parapets, infill witht openings) where the lengths of contact are not known in advance, it may be necessary to determine them by iteration.

Some mathematical models, reported in the literature, consider the influence of sliding at the connection between frame and infill. In principle, this influence can be easily included in the presented method. It requires, however, an iterative procedure and has only a small effect on results, with the exception of the stresses in the infill in the vicinity of the contact ((Riddington and Stafford Smith 1977). For these reasons sliding at the contact zones was not included in the mathematical model.

In a finite element analysis the accuracy of results depends on the density of the mesh of finite elements. It has been found that an accuracy appropriate for design purposes can be obtained by using a 8 x 8 mesh in the panel. In many cases even a 4 x 4 mesh may yield acceptable results for displacements.

METHOD OF ANALYSIS

An analysis of a mathematical model of a real multistory building with infilled frames, which may have several thousands degrees of freedom, is computationaly demanding and its results are hard to be looked over. Computational efficiency and survailance of results will be greatly improved if substructuring technique is used. This technique is ideally suited for the analysis of infilled frames where typically identical infill walls are provided in different stories and different bays.

Each infill is treated as one substructure. The stiffness matrix for each substructure is first formulated by the finite-element approach. Then, all degrees of freedom with the exception of those belonging to the nodes at the contact with the frame are eliminated from the stiffness matrix by the static condensation procedure (e.g. Przemieniecki 1968). Finally, the stiffness matrix of the whole structure is formed. Only degrees of freedom which correspond to the contact points between frame and infill are included. The number of degrees of freedom is larger than in the case of a bare frame or in the case of an infilled frame modeled with equivalent diagonals, but it is an order of magnitude lower than in the case of the classical finite element approach. The same model and the same substructuring procedure can be used for the static and free-vibration analysis.

COMPARISON WITH EXPERIMENTS

In the Institute for Testing and Research in Materials and Structures in Ljubljana a series of experiments on 1:3 models of infilled reinforced concrete single-story single bay frames has been made (Žarnić 1990). The test structure is shown in Fig. 2. The observed horizontal force - top displacement relation is shown in Fig. 3. Relatively large standard deviation from the mean value can be observed even under laboratory conditions mainly due to the variation in the quality of workmanship. Two different mathematical models were used in the analysis. The first model was intended to reproduce the initial behaviour of the structure under small loading. A rigid contact between frame and brick masonry infill was assumed on all sides. The shear modulus of the infill G_p was determined according to the theory of elasticity, using Poisson's coefficient $\nu = 0.1$. As shown in Fig. 3, the stiffness of the mathematical model corresponds very well to the initial mean stiffness of the test models.

The second model was designed to simulate the state of the structure after the separation of the infill from the frame. The contact between the frame and the infill was provided only in the contact zone determined according to Eq. 1. A lower shear modulus, based on experiments on brick masonry walls subjected to large horizontal loading, was used. The stiffness of the mathematical model corresponds to the stiffness of the test models at approximately 50% to 60% of the ultimate loading, as shown in Fig. 3.

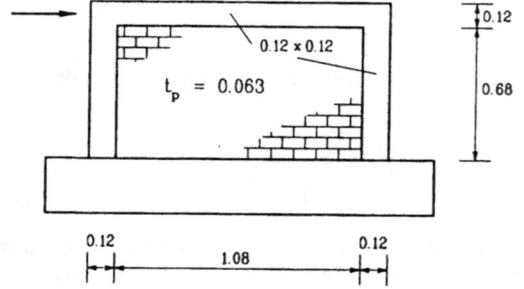

CONCRETE: $E = 1.1 \times 10^7$

$G = 4.4 \times 10^6$

INFILL: $E_p = 3.857 \times 10^6$

$G_p = 1.753 \times 10^6$ (model 1)

$G_p = 0.505 \times 10^6$ (model 2)

Figure 2. Test structure (units are meters and kilonewtons)

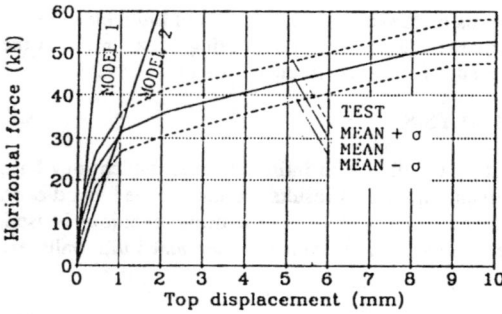

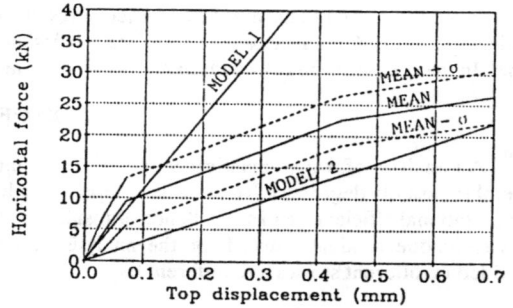

Figure 3. Horizontal force - top displacement relationships obtained in tests and analyses. The figure on the right side represents a detail of the figure on the left side

PARAMETRIC STUDY OF AN INFILLED FRAME WITH OPENINGS

The advantage of the proposed procedure in comparison to the classical finite element approach can be fully appreciated if analysing multi-story multi-bay infilled frames. Due to the space limitation, however, in this paper only some results of a parametric study of a simple single-story single-bay infilled frame with different types of openings are presented as an example of the application of the analysis procedure. The program SAP84, which includes the substructuring option, was used for the analysis. SAP84 has been originally developed at Peking University. It has been partially modified and extended in Ljubljana (Yuan and Peruš, 1988). The U.S. version of the program is called MICSAP (Yuan and Chang 1990).

Two horizontal forces (50 kN each) act in the upper two corners of the structure. The structural characteristics of the infilled frame are as follows (units are meters and kilonewtons)

h = 4, l = 4, λh = 6.2
Columns and beams 0.40/0.40
Panel t_p = 0.20
E = 3.2 x 10^7 G = 1.37 x 10^7
E_p = 3.2 x 10^6 G_p = 1.37 x 10^6

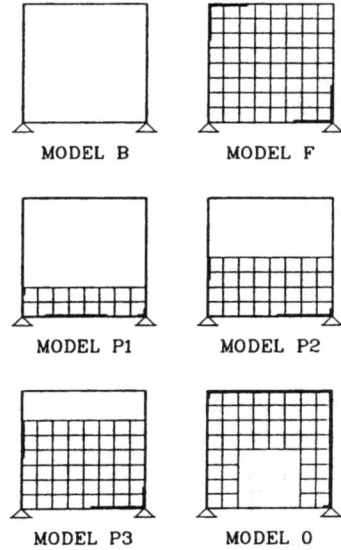

In addition to the bare frame (Model B), five different infilled frames have been studied (Fig. 4). The length of contact in the model F was determined according to Eq. 1. In the case of other models the lenghts of contact were defined using an iterative procedure.

Some typical results are shown in Figs. 5 to 7. Comparison of displacements and internal forces in the left column of the investigated models are shown in Fig. 5. The stiffness of infilled framed is up to ten times greater than the stiffness of the bare frame. Maximum tensile axial force in the left column of infilled frames is larger than in the column of the bare frame. Maximum shear force in the left column in the case of parapet infill is larger than in the column of the bare frame. Maximum values of bending moments in the column are observed at the top of parapets. It should be noted that the comparison in Fig. 5 was made at the same horizontal loading. In reality, however, usually much larger forces will be attracted to infilled frames due to their much larger stiffness. Maximum compresive stresses in the infill of four models are shown in Fig. 6 and the principal stresses are shown in Fig. 7 (tensile stresses are marked with arrows). It should be noted that the details of the stresses in the corners of the completely filled frame are strongly influenced by

Figure 4. Models analysed in the parametric study. The mesh of finite elements in the infill and the length of contacts are shown in each model

the parameter λh and might be numerically sensitive. High stress concentrations can be observed near to the corner of the opening in Model O and at the left upper corner of the parapet walls. A very distinctive diagonal path of compressive stresses can be observed in the case of the completely filled frame (Model F). Diagonals can be observed also in the case of parapet walls. However, they do not reach the bottom corner on the right side. A distorted diagonal path can be seen in the infill with an opening (Model O).

CONCLUSIONS

The main findings of the research reported in (Reflak 1990), and partly summarized in this paper, are as follows.

Masonry infill can drastically alter the structural response of frames and of whole structures. Thus, for a realistic simulation of the behaviour of infilled frames subjected to horizontal loading infill should be included in the mathematical model. A model with an equivalent diagonal strut is the most usual model for the elastic analysis. It can predict the stiffness of a completely infilled frame with resonable accuracy. It may fail, however, in the case of the infill with openings. It also usually does not provide correct results for shear forces and bending moments in the columns of the frame. (Shear failure of columns is a frequent failure mode for masonry infilled frames.)

A finite element model can be easily applied for any type of infill. Partial separation of frame and infill is usually assumed in the case of an elastic analysis. The lenghts of contact of infill with the surrounding frame can be determined by Eqs. 1 and 2, or, in the case of an infill with openings and unknown behaviour, by an iteration procedure. Provided that an appropriate mesh is used, a finite element model yields correct results for

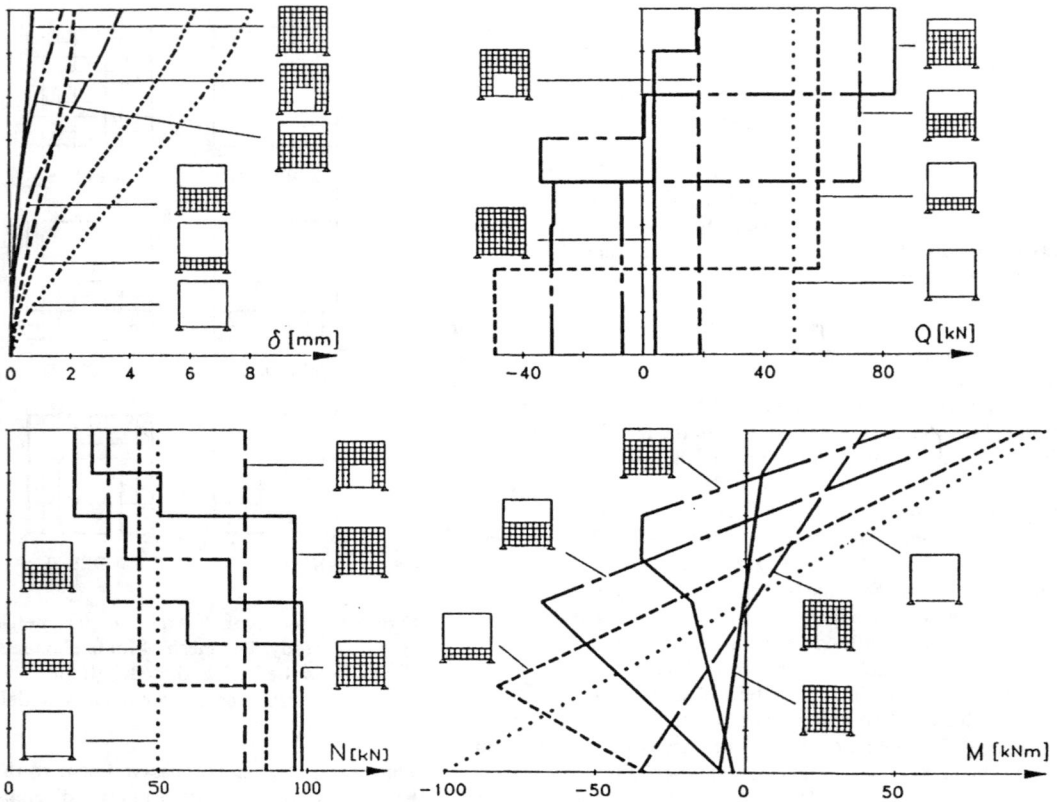

Figure 5. Displacements δ and internal forces (shear force q, axial force n and bending moment m) in the left column

displacements, internal forces in frame and stresses in infill, except of the local stresses in the vicinity of the contact between frame and infill, especially in the corners of the infill. These stresses are extremely dependent on many details of the mathematical model. Some of the details (e.g. the exact length of contact between frame and infill and the detailed behaviour of the contact) depend a lot on the quality of the workmanship and cannot be accurately predicted even in the case of laboratory conditions. For these reasons, any calculated local stress in the vicinity of the contact can be only a very rough approximation. The shear modulus of infill G_p may have a relatively important influence on the behaviour of infilled frames. At higher loads it does not follow the theory of elasticity. It should be based on the experiments on masonry walls, if available.

A finite element model may have in case of a multistory multibay infilled frame several thousands degrees of freedom and be prohibitively restrictive for the analysis of an ordinary building in a design office. However, when using the substructuring technique, treating each infill as one substructure, and eliminating all degrees of freedom cor-responding to the infill but those in the contact joints with the frame, the feasibility of the analysis will be largely improved.

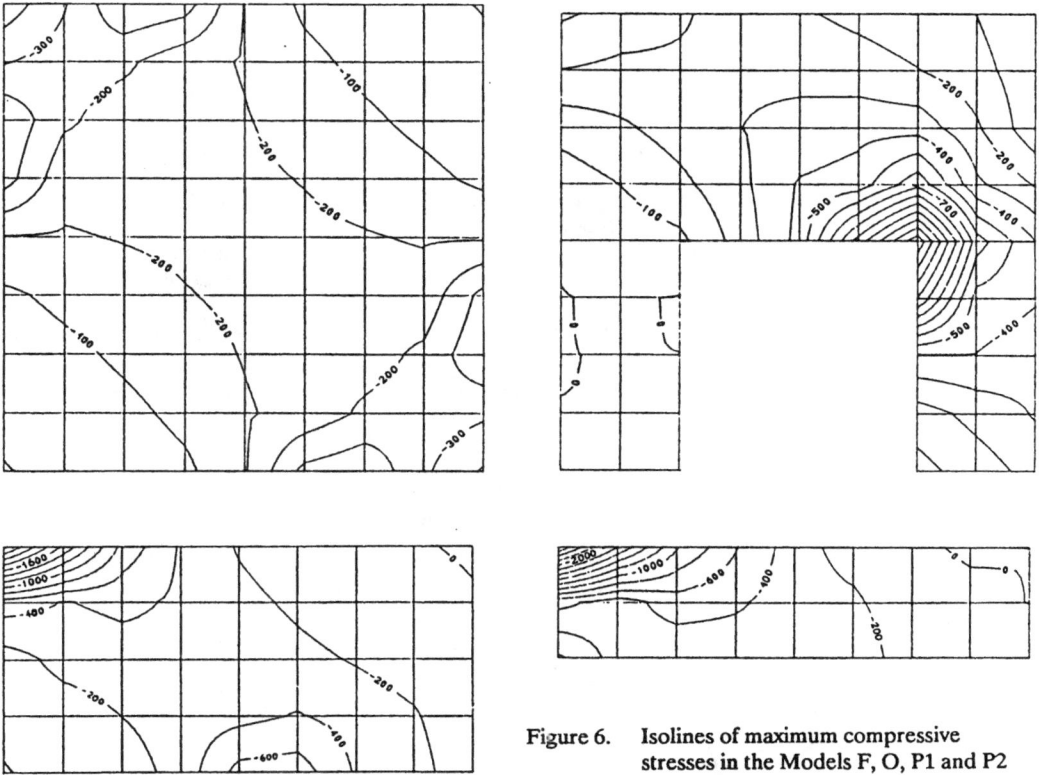

Figure 6.   Isolines of maximum compressive
            stresses in the Models F, O, P1 and P2

ACKNOWLEDGEMENTS

The authors would like to express their gratitude to the Ministry for Research of Slovenia for the partial support, to Mr. Iztok Peruš for assistance with plotting results and to Prof. Mingwu Yuan from Peking University for providing the SAP 84 program.

REFERENCES

Axley, J.W. and Bertero, V.V. 1979. Infill panels: their influence on seismic response of buildings. Report No. UCB/EERC-79/28, University of California, Berkeley.

King, G.J. and Panday, P.C. 1978. The analysis of infilled frames using finite elements. Proceedings of the Institution of Civil Engineers, 65, Part 2, 749-760.

Mainstone, R.J. 1971. On the stiffness and strength of infilled frames. Proceedings of the Institution of Civil Engineers, Supplement IV, 57-90.

Mallick, D.V. and Garg, R.P. 1971. Effect of openings on the lateral stiffness of infilled frames. Proceedings of the Institution of Civil Engineers, 40, 193-210.

Mallick, D.V. and Severn, R.T. 1967. The behaviour of infilled frames under static loading. Proceedings of the Institution of Civil Engineers, 38, 639-656.

Moss, P.J. and Carr, A.J. 1971. Aspects of the analysis of frame-panel interaction. Bulletin of the New Zealand Society for Earthquake Engineering, 4, 126-144.

291

Polyakov, S.V. 1957. Masonry in framed buildings; An investigation into strength and stiffness of masonry infilling (English translation), Moscow.

Przemieniecki, J.S. 1968. Theory of matrix structural analysis, Mc Grow-Hill, London.

Reflak, J. 1990. The influence of infill on linear static and dynamic behaviour of frames, Ph. D. Dissertation, University of Ljubljana, Department of Civil Engineering (in Slovene).

Riddington, J.R. and Stafford Smith, B. 1977. Analysis of infilled frames subject to racking with design recommendations. The Structural Engineer, 55, 263-268.

Schmidt, T. 1990. An approach of modeling masonry infilled frames by the FE-method and a modified equivalent strut method. Darmstadt Concrete, 5, 171-180.

Stafford Smith, B. 1966. Behaviour of square infilled frames. Proceedings ASCE, 92, ST1, 381-403.

Stafford Smith, B. and Carter, C. 1969. A method of analysis for infilled frames. Proceedings of the Institution of Civil Engineers, 44, 31-48.

Thiruvengadam, V. 1985. On the natural frequencies of infilled frames. Earthquake Engineering and Structural Dynamics, 14, 401-419.

Yuan, M. and Peruš, I. 1988. SAP 84, User's manual, IKPIR Publication No. 30. University of Ljubljana (in Slovene).

Yuan, M.W and Chang, T.Y. 1990. MICSAP, A general purpose static and dynamic structural analysis program. Univesity of Akron, Department of Civil Engineering.

Žarnić, R. 1990. Masonry infilled reinforced concrete frames as subassemblages of earthquake resistant buildings. Earthquake damage evaluation & vulnerability analysis of buildings structures (A. Koridze, editor), UNESCO INEEC, Omega Scientific, Wallingford, 79-100.

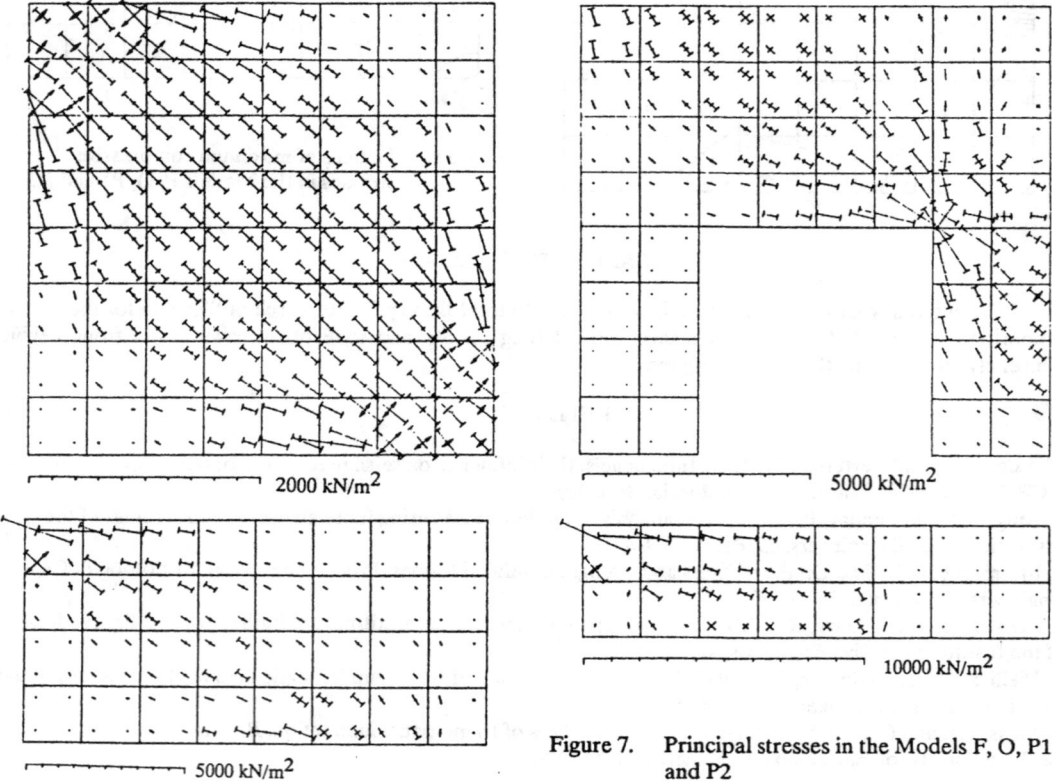

2000 kN/m^2

5000 kN/m^2

5000 kN/m^2

10000 kN/m^2

Figure 7. Principal stresses in the Models F, O, P1 and P2

Uncertainty in system modeling for seismic performance of structural systems

Sharif Rahman[1]

ABSTRACT

This paper examines the effects of uncertainty in structural and material models on seismic response and reliability of structural systems. The analysis involves hysteretic constitutive laws commonly used in earthquake engineering to model restoring forces and extensive Monte Carlo simulation for obtaining probabilistic characteristics. Several numerical examples based on single- and multi-degree-of-freedom systems are presented.

INTRODUCTION

It is widely recognized that the seismic performance of structural systems is overwhelmingly dominated by the uncertainty in seismic load processes. Accordingly, reliability analysis is carried out by assuming deterministic structural and material characteristics thus ignoring their inherent stochasticity. The uncertainty in system modeling is usually present due to the variabilities in (*i*) mathematical idealization of structural system, (*ii*) mathematical representation of hysteretic restoring forces, and (*iii*) the parameters of restoring force characteristics given a hysteretic model [3,4].

This paper conducts a systematic investigation to determine the effects of uncertainty in system modeling on seismic response and reliability of structural systems. The method of analysis is based on (*i*) common hysteretic constitutive laws for material models and (*ii*) extensive Monte Carlo simulation for performance evaluation. Several numerical examples on single- and multi-degree-of-freedom systems are presented.

STRUCTURAL SYSTEMS

In predicting the response and damage of actual structures, modeling of structural systems is an essential task. A model of a structure is defined as a mathematical representation of the behavior of the structure in its environment. The accuracy of response prediction depends on how well the models approximate the actual behavior of the structure.

---

[1]Research Scientist, Battelle Laboratories, Columbus, OH 43201

While derivation of the governing field equations of continuous models is not unduly diffi-cult, the attainment of general solution is a formidable task. To date, analytic solutions are known only for a few relatively simple continuous systems with linear elastic constitutive law, e.g., uniform beams, strings, plates and shells with simple boundary conditions. For the dynamic analysis of skeletal structures like frames, the continuous models becomes extremely complex and have thus found limited use in practice.

Consider a discrete, nonlinear structural system with critical cross-sections associated with appropriate restoring forces. The seismic modeling of this multi-degree-of-freedom, hysteretic system leads to the matrix differential equations of the form [3]

$$\mathbf{m}\ddot{\mathbf{X}}(t) + \mathbf{g}\left(\{\mathbf{X}(s), \dot{\mathbf{X}}(s), 0 \le s \le t\}; t\right) = -\mathbf{m}d\mathit{W}(t) \tag{1}$$

with the initial conditions

$$\mathbf{X}(0) = \mathbf{0}, \quad \text{and} \quad \dot{\mathbf{X}}(0) = \mathbf{0} \tag{2}$$

where t is the time coordinate originating at the beginning of seismic event $W(t)$, $\mathbf{X}(t)$ is a vector of generalized displacements, $\mathbf{g}$ is the vector functional representing general nonlinear hysteretic restoring forces, $\mathbf{m}$ is the constant mass matrix, and $\mathbf{d}$ is a vector of influence coefficients. In earth-quake engineering, the total restoring force $\mathbf{g}$ is usually allowed to admit an additive decomposition of a nonhysteretic component

$$\mathbf{g}_{nh} = \mathbf{c}\,\dot{\mathbf{X}}(t) + \mathbf{k}_{nh}\left(\mathbf{X}(t)\right)\mathbf{X}(t) \tag{3}$$

and a hysteretic component

$$\mathbf{g}_h = \mathbf{k}_h\left(\mathbf{Z}(t)\right)\mathbf{X}(t) \tag{4}$$

where $\mathbf{c}$ is the constant viscous damping matrix, $\mathbf{k}_{nh}$ is the nonhysteretic part of stiffness matrix, $\mathbf{k}_h$ is the hysteretic part of stiffness matrix, and $\mathbf{Z}(t)$ is the vector of additinal hysteretic variables the time evolution of which can be modeled by a set of general nonlinear ordinary differential equations

$$\dot{\mathbf{Z}}(t) = \mathbf{F}\left(\mathbf{X}(t), \dot{\mathbf{X}}(t), \mathbf{Z}(t); t\right) \tag{5}$$

in which $\mathbf{F}$ is a general nonlinear vector function the explicit expression of which depends on the hysteretic rule governed by a particular constitutive law. Following the state vector approach with the designation of $\theta_1(t) = \mathbf{X}(t)$, $\theta_2(t) = \dot{\mathbf{X}}(t)$, and $\theta_3(t) = \mathbf{Z}(t)$ the equivalent system of first-order nonlinear differential equations in state variables become

$$
\begin{aligned}
\dot{\theta}_1(t) &= \theta_2(t) \\
\dot{\theta}_2(t) &= -\mathbf{m}^{-1}\left[\mathbf{c}\,\theta_2(t) + \mathbf{k}_{nh}\left(\theta_1(t)\right)\theta_1(t) + \mathbf{k}_h\left(\theta_3(t)\right)\theta_1(t)\right] - \mathbf{d}\,W(t) \\
\dot{\theta}_3(t) &= \mathbf{F}\left(\theta_1(t), \theta_2(t), \theta_3(t); t\right)
\end{aligned}
\tag{6}
$$

which can be recast in a more compact form

$$\dot{\theta}(t) = \mathbf{h}\left(\theta(t); t\right) \tag{7}$$

with the initial conditions $\theta(0) = 0$ where $h(\cdot)$ is a vector function and $\theta(t) = \{\theta_1(t), \theta_2(t), \theta_3(t)\}^T$ is the response state vector. The nonlinear system of first-order ordinary differential equations in the initial-value problem of Eq. 7 can be solved by using step-by-step numerical integration such as fifth- and sixth-order Runge-Kutta integrators. When the excitation and/or the structural and material characteristics are random, $\theta(t)$ becomes a vector stochastic process which characterizes state of structural system.

UNCERTAINTY IN STRUCTURAL MODELS

A major source of seismic risk in New York City relates to the hundreds of flat slab apartment buildings constructed during the past 40 years. These buildings house thousands of people and were designed primarily for gravity loads. Fig. 1 shows a floor plan of a 24-story R/C flat-slab building.

The structure is modeled as a two-dimensional frame-shear wall type building based on the assumption that the floors have perfect in-plane rigidity. Moment of inertia of all the columns are lumped into columns of a 3-bay planar frame (System-A). For the shear walls, the moment of inertia are lumped into two separate walls (System-B and -C) corresponding to contributions from small and large walls. Hinged links are then used to transfer the axial loads from System-A to System-B and then from System-B to System-C. The simplified idealized structure is shown in Fig. 2.

Out-of-plane bending of floor slabs are considered by idealizing slabs into equivalent beams [5] of same depth with effective width being some fraction λ_w (effective width coefficient) of slab panel width. Using the chart used in Ref. 5 with proper regard to the irregularity of plan, a lower bound of $\lambda_w = 0.35$ and an upper bound of $\lambda_w = 1.0$ are obtained.

The variability of effective width coefficient λ_w may incorporate substantial amount of uncertainty in the response of structure due to earthquake loads. For example, the initial fundamental natural period T_0 of the building is 2.9 s for $\lambda_w = 0.35$ and 2.3 s for $\lambda_w = 1.0$. Fig. 3 shows a plot of top displacement of the building versus seismic base shear coefficient obtained from nonlinear static analysis based on a bilinear force-deformation model [5]. Significant differences are noticed in the values of maximum base shear coefficients, $e.g.$ 0.045 and 0.08 when calculated for $\lambda_w = 0.35$ and $\lambda_w = 1.0$,   respectively.

UNCERTAINTY IN MATERIAL MODELS

The variability in material models constitutes another major source of uncertainty in the evaluation of seismic performance of structural systems. Two sources can be identified and they correspond to the uncertainty in (i) the mathematical idealization of hysteretic constitutive law and (ii) the parameters of restoring force characteristics given a hysteretic model.

Single-Degree-of-Freedom Systems

Consider a nonlinear hysteretic oscillator with mass $m = 1.0$ $kips$ s^2 in^{-1}, damping coefficient $c = 0.06$ $kips$ s in^{-1}, initial stiffness $k = 1.0$ $kips$ in^{-1}, yield strength $F_y = 1.0$ $kips$ which is subjected to a $zero$-mean stationary Gaussian random process $W(t)$ with one-sided power spectral density $G(\omega) = G_0$ for $\omega \leq \bar{\omega} = 3$ rad s^{-1} and $zero$ otherwise. The duration of motion is assumed to be $t_d = 40$ s.

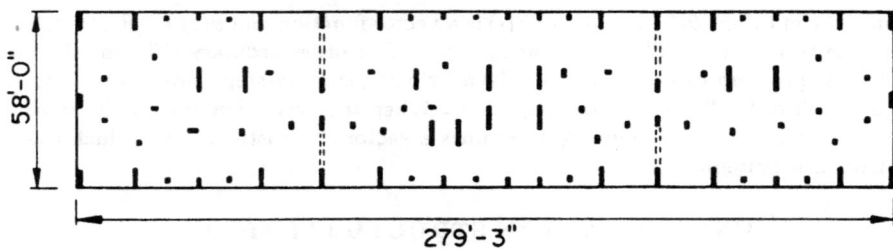

Fig. 1: Typical Floor Plan of Building System

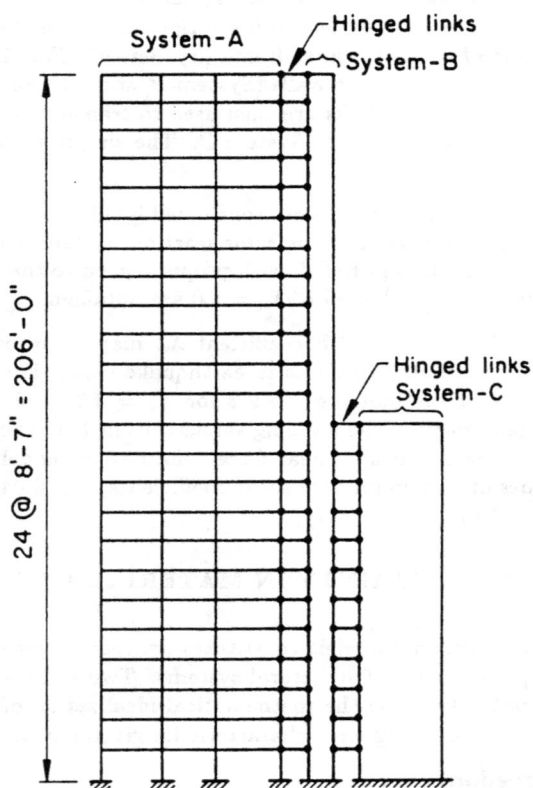

Fig. 2: Idealized Frame-Wall System

Three nondegrading restoring force models are considered in the sensitivity study: the ideal elasto-plastic, bilinear and Bouc-Wen. Details of these hysteretic models are available in the current literature [1,3,7]. Explicit description of the associated F functions in Eq. 5 can be obtained from Ref. 3. The models are equivalent in the sense that they have the same initial stiffness and strength characteristics and the parameters of each model are assumed to be deterministic.

Seismic performance of structural systems can be evaluated in terms of the condition that a specific response or damage level is exceeded during seismic ground motion. One such response quantity of interest is the ductility ratio μ given by

$$\mu = \frac{1}{X_y} \left[\max_t X(t) \right] \tag{8}$$

where X_y is the yield displacement and $X(t)$ is the relative displacement response of the oscillator. Since the ground motion is modeled as random process, a realistic assessment requires computation of the the probability $p = \Pr(\mu > \mu_0)$ with μ_0 representing the ductility threshold. This probability is estimated in the paper by Monte Carlo simulation with 3000 samples.

Fig. 4 shows the probability $p = \Pr(\mu > \mu_0)$ for several intensities of the input. For weak noise ($G_0 = 0.005 \ in^2 \ s^{-3}$), the exceedance probability of μ for Bouc-Wen hysteresis is considerably smaller than that for either elasto-plastic or bilinear models which exhibit identical behavior due to mostly linear response. For strong noise ($G_0 = 0.5 \ in^2 \ s^{-3}$), the probabilities $\Pr(\mu > \mu_0)$ becomes similar for bilinear and Bouc-Wen models both of which show smaller values of above probability than that for the elasto-plastic model. When the strength of noise is somewhat intermediate ($G_0 = 0.5 \ in^2 \ s^{-3}$), all the hysteretic models exhibit practically similar behavior.

Multi-Degree-of-Freedom Systems

Consider a 10-story steel frame building in Ref. 2 which is idealized here as a 10-degree-of-freedom shear beam system (stick model) with one degree of freedom per story. The mean values for the physical properties of stick model obtained from Refs. 2 and 4. Mass, damping coefficient, stiffness, and strength are treated as independent lognormal random variables with coefficients of variation 11%, 65%, 13%, and 23%, respectively. The coefficients of variation account for uncertainty in both mathematical idealization of hysteretic models and model parameters and are obtained from Ref. 6. Due to common construction and workmanship, each of these random variables are assumed to be perfectly correlated among all the stories. The restoring force at each story k is modeled by the nondegrading Bouc-Wen hysteresis with the hysteretic parameters defined in Ref. 4.

The ground motion is modeled as a uniformly modulated random process $W(t) = \psi(t)\widetilde{W}(t)$ where the modulation function

$$\psi(t) = \begin{cases} \left(\frac{t}{t_1}\right)^2, & 0 \leq t \leq t_1 \\ 1, & t_1 \leq t \leq t_2 \\ \exp[-c_\psi(t - t_2)], & t_2 \leq t \end{cases} \tag{9}$$

and the $\widetilde{W}(t)$ is a zero-mean stationary Gaussian colored process with one-sided power spectral density

$$\tilde{G}(\omega) = G_0 \frac{1 + \left[2\zeta_g(\frac{\omega}{\omega_g})\right]^2}{\left[1 - (\frac{\omega}{\omega_g})^2\right]^2 + \left[2\zeta_g(\frac{\omega}{\omega_g})\right]^2} \tag{10}$$

in which $t_1 = 1.5\ s$, $t_2 = 8.5\ s$, $c_\psi = 0.18\ s^{-1}$, $\omega_g = 16.5\ rad\ s^{-1}$, $\zeta_g = 0.8$, and $G_0 = 130.0\ in^2\ s^{-3}$. The duration of motion is assumed to be $t_d = 15\ s$.

Table 1 shows the exceedance probability of story level ductility ratio μ_k $(k = 1, 2, \cdots, 10)$ for several thresholds $\mu_0 = 3, 4, 5, 6$ at different stories of the 10-story steel frame structure. Two cases are considered. In the first case, the analysis is based on deterministic structural and material characteristics obtained from the mean values in Ref. 2. In the second case, the analysis accounts for the uncertainty in structural system with its probabilistic characteristics mentioned earlier. In both cases, Monte Carlo simulation is performed with 3000 samples. Results from Table 1 suggest that the uncertainty in system modeling can increase significantly the exceedance probability of story ductility.

CONCLUSIONS

A systematic investigation is conducted to study the sensitivity of seismic performance to the uncertainty in structural and material characteristics. The analysis is based on commonly used hysteretic constitutive laws for modeling restoring forces and extensive Monte Carlo simulation for seismic performance evaluation.

Results from a 24-story R/C flat-slab building suggest that the variability in structural model itself can significantly alter dynamic characteristics of structural system. Reliability analysis performed on a single-degree-of-freedom system and a 10-story steel frame building reveal that the material uncertainty can also have significant effect on seismic response and reliability.

REFERENCES

1. Bouc, R., "Forced Vibration of Mechanical Systems with Hysteresis," Abstract, Proceedings of the 4th Conference on Nonlinear Oscillation, Prague, Czechoslavakia, 1967.

2. Lai,P. S-S., "Seismic Safety: 10-Story UBC Designed Steel Building," *Journal of Engineering Mechanics*, ASCE, Vol. 109, No. 2, April 1983.

3. Rahman, S., "A Markov Model for Local and Global Damage Indices in Seismic Analysis," Dissertation presented to the Graduate School of Cornell University, in partial fulfillment of the requirements for the degree of Doctor of Philosophy, 1991.

4. Rahman, S., and Grigoriu, M., "Effects of Model Uncertainty on Seismic Response and Reliability of Structural Systems" *Proceedings of the 9th Structures Congress*, Indianapolis, Indiana, 1991.

5. Rahman, S., Turkstra, C., Grigoriu, M., and Kim H-J, "Seismic Performance of Existing Building in New York City," Proceedings of the *7th Structures Congress*, San Francisco, California, 1989.

6. Sues, R. H., Wen, Y. -K., and Ang, A. H-S., "Stochastic Seismic Performance Evaluation of Buildings," Civil Engineering Studies, *Structural Research Series No. 506*, University of Illinois at Urbana-Champaign, Urbana, Illinois, May 1983.

7. Wen, Y. -K., "Method for Random Vibration of Hysteretic Systems", *Journal of Engineering Mechanics*, ASCE, Vol. 102, pp. 249-263, April 1976.

BASE SHEAR COEFFICIENT

$\lambda = 0.35$
($T_0 = 2.9$ s)

$\lambda = 1.00$
($T_0 = 2.3$ s)

TOP DISPLACEMENT (PERCENT OF HEIGHT)

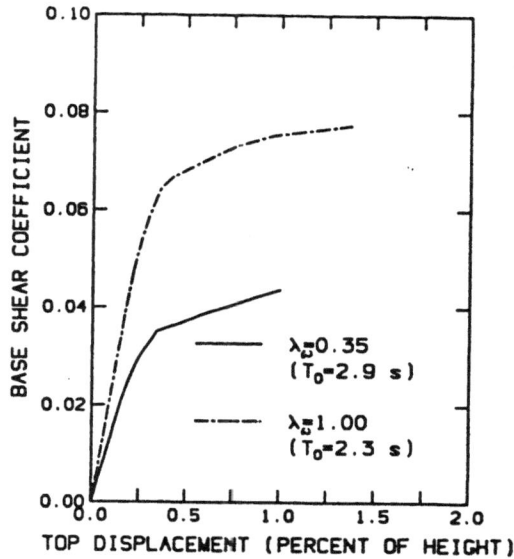

Fig. 3: Base Shear Coefficient Versus Top Displacement

Table 1: Exceedance Probability of Ductility Ratio

Cases	story k	$\Pr(\mu_k > \mu_0)$			
		$\mu_0 = 3$	$\mu_0 = 4$	$\mu_0 = 5$	$\mu_0 = 6$
Deterministic	1	0.383000	0.151000	5.666×10^{-2}	2.233×10^{-2}
System	2	0.201333	6.266×10^{-2}	1.566×10^{-2}	5.333×10^{-3}
	3	0.121000	2.400×10^{-2}	6.000×10^{-3}	1.666×10^{-3}
	4	7.633×10^{-2}	1.166×10^{-2}	2.000×10^{-3}	3.333×10^{-4}
	5	4.500×10^{-2}	5.666×10^{-3}	0.000	0.000
	6	4.400×10^{-2}	3.000×10^{-3}	0.000	0.000
	7	3.133×10^{-2}	9.999×10^{-4}	0.000	0.000
	8	0.000	0.000	0.000	0.000
	9	0.000	0.000	0.000	0.000
	10	0.000	0.000	0.000	0.000
Uncertain	1	0.452666	0.269000	0.154333	9.099×10^{-2}
System	2	0.312666	0.153000	7.733×10^{-2}	4.133×10^{-2}
	3	0.223000	9.200×10^{-2}	4.066×10^{-2}	2.200×10^{-2}
	4	0.161000	5.700×10^{-2}	2.433×10^{-2}	1.033×10^{-2}
	5	0.109333	3.466×10^{-2}	1.066×10^{-2}	5.000×10^{-3}
	6	0.114000	3.066×10^{-2}	7.333×10^{-3}	3.666×10^{-3}
	7	0.104666	2.133×10^{-2}	5.000×10^{-3}	1.333×10^{-3}
	8	1.333×10^{-3}	0.000	0.000	0.000
	9	9.999×10^{-4}	0.000	0.000	0.000
	10	0.000	0.000	0.000	0.000

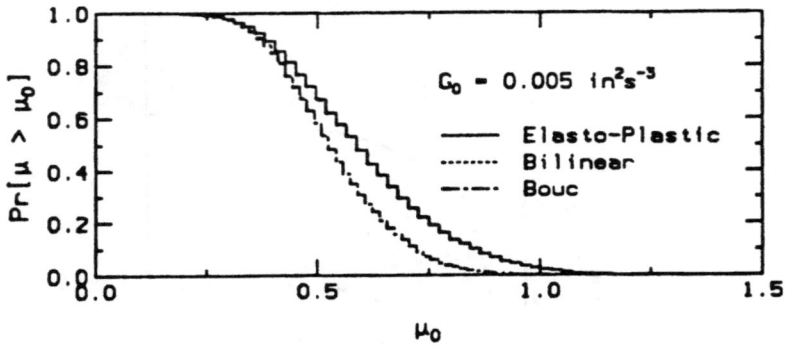

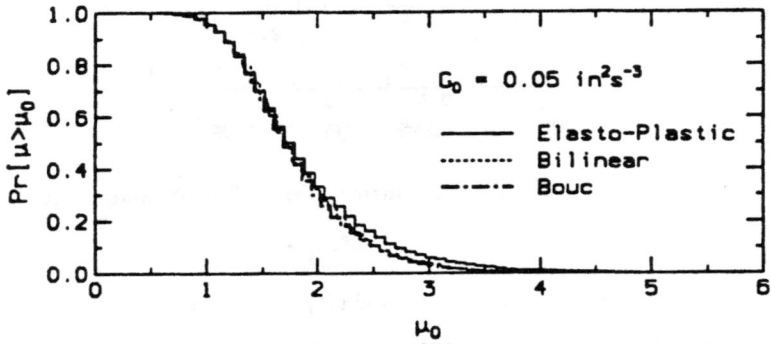

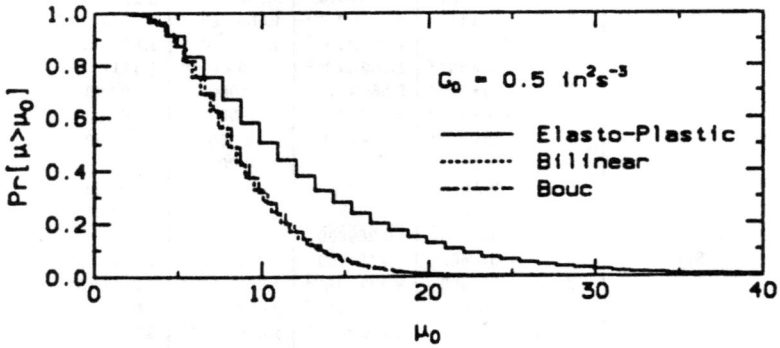

Fig. 4: Exceedance Probability of Ductility Ratio

Spatial correlation effects on seismic response of structures

O. Ramadan[I] and M. Novak[II]

ABSTRACT

The effects that spatial randomness of ground motion may have on seismic response of large structures to earthquakes are discussed. It is outlined how spatial randomness of this motion can be incorporated in the analysis. Spatial correlation of seismic ground motion may have important consequences for extensive structures such as long dams, pipelines, large buildings etc. Usually, such structures are examined using a two-dimensional finite element analysis of its "slice". However, lack of spatial correlation of the ground motion along the longitudinal axis of the structure may result in bending and shear stresses that are very significant. This is demonstrated using the example of the horizontal seismic response of a long, concrete gravity dam. Soil-structure interaction is accounted for in the analysis.

INTRODUCTION

The performance and safety of earthquake-resisting structures can be enhanced by improving the understanding and representation of earthquake ground motions. One aspect of these motions, relevant for the analysis of extended structures such as tunnels, pipelines, large dams and multi-supported long bridges, is spatial variability. This means that the ground motion and the resulting dynamic loads may not be perfectly coherent (synchronized) in space. This property reduces the total load in the structure in some cases while in others it can produce a type of response that remains completely unforeseen if spatial correlation is not accounted for. This paper outlines a procedure that makes it possible to account for spatial correlation of ground motions and soil-structure interaction, using an example of the horizontal response of a large gravity dam. The paper complements an earlier study by Novak and Suen (1987) which was limited to the vertical response and used a technique different from the one employed here.

[I] Research Assistant, Faculty of Engineering Science, The University of Western Ontario, London, Ontario N6A 5B9

[II] Professor, Faculty of Engineering Science, The University of Western Ontario, London, Ontario N6A 5B9

GROUND MOTION REPRESENTATION

While structural response to incoherent ground motions can be conveniently analyzed in terms of random vibration (Novak & Hindy, 1979; Hindy & Novak,
1980; Novak and Suen, 1987), the deterministic approach is more common. For
the latter approach, time histories of ground motions are needed. These can
either be actual motions recorded in seismic events or digitally simulated
motions generated to fit specified spectra. In this study, horizontal ground
motions are generated using the technique due to Shinozuka et al. (1988). The
motions generated are stationary, spatially two-dimensional and homogeneous
with zero mean and match a prescribed power spectral density. The motion
components feature two independent random phase angles uniformly distributed
between 0 and 2π. The time envelope function is applied on the simulated
stationary ground motions to convert them into nonstationary ones.

The stationary time histories are generated to fit the local power
spectral density represented by the modified Kanai-Tajimi acceleration
spectrum (Clough & Penzien, 1975), i.e.

$$S_{u_g}(\omega) = S_0 \frac{[1+4\xi_g^2(\frac{\omega}{\omega_g})^2]}{[1-(\frac{\omega}{\omega_g})^2]^2+4\xi_g^2(\frac{\omega}{\omega_g})^2} \cdot \frac{(\frac{1}{\omega_f})^4}{[1-(\frac{\omega}{\omega_f})^2]^2+4\xi_f^2(\frac{\omega}{\omega_f})^2} \quad (1)$$

in which S_0 is the white-noise bedrock acceleration spectrum, ω_g, ω_f are
resonance frequencies, and ξ_g, ξ_f are damping ratios. One of the advantages of
this spectrum is that the corresponding spectrum of ground displacements,
which will be needed, does not feature a singularity at frequency $\omega = 0$.

The power spectral density by Eq. 1 is used in the double series
expression for the generated ground motion with the frequency ω replaced by

$$\omega = g(k_1, k_2) = \alpha\sqrt{k_1^2 + k_2^2} \quad (2)$$

in which k_1, k_2 are the wave numbers of non-dispersive waves in the directions
x and z, respectively, and α is the wave phase velocity.

MATHEMATICAL MODEL

The Dam

The dam is assumed to have a large aspect ratio (length/width), and is
modelled by N beam elements (Fig. 1). The mass, m, and the polar mass moment
of inertia, I, associated with rocking are lumped at the element nodes. Water
behind the dam is not considered. Horizontal translations perpendicular to
the dam longitudinal axis and rotations about this axis (rocking) are allowed,
resulting in bending of the dam and its twisting. The horizontal joint
rotations are eliminated through static condensation leaving the model with
2(N+1) degrees of freedom. In the example, the dam is assumed to be of
constant cross-section as may be adequate for a rectangular canyon.

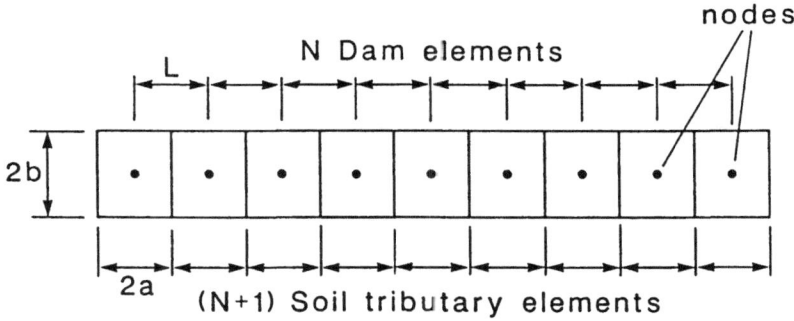

Figure 1. Schematic of dam elements and soil tributary elements

Material damping of the dam is hysteretic, i.e. frequency independent, and is introduced using complex Young's and shear moduli with material damping ratio β.

Soil Reactions

Soil-structure interaction is incorporated in the analysis through the complex, frequency dependent soil stiffness matrix, established for a sequence of rectangular tributary areas indicated in Fig. 2. The soil stiffness matrix associated with horizontal translations is obtained by inversion of the dynamic flexibility matrix which describes the displacements of the tributary areas due to unit harmonic loads distributed over the areas 2a x 2b on the surface of a viscoelastic halfspace. Coupling is considered. The horizontal displacements so generated can be written as

$$u(x,y) = \frac{F}{2G} [f(x,y)+i \ g(x,y)] \tag{3}$$

in which F = amplitude of the harmonic load taken as unity, G = shear modulus of the soil, and f,g = complex compliance functions evaluated using the solution of the stress boundary value problem. These functions were calculated following Gaul's (1977) solution. The part of the soil stiffness matrix stemming from horizontal translations is first calculated at the dam base and then transferred to the nodal points of the dam elements placed on the longitudinal axis passing through the dam centre of gravity. This couples the horizontal translation and rocking of the dam elements. The rocking soil stiffness matrix linking the rotations of individual soil elements was investigated in an analogous way. It turns out that the soil rotations do not propagate from the source very far but diminish quickly with distance. Therefore, the rocking soil stiffness matrix was evaluated in a simpler way using the well known solution for rocking of a strip footing. The torsional soil stiffness associated with the rotation in the horizontal plane is neglected because it is adequately accounted for by the horizontal translations when the dam elements are sufficiently small relative to the dam length.

303

Governing Equations of Dam Response

Denote the horizontal ground motion at node i, u_{g_i}, the relative motion between the ground and the dam, u_i, the absolute horizontal motion of the dam $U_i = u_i + u_{g_i}$ (Fig. 2); finally, rocking at node i is ψ_i. The dam structural stiffness matrix is related to the absolute translations U_i and rotations ψ_i, while the soil resistance derives from the relative horizontal displacements u_i and rocking ψ_i.

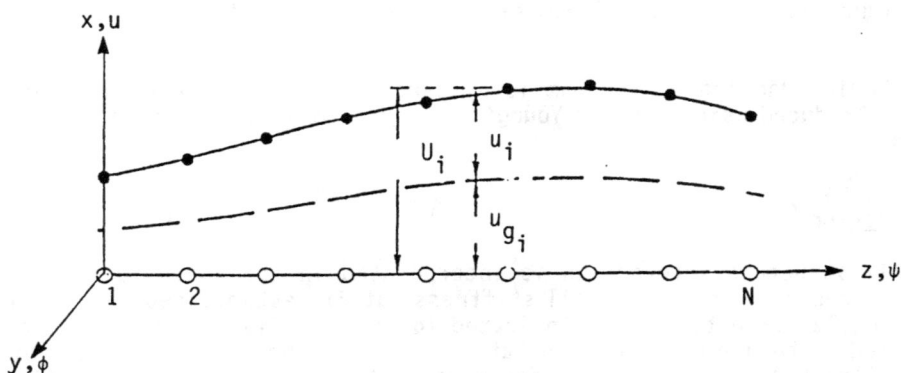

Figure 2. Dam displacements

Using the lumped mass matrix, the governing equations are

$$[M]\{\ddot{Y}\} + [k]_s\{Y\} + [k]_f\{y\} = \{0\} \tag{4}$$

in which the displacement vectors are written as

$$\{Y\} = [U_1 U_2 \ldots U_{N+1} \; \psi_1 \psi_2 \ldots \psi_{N+1}]^T = [\{U\} \; \{\psi\}]^T \tag{5}$$

$$\{y\} = [u_1 u_2 \ldots u_{N+1} \; \psi_1 \psi_2 \ldots \psi_{N+1}]^T = [\{u\} \; \{\psi\}]^T \tag{6}$$

The mass matrix is diagonal and lists all lumped masses, m, and mass moments of inertia, I, and the structural stiffness, $[k]_s$, condensed with regard to rotation, ϕ, is standard. The foundation stiffness matrix can be written as

$$[k]_f = \begin{bmatrix} [k_{uu}]_f & [k_{u\psi}]_f \\ [k_{\psi u}]_f & [k_{\psi\psi}]_f \end{bmatrix} \tag{7}$$

304

Denoting the total stiffness matrix

$$[k] = [k]_s + [k]_f \qquad (8)$$

the governing equation in terms of absolute motion, preferred here because of subsequent stress calculations, becomes

$$[\,M\,]\{\ddot{Y}\} + [k]\{Y\} = [k]_f\{y_g\} \qquad (9)$$

where the vector of the input ground motion, featuring zero rotations, is

$$\{y_g\} = [u_{g_1} \; u_{g_2} \cdots u_{g_{N+1}} \; 0 \; 0 \ldots 0_{N+1}]^T = [\{u_g\} \; \{0\}]^T \qquad (10)$$

With the notations by Eqs. 5 to 10, and partitioning Eq. 8 as in Eq. 7, the governing equation, Eq. 9, can be rewritten in a partitioned form, i.e.

$$
\begin{bmatrix} [m] & \vdots \\ ---\vdots--- \\ \vdots & [I] \end{bmatrix}
\begin{Bmatrix} \{\ddot{U}\} \\ -- \\ \{\ddot{\psi}\} \end{Bmatrix}
+
\begin{bmatrix} [k_{uu}] & \vdots & [k_{u\psi}] \\ ---\vdots--- \\ [k_{\psi u}] & \vdots & [k_{\psi\psi}] \end{bmatrix}
\begin{Bmatrix} \{U\} \\ -- \\ \{\psi\} \end{Bmatrix}
=
\begin{bmatrix} [k_{uu}]_f & \vdots & [k_{u\psi}]_f \\ ---\vdots--- \\ [k_{\psi u}]_f & \vdots & [k_{\psi\psi}]_f \end{bmatrix}
\begin{Bmatrix} \{u_g\} \\ -- \\ \{0\} \end{Bmatrix}
$$
(11)

Because of the frequency dependence of the soil stiffness matrix these equations are solved by Fast Fourier Transform in the frequency domain (complex response analysis), transforming first the simulated ground motions generated digitally. For each harmonic component further condensation with regard to $\{\psi\}$ is implemented to save on computing time.

EXAMPLE

To illustrate the type of response the above procedure yields, a long, concrete gravity dam with a cross-section similar to that of the Koyna Dam in western India is analyzed. The dam is 103 m high with a crest length of 853 m. Its cross-section, assumed to be constant, is shown in Fig. 3. The dam specific mass and Young's modulus are 2300 kg/m^3 and 30000 MPa, respectively. The Poisson's ratio is 0.2. The foundation rock is basalt with a shear wave velocity of 1218 m/s (4000 ft/s) and Poisson's ratio equal to 0.3. Further data on this dam are given in Novak and Suen (1987). The ground motion is generated by Eq. 1 with $\omega_f = 0.1\,\omega_g$, $\xi_f = \xi_g = 0.6$, $\omega_g = 15.71\ s^{-1}$, and the envelope function shown in Fig. 4. The dam is divided into ten equal elements.

The ground motions are simulated at the eleven nodes in the direction X with the dam lying along the axis Z and the wave phase velocity of 1400 m/s. The duration of the ground motions is 10 s. The simulated ground motions are shown in Fig. 5. The maximum ground acceleration is 0.15g. The correlation

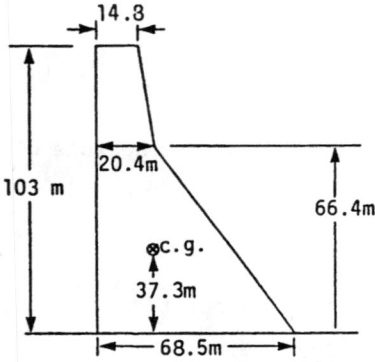

Figure 3. Dam cross-section

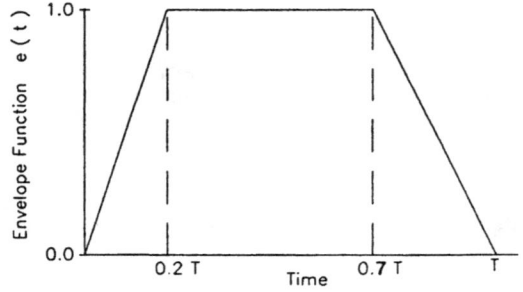

Figure 4. Envelope function

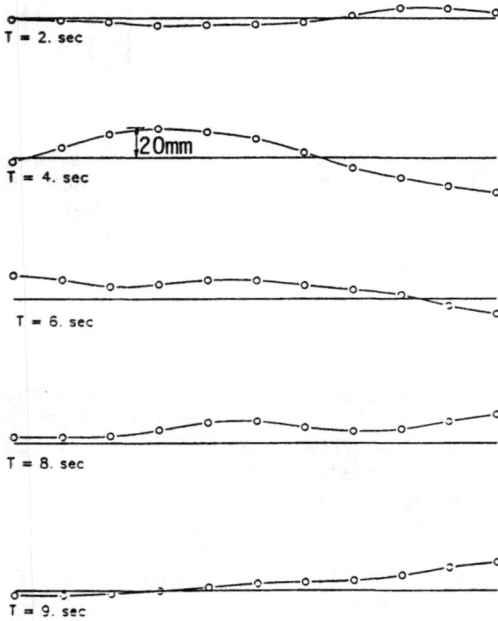

Figure 5. Simulated ground motions

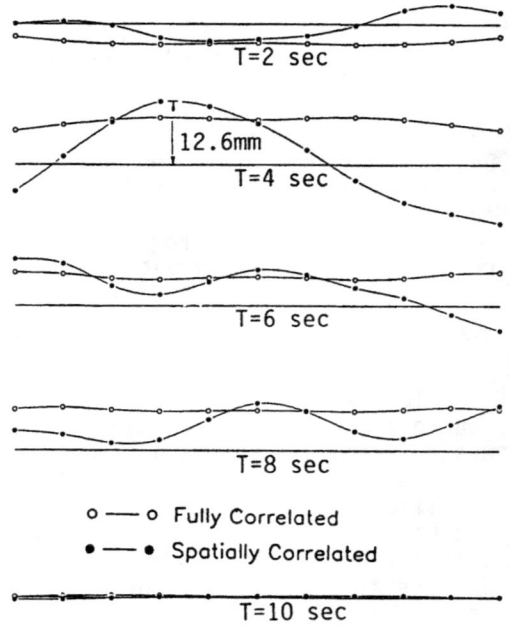

o —— o Fully Correlated

• —— • Spatially Correlated

Figure 6. Dam horizontal response

306

coefficients between the node 1 and the nodes 2 to 11 are 0.96, 0.85, 0.73, 0.60, 0.48, 0.35, 0.22, 0.10, 0.0 and -0.08. These correlation coefficients suggest a correlation length close to the dam length. The dam response to this ground motion is shown in Fig. 6 in which the response to fully correlated ground motions is also plotted for comparison, as in the subsequent figures. The bending of the dam due to the lack of ground motion coherence is obvious. For fully correlated ground motions, the dam moves almost as a rigid body. Notice also that the maximum dam displacement due to incoherent ground motions exceeds the response to fully correlated excitation. The torsional moments associated with the dam response are displayed in Fig. 7 while the bending moments are shown in Fig. 8. These moments are much higher for the incoherent ground motion than for the fully correlated one. Under the usual plane strain assumptions, these moments would be all equal to zero. The maximum bending and combined shear stresses are 1.24 and 0.36 MPa for the incoherent ground motions and only 0.26 and 0.09 MPa for the fully correlated ground motions, respectively. The actual level of the stresses depends, of course, on the actual ground motion intensity, frequency content and degree of coherence.

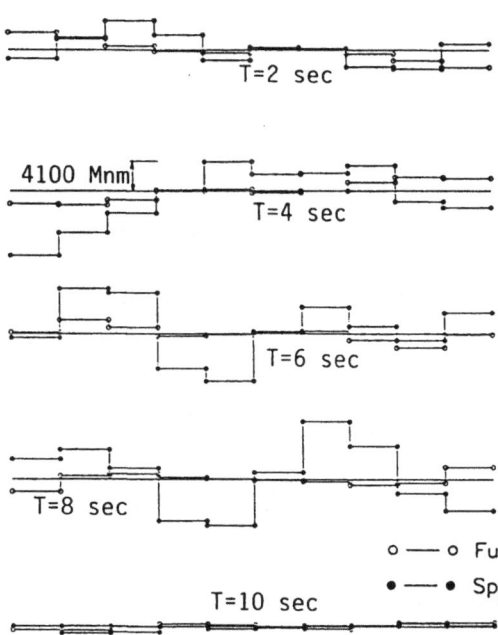

Figure 7.   Dam torsional moments

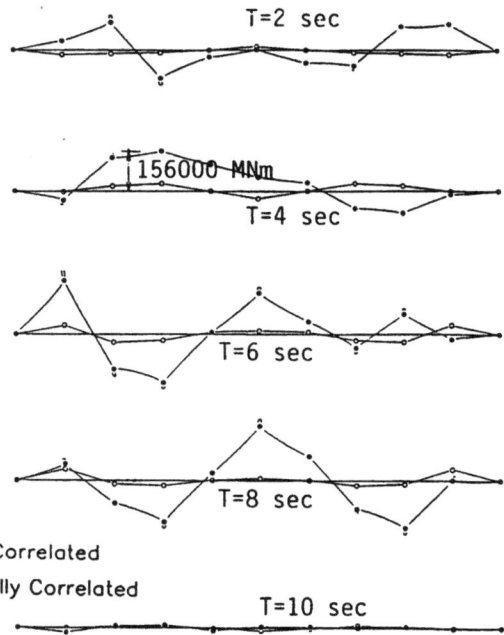

Figure 8.   Dam bending moments

307

The results shown are affected by soil-structure interaction. This can be recognized from the ratio of the Fourier spectrum of the dam horizontal response to the spectrum of the ground motion.

CONCLUSIONS

A solution which accounts for both spatial correlation of ground motion and soil-structure interaction in the seismic response analysis of large structures is presented and used to analyze a long concrete gravity dam. The following conclusions emerge:

1. While the dam responds almost as a rigid body to fully correlated ground motions, it bends and twists significantly due to ground motion incoherence.
2. The dam stresses are small under fully correlated motions but can be quite high even under moderately incoherent motions.
3. Absolute dam displacements may be increased by seismic motion incoherence.
4. Significant stresses in large structures may remain unforeseen if the ground motion is assumed to be fully correlated.

ACKNOWLEDGEMENT

The research reported here was supported by a grant from the Natural Science and Engineering Research Council of Canada.

REFERENCES

Clough, R.W. and Penzien, J. 1975. Dynamics of Structures. New York: McGraw-Hill, Inc..

Gaul, L. 1977. Dynamic interaction of a foundation with viscoelastic halfspace. Proc. Dyn. Methods in Soil & Rock Mech., Karlsruhe, Germany, 1: 167-183.

Hindy, A. and Novak, M. 1980. Response of pipelines to random ground motion. J. Eng. Mech. Div. ASCE, April, Vol. 106, No. EM2, pp. 339-360.

Novak, M. and Hindy, A. 1979. Seismic response of buried pipelines. Third Canadian Conference on Earthquake Engineering, Montreal, Quebec, June 4-June 6, Vol. 1, pp. 177-203.

Novak, M. and Suen, E. 1987. Dam-foundation interaction under spatially correlated random ground motion. Soil-Structure Interaction, Elsevier, 25-33.

Shinozuka, M. and Deodatis, G. 1988. Stochastic wave models for simulation of seismic ground motion. Proc. of the Workshop on Spatial Variation of Earthq. Ground Motion, Dunwalke, Princeton Univ., Nov. 7-9.

Nondeterministic inelastic response spectra

Ashok K. Jain[I]

ABSTRACT

The present study aims at generation of probabilistic inelastic response spectra using the analytical probability distributions of the strength and reduction factors. These spectra consists of constant ductility, constant strength, reduction factor and acceleration spectra. Elasto-plastic and stiffness degrading hysteresis models are used to generate these spectra for two ensembles of the artificial earthquakes for different confidence levels. It is concluded that for strength and reduction factors, the extreme value type I and II distributions are within 10% of the analytical distribution upto a confidence level of 85% beyond which the difference is 15%. The Weibull distribution does not give satisfactory results. It is shown that the elastic forces can be reduced upto a factor of 10 for confidence level of about 85%. For lower confidence levels, the reduction factors are about 24.

INTRODUCTION

Riddell and Newmark (1979) studied the statistical response of single degree of freedom systems subjected to ten real earthquakes and proposed amplification and deamplification factors to construct trapezoidal inelastic response spectra. Briseghella, Zaccaria and Guiffre (1982) also proposed reduction factors, that is, deamplification factors, to generate response spectra. Jain (1985) and Pal (1987) carried out statistical analysis and presented constant strength, constant ductility, reduction factor and inelastic spectra. These analyses were based on the assumption that the inelastic response follows Gaussian probability distribution.

This paper aims at the generation of probabilistic inelastic response spectra using the analytical probability distribution for viscously damped single degree of freedom systems. The amplification factors were determined by fitting trapezoidal lines to each accelerogram rather than to the mean normalized spectra of the ensemble since the variation in the knee periods was enormous. In this study, two ensembles of fifty records each were

---

[I] Professor of Civil Engg., University of Roorkee, Roorkee, India

generated using the nonstationary shot noise modelling using the intensity curves shown in Fig. 1 (Murakami and Penzien 1975). Earthquake E1 was of 5 sec duration and peak ground accelerations varied between 0.15 g and 0.3 g. It simulates a shallow ground motion of magnitude 4.5 to 5.5. Earthquake E2 was of 30 sec duration and peak ground accelerations varied between 0.25 g and 0.4 g. It simulates a motion of magnitude 7 close to a fault.

A computer code IRS was written to generate the various response spectra and PDF was written to compute the various probability distributions. The equation of motion was integrated using a variable time step so as not to miss any peak or trough. The maximum time step was 0.01 sec. These inelastic spectra were generated using 27 values of strength factors ranging from 0.001 to 4. The displacement ductility ratios were 1,1.5,2,2.5,3,4.5, 6,7, and 8. The response was computed for fifty six values of the time periods ranging from 0.05 sec to 10 sec. A larger number of ordinates were selected in the shorter time period range since the response spectra is very sensitive to the system characteristics in this range.

AMPLIFICATION FACTORS

The trapezoidal lines were fitted to the elastic spectra and amplification factors were determined. The procedure of fitting the trapezoidal lines is adopted after Riddell and Newmark (1979) and is as follows :

(i)   Initialize values of knee periods T_{av} and T_{vd}, where, T_{av} is the knee period at the junction of acceleration and velocity regions, and T_{vd} is the knee period at the junction of velocity and displacement regions.

(ii) Compute average values of acceleration, velocity and displacement for the spectral regions as follows :

$$S_a^i = \frac{\int\limits_{0.05}^{T_{av}^i} S_a(T)dT}{T_{av}^i - 0.05} \quad ; \quad S_v^i = \frac{\int\limits_{T_{av}^i}^{T_{vd}^i} S_v(T)dT}{T_{vd}^i - T_{av}^i} \quad ; \quad S_d^i = \frac{\int\limits_{T_{vd}^i}^{10} S_d(T)dT}{10 - T_{vd}^i} \quad (1)$$

where, subscript i denotes the ith iteration, T is time period, and values 0.05 and 10 are the smallest and largest time periods for the spectra.

(iii) Compute new values of the knee periods using the relation as :

$$T_{av}^{i+1} = 2\pi \frac{S_v^i}{S_a^i} \quad ; \quad T_{vd}^{i+1} = 2\pi \frac{S_d^i}{S_v^i} \quad (2)$$

(iv) Repeat steps (ii) and (iii) till $T_{av}^{i+1} = T_{av}^i$, and $T_{vd}^{i+1} = T_{vd}^i$ within a specified tolerance.

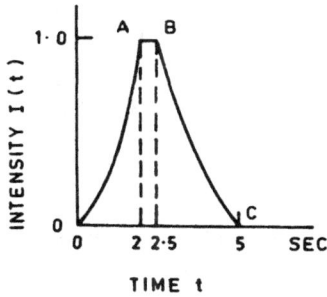

OA: $I(t) = t^3/8$

BC: $I(t) = EXP\left[-1.606(t-2.5)\right]$

(a) EARTHQUAKE E 1

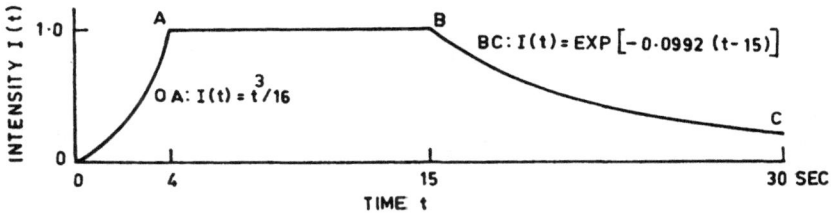

BC: $I(t) = EXP\left[-0.0992(t-15)\right]$

OA: $I(t) = t^3/16$

(b) EARTHQUAKE E 2

FIG.1 TIME INTENSITY FUNCTION CURVES FOR ARTIFICIAL EARTHQUAKES

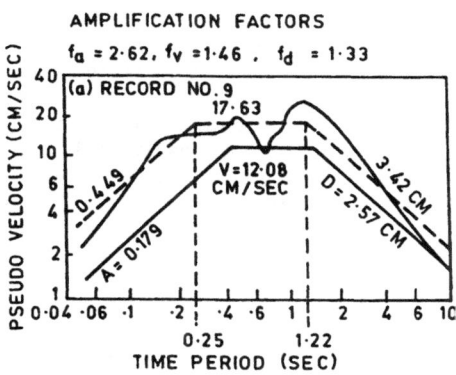

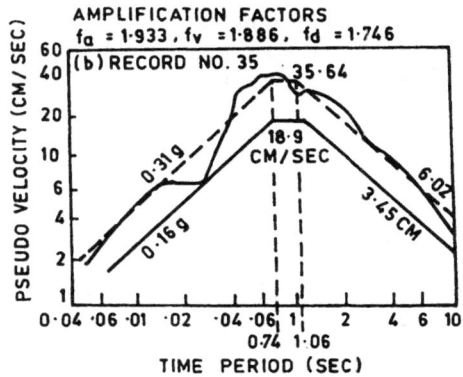

AMPLIFICATION FACTORS
$f_a = 2.62$, $f_v = 1.46$, $f_d = 1.33$

(a) RECORD NO. 9

17.63
0.449
V = 12.08 CM/SEC
A = 0.179
D = 2.57 CM
3.42 CM

0.25   1.22

AMPLIFICATION FACTORS
$f_a = 1.933$, $f_v = 1.886$, $f_d = 1.746$

(b) RECORD NO. 35

35.64
0.319
18.9 CM/SEC
0.169
3.45 CM
6.02

0.74 1.06

FIG.2 — ELASTIC RESPONSE SPECTRA - EARTHQUAKE E1

There was considerable dispersion in the knee period values. For earthquake E1, the knee period T_{av} varied from 0.2 sec to 1 sec, and T_{vd} varied from 0.4 to 1.7 sec. For E2 earthquake the values were 0.35 to 0.9 sec and 1.8 to 9 sec., respectively. The trapezoidal lines fitted to the mean spectra of E1 earthquake gave the knee periods as 0.31 sec and 0.95 sec. Figure 2 shows typical variation in the knee periods for record numbers 9 and 35.

The amplification factors proposed by Riddell and Newmark (1979) are in close agreement in the acceleration and displacement regions. In the velocity region, however, there is appreciable difference (Jain and Pal 1991). Riddell and Newmark obtained lower values in the velocity region because they used the mean spectra of ten records and amplification factors were computed by averaging the normalized spectra of each earthquake within each interval. This results in lowering the trapezoidal line in the velocity region. As the interval between the knee periods increases, the velocity region ordinates decrease. Figure 3 shows the distribution of amplification factors for different confidence levels.

INELASTIC SPECTRA

Four probabilistic models were employed to examine their suitability in representing the random behaviour of the inelastic response, viz : Extreme value distribution type 1, type 2, type 3 and analytical probability distribution models. The probability density function of the analytical model is computed using the theory of curve fitting and interpolation on the available data. More details can be seen in Pal (1989) and Siddal (1983).

Elasto-plastic and stiffness degrading hysteresis models were used to generate constant ductility, constant strength, reduction factor and inelastic response acceleration spectra. The stiffness degrading model is shown in Fig. 4 and its salient features are given elsewhere (Jain 1985). The various probability distributions of the strength factors are shown in Fig. 5. It can be seen that the type I and type II distributions are fairly close to the analytical distribution till about 85% confidence level beyond which the difference is quite large. Similarly, the various probability distribution for reduction factors are shown in Fig. 6. The type I and type II distributions are very close to the analytical distribution except in the lower confidence level range (less than about 20%) which is practically of not much relevance. It may be noted that mean + standard deviation curve corresponds to 84.1% confidence level for strength factors while mean - standard deviation curve corresponds to 84.1% confidence level for reduction factors.

The confidence level spectra of the strength factors for a ductility of 4 for different times periods is shown in Fig. 7. It also shows mean + standard deviation curve corresponding to 84.1% confidence level. It can be seen that the statistical curve is close to the probabilistic curve within about 10%. The probability distributions of the reduction factors for earthquake E1 and time period of 1.0 sec for all ductility ratios

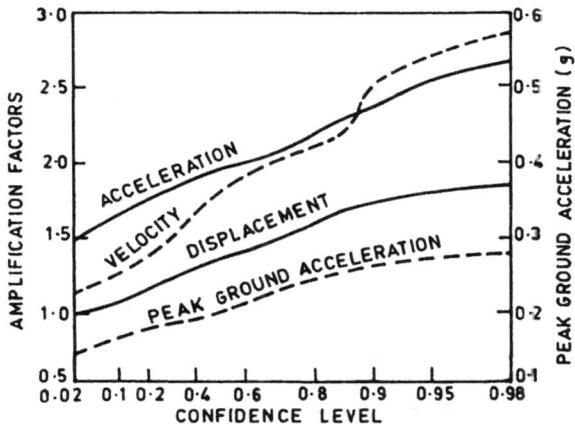

FIG.3 AMPLIFICATION FACTORS FOR
ELASTIC SPECTRA EARTHQUAKE E1

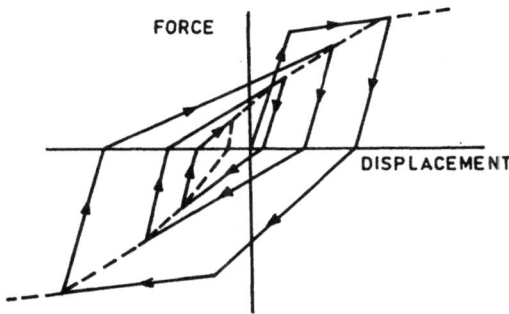

FIG.4- STIFFNESS DEGRADING MODEL

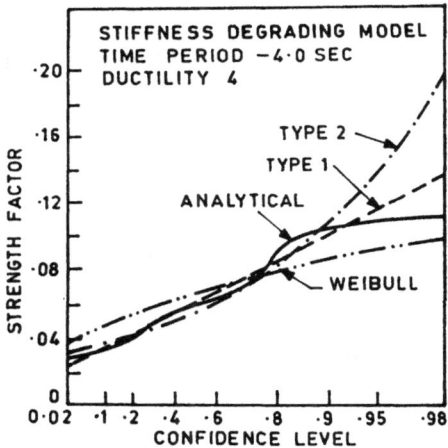

FIG.5 COMPARISON OF PROBABILITY
DISTRIBUTION OF STRENGTH
FACTORS-EARTHQUAKE E 2

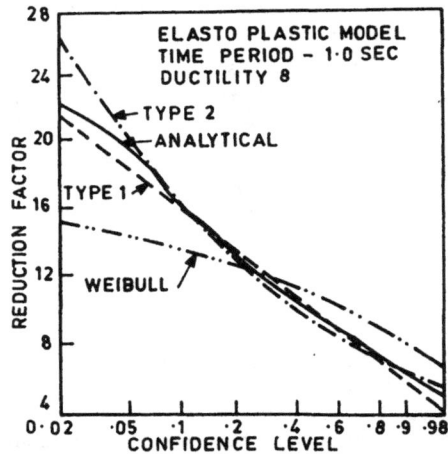

FIG.6. COMPARISON OF PROBABILITY
DISTRIBUTIONS OF REDUCTION
FACTORS-EARTHQUAKE E 1

313

and confidence levels are shown in Fig. 8. The type I distribution compares very well with the analytical probability distribution. For very low confidence levels, the reduction factors were as high as 24 (Fig. 6). For a confidence level of 85%, the reduction factors were up to 10 depending upon the ductility ratio, hysteresis model and the ground motion. Figure 9 shows confidence level spectra for the two hysteresis models for a ductility ratio of 4 and earthquake E1. It may be seen that the reduction factors decrease with the increase in confidence levels. It is because the inelastic acceleration spectra is inversely proportional to the reduction factors. Hence, the accelerations increase with the increase in confidence levels.

Inelastic acceleration spectra for earthquake E2 for different ductility ratios and 85% confidence level are shown in Fig. 10. The trapezoidal elastic response spectra can be obtained from the mean peak ground motion parameters by first applying the amplification factors obtained from the elastic analysis for different confidence levels to get the elastic spectra. Next, the reduction factors may be applied to the smoothened trapezoidal response spectra for different ductility ratios and confidence levels.

CONCLUSIONS

Based on the results presented in this paper, the following significant conclusions can be made :

1. The knee periods of each spectra of the ensemble varied considerably from those of the mean spectra. This results in considerable difference in the amplification factors in the velocity region.

2. For the strength and the reduction factors, the extreme value type I and II distributions are within 10% of the analytical probability distribution upto a confidence level of 85% beyond which the difference is about 15%. The Weibull distribution does not give satisfactory results.

3. The (mean + standard deviation) curve in the case of strength factors and (mean - standard deviation) curve in the case of reduction factors are within 15% of the respective 85% confidence level curves.

4. For 85% confidence level, the reduction factors vary up to 10 for various ductility ratios, hysteresis models and ground motions. For lower confidence levels, the reduction factors are as high as 24.

ACKNOWLEDGEMENTS

The research reported in this paper was partly supported by the University Grants Commission, New Delhi under the Career Award Scheme. The results and conclusions do not necessarily reflect the views of the sponsor. Much of the results summarized were obtained by Shri Pal in conjunction with his Ph.D. studies at the University of Roorkee. His originality and contribution to this effort are gratefully acknowledged.

314

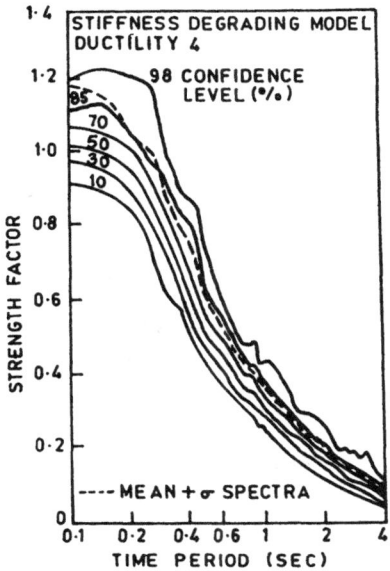

FIG.7 CONFIDENCE LEVEL SECTRA OF
STRENGTH FACTORS - EARTHQUAKE E.2

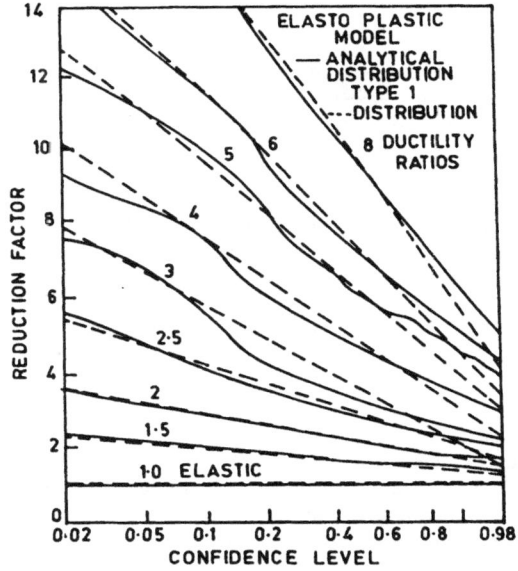

FIG.8 PROBABILITY DISTRIBUTION OF REDUCTION
FACTORS - EARTHQUAKE E 1
(TIME PERIOD = 1·0 SEC)

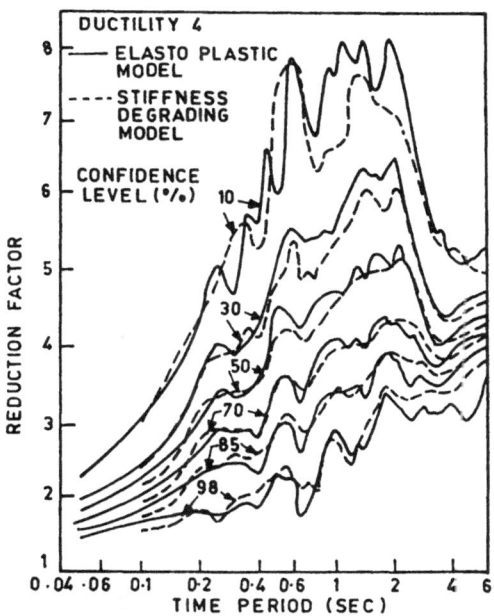

FIG. 9 CONFIDENCE LEVEL SPECTRA OF REDUCTION
FACTORS - EARTHQUAKE E 1

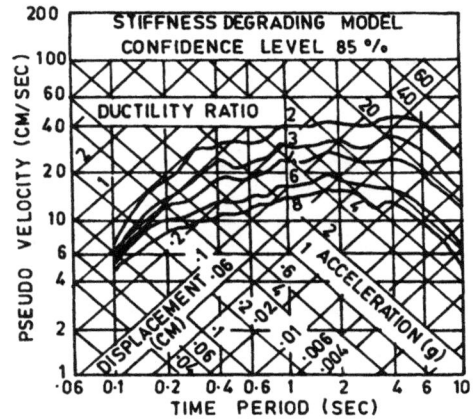

FIG.10 - INELASTIC RESPONCE SPECTRA
FOR 85 % CONFIDENCE LEVEL -
EARTHQUAKE E 2

315

REFERENCES

Briseghella, L., Zaccuria, P.L. and Guiffre, A. 1982 . Inelastic Response Spectra, Proc. 7th Symposium on Earthquake Engineering, University of Roorkee, Roorkee, 1, 159-162.

Jain, A.K. 1985. Inelastic Response Spectra, Research Report, Deptt. of Civil Engg., University of Roorkee, Roorkee.

Jain, A.K. and Pal, S. 1991. Probabilistic Amplification Factors for Response Spectra, J. Struct. Engg., ASCE, (accepted for publication).

Murakami, M. and Penzien, J. 1975. Nonlinear Response Spectra for Probabilistic Seismic Design and Damage Assessment of R.C. Structures, EERC Report No. 75-38, University of California, Berkeley.

Pal, S. 1989. Inelastic Response Spectra for Dynamic Excitations Using Nondeterministic Approach, Ph.D. Thesis, College of Engineering, University of Poona, Pune.

Pal, S., Dasaka, S.S. and Jain, A.K. 1987. Inelastic Response Spectra, Computers and Struct., 25, (3), 335-344.

Riddell, R. and Newmark, N.M. 1979. Statistical Analysis of the Response of Nonlinear Systems Subjected to Earthquakes, Report No. 468, University of Illinois, Urbana.

Siddal, J.N. 1983. Probabilistic Engineering Design, Marcel Decker Inc., New York.

Stochastic response of large structures to multiple excitations

Hong Hao[I]

ABSTRACT

A multiply-supported rigid plate to spatially correlated ground excitations are analysed. Quasi-static, dynamic and total structural responses are calculated. Different ground motion assumptions are: general case, neglecting phase shifts, neglecting coherency losses and neglecting propagation effects (single input). The results are compared. Some general conclusions on structural responses to correlated multiple ground excitations are drawn.

INTRODUCTION

large structures, such as bridges, pipelines, will be affected by ground motion propagation. The properties of ground motion propagation have been studied based on the actual recorded data at a high density earthquake accelerometer array, SMART-1. Among those studies, Harichandran and Vanmarcke (1984) proposed a one-dimensional coherency model. Hao, et al. (1989) proposed a two-dimensional coherency model. Using the coherency model (Harichandran and Vanmarcke 1984), Harichandran and Wang (1988; 1990) calculated the responses of a single-span beam and a double-span beam to spatially correlated multiple excitations. Zerva (1990) calculated the responses of a continuous beam using an assumed coherency model. Using the coherency model (Hao, et al. 1989), Hao (1991) analysed a two-dimensional multiply-supported rigid plate to multiple excitations by assuming the ground motion propagating along x direction, Fig. 1. Hao (1989) also simulated spatially correlated ground motion time histories based on the both one- and two-dimensional coherency models mentioned above, and calculated the structural response time histories by using those simulated ground motions as multiple inputs.

In this paper, a multiply-supported rigid plate to multiple excitations is analysed. The two-dimensional coherency model (Hao, et al. 1989) is used. The ground motion propagation direction is arbitrary. Quasi-static, dynamic and total structural responses are calculated. Three cases of ground motion inputs are: Case 1, multiple inputs with both phase shifts and coherency losses, Case 2, multiple inputs with coherency losses only, and Case 3, multiple inputs with phase shifts only. The results from these three cases are normalized by the corresponding results from single input. The normalized results are compared. Some general conclusions on the effects of multiple inputs on structural responses are obtained.

[I] Lecturer, School of Civil and Structural Engrg., Nanyang Technological Institute, Nanyang Ave., Singapore 2263

GROUND MOTION MODEL

Assume earthquake ground motions are stationary and ergodic, the power spectral density function of ground accelerations can be expressed as (Hao 1991)

$$S_{kl}(i\bar{\omega}) = |H_1(i\bar{\omega})|^2 \, S_0(\bar{\omega}) \, |\gamma_{kl}(\bar{\omega}, d_{kl}^l, d_{kl}^t)| \, exp(i\bar{\omega}d_{kl}^l/v) \tag{1}$$

where $\bar{\omega}$ is circular frequency, v is apparent velocity, $|H_1(i\bar{\omega})|^2$ is a highpass filter, which has the form (Ruiz and Penzien 1969)

$$|H_1(i\bar{\omega})|^2 = \frac{\bar{\omega}^4}{(\omega_1^2 - \bar{\omega}^2)^2 + 4\xi_1^2\omega_1^2\bar{\omega}^2} \tag{2}$$

where the optimal values are $\omega_1 = 1.636 \; rad/s$ and $\xi_1 = 0.619$.

$S_0(\bar{\omega})$ is a Kanai-Tajimi power spectral density function given as (Tajimi 1960)

$$S_0(\bar{\omega}) = \frac{1 + 4\xi_g^2\frac{\bar{\omega}^2}{\omega_g^2}}{(1 - \frac{\bar{\omega}^2}{\omega_g^2})^2 + 4\xi_g^2\frac{\bar{\omega}^2}{\omega_g^2}} S \tag{3}$$

where ξ_g is damping ratio, ω_g is central frequency, and S is a scale factor.

The coherency loss function has the form (Hao, et al. 1989)

$$|\gamma_{kl}(\bar{\omega}, d_{kl}^l, d_{kl}^t)| = exp(-\beta_1|d_{kl}^l| - \beta_2|d_{kl}^t|)exp\left\{-\left[\alpha_1(\bar{\omega})\sqrt{|d_{kl}^l|}\right.\right.$$
$$\left.\left. + \alpha_2(\bar{\omega})\sqrt{|d_{kl}^t|}\right](\frac{\bar{\omega}}{2\pi})^2\right\} \tag{4}$$

where d_{kl}^l and d_{kl}^t are projected distances between the two stations to ground motion propagation direction and its transverse direction, respectively; β_1 and β_2 are constants; $\alpha_1(\bar{\omega})$ and $\alpha_2(\bar{\omega})$ are given as (Hao 1989)

$$\begin{aligned}\alpha_1(\bar{\omega}) &= \frac{2\pi a}{\bar{\omega}} + \frac{b\bar{\omega}}{2\pi} + c \\ \alpha_2(\bar{\omega}) &= \frac{2\pi d}{\bar{\omega}} + \frac{e\bar{\omega}}{2\pi} + g\end{aligned} \qquad 0.314 \le \bar{\omega} \le 62.83 \tag{5}$$

where a, b, c, d, e and g are constants.

By processing the recorded horizontal motions of Event 40 at SMART-1 array, the above constants obtained are (Hao 1989): $\beta_1 = 9.323 \times 10^{-5}$, $\beta_2 = 1.421 \times 10^{-4}$, $a = 1.037 \times 10^{-2}$, $b = 9.33 \times 10^{-5}$, $c = -1.821 \times 10^{-3}$, $d = 8.09 \times 10^{-3}$, $e = 4.083 \times 10^{-5}$ and $g = -1.007 \times 10^{-3}$.

DYNAMIC RESPONSE EQUATIONS

The equations of motion of the rigid plate shown in Fig. 1 can be derived (Hao 1991). The total structural response equations are

$$\mathbf{M}_{ss}\ddot{\mathbf{U}}^t + \mathbf{C}_{ss}\dot{\mathbf{U}}^t + \mathbf{K}_{ss}\mathbf{U}^t = -\mathbf{K}_{sb}\mathbf{V}_g \tag{6}$$

318

The dynamic response equations are

$$\mathbf{M}_{ss}\ddot{\mathbf{U}} + \mathbf{C}_{ss}\dot{\mathbf{U}} + \mathbf{K}_{ss}\mathbf{U} = \mathbf{M}_{ss}\mathbf{K}_{ss}^{-1}\mathbf{K}_{sb}\ddot{\mathbf{V}}_g \tag{7}$$

and the quasi-static response equations are

$$\mathbf{U}^{qs} = -\mathbf{K}_{ss}\mathbf{K}_{sb}\mathbf{V}_g \tag{8}$$

where $\mathbf{C}_{ss}$ is a proportional damping coefficient matrix, and

$$\mathbf{M}_{ss} = \begin{pmatrix} m & 0 & 0 \\ 0 & m & 0 \\ 0 & 0 & I \end{pmatrix} \tag{9}$$

$$\mathbf{K}_{ss} = \begin{pmatrix} 4k & 0 & 0 \\ 0 & 4k & 0 \\ 0 & 0 & 4k \end{pmatrix} \tag{10}$$

$$\mathbf{K}_{sb} = \begin{pmatrix} -k & 0 & -k & 0 & -k & 0 & -k & 0 \\ 0 & -k & 0 & -k & 0 & -k & 0 & -k \\ \frac{1}{2}kd & -\frac{1}{2}kd & \frac{1}{2}kd & \frac{1}{2}kd & -\frac{1}{2}kd & \frac{1}{2}kd & -\frac{1}{2}kd & -\frac{1}{2}kd \end{pmatrix} \tag{11}$$

and

$$\mathbf{V}_g = (v_{1x} \; v_{1y} \; v_{2x} \; v_{2y} \; v_{3x} \; v_{3y} \; v_{4x} \; v_{4y})^T \tag{12}$$

where m is lumped mass, I is polar moment of inertia, k is column stiffness, d is structural dimension, and v_{ix}, v_{iy} are the ground displacement in x and y directions at support i, Fig. 1.

STOCHASTIC RESPONSE FORMULATION

Assume ground motion propagating along $\bar{x}$ direction, Fig. 1, then ground motions in $\bar{x}$ and $\bar{y}$ directions can be considered statistically independent and the power spectral density functions of ground motions in $\bar{y}$ direction are approximately 0.7 of those in $\bar{x}$ direction (Penzien and Watabe 1975).

Ground accelerations in x and y directions can be obtained by transformation,

$$\ddot{v}_x = \ddot{v}_{\bar{x}}\cos\alpha - \ddot{v}_{\bar{y}}\sin\alpha$$
$$\ddot{v}_y = \ddot{v}_{\bar{x}}\sin\alpha + \ddot{v}_{\bar{y}}\cos\alpha \tag{13}$$

where α is the angle between x and $\bar{x}$ axes, and is defined as the ground motion incident angle.

By some tedious but otherwise straightforward derivation, the power spectral density function of the quasi-static responses in x direction can be obtained as

$$S_{x^{qs}}(\bar{\omega}) = \frac{1}{\bar{\omega}^4} \mid H_1(i\bar{\omega}) \mid^2 S_0(\bar{\omega}) RT_x(\bar{\omega}, d_{kl}^l, d_{kl}^t) \tag{14}$$

where $RT_x(\bar{\omega}, d_{kl}^l, d_{kl}^t)$ is a factor function for translational responses in x direction, it has the form

$$RT_x(\bar{\omega}, d_{kl}^l, d_{kl}^t) = \frac{\cos^2\alpha + 0.7\sin^2\alpha}{8} \left[2 + \sum_{k=1}^{3}\sum_{l=k+1}^{4} \mid \gamma_{kl}(\bar{\omega}, d_{kl}^l, d_{kl}^t) \mid \cos(\bar{\omega}d_{kl}^l/v) \right] \tag{15}$$

319

The power spectral density functions for responses in y direction have the same form as Equation (14), the only difference between them is an α dependent constant factor. Therefore, only responses in x direction are discussed.

The rotational response power spectral density function is obtained as

$$S_{\theta q \cdot}(\bar{\omega}) = \frac{1}{\bar{\omega}^4} \mid H_1(i\bar{\omega}) \mid^2 S_0(\bar{\omega}) RR(\bar{\omega}, d^l_{kl}, d^t_{kl}) \tag{16}$$

where

$$RR(\bar{\omega}, d^l_{kl}, d^t_{kl}) = \frac{1}{8d^2} \left[3.4 + 0.3(cos^2\alpha - sin^2\alpha)(r_{12}cos\phi_{12} - r_{14}cos\phi_{14} - r_{23}cos\phi_{23} + \right.$$
$$\left. r_{34}cos\phi_{34}) - 1.7(r_{13}cos\phi_{13} + r_{24}cos\phi_{24}) + 0.6sin\alpha cos\alpha(r_{13}cos\phi_{13} - r_{24}cos\phi_{24}) \right] \tag{17}$$

where $r_{kl}cos\phi_{kl} = \mid \gamma_{kl}(\bar{\omega}, d^l_{kl}, d^t_{kl}) \mid cos(\bar{\omega}d^l_{kl}/v)$.

The power spectral density functions of the dynamic and total structural responses in x direction can be derived as

$$S_{x^d}(\bar{\omega}) = \bar{\omega}^4 \mid H_x(i\bar{\omega}) \mid^2 S_{x q \cdot}(\bar{\omega}) \tag{18}$$

and

$$S_{x^t}(\bar{\omega}) = \omega_0^4 \mid H_x(i\bar{\omega}) \mid^2 S_{x q \cdot}(\bar{\omega}) \tag{19}$$

where $\mid H_x(i\bar{\omega}) \mid^2$ is a transfer function for translational response mode

$$\mid H_x(i\bar{\omega}) \mid^2 = \frac{1}{(\omega_0^2 - \bar{\omega}^2)^2 + 4\xi^2\omega_0^2\bar{\omega}^2} \tag{20}$$

where $\omega_0 = \sqrt{\frac{4k}{m}}$ is natural frequency, ξ is damping ratio.

The dynamic and total rotational response power spectral density functions can be derived as

$$S_{\theta^d}(\bar{\omega}) = \bar{\omega}^4 \mid H_\theta(i\bar{\omega}) \mid^2 S_{\theta q \cdot}(\bar{\omega}) \tag{21}$$

and

$$S_{\theta^t}(\bar{\omega}) = \omega_\theta^4 \mid H_\theta(i\bar{\omega}) \mid^2 S_{\theta q \cdot}(\bar{\omega}) \tag{22}$$

where $\omega_\theta = \sqrt{\frac{2kd^2}{I}}$ is natural frequency for rotational mode, $\mid H_\theta(i\bar{\omega}) \mid^2$ is corresponding transfer function.

The power spectral density functions of the acceleration responses can be obtained by multiplying the corresponding displacement response spectra by $\bar{\omega}^4$. The variances of the responses can be obtained by integrating the power spectral density functions.

NUMERICAL RESULTS

Structural response variances are calculated. The parameters used are $\xi_g = 0.6$, $\omega_g = 5\pi$ rad/s, $d = 100m$, $v = 1000m/s$, $S = 10^7$ cm^2/s^2 and $\alpha = 30^0$. The translational responses are normalized by the corresponding responses from single input. Rotational responses to single ground motion input are zero. The results for the three cases are presented and compared.

Fig. 2 shows quasi-static responses with respect to d/v. It can be seen that the ratios of all the responses decreases as d/v increases. The multiple input effect to translational responses are dominated by ground motion phase shift effects, while coherency loss effects are more critical to rotational displacement responses.

Fig. 3 shows dynamic responses with respect to a dimensionless parameter f_0/f_d, where f_0 is natural frequency of the system and $f_d = v/d$ is wave frequency of wavelength d. It can be seen that single input assumptions overestimate translational responses and underestimate rotational responses. By comparing with the general input results (Case 1), it can be concluded that the translational responses are always overestimated while the rotational responses are always underestimated by neglecting phase shifts effects, and the responses are sometimes underestimated and sometimes overestimated by neglecting coherency loss effects.

Fig. 4 shows total responses with respect to f_0/f_d. It can be seen that the total displacements are overestimated by either neglecting the phase shift or coherency loss effects, while accelerations are underestimated by neglecting phase shift effects, but sometimes overestimated and sometimes underestimated by neglecting coherency loss effects.

From Equations (14) to (22), it can be noticed that the critical incident angle α to translational responses is either 0^0 or 90^0. But the critical α to rotational responses depends on the coherency properties. Fig. 5 shows the comparisons between the total responses from the ground motions with different incident angles α. It can be seen that all the responses, except rotational accelerations, are reduced if $\alpha \neq 0$. For rotational accelerations, the total responses vary with the incident angles.

CONCLUSIONS

Single input representations of the ground motions always overestimate translational responses but underestimate rotational responses. By considering the ground motion phase shifts only, structural responses are sometimes overestimated and sometimes underestimated. By considering ground motion coherency loss effects only, translational responses are overestimated and rotational responses are underestimated. The ground motion incident angles also affect structural responses. The total responses are generally reduced by a non zero incident angle except for the responses of rotational accelerations.

REFERENCES

Harichandran, R.S. and Vanmarcke, E. 1984. Space-Time Variation of Earthquake Ground Motion. Research Report R84-12, Dept. of Civil Engrg., Massachusetts Institute of Technology.

Hao, H., Oliveira, C.S. and Penzien, J. 1989, Multiple-Station Ground Motion Processing and Simulation Based on SMART-1 Array Data. Nuclear Engineering and Design, Vol. 111, 293-310.

Harichandran, R.S. and Wang, W. 1988, Response of Simple Beam to Spatially Varying Earthquake Excitation. J. of Engrg. Mech., ASCE, Vol. 114, No. 9, 1526-1541.

Harichandran, R.S. and Wang, W. 1990, Response of Indeterminate Two-Span Beam to Spatially Varying Seismic Excitation. Earthq. Engrg. and Str. Dyn., Vol. 19, 173-187.

Zerva, A. 1990, Response of Multi-Span Beams to Spatially Incoherent Seismic Ground Motions. Earthq. Engrg. and Str. Dyn., Vol. 19, 819-832

Hao, H. 1991, Response of Multiply-Supported Rigid Plate to Spatially Correlated Seismic Excitations. Submitted for Review for Publication in Earthq. Egnrg. and Str. Dyn..

Hao, H. 1989, Effects of Spatial Variation of Ground Motions on Large Multiply-Supported Structures. Report No. EERC 89-06, Earthquake Engineering Research Center, University of California, Berkeley.

Ruiz, P. and Penzien, J. 1969, Probabilistic Study of the Behavior of Structures during Earthquakes. Report No. EERC 69-03, Earthquake Engineering Research Center, University of California, Berkeley.

Tajimi, H. 1960, A Statistical Method of Determining the Maximum Response of a Building Structure During an Earthquake. Proc. 2WCEE, Vol. 2, Tokyo, 781-797.

Penzien, J. and Watabe, M. 1975, Characteristics of 3-Dimensional Earthquake Ground Motions. Earthq. Engrg. and Str. Dyn., Vol. 3, 365-373.

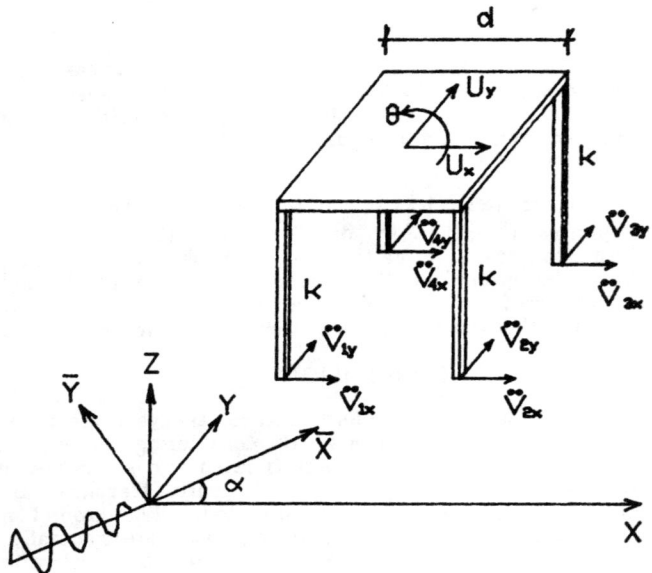

Figure 1. Multiply-supported rigid plate

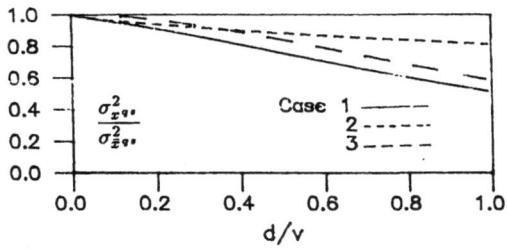

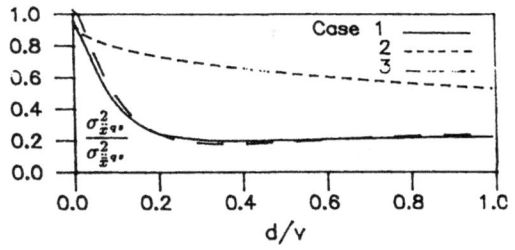

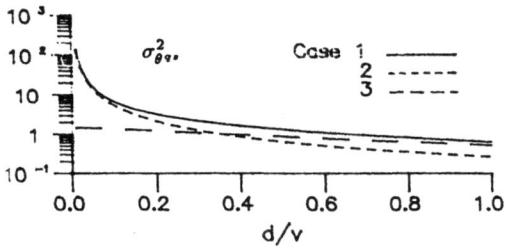

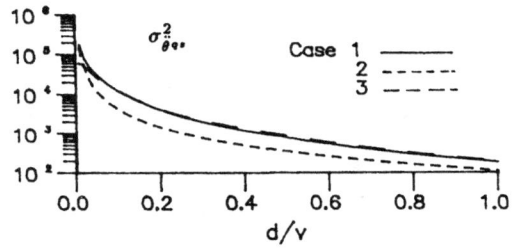

Figure 2. Quasi-static response variances

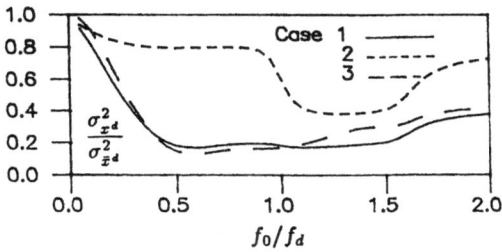

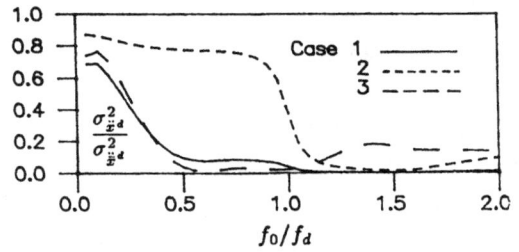

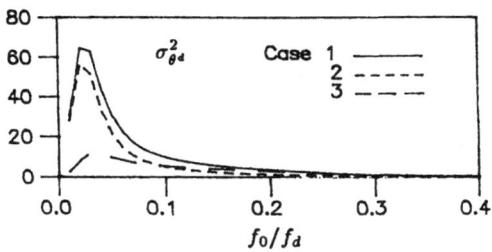

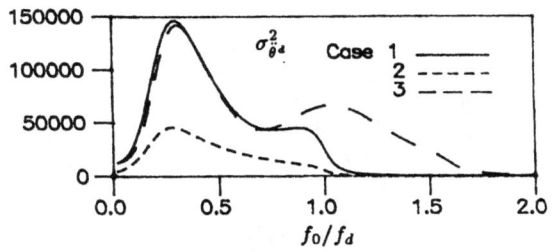

Figure 3. Dynamic response variances

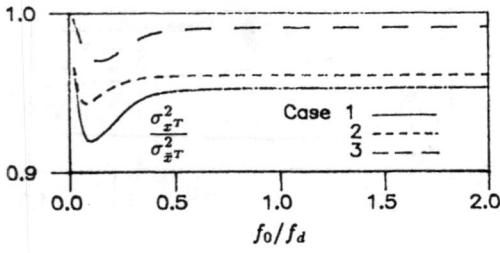

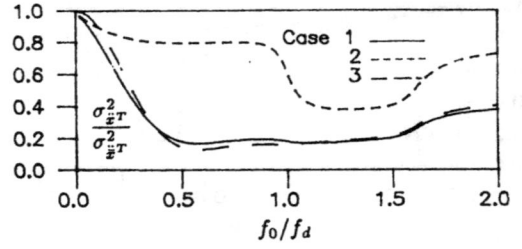

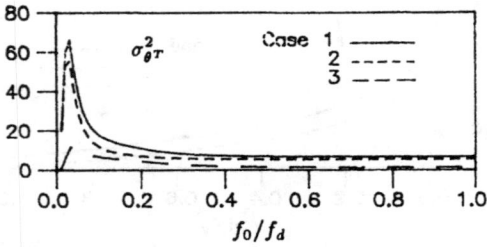

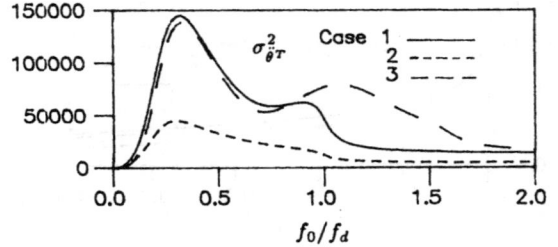

Figure 4. Total structural response variances

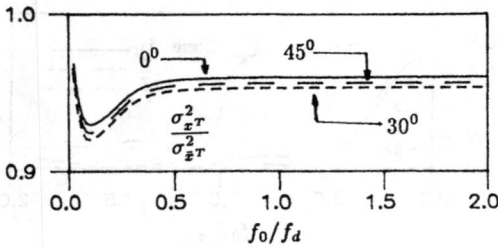

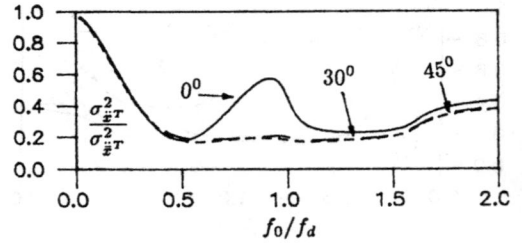

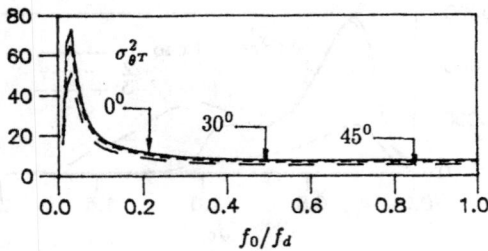

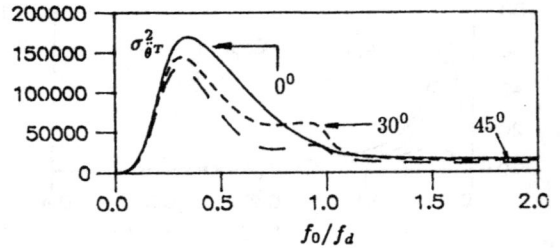

Figure 5. Total response variances for different incident angles

2 Seismicity, mitigation, soil response, and soil-structure interaction
2 Sismicité, mitigation, réponse des sols, et interaction sol-structure

Shake table studies of seismic response of single partially supported piles

A.J. Valsangkar[I], J.L. Dawe[I], and K.A. Mita[II]

ABSTRACT

An experimental investigation into the seismic response of soil-pile interaction has been carried out. Shake table tests have been conducted on single flexible model pile embedded in loose dry sand. Several parameters such as unsupported pile length, supported mass at the pile head, and intensity of applied base motion were examined experimentally in order to study their effects on the distribution and intensity of bending moments induced in each model pile. The natural frequency of the model pile was determined using two different techniques.

INTRODUCTION

Piles vibrate when supported structures are exposed to dynamic forces resulting from wind, waves, earthquakes, etc. Earthquake vibrations are transmitted through soil or rock into the structure generating dynamic stresses and displacements in structure and ground resulting in an extremely complex interaction problem. Use of pile foundations for supporting machines, off-shore platforms, and structures built in earthquake zones, requires a better understanding of dynamic behaviour of piles in order to produce more accurate and effective designs.

Over the past years, several analytical models have been developed. Many of these are based on either visco-elastic or Winkler type models (Flores-Berrones and Whitman 1982; Novak and Nagomai 1977; Novak et al. 1978; Kagawa and Kraft 1981). Some researchers have used three dimensional analysis to obtain dynamic pile response (Nogami et al. 1976; Baba et al. 1977; Neilson 1982). Others have adopted various numerical methods to treat the soil-pile interaction problem using discretized models (Kuhlemeyer 1976; Emery and Nair 1977; Baguelin and Frank 1980). While there have been a large number of analytical studies on dynamic response of piles, published records of experimental data are somewhat scarce (Finn and Gohl, 1987; Gohl and Finn, 1987; Steedman and Maheetharan, 1989; Stanton et. al 1988). Detailed measurements of the behaviour of prototype piles subjected to strong earthquake loading are also not yet available. The present experimental research therefore was undertaken to obtain experimental data against which the performance of theoretical models for predicting seismic response of single piles could be calibrated.

I   Professor, Department of Civil Engineering, P.O. Box 4400, University of New Brunswick, Fredericton, New Brunswick, E3B 5A3, Canada.

II   Graduate Student, Department of Civil Engineering, P.O. Box 4400, University of New Brunswick, Fredericton, New Brunswick, E3B 5A3, Canada.

EXPERIMENTAL PROGRAM

In the present experimental program, single flexible model piles embedded in loose dry sand were tested on a shake table under sinusoidal motion to simulate seismic loading on the soil-pile system. Bending strains induced in each model pile and accelerations at the base and at the pile head were recorded. Piezoceramic bender elements were used to monitor the distribution of shear wave velocity through the sand medium. Natural frequency of the model pile was also measured experimentally using techniques described by Gohl and Finn (1987).

The experimental study consisted of twelve small scale tests comprising four test series. Details of these tests are given in Table I. The objectives were to study the effects of the following parameters on dynamic response of single piles:

1.   unsupported length of the model pile
2.   supported mass at the pile head
3.   magnitude and frequency of applied base acceleration

Test set-up

A schematic representation of the test set-up is shown in Figure 1. Motion of the shake table was controlled by a hydraulic actuator and feedback control. A rigid wooden sand container (495 mm x 1016 mm), diagonally braced to prevent any movement other than that of the shake table, was bolted to the shake table. Two 25mm thick styrofoam pads were placed at each end of the container to reduce effects of wave reflection from the sides of the box perpendicular to the direction of the base motion. An accelerometer was attached at the base of the sand container to measure input acceleration.

The model piles used in this study were made from hollow aluminium tubing with a wall thickness of 1 mm and an outside diameter of 6.35 mm. Ten calibrated single element gauges with gauge lengths of 3.175 mm (0.125 in) were installed at various elevations on the outside of the model pile. The flexural rigidity (EI) of the model pile was measured to be approximately 4.5×10^6 N-mm^2 and its mass per unit length including the contributions of strain gauges and lead wires was found to be about 0.00028 kg/mm. The head mass with an attached accelerometer was clamped at the pile head. Strain gauge, accelerometer, and bender element signals were stored in a microcomputer using a data acquisition interface.

The instrumented pile was placed inside the model sand container and the foundation prepared using dry silica sand with an average void ratio of 0.65. A raining technique was used for sand placement. The sand container was filled several times prior to shake table studies to ensure that uniform and consistent sand beds were prepared.

Shear Wave Velocity of Sand Medium

Prior to the main series of tests on model piles using a shake table, shear wave velocity transmission in model sand bed was studied using bender elements. Three bender elements were placed inside the prepared sand bed at depths of 127 mm, 254 mm and 381 mm below finished sand surface to measure shear wave velocity. Each bender element consisted of a thin metal disc with a ceramic coating at the central area. The coating, which is extremely sensitive, generated electrical charge when excited by any kind of vibration. A shear wave vibration was generated by applying a horizontal shock load to a shear plate placed on top of the soil surface.

The average shear wave velocity for the top 381 mm sand layer was calculated to be 254 m/sec. This agrees with the data from Gohl and Finn (1987) who reported an average shear wave velocity of 211 m/sec for 300 mm thick dense Ottawa sand with an average void ratio of 0.57.

Natural Frequency of the Model Pile

Ringdown technique (Gohl and Finn 1987) was used to measure natural frequency of the model pile. In this method, the pile head was displaced a certain amount and then released so that it vibrated freely. Pile head acceleration was recorded as shown in Figure 2(a) for pile test C1 (Table 1). A Fourier spectrum shown in Figure 2(b) indicates a fundamental frequency of the pile to be about 4.2 Hz.

The same model pile (test C1) was subjected to a base acceleration of about 0.1g at different excitation frequencies. Maximum bending moment plotted against the different frequencies of base excitation is shown in Figure 2(c). This figure indicates that the pile has a resonant frequency of about 3 Hz. This is smaller than the value of natural frequency obtained from the ringdown test. It appears that when the pile was subjected to shake table motion, there was greater strain softening in sand caused by the high steady state response amplitudes thereby reducing the value of resonant frequency. Similar observations have been reported by Gohl and Finn (1987).

Test Procedure for Dynamic Pile Response

The shake table assembly was subjected to a sinusoidal base motion of approximately 0.1g at a starting frequency of 1 Hz which was gradually increased up to 9 Hz in test series A, B and C (Table 1). This procedure was adopted as only large amplitude shaking with peak acceleration amplitude of 0.5g with frequencies between 20 to 30 Hz resulted in significant sand densification. The base acceleration was maintained for 12 seconds at each frequency during which time signals from strain gauges and accelerometers were recorded at an interval of 0.01 second. This procedure was repeated in series D for base accelerations of 0.25g and 0.34 g.

EXPERIMENTAL RESULTS

Time history of acceleration applied at the base and recorded at the pile head was monitored in all the tests performed. The data indicated that the base acceleration was amplified by a factor of about 2 at the pile head. In the experimental data reported by Gohl and Finn (1987) on model piles in dense sand, the base acceleration of 0.6g was amplified to 3.5g at the pile head. Time history of bending moments at various depths along the pile is shown in Figure 3.

Distribution of bending moment along the pile length for different unsupported pile lengths and head masses are shown in Figures 4 (a) and (b) respectively. These figures indicate that pile bending moment increases linearly from the top of the pile to the soil surface and then from there decreases nonlinearly to zero at greater depths. Maximum bending moments occurred approximately 12 pile diameters below the soil surface when the pile was almost fully embedded. Gohl and Finn (1987) have reported under similar loading conditions that maximum bending moment occurs approximately 13 pile diameters below the soil surface. The common trend of distribution of bending moment along pile length indicated that the pile displacements and bending moments reduce to negligible values after a depth of about 30 pile diameters below the soil surface. Similar observations have also been reported by Gohl and Finn (1987) for fully embedded piles.

A summary of experimental results of the study presented herein is given in Table 2. The results show that maximum bending moments induced in the pile increase with increased unsupported length of the pile, head mass at the pile top, and the base accelerations. Within the range of parameters studies the magnitude of head mass at the pile top appears to have maximum effect on the bending moments in the pile.

CONCLUSIONS

The magnitude and distributions of shear moduli in the sand medium was determined using piezoceramic bender elements. This procedure was found to produce reliable results which can be used in calibrating the present experimental work with analytical predictions.

The procedures proposed by Gohl and Finn (1987) were used for the determination of fundamental natural frequency of the pile. The results confirmed the earlier findings by others that the frequency sweep tests yield reduced natural frequency values due to cyclic strain sorting of the soil around the pile.

Lateral pile response to sinusoidal base accelerations indicated that the model piles vibrated mainly in the first mode. The parametric study of piles subjected to sinusoidal base motion input showed that the peak bending moments increased with increase in unsupported length, head mass at the pile top, and the amplitude of base accelerations.

ACKNOWLEDGEMENTS

Authors are thankful to Natural Science and Engineering Research Council of Canada for financial assistance. Donation of the bearings for shake table by L.E. Shaw Ltd. is greatly appreciated. Special thanks are due to Dr. A.B. Schriver of the Dept. of Civil Engineering, UNB, for his valuable contributions in this research.

REFERENCES

Baba, K., Kobori, T., and Minai, R. (1977). "Dynamic Behaviour of a Laterally Loaded Pile". Proceedings of Speciality Session No. 10, IX International Conference on Soil Mechanics and Foundation Engineering, Tokyo.

Baguelin, F. and Frank, F. (1980). "Theoretical Studies of Piles Using the Finite Element Method". Chapter 11 in Numerical Methods in Off-Shore Piling, ICE, London, pp. 83-91.

Emery, J.J. and Nair, G.P. (1977). "Dynamic Response of a Single Pile". Proceedings Speciality Session No. 10. IX International Conference on Soil Mechanics and Foundation Engineering, Tokyo, pp.

Flores-Berrones, J.R. and Whitman, R.V. (1982). "Seismic Response of End-Bearing Piles". Proceedings, ASCE Journal of the Geotechnical Engineering Division, 103 (GT4), pp. 554-569.

Finn, W.D.L., and Gohl, W.B. (1987). "Centrifuge Model Studies of Piles under Simulated Earthquake Loading", ASCE Convention, Speciality Session on Dynamic Behaviour of Pile Foundations, Atlantic

City, NJ., pp. 21-38.

Gohl, W.B. and Finn, W.D.L. (1987). "Seismic Response of Single Piles in Shake Tables Studies". 5th Canadian Conference on Earthquake Engineering, Ottawa, pp. 435-444.

Kagawa, T., and Kraft, L.M. (1981). "Lateral Pile Response During Earthquakes". Proceedings, ASCE Journal of Geotechnical Engineering Division, 107 (GT12), pp. 1713-1731.

Kuhlemeyer, R.L. (1976). "Static and Dynamic Laterally Loaded Piles". Research Report No. CE 76-9, Department of Civil Engineering, University of Calgary, pp. 48-61.

Neilson, M.T. (1982). "Resistance of Soil Layer to Horizontal Vibration of a Pile". Earthquake Engineering and Structural Dynamics, 10, pp. 497-510.

Novak, M. and Nogami, T. (1977). "Soil-Pile Interaction in Horizontal Vibration". Earthquake Engineering and Structural Dynamics, 5, pp. 263-281.

Novak, M., Nogami, T., and Aboul-Ella, F. (1978). "Dynamic Soil Reaction for Plane Strain Case". Proceedings, ASCE Journal of Engineering Mechanics Division, 104 (EM4), pp. 953-959.

Stanton, J.F., Banerjee, S. and Hasayan, I., (1988), "Shaking Table Tests on Piles". Final Report, Washington State Department of Transportation, pp. 97.

Steedman, R.S. and Maheetharan, (1989). "Modelling the Dynamic Response of Piles in Dry Sand." Proceedings, Twelfth International Conference on Soil Mechanics and Foundations Engineering, Rio de Janeiro, Vol. 2, pp. 983-986.

TABLE I
Summary of Experimental Program

Test Series	Test Number	Pile Length (mm)	Unsupported Length (mm)	Head Mass (kg)	Exciting Frequency (Hz)	Base Acc. (g)
A	A1	610	40	0.79	1 to 9	0.14
	A2	610	152	0.79	1 to 9	0.12
	A3	610	304	0.79	1 to 9	0.14
C	B1	610	40	0.45	1 to 9	0.12
	B2	610	40	0.79	1 to 9	0.14
	B3	610	40	1.13	1 to 9	0.14
C	C1	305	40	0.79	1 to 9	0.13
	C2	610	40	0.79	1 to 9	0.14
	C3	915	40	0.79	1 to 9	0.15
D	D1	305	40	0.79	3	0.16
	D2	305	40	0.79	3	0.25
	D3	305	40	0.79	3	0.34

Table 2: Summary of Experimental Results

Test Number	Pile Length (mm)	Unsupported Length (mm)	Head Mass (kg)	Exciting Frequency (Hz)	Base Acc. (g)	Maximum B.M. (N-mm)
(a) Effect of unsupported length						
A1	610	40	0.79	2.7	0.18	211.1
A2	610	152	0.79	2.7	0.11	283.6
A3	610	304	0.79	2.7	0.15	300.7
(b) Effect of Magnitude of Head Mass:						
B1	610	40	0.45	2.7	0.13	174.8
B2	610	40	0.79	2.7	0.13	310.0
B3	610	40	1.13	2.7	0.15	534.8
(c) Effect of Base Acceleration:						
D1	305	40	0.79	3.0	0.16	300.2
D2	305	40	0.79	3.0	0.25	352.9
D3	305	40	0.79	3.0	0.34	451.8

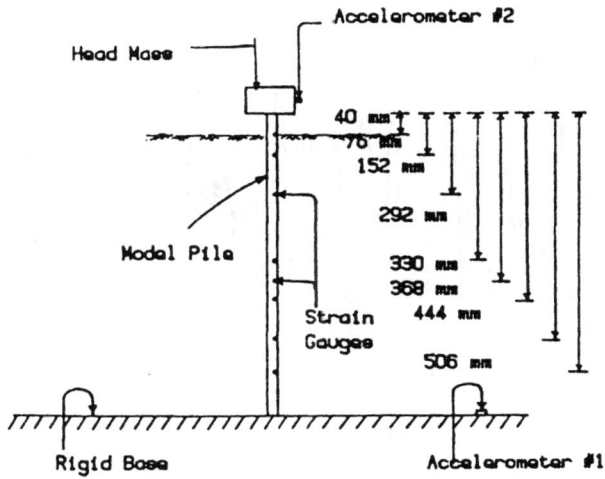

Fig. 1. Experimental Set-Up

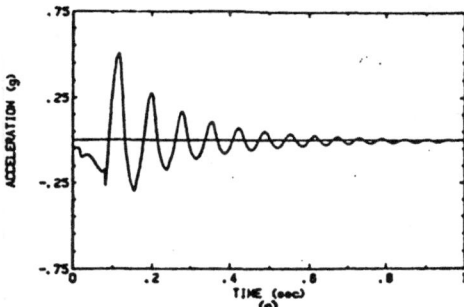

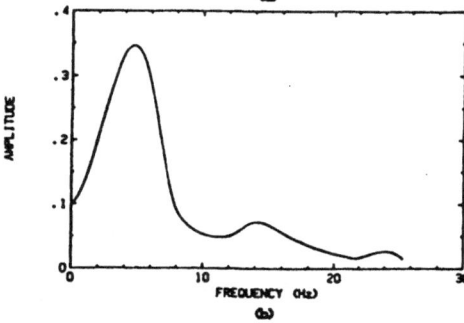

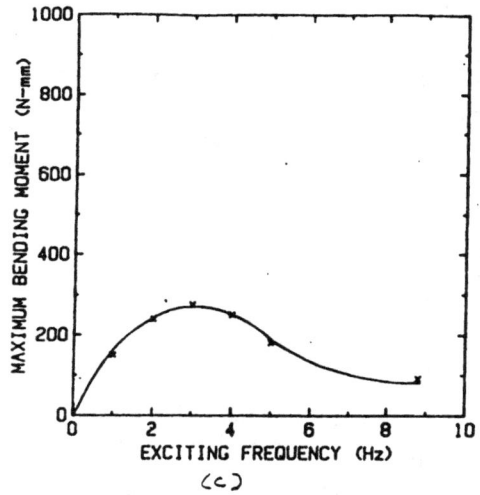

Fig. 2. Natural Frequency of Model Pile, Test C1
(a) and (b) Ring Down Test (c) Frequency Sweep Test

333

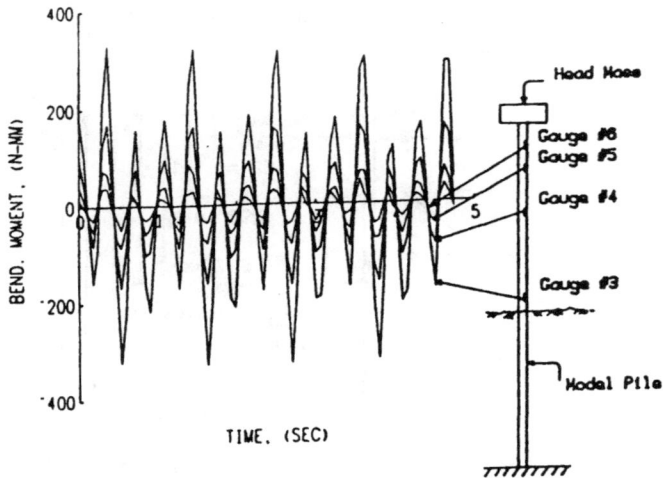

Fig. 3. Time History of Bending Moments (Test A3, Exciting Frequency 1Hz)

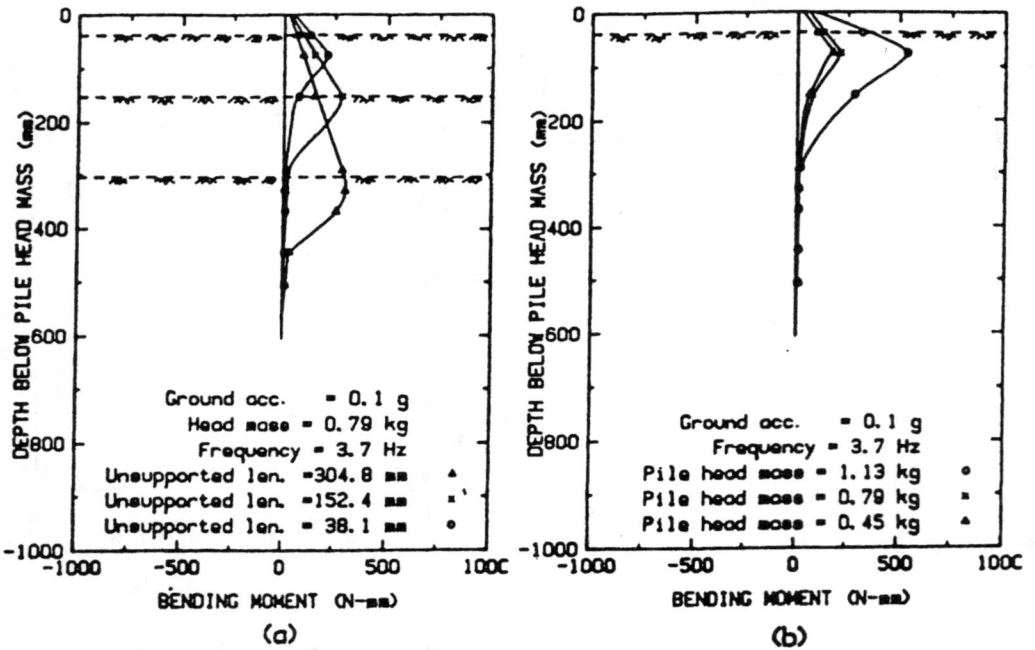

Fig. 4. Bending Moments Along the Pile Length (a) Effect of Unsupported Length
(b) Effect of Pile Head Mass

334

Model studies of dynamic pile response using hydraulic gradient shaking table tests

Li Yan[1], Peter M. Byrne[2], and Huaren Dou[2]

ABSTRACT

A new method of shake table model testing of soil-pile interaction is presented. A unique feature of this method is the use of hydraulic gradient to increase the stress level in the models, as opposed to the centrifuge technique. The testing principle and procedure are discussed. A series of free and forced vibration tests on single piles are presented. It is shown that this new testing technique provides an easy and inexpensive way to model the seismic response of piles at the field stress condition using a conventional shake table.

INTRODUCTION

Pile foundations have suffered severe damage in past earthquake events (Fukuoka 1966, Margason 1975, Sugimura 1981). Several analytical methods have been proposed to analyze the seismic response of piles under earthquake loading. These methods vary from linear elastic (e.g. Novak et al. 1978) to nonlinear solutions (e.g. Matlock et al. 1978). However, because of the scarcity of well documented field pile response data under real earthquake events, the actual performance of a pile foundation under seismic loading is still poorly understood. Thus, model tests have often been performed to study the seismic response of pile foundation, and these tests form a data base against which analytical methods can be evaluated.

The shaking table has been by far the most commonly used equipment in examining earthquake effects on soils and soil-structure systems. Many research institutes and universities have used shaking table facilities to simulate earthquake loading. However, the application of shake table tests in soil-structure interaction, especially for pile foundations, is severely limited mainly due to the small size of the model which results in low stresses in the soil. Because the stress-strain response of soil is highly dependent on the stress level, the response of the prototype will be quite different than the model. Consequently, this type of modelling test is not appropriate, and has been severely criticized. Finn and Gohl (1987) have reported the results of dynamic model tests on pile in which the centrifuge was used to increase the self stresses in the model to the field condition. However, this is an expensive approach, requiring costly

---

[1] Geotechnical Engineer, Klohn Leonoff Consultants, 10200 Shellbridge Way, Richmond, B.C.

[2] Prof. and Grad. Student, Respectively, Department of Civil Engineering, University of British Columbia, Vancouver, B.C.

equipment and well trained personnel. In addition, the application of seismic loading during centrifugal flight is not that easy, involving specially built facilities (Whitman 1984). Thus, the existing body of test data available on seismic soil-pile interaction at field stress condition is still meagre. Generation of such data is very much needed to validate numerical procedures.

In this paper, an inexpensive alternative method to test models on conventional shake tables but at field stress level is presented. This method, called the Hydraulic Gradient Similitude method (HGS), employs a high hydraulic gradient which causes a seepage force within the granular soil and creates a high body force. This body force results in high stresses simulating field conditions. The testing principle and its application to seismic pile response are presented.

PRINCIPLE OF HYDRAULIC GRADIENT SHAKING TABLE TEST

Similar to the centrifuge modelling technique, the HGS method is just another way of increasing soil stresses in the model. The only difference is that the body force of the model soil is effectively increased by the seepage force through the porous material rather than by centripetal acceleration. This offers the advantage that the models with escalated stress field can be mounted on the usual shake table and readily subjected to prescribed input excitation, rather than in the seismic centrifuge testing where seismic inputs have to be supplied within the high stress environment during the centrifugal flight.

For a model test subjected to a controlled downward hydraulic gradient, seepage force will increase the unit volume body force of a soil element by an amount of $i\gamma_w$. This is equivalent to increasing the unit weight of the material by $i\gamma_w$. Hence the effective unit weight, γ_m, of the model soil is:

$$\gamma_m - i\gamma_w + \gamma' \tag{1}$$

where i is the applied downward hydraulic gradient, γ_w is the unit weight of water if water is used in the test, and γ' is the submerged unit weight of soil. Thus, the vertical effective stress in the model soil has increased at any depth, z, below the surface is given by:

$$\sigma'_v - \gamma_m \cdot z \tag{2}$$

and the HGS scale factor, N, is defined as:

$$N - \frac{\gamma_m}{\gamma_p} - \frac{i\gamma_w + \gamma'}{\gamma_p} \tag{3}$$

where γ_p is the effective unit weight of the soil in the prototype, which could either be total or submerged unit weight depending upon the ground water conditions in the prototype soil. Thus, when the $1/n$ scaled model test is performed under a hydraulic gradient scale factor $N=n$, the stresses due to the self-weight of soils at homologous points of model and prototype will be the same, i.e., the scale factor for stress is unity. If the same soil is tested and the same stress path is followed in the model as in the prototype, the strains in the model and prototype will be the same (Roscoe 1968), i.e., the scale factor

336

for the strain is unity, while the displacements of the prototype will be larger than the model by the factor n=N. Thus, the scaling laws for the HGS tests are expected to be the same as in the centrifuge tests.

However, in the actual testing the scaling laws related to the problems studied have to be verified experimentally, as many factors may not be scaled due to technical limitations. In centrifuge tests, the "modelling of models" technique is often used, in which an assumed prototype behaviour is simulated with different scaled models under different stress fields. In this paper, this same technique will be employed to examine the HGS scaling laws for testing dynamic pile response.

HGS has been successfully applied in some model testings (Yan and Byrne 1989, 1990, 1991). Herein, a model study of seismic response of single piles to the simulated earthquake loading is presented to illustrate the application of HGS technique to dynamic testing. Test program consists of free and forced vibration tests of single piles in dense sand.

TEST SET-UP AND PROCEDURES

A test device using HGS testing principle has been developed at the University of British Columbia. A schematic of the device is shown in Fig.1. Detailed description of HGS device is given by Yan (1990).

During a test, water is continuously pumped to the sand surface. The given hydraulic gradient is obtained by controlling the air pressure in air chamber and draining the water to a low pressure at the base. Thus, pore water pressure in the soil decreases with depth, giving escalated effective stresses that increase linearly with depth. This test device is mounted on the normal shaking table, and the model tests performed as usual shaking table tests.

The soil deposit is formed of uniform fine Ottawa sand using the "quick sand" sample preparation technique (Yan and Byrne 1989). The sand deposit is 323.6 mm in height, and 404x190 mm in plan with the larger dimension in the shaking direction. No soft material is used at soil container walls to simulate the free field condition as it is found that the "soft" boundary is not sufficient to simulate the simple shear mode of soil motion, rather it introduces active soil failures at the boundaries when the stress in the soil is increased by the hydraulic gradient, thus violating zero strain boundary conditions before earthquake loading. The effect of rigid boundary will be discussed later in light of experimental data.

Three model piles made of 6.35, 9.53, 12.7 mm O.D. alum. tubing were used in the test program. The 6.35 mm O.D. pile is instrumented with 8 pairs of foil type strain gauges along its length to measure the bending moments. Brass masses of different weights are clamped at the pile head to simulate different structure masses.

In the free vibration tests, after a given soil stress condition is established, the pile is pushed to a given displacement at the pile head, and then released quickly to undergo free vibration. Pile head acceleration and lateral displacement are measured respectively by a miniature accelerometer at the mass centre and two LVDTs. For the forced vibration tests, a sinusoidal

337

motion is input through the shake table at the sand base to simulate earthquake excitation. A miniature accelerometer is installed at the surface about 16 pile diameter away from both boundary and pile to measure the free field response. Measurement is also made of the base input acceleration.

RESULTS AND DISCUSSIONS

Free Vibration Test

SYSTEM STIFFNESS AND DAMPING. Fig.2 shows a typical pile head acceleration response in the free vibration test at N = 30. This is a typical response of a under-damped system. The natural frequency of soil-pile system can be obtained from the period between acceleration peaks or FFT analysis of the acceleration response. Both methods have been used and found to give very similar results. The decay in the acceleration amplitude is a result of system damping, and an equivalent viscous damping can be obtained from the logarithmic decrement of the amplitude.

A series of pile free vibration tests were performed at different hydraulic gradients but at the same initial lateral displacement to evaluate the stress level dependency of the soil-pile natural frequency and damping. Fig.3 shows the relation between natural frequency, f_n, of the soil-pile system and the HGS scale factor, N. It is shown that as the hydraulic gradient increases the natural frequency of soil-pile system increases, and can be expressed as a power function of soil stresses. If the natural frequency, f_n, of the pile is normalized by the 1st natural frequency of the soil deposit, i.e. $f_n = V_s/4H$, where V_s and H are respectively the shear wave velocity and depth of soil deposit, the normalized soil-pile natural frequency appears to be independent of soil stress level, as also shown in Fig.3.

Fig.4 shows the equivalent viscous damping of soil-pile system determined at different soil stress levels. It is shown that the damping is nearly independent of soil stress levels, but appears to be a function of vibration amplitude. At the 1st cycle, large vibration amplitude gives a damping of about 8.6%, while at the 7th cycle smaller amplitude gives a damping of about 4%. The equivalent damping obtained from the logarithmic decrement of amplitude represents total damping of the system including material and geometric damping. The strong dependency of the measured damping with the vibration amplitude suggests that the major component of the measured damping results from material damping rather than geometric damping. Theoretical studies (Novak and Nogami 1977) have shown that material damping is the major source of system damping when the pile is vibrated in a frequency lower than the natural frequency of the soil deposit, as is the case for these tests (Fig.3). In the absence of radiation damping, the rigid boundary would have little effect on the free vibration test results.

EVALUATION OF SCALING LAWS. The scaling laws implied in the HGS test can be evaluated using the "modelling of models" technique. Three geometrically similar piles of different sizes were made, and tested in free vibration to examine the scaling law. These models were tested at the appropriate HGS scale factor, N, to produce the same prototype condition. According to the scaling laws, the natural frequencies measured in model tests, $(f_n)_m$, are related to the natural frequency of the assumed prototype, $(f_n)_p$, as in Eq.(4). This implies that the measured model frequency is proportional to HGS scale factor, N. As

$$(f_n)_m = N \cdot (f_n)_p \qquad \qquad (4)$$

shown in Fig.5, the measured model frequency does vary linearly with N, indicating that in HGS tests the scaling laws are satisfied. This result also indicates that for our test conditions the effects of lateral boundaries are not significant.

Forced Vibration Test

Forced vibration tests on a given model pile were performed at a HGS scale factor of 60. The natural frequency of this model is about 18.5 Hz determined from a free vibration test. The peak base accelerations used are 0.51g and 0.43g at input frequency of 10 and 20 Hz, respectively. The input vibration frequency is changed to examine the pile response under or near the pile natural frequency.

Table 1 give a summary of peak base, free-field, and pile head acceleration values. It can be seen that for both input base accelerations a similar small amount of amplification occurs between the base and sand surface. However, this is not the case for pile head acceleration.

When the pile is vibrated under a base input frequency significantly less than its natural frequency, only a small amount of amplification from free field to pile head occurs. On the other hand, when the pile is vibrated under a base input frequency close to its natural frequency, significant amplification occurs at the pile head acceleration. At this resonant condition, the acceleration at the pile head is about 3 times higher than that in free field.

Such a high pile head acceleration produces a significant increase in the pile bending moment. Fig.6 shows a comparison of pile bending moment distributions between resonant and non resonant conditions. This bending moment distribution is very similar to that observed from centrifuge test (Finn and Gohl 1987). The maximum bending moment occurs at a depth of 3.5 pile diameter below the surface. It is seen from this figure that the maximum bending moment for the near resonant condition is about 4 times higher than that for the non resonant condition. Thus, it is important in the design to avoid resonant condition, and provide enough damping and ductility to control the amplification and prevent pile bending damage should resonance occur.

SUMMARY AND CONCLUSION

In this paper, a new method of performing seismic shake table tests at a field stress condition is presented. The unique feature of this method is the use of the hydraulic gradient to increase the stress level in the models. Scaling laws implied in dynamic HGS tests have been evaluated and found to be satisfied. A series of free vibration and forced vibration tests have been presented to illustrate the application. Relations between pile stiffness and damping with the soil stress level have been evaluated, and different pile response at resonant and non resonant conditions has been clearly demonstrated. From these test results it is shown that a conventional shake table in combination with HGS technique can provide a simple and inexpensive way of seismic model testing at the field stress condition. Such tests can enrich our data base from which the analytical methods can be checked.

REFERENCES

Finn, W.D.L. and Gohl, W.B. 1987. Centrifuge Model Studies of Piles under Simulated Earthquake Lateral Loading. Proc. Geot. Spec. Pub. 11, Atlantic City, 21-39

Fukuoka, M. 1966. Damage to Civil Engineering Structures. Soils and Fdns., 6(2), 45-52

Margason, E. 1975. Pile Bending During Earthquakes, Design, Construction and Performance of Deep Foundations, ASCE Continuing Education Committee, San Francisco, California.

Matlock, H., Foo, S. and Bryant, L. 1978. Simulation of Lateral Pile Behaviour Under Earthquake Motion. Proc. Earthquake Engng. and Soil Dynamics, ASCE Spec. Conf. Pasadena, Calif., 601-619

Nogami, T. and Novak, M. 1977. Resistance of Soil to a Horizontally Vibrating Pile. J. of Earthquake Engng & Struct. Dynamics., 5, 249-261

Novak, M., Nogami, T. and Aboul-Ella, F. 1978. Dynamic Soil Reactions for Plane Strain Case. ASCE, J. Engng. Mech. Div., 104(4) 953-959

Roscoe, K.H. 1968. Soils and Model Tests, J. of Strain Analysis, 3(1) 57-64

Sugimura, Y. 1981. Earthquake Damage and Design Method of Piles. Proc. 10th ICSMFE, 2, 865-868

Whitman, R.V. 1984. Experiments with Earthquake Ground Motion Simulation. Proc. Sym. on Application of Centrifuge Modelling to Geotechnical Design. Ed. W.H. Craig, Manchester, 281-300

Yan, L. and Byrne, P.M. 1989. Application of Hydraulic Gradient Similitude Method to Small-scale Footing Tests on Sand, Can. Geot. J., 26(2), 246-259

Yan, L. and Byrne, P.M. 1990. Simulation of Downhole and Crosshole Seismic Tests on Sand using The Hydraulic Gradient Method. Can. Geot. J. 27(4).

Yan, L. 1990. Hydraulic Gradient Similitude Method for Geotechnical Modelling Tests with Emphasis on Laterally Loaded Piles, Ph.D. Thesis, Dept. of Civil Engng., Univ. of British Columbia, Vancouver, B.C.

Yan, L. and Byrne, P.M. 1991. Laboratory Small Scale Modelling Tests using the Hydraulic Gradient Similitude Method. Accepted for the 1991 Geot. Congress, ASCE, Boulder, CO.

Table I   SUMMARY OF FORCED VIBRATION ACCELERATION RESULTS

Base Motion	Free Field	Pile Head	Remarks
10Hz, 0.51g	0.69g	0.75g	Non Resonant
20Hz, 0.43g	0.53g	1.72g	Resonant

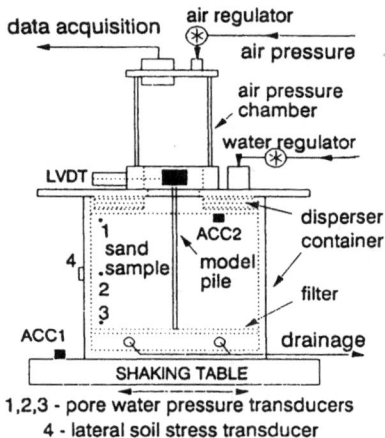

Figure 1 Schematic of HGS Shake Table Device

1,2,3 - pore water pressure transducers
4 - lateral soil stress transducer

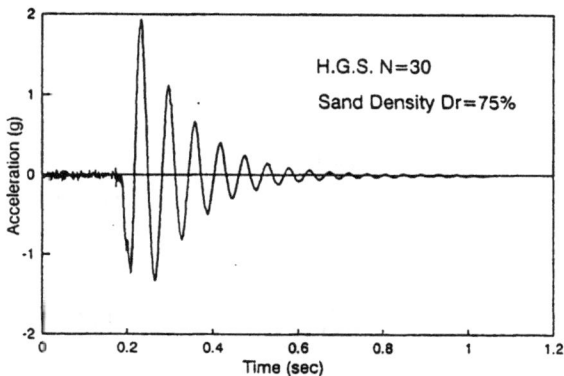

Figure 2 A Typical Acceleration Response of
Free Vibration at Model Pile Head

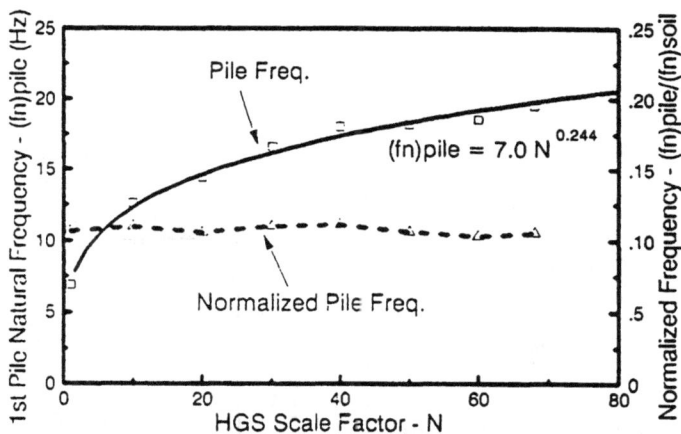

Figure 3 Soil Stress Level Effect on 1st Pile Natural Frequency

$(fn)pile = 7.0 \, N^{0.244}$

341

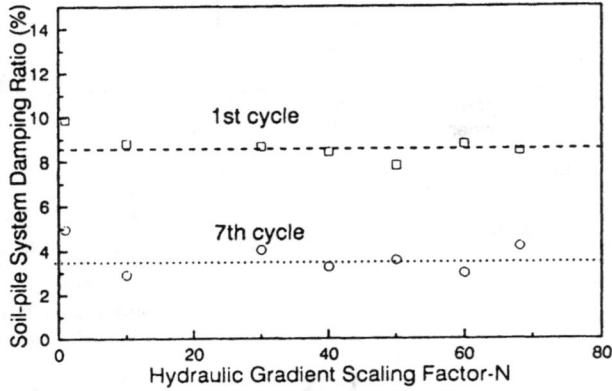

Figure 4 Soil Stress Level Effect on Equivalent Viscous Damping Ratio

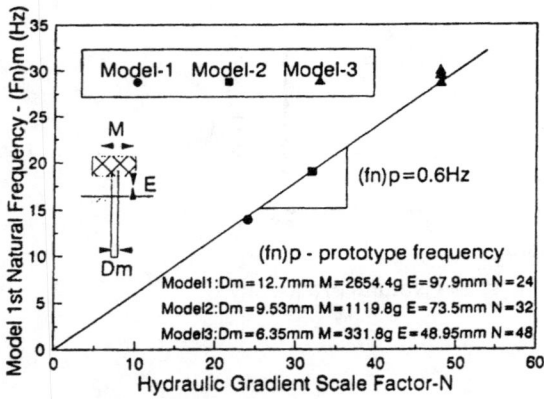

Figure 5 Evaluation of HGS Scaling Law

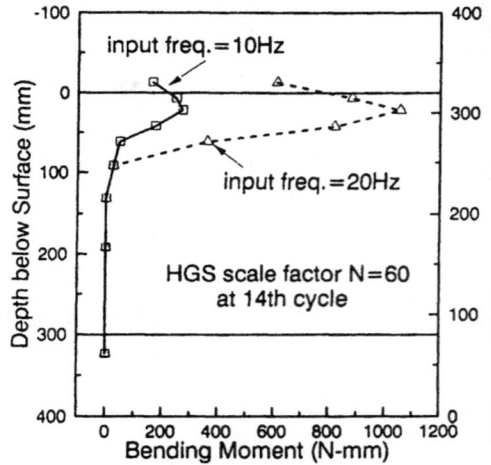

Figure 6 Comparison of Pile Bending Moment at Different Base Input Frequencies

342

A study of shield tunnel's earthquake prevention under a big earthquake

Koichi Murakami[1], Masahiro Nakano[1], and Yutaka Horibe[1]

1. Introduction

To date, the aseismicity of a shield tunnel is usually determined by the "Response Displacement Method" from analysed results of tests performed during an earthquake of magnitude equal to that indicated by "Criterion Utility Tunnel Design". However, by reason as follows, this shield tunnel is also thoroughly checked for aseismicity under a big earthquake by the "Dynamic Analysis Method".
① As part of Telecommunication Disaster Prevention Project in Tokyo Metro Politan Area, it is designed to be a highly reliable cable tunnel capable of immediate and accurate data transmission during a disaster.
② It crosses Class A rivers, and when destruction results, has significant effects on important structures like embankments.
③ The soil in which the shield tunnel drives changed from diluvial clay to sand.
This reports on the results of studies performed to check the aseismicity of the shield tunnel under a large earthquake.

2. Outline of Construction

This shield tunnel was constructed on behalf of Telecommunication Disaster Prevention Project in Tokyo Metropolitan Area. From a starting shaft in Katsushika ward, the tunnel drives under two Class A Rivers, the Nakagawa and the Arakawa, which are 150 m and 50 m wide, to an underground connection point in Edogawa Ward, a distance of 1.6 km.
The construction outlines of the shield tunnel are shown in table 1.

Table 1   Construction Outline

Length	1586 m	Segment Diameter	3.60 m
Horizontal Alignment	R=40 m, 65 m and 250 m	Machine Diameter	3.73 m
Earth Cover	38.6 m ~ 33.7 m	Machine Type	Slurry Shield Machine

3. Soil Conditions

The shield tunnel was constructed within the diluvium stratum called the Nanago stratum which is stable after considerations given to the effects of ground subsidence as a result of either a tail void created during excavation. Figure 1 shows the shield tunnel and the soil strata.

I Nippon Telegraph and Telephone Corporation

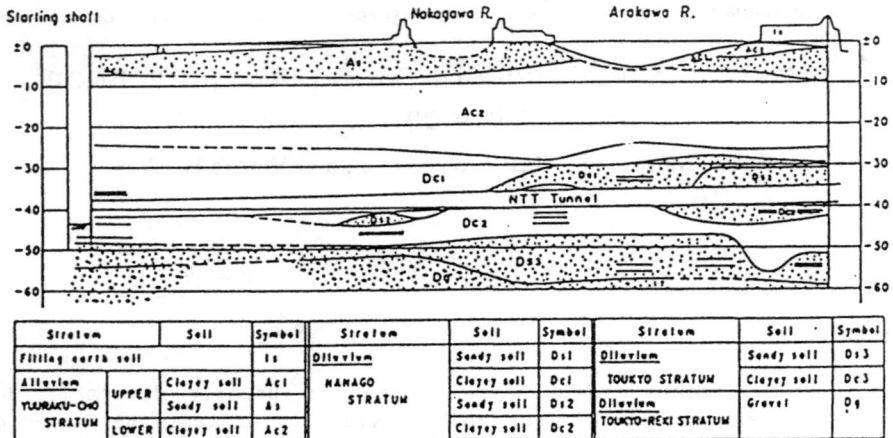

Fig.1 Geological Cross Section

Stratum		Soil	Symbol	Stratum	Soil	Symbol	Stratum	Soil	Symbol
Fillilag earth soil			Is	Diluvium	Sandy soil	Dsl	Diluvium	Sandy soil	Ds3
Alluvium YUURAKU-CHO STRATUM	UPPER	Clayey soil	Acl	NANAGO STRATUM	Clayey soil	Dcl	TOUKYO STRATUM	Clayey soil	Dc3
		Sandy soil	As		Sandy soil	Ds2	Diluvium TOUKYO-REKI STRATUM	Gravel	Dg
	LOWER	Clayey soil	Ac2		Clayey soil	Dc2			

4. Investigations on the Aseismicity of the Shield Tunnel

4.1 Assumptions of the analysis

The characteristics of the shield tunnel behavior as an underground conduit are as follows.
① Since the shield tunnel has a cavity cross-section, its mass is comparatively smaller than the surrounding soil.
② Since the shield tunnel is in contact with the ground on all sides, its vibration decays easily.
The shield tunnel does not vibrate during an earthquake for itself and is indicated by the displacement of the soil around the tunnel as proved by past vibration examinations. This analytical method therefore assumes
① that the shield tunnel vibrates together with the surrounding soil during an earthquake.
② that the relationship between the displacement of the shield tunnel and the

surrounding soil is similar to that as if the tunnel is a beam on an elastic floor.
③ that it is sufficient for considering the displacement of the shield tunnel to be that of the surrounding soil.

4.2 Structural characteristics of the shield tunnel

The shield tunnel is structurally different from the singlebodied circular tube since its segments are joined together by nuts and bolts.
Various models like "Equivalent rigid beam", "Finite Elements Method" and "Bone Structures" models can be considered for analysis.
This test however, uses the "Equivalent rigid beam" model which are applied for many cases. This model can analyze along the length of the tunnel well.

344

4.2.1 Segment structures

The dimensions of the standard segment used in this analysis and the allowable stress force and modulus of elasticity for each type of material used in the segments are listed in tables 2 and 3.

Table 2. Segment dimensions

· Segment	outer diameter	$D_s = 3750$ mm
	width	$l_s = 900$ mm
	height	$h = 250$ mm
· Bolts joining the segments		
	diameter	M24 $lb = 80$ mm
	number used	$n = 36$
	efective sectional area	3.53 cm²
· Plate thickness of each segment		
	surface plate	$t1 = 3.2$ mm
	main beam	$t2 = 22$ mm
	vertical rib	$t3 = 8$ mm
	connecting plates	$t4 = 22$ mm
	reinforcing plates	$t5 = 22$ mm

Table 3. Moduli of eelasticity. Allowable stress

Material	Standard	Modulus of elasticity (kg/cm²)	Allowable stress (kg/cm²)		
			Force type	Normal condition	During earthquake
Stell	SM50	2.1×10^6	tention	1.900	2.850
			comprressive	1.900	2.850
			bending	1.900	2.850
			shoar	1.100	1.650
Bolt	8·8	2.1×10^6	tention	2.400	3.600
			shoar	1.500	2.200

* Allowable stress during an earthquake is 50% higher than normal conditions.

4.2.2 Equivalent rigidity used in analysis

The equivalent rigidity of shield tunnels have been reported in past researches on dynamic vibrations along the length of the tunnel.

However in this case, the equivalent rigidity is assumed to be the elastic constant of the ring joint calculated from the bending rigidity of the main girder of segments within the limits of elasticity.

The bending rigidity is assumed to be different from that in the compressive or tensile direction. Consequently,

$$(E A)^T_{eq} / (E A)^C_{eq} = 15\%$$

$(E A)^T_{eq}$: equivalent compressive rigidity
$(E A)^C_{eq}$: equivalent tensile rigidity

Moreover, the bending rigidity is equal to the resistance in the segmental section when compressing or the resistance in the ring joint bolts when pulling.
Then

$$(E I)_{eq} / (E s \cdot I s) = 30\%$$

$(E I)_{eq}$: Equivalent bending rigidity
$E s$: Segment's modulus of elasticity
$I s$: Segment's moment of the second order

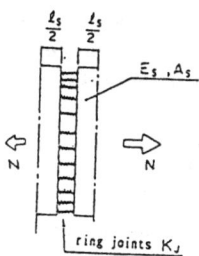

Fig. 2. Axial rigidity of segment ring joints

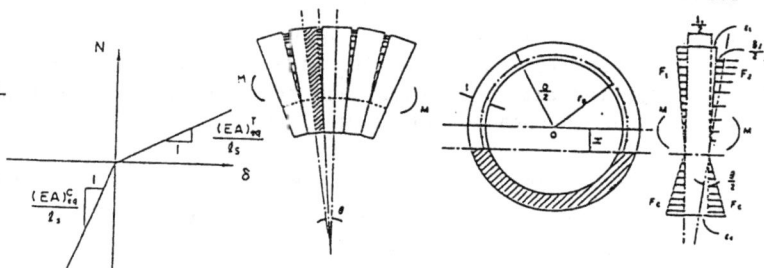

Fig. 3. Equivalent flexural rigidity of segment ring joint

4.3 Determination of dynamic characteristics of soil

Soil constants used in dynamic analysis are based on results of soil tests. However, since it is believed that the speed of shearing greatly affects the analysed results, boring test of PS logging was done. To determine the decrease in rigidity during an earthquake, dynamic strain curves at 6 places within the areas of displacement in the soil strata are determined by the theory of multiple reflection.

The soil constants used in this analysis is shown in Table 4.

Figure 4 shows an example of a dynamic strain curve measured within the Dc1 soil layer of the shield tunnel.

Table 4. Soil types used for analysis

Stratum		soil	symbol	Thickness (m)	N value	Unit weight (t/m)	Shear force speed (m/sec)	poisson's ratio
Stratum Top stratum			t s	1.2~2.7	4	1.77	135	0.49
Silt stratum			Ac1	2~2.5	0	1.4	50	0.50
Alluvial YURAKU-CHO STRATUM	UPPER	Clayey soil	As	4~3	10	1.93	120	0.49
		Sandy soil	Ac2	0.7~2.9	1	1.6	100	0.49
	LOWER	Clayey soil	Ac3	17~23	1	1.65	140	0.49
Diluvium KASAGO STRATUM		Sandy soil	Ds1	4~10	15	1.96	200	0.49
		Clayey soil	Dc1	2~13	3	1.75	200	0.43
Diluvium TOKYO STRATUM		Sandy soil	Ds2	11~13	30	1.93	240	0.43
		Clayey soil	Dc2	3~10	23	1.77	225	0.43
		Sandy soil	Ds3	2~11	60	2.17	240	0.47
		Clayey soil	Dc3	1~1.9	17	1.86	225	0.43
Diluvium TOKYO-REKI STRATUM			Dg	—	119	—	—	0.46

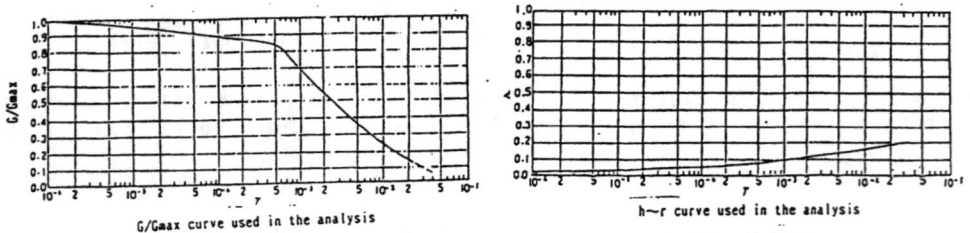

G/Gmax curve used in the analysis

h~r curve used in the analysis

Fig. 4. Dynamic strain curve measured within the Dc1 soil layer

4.4 Determination of elastic constant between tunnel and soil

The elastic constant is determined by using FEM models at 6 places with soil changes and calculated from the displacement at each intersection where unit weights are applied along the axial and perpedecular direction of the tunnel as well as directions perpendicular to the axis.

4.5 Basic policy in determination of input earthquake magnitude and aseismicity

Two aseismic standards were determined in proportion to the two input earthquake magnitude. Safety of the shield tunnel was tested for each earthquake magnitude. Table 5 shows the basic principles of the test and the input standards which are targetted to be the aseismic standards.

Table 5. Input standards which are the aseismic targets

Input standards	Earthquake magnitude	Basic principle of aseismic tests	Aseismic tests
L-1	Common channel spectrum (equivalent to M=7.0, Δ=50km) ★That whose response spectrum maximum.	Ensure that structures are sufficientry safe during the occurence of a relatively major earthquake. these same structures should also be safe to use after an earthquake.	Stresses generated in bolt joining the segments or the joints (surface plates) should be below the allowable stress value.
L-2	Major earthquake (Kanto Earthquake) (equivalent to M=8.0, Δ=50km) ★Imagine an earthquake of the maximum possible magnitude in this area.	Earthquakes of this magnitude are quite rare in this area. Structures should not be functionally damaged and should be able to withstand further damage. Above all, the structures should be safe.	Stresses generated in bolts joining the segments or the joints (surface plates) should be below the yield stress value.

4.5.1 Input earthquake magnitude

① The seismic movements of magnitude L1 are generated by the common channel method (equivalent to M=7, Δ =50 km) whose waveform is very similar to that the 1983 Nihonkai Chubu earthquake observed near a place called Tsugaru Ohashi. The amplitude of the dynamically analysed acceleration response spectrum is adjusted to match that recorded on the actual quake.
② The seismic activity of magnitude L2 is assumed to be a major earthquake (Kanto earthquake class)(M=8.0, Δ =50 km). The forces generated on the cross-section of shield tunnel are estimated from the design results at the L1 level. The estimation is described below.

4.5.2 Estimation of response values to major earthquake

The response of subterranean structures during seismic activity is similar to the shield tunnel and can be expressed as follows :

$$S_\gamma(\omega) = S_1(\omega) \cdot S_2(\omega) \cdot S_3(\omega)$$

$S_\gamma(\omega)$: tunnel response
$S_1(\omega)$: response of input seismic force
$S_2(\omega)$: response of surface soil
$S_3(\omega)$: tunnel's vibration coefficient

Previous experiments have proved that the tunnel does not vibrate for itself, $S_3(\omega)$ can be assumed to be a constant independent of frequency. $S_2(\omega)$ is not expected to change very much when the shearing strain of the surface soil during an earthquake is only a few percent.
Therefore, the tunnel response $S_\gamma(\omega)$ will mainly be affected by the response characteristics of the applied seismic force $S_1(\omega)$. Once the ratio between the characteristic frequency of the soil to the strength of the seismic force can be determined, seismic response values at other areas can be deduced if those values at any one place is known.
The deduction of response values during a major earthquake, uses the attenuation equation in the acceleration response spectrum suggested by Kawashima et.al.. By inserting into his equation, the values of soil type at the construction site, distance from the epicenter and the magnitude of the earthquake, the acceleration response spectrum can be determined.

347

Next, existing records of major earthquakes are checked for something whose waveform matches that of the spectrum determined theoretically. The amplitude within the band width is then ajusted. The imaginary surface earthquake can be altered to the applied seismic waveform at the foundation by using the multiple reflection theory. The response spectrum at the foundation can then be determined.

By comparing this spectrum and that of the common channel, the response values during a major earthquake can be deduced from calculations performed on the common channel spectrum if the ratio near the characteristic frequency of the soil is determined.

Figure 5. shows the flowchart for deducing tunnel response during a major earthquake.

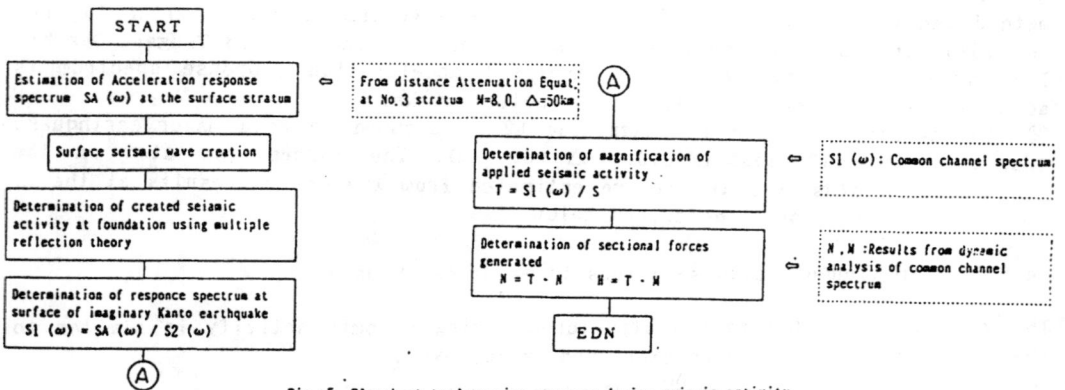

| START |

| Estimation of Acceleration response spectrum SA (ω) at the surface stratum | ⇐ | From distance Attenuation Equat. at No.3 stratum M=8.0. △=50km (A) |

| Surface seismic wave creation |

| Determination of created seismic activity at foundation using multiple reflection theory |

| Determination of magnification of applied seismic activity T = S1 (ω) / S | ⇐ | S1 (ω): Common channel spectrum |

| Determination of responce spectrum at surface of imaginary Kanto earthquake S1 (ω) = SA (ω) / S2 (ω) |

| Determination of sectional forces generated N = T · M   H = T · M | ⇐ | N,M :Results from dynamic analysis of common channel spectrum |

(A)

| EDN |

Fig. 5. Flowchart to determine stresses during seismic activity

4.5.3 Results of an deducing major earthquake

This analysis assumes a major earthquake of magnitude equivalent to Kanto Earthquake (M=8, △ =50 km).

Figure 6 shows the acceleration response spectrum during the occurrence of a major earthquake.

Figure 7 shows the amplitude adjusted waveforms for the acceleration response spectrum by matching them with records of the Nihonkai Chubu earthquake observed at a place called Tsugaru Ohashi. This waveform is then transformed into a waveform at basic stratum by using SHAKE formula to give the results shown in figure   .

For the characteristic cycle of the relative stratum is 1.0 ~ 2.0 seconds, the seismic activity is amplified by least 1.8 times. In other words, the response magnification generated in the tunnel during the Kanto earthquake is about 1.8 times that of the applied seismic activity determined by the common channel spectrum.

Figure 9 shows the acceleration response magnification at the basic stratum within a cycle of the relative stratum.

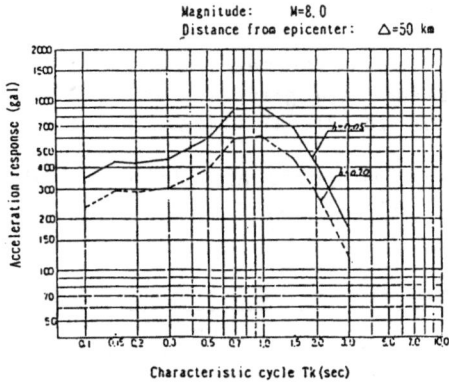

Magnitude: M=8.0
Distance from epicenter: Δ=50 km

Fig. 6. Acceleration response spectrum of imaginary Kanto Earthquake

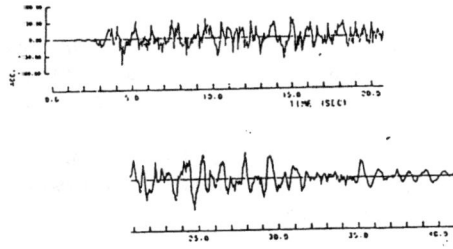

Fig. 7. Acceleration waveform generated during imaginary Kanto Earthquake

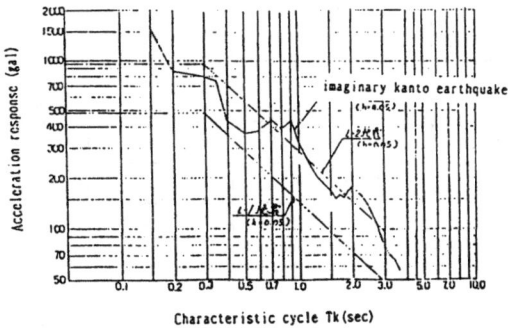

Fig. 8. Acceleration response spectrum at the basic stratum

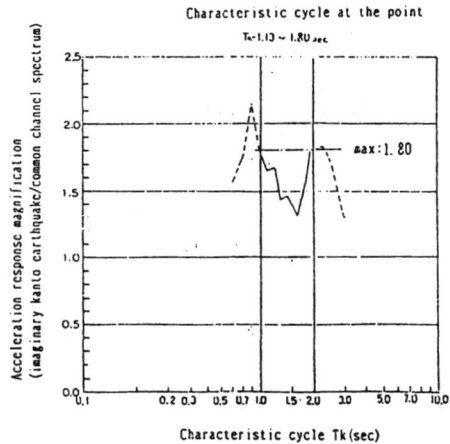

Fig. 9. Acceleration response magnification at basic stratum

4.6 Results of aseismicity tests

The places of testing is as follows :

① Response displacement method (tested at shield's horizontal cross-section)
  • 4 places (a) vicinity of starting vertical shaft (b) left levee of Arakawa river (c) riverbed of Arakawa river (d) right levee of Arakawa river

② Response displacement method (tested at shield's vertical cross-section)
  • 5 places (a) arc with radius of curvature R=45 m (b) arc with radius of curvature R=65 m (c) vicinity of left banks of Arakawa river (d) Arakawa riverbed (e) vicinity of right banks of Arakawa river

③ Dynamic analysis method
  • A model with 60 test points each 20m apart along a total distance of 1200m from the vicinity of the right banks of the Arakawa river

4.6.1 Test results of L1 magnitude earthquake

During a relatively frequently occurring earthquake of magnitude L1 (M=7, Δ =50 km), similar in magnitude to that created by the common channel method, the stresses generated in the various segments, bolts and surface plates of structures supposed to be still durable were all within the allowable stress limit.

349

4.6.2 Test results of L2 magnitude earthquake

In this area, the rarely occurring major earthquake of magnitude L2 (M=8, Δ =50 km) creates stresses in segments, bolts and main girders which are within the yield limit. Structures do not receive any functional damage and can still withstand further damage.
The results obtained by the response displacement method are shown in Table 6 while those by the dynamic analysis method are in Table 7 under L1 or L2 magnitude earthquake.

Table 6. The results obtained by the response displacement method

stress (kg/cm)	L1 level		L1 level	
	generated stress	allowable stress	generated stress	yield stress
[Perpendicular cross-section] Main girder (SM50)	2. 210	2. 850	2. 432	3. 200
[Axial cross-section] Bolts (8.8) Surface pleta (SM50)	2. 293 1. 575	3. 600 2. 850	4. 127 2. 835	6. 400 3. 200

Table 7. The results obtained by the dynamic analysis method

stress (kg/cm)	L1 level		L1 level	
	generated stress	allowable stress	generated stress	yield stress
Bolts (8.8) Surface pleta (SM50)	2. 293 1. 575	3. 600 2. 850	4. 127 2. 835	6. 400 3. 200

5. Afterword

This shield tunnel's construction (primary lining) was completed in November 1989. After completion of the secondary lining, measuring instruments will be installed and seismic tests will be performed.
The necessity for a highly reliable telecommunications network system in this age of information is increasing.
NTT recognizes this need and will continue performing tests on how to improve the quake resistance of cable tunnels with due considerations given to the results of this report.

6. References

(1)Kawashima, Aizawa :*Attenuation of Earthquake Response Spectra Based on Multiple Regression Analysis of Japanese Strong Motion Data*, Proceedings of Japan Society of Civil Engineers No.350.
(2)Suzuki, Yagi, Kobayasi :*An Aseismatic Study on Tunnels Under an Imaginary Major Earthquake*, Proceedings of the 44th Annual Conference of the Japan Society of CIvil Engineers, 3, 1989.

A microcomputer software package for shake table testing

M. Penn[I], A. Filiatrault[II], R.O. Foschi[III], and S. Cherry[IV]

ABSTRACT

A new microcomputer data acquisition and shake table control system has been developed for the Earthquake Engineering Research Laboratory at the University of British Columbia. The new system utilizes an IBM/AT compatible microcomputer for acquiring data and controlling the motion of the shake table. Two Metrabyte DAS20 boards attached to a microcomputer provide 32 channels (Analog to Digital) for data acquisition and 4 channels (Digital to Analog) for control. The software is capable of simultaneously acquiring data at a rate of 500 Hz per channel (16 kHz total), saving this data directly on the hard disk and controlling the shake table. The software, written in C language, was designed to be user friendly. Setting up a multi channel experiment, controlling the shake table, checking the data acquired and saving the data from all the channels in ASCII single column files is achieved through very simple window menus. Special features, such as automatic file naming, automatic parameter setting, translation functions and graphics enable the completion of an experiment within minutes.

INTRODUCTION

Shake table testing is employed extensively to demonstrate the operability of vital equipment during a seismic event, to verify the accuracy of analytical studies and to experimentally qualify and study systems which are not readily amenable to mathematical analysis. Shake tables can be used to investigate the dynamic behaviour of diverse and complex structural and mechanical systems, such as nuclear reactor components, submerged structures, piping networks, bridges, dams, buildings, tanks, transformers and circuit breakers, hospital equipment etc.

[I]Graduate Research Assistant, Dept. Civil Eng., Univ. of B.C., Vancouver, B.C.;

[II]Assistant Professor, Dept. Civil Eng., Ecole Polytechnique, Montréal, QC;

[III], [IV]Professors, Dept. Civil Eng., Univ. of B.C., Vancouver, B.C.

The University of British Columbia (UBC) Earthquake Engineering Research Laboratory offers comprehensive facilities for seismic research and qualification testing. The central feature of the laboratory is an advanced, closed—loop, servo—controlled electro—hydraulic shake table. The shake table is controlled by an electronic feedback control system for simulation of single axis horizontal ground motion. This system is in the process of being upgraded for multi—axis excitation.

Recently, a new data acquisition and control system has been incorporated into the UBC Earthquake Shake Table Facility. This involves a feedback control system, signal conditioning cards and a microcomputer with the addition of two Analog/Digital cards, as shown in Fig. 1.

UBC SHAKE TABLE CONTROL SYSTEM

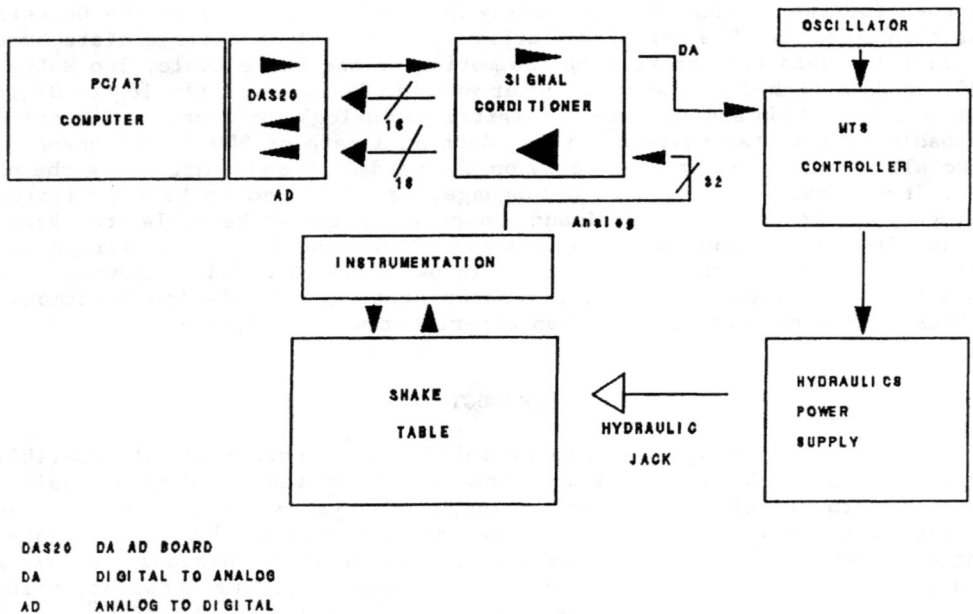

DAS20   DA AD BOARD
DA   DIGITAL TO ANALOG
AD   ANALOG TO DIGITAL

Figure 1 — Block diagram of the control system

The UBC microcomputer system was designed to deal with up to 32 channels having sampling rates of up to 500 Hertz per channel (16 kHz total) and to have the ability to simultaneously control the shake table and transfer the acquired data directly to the computer's hard disk. In addition, special features were added to help the user in defining active channels, defining names of result files and in obtaining result data that are in physical units and in ASCII formatted files ready to be imported to any data analysis software package.

GENERAL DESCRIPTION OF THE SYSTEM

The Shake Table

The shake table, located in the Earthquake Engineering Research Laboratory at the University of British Columbia, is a 3m x 3m cellular aluminium construction weighing 20 kN. The table is driven by an uniaxial 135 kN hydraulic actuator, can support a payload of 155 kN and is mounted on four vertical posts with swivel end bearings located in an isolated concrete pit foundation. In normal operation the shake table can achieve a maximum acceleration of 2.5 g, a maximum velocity of 130 cm/s and maximum displacements of ±7.5 cm.

The MTS Control system

The movement of the hydraulic actuator is controlled by a MTS (MTS Systems Corporation) system which consists of a MTS 443 Controller with the addition of control and selector cards. Control of the servo loop is accomplished with a displacement transducer.

The Signal Conditioner Cards

The signal from each recording instrument goes through an analog filter and an amplifier before reaching the AD board in the computer. This achieves two functions; it serves mainly as a signal conditioner but also protects the microcomputer system. The filters are low pass filters with settings from 2.5 to 100 Hz. The amplifiers allow amplification by factors of 0.5 to 10. Further amplification is also obtainable from the amplifier in the DAS20 analog/digital board.

Table I Microcomputer System Specifications

Number of channels	up to 32 channels.
Sampling rate	from 1/3600 to 500 Hz.
DA output rate	from 1/3600 to 500 Hz.
Data acquired	up to 5 Mbyte (can be expanded).
Duration of a test	from one second to 27 hours.

The Microcomputer System

Table 1 contains a condensed specification list of the microcomputer system. A 10 mHz IBM AT compatible computer (an AST product) with the addition of two Metrabyte DAS20 analog/digital converters are used for controlling the shake table as well as acquiring all the data from the instruments. The software is written mostly in the C language; only the drivers of the DAS20 boards are written in assembly. The software was developed under Microsoft's C Compiler and Assembler. The user interface is fully menu driven using windows for the display of each menu. The program accepts standard earthquake files in ASCII format, or any ASCII formatted single column file, as a DA control file (this file controls the shake table). The program offers only basic post-analysis options; excellent data analysis programs exist on the market and can be used more efficiently than home written packages. The physical measurement of each instrument is produced in a single column ASCII file. The program can display the maximum value (and the time it occurred) for each channel. The user can also view a graph of up to eight channels at a time on the screen. These features are given mainly to allow the user a quick view of the data recently acquired. A comments file contains a summary of all the parameter settings of the experiment together with the user's comments. In the CALIB menu it is possible to display the value of each channel in real time. This feature enables the user to calibrate the instrumentation and also to read the initial offsets of the instruments before the experiment. Finally, the user may save the settings of the experiment in a file which can be reloaded later.

Since an experiment could produce more than 32 files, a special feature allows automatic file naming. The user defines a name for the binary result file. The program will use this name to automatically create names for the 32 result ASCII files, the statistics file and the comments file. In addition, the next time the experiment is run the program automatically modifies the name of the previous binary file. Therefore, the user can run any number of experiments (up to 100) while having to define only one name. In addition, there is no danger of overwriting a previous file.

The source earthquake files are received in an ASCII format. To use these files an ASCII to binary translation is necessary. In addition the user may wish to create his own "earthquake" file or to use a waveform created in a data analysis package. The BUILD menu allows these features.

SOFTWARE MENUS DESCRIPTION

Menu Display

Figure 2 illustrates a typical screen menu display. The main menu always appears at the top of the screen; this menu allows the user to choose one of the many available functions. These functions are displayed by overlapping windows as shown for the BUILD option in fig 2.

A typical experiment may involve the following steps :

1)   Choose the OUTPUT menu to modify any of the parameter fields (Earthquake file name, DA rate, DA duration etc.).

2)   Choose the INPUT menu to select the active channels and to change the sampling parameters (AD rate, AD duration, etc.).

3)   Choose the CALIB menu to calibrate the instrumentation and to read the initial value of each channel.

4)   Choose the RUN menu to proceed with the experiment. At the end of the experiment the user may change the name of the AD binary file which contains the raw data of the experiment.

5)   Choose the RESULT menu to obtain the physical data in ASCII files. The user may display the results for a quick check.

```
┌─────────┬──────┬─────────┬─────────┬─────────┬─────────┬─────────┬──────┐
│ SETUP   │ RUN  │ OUTPUT  │ INPUT   │ RESULT  │ BUILD   │ CALIB   │ DOS  │
└─────────┴──────┴─────────┴─────────┴─────────┴─────────┴─────────┴──────┘
```

```
═══════════ BUILD DA File ═══════════
     Generate wave
     ═════════ Wave Form ═════════
                    ════════ Sin Sweep ════════

     Name of file            d:\eq\data
     ASCII, Binary (A,B)     B
     Max value (%)           100
     First freq (Hz)         1
     Last freq (Hz)          3
     Freq step (Hz)          0.5
     No. cycl/freq           3
     DA Rate (sec)           0.01

     EXECUTE (Y/N)           Y
```

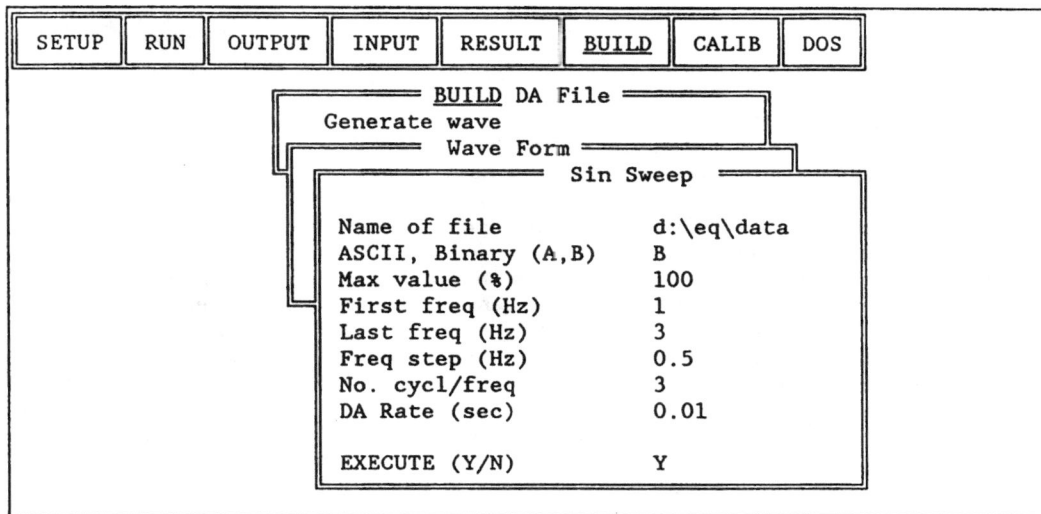

Figure 2 — Window Menu Display

6) Choose the DOS menu to exit, or the RUN menu to start another experiment, or the SETUP menu to reload the parameters with a previously defined experiment.

SETUP Menu

This menu allows the user to choose a setup file which contains the parameters of a previously defined experiment.

RUN Menu

This menu gives a check list for starting the hydraulics system and prompts the user to press the start/stop switch. The bottom of the screen contains the main settings of the experiment for a last check. At the completion of the experiment the user may choose a name for the binary file. This file contains all the raw data of all the active channels.

OUTPUT Menu

The user may choose a DA file for controlling the shake table. Then the DA output rate and the duration of the output are selected. An additional feature allows the user to select a portion of the DA file for output.

INPUT Menu

The user selects any subset of the 32 available AD sampling channels. The sampling rate (AD Rate) and duration for sampling are selected. An additional feature allows the sampling to be delayed relative to the DA output (or vice versa). The user may also set the programmable gain for each of the 32 channels separately.

355

RESULT Menu

This menu contains several sub-menus namely : SET SCALE/CHANNEL, STATISTICS, TRANSFER DATA TO ASCII, COMMENTS and DISPLAY DATA. At the completion of a test the user may decide to run another test or to view the results of the current test. To obtain the results of the test the first step is to use the SET SCALE/CHANNEL option to set the scales for each channel separately. This includes a multiplication factor as well as an addition factor for translating the sampled data into physical units. The STATISTICS menu creates an ASCII file of the maximum values for each channel and the time of these results. In addition these results are displayed on the screen for a quick review. The TRANSFER TO ASCII function allows the creation of the final result files. These files can later be used in any available data analysis package. The COMMENTS function creates an ASCII file which contains all the parameter settings of the test. In addition, the user may add any of his own comments to the end of the file. There are no limits to the amount or format of the comments. The DISPLAY DATA function presents a plot of any of the sampled channels. The user may choose up to eight channels at a time for display.

BUILD Menu

This menu contains some utilities to create DA binary output files for controlling the shake table. Functions are available to generate a waveform, translate a standard earthquake ASCII file to a binary DA file, translate a single column ASCII file to a DA binary file or to translate a binary DA file into a single column ASCII file.

The waveform generation function allows the user to generate a ramp waveform (i.e. a linear increasing or decreasing function) or a sine sweep function. The parameters (amplitude, DA Rate, duration etc.) are set interactively by the user. For the sine sweep function, the first and last frequencies and number of cycles per frequency must be defined. For clarity, an example of a sine sweep wave is presented together with measurements of the reaction of a model structure to this waveform.
First frequency = 2 Hertz
Last frequency = 5 Hertz
Frequency step = .5 Hertz
Number of cycles per frequency = 10
DA Rate = .005 second.

Figure 3 presents the waveform created by the BUILD menu which will be transmitted to the hydraulics control system as a DA output file — an acceleration file in this case. Figure 4 shows the acceleration measured by an accelerometer attached to a model structure, having a fundamental natural frequency of about 4Hz, which was mounted on the shake table.

CALIB Menu

This menu displays, in real time, the sampled data of all the selected channels. This menu serves two purposes. First, the user may use this menu to calibrate the instrumentation. Secondly, the data displayed on the screen can be used in the RESULT menu to translate the sampled data into physical units.

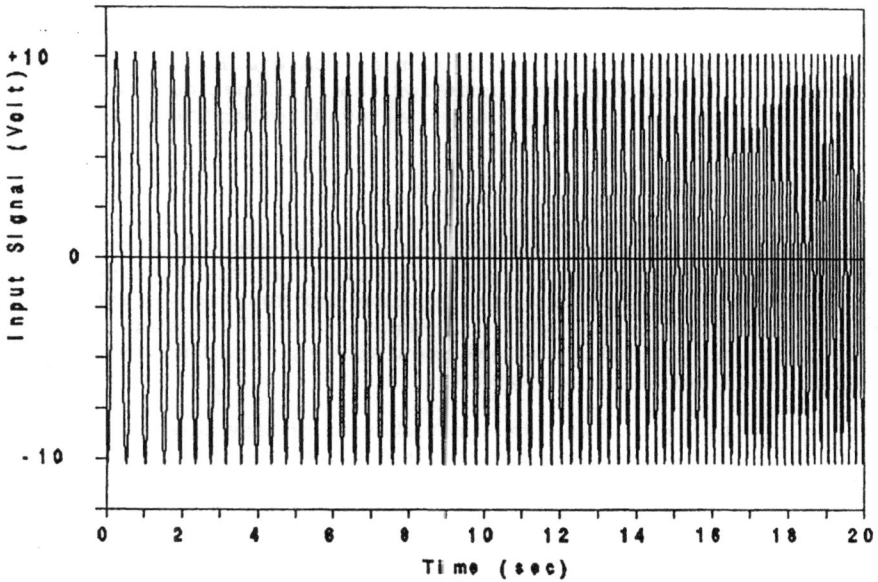

Figure 3 — Acceleration sine sweep file.

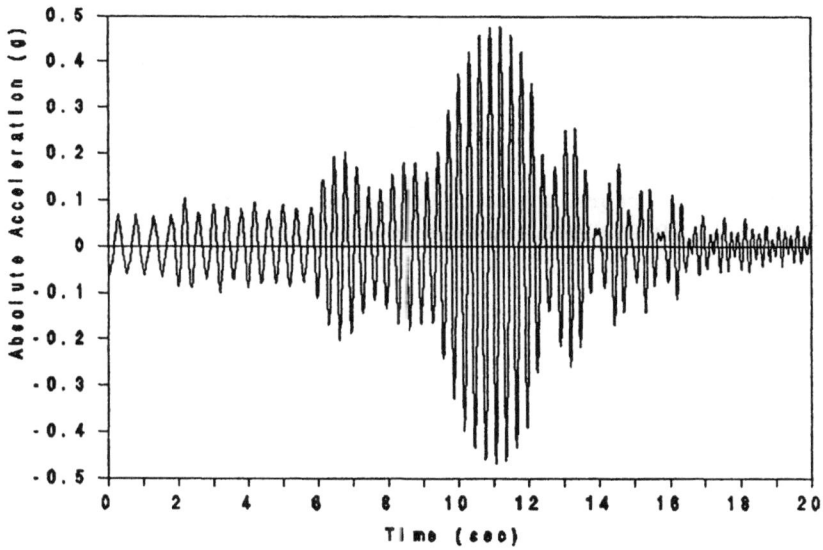

Figure 4 — Acceleration of a model structure under an acceleration sine
sweep DA file.

DOS Menu

Before exiting the program, the user may save all the parameter settings of the experiment. The user gives a name for the file that contains the settings. There is no limit on the number of files allowed (other than DOS limitations).

CONCLUSION

The data acquisition and control system developed for shake table testing in the Earthquake Engineering Research Laboratory at the University of British Columbia meets all the specifications required for running a seismic experiment. Because it was developed and designed in—house, the software enables the user to set up an experiment for multi channel acquisition, view the results, and obtain ASCII result files within a few minutes.

The UBC software package does not depend on any special feature of the microcomputer; it can therefore be installed on any IBM compatible machine. Since the control and acquisition parameters are independent, the software can also be utilized strictly as a data acquisition system for general dynamic testing.

ACKNOWLEDGEMENTS

The authors acknowledge the support of the National Sciences and Engineering Research Council of Canada, which provided operating and strategic grants in support of this project.

DISCLAIMER

The use of specific manufacturer names, model numbers, and other specifications do not represent an endorsement of the products by the authors, but are given only to provide adequate understanding of the subject.

Studies on friction base isolators

X. Wang[I], C. Marsh[II], and A. Pall[III]

ABSTRACT

This paper deals with friction type base isolators including the pure-friction system (P-F) and resilient-friction system (R-FBI). Using the computer program DRAIN-2D for time-history dynamic analysis, various intensities and frequencies of the excitation were employed in the simulation. The peak relative displacements and the maximum accelerations of the mass were determined. Particular attention is given to the effect of the elastic stiffness of the R-FBI system. It is shown that the slip displacement decreases sharply as the elastic restoring force of the isolator increases. Several sensitivity analyses for variations in the type and intensity of the excitation were carried out, and parametric studies were performed. Preliminary laboratory tests were conducted to check the theoretical prediction.

INTRODUCTION

Base isolators can effect useful reductions in the accelerations developed in buildings due to seismic ground motions. The isolators may operate in any of several manners, or combinations thereof, which include:

a) By introducing a horizontally acting spring between the ground and the building, the natural frequency can be reduced to a value below the range of frequencies of the ground motion, thereby avoiding resonance.
b) By introducing material with a high hysteretic damping capacity, between the ground and the building, excessive motion can be prevented by damping.
c) By introducing rollers between the ground and the building, the

----

[I] Graduate student, CBS, Concordia University, Montreal, Quebec.

[II] Professor, CBS, Concordia University, Montreal, Quebec.

[III] President, Pall Dynamics Ltd., Pierrefonds, Quebec.

transfer of ground motion to the building is prevented. By resting the rollers on dished surfaces, the action of gravity can control the displacements.
d) By supporting the building on surfaces which can slide when the earthquake intensity overcomes the friction, the force transferred, and hence the acceleration of the building, can be limited.

By combining these ideas it is possible to obtain results which are closer to some specified ideal than can be achieved by one idea alone. The aim is to reduce the maximum acceleration suffered by the building. This is evidently best achieved by introducing rollers. However, the building must resist wind forces, so a minimum lateral resistance is needed. This is most readily provided by a sliding surface with the desired coefficient of friction, but the value required tends to be low (< 0.1) with the result that the motion of the building relative to the ground during earthquake may be large, influencing the design of service connections. To reduce the displacements, the friction coefficient must be higher, but this leads to greater accelerations being transferred to the building. By adding resilient restraints as a means of controlling the relative lateral displacement, in conjunction with a low value for the friction coefficient, both the acceleration and the relative displacement can be minimized. When a spring is combined with friction, the natural frequency changes with amplitude thus resonance is not possible. Because of the low restraining force required, the natural frequency of the building is low and hence outside the range of exciting frequencies. Once sliding starts the rate of energy dissipation is high, and there is no need for any hysteretic damping from the resilient restraints.

PURE-FRICTION SYSTEM

The pure-friction system is used to support the weight of the structure and provides sliding friction, sufficiently high to withstand the wind load. For harmonic excitation, there are three ranges of building response[1]:

a) There is no sliding when $\mu g/A < 1$
b) There is a stick-slip behaviour when $0.537 < \mu g/A < 1$
c) There is continuous sliding when $\mu g/A < 0.537$
where: μ = the coefficient of friction
A = the maximum harmonic acceleration of the ground

When a body can slide on the supporting surfaces, the equation of motion during sliding, (case c), can be written as:

$$\ddot{u}_n(t) = (-1)^{n+1} \mu_d g - \ddot{x}_g(t) \qquad (1)$$

for the different time intervals, t_{n-1} to t_n, n =1,2,3,4,.., at which times, $\dot{u}$ changes sign, where:

u = displacement of the body relative to the ground = $x-x_g$
μ_d= sliding friction coefficient
g = acceleration due to gravity
x_g= horizontal ground displacement
x = absolute displacement of the body

For the initial condition at $t_0 = 0$, the body is assumed to be at rest and the ground acceleration is assumed to exceed μg, thus:

$$u(0) = -x_g(0) \, , \quad \dot{u}(0) = \dot{x}_g(0) \, , \quad \ddot{x}(0) = \mu g$$

The absolute displacement x_1 in the first time interval is then:

$$x_1 = \frac{1}{2}\mu_d gt^2 \qquad\qquad (2)$$

where $0 \leq t \leq t_1$

At subsequent times t_{n-1}, n=2,3,4..., the relative velocity between body and ground is taken as zero, thus the initial conditions for each time interval are:

$$\dot{u}_n(t_{n-1}) = 0 \, , \quad u_n(t_{n-1}) = u_{n-1}(t_{n-1})$$

Solving equation (1) by incorporating these initial conditions leads to an expression for the absolute displacement of the body at time t, for continuous sliding:

$$x = \frac{(-1)^{n+1}}{2}\mu_d gt^2 + \mu_d g[(-1)^{n}{\cdot}t_{n-1} + \sum_{k=1}^{n-1} (-1)^{k+1}{\cdot}(t_k - t_{k-1})]\,t - \dot{x}_g(t_1){\cdot}t.$$

$$+\; \mu_d{\cdot}g{\cdot}\sum_{k=2}^{n-1} (-1)^{K}{\cdot}t_k^2; \qquad for \quad t_{n-1} < t < t_n; \quad n = 2,3,4,$$

This is represented in Fig.1 (a), where the velocities of the ground and the body are plotted against time.
To include stick-slip behaviour, an additional term is required and the final expression becomes:

$$x = \frac{(-1)^{n+1}}{2}\mu_d gt^2 + \mu_d g[(-1)^{n}t_{n-1} + \sum_{k=1}^{n-1} (-1)^{k+1}(t_k - t_{k-1})\,t - \dot{u}_g(t_1)\,t_1$$

$$+\; \mu_d g\sum_{k=2}^{n-1} (-1)^{k}t_k^2 + \sum_{s=1}^{m} \int_{t_s}^{t_s{'}} \dot{u}_g(t)\,dt \qquad\qquad (3)$$

where t_{n-1} is the time when stick starts in each cycle, s and s' are the points at the beginning and end of the stick interval, respectively, and m is the number of reattachments that have taken place (Fig.1 (b)).

Computer simulation

A program of tests on a shaking table is in progress. The model is a 2025kg concrete block resting on four sliding bearing plates, on which different materials can be mounted. The analyses have been conducted for this model.

The computer program DRAIN2D was used to study the dynamic behaviour of the body supported on a sliding surface. The uppermost curve in Fig.2 is the relationship between the maximum relative displacement and the friction coefficient, for EL-CENTRO 1940 NS ground motion. It shows how the increase of the friction coefficient has a significant effect in reducing the peak relative displacement in the range of higher friction coefficients.

The lowest curve in Fig.3 represents the maximum acceleration of the body vs. the friction coefficient. The results show that the maximum acceleration increases as the friction coefficient increases, eventually the acceleration approaches that of the fixed-base case.

Relative displacement is quite sensitive to variations in the intensity of excitation for the pure friction system, especially for lower friction coefficients, becoming less sensitive as the friction coefficient increases. This is shown in Fig.4.

The frequency content was varied by changing the duration of the earthquake record. Fig.5 illustrates the sensitivity of the displacement to this variation, and how the sensitivity is less as the friction coefficient increases.

R-FBI SYSTEM

In R-FBI systems[2], a resilient restraint is added to a friction isolator. The resilient elements provide a horizontal restoring force but carry no gravity load. The sliding surface supports the vertical load, and the interfacial friction acts both as the structural fuse and as an energy absorber.

The resilient element has a spring constant, k, giving a natural frequency, for a mass M, of $f = \omega/2\pi$, where $\omega^2 = k/M$. The acceleration of the mass due to the action of the spring for a relative displacement u is: $ku = M\omega^2u$. In a non-sliding phase, $(\dot{u} = \ddot{u} = 0)$:

$$\mu g - |\ddot{x}_g + \omega^2u| > 0$$

At the start of a sliding phase:

$$\mu g - |\ddot{x}_g + \omega^2 u| = 0$$

In a sliding phase ($\dot{u} \neq 0$):

$$\mu g - |\ddot{x}_g + \omega^2 u| \leq 0$$

When sliding, the equation of motion of the body can be expressed as:

$$\ddot{u}(t) + \omega^2 u(t) = (-1)^{n+1}\mu_d g - \ddot{x}_g(t) \qquad (4)$$

Using the same initial conditions as before leads to the expressions of u_1 and u_n:

$$u_1(t) = \int_0^t \omega^{-1}\sin\omega(t-\tau)[\mu_d g - \ddot{x}_g(\tau)]d\tau - x_g(0)\cos\omega t - \dot{x}_g(0)\frac{\sin\omega t}{\omega} \qquad (5$$

where $0 \leq t \leq t_1$; $0 \leq \tau \leq t$, and:

$$u_n(t) = \int_{\tau - t_{n-1}}^t \omega^{-1}\sin\omega(t-\tau)[(-1)^{n+1}\mu_d g - \ddot{x}_g(\tau)]d\tau + u_{n-1}(t_{n-1})\cos\omega(t-t_{n-1}$$

$$(6)$$

where $t_{n-1} \leq t \leq t_n$; $t_{n-1} \leq \tau \leq t$; $n = 2,3,4,\ldots$

This relationship is used to provide a check of the results given by DRAIN-2D.

Computer simulation

Time-history dynamic analyses were made for a body on a sliding surface with a spring restraint. Varying the friction coefficient, μ, and the spring constant, k, represented by the natural frequency, f, the behaviour was determined in each case.

Fig.2 represents the relation between the peak relative displacement and the friction coefficient for different spring constants. The results show that only in the range of lower friction coefficients, does the addition of a spring usefully decrease the peak relative displacement.

Fig.3 represents the relationship between the maximum acceleration of the body and the friction coefficient, showing that an increase in the spring constant leads to an increase in the acceleration of the body. The greater the friction coefficient, the less sensitive is the acceleration of the body to a change in the spring constant.

Fig.6 shows that the acceleration response for various intensities of excitation. After sliding has started, the R-FBI system is not very sensitive to variations of intensity.

To study the dynamic behaviour of the isolator for different frequency contents, the accelerogram of the EL-CENTRO 1940 earthquake was modified by multiplying the time scale by factors of 0.5 and 2.0. Fig.7 show that the relative displacement is somewhat insensitive to the frequency content.

The time-history analysis of the R-FBI system with a weak restoring force shows, for a given excitation, that the spring force after the start of motion is smaller than the friction force in almost the whole time history. This means that the R-FBI system is still a friction-type isolator, and that the energy dissipation is a primary action.

Laboratory test
In a typical test on the shaking table, the concrete block(2025kg) rested on four bearings of unfilled Teflon sliding on polished stainless steel (μ=0.06), with resilient restraints of rubber giving a natural frequency of 0.4 Hz. Sinusoidal signals sweeping through frequencies from 1 to 9 Hz, in varying times, were used as excitation. After the start of motion (>3 Hz), the block slides continuously at all frequencies, and the relative displacement, peak to peak, is less than 0.03" (Fig.8). Results of the tests will be reported later.

CONCLUSION

For a pure friction base isolator, μ = 0.2 provides a balance between acceleration and displacement. To reduce acceleration, a low coefficient of friction (<0.1) is used, with a relatively weak elastic restoring force (<0.4 Hz), to prevent excessive displacement between the building and the ground. Such a system resists resonance and can be designed to respond in a predictable manner, for a wide range of earthquake motions.

REFERENCES

[1] B. Westermo and F. Udwadia, 1983, "Periodic Response of a Sliding Oscillator System to Harmonic Excitation", Earthquake Engineering and Structural Dynamics, VOL.11.
[2] Mostahgel, N.,1986, "Resilient-friction Base Isolator", Proceeding Base Isolation and Passive Energy Dissipation, ATC-17, p.221-230.

ACKNOWLEDGEMENTS
This research is conducted under a grant from the Natural Sciences and Engineering Research Council of Canada.

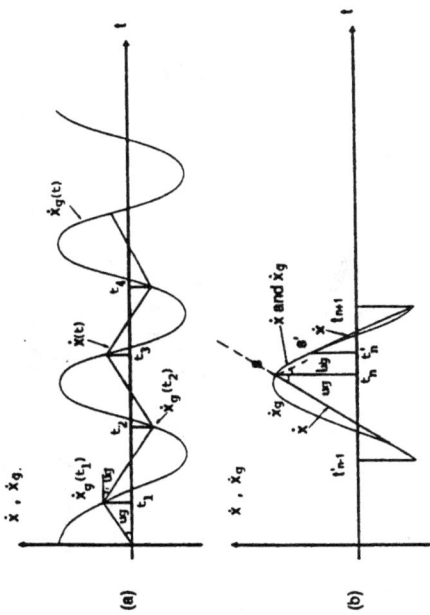

Fig.1 Ground Motion and Response

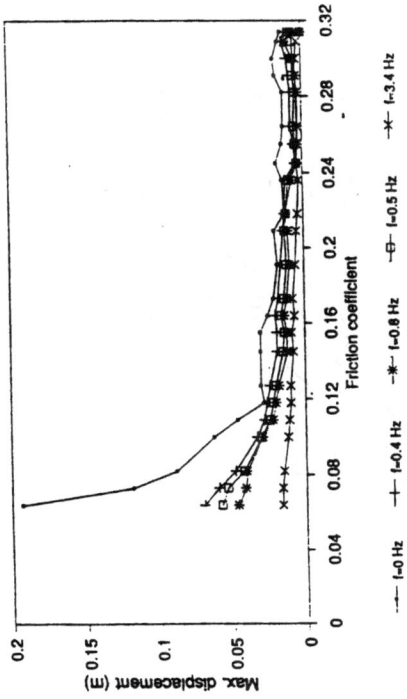

Fig.2 Displacement Responses of P-F and R-FBI Systems (EL-CENTRO Earthquake)

f=0 Hz   f=0.4 Hz   f=0.8 Hz   f=0.5 Hz   f=3.4 Hz

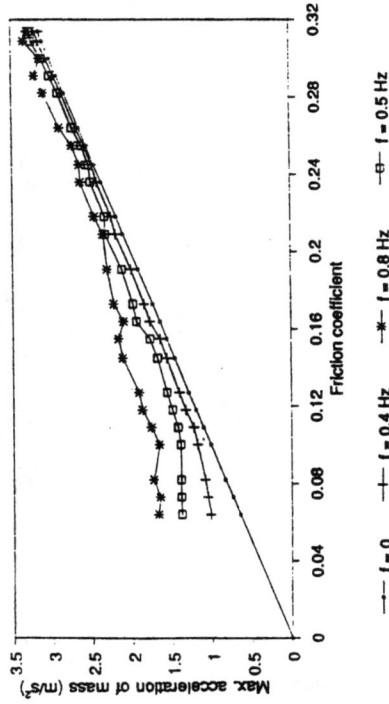

Fig.3 Acceleration Responses of P-F and R-FBI Systems (EL-CENTRO Earthquake)

f = 0   f = 0.4 Hz   f = 0.8 Hz   f = 0.5 Hz

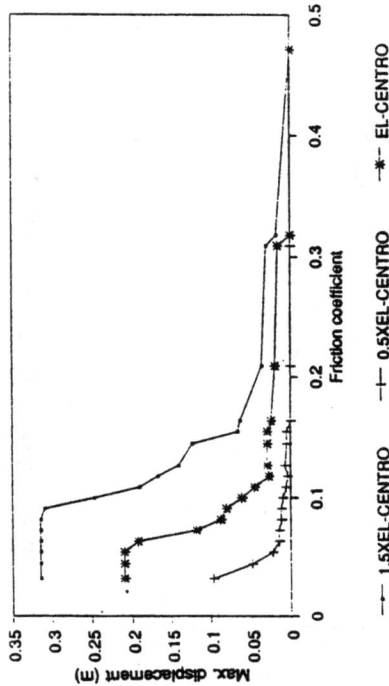

Fig.4 Displacement Responses for Different Intensities (Pure-friction System)

1.5XEL-CENTRO   0.5XEL-CENTRO   EL-CENTRO

365

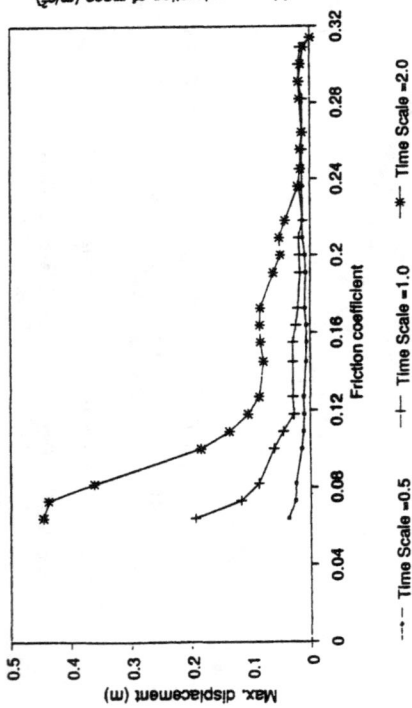

Fig.5 Displacement Responses for Different Frequency Contents (Pure-friction System for EL-CENTRO Record)

Max. displacement (m)

Friction coefficient

-·- Time Scale =0.5   -+- Time Scale =1.0   -*- Time Scale =2.0

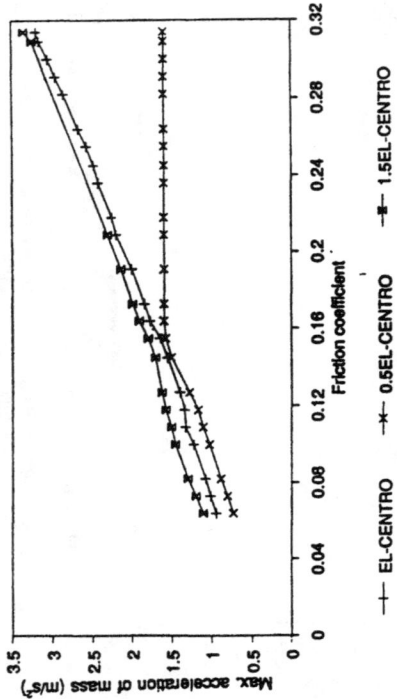

Fig.6 Acceleration Responses of R-FBI System (Natural Frequency = 0.4 Hz)

Max. acceleration of mass (m/s²)

Friction coefficient

-+- EL-CENTRO   -*- 0.5EL-CENTRO   -■- 1.5EL-CENTRO

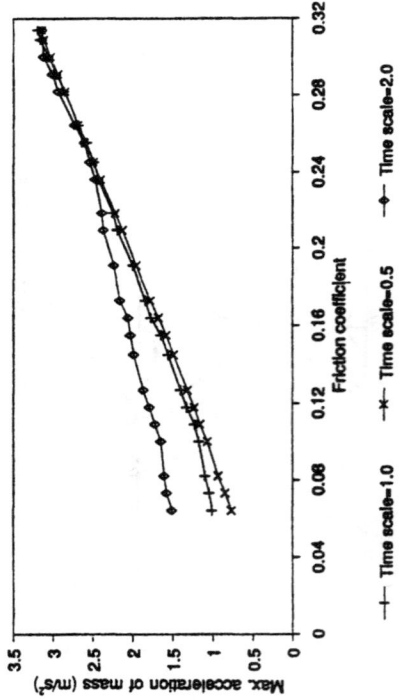

Fig.7 Acceleration Responses (EL-CENTRO Earthquake) (Natural Frequency = 0.4 Hz)

Max. acceleration of mass (m/s²)

Friction coefficient

-+- Time scale=1.0   -*- Time scale=0.5   -♦- Time scale=2.0

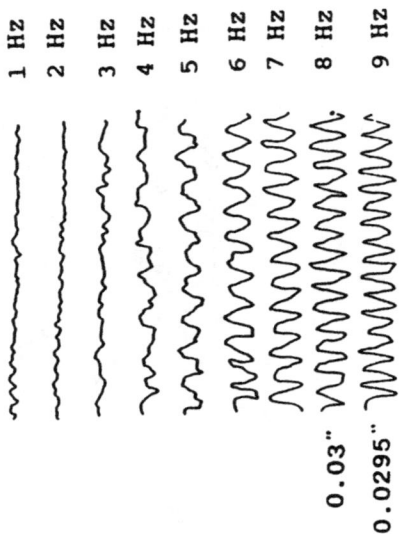

1 Hz
2 Hz
3 Hz
4 Hz
5 Hz
6 Hz
7 Hz
8 Hz
9 Hz

0.03"
0.0295"

Fig.8 Displacement Time Histories of the Block
For Different Frequencies of the sine signal
(Natural Frequency = 0.4 Hz)

366

Actively regulated friction slip braces

Zekai Akbay[1] and Haluk M. Aktan[2]

ABSTRACT

This paper describes use of a hybrid energy dissipation device within the structural framework of buildings to regulate its energy dissipation characteristics during seismic action. The bracing system with these energy dissipation devices called Actively Regulated Friction Slip Braces, and abbreviated as ASB. These braces are designed to monitor and regulate the preload along a frictional interface during seismic action. The ASB concept is derived from friction braces implemented as passive energy dissipation devices where the preload is kept constant. The passive friction braces in this text will be referred to as CSB (Constant Strength Friction Slip Braces). The effectiveness of the ASB system will be demonstrated by the comparative response of CSB simple frame structure under impulsive and harmonic actions. Principle of operation and key design parameters of ASB are evaluated on the response of ASB simple frame.

INTRODUCTION

A recent trend in the structural research environment in dealing with seismic design is to use systems such as energy dissipation devices that defend the building against the effects of seismic actions in order to prevent or reduce the damage [Kobori]. In all applications the energy dissipater exhibits a stable nonlinear restoring force hysteresis that enhances the damping capacity of the structure.

The energy dissipation devices are in some cases designed to operate under low vibration amplitudes during the seismic action in order to delay the built of vibratory energy [Pall]. However, in most cases the energy dissipation devices are designed to supplement the ductility supply of the building and will only be operational during the large amplitude response segments to ground excitation [Filiatrault, Whittaker]. The principle is to complement the energy dissipation capacity of the building by increasing the ductility supply which will reduce the strength demand. In other words, the two groups of energy dissipater either supplement the structural stiffness or structural strength. The hybrid system we will be describing here is a unique energy dissipater that operates during all response phases and complements both structural stiffness and strength.

The purpose of this paper is to present a hybrid energy dissipation device called Actively Regulated Friction Slip Braces, ASB, to provide solutions to the problems encountered with the applications of conventional energy dissipaters. This bracing system includes an innovative device [Akbay and Aktan] which activates the energy dissipation mechanism at all stages of the building response. The energy dissipation is accomplished along a friction interface under a clamping load on the structural bracing. The slippage along the friction

---

1 Ph.D. Candidate, Wayne State University, Detroit, MI 48202 USA

2 Associate Professor, Wayne State University, Detroit, MI 48202 USA

interface is observed during the inter-story response of the building. The clamping force on the friction interface is monitored and regulated during the response of the building to ground excitation.

METHOD

Overview:

The Actively Regulated Friction Slip Bracing was developed from a friction type energy dissipater that was in part developed by one of the authors. The early device allowed energy dissipation by slippage under a constant clamping force. The hybrid device (ASB) that is described here includes a mechanism that can regulate the clamping force on the friction interface upon demand. The decision on the clamping force regulation is based on a feedback response parameter.

In the following sections the friction energy dissipater with constant preload will be described first. The advantages of actively regulating the clamping force will be described following the passive device.

Friction Slip Braces:

Friction energy dissipation devices with constant clamping forces are installed on the structural bracing and designed to slip at an inter-story drift that is below the damage threshold of the building structure. The corresponding brace axial load at that drift level during slippage is designed to be lower then buckling load of the bracing. The energy dissipation is observed during the slippage along the friction interface at predefined brace axial loads. The lateral load vs. inter-story drift response of the structure appears as nondegrading elasto-plastic where, the yielding of the building is simulated by the slippage of the braces along the friction interface. The advantage of simulating structural yield before the damage threshold story drift is the increase to the ductility of the structure.

The stiffness, strength, and ductility of the building with friction bracing are uncoupled and are designed independently to provide the required balance. For example, the brace stiffness is designed by selecting a specific brace configuration and cross-sectional area. Brace strength is designed by selecting the clamping force on the friction interface. The operation of the passive energy dissipation devices are verified in scaled model of a multistory building experiments on the earthquake platform [Whittaker]. Experimental results showed significant reduction in the building response under ultimate limit state ground motion.

In multistory buildings under lateral load the slip strength of the structural bracing with friction interface also defines the yielding strength of that particular story. The clamping force on the friction interface is computed from an assumed lateral load distribution along the building height. The uncertainties in establishing the lateral load distribution will dictate whether slippage along the friction interface will initiate at a particular story. For example, soft story may form if clamping forces at all storeys except one are overestimated. In summary, the use of friction energy dissipaters in seismic defence of buildings requires a very accurate estimation of ground motion parameters. It should be also reiterated that, the friction energy dissipater is implemented to defend the building during the ultimate limit state seismic event and there is no functional expectation from these devices during the serviceability limit state lateral load demands. One may point out a practical concern on the operational reliability of these devices due to very infrequent service expectation.

368

As it is described in the previous paragraph, predefined level of brace axial load sets the strength of the brace which remains constant during the operation. Therefore, in this study, these brace components to implement the passive energy dissipation concept are referred as Constant Strength Friction Slip Braces, CSB.

Actively Regulated Friction Slip Braces, ASB:

Author's recent research to overcome the difficulties in the implementation of CSB in buildings led to the development of a hybrid energy dissipation device which will monitor and alter the clamping force on the friction interface thus, the energy dissipation characteristics of the structural system during the seismic action [Akbay and Aktan]. The structural bracing system having this new hybrid energy dissipation devices is called Actively regulated friction Slip Brace, and abbreviated as ASB.

Contrary to the conventional applications where the accumulated vibrational energy of the building is being dissipated near or at the ultimate level of the seismic response, the use of ASB will prevent the build-up of the vibrational energy in the early stages of seismic response. Energy dissipation is monitored and actively corrected to retain the building response below damage threshold. The ASB device can be installed to existing bracing or can be implemented with new bracing in different configurations such as chevron, X and K. The eccentric bracing could also be used so that a staggered seismic defence system is used where the yielding beams provide the last line of defence.

The prototype design of the active energy dissipation device, Fig. 1, is based on a passive device developed in part by one of the authors and reported as "Friction Slip Braces" [Whittaker]. The mounting of the device to structural bracing within a simple frame is demonstrated in Fig. 2. The main alteration in the design of the device is: its capability of controlling the pressure on the frictional interface to regulate the strength of the brace component.

OPERATIONAL FEATURES

Operation:

Basic operation scheme of the ASB is controlled by its energy dissipation device. The device is, like the one in CSB, a preloaded friction shaft and rigidly connected to the bracing. Brace is allowed to slip axially through the friction interface when the axial load exceeds the friction force developed under the regulated clamping force. Thus, when the slippage occurs, an amount of energy (equals to brace axial load times the slip displacement) is dissipated prior to any structural damage. Under increasing ground pulses, the clamping force on the friction interface is increased to give an additional strength to the brace.

The slippage along the friction interface is coerced during the early stages of the building's response by reducing the clamping force on the friction interface when the response amplitudes are small. Early initiation of energy dissipation prevents the build-up of vibrational energy and reduces the strength demand under large ground pulses during the middle portion of seismic motion.

A simple operational algorithm is incorporated in this study. The clamping force mechanism is changed at single fixed increments. The clamping force is lowered one increment if the slippage is not taking place. The clamping force is raised if the device is currently at slip state. The monitoring is performed at fixed time intervals. The design parameters of the active clamping mechanism are specified as time interval between decisions and clamping force increment. The effect of these parameters on the performance of the ASB device is studied in next section.

Design parameters:

Equation of motion for a simple frame structure modelled as a single degree-of-freedom system and subjected to a ground acceleration is written as:

$$m\ddot{x} + c\dot{x} + R(x) = -m\ddot{x}_g \qquad \qquad \dots (1)$$

where m is the total mass of the frame; c is the coefficient of inherent viscous damping; $x, \dot{x}$, and $\ddot{x}$ are displacement, velocity and acceleration of the mass respectively; $R(x)$ is the resistance force function, and $\ddot{x}_g$ is the ground acceleration which acts at the base of the system. The resistance function $R(x)$ becomes equal: to kx when structure is linearly-elastic where k is the stiffness of the structure, and to R_y when the yield (slip) strength of the structure is reached. The R_y corresponds to the lateral load at which the slippage at the frictional interface initiates.

The operation of ASB is implemented in a discrete manner at fixed time intervals of t_s. The status of the ASB is monitored only at each t_s and decision is made on changing the clamping force. Eq. (1) is solved at fixed time increments and the ASB action is taken at fixed time intervals and the change in the strength of the brace is defined as a fraction of its maximum strength R_y. Eq. (1) is then written in the incremental form as follows:

$$m\Delta\ddot{x} + c\Delta\dot{x} + \Delta R(x) = -m\Delta\ddot{x}_g \qquad \qquad \dots (2)$$

Two important design parameters of ASB are: a) decision time interval t_s, and b) clamping force increment. The parameter t_s is expressed in a nondimensional manner by normalizing it with the fundamental period of the frame, T, as $\frac{t_s}{T}$. The clamping force increment is the change to $\Delta R(x)$ at each $\frac{t_s}{T}$ in Eq. (2).

At each time interval t_s a decision is made on the clamping force. For example, if there the friction interface is currently not at slip state, the clamping force thus, the brace strength is reduced to a level which will invoke the energy dissipation immediately. On the other hand, if the friction interface is at slip state, the clamping force thus, the brace strength is increased one increment. For the purposes of this study the clamping force increments are retained constant. The clamping force increment size effects are studied by a parameter describing the incremental change in slip strength $\Delta R = \alpha R_y$, where α is a coefficient defining the strength increment ratio to maximum strength.

NUMERICAL EXAMPLE

General:

The effects of the two design parameters on the response of ASB is studied using a simple framed structure with one diagonal bracing scheme with a fundamental period of 0.5 seconds. In addition, response comparisons are carried out on the same frame one with an elastic bracing and another with a CSB to demonstrate the effectiveness of ASB. The elastic braced frame is assumed undamped to explicitly demonstrate the effect of ASB. Other members of the frame (the columns and the beam) are retained elastic.

370

The primary difficulty in the analytical simulation of ASB response is the unavailability of tools. A significant effort was made to append the ABAQUS finite element analysis program to allow the testing of active control algorithms in large non-linear structural systems [Akbay]. An element subroutine describing the ASB and clamping force regulation algorithm is developed and appended to the finite element analysis software ABAQUS [Users Manual].

The ASB operation is demonstrated on the simple frame subjected to a ground impulse. The effects of ASB parameters are evaluated using the response of the simple frame to a ground impulse as well as a simple harmonic ground acceleration.

ASB Frame under Impulsive Excitation:

The frame response is obtained under a ground acceleration impulse with an amplitude of 1 g and duration of $\frac{T}{40}$ where T is the fundamental period of the frame. The maximum slip strength of the frame is described as the base shear coefficient of 0.3. Three different values of clamping force increments of 0.04, 0.08, and 0.16 α are compared. The frame response is computed for a total time duration of $10T$.

The response of ASB building with a decision time interval ratio of $\frac{t_s}{T} = 0.05$ and a clamping force increment ratio of $\alpha = 0.04$ is compared with the response of the CSB and elastic frame in Fig. 3. ASB force vs. displacement hysteresis is shown in Fig. 4. Under a short duration impulse the maximum displacement during the first cycle by the ASB frame is greater than the other frames. The next response cycle however, the feature of ASB is demonstrated. The early dissipation of the energy through Coulomb damping significantly reduces the response amplitudes of the frame.

The equivalent viscous damping coefficients of the ASB frame during the first response cycle is computed and plotted in Fig. 5. Equivalent viscous damping coefficient of around 10% is computed for average values of $\frac{t_s}{T}$. In other words, under seismic excitations, the input energy will be dissipated immediately upon the application of the pulse and the build-up of vibration energy is prevented.

ASB Frame under Simple Harmonic Excitation:

The ASB frame response is computed under a harmonic ground acceleration with an amplitude of 0.01 g, and its frequency matching the initial frequency of the frame. The response of ASB building with a decision time interval of $\frac{t_s}{T} = 0.05$ and a clamping force increment ratio of $\alpha = 0.05$ is compared with CSB and elastic frame in Fig. 6. The maximum slip strength of the ASB frame and the slip strength of the CSB frame are taken equal and defined as the base shear coefficient of 0.3. It is clearly observed in Fig. 6 that the response of ASB is bounded at a limit described by its strength characteristics. The effect of early slippage is demonstrated in the ASB axial force-axial displacement relationship shown in Fig. 7. The early energy dissipation resulted in a 40% reduction of maximum displacements when compared to the CSB frame.

Effect of the decision time interval t_s is studied in Fig. 8. Maximum displacement amplitude of ASB, CSB and elastic frame are compared for a time duration of $10T$ in Fig. 8. The ASB response characteristics are near optimum for the range of $\frac{t_s}{T}$ between 0.05 and 0.25 for all α values.

Maximum response amplitudes of ASB under a varying amplitude harmonic ground motion at resonant frequency with the ASB frame is compared in Fig. 9. In this analysis, the response of the ASB building with various decision time intervals $\frac{t_s}{T}$ are compared with the responses CSB and elastic building. The slip strength of the CSB frame and maximum clamping force of the ASB frame is described as the base shear coefficient and retained constant under all excitation amplitudes. The clamping force increment of the ASB building is also retained constant. Response of ASB building is computed to be the lowest for all cases analyzed. Also, the reduction of maximum response is observed in Fig. 9 as increased apparent stiffness of the ASB frame.

CONCLUSION

The concept and development presented here is a hybrid energy dissipation device that can actively regulate its energy dissipation characteristics during seismic action. The device that is presented as a bracing system (called Actively regulated friction Slip Braces, and abbreviated as ASB) can be implemented to existing bracing or incorporated as new bracing in building structural systems.

Response of ASB building is presented under impulsive and harmonic actions. A very simple and preliminary operation algorithm is adopted for the implementation of ASB. Comparisons with other conventional frame types indicate very favorable influence of ASB to reduction of response amplitudes.

Further research will include multistory building response simulations with ASB and search for other operation algorithms to establish a design guideline for upgrading of seismic deficient buildings using ASB.

REFERENCES

ABAQUS, Theory and User Manuals, Hibbit, Karlsson & Sorensen, Inc., 4.8, 1989. 100 Medway Street, Providence, R.I. 02906.

Akbay, Z., "ASB Element Users Manual: ABAQUS User Subroutine for ASB" In preparation, Wayne State University, Civil Engineering Department, Detroit, MI, (1991)

Akbay, Z., and Aktan, H. M., "Intelligent Energy Dissipation Devices", Proc. Fourth National Conference on Earthquake Engineering, EERI, May 20-24, Palm Springs, III, 427-435 (1990).

Filiatrault, A., and Cherry, S., "Comparative Performance of Friction Damped Systems and Base Isolation Systems for Earthquake Retrofit and Aseismic Design", Earthquake Engineering and Structural Dynamics, Vol. 16, 389-416 (1988).

Kobori, T., "State-of-the Art Report. Active Seismic Response Control", Proc. Ninth World Conference on Earthquake Engineering, Aug. 2-9, Tokyo-Kyoto, VIII, (1988).

Pall, A. S., And Marsh, C., "Response of Friction Damped Braced Frames", J. Str. Eng., ASCE, 108(ST6), 1313-1323 (1982).

Whittaker, A. S., Bertero, V. V., Aktan, H. M., and Giacchetti, R., "Seismic Response of a DMRSF Retrofitted with Friction-Slip Devices", Proc. EERI Annual Conference, Feb. 9-12, San Francisco (1989).

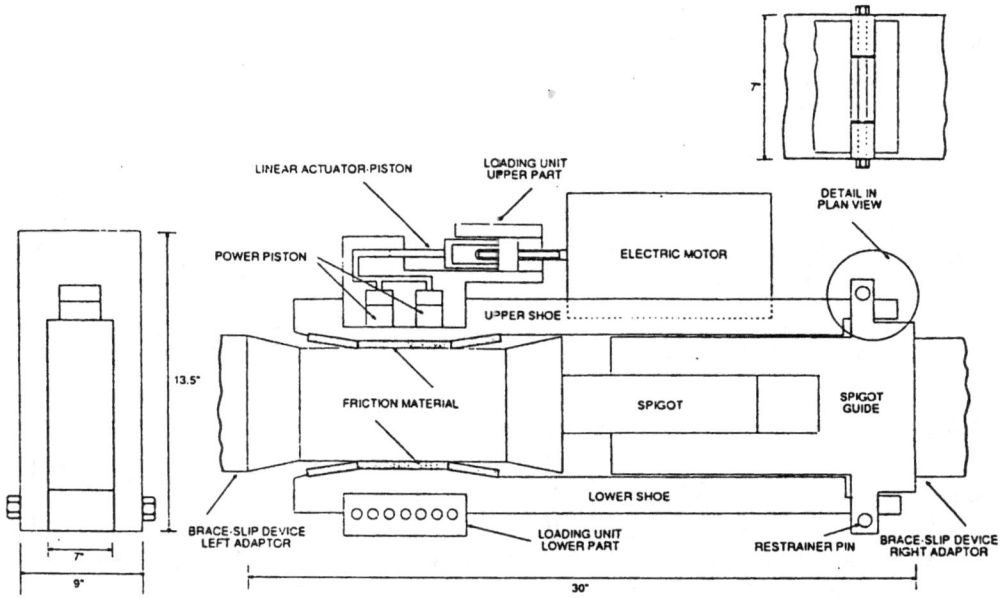

Figure 1. Prototype design of the active energy dissipation device.

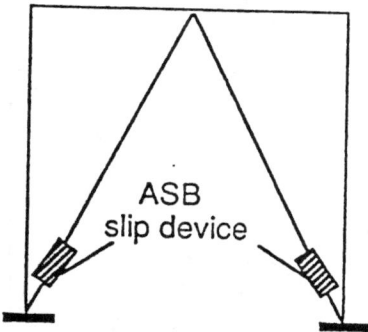

Figure 2. The mounting of the device to structural bracing.

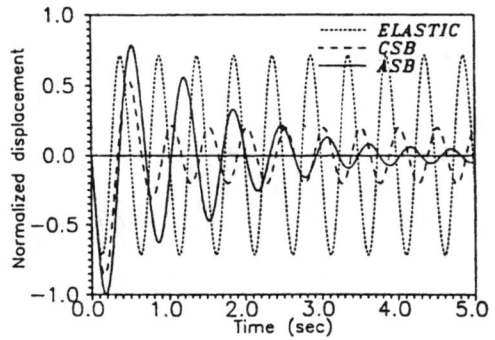

Figure 3. Displacement response under impulsive excitation.

373

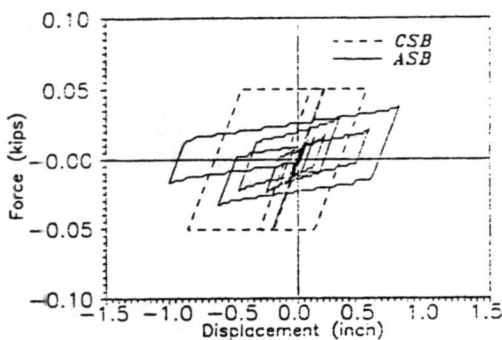

Figure 4. Force displacement relation of ASB under impulsive excitation.

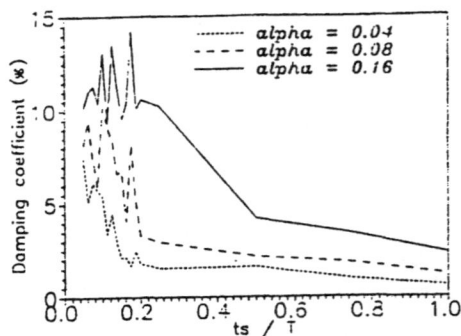

Figure 5. First cycle equivalent viscous damping coefficient of ASB under impulsive excitation.

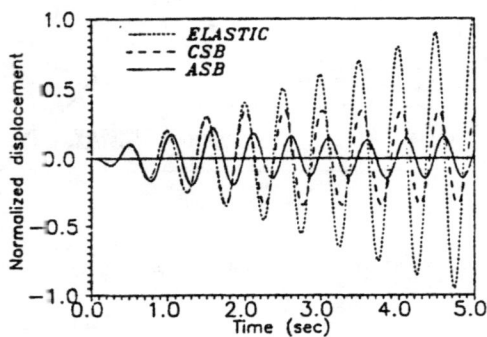

Figure 6. Displacement response under harmonic excitation.

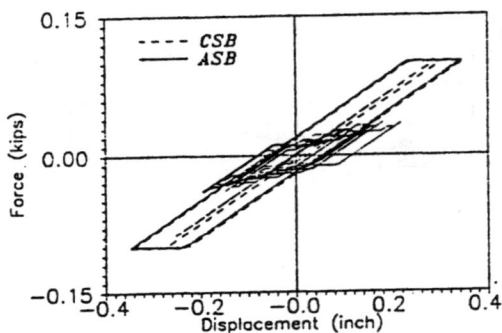

Figure 7. Force displacement relation of ASB under harmonic excitation.

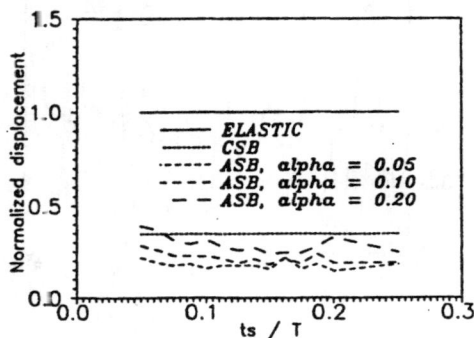

Figure 8. Effect of decision time interval.

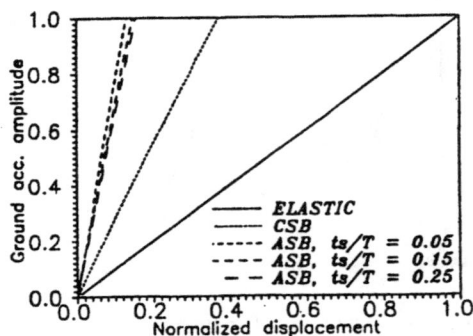

Figure 9. Max. response amplitudes under varying amplitude harmonic excitation.

374

Seismic response of a friction-base-isolated house in Montreal

Avtar S. Pall[1] and Rashmi Pall[1]

ABSTRACT

A two storey residential house, incorporating friction-base isolators, has been built in Montreal. Three-dimensional nonlinear time-history dynamic analysis was chosen to determine the seismic response of the structure. Compared to conventional construction, the stresses and accelerations in a friction-base isolated building are dramatically reduced, thereby, the damage to the building and its contents is minimized. The friction-base isolators are simple in construction and need no maintenance, repair or replacement over the life of building. The low cost of friction-base isolators suggests wide application in low-rise construction including residential houses.

INTRODUCTION

During a major earthquake, a large amount of energy is released in rapid ground motion. The amount of energy fed into the structure depends on the relationship between the frequency content of the ground motion and the natural frequency of the building. When the two frequencies are in close proximity, the building resonates and shakes violently. Unfortunately , this is the case in most low-rise to medium height buildings. The amplified accelerations can cause severe damage to the contents of the building even when the structure itself does not suffer any damage. All building codes, including National Building Code of Canada, recognize that it is economically not feasible to reconcile the seismic energy within the elastic capacity of structure. The code philosophy is to design structures to resist moderate earthquakes without significant damage and to resist major earthquakes without structural collapse. In general, reliance for survival is placed on the ductility of the structure to dissipate energy while undergoing inelastic deformations. This assumes permanent damage, after repair costs of which could be economically as significant as the collapse of the structure.

The problems created by the dependence on ductility of the structure can be reduced if the amount of seismic energy getting into the structure can be controlled and a major portion of the energy can be dissipated independently from the primary structure. In low-rise buildings, where overturning moments are not significant, the superstructure is decoupled from the forcing ground motion by providing base isolators. The introduction of supplemental damping in framed buildings is more convenient and economical. With the emergence of new techniques like base-isolation and friction-damping devices, it has become economically feasible to design damage free structures. In September 1985, the State of California passed an Assembly Resolution 'ACR 55 - Seismic Safety' that all publicly owned buildings must incorporate new seismic technology

I. Pall Dynamics Limited, 100 Montevista, D.D.O., Montreal, Quebec, H9B 2Z9, Canada.

like a ficticious truss element, having an elasto-plastic behaviour. Hysteretic behaviour of the friction base-isolator is shown in Fig. 4. The dynamic coeffigent of friction of the isolator is 0.2. The effect of 20% accidental eccentricity of mass on the overall seismic response of the building was also studied.

In order to compare the seismic response of the friction-base isolation, analysis were also carried out for this house with a fixed base.

Results of Analysis

1. At earthquake intensity of 0.10g, there is no slippage in the isolator. The friction-base isolated house behaves like a fixed base house.
2. At earthquake intensity of 0.18g, the maximum slippage in the isolator is 3 mm. The accelerations at top of friction-base isolated house and fixed base house are 0.17g and 0.29g respectively - a reduction of 42%. After the earthquake, the permanent offset is 1.5 mm (Fig. 6).
3. At earthquake intensity of 0.33g, the maximum slippage in the isolator is 11mm. The accelerations at top of the friction-base isolated house and fixed base house are 0.21g and 0.55g respectively - a reduction of 62%. After the earthquake, the permanent offset is 4 mm.
4. At earthquake intensity of 0.50g, the maximum slippage in the isolator is 18mm. The accelerations at top of the friction-base isolated house and fixed base house are 0.25g and 0.86g respectively - a reduction of 70% (Fig. 7). After the earthquake, the permanent offset is 7 mm.
5. The effectiveness of friction-base isolation increases with the severity of earthquake (Fig. 6).
6. The friction-base isolated building is not very sensitive to accidental eccentricity.
7. Although there is significant reduction in stresses when compared to fixed base house, it was not possible to save in material cost as standard size materials available in the market are used. However, in larger buildings of 3 storey height or more, the additional cost of isolators will be more than offset by the savings in material cost.

Friction-damped Buildings

Base isolation is not the only technique for protection against earthquakes. Passive energy dissipators provide protection by absorbing earthquake energy while the structure is deforming. Several types of inexpensive friction-damping devices suitable for different construction techniques have been developed by Pall (1980,81,82,84,86,89). These devices are for: large panel precast concrete construction; cast-in-place concrete shearwalls; braced frames; and for connecting precast cladding to frames. Cyclic dynamic laboratory tests have been conducted on specimen devices (Pall 80, Filiatrault 1986). The performance is reliable, repeatable and possesses large rectangular hysteresis loops with negligible fade over several cycles of reversals that can be encountered in successive earthquakes. Much greater quantity of energy can be disposed of in friction than any other method involving the damaging process of yielding of steel. Unlike visco-elastic materials, their performance is not affected by temperature, velocity and stiffness degradation due to aging. Furthermore, these friction-damping devices need no maintenance or replacement over the life of building and are always ready to do their job regardless of how many times they have performed. (Aiken 1988, Baktash 1986, Filiatrault 1986,1988). Shake table studies at the University of British Columbia in Vancouver, the University of California at Berkeley and Imperial College in London have successfully demonstrated the superior seismic performance of friction-damped frames. Friction-damping devices are now finding practical application in new construction as well as retrofitting of existing buildings (Pall 87,91). Their use has resulted in significant savings in the initial cost of construction while the earthquake resistance of the building has increased considerably.

and existing buildings be retrofitted to increase their earthquake resistance. This resolution is based on the consideration that while the past building code philosophy was only concerned with the avoidance of collapse of structure, the modern buildings have expensive finishes and contain extremely sensitive and costly equipment which must be protected. The National Building Code of Canada 1990, Clause 83 of Commentary-J of the Supplement, allows the use of new technology.

In 1988, a residential house incorporating friction-base isolators was built in Montreal. This paper discusses the results of the analytical studies and describes the construction details of its practical application.

STATE-OF-THE-ART

The concept of base isolation is not new. Mechanical engineers have used this concept for centuries to isolate the transmission of vibration of machinery to the foundation or vice-versa. In recent years, the use of base isolation system as a mean of aseismic design of structures has attracted considerable attention. The reviews on its historical and recent developments have been extensively provided by Kelly 1986, Tarics 1987 and Buckle 1990. Now, the concept of base isolation has matured into a practical reality and is taking its place as a viable alternative to conventional fixed base seismic resistant construction. In the past decade about 40 base isolated buildings have been built or retrofitted in the U.S and Japan. The most commonly used isolators are of laminated rubber bearings with or without lead core. Friction type base isolators have been used by Electricite de France in nuclear power plants (Vaidya, Plichon 1986). All base isolators have certain features in common. These are: horizontal flexibility, energy dissipation capacity. The purpose of the horizontal flexibility is to shift the natural frequency of the structure to a lower value and away from the energy containing frequencies of the earthquake. However, low flexibility could result in excessive displacements relative to ground. Energy absorbing capacity reduces both base displacement and the transmission of the seismic energy to the building. The isolators should also have some rigidity against wind and low earthquakes.

Friction-Base Isolators

In low-rise structures, where overturning moments are not significant, the friction-base isolators are located horizontally between the foundation and the superstructure to partly isolate it from the forcing ground motion (Pall 1981 ii, 1986). Ideally, frictionless joints will allow the foundation to move without exerting any force on the building, but the displacements of the building relative to ground will be very large. A friction force is therefore required, sufficient to react to wind and small earthquakes. During a severe earthquake, the magnitude of lateral force that the building can experience is limited to the slip load. The slip load or the coefficient of friction is so selected that the stresses in the materials do not exceed the permissible stresses of the materials and that the relative displacements are limited to an acceptable value of say 25 mm. In friction-base isolated buildings, the displacement at the end of an earthquake is a permanent offset as there is no restoring force. In order to provide restraint on the total movement during catastrophic conditions, increasing resistance to sliding is provided by the ramp shape in steel plates or by providing an elastic pad at the end of travel (Fig. 3). The slippage of device acts like a safety valve to limit the forces exerted and as a damper to limit the amplitude of vibration. Another interesting feature of the friction-base isolated building is that the natural period of the structure varies with the amplitude of vibration i.e. the severity of earthquake. Hence the phenomenon of resonance is difficult to establish. Some of the many technical and economic advantages of friction-base isolators are:

1. Friction-base isolators provide high damping by dissipating energy in friction during slippage.

2. The relative displacement of rubber pad base isolators is in the order of 150-180 mm. This involves expensive detailing of service connections. In friction-base isolators, the maximum displacements are less than 25 mm.
3. Rubber pad base isolators have a natural period of about 2 seconds. Hence, these are suitable for regions with high frequency ground motions. In case of earthquakes similar to 1985 Mexico earthquake (0.5 Hz.), the buildings with rubber pad base isolators would have resonated and damaged. Friction-base isolators are suitable for any type of future earthquake.
4. The cost of rubber pad base isolators is high and thus has so far found application only in government buildings of national or historic importance. The low cost of friction-base isolators is very appealing and it is therefore possible to extend the benefits of this concept to all low-rise construction including residential houses.

Description of the Friction-Base Isolated House in Montreal

The residential house is located in the Dollard-des-Ormeaux, a suburb of Montreal. The house has two storeys above grade and one basement below grade. It is about 20 m x 15 m in plan and has a total living area of about 700 m^2 including the basement. The basement walls are of reinforced concrete upto the ground floor. The superstructure is of typical Canadian construction of wood stud wall framing with brick veneer. The front view of the house is shown in Fig. 1. The superstructure is actually floating over the foundation/basement wall. A continuous white line, just below the brick masonry, is a strip of flashing on the outside of the isolation joint to protect against rain or moisture penetrating the jont.

A total of 15 friction-base isolators are provided all along the outer basement wall. The location of the isolators is shown in plan in Fig. 2. The cross-section of the wall shows the location of friction-base isolators. The reinforced concrete wall above the isolators is a continuous tie-beam. The floor acts as a rigid diaphragm and is bolted to the tie-beam. The interior columns have pinned connection at the top and bottom and can rotate to accomodate a displacement of 25 mm. During construction, to avoid blowing of the tie beam and light weight wooden shell structure in wind storms, 6 low yielding anchor bars of 10 mm diameter were provided at the corners. These anchor bars are redundant after the outside masonry veneer is constructed. These bars will yield during a major earthquake and allow the base isolators to slide.

The cost of 15 friction-base isolators was only $8000. Patented (Pall 1980) friction-base isolators were designed and supplied by Pall Dynamics Limited.

Nonlinear Time-History Dynamic Analysis

Three-dimensional nonlinear time-history dynamic analysis was carried out by using the computer program DRAIN-TABS, developed at the University of California, Berkeley. This program consists of series of subroutines that carry out a step by step integration of the dynamic equilibrium equations using a constant acceleration within any time step. As future earthquakes may be erratic in nature, an artificial earthquake record generated to match the design spectrum of Newmark-Blume-Kapur, which is an average of many earthquake records, has been used (Fig.5). This earthquake record forms the basis of the NBC response spectrum. For Montreal, the ground accelerations of this earthquake record were scaled to 0.18g. Analysis were also carried out for intensities of 0.10g, 0.33g and 0.5g accelerations. The integration time step was 0.005 second. Analysis were done for the earthquake acting in directions x, y and 45 degrees axis.

The superstructure is assumed to be rigid elastic. Nonlinearity is assumed in the friction-base isolator only. Viscous damping of 3% of critical was assumed in the initial elastic stage to account for presence of non-structural elements. Hysteretic damping due to slipping of friction base-isolator is automatically taken into account by the program. Friction-base isolator is modelled

CONCLUSIONS

The use of friction-base isolators have shown to provide a practical, economic and effective new approach to design low-rise buildings to resist future earthquakes. Whereas the use of base isolation has so far been limited to only government buildings of national or historic importance, the low cost of friction-base isolators suggest their wide application for all low-rise buildings including residential houses.

REFERENCES

Aiken, I.D., Kelly, J.M., Pall, A.S., 1988 "Seismic Response of a Nine-Story Steel Frame with Friction Damped Cross-Bracings", Report No. UCB/EERC-99/17, Earthquake Engineering Research Center, University of California, Berkeley. pp. 1-7.

Baktash, P., Marsh, C. 1986, "Comparative Seismic Response of Friction-Damped Braced Frames", ASCE Proceedings Dynamic Response of Structures, Los Angeles, pp. 582-589.

Buckle, I.G., Mayes, R.L. 1990, " Seismic Isolation: History, Application and Performance - A World View", Earthquake Spectra, EERI, Vol.6, No.2

Filiatrault, A., Cherry, S. 1986, "Seismic Tests of Friction Damped Steel Frames", Proceedings-third conference on Dynamic Response of Structures, ASCE, Los Angeles.

Filiatrault, A., Cherry, S. 1988, "Comparative Performance of Friction-Damped System and Base-Isolation System for Earthquake Retrofit and Seismic Design", Earthquake Engg. and Structural Dynamics, Vol. 16, pp. 389-416.

Kelly, J.M. 1986, "Progress and Prospects in Seismic Isolation", Proceedings, Seminar and Workshop on Base Isolation and Passive Energy Dissipation, ATC-17, San Francisco, pp. 29-35.

Pall, A.S., Marsh, C., Fazio, P. 1980, "Friction Joints for Seismic Control of Large Panel Structures", Journal of Prestressed Concrete Institute,Vol. 25, No. 6, pp. 38-61.

Pall, A.S., Marsh, C. 1981 i, "Friction Damped Concrete Shearwalls", Journal of American Concrete Institute, No. 3, Proceedings, Vol. 78, pp. 187-193.

Pall, A.S., Marsh, C. 1981 ii, "Friction Devices to Control Seismic Response", Proceedings of the Second Specialty Conference on Dynamic Response of Structures, ASCE, Atlanta. pp. 808-819.

Pall, A.S., Marsh, C. 1982, "Seismic Response of Friction Damped Braced Frames", ASCE, Journal of Structural Division, Vol. 108, St. 9, June 1982, pp. 1313-1323.

Pall, A.S. 1984, "Response of Friction Damped Buildings", Proceedings Eighth World Conference on Earthquake Engineering, San Francisco, Vol. V, pp. 1007-1014.

Pall, A.S. 1986, "Energy-Dissipating Devices for Aseismic Design of Buildings", Proceedings of the Seminar and Workshop on Base Isolation and Passive Energy Dissipation, ATC-17, San Francisco, pp. 241-250.

Pall, A.S., Verganelakis, V., Marsh, C. 1987, "Friction-Dampers for Seismic Control of Concordia University Library Building", Proceedings of Fifth Canadian Conference on Earthquake Engineering" Ottawa, Canada, pp. 191-200.

Pall, A.S. 1989, "Friction-Damped Connections for Precast Concrete Cladding", Proceedings PCI - Architectural Precast Concrete Cladding and its Lateral Resistance, pp. 300-309.

Pall, A.S., Ghorayeb, F., Pall, R. 1991, "Friction-Dampers for Rehabilitation of Ecole Polyvalente at Sorel, Quebec", Proceedings, Sixth Canadian Conference on Earthquake Engineering, Toronto.

Tarics, A.G., 1987, "Earthquakes - Are we Ready", The Journal of The Society of American Military Engineers, Vol. 79, Number 517, Proceedings of the Seminar and Workshop on Base Isolation and Passive Energy Dissipation, ATC-17, San Francisco, pp.486-490.

Vaidya, R.N., Plichon, C.E., "On the Concept of Base Isolation Design in France", Proceedings of the Seminar and Workshop on Base Isolation and Passive Energy Dissipation, ATC-17, San Francisco pp. 175-184.

Figure 1. Front View of House

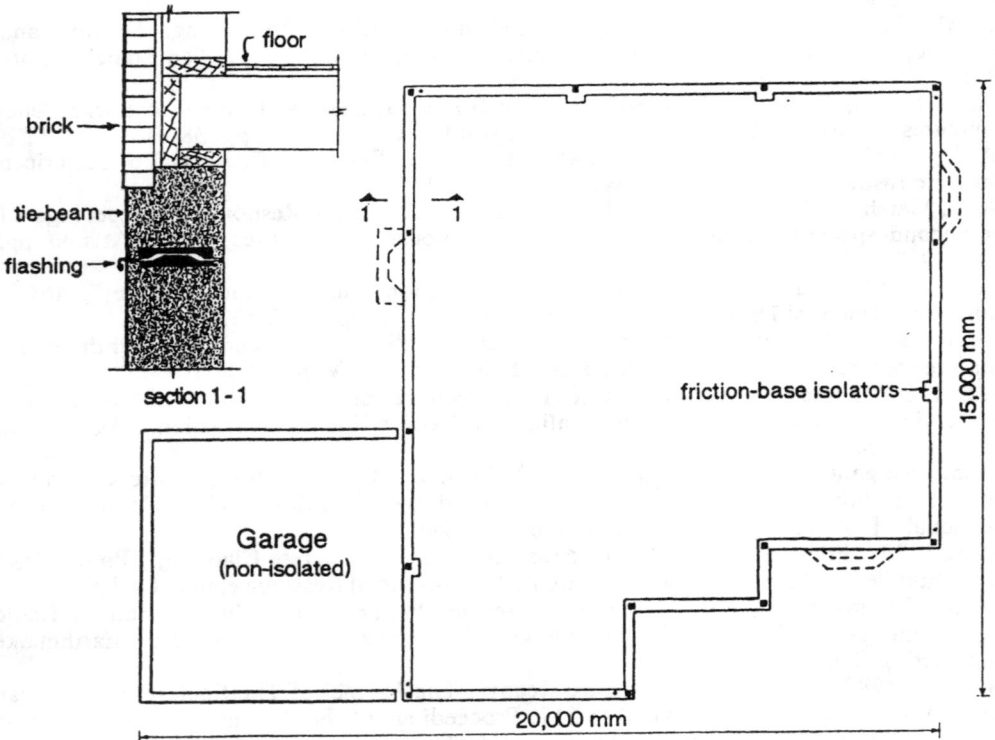

Figure 2. Location of Friction-Base Isolators - Plan

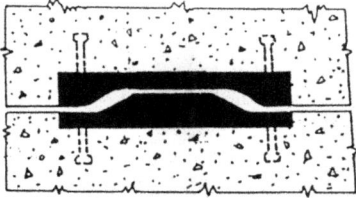

a) with ramped surface   b) with rubber cushion

Figure 3. Detail of Friction-Base Isolator

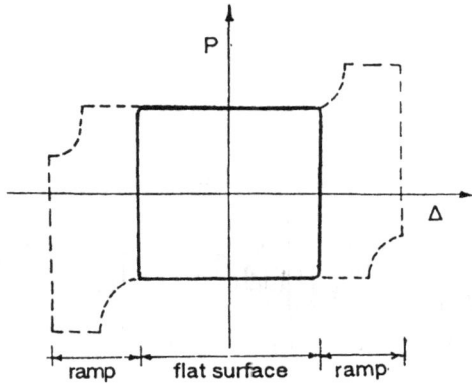

Figure 4. Hysteresis Loop of Friction-Base Isolator (with ramped surface)

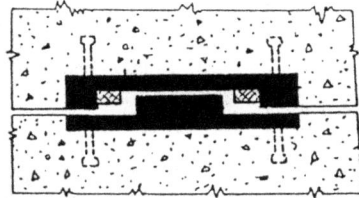

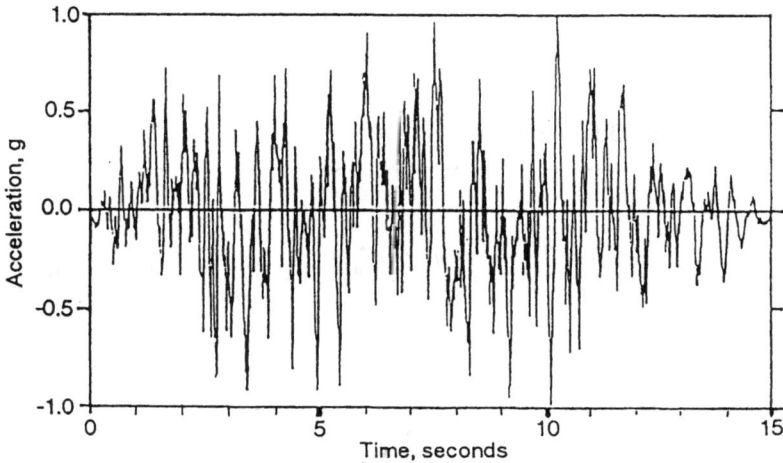

Figure 5. Time-Histories of Artifical Earthquake (Newmark, Blume & Kapur)

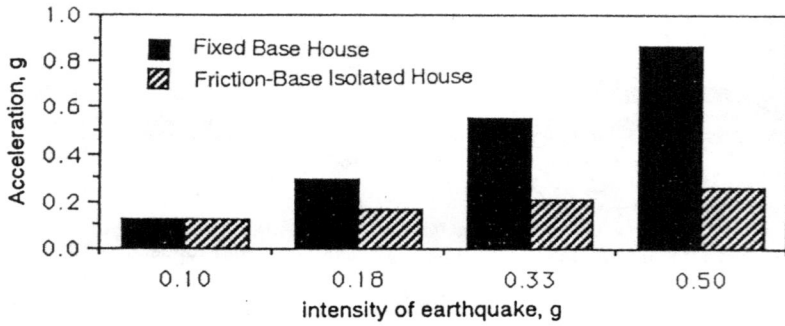

Figure 6. Comparative Seismic Performance

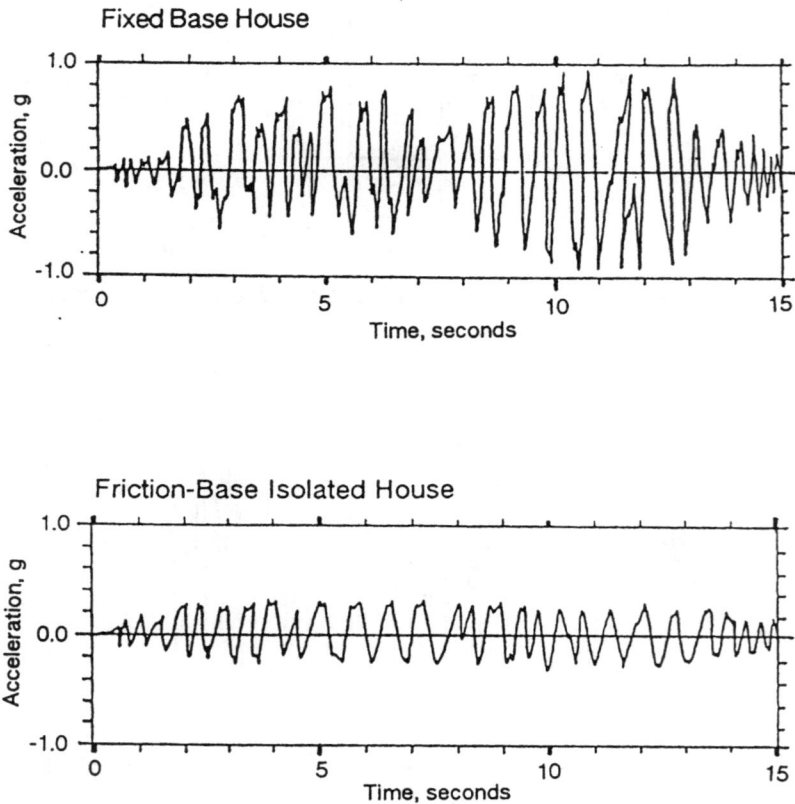

Figure 7. Time-Histories of Accelerations at Top

New approach to seismic base isolation

Valentin Shustov

ABSTRACT

A flexible mounting and an energy dissipation mechanism are considered to be the two basic elements in most of practical base isolation systems. At the same time an effect of energy dissipation in the frequency spectrum area that is rather distant from resonant frequencies, what is just typical for flexible mounting, is actually negligible. Besides, a reduction of relative displacement of a superstructure with respect to its foundation, which is proclaimed as a goal of said dissipation is none other than a result of transmission of earthquake movement into the superstructure and therefore should be avoided.

A new concept embodied in the Antifriction and Multi-Step Base Isolation (AF&MS BI) is an alternative approach [3, 4, 5]. The global strategy of the AF&MS BI may be set out as follows: there is no sense both in trying to resist violent tremors and to damp them, it is much easier to escape.

The main advantage of the AF&MS BI lies in its double level protection. A mechanism of the first level lessens an intensity of the earthquake input itself. The second level reduces dynamic responses of the structure.

AF&MS BI REALIZATION

To minimize transmission of destructive earthquake ground motion into a structure, to prevent permanent horizontal post-earthquake offsets, to keep the system's ability to withstand wind pressure as well as minor earthquakes without being decoupled from its foundation, the AF&MS BI consists (Fig.1) of a ball transfer unit (1) supporting a superstructure (2) and resting on a depression (3) of a pedestal plate (4). The depression is shaped in compliance with the configuration of the contacting surface of the ball and is centered at the lowest point of the pedestal plate (4) having a concave upper surface (5) and resting on a foundation (6). The depth of the depression at given radius of the ball is governed by weight of the superstructure and by design wind load. The force of gravity will keep the superstructure in a steady position on the pedestal plate both at any wind at a slight earthquakes. When

---

Principal, Seismic Risk Evaluation, Scientific Consulting Co, Los Angeles, California, 90036.

magnitude of the earth movement exceeds a certain threshold, the ball gets out of the depression, any transfer of horisontal movement to the superstructure considerably decreases.

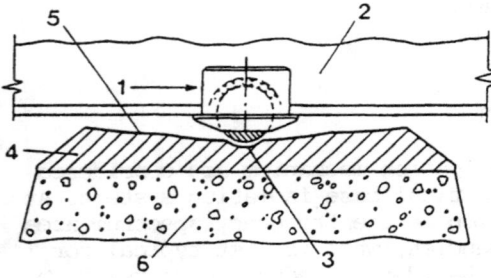

Figure 1. AF&MS BI unit

base isolation by means of successive tem, thus protecting it from resonant ty in the pedestal plate is not to be

To confine the base shear as well as the travel of the superstructure by an acceptable leval, the upper surface of the pedestal plate is shaped as a combination of concentric spherical surfaces with successively increasing radii of curvature which are continuously transforming into each other. Maximum vertical grade of every component surface is the same and approximately equals to the ratio of the design wind load to the weight of the superstructure (Fig.2). This design of upper surface provides a multi-step tuning-out the forcibly vibrating system, thus protecting it from resonant amplification. The diameter of the cavity in the pedestal plate is not to be less than a double maximum amplitude

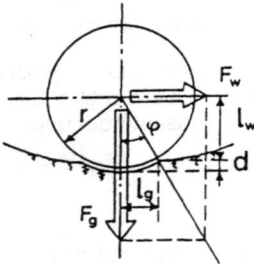

Figure 2. Diagram of forces
for ball in central depression

of earth displacement during a strong earthquake. The radius of vertical curvature of the central sphere of the upper surface of the pedestal plate is designed as big as to provide a proper tuning-out natural frequencies of the base-isolated from the fix base structure.

One of the most critical characteristics of any seismic isolator is its load-deflection curve. Subject to a particular material and design, those curves may differ in details but in the main thing they are very

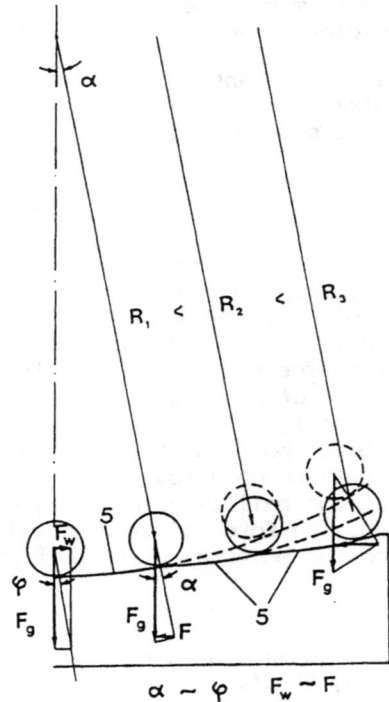

Figure 3. Fragment of multi-curved surface with balls in critical positions

384

much the same: with an increase of deflection the corresponding horizontal reaction in the system steadily builds up (Figs.3,4). The case is entirely different in the event of the AF&MS BI: with this concept now it becomes possible to create isolators of any set properties simply by changing their pedestal plates' configuration. Thus, the AF&MS BI is not only the next but the simplest and the most powerful seismic-isolation technology.

By the way, the main component of the AF&MS BI, a Ball Transfer Unit is widely used in stationary and mobile transport systems intended for heavy duty and extreme condition environments.

Another advantage of the AF&MS BI in comparison, for example, with rubber bearings is an absence of alternating eccentrically applied vertical base reactions which can excite damaging flextural stress waves.

The AF&MS BI can be easily applied both to new constructed and to retrofitting existing structures. It is incomparable for one-two story buildings which cannot be effectively equipped with any other type of base isolators.

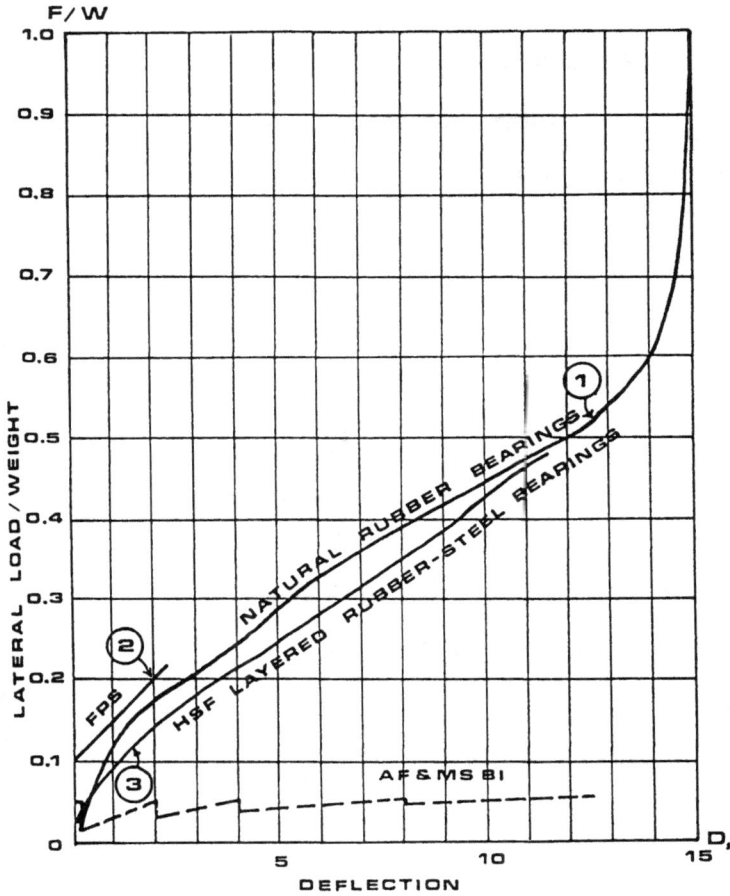

Figure 4. Load-deflection curves of different types of base isolators, investigated by:

1 - T. Anderson [1]; 2 - V. Zayas et al [7];
3 - F. Tajirian et al [6]

CONDENSED PHILOSOPHY OF SEISMIC ISOLATION

Leave buildings alone, do not tie them to the ground and decrease to the utmost extent any existing lateral ties, or briefly: let the earth move its way.

385

Remember: it is not the building, it is the earth which is vibrating when the building is supported on an adequate isolation system. Any attempt to reduce a relative displacement of the superstructure with respect to the foundation inevitably results in an additional transmission of earthquake energy into the building.

MATHEMATICAL MODEL

Basic features of the AF&MS BI systems can be received from a two-degree-of-freedom model shown in Fig.5 where parameter u stands for absolute, and parameter v for relative displacements.

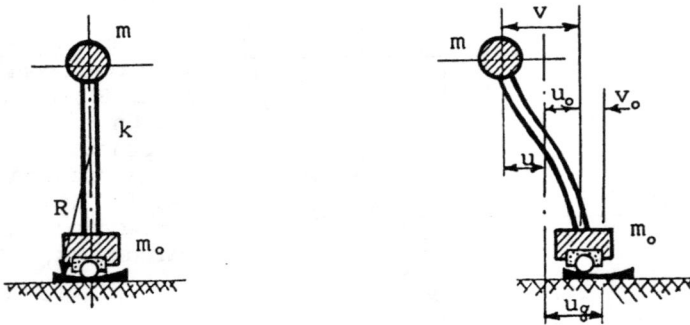

Figure 5. Theoretical model of AF&MS BI system

The equation of motion can be written in the form:

$$\begin{cases} m\ddot{v} + r\dot{v} + kv = - m\ddot{u}_o \\ M\ddot{u}_o + m\ddot{v} + (g/R)\, Mu_o = (g/R)\, Mu_g + fgM\, (\dot{u}_g - \dot{u}_o)|\dot{u}_g - \dot{u}_o|^{-1} \end{cases}$$

where $M = m + m_o$

    f is the friction factor,

    R the radius of curvature of the pedestal plate in the point of the ball's instant contact,

    $v = u - u_o$

From the first equation it is easy to see that deformations of a superstructure are fully governed by an external force which is equal to the inertia force of the superstructure as a rigid body rocking on isolators.

The second equation shows that an external force acting on a base-isolated structure consists of two components. One of them is proportional to the ground displacement and inversely proportional to the radius of curvature R. another is proportional to the friction factor f and is acting opposit to the sense of relative velocity $\dot{v}_o = \dot{u}_o - \dot{u}_g$. As long as $(\dot{u}_o)_{amp} < (\dot{u}_g)_{amp}$

the so called "damping force" is more exciting than damping. That is a reason for the "antifriction approach". If $(\dot{u}_o)_{amp}$ achieves the value of $(\dot{u}_g)_{amp}$, the further rate of excitation is mostly controlled by the "magic" R: any time the relative displacement v_o riches some predetermined value, the radius R sharply increases thus correspondently decreasing the external force as well as changing the natural period of the isolated structure ("multi-step approach").

CONCLUSIONS

Damping mechanism of any kind under kinematic exitation is simultaneously a driving one. Its "negative", pushing effect is immediate, where as its "positive", dissipating effect needs more time to fully develop. Besides, the frequency spectrum area of the superstructure and that of the isolated system are supposed to be well separated, so usefulness for the superstructure of adding some damping to the isolators is rather doubtful. And what is more, there is no need to confine the superstructure's displacement relative to the earth, or better to say the earth's displacement relative to the superstructure. The deformation of the structure itself is what matters. Therefore, low friction base isolation in combination with progressive (multi-step) frequency separation appears to be a much more fruitful goal.

REFERENCES

[1]   Anderson, T.L., "Seismic Isolation Design and Construction Practice", Proc. 4th Nat'l Conf. on Earthquake Engineering, Palm Springs, California, Vol.3, pp.519-528, 1990.

[2]   Kelly, J., "Base Isolation: Linear Theory and Design", Journal "Earthquake Spectra", Vol.6, No.2, pp.223-244, 1990.

[3]   Shustov, V., "Earthquake Shelter with Bed Support and Canopy", US Patent 4,965,895, 1990.

[4]   Shustov, V., "Seismic-Isolator", US Patent 4,974,378, 1990.

[5]   Shustov, V., "Antifriction and Multi-Step Base Isolator", US Patent pending.

[6]   Tajirian, F., Kelly, L., Aiken, I., "Seismic Isolation for Advanced Nuclear Power Stations", Journal "Earthquake Spectra", Vol.6, No.2, pp. 371-401, 1990.

[7]   Zayas, V., Low, S., Bozzo, L., Mahin, S., "Feasibility and Performance Studies on Improving the Earthquake Resistance of New and Existing Buildings Using the Friction Pendulum System", Report to NSF No. UCB/EERC-89/09, Univ. of Calif., Berkley, 1989.

Friction-dampers for rehabilitation of Ecole Polyvalente at Sorel, Quebec

Avtar S. Pall[I], Fadi Ghorayeb[II], and Rashmi Pall[I]

ABSTRACT

An innovative structural system, which combines the strength and stiffness of a braced frame and high energy dissipation capacity of the friction-dampers, has been adopted to rehabilitate the school buildings damaged during the 1988 Saguenay earthquake. The existing structure, built in 1967, lacked in lateral resistance and ductility requirements of the new building code. The introduction of supplemental damping provided by the friction-dampers reduced the force level and eliminated the necessity of dependence on the ductility of structure. Nonlinear time-history dynamic analysis was chosen to determine the seismic response of the structure. The conventional method of retrofitting with concrete shearwalls involved extensive foundation work which was very expensive and time consuming. The new method of retrofitting, while significantly lower in initial cost of construction, offers greater savings in the life cycle cost as damage to the building and its contents is minimized. The retrofitting work was completed in a record time during the summer vacation of 1990.

INTRODUCTION

On November 25, 1988 an earthquake of magnitude 6.2 on the Richter scale occured in the Saguenay region approximately 36 km south of Chicoutimi, close to the northern boundary of Parc Laurentides in the province of Quebec (Mitchell et al. 1989). The focal depth was 28 km. The peak horizontal and vertical accelerations at the Chicoutimi station were 13.1% and 10.2% of gravity with a frequency content of 13.3 Hz and 18.2 Hz respectively. The earthquake was felt over an extremely large area, as far south as New York and as far west as Toronto. The seismologists of the Geological Survey of Canada have warned that an earthquake worse than the above could well occur in the St. Lawerence valley before the end of the century.

After the earthquake, some minor structural and non-structural damage was noticed in the school buildings. The school authorities, La Commission Scolaire de Sorel, retained the services of Les Consultants Dessau Inc. to investigate the extent of damage and to suggest strengthening measures to rehabilitate the buildings. The existing three building complexes are of precast concrete construction and were built in 1967. Based on the detailed site inspection and analytical studies, the structural engineers concluded that:

I. Pall Dynamics Limited, 100 Montevista, D.D.O., Montreal, P.Q., H9B 2Z9, Canada.

II. Les Consultants Dessau Inc., 370 Chemin Chambly, Longueuil, P.Q., J4H 3Z6, Canada.

1. The lateral earthquake resistance of the structures is not adequate and that it is only relying on the resistance of unreinforced masonry infilling which is highly vulnerable to damage.
2. The precast concrete frame structure lacks ductility - an essential prerequisite for the survival of structures during a major earthquake.
3. The roof elements are not properly tied to offer adequate diaphragm action for the transfer of lateral inertial forces to the shear resisting arrangement.
4. The damages are repairable but the structure must be strengthened to meet the requirements of the latest National Building Code of Canada.

The structural engineers studied two alternative schemes to rehabilitate the structures. These were: a) a conventional method of strengthening with cast-in-place concrete shearwalls, and b) an innovative technique of introducing supplemental damping by installing friction-dampers.

The conventional method of introducing concrete shearwalls involved extensive new foundation work. This was very expensive and time consuming. The school authorities could not afford to close the school for a long duration. Also, it was felt that any benefits gained from the stiffening with shearwalls could be negated by the shift of the natural frequency of the structure closer to the resonant frequency content of the ground motion, typically experienced during the Saguenay earthquake.

The innovative technique of introducing supplemental damping in conjunction with appropriate stiffness was considered to be the most effective, practical and a smart hi-tech solution for the seismic upgrading of these buildings. This could be conveniently implemented by incorporating friction-dampers in steel cross-bracings in precast concrete frames and connectors on the vertical joints in precast wall panels. It was possible to stagger the bracings at different storeys to avoid interference with the services and avoid overloading of columns/foundations. Flexibility in the location of braces resulted in better space planning. This was not possible with the use of shearwalls. Finally, the client's criteria of shorter construction time and lower initial construction cost, overwhelmingly decided in the favour of this technique. The use of new technology resulted in a saving of 40% in construction cost and 60% in construction time. The roof diaphragm was strengthened with conventional steel plate bracing under the roof insulation.

This paper will discuss the analysis, design and construction details of the chosen system.

STATE-OF-THE-ART

Based on economic considerations, the building code philosophy is to design structures to resist moderate earthquakes without significant damage and to resist major earthquakes without structural collapse. In general, reliance for survival is placed on the ductility of the structure to dissipate energy while undergoing inelastic deformations. This assumes permanent damage, after repair costs of which could be economically as significant as the collapse of the structure.

The problems created by the dependence on ductility of the structure can be reduced if a major portion of the seismic energy is dissipated independently from the primary structure. With the emergence of new techniques like friction-dampers and base-isolation, it has become economically possible to design damage free structures. In September 1985, the State of California has passed an Assembly Resolution 'ACR 55 - Seismic Safety' that all publicly owned buildings must incorporate new seismic technology and existing buildings be retrofitted to increase their earthquake resistance. This resolution is based on the consideration that while the past code philosophy was concerned with the avoidance of collapse of the structure, the modern buildings have expensive finishes and contain extremely sensitive and costly equipment which must be protected. The National Building Code of Canada 1990, Clause 83 of Commentary-J of the Supplement, allows the use of new technology.

Friction-Damped Buildings

Of all the methods available to extract kinetic energy from a moving body, the most widely adopted is undoubtedly the friction brake. It is the most effective, reliable and economical mean to dissipate energy. For centuries, mechanical engineers have successfully used this concept. This concept has been extended to building construction to control their vibratory motion caused by the lateral inertial forces of an earthquake.

Several types of inexpensive friction-damping devices suitable for different construction techniques have been developed by Pall (1980,81,82,84,89,91). The devices are for: large panel precast concrete construction; cast-in-place concrete shearwalls; braced frames, friction base-isolators for low-rise buildings; and for connecting precast cladding to frames. Cyclic dynamic laboratory tests have been conducted on specimen devices (Pall 80, Filiatrault 86). The performance is reliable, repeatable and possesses large rectangular hysteresis loops with negligible fade over several cycles of reversals that can be encountered in successive earthquakes. Much greater quantity of energy can be disposed of in friction than any other method involving the damaging process of yielding of steel or cracking of concrete. Unlike visco-elastic materials, their performance is not affected by temperature, velocity and stiffness degradation due to aging. Furthermore, these friction-damping devices need no maintenance or replacement over the life of building and are always ready to do their job regardless of how many times they have performed.

Friction-dampers are designed not to slip under normal service loads, wind storms or moderate earthquakes. During a major earthquake, the dampers slip at a predetermined load, before yielding occurs in the other structural elements. This allows the building to remain elastic or at least the yielding is delayed to be available during catastophic conditions. Another interesting feature of friction-damped buildings is that their natural period varies with the amplitude of vibration, i.e. the severity of earthquake. Hence the phenomenon of resonance or quasi-resonance is avoided. Parametric studies have shown that optimum slip load of the structure is independent of the character of future earthquakes and is rather a structural property. Also, within a variation of ±20% of the optimum slip load, the response is not significantly affected.

In 1985, a large scale 3-storey friction-damped braced frame was tested on a shake table at the University of British columbia, Vancouver (Filiatrault 1986). The response of the friction damped braced frame was much superior to that of moment-resisting frame and braced frame. Even an earthquake record with a peak acceleration of 0.9g did not cause any damage to friction-damped braced frame, while the other two frames suffered large permanent deformations. In 1987, a 9-storey three bay frame, equipped with friction-dampers, was tested on a shake table at the Earthquake Engineering Research Center of the University of California at Berkeley (Aiken 1988). All members of friction-damped frame remained elastic for 0.84g acceleration - maximum capacity of the shake table, while the moment-resisting frame would have yielded at about 0.3g acceleration. In 1988, a single storey friction-damped frame was tested on a shake table at the Imperial college in London. Here again, the performance of the friction-damped braced frame was superior to the conventional moment-resisting frame.

Other researchers have investigated the seismic response of friction damped frames and reported on the superior performance of friction-damped frames (Austin 1985, Baktash 1986, Filiatrault 1986, 1988, Aiken 1988). In Montreal, a 10-storey Concordia University library building has been recently completed (Pall 1987). Use of steel bracing in concrete frames has eliminated the need for expensive shearwalls and the use of friction-dampers has eliminated the need of dependence on the ductility of structural components. Use of this system has resulted in a net saving of 1.5% of the total building cost while its earthquake resistance and damage control potential has significantly increased.

391

REHABILITATION OF ECOLE POLYVALENTE

Description of the Structure

The school complex consists of three blocks of buildings interconnected to each other. A typical plan view of the buildings at the first floor is shown in Fig. 1. Block-A is of two storey height and Blocks-B & C are of three storey height above the basement. Front view of the Block-A is shown in Fig. 2. Total covered area of the three buildings is approximately 40,000 m. sq. The structural frames are made up of precast concrete columns or wall panel units and beams tied together with welded connections. The columns or wall panel units are of full height of the building. The floor over the basement is of cast-in-place concrete. The upper floor and roof units are of prestressed precast concrete single or multiple T units. Floor units have a concrete overlay of 75mm thickness. Roofing units have standard insulation with tar and gravel finish. Floor and roofing units are placed on the concrete beams and tied with welded connections. These units are also tied to each other with welded connections at 3-4 m spacing. However, site inspections revealed that the connections at the roof are inadequate to offer diaphragm action. All partition walls in corridors, class rooms and around staircases are of unreinforced hollow concrete or terracotta masonry "stack-block" construction.

The location of friction-dampers at the first floor are shown in Fig. 1. The typical details of friction-damper for steel cross-bracing are shown in Fig. 3. Epoxy grouted anchor bolts were used to connect the steel bracings and wall connectors to the precast concrete units.

Nonlinear Time-History Dynamic Analysis

Non-linear time-history dynamic analysis was carried out by using the computer program DRAIN-2D, developed at the University of California, Berkeley. This program consists of series of subroutines that carry out a step by step integration of the dynamic equilibrium equations using a constant acceleration within any time step. As future earthquakes may be erratic in nature, an artificial earthquake record generated to match the design spectrum of Newmark-Blume-Kapur, which is an average of many earthquake records, has been used (Fig. 4). This earthquake record forms the basis of the NBC response spectrum. For Sorel, the peak ground accelerations of this earthquake record were scaled to 0.18g. The integration time step was 0.002 second.

Viscous damping of 5% of critical was assumed in the initial elastic stage to account for the presence of non-structural elements. Hysteretic damping due to inelastic action of structural elements and slipping of friction-damped connections is automatically taken into account by the program. Interaction between axial forces and moments for columns and P-Δ effect were taken into account by including geometric stiffness based on axial force under static loads.

Actual capacities of the precast concrete beam/column or panel connections and masonry infill panels were included in the analysis. The infill panels are assumed to have shear stiffness only. Cracking of masonry within repairable limits was considered acceptable. The ratio of strain at complete failure to strain at yield was taken to be 5. On the failure of the infill, the forces being resisted by the element immediately prior to failure are transfered to the structure as a shock loading. Strength and stiffness are reduced to zero after the failure.

Results of Analysis

1. The existing structure could not withstand peak ground accelerations of 5% of gravity.
2. The slip load of the dampers was governed by the safe load carrying capacity of the precast units, connections and foundations. The slip load for dampers in cross-bracings and panel connectors varried between 225-355 kN and 30-55kN respectively depending on the location.

3. A total of 64 dampers in cross-bracings and 388 panel connectors were required in all the three buildings to provide the desired energy dissipation.
4. The time-histories of deflection at the roof of a three-storey building are shown in Fig. 5. Maximum deflection was 15mm. The storey drift did not exceed 6mm in any building.
5. Typical time-histories of dampers at the first storey are shown in Fig. 6.
6. The introduction of supplemental damping by the friction-dampers reduced the forces in the structure. Strengthening of the existing members or foundations was not necessary.

CONCLUSIONS

The use of friction-dampers has shown to provide a practical, economic and effective new approach to rehabilitate the existing buildings to resist future earthquakes. The use of new technology resulted in significant savings in the initial construction cost and construction time.

ACKNOWLEDGEMENTS

The authors gratefully acknowledge the cooperation of Daniel Chartrand, Director of Commission Scolaire de Sorel; Marcel Dubois, Vice-President and Dominique Trottier, Director of structures of Dessau Inc. The assistance of Tom Tan and Jeff Matheuszik, engineers at Pall Dynamics Ltd. and Maurice Simard of Dessau Inc. is greatly appreciated. Patented Friction-Dampers were designed and supplied by Pall Dynamics Limited.

REFERENCES

Aiken, I.D., Kelly, J.M., Pall, A.S., 1988 "Seismic Response of a Nine-Story Steel Frame with Friction Damped Cross-Bracings", Report No. UCB/EERC-99/17, Earthquake Engineering Research Center, University of California, Berkeley. pp. 1-7.

Austin, M.A., Pister, K.S. 1985, "Design of Seismic-Resistant Friction-Damped Braced Frames", Journal of Structural Division, ASCE, Vol. III, No. 12, December 1985, pp. 2751-69.

Baktash, P., Marsh, C. 1986, "Comparative Seismic Response of Friction-Damped Braced Frames", ASCE Proceedings Dynamic Response of Structures, Los Angeles, pp. 582-589.

Filiatrault, A., Cherry, S. 1986, "Seismic Tests of Friction Damped Steel Frames", Proceedings-third conference on Dynamic Response of Structures, ASCE, Los Angeles.

Filiatrault, A., Cherry, S. 1988, "Comparative Performançe of Friction-Damped System and Base-Isolation System for Earthquake Retrofit and Seismic Design", Earthquake Engg. and Structural Dynamics, Vol. 16, pp. 389-416.

Mitchell, D., Tinawi, R., and Law, T. 1989. "The 1988 Saguenay Earthquake - A Site Visit Report ", Open Report # 1999, Geological Survey of Canada.

Pall, A.S., Marsh, C., Fazio, P. 1980, "Friction Joints for Seismic Control of Large Panel Structures", Journal of Prestressed Concrete Institute,Vol. 25, No. 6, pp. 38-61.

Pall, A.S., Marsh, C. 1981, "Friction Damped Concrete Shearwalls", Journal of American Concrete Institute, No. 3, Proceedings, Vol. 78, pp. 187-193.

Pall, A.S., Marsh, C. 1982, "Seismic Response of Friction Damped Braced Frames", ASCE, Journal of Structural Division, Vol. 108, St. 9, June 1982, pp. 1313-1323.

Pall, A.S. 1984, "Response of Friction Damped Buildings", Proceedings Eighth World Conference on Earthquake Engineering, San Francisco, Vol. V, pp. 1007-1014.

Pall, A.S., Verganelakis, V., Marsh, C. 1987, "Friction-Dampers for Seismic Control of Concordia University Library Building", Proceedings of Fifth Canadian Conference on Earthquake Engineering" Ottawa, Canada, pp. 191-200.

Pall, A.S. 1989, "Friction-Damped Connections for Precast Concrete Cladding", Proceedings PCI - Architectural Precast Concrete Cladding and its Lateral Resistance, pp. 300-309.

Pall, A.S., Pall, R. 1991, "Friction Base-Isolated House in Montreal", Proceedings, Sixth Canadian Conference on Earthquake Engineering, Toronto.

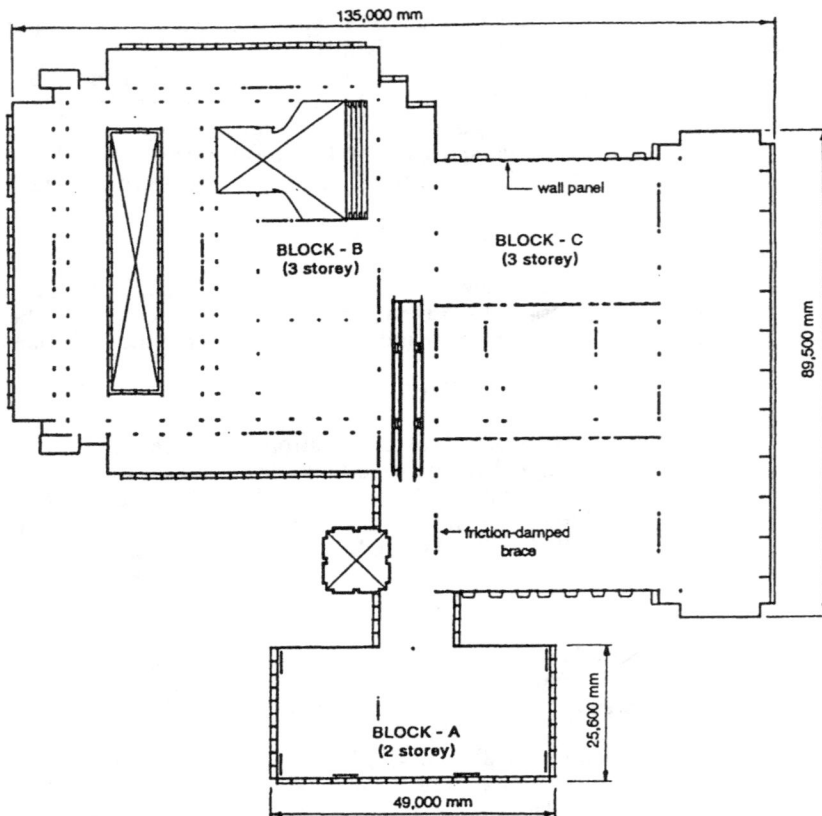

Figure 1. Plan View of School Complex

Figure 2. Front View of Block - A

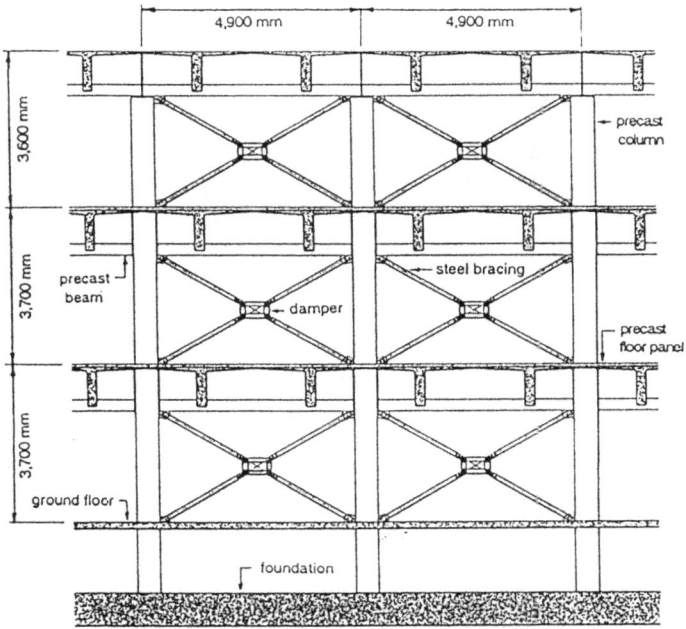

a) Location of Friction-Damper in Frame

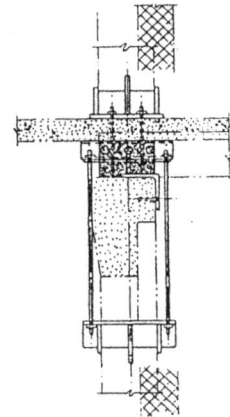

Section 1 - 1

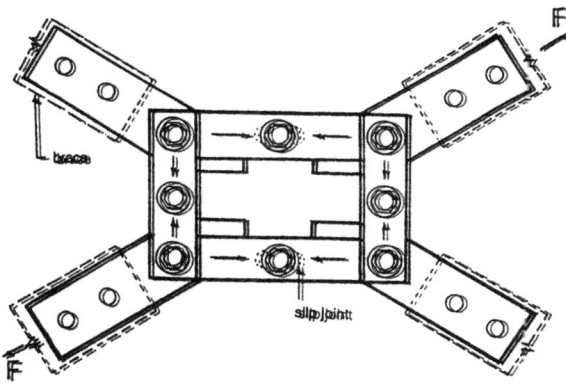

b) Typical Friction-Damper

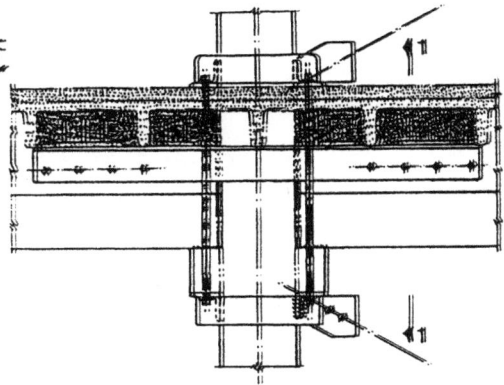

c) Typical Detail of Fixing Steel Bracing to Precast Concrete Columns

Figure 3. Typical Detail of Friction-Damper in Steel Cross-Bracing

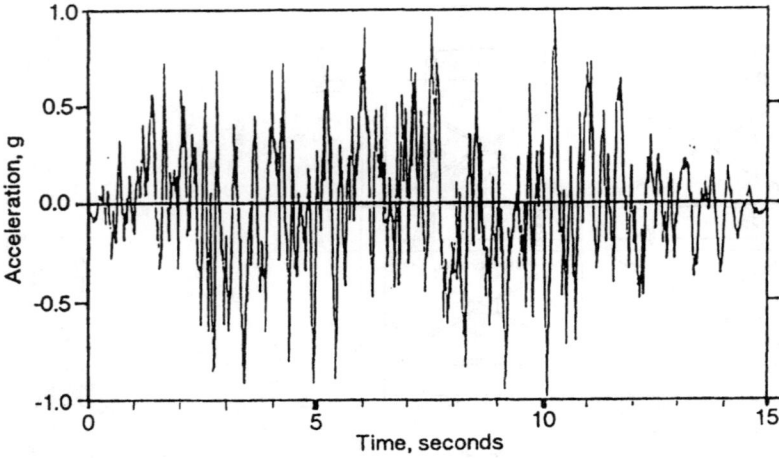

Figure 4. Time-Histories of Artifical Earthquake (Newmark, Blume & Kapur.)

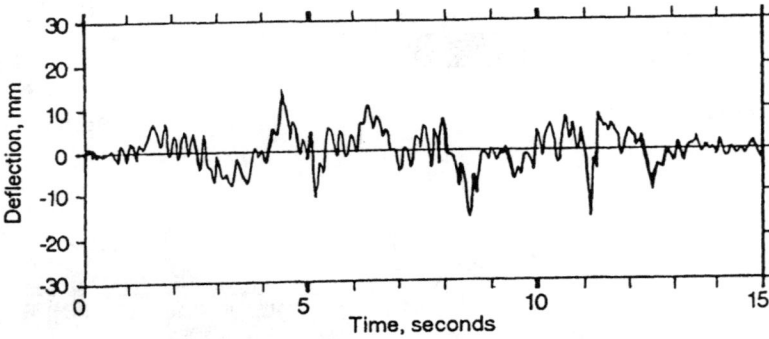

Figure 5. Time-Histories of Deflection at Roof

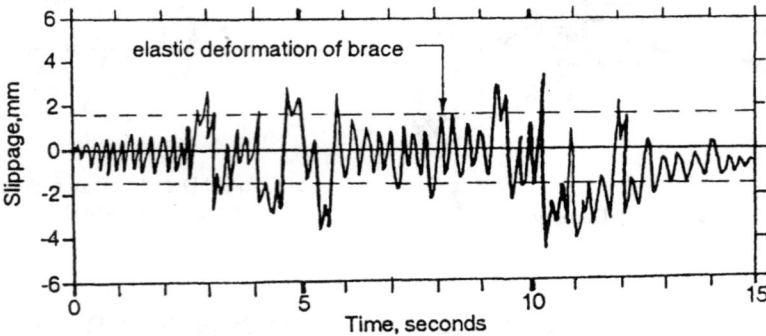

Figure 6. Time-Histories of Slippage in Friction-Damper in Cross-Brace

1988 Saguenay earthquake and design of rigid underground walls for Southeastern Canada seismicity

Cetin Soydemir[I]

ABSTRACT

The classic Mononobe-Okabe formulation has been widely adopted in seismic design of retaining walls for several decades. Although the formulation has proved to be quite appropriate for retaining walls which experience an active yielding condition during ground shaking, it is suggested that the formulation would underestimate the induced dynamic soil pressure increment for rigid, non-yielding walls of underground structures. Based on the results of more recent analytical work by Wood (1973) and model wall tests undertaken at the Central Laboratories of New Zealand (Yong 1985; Elms and Wood 1987), alternate recommendations are proposed for the seismic design of rigid, non-yielding walls in southeastern Canada in the wake of the well documented 1988 Saguenay earthquake.

INTRODUCTION

The current practice of seismic design of rigid, non-yielding retaining walls of underground structures such as multi-level building basements, box-like transportation and hydraulic modules, monolithic bridge abutments, power plants, pumping stations, etc., generally uses the well-known Mononobe-Okabe (MO) formulation developed over 60 years ago, or a simplified version of it proposed by Seed and Whitman (1970) (MO-SW). Although the MO formulation has stood the test of time quite well for walls which yield adequately under ground shaking, it is often overlooked that the formulation would underestimate the earthquake induced soil pressure increments against rigid, non-yielding walls as described above, and may lead to unsafe designs. Accordingly, the same argument is also valid for the MO-SW formulation, which is given for a typical cohesionless soil backfill with horizontal surface as:

I Vice President, Haley & Aldrich, Inc., Cambridge, Mass., U.S.A.

$$\Delta P_{AE} = \frac{3}{8} K_h \gamma H^2 \qquad\qquad (1)$$

where ΔP_{AE} is dynamic force increment due soil/backfill, K_h g is horizontal ground acceleration at the base of wall, γ is unit weight of the soil, and H is height of soil/backfill against the wall.

SEISMICITY OF SE CANADA

For earthquake resistant design of buildings seismic risk in Canada has been prescribed by probabilistic strong ground motion maps in the National Building Code of Canada (NBCC). Background information on the progressive development of these maps, the methodology followed and the uncertainties involved were reported by Basham et al (1985). 32 earthquake source zones have been defined within the boundaries of Canada based on historical seismicity, regional seismotectonics and geology as understood by early 1980's. With this source model Cornell-McGuire method of probabilistic ground motion estimation was used to develop the maps in the Code (Basham et al. 1985; Basham 1987).

The earthquake source zones defined for the general region of southeastern Canada are shown in Fig. 1, including the subregions of Charlevoix (CHV), western Quebec (WBU), northern Appalachians (NA), Laurentian slope (LSP) and Lower St. Lawrence (LSL) (Basham et al. 1985). Charlevoix subregion has been the most active seismic source in eastern North America with five or six earthquakes of magnitude 6 or greater having occurred in 1638, 1663, 1791, 1860, 1870 and 1925 (Basham 1987). In the western Quebec subregion documented large seismic events have been the M6.2 Montreal earthquake of 1732, M6.2 Temiskaming earthquake of 1935 and the M5.6 Cornwall earthquake of 1944 (Adams and Basham 1987).

The 1985 NBCC includes a peak horizontal ground acceleration map and a peak horizontal ground velocity map with probability of exceedance of 10 percent in 50 years. The acceleration and velocity maps provide independent ground motion reference levels in the design of squatty rigid structures with shorter periods and tall flexible structures with longer periods, respectively (Basham et al. 1985; Ishiyama and Rainer 1987). Typical underground structures described earlier belong to the former group. The peak horizontal acceleration map for the portion of southeastern Canada is reproduced in Fig. 2 (Basham et al. 1985).

1988 SAGUENAY EARTHQUAKE

An earthquake of magnitude 6.0 occurred in the Saguenay region of the
province of Quebec on 25 November 1988. It has been the largest earthquake
in eastern North America since the 1935, M6.2 earthquake near Temiskaming,
Quebec. The epicenter of the Saguenay earthquake, which is depicted with
an encircled star in Fig. 3 was located at approximately 35 km south of the
cities of Chicoutimi and Jonquiere and about 100 km northwest of the active
Charlevoix earthquake zone. There has been no previously known significant
earthquake activity in the Saguenay region. The focal depth of the
earthquake was determined by Geological Survey of Canada at 29 km, twice as
deep as compared to most of the eastern Canadian earthquakes previously
studied (EERI 1989; Adams and Basham 1989).

The seismic activity generated by the Saguenay earthquake was recorded
by the eastern Canadian strong-motion seismograph network shown in Fig. 3
(EERI 1989). All instruments were installed on bedrock except Baie-St.
Paul. Accelerographs at 15 of the 22 sites triggered on the main shock of
25 November 1988. In Fig. 3 the triggered sites are depicted with solid
circles and the maximum horizontal acceleration recorded in (g) by each
instrument are given (Munro and Weichert 1989). The largest maximum
horizontal acceleration, 0.174 g, was recorded at Baie-St. Paul on
alluvium. The largest maximum horizontal acceleration on bedrock, 0.156 g,
was measured at St. Andre. The seismic activity was also measured by the
U.S. National Center of Earthquake Engineering Research strong-motion
stations. The maximum horizontal acceleration of 0.09 g was recorded at
Dickey, Maine about 200 km from the epicenter (EERI 1989).

Much of the damage caused by the Saguenay earthquake in the epicenter
region and at distances of up to 350 km was due in whole or in part to soft
subsoil conditions or to poor performance of unreinforced masonry (EERI
1989, EQE 1989). Amplification of rock motion through soft soil deposits
was demonstrated at Baie-St. Paul and elsewhere. The lack of soft soils
overlying the bedrock may account for the modest level of damage observed
in the populated areas near the epicenter (EQE 1989).

DESIGN CRITERIA FOR NON-YIELDING RIGID WALLS

Design earthquake for the study

Based on an overall assessment of the peak horizontal ground
acceleration zoning map for SE Canada (Fig. 2), and maximum horizontal
ground accelerations recorded during the 1988 Saguenay earthquake (Fig. 3),
a design horizontal ground acceleration of 0.16 g has been selected to be
representative of the seismic risk in SE Canada within the context of this
study. An exception has been made for the highly seismic Charlevoix

subregion encircled by the 0.32 g contour in Fig. 3, for which a design horizontal ground acceleration of 0.32 g has been assigned.

Wood's analytical work

Wood (1973, 1975) considered the rigid, non-yielding retaining wall problem as depicted in Fig. 4. The soil/backfill is assumed to be homogeneous, isotropic and elastic. A uniform body force field is considered representative of the earthquake induced inertia forces on the soil. The lower boundary consists of competent ground along which no soil displacement occurs. Wood (1973), utilized the finite element technique and obtained static elastic solutions for the dynamic soil pressure increments against the rigid, non-yielding wall. Solutions for the case of the soil shear modulus increasing linearly with depth have been generated by the author for a horizontal acceleration of 0.16 g in Fig. 5. The actual value of the modulus does not affect the results. The solutions indicate the significant influence of the problem geometry defined by the L/H ratio. The pressure distribution is capped by the passive pressure envelope for the particular soil.

Elastic solutions in close agreement with Wood's (1973) were also obtained by Scott (1973) who used a one-dimensional shear beam analogy to model the soil retained by the rigid non-yielding wall, as demonstrated by Soydemir (1991).

In Fig. 6, the dynamic force increment calculated both by Wood's (1973) model, and the MO-SW formulation for a horizontal ground acceleration of 0.16 g, are presented for comparison. It is observed that the dynamic thrust on rigid, non-yielding walls may be underestimated by about 2.5 times by the MO-SW formulation. Whitman (1990) has also drawn attention to this condition.

Model tests by the Central Laboratories of New Zealand

Since the early 1980's, a comprehensive model testing program has been undertaken at the Central Laboratories, Ministry of Works and Development in New Zealand (CLNZ) to study the dynamic and static earth pressure increments on non-yielding and forced-displaced rigid walls (Elms and Wood 1987). A large number of tests for model retaining walls built on a shaking table have been performed (Yong 1985) to provide experimental data to assess the validity of the Wood's (1973) analytical results which had been the basis in developing recommendations for seismic design practice in New Zealand (Matthewson et al. 1980).

Results of Yong's (1985) tests for a non-yielding rigid model wall are summarized in Fig. 7, which depicts the measured maximum dynamic force increments by a cohesionless soil backfill as a function of the applied horizontal base acceleration. Dynamic force increments calculated from Wood's (1973) analytical solutions are included in the figure for comparison. The author has also incorporated in the figure the force increments calculated using the MO-SW formulation. It is observed that the experimental results are in close agreement with Wood's (1973) analytical solutions, however, they are significantly greater than the dynamic force increments given by the MO-SW formulation.

Recommendations for SE Canada

Based on the results of analytical and experimental works presented above, it is proposed that recommendations provided in Fig. 8 be adopted in seismic design practice of rigid, non-yielding retaining walls in SE Canada. Dynamic soil pressure increments in Fig. 8 correspond to a horizontal ground acceleration of 0.16 g and are believed to be applicable to the region in general. For the highly seismic Charlevoix subregion, as defined earlier, the pressure increments in Fig. 8 should be multiplied by two in compliance with the 0.32 g design horizontal ground acceleration selected for the subregion. Linear proportionality is acceptable since the soil is assumed to be elastic in Wood's (1973) solutions. Fig. 8 also includes the dynamic soil pressure increments calculated from the MO-SW formulation for comparison, and the criteria currently used in New Zealand design practice (NZP) for the regions represented by a horizontal ground acceleration level of 0.16 g.

ACKNOWLEDGEMENT

The author is grateful to Dr. J. Wood and Mr. P.M.F. Yong for their illuminating correspondence on the subject which they have provided graciously. Haley & Aldrich, Inc. Professional Development Program has provided funds for the preparation of the paper. Ms. A. Welch has drafted the figures, and Ms. D. Correia has typed the text. Their contributions are greatly appreciated.

REFERENCES

Adams, J. and Basham, P.W. 1987. Seismicity, crustal stresses and seismo-tectonics of eastern Canada. Natl. Cent. Earthquake Engr. Res., Tech. Rept. NCEER-87-0025, 127-142.
Adams, J. and Basham, P.W. 1989. Implications of the 1988 Saguenay earth-quake for seismic hazard zoning of southeastern Canada. Eastern Sect. Seism. Soc. Am., Seism. Res. Let., 60(1).

Basham, P.W., Weichert, D.H., Anglin, F.M. and Berry, M.J. 1985. New probabilistic strong seismic ground motion maps of Canada. Bull. Seism. Soc. Am., 75(2), 563-595.

Basham, P.W. 1987. Seismic hazards assessment and seismic codes for eastern Canada. Natl. Cent. Earthquake Engr. Res., Tech. Rept. NCEER-87-0025, 1-15.

EERI. 1989. The Saguenay, Quebec, Canada, earthquake of 25 November 1988. EERI Newslet., 23(5).

Elms, D.G. and Wood, J.H. 1987. Earthquake induced displacements and pressures on retaining walls and bridge abutments. Proceed. New Zealand Roading Symp., 4, 809-820.

EQE Engr. 1988. The Saguenay earthquake of November 25, 1988, a preliminary summary, San. Fran., CA.

Ishiyama, Y. and Rainer, J.H. 1987. Comparison of seismic provisions of 1985 NBC of Canada, BSC of Japan and 1985 NEHRP of the USA. Proceed. 5th Can. Conf. Earthquake Engr., 747-756.

Matthewson, M.B., Wood, J.H., and Berrill, J.B. 1980. Seismic design of bridges - earth retaining structures. Bull. New Zealand Soc. Earthquake Engr., 13(3), 280-293.

Munro, P.S. and Weichert, D. 1989. The Saguenay earthquake of Nov. 25, 1988 - processed strong-motion records. Geol. Surv. Can., Open File Rept. No. 1996.

Seed, H.B. and Whitman, R.V. 1970. Design of earth retaining structures for dynamic loads. Proceed. ASCE Spec. Conf. Cornell Univ., 817-842.

Soydemir, C. 1991. Seismic design of rigid underground walls in New England. Proceed. 2nd Intl. Conf. Geotech. Earthquake Engr. Soil Dynamics, St. Louis, Missouri.

Whitman, R.V. 1990. Seismic design and behavior of gravity retaining walls. Proceed. ASCE Conf. Design Perform. Earth Retain. Struct., Cornell Univ., 817-842.

Wood, J.H. 1973. Earthquake-induced soil pressures on structures. Earthquake Engr. Res. Lab., Calif. Inst. Tech., Rept. EERL 73-05.

Wood, J.H. 1975. Earthquake induced pressures on rigid wall structure. Bull. New Zealand Soc. Earthquake Engr. 8(3), 175-186.

Yong, P.M.F. 1985. Dynamic earth pressures against a rigid earth retaining wall. Central Lab., Min. Works and Devel., New Zealand, Rept. 5-85/5.

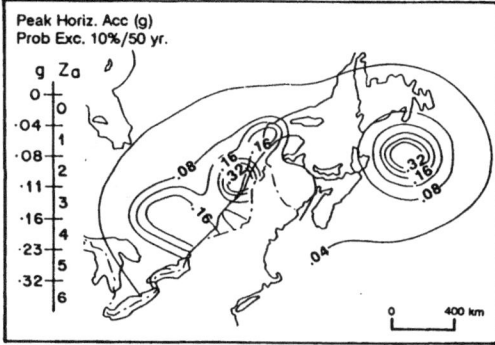

Figure 1. Earthquake source zones in SE Canada (Basham et al. 1985)

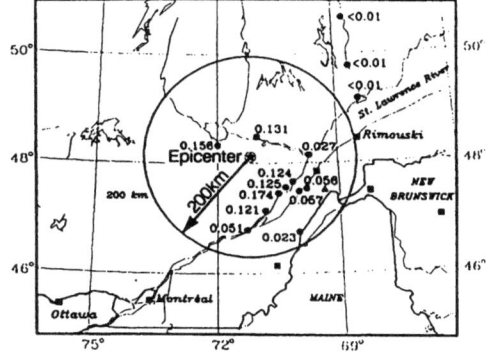

Figure 3. Maximum horizontal ground acceleration in (g), Saguenay, 25 Nov. 1988 (Munro and Weichert 1989; EERI 1989)

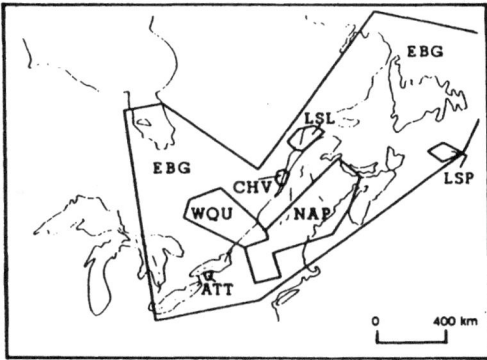

Figure 2. Peak horizontal acceleration zoning map for SE Canada (Basham et al. 1985)

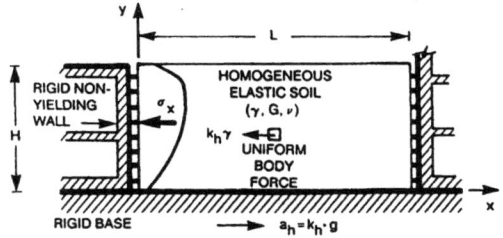

Figure 4. Wood's (1973) analytical problem

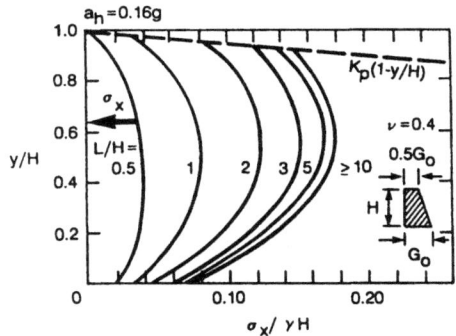

Figure 5. Seismic soil pressure increments on rigid non-yielding wall for $a_h = 0.16 \cdot g$ by Wood's (1973) solution

403

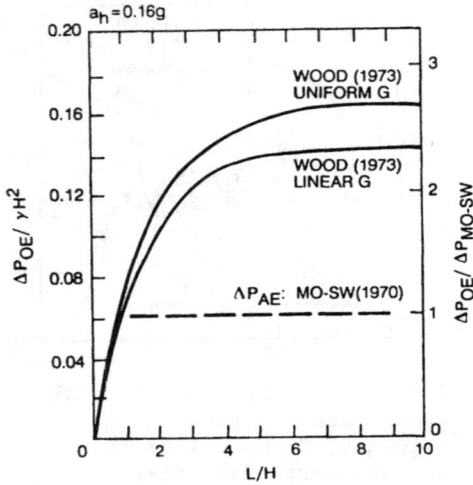

Figure 6. Analytical results for seismic pressure increment induced by soil on rigid yielding and non-yielding walls

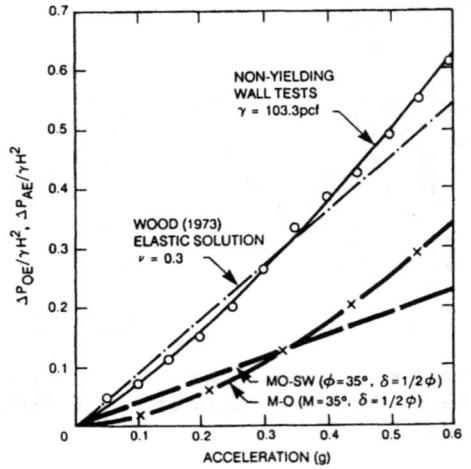

Figure 7. Dynamic soil pressures on yielding and non-yielding walls (after Yong 1985, Elms and Wood 1987)

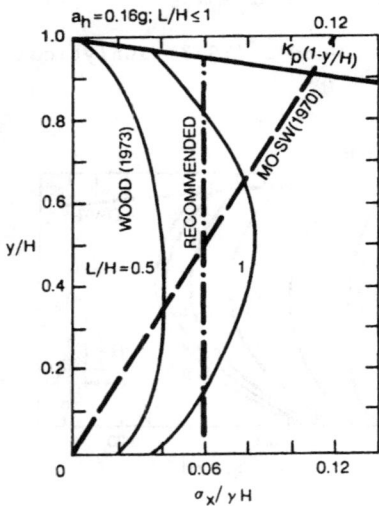

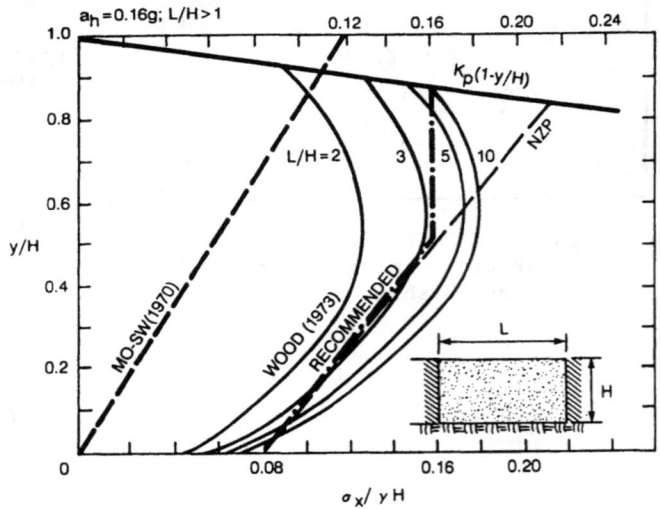

Figure 8. Recommended seismic soil pressure increment against rigid non-yielding walls in SE Canada

404

Effects of soil-structure interaction on the seismic performance of a concrete frame-wall structure

Patrick Paultre[I] and Maryse Lavoie[II]

ABSTRACT

The effects of the soil and of the soil-structure interaction on the seismic performance of a six-storey reinforced concrete office building is studied. The frame-wall structural system is designed according to the latest Canadian codes to possess nominal ductility. Three cases are studied corresponding to three different thicknesses of the underlying soil. The finite element soil-structure system is first reduced by means of derived Ritz vectors. Step-by-step analyses of the reduced linear system are then performed to obtain horizontal and vertical acceleration responses at the soil-structure interface. These modified motions are then used as input motions for the non-linear analyses of the frame-wall structure alone. The performance of the structure is then evaluated in terms of lateral displacement, drift index and base shear.

INTRODUCTION

Soil effects and soil-structure interaction are important parameters on the behaviour of building structures under seismic loading. The true response of the structure can only be obtained if the interaction of the vibrating structure and the underlying soil is considered and properly modelled. There is a clear distinction between site effects and soil-structure interaction. Site effects are the modification of the motion at the free surface after propagation through soil layers. Soil structure interaction depends not only on the free surface motion but on the mechanical properties of the structure and the underlying soil. Seed (1975) has shown that a quasi-resonance between the structure and the soil layer can increase significantly the seismic forces acting on the structure and therefore their damage potential. Seed (1975), Balendra and Heidebrecht (1987), among others, suggested the use of a period-dependent foundation factor to account for these effects.

The Uniform Building Code (UBC 1988) includes the soil effect in the minimum lateral load through the factor S, which depends on the height and the type of soil only. In the 1982 edition (UBC 1882) however, the factor S was also a function of the ratio of the period of the structure to the period of the soil layer.

[I]Associate Professor, Département de génie civil, Faculté des sciences appliquées, Université de Sherbrooke

[II]Graduate Student, Département de génie civil, Faculté des sciences appliquées, Université de Sherbrooke

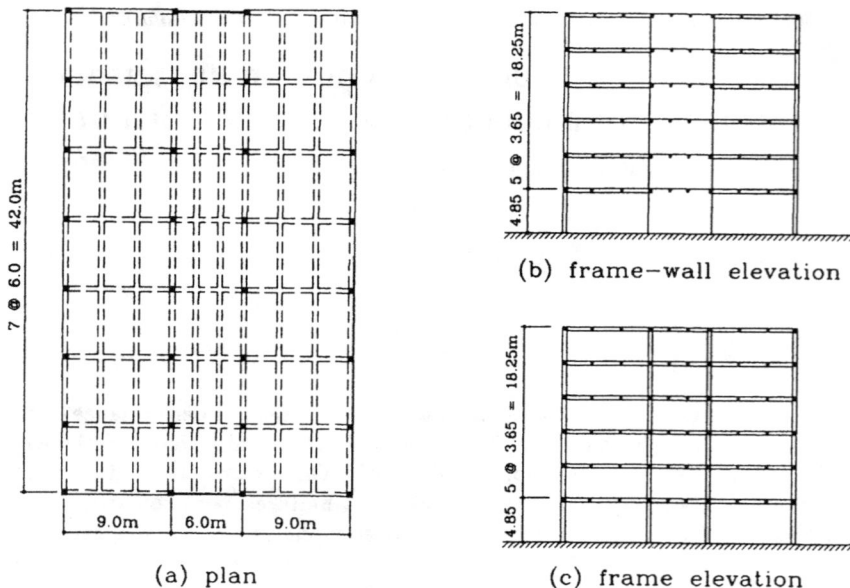

(b) frame—wall elevation

(a) plan

(c) frame elevation

Figure 1. Plan view of six-storey frame-wall structure.

The NEHRP (NEHRP 1986) accounts for the height and the type of soil in the equation of the minimum lateral load. Supplementary equations are given to account for soil-structure interaction.

The National Building Code of Canada (NBC 1990) includes a foundation factor in the equation of the minimum lateral load. This factor, F, depends on the thickness and type of soil. No particular procedure is recommended to account for soil-structure interaction.

DESCRIPTION OF STRUCTURE AND SUPPORTING SOIL

In order to study the adequacy of the F factor in the National Building Code of Canada, a six-storey office building was designed for a Montréal area. The plan view of the building is shown in Fig. 1. The reinforced concrete frame-wall building has 7 - 6 m bays in the longitudinal (N-S) direction and 3 bays in the transverse (E-W) direction, consisting of a central corridor bay and 2 - 9 m external office bays. The storey height is 4.85 m for the ground level and 3.65 m for all the other levels. The structural system is made of 6 internal moment-resisting frames and two external bents consisting of two perimeter frames attached to a central wall. The design is according to the 1990 NBC and the 1984 CSA Standard for the design of concrete structures for buildings with $R = 2.0$. This implies that the walls possess nominal ductility and take 100% of the lateral load and the frames carry only gravity loads. It is assumed that the central roof bay supports machinery. The specified yield stress for the steel is 400 MPa and the concrete strength is 30 MPa. The calculated natural period of the building is 1.1 s.

The site has been chosen from boring data on the island of Montréal to correspond to a foundation factor $F = 1.3$. The soil medium is made up of granular material of average density.

406

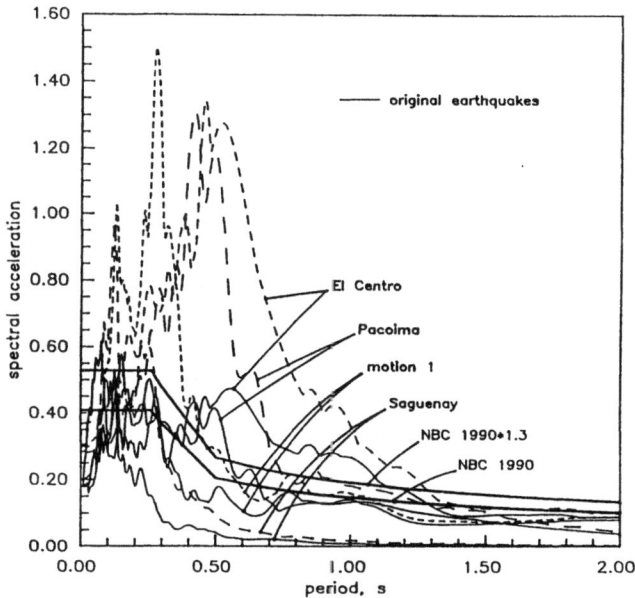

Figure 2. Response spectra of earthquake motions used.

The properties of the soil are derived after equations proposed by Seed et al. (1986). The initial values for the shear modulus, the mass density and the Poisson ratio, are 128 MPa, 2080 kg/m^3 and 0.35, respectively. Three different thicknesses of soil are studied to appreciate the site effect and the effects of soil-structure interaction on the nonlinear response of the building.

EARTHQUAKE MOTIONS

The use of several accelerograms is necessary to assess adequately the nonlinear performance of structures under seismic loading. In this study a set of four earthquakes is used, consisting of three historic recordings and a motion artificially generated with SIMQKE (1976). The artificially-generated accelerogram has a spectrum compatible with the spectra given in the Commentary of NBC (1980) with the velocity bound adjusted by multiplying by the ratio of the peak horizontal ground velocity, v, over the peak horizontal ground acceleration, a. The characteristics of the four earthquakes are presented in Table 1. Figure 2 compares the response spectra for the four motions with the NBC 1990 equivalent spectral acceleration.

Table 1. Earthquakes selected

Earthquakes	Date	Location	Component	PGA, g	PGV, m/s	a/v
Imperial Valley	05/18/40	El Centro	S00E	0.348	0.335	1.04
San Fernando	02/09/71	Pacoima Dam	S74W	1.075	0.577	1.86
Saguenay	11/25/88	Chicoutimi	Long.	0.106	0.015	7.02
Motion 1				1.000	0.857	1.17

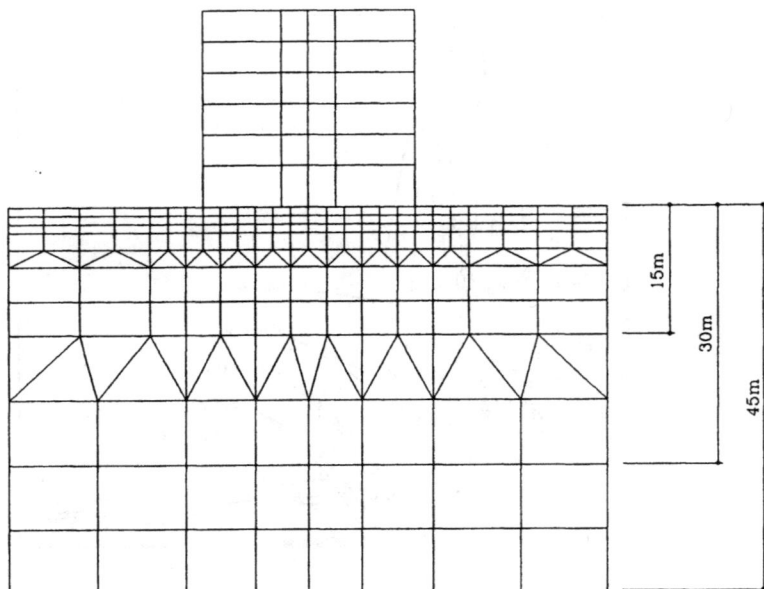

Figure 3. Finite element model of soil and structure.

All the ground motions were scaled to the desired maximum peak ground accelerations of 0.078 g, 0.18 g and 0.27 g and are referred to as "low", "intermediate " and "high", respectively. The "low level" earthquakes were assumed to have a maximum acceleration having a probability of exceedance of about 40% in 50 years (100-year return period). The "intermediate level" earthquakes correspond to peak horizontal acceleration having a probability of exceedance of 10% in 50 years (about 500-year return period). The peak horizontal acceleration used for the "high level" earthquakes corresponds to a probability of exceedance of about 5% in 50 years (1000-year return period) as determined from the data base of the Geological Survey of Canada. The Saguenay earthquake was also scaled according to the velocity because the period of the structure studied was in the velocity bound of the spectrum. This scaling resulted in maximum peak ground accelerations of 0.22 g, 0.68 g and 1.02 g for the "low", "intermediate" and "high" levels of earthquake, respectively.

ACCOUNTING FOR SOIL-STRUCTURE INTERACTION

It is known that the predicted nonlinear response of certain classes of structures is approximately equal to the predicted linear response for the same motion. Based on that fact, the effects of soil-structure interaction on the nonlinear seismic response of a structure can be studied in two steps. First, a linear analysis of the soil and the structure under seismic loading is carried out. The resulting motion at the soil-structure interface, which includes the interaction effects, is then used as the input motion in a nonlinear analysis of the structure alone.

The program SHAKE (Schnabel et al. 1972) was used to analyze the response of the soil layers. This program calculates the response of a layered infinite soil medium and accounts for the nonlinearity of the shear modulus and the damping. Response curves given by Seed et al. (1986)

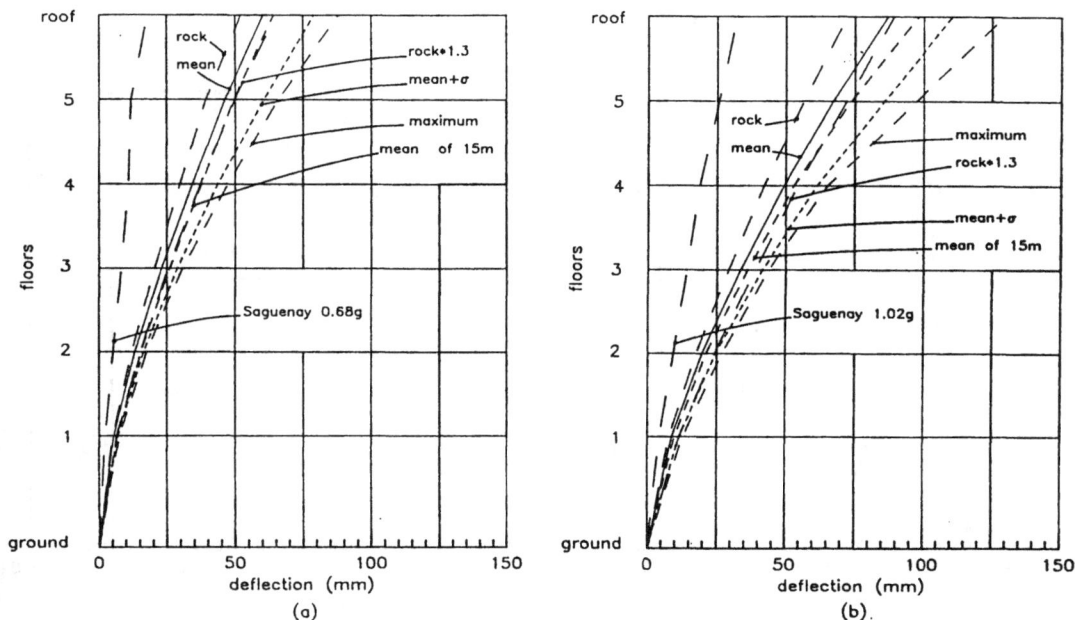

Figure 4. Maximum lateral displacements for structure subjected to ground motions with maximum accelerations of (a) 0.18 g and (b) 0.27 g.

were used to find the variation of the shear modulus and the damping with respect to the rate of shear deformation.

To study the soil-structure interaction, different models can be used. In this study the structure and the soil were modelled with beam and plane strain finite elements respectively, and analysed with the program CAL (Wilson 1986). The rigid foundation was modelled through the use of transfer matrices. The extent of soil modelled was approximately three times the width of the building with viscous boundaries, as proposed by Lysmer and Kuhlemeyer (1969) (see Fig 3).

The variations of the shear modulus and the damping of the soil were determined with the program SHAKE for each level of excitation. The shear modulus varied from 14 MPa to 128 MPa, with an average value of 72 MPa. The average damping for the soil was 11.5% of critical. The period of the combined soil-structure varied from 1.12 s to 1.34 s for a soil thickness of 15 m to 45 m.

The largest combined soil-structure finite element model had 411 degrees of freedom. These models were reduced with 35 load dependent Ritz vectors. The response of the reduced model was obtained by step-by-step integration. A measure of the soil-structure interaction is the acceleration time histories obtained at the soil-structure interface. Figure 2 shows the soil-structure interaction effects on the original motions for a peak ground acceleration of 0.18 g and 5% of critical damping. Also shown in Fig. 2 is the NBC 90 equivalent spectral acceleration times the foundation factor $F = 1.3$.

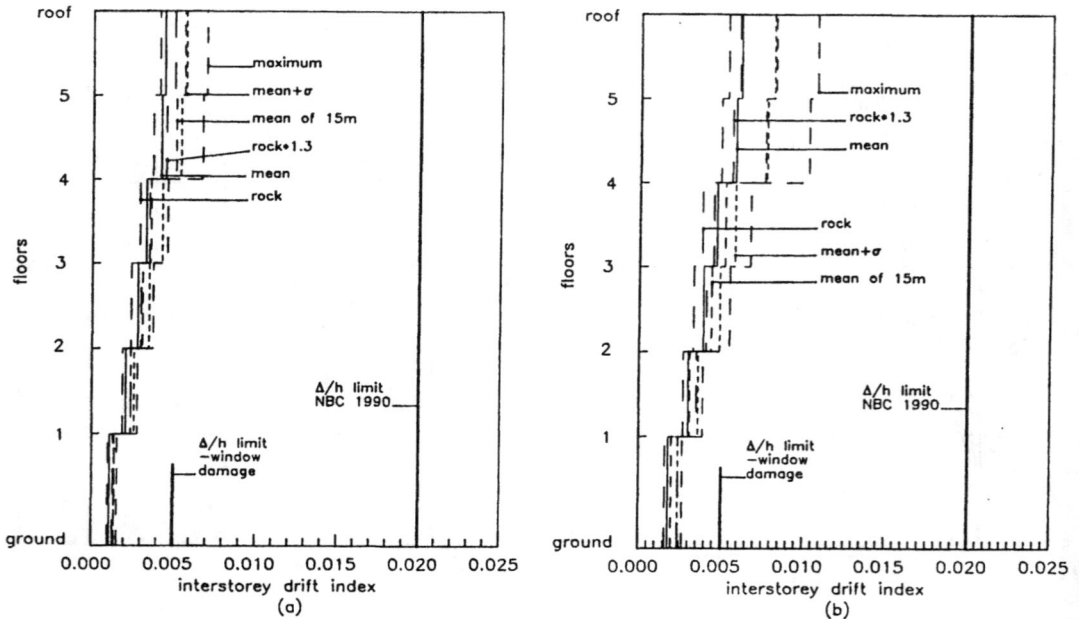

Figure 5. Maximum interstorey drift indices for structure subjected to ground motions with maximum accelerations of (a) 0.18 g and (b) 0.27 g.

PREDICTED NONLINEAR DYNAMIC RESPONSES

The resulting horizontal acceleration time histories at the soil-structure interface was used as input motion in the nonlinear analyses of the structure alone. To predict the nonlinear dynamic responses of the six-storey frame-wall structure, the general purpose nonlinear dynamic analysis program DRAIN-2D (Kanaan and Powell 1975) was used. It is interesting to note that the displacements calculated in the nonlinear dynamic analyses were similar to those calculated in the corresponding dynamic linear analyses. This justifies the procedure that was followed to study the dynamic soil-structure interaction effects.

Figure 4 shows the maximum lateral displacements for the four different earthquakes scaled to 0.18 g and 0.27 g. It is interesting to note that the predicted deflections for the "base rock" motion multiplied by 1.3 is approximately equal to the average maximum deflection predicted with 15 m of soil. As can be seen from Fig 4 b, the Saguenay earthquake, even scaled to 1.02 g, causes very small deflections due to its high frequency content.

Figure 5 shows the averages of the maximum interstorey drift indices (i.e., the ratio of the interstorey drift, Δ, to the storey height, h) for the "intermediate level" and the "high level" earthquakes. The interstorey drift ratio, $\Delta/h = 0.005$, corresponding to likely window damage, is also shown. The maximum predicted interstorey drift indices are well below the 0.02 limit given in the NBC (1990) for structures other than post-disaster buildings. However, due to large interstorey drift ratios, significant nonstructural damage is expected in the higher storeys.

410

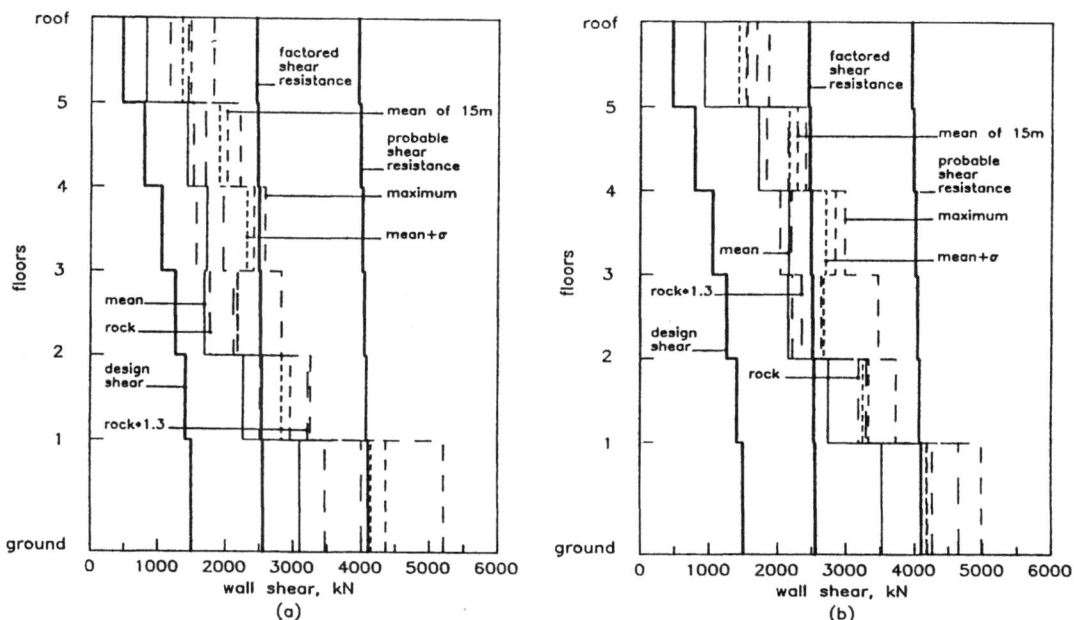

Figure 6. Maximum predicted wall shear with code design shear and shear resistances for structure subjected to ground motions with maximum accelerations of (a) 0.18 g and (b) 0.27 g.

Figure 6 illustrates the predicted wall shear for the "intermediate level" and the "high level" earthquakes. Also shown are the design wall shear. The factored and the probable resistance shown in Fig. 6 illustrate the range of wall shear capacities available. The shear capacities of the walls were calculated assuming axial loads corresponding to 0.85% of dead load. As expected, the nonlinear dynamic analyses give wall shears larger than the computed design shears for the "intermediate level" and "high level" earthquakes. An important fact is that the predicted shears at the first storey are in the majority of the cases larger than the probable resistance of the walls. The underestimation of the maximum shear is due to the fact that the same lateral force distribution is used for both flexural and shear design. In fact, in this case, the maximum moment and the maximum shear in the wall do not occur simultaneously. It should be noted that the predicted column shears are well below their nominal capacities; the lateral load is almost entirely taken by the walls.

CONCLUSION

Predicted displacements using the foundation factor are close to the ones obtained from a full soil-structure interaction study. This can be explained from the good agreement between the different spectra and the Code spectra in the frequency range of interest. It is also evident that the code procedure can significantly underestimate the response in the short period range. It is not clear, at least with the motions used, whether the large differences observed in the short period range can be attributed to the interaction problem or to an underestimation of the seismic response.

411

ACKNOWLEDGMENTS

The financial assistance provided by the Natural Science and Engineering Research Council of Canada and by the Fonds pour la Formation de Chercheurs et l'Aide à la Recherche of the Government of Québec is gratefully acknowledged.

REFERENCES

Balendra, T. and Heidebrecht, A. C. 1987. A Foundation Factor for Earthquake Desing Using the Canadian Code of Practice. Canadian Journal of Civil Engineering, 14(4), 498–509.

CSA Standard. 1984. CSA Standard CAN3-A23.3-M84 - Design of Concrete Structures for Buildings. Canadian Standard Association, Rexdale, Ontario, 282 p.

Kanaan, A.E. and Powell, G. H. 1975. DRAIN-2D - A General Purpose Computer Program for Dynamic Analysis of Inelastic Plane Structures. EERC Report No. 73-22. Earthquake Engineering Research Center, University of California, Berkeley, California, 138 p.

Lysmer, J. and Kuhlemeyer R.L. 1969. Finite Dynamic Model for Infinite Media. Journal of Engineering Mechanics Division, ASCE, 95(EM14), 859–877.

NBC. 1980. National Building Code of Canada 1980 and Supplement to the National Building Code of Canada 1980. National Research Council of Canada, Ottawa, Ontario.

NBC. 1990. National Building Code of Canada 1990 and Supplement to the National Building Code of Canada 1990. National Research Council of Canada, Ottawa, Ontario.

NEHRP. 1986. National Earthquake Hazards Reduction Program, Recommended Provisions for the Development of Seismic Regulations for New Buildings. Building Seismic Safety Council, Washington.

Robert, Y. 1980. Etude géotechnique des sols de l'île de Montréal - Prédiction de comportement pendant les séismes. M.A.Sc. Thesis, University of Montreal, Montreal, Quebec.

Schnabel, B., Lysmer, J. and Seed, B.1972. SHAKE - A Computer Program for Earthquake Response Analysis of Horizontally-Layered Sites. EERC Report No. 72-12. Earthquake Engineering Research Center, University of California, Berkeley, California, 88 p.

Seed, H. B. 1975. Design Provision for Assessing the Effects of Local Geology and Soil Conditions on Ground and Building Response During Earthquakes, New Earthquake Design Provisions, Seminar Papers from ASCE/SEAONC Professional Development Committee Series.

Seed, H. B., Wong, R. T., Idriss, I. M. and Tokimatsu, K. 1986. Moduli and Damping Factor for Dynamic Analyses of Cohesionless Soils. Journal of Geotechnical Engineering Division, ASCE, 112(GT11), 1016–1032.

SIMQKE. 1976. A Program for Artificial Motion Generation. User's Manual and Documentation. Massachusetts Institute of Technology, Department of Civil Engineering, Cambridge, Massachusetts, 32 p.

UBC. 1988. Uniform Building Code 1988. International Conference of Building Officials, Whittier, California.

Wilson, E. L. 1986. CAL-86 Computer-assisted learning of structural analysis and the CAL/SAP development system. Report UCB/SESM-86/05, U.C. Berkeley, California.

Soil-structure interaction effects on seismic response of elevated, ground-based, and buried tanks

Medhat A. Haroun[I] and Mohamed K. Temraz[II]

ABSTRACT

Effects of soil-structure interaction on the seismic response of elevated, ground-based, and buried, liquid storage tanks are assessed. Two-dimensional models of x-braced elevated tanks supported on isolated footings and subjected to horizontal ground motions were analyzed under static, dynamic elastic, and dynamic inelastic conditions. Further, the interaction between ground-supported, flexible-base, cylindrical tanks and the underlying soil under vertical ground excitations was examined. Flexibility of the shell and the base plate as well as stiffness and damping characteristics of the foundation soil were included. Lastly, stresses developed in the walls of underground rectangular tanks subjected to three components of ground motion were computed. Wall inertia, liquid hydrostatic and hydrodynamic pressures, in addition to static and dynamic earth pressures were considered.

ELEVATED TANKS

A common configuration of storage tanks in municipal water supply systems is the elevated type to provide the required head for water distribution. These structures consist of two main parts: a tower and a vessel. The former can be a steel braced frame, a multi-column assembly, or an axisymmetric pedestal shell, whereas the vessel comes in a variety of geometric shapes. Many elevated tanks are regarded as essential facilities, yet they are vulnerable to earthquakes mainly because of the relatively small resistance that the supporting system offers. This has been demonstrated by incidents of seismic damage which have been mostly confined to the supporting system rather than the vessel.

An extensive study of the earthquake response of elevated tank systems was conducted by Ellaithy [3]. It evaluated liquid-induced forces on top of the supporting structures, and estimated the behavior of x-braced towers in both 2- and 3-dimension spaces as well as the response of axisymmetric pedestal towers. The present analysis further examines the effects of soil-foundation-tower interaction on the seismic response of x-braced elevated tanks. Lumped-parameters method is used to model the soil-foundation system, and coupled with a two-dimensional model of a x-braced tower, the static, dynamic elastic, and dynamic inelastic responses of the overall system were evaluated.

[I]Professor and Chairman, Department of Civil Engineering, University of California, Irvine, CA 92717, and [II]Senior Geotechnical Engineer, ICG, Inc., Irvine, CA 92718.

METHOD OF ANALYSIS

A typical tower is essentially a three dimensional structure; however, Ellaithy [3] showed that the difference in response between three- and two-dimensional models ranges from 5% to 10%. Therefore, it was decided to employ a two-dimensional idealization of the tower as shown in Fig. (1). All members numbered 1 to 9 are designed as frame members while the rest are bracing members; their properties and dimensions can be found in [8]. Footings are designed to resist developed seismic uplift forces by both their own weight and the friction forces mobilized along their sides.

In current design standards, soil effects are explicitly represented by coefficients which reflect soil type and amplification of ground motion, but the degree of fixation of the tower as well as the foundation characteristics are not considered. In this study, the foundation system is represented by a discrete three degree of freedom model undergoing vertical, horizontal, and rocking motions, and effects of footings embedment are accounted for by modifying the spring constants.

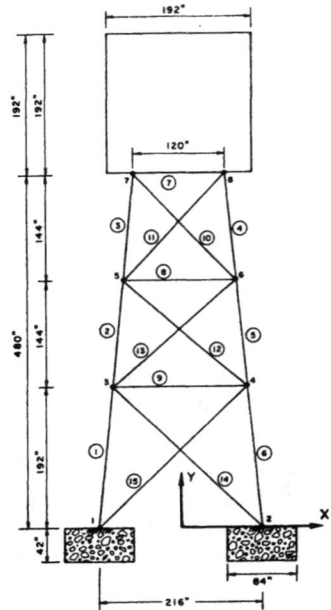

Figure 1: X-braced tower.

UBC-Based Loads

The Uniform Building Code (UBC) [9] and the American Water Works Association Standard D100 (AWWA) [2] are widely used for evaluating equivalent static lateral forces on towers of elevated tanks. The AWWA standard adopts a formula for calculating the total lateral force which is similar in form to that used in the earlier UBC versions prior to 1988. It is noted that the value of the structural system factor for a cross-braced tower supporting an elevated tank was among the highest for all types of structures, reflecting a perceived absence of hidden reserve strength. The UBC no longer requires that supporting towers be designed to resist torsional effects, not to compensate for a loading condition which occurs during failure. The 1988 UBC revised its formula for calculating the lateral shear force and uses a newly-defined structural system factor R_w.

Mechanical Model-Based Loads

In calculating lateral static loads according to current codes, the tank is assumed totally filled with water whose entire weight is augmented with the weight of container and tower to yield the total weight of the structure. As proposed by Haroun, only the weight of liquid accelerating with the vessel shall be considered as an "effective weight" contributing to lateral load computations. These loads are magnified by the ratio of the spectral acceleration, obtained from a response spectrum, divided by peak ground acceleration. Moreover, the additional overturning moment of the vessel must also be considered by equating it to two opposite vertical forces acting at the tower top.

Soil-Dependent Lateral Forces

Soil-tower interaction affects not only the kinematics of the supporting system but also

414

the values of the equivalent lateral static loads. Changes in the horizontal displacement of the tower in addition to its rocking motion imply a corresponding change in the inertia forces applied at the top of tower. This could be accounted for by evaluating the horizontal and angular accelerations at the tower top through a simplified dynamic analysis of the system, from which the equivalent lateral static forces can be computed.

Dynamic Elastic Analysis

The response to earthquake excitations was determined by direct time-integration of the matrix equation of motion of the overall system. A Rayleigh damping model was used for the tower whereas the damping matrix of the soil-foundation system was defined by lumped parameters. The water tower was investigated under the action of a horizontal component of ground acceleration of the 1940 El Centro earthquake.

Dynamic Inelastic Analysis

It is known that bracing members of x-braced towers undergo inelastic deformations. Therefore, the model developed by Ellaithy [3] which incorporates expected theoretical and observed experimental inelastic behavior of such members was used to reproduce the inelastic deformations. The overall stiffness matrix of the supporting cross-braced steel frame and the soil-foundation system was reformulated, but the overall mass and damping matrices remained similar to those used in the elastic analysis.

ILLUSTRATIVE NUMERICAL EXAMPLES

For the foundation-tower system shown in Fig. (1), the total lateral 1988 UBC-based shear force was found to be 104.2 kips, whereas Haroun's model when used with the ground motion of the 1940 El Centro earthquake (2% damping) yields a total lateral force of 99.2 kips, in addition to two vertical forces, each 40.5 kips, resulting from the rocking motion of the vessel. Figure (2) shows the variation of the horizontal displacement at the top of tower with the shear wave velocity of the soil. It is seen that the UBC and Haroun's model follow the same general trend but have different response values. This is because the UBC formula does not account for the additional overturning moment exerted on the tower by the vessel nor the influence of the height-to-radius ratio of the container on the effective mass of liquid.

Figure 2: Tower displacement.

Whereas the absolute lateral translations of towers supported on flexible soils are higher than those supported on infinitely-stiff soil, most member end actions are hardly affected as their values depend on the relative displacements between joints. Only bending moments developed at the connection between tower leg and footing were magnified for soft soils.

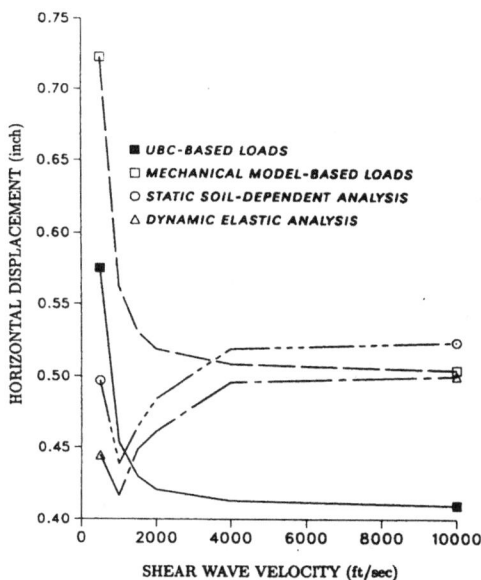

415

It was generally noted that employing the mechanical model to evaluate the tower's static response, the resulting displacements and member end actions were usually on the conservative side, especially for soft soils. For stiff soils, the mechanical model reliably predicted the dynamic response of the tower. It was also observed that the inelastic behavior redistributed the internal forces in the tower. Most notably, there was an increase in the member end actions of those members located at or near the top of tower where braces experienced significantly higher axial forces than those obtained from an elastic analysis, and this in turn produced higher end moments in the elastic members. Figure (3) shows that inelastic deformations in braces increase the maximum axial force in bracing member no. 10 whereas they reduce it in member no. 1 for all values of soil shear wave velocity.

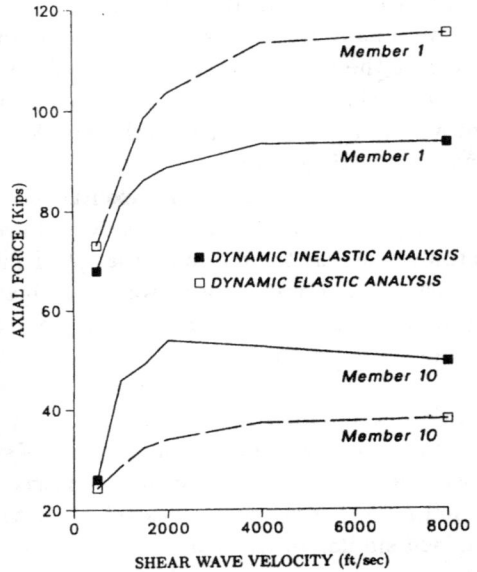

Figure 3: Axial forces.

GROUND-BASED TANKS UNDER VERTICAL EXCITATIONS

Whereas the response of ground-based tanks to horizontal earthquake excitations was thoroughly investigated, their response to vertical excitations has received little attention. The relatively weak state of knowledge of this behavior is reflected in current standards; for example, the API standard [1] neglects entirely the effects of vertical excitations on design forces whereas the AWWA standard [2] crudely considers such effects, and only in calculating hoop tensile stress in tank shell. Research into effects of vertical excitations was mainly conducted in the 80's. Haroun and Abdel-Hafiz [5] used a simplified two-degree-of-freedom system to evaluate such effects on rigid-base tanks considering soil-structure interaction. The main objective herein is to accurately evaluate the response to a vertical seismic excitation of a tank whose flexible base plate rests directly on a compacted soil.

METHOD OF ANALYSIS

A ground-based, cylindrical, thin-walled container supported on flexible soil is subjected to a vertical excitation. The tank and the soil were treated as substructures [10] of the overall system. At first, equations of motion of the soil were formulated to obtain its dynamic stiffness matrix; the elements of which represent the dynamic stiffness corresponding to those degrees of freedom located at the interface between soil and tank. Equations of motion of the liquid-filled tank were separately formulated and the dynamic stiffness matrix of soil was then added to the structural stiffness matrix for the degrees of freedom at the interface only. Finally, only the equations of motion of the tank (including soil effects) were solved in the

416

frequency domain to yield its response. Since the free-field earthquake motion is specified at structure-soil interface, this procedure eliminates deconvolution calculations.

In modeling the soil, two types of damping were considered: radiation and material. The former is produced by energy radiation due to propagation of waves away from the structure, whereas the latter is a frictional loss of energy which produces a damping force proportional to displacement but in phase with velocity. As for the other major substructure, the shell was analyzed by axisymmetric ring-shaped finite elements and the base plate was represented by annular elements. Liquid effects were in the form of hydrodynamic added mass on tank shell and base plate, and in this analysis, a novel approach for its computation was adopted [8].

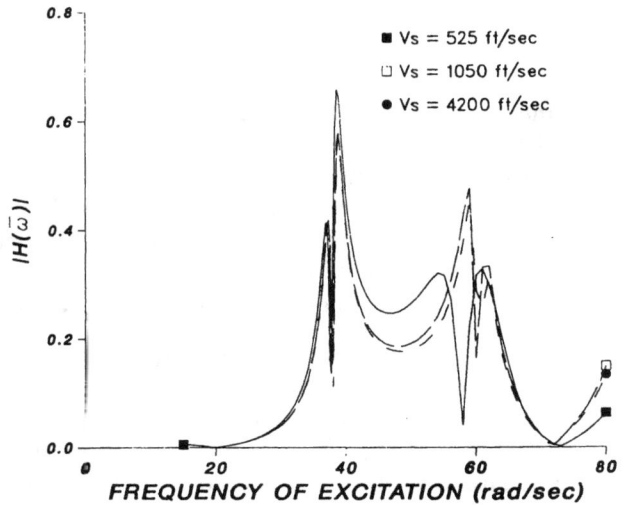

Figure 4: Frequency response functions.

ILLUSTRATIVE NUMERICAL EXAMPLES

Two tanks representing the class of broad and tall tanks, and three shear wave velocities representing soft, medium, and stiff soils were analyzed, subject to the vertical component of the 1940 El Centro earthquake as free field motion. Figure (4) plots the frequency of excitation versus maximum amplitude of the broad tank's radial displacement. In the case of a tall tank, the effect of changing the stiffness of soil was less pronounced. By employing these transfer functions and applying Fourier Transform, the time history of shell radial displacement can be determined. It was observed that soil-tank interaction, for same material damping ratio, reduces the maximum shell radial displacement on the softer soil to about 60% of that on the stiffer soil. It was also noted that the general shape of time history produced by other vertical components remains similar to that obtained using El Centro record.

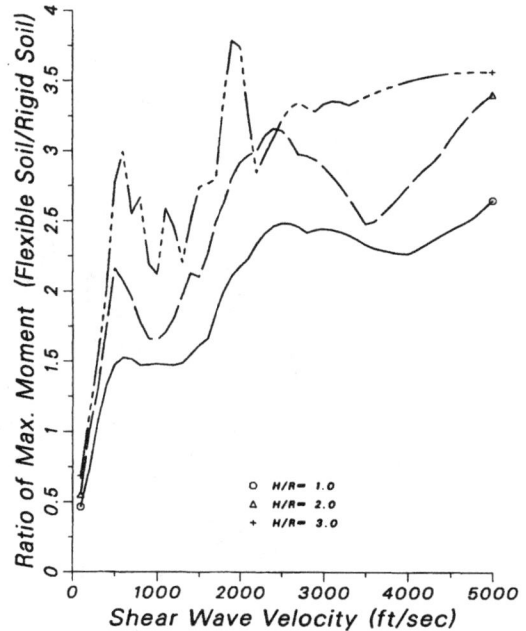

Figure 5: Horizontal excitation.

417

Observed reduction in tank displacements under vertical excitation is in contrast to the amplification [6] of tank response exhibited under horizontal ground motion as demonstrated in Fig. (5).

UNDERGROUND RECTANGULAR TANKS

Underground tanks for storage of liquids have beneficial characteristics such as environmental acceptability and effective use of land area. Yet, only a few analyses are available in the literature to design such tanks. In the present study, Haroun's analytical model [4] for the analysis of above ground, rectangular walls under seismically induced hydrodynamic loads was coupled with models for evaluating lateral earth pressures due to Mononabe-Okabe and Scott [7] to compute the response of the tank-soil system. Linearity of the problem is retained, and thus, laws of superposition remain applicable.

METHOD OF ANALYSIS

System under consideration is a rectangular tank, completely buried, and subjected to simultaneous action of three components of earthquake acceleration. Static and pseudo-dynamic lateral pressures, produced by both soil and liquid, were modeled. A typical wall would be subjected to a system of loads as shown in Fig. (6). Analytical as well as finite element solutions for the deflections and moments were evaluated at discrete points on the wall for each of the loading systems under consideration.

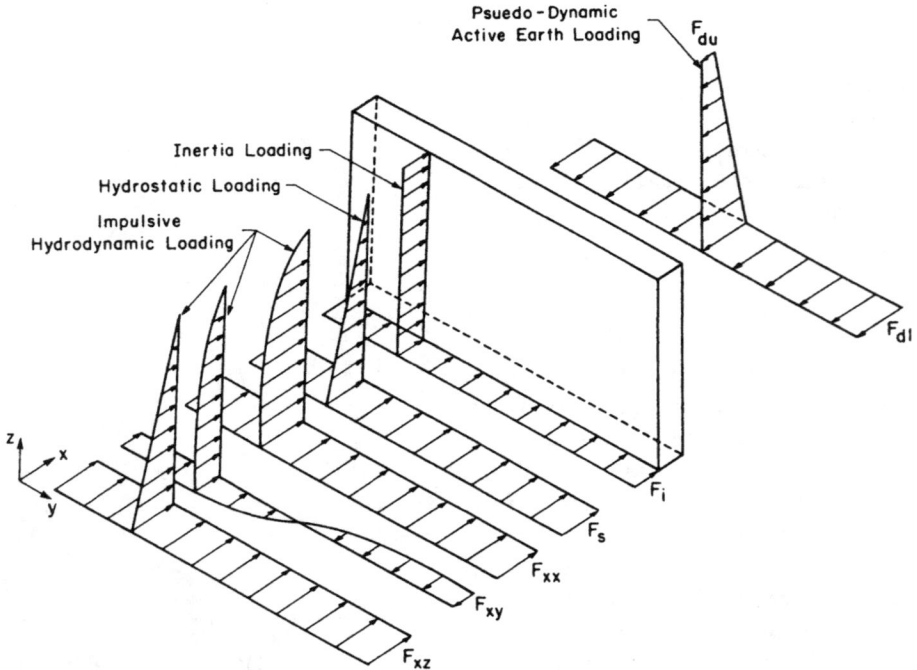

Figure 6: Seismic loads on a typical wall.

Mononobe and Okabe pressure distribution on the wall increases linearly from top to bottom and its value depends on the horizontal and vertical accelerations of an earthquake record. This pressure can also be presented in a convenient way for direct application at the design stage using seismic design coefficients [8]. In an effort to devise a more accurate method to calculate the dynamic earth pressure, Scott presented a model for a soil-retaining wall system in which the soil was considered as one-dimensional shear beam attached to the wall by springs representing soil-wall interaction.

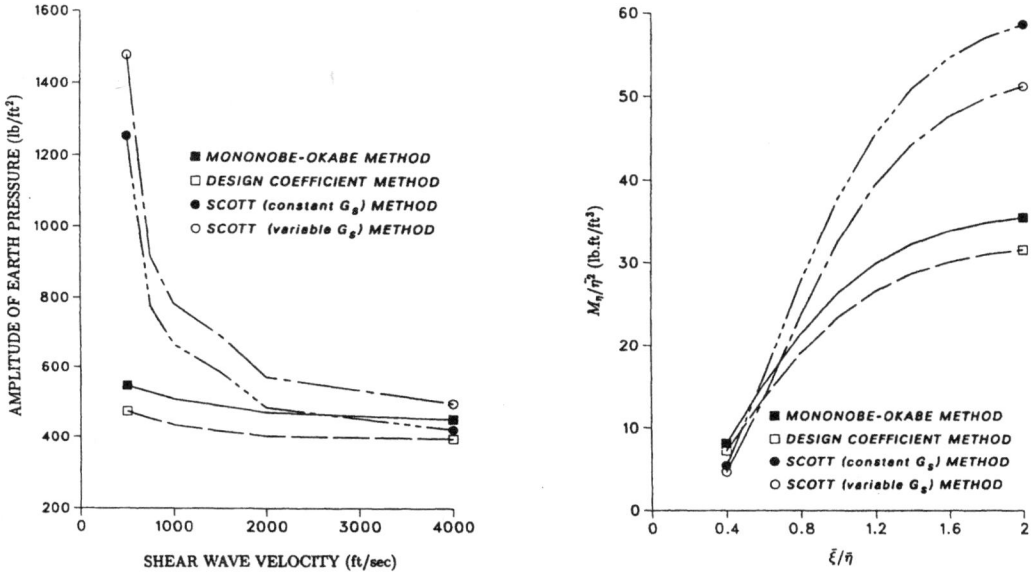

Figure 7: Soil pressure effects on tank wall.

ILLUSTRATIVE NUMERICAL EXAMPLES

A typical wall is divided into a rectangular mesh, and deflections and bending moments in both directions of the wall are computed at each nodal point. Effects of soil stiffness, expressed by its shear wave velocity, on the amplitudes of dynamic earth pressure are illustrated in Fig. (7). It is seen that soft soils produce higher amplitudes of dynamic earth pressure than stiff soils. Furthermore, discrepancy in evaluating the earth pressures is much larger for soft soils. To depict effects of the wall width-to-height ratio on internal moments, relationships between the wall aspect ratio and the bending moment are also plotted in Fig. (7) for each dynamic earth pressure distribution. Since the design coefficient method has similar pressure distribution as that of Mononobe-Okabe method, results of both methods follow the same trend as far as effects on the wall moment are concerned except for moment amplitudes. In general, one must consider three loading combinations for wall design. In an empty tank, acting loads on the wall are the earth pressure plus wall inertia. For a full tank, an extreme situation occurs when surrounding soil is separated from tank wall. Acting forces, all in same direction, are due to inertia loads, hydrostatic pressure,

419

and hydrodynamic pressure. The last case is for a full tank where acting forces, including opposing forces, are considered. It was observed that, in all possible cases of loading, the hydrostatic pressure was dominant, and therefore, the exclusion of the earth pressure results in the most critical loading case.

CONCLUSIONS

Soil-tower interaction reduces, in general, the member-end actions in x-braced elevated tanks except near the base of the tower. It was also confirmed that ground-based tank interaction with the foundation soil under vertical excitations reduces shell radial displacement contrary to the amplification exhibited under horizontal earthquake motions. For underground tanks, it was concluded that hydrodynamic and hydrostatic liquid pressures are generally more critical design load components than the dynamic earth pressure.

REFERENCES

[1] American Petroleum Institute, "Welded Steel Tanks for Oil Storage," API Standard 650, 7th Edition, Washington, D.C., 1980.

[2] American Water Works Association, "AWWA Standard for Welded Steel Tanks for Water Storage," AWWA-D100, Denver, Colorado, 1984.

[3] Ellaithy, H.M., "Dynamic Analysis and Computer Modeling of Elevated Tanks and Three-Dimensional Supporting Towers," Ph.D. Dissertation, University of California, Irvine, 1986.

[4] Haroun, M.A., "Stress Analysis of Rectangular Walls Under Seismically Induced Hydrodynamic Loads," Bulletin of the Seismological Society of America, Vol. 74, 1984, pp. 1031-1041.

[5] Haroun, M.A., and Abdel-Hafiz, E.A., "A Simplified Seismic Analysis of Rigid Base Liquid Storage Tanks Under Vertical Excitation With Soil-Structure Interaction," Journal of Soil Dynamics and Earthquake Engineering, Vol. 5, No. 4, October 1986, pp. 217-225.

[6] Haroun, M.A., and Abou-Izzeddine, W., "Parametric Study of Seismic Soil-Tank Interaction: Horizontal Excitation," submitted to the Journal of Structural Engineering, ASCE, 1990.

[7] Scott, R.F., "Earthquake Induced Earth Pressures on Retaining Walls," Proceedings of the 5th World Conference on Earthquake Engineering, Vol. 3, Rome, 1973.

[8] Temraz, M.K., "Dynamic Soil-Structure Interaction with Application to Liquid Storage Tanks," Ph.D. Dissertation, University of California, Irvine, 1989.

[9] International Conference of Building Officials, Uniform Building Code (UBC), Whittier, California, 1988.

[10] Wolf, J.P., Dynamic Soil-Structure Interaction, Prentice-Hall, New Jersey, 1984.

Earthquake observation of deeply embedded building structure

H. Matsumoto[I], K. Ariizumi[I], K. Yamanouchi[II], H. Kuniyoshi[II],
O. Chiba[III], and M. Watakabe[III]

ABSTRACT

Earthquake observation of deeply embedded building structures in the suburbs of Tokyo has been continued on a large scale for the purpose of investigation of dynamic soil-structure interaction behavior. Many earthquake records have been obtained since June, 1985.

This paper introduces the outline of our earthquake observation system and the results obtained from the investigation of earthquake records observed in and around deeply embedded building structures and of dynamic lateral pressure acting on the basement walls.

(1) The soil model for dynamic soil-structure interaction behavior subjected to earthquake ground motions is estimated by an identified technique which minimize the square sum of the residuals between the observed records and a simulated model. This soil model presented herein, closely simulates the observed records.

(2) It is found that by comparison between the normalized spectra of the dynamic lateral pressure and those of the velocities at approximately the same depth, the both shapes of the spectra closely coincide.

INTRODUCTION

Earthquake observation has been carried out at many building structures to investigate the dynamic soil-structure interaction behavior in Japan(Tsubokura et al. 1983, etc). Up till now, however, there has been little research related to the behavior of dynamic lateral pressures acting on the basement walls of deeply embedded building structures. In this paper, the authors wish to introduce the investigation of the fundamental characteristics of earthquake records observed in and around deeply embedded building structures and of the dynamic lateral pressure acting on the basement walls during earthquakes.

OUTLINE OF EARTHQUAKE OBSERVATION SYSTEM

Outline of the ground , the building and observation system

Ground conditions around the building and cross-section of building are as

---

I   Engineering Research Center, Tokyo Electric Power Company, Tokyo, Japan
II Architecture & Structural Technical Dept.,
    Tokyo Electric Power Services Co., Ltd.,Tokyo,Japan
III Institute of Construction Technology, Toda Corporation,Tokyo,Japan

summarized in Fig. 1(a),(b),respectively. Geological structure consists of soft
alluvial deposits above thirty two meters and underlying diluvial deposits. About
forty two meters below the ground surface, there exists a firm diluvial
deposit,generally called the "Upper Tokyo Formation", which is selected as the bearing
stratum of the building.

Earthquakes were observed with arrays set up inside the building and those set up
vertically in the ground about one meter and about 12 meters from the building,
respectively. Seismometers were installed in the building (4 sets, 12 components),
in the ground about one meter from the basement wall (panel A) (3 sets, 9 components)
, and in the ground about 12 meters from the basement wall (panel D) (4 sets, 12
components). See Fig. 1(b). Dynamic earth and water pressures were observed during
earthquake observation using 20 gauges (20 components) installed on the basement
walls (See Fig. 1(b)).

Observed earthquakes

As of March 1990, many earthquakes have been observed since June 1985 when
earthquake observation began. The locations of the epicenters of these earthquake are
as shown in Fig. 2. For the purposes of this study, 21 earthquakes (Sakai et al.
1989) were chosen from earthquakes with an intensity of 2 or above in Tokyo(Japanese
Meteorological Agency Scale) and a maximum peak acceleration value of $3cm/sec^2$ or
more observed at GL-142m.

ANALYSIS OF OBSERVED RECORDS AND DISCUSSION

Observed records and maximum velocity distribution

The example of observed records of the East Off Chiba Prefecture Earthquake which
occurred on December 17,1987 is as shown in Fig. 3. The records observed in ground
show a tendency that the velocity is hardly amplified between GL-142m and GL-42m, but
it is amplified near the ground surface. Little amplification is observed between
B6F and 1F of the building.

On the basis of observed records, the ratios of the maximum velocity at each
observation point in the soil and in the building to the maximum velocity at GL-142m
were determined. The mean values and standard deviations of the ratios obtained are
as shown in Fig. 4. Both the horizontal and vertical components of velocity at GL-
25.9m in the ground about 12 meters from the building were amplified by as low as 1.5
times on average. By contrast, they were more amplified above this level. Also, near
the ground surface, the horizontal component was amplified by about 2.3 times on
average and the vertical component by about 3.7 times on average. In the building,
although amplification was observed at PH2, the horizontal component at B6F and 1F was
amplified by as low as about 1.3 times and 1.5 times on average, respectively,
indicating the effects on the embedment of the building.

Amplitude characteristics of the soil and the building

Correlation analysis was conducted on combinations of observed values obtained
at two points in Array 1 as shown in Fig. 1, which was vertically set in the soil.
The mean spectral ratios and standard deviations of GL-42m/GL-142m and GL-1.5m/GL-
142m (horizontal component) were obtained as shown in Fig. 5(a) and (b),
respectively, which represent transfer characteristics. Fourier spectra on smoothed
by Parzen window having a bandwidth of 0.2 Hz are employed in this analysis. Fig.

5(a) shows that the primary frequency was predominant at around 0.71 Hz, while no predominance was observed in the secondary or higher mode of oscillation. In Fig. 5(b), predominance was observed at frequencies 0.71 Hz, 1.6 Hz, 2.56 Hz and 3.2 Hz, which indicated that predominant frequencies in higher modes of oscillation having a frequency of 1.60Hz or above were produced in shallow layers at GL-42m or above.

In order to investigate transfer characteristics between the ground and the building, the spectral ratio at each level was calculated using data observed in the ground at levels roughly corresponding to those observed at B6F and 1F. The results of this calculation are as shown in Fig. 6 (a) and (b), respectively. From the comparison between the spectrum at basement (B6F) and that in the soil layer at the depth of 25.9m, it is found that in the frequency range higher than the first natural frequency of the structure, the amplitude in the structure decreases to about half of that in the soil, and is accordance to the frequencies. The spectral ratio of 1F to GL-1.5m showed a similar tendency as estimated above. These phenomena indicated that dynamic soil-structure interaction occurred.

ESTIMATION OF SOIL MODEL BASED ON OBSERVATION RECORDS

The soil profile based on identified technique

A technique to minimize the square sum of the residuals between the observed records and the simulated model is employed, in which the shear wave velocities and the frequency-depentant damping factors are selected in accordance to the parameters, and the density and the layer thickness are set in accordance to the constant values. In the analysis, on the assumption that the soil consists of a multiple-layer horizontal formation, the amplitude characteristics were calculated by the one dimensional wave propagation theory. The soil constants for the initial soil model and the soil model estimated by the identified technique are as shown in Table 1. Example of the transfer characteristics obtained from the identified soil model and observed records are as shown in Fig. 7. As shown in the figure, predominant frequencies and amplitude characteristics of the soil model estimated by the identified technique showed close agreement with the observed ones.

Comparison of the results of observation and simulation

In order to evaluate the validity of the soil model estimated by the identified technique, simulation analysis was conducted. In the simulation analysis, the wave records observed at GL-142m were used as input. The example of comparisons of waveform and response spectrum(with a damping factor of 5%) between the observed and simulated waveforms are as shown in Fig. 8(a) and (b), respectively. In these figures, the simulated wave showed good agreement with the observed one. Although response spectrum values of the simulated wave at about 1.0 second were greater than the observed one, the simulated response spectrum showed good agreement with the observed one. Similar simulations were carried out for the remaining 20 earthquakes, and the ratios of response spectrum values of the simulated waves to those of the observed ones were calculated. The mean values and standard deviations are as shown in Fig. 9. It was confirmed that the average value of the simulations well agreed with the observed ones, despite greater standard deviations for periods of two seconds and more.

FUNDAMENTAL CHARACTERISTICS OF DYNAMIC LATERAL PRESSURE ON BASEMENT WALLS

Distribution of maximum dynamic lateral pressure

The mean values and standard deviations of the maximum dynamic lateral pressures on the basement walls are as shown in Fig. 10, in which the dynamic lateral pressure herein described signifies the fluctuation of earth pressure due to earthquake. The maximum dynamic lateral pressures shown in Fig. 10 are normalized by the maximum velocities at GL-142m. Fig. 11 shows variation coefficients between the maximum dynamic lateral pressures normalized by the maximum velocities and by the maximum accelerations at GL-142m, respectively. The variation coefficients of the maximum dynamic lateral pressures normalized by the maximum velocities are lower than those normalized by the maximum accelerations, and the variations in the vertical direction, smaller.

Time history and frequency characteristics of dynamic lateral pressure

The time history and frequency characteristics of dynamic lateral pressure which acts on the basement walls were investigated. The examples of dynamic lateral pressure at GL-3.2m on the panel A side and GL-3.7m on the panel D side, which were observed during the East Off Chiba Prefecture earthquake on December 17,1987, are as shown in Fig. 12(a). The time history of dynamic lateral pressure is as shown in Fig. 12(b), in which the time history was passed through a band-pass filter (0.7 Hz-1.7 Hz) having a frequency bandwidth close to the predominant frequency of the Fourier spectra of dynamic lateral pressures as shown in Fig. 13. The time history obtained above (20-40 second in Fig. 12(a)) and an example of particle orbits are as shown in Fig. 12(b). Some phase difference was observed in the changes in dynamic lateral pressure acting on panels A and D.

Fourier spectra was normalized by the maximum values. These spectra consist of dynamic lateral pressures acting on panels A and D at the ground surface level and the foundation level, and of observed velocities at the corresponding levels of the soil. Their mean values and standard deviations are as shown in Fig. 13(a), (b) and (c),respectively. Spectra determined from observed dynamic lateral pressures showed good agreement with spectra of velocities observed at almost the same levels. This result indicates that observed dynamic lateral pressures and velocities were interrelated, though qualitatively.

CONCLUSIONS

Analysis on the fundamental characteristics of earthquake records observed in and around deeply embedded building structures was carried out and the following results have been obtained.

(1) The amplification factor of the ground against GL-142m was as low as 1.5 or so on average (both horizontal and vertical) at GL-25.9m, while that at the ground surface level was about 2.3 (horizontal) and 3.7 (vertical) on average. The amplification factor of the building at the 6th basement level was about 1.3 (horizontal) on average, and that at the 1st floor level, about 1.5 on average. These results indicate the effects on the embedment of the building.

(2) Spectrum ratios between the soil and the building at the levels of (1F/GL-1.5m) and (B6F/ GL-25.9m) were examined. From comparison between the spectrum at basement (B6F) and that in the soil layer at the depth of 25.9m, it is found that in a frequency range higher than the first natural frequency of the structure,the amplitude in the structure decreases to about half of that in the soil, and is in accordance to the frequencies. The spectral ratio of 1F to GL-

1.5m showed a similar tendency as estimated above. These phenomena indicated that dynamic soil-structure interaction occurred.

(3) It was confirmed that the soil model estimated by an identified technique assuming a frequency-dependent damping factor closely simulates the observed record. Therefore, it was confirmed that the soil model could be applied for further investigation of dynamic soil-structure interaction behavior.

(4) The maximum dynamic lateral pressure requires a smaller variation coefficient and a smaller vertical variation when normalized by the maximum velocity than by the maximum acceleration. Examination of the time history and frequency characteristics of dynamic lateral pressure acting on basement walls revealed that there was some phase difference in dynamic lateral pressure acting on panels A and D at the ground surface level. Spectra determined from observed dynamic lateral pressures showed good agreement with spectra of velocities obsbserved at almost the same levels. This result indicates that observed dynamic lateral pressures and velocities were interrelated, though qualitatively.

REFERENCES

Sakai,Y.,et al.1989. Earthquake Observation of Deeply Embedded Building(in Japanese). Annual Meeting of A.I.J.,267-268.
Tsubokura,H.,et al.1983. Observation and Analysis of Earthquake Motions to Deeply Embedded Underground Structure(in Japanese). Annual Meeting of A.I.J.,779-780.

Table 1 Data of soil profile

Observation points(m)	Thickness (m)	Density (g/cm³)	Initial model		Identified model		
			V s (m/s)	h(%)	V s (m/s)	h_o* (%)	α
GL-1.5 ●	1.5	1.80	105.0	5.0	105.0	9.1	0.48
	2.3	1.80	180.0	5.0	147.0	9.1	0.48
	1.5	1.80	125.0	5.0	126.0	9.1	0.48
	7.6	1.58	100.0	5.0	131.0	9.1	0.48
	8.1	1.64	170.0	5.0	195.0	9.1	0.48
GL-25.9 ●	4.9	1.68	225.0	5.0	239.0	9.1	0.48
	6.2	1.69	262.0	5.0	230.0	9.1	0.48
GL-42.0 ●	9.9	1.80	490.0	5.0	410.0	10.3	0.66
GL-142.0●	100.0	1.97	420.0	5.0	431.0	10.3	0.66
	───	2.00	500.0	5.0	500.0	10.3	0.66

● Observation points   *:$h(f)=h_o \cdot f^{-\alpha}$

where, α :Constant expressing the degree of dependence on frequency
h_0 :Damping factor at 1 Hz , f :Frequency (Hz)

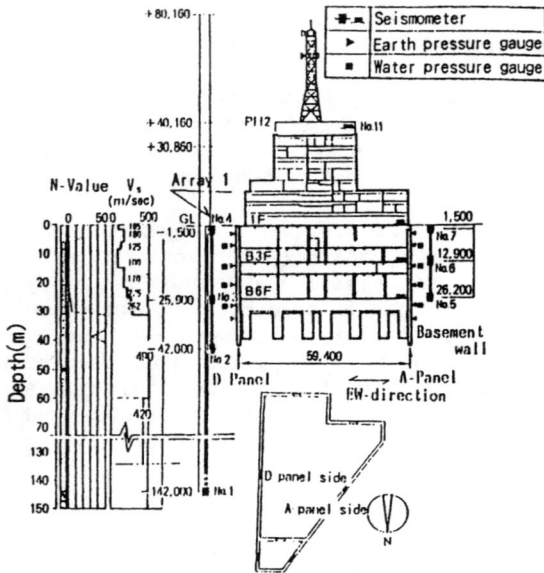

(a) Soil profile (b) Location of observation
                             points
Fig.1 Outline of soil profile and building
     structure

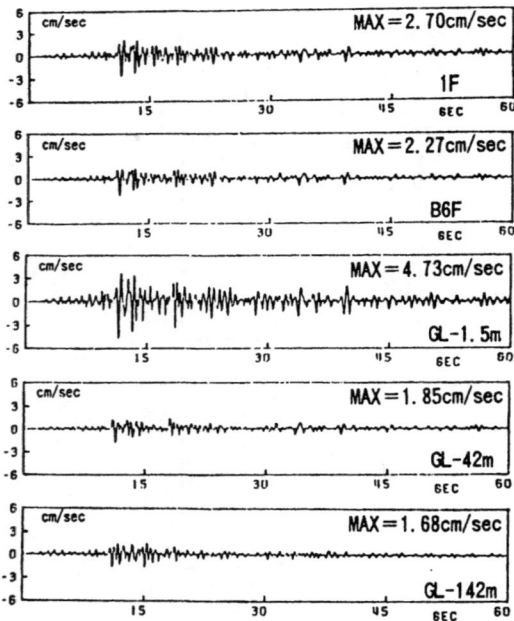

Fig.2 Location of epicenter and magnitude for
     observed earthquakes

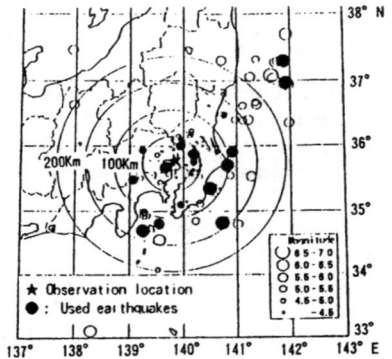

Fig.3 Time histories observed during 1987 E-Off
     Chiba Prefecture Earthquake (NS component)

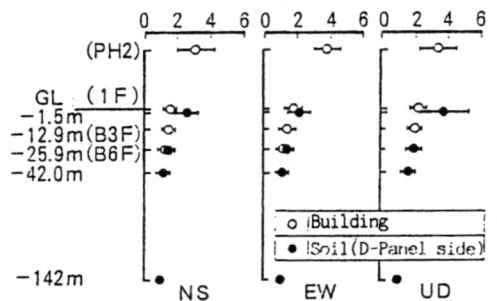

Fig.4 Distributions of mean values and standard
     deviations of maximum velocity(Normalized
     by the maximum velocity at GL-142m)

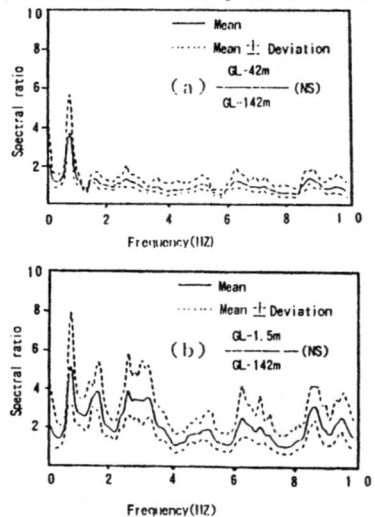

Fig.5 Averaged spectral ratio of soil layer
     (NS-direction)

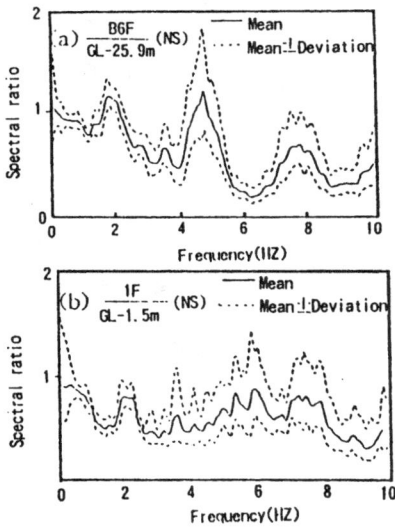

Fig.6 Averaged spectral ratio obtained from observed records in the building and the surrounding soil

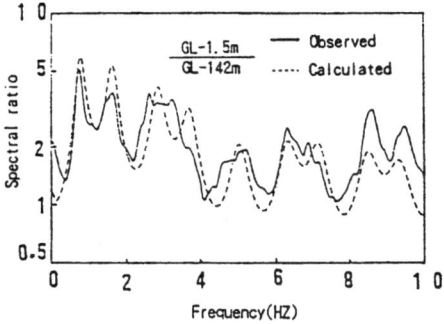

Fig.8 1985 Southern Ibaraki Prefecture Earthquake
(M=6.2,Focal depth=76km,Epicental distance=38km)

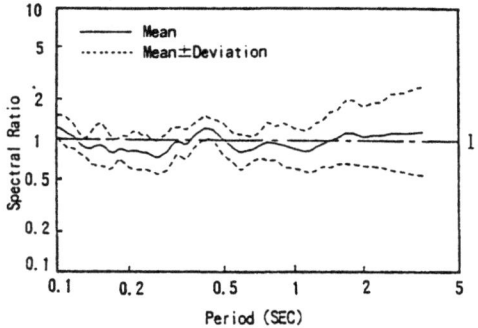

(a) Simulated wave and observed wave at GL-1.5m
(b) Velocity response spectra of simulated wave and observed record at GL-1.5m

Fig.7 Comparison of observed transfer function and theoretical one

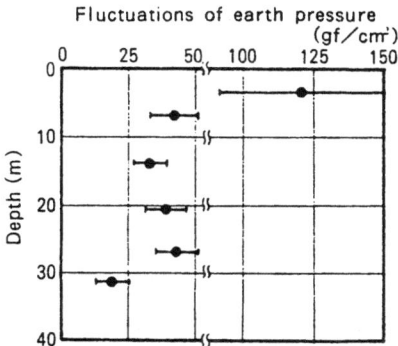

Fig.9 Averaged velocity response spectral ratio at GL-1.5m (simulated/observed)

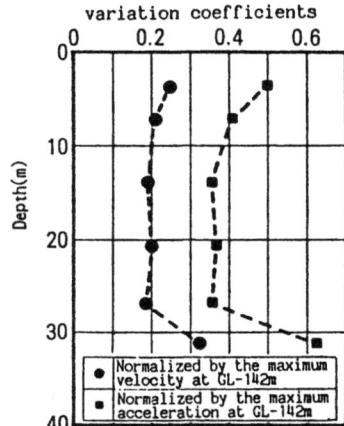

Fig.10 Maximum dynamic lateral pressure distribution(mean and deviation)

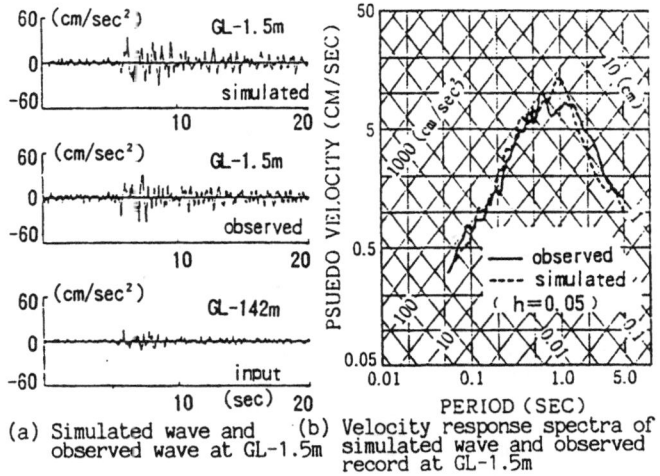

Fig.11 Distribution of variation coefficients

427

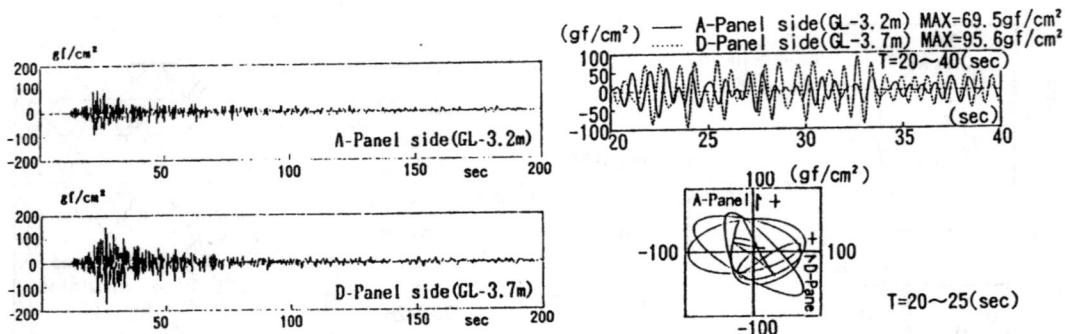

(a) Dynamic lateral pressure records of approximately the same observed points

(b) Records of major tremor and orbit (band pass filter, f:0.7Hz ~1.7Hz)

Fig.12 Dynamic lateral pressure records and its orbit during 1987 E-Off Chiba Prefecture Earthquake (M=6.7, Focal depth=58km, Epicental distance=75km)

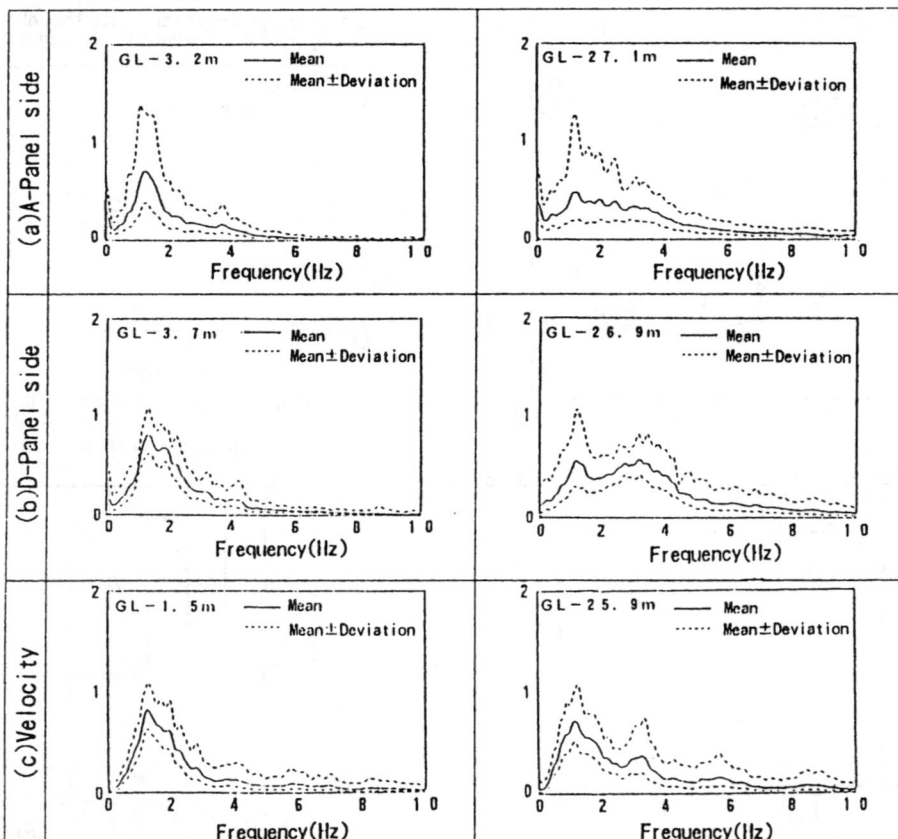

Fig.13 Comparison between normalized spectra obtained from dynamic lateral pressure records and that obtained from velocity records

Seismic liquefaction and flow deformations

W.D. Liam Finn[I], R.H. Ledbetter[II], and M. Yogendrakumar[III]

ABSTRACT

This paper presents a nonlinear finite element method for calculating the displacements in earth structures caused by seismic liquefaction. The method is used to evaluate post-liquefaction behaviour of Sardis Dam.

INTRODUCTION

One of the most challenging problems of seismic liquefaction is calculating the seismic response of earth structures which include zones of potentially liquefiable soils.

In current practice the post-liquefaction behaviour of an earth structure is assessed by means of static equilibrium analysis of the undeformed structure using the steady state or residual strength as the operating strength in the liquefied zones.

An alternative approach, more in keeping with the concept of designing dams for acceptable deformations proposed by Newmark (1965), is to evaluate the safety of the dam and the extent of necessary remedial measures on the basis of a tolerable amount of deformation for the low probability event specified by the design earthquake. The potential post-liquefaction deformations may be estimated using the computer program, TARA-3FL (Finn and Yogendrakumar, 1989), which is a specialized derivative of the general program TARA-3 by Finn et al. (1986).

STEADY STATE STRENGTH

The undrained steady state strength, S_{us}, is the crucial element controlling post-liquefaction deformations. There are two methods for

---

[I]Professor, Civil Engineering Dept., University of British Columbia, Vancouver, B.C., Canada
[II]Research Associate, Royal Military College, Kingston, Ontario, Canada
[III]U.S. Army Corps of Engineers, Waterways Experiment Station, Vicksburg, MS, U.S.A.

obtaining S_{us}; laboratory tests on good quality samples from the field (Poulos et al., 1985), and from correlations between normalized standard penetration resistance $(N_1)_{60}$ and S_{us} determined by analysis of liquefaction failures in the field (Seed, 1987; De Alba et al., 1987). Great uncertainties are associated with estimates in S_{us} by either method. The causes of uncertainty associated with these approaches have been discussed critically by Seed et al. (1988) and Finn (1989). Marcuson et al. (1990) also reviewed the problems associated with the above methods and concluded "the empirical correlation relating penetration resistance to field performance is adequate and may even be preferable to the testing of undisturbed samples". The most recent version of the Seed correlation is shown in Fig. 1.

STRUCTURE OF THE PROGRAM TARA-3FL

The basic theory of the finite element program TARA-3 has been reported by Finn (1985,1990) and Finn et al. (1989). So only procedures specific to TARA-3FL will be described here.

In a particular element in the soil structure, the shear stress-shear strain state which reflects pre-earthquake conditions is specified by a point P_0 on the stress-strain curve as shown in Fig. 2. When liquefaction is triggered, the strength will drop to the steady-state value. The post-liquefaction stress-strain curve cannot now sustain the pre-earthquake stress-strain condition and the unbalanced shear stresses are redistributed throughout the dam. In the liquefied elements, the stresses are adjusted according to the following equation,

$$d\tau = \frac{\partial f}{\partial \sigma'_m} d\sigma'_m + \frac{\partial f}{\partial \gamma} d\gamma \qquad (1)$$

where $\tau = f(\sigma'_m, \gamma)$. This process leads to progressive deformation of the dam until equilibrium is reached at the state represented by P_2.

Since the deformations may become large, it is necessary to update progressively the finite element mesh. Each calculation of incremental deformation is based on the current shape of the dam, not the initial shape as in conventional finite element analysis.

DEFORMATION ANALYSIS OF SARDIS DAM, MISSISSIPPI

Preliminary deformation studies of Sardis Dam have been conducted using TARA-3FL for various assumptions about the magnitudes and distributions of the residual strengths in the liquefied zones. Evaluations of dam performance during seismic shaking and potential remedial measures are still underway, so final results cannot be presented here. Instead some results are presented to show the type of information provided to the design engineer by deformation analyses.

The general configuration of the Sardis Dam is shown in Fig. 3. During the design earthquake, liquefaction is predicted to occur in the core and in a thin seam of clayey silt or silty clay in the top stratum clay in the foundation. The thin layer may be seen clearly in Fig. 4. The liquefaction potential of the sands and silts was evaluated using Seed's liquefaction assessment chart (Seed et al., 1985) and the Chinese criteria for soils with plastic fines by Wang (1979).

The residual strength (steady state strength) in the core was assumed to be 5 kPa (100 psf) based on Seed's correlation between corrected standard penetration resistance $(N_1)_{60}$ and residual strength shown in Fig. 1 (Seed and Harder, 1990). From a variety of studies, the residual strength in the thin layer in the foundation was assumed to be $S_r = 0.075 \, \sigma'_{vo}$ (Woodward Clyde Consultants, 1989). The original strength of the thin layer was taken as 100 kPa (2000 psf).

The large differences between the initial and post-liquefaction strengths in Sardis Dam resulted in major load shedding from liquefied elements. This put heavy demands on the ability of the program to track accurately what was happening and on the stability of the algorithms. Therefore it was imperative to have an independent check that the computed final deformed positions were indeed equilibrium positions. The most direct check is to run a conventional stability analysis on the deformed position. If the major deformations occur during the earthquake, the resulting factor of safety should be greater than unity because some of the deformation field is driven by the inertia forces. If the major deformations occur relatively slowly after the earthquake, the factor of safety should be close to unity.

In one analysis of Sardis Dam, the residual strength values specified above were used, but assuming a minimum value of 17.5 kPa (350 psf) in the thin liquefied layer in the foundation. The initial and final deformed shapes of the dam for this case are shown in Fig. 4. Very substantial vertical and horizontal deformations may be noted, together with intense shear straining in the weak thin layer. The static stability of the deformed shape was analyzed using the program UTEXAS2 (USACE, 1989). This program uses Spencer's method (1973) which satisfies both moment and force equilibrium. The factor of safety was found to be close to 1.0. It is also interesting to note that the critical slip surface for UTEXAS2 analysis exited the slope near the location suggested by the finite element analysis using TARA-3FL.

The reliability of TARA-3FL in predicting stable deformed shapes was tested by parametric studies with various assumptions about the steady state strength in the thin liquefiable layer. The factors of safety of the undeformed and deformed dam cross-sections were determined using UTEXAS2. The factors of safety for the undeformed dam are given by the solid sloping line in Fig. 5. The points give the factors of safety of the post-liquefaction deformed sections. The steady state strengths in Fig. 5 represent either a constant value for the layer or an imposed minimum value when $S_{us} = 0.075 \, \sigma'_{vo}$. In the clearly unstable region defined by a factor of safety less than one for the undeformed section, the computed factors of safety for the deformed sections were in the range of 1 ± 0.05. This is the theoretical error band associated with UTEXAS2. Many results of this type, for different assumptions about the residual strengths, suggest that the TARA-3FL analysis does indeed achieve equilibrium positions even for large drops in strength due to liquefaction.

Studies were made of the sensitivity of displacements to various levels of residual strength in the thin layer. The variations in the vertical displacements at the upstream edge of the crest (curve 1) and in the horizontal deformations at the midpoint of the upstream slope (curve 2) are shown for various levels of constant residual strength in the thin liquefied layer in the foundation in Fig. 6. The increase in displacement is gradual with decrease in residual strength until the strength drops to about 20 kPa (400 psf) when the displacements begin to increase very rapidly. The variation in vertical displacements (curve 3) is also shown in Fig. 6, for residual strengths $S_r = 0.075 \; \sigma'_{vo}$. For variable minimum residual strengths, the displacements increase rapidly when the minimum strength is about 15 kPa (300 psf). This information is critically important when choosing an appropriate residual strength for design purposes from the usually very scattered residual strength data derived from laboratory tests or from field penetration test data.

It is also possible to determine the loss in freeboard associated with various factors of safety based on the original configuration of the dam for various residual strengths. The variation of vertical crest displacement (loss of freeboard) with factors of safety of the undeformed dam are shown for various values of residual strength in Fig. 7. For the first time a designer has available the deformation fields associated with different factors of safety for a particular dam. This information is helpful in interpreting the implications of a given factor of safety.

CONCLUSIONS

Deformation analysis of earth structures with liquefied zones using TARA-3FL is a very useful complement to conventional slope stability analysis for studying the effects of liquefaction and designing remedial measures. It provides the designer with additional useful information which assists him in exercising his judgement effectively. It is particularly useful when only partial remedial measures are employed and liquefaction is still permitted to occur in other sections of the structure. In these cases not only the global stability guaranteed by conventional stability analysis is important but also the localized displacements that occur in the liquefied sections.

TARA-3FL analysis also permits magnitudes of crucial deformations such as crest subsidence to be associated with factors of safety of the undeformed structure. This capability provides an essential link between current design practice and the proposed new procedure based on tolerable deformations.

ACKNOWLEDGEMENTS

Initial development of TARA-3FL was supported by Sato-Kogyo Co., Tokyo, Japan. Later developments were funded by the National Science and

Engineering Council of Canada under Grant No. 81948. Permission by Chief of Engineers, U.S. Army Corps of Engineers to present data from the Sardis Dam study is gratefully acknowledged.

REFERENCES

De Alba, P., H.B. Seed, E. Retamal, and R.B. Seed (1987), "Residual Strength of Sand from Dam Failures in the Chilean Earthquake of March 3, 1985," Earthquake Engineering Research Center, Report No. UCB/EERC-87-11, University of California, Berkeley, September.

Finn, W.D. Liam. 1985. "Dynamic Effective Stress Response of Soil Structures; Theory and Centrifugal Model Studies," Proc. 5th Int. Conf. on Num. Methods in Geomechanics, Nagoya, Japan, Vol. 1, 35-46.

Finn, W.D. Liam 1989. "Analysis of Post-Liquefaction Deformations in Soil Structures", Proceedings, H. Bolton Seed Memorial Symposium, Editor J. Michael Duncan, Bi-Tech Publishers, Vancouver, Vol. 2, May, pp. 291-311.

Finn, W.D. Liam 1990. "Seismic Analysis of Embankment Dams," Dam Engineering, Vol. 1, Issue 1, pp. 59-75.

Finn, W.D. Liam, Yogendrakumar, M., Yoshida, N. and Yoshida, H. 1986. "TARA-3: A Program for Nonlinear Static and Dynamic Effective Stress Analysis," Soil Dynamics Group, University of British Columbia, Vancouver, B.C.

Finn, W.D. Liam and Yogendrakumar, M. 1989. "TARA-3FL - Program for Analysis of Liquefaction Induced Flow Deformations," Dept. of Civil Engineering, University of British Columbia, Vancouver, B.C., Canada.

Finn, W.D. Liam, Yogendrakumar, M., Lo, R.C. and Ledbetter, R.H. 1989. "Seismic Response of Tailings Dams," State of the Art Paper, Proc., Int. Symposium on Safety and Rehabilitation of Tailings Dams, International Commission on Large Dams, Sydney, Australia, May, pp. 7-33.

Marcuson W.F. III, M.E. Hynes and A.G. Franklin (1990), "Earthquake Spectra," Vol. 6, No. 3, August, pp. 529-572.

Newmark, N.M. 1965. "Effects of Earthquakes on Dams and Embankments," 5th Rankine Lecture, Geotechnique, Vol. 15, No. 2, June, pp. 139-160.

Poulos, S.J., G. Castro and J.W. France (1985), "Liquefaction Evaluation Procedures," Journal of the Geotechnical Engineering Div., ASCE, Vol. 111, No. 6, June, pp. 772-792.

Seed, H.B. 1987. "Design Problems in Soil Liquefaction," Journal of Geotechnical Engineering, ASCE, Vol. 113, No. 7, August, pp. 827-845.

Seed, H. Bolton, Tokimatsu, K., Harder, L.F. and Chung, R.M. 1985.
"Influence of SPT Procedures in Soil Liquefaction Resistance
Evaluations", Journal of the Geotechnical Eng. Div., ASCE, Vol. 3, No.
12, December.

Seed, H.B., R.B. Seed, L.F. Harder and H.-L. Jong (1988), "Re-evaluation
of the Slide in the Lower San Fernando Dam in the Earthquake of
February 9, 1971," Report No. UCB/EERC-88/04, University of
California, Berkeley, April.

Seed, R.B. and Harder Jr., L.F. 1990. "SPT-Based Analysis of Cyclic Pore
Pressure Generation and Undrained Residual Strength," Proceedings, H.
Bolton Seed Memorial Symposium, J. Michael Duncan ed. , Vol. 2, May,
pp. 351-376.

Spencer, E. 1973. "Thrust Line Criterion in Embankment Stability
Analysis," Geotechnique, Vol. 23, No. 1, pp. 85-167.

USACE 1989. "User's Guide: UTEXAS2 Slope Stability Package, Vol. II.
Theory by Task Group on Slope Stability," Instruction Report GL-87-1,
Final Report, U.S. Army Corps of Engineers, Washington, D.C.

Wang, W. 1979. "Some Findings in Soil Liquefaction," Water Conservancy
and Hydroelectric Power Scientific Research Institute, Beijing, China,
August.

Woodward Clyde Consultants. 1989. Private Communication.

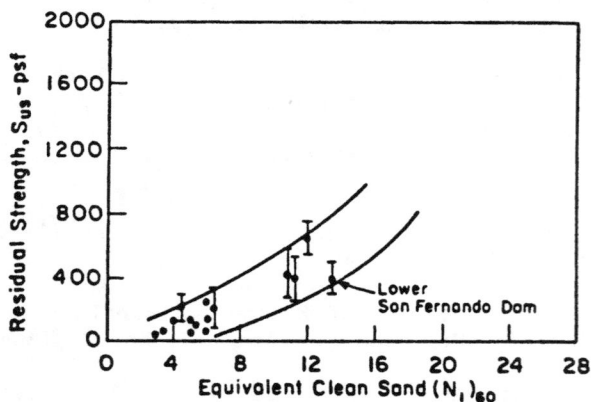

Figure 1.   Tentative Relationship Between Residual Strength and
Standardized SPT N Values for Sands (Seed et al., 1988).

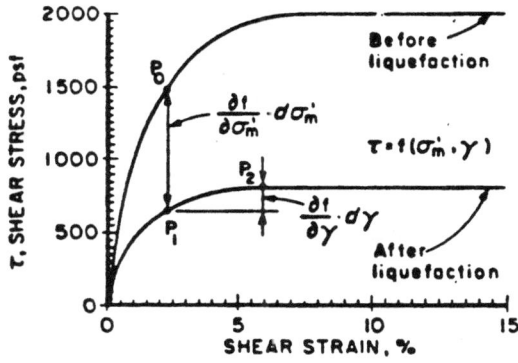

Figure 2.   Adjusting Stress-Strain State to Post-Liquefaction
Conditions.

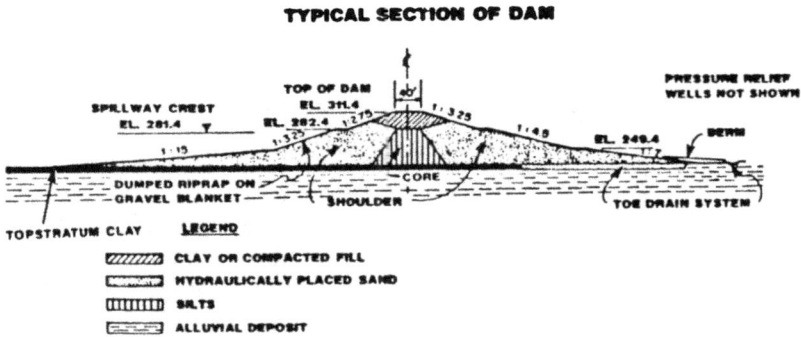

Figure 3.   Typical Section of Sardis Dam.

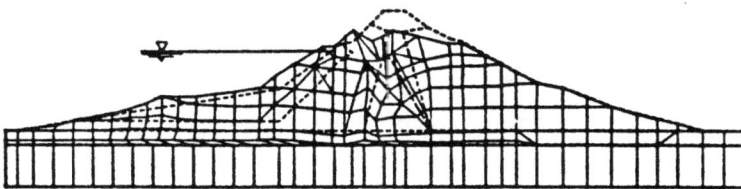

Figure 4.   Initial and Post-Liquefaction Configurations of Sardis Dam.

435

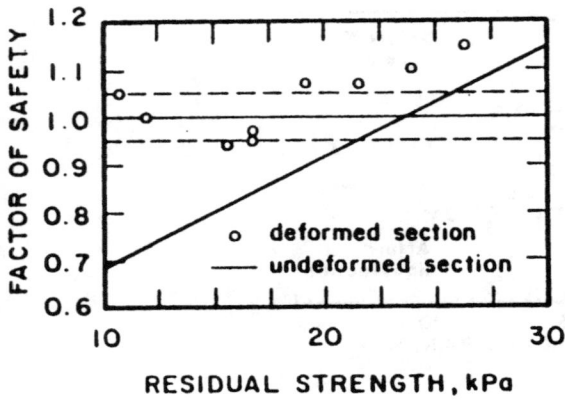

Figure 5. Variations in the Factor of Safety with Residual Strength.

Figure 6. Variation in Displacement with Residual Strength

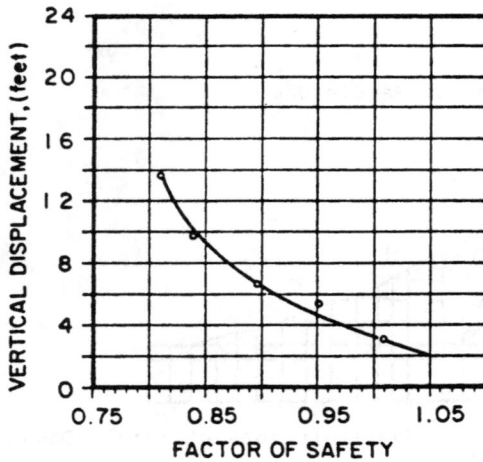

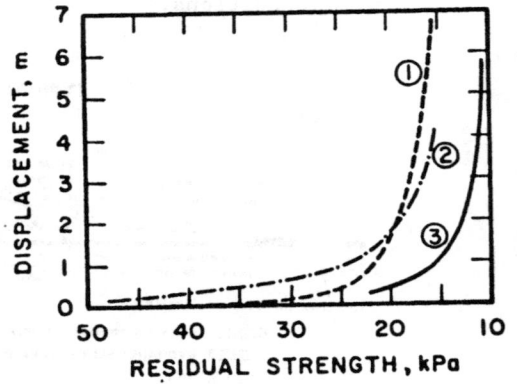

Figure 7. Variation of Vertical Displacement with Factor of Safety.

436

Seismic response, liquefaction, and resulting earthquake induced displacements in the Fraser Delta

Peter M. Byrne[I], Li Yan[II], and M. Lee[III]

ABSTRACT

Seismic response analyses carried out in the Fraser Delta indicate higher amplifications of accelerations than have previously been considered. The analyses are based on recent information on soil properties and layer thicknesses and the results are in accord with amplifications recorded in Mexico City, 1985, and San Francisco, 1989. The computed cyclic stresses for the NBCC design earthquake would trigger widespread liquefaction in the delta and result in large vertical and horizontal displacements that would likely cause severe damage to lifeline facilities.

INTRODUCTION

The coastal area of mainland Southern British Columbia lies within a highly active seismic region. Seven earthquakes in the magnitude range M5-M7 have occurred in the recorded past 100 years. Geological evidence suggests that very large subduction earthquakes of the order M9 have occurred in the past. The recurrence period of these earthquakes is thought to be about every 700 years on average, with the most recent such event having occurred about 300 years ago.

The Fraser Delta lies within this region and is particularly prone to damage in the event of a major earthquake. This is because it is underlain by deep deposits of relatively loose or soft soils. The presence of such soils can: amplify the intensity of shaking; lengthen the predominant period of the motion, and cause strength loss or liquefaction of saturated sandy soils.

Experience at Mexico City during the 1985 earthquake showed that a major cause of damage was the very high amplification of acceleration that occurred as the motion propagated upward through the soft clay lakebed deposits. A similar amplification occurred in the San Francisco bay muds and caused much of the damage in San Francisco and Oakland during the 1989 Loma Prieta earthquake (Idriss, 1990). In addition, liquefaction of loose sand fill placed on top of the Bay mud greatly added to the damage where it was present.

In much of the Fraser Delta, natural deposits of loose to medium dense sands overlie deep silt and clayey deposits, so that the combined effects of both amplification and liquefaction are a possibility in the event of a major earthquake, and should be considered in design. The major design considerations from the foundation point of view are the depth of liquefaction and the resulting displacements. These aspects were discussed in

---

[I]Professor, Department of Civil Engineering, University of British Columbia, Vancouver, B.C.
[II]Klohn Leonoff Consultants, 10200 Shellbridge Way, Richmond, B.C.
[III]B.C. Hydro & Power Authority, 970 Burrard Street, Vancouver, B.C.

detail by Byrne and Anderson (1987). However, the larger than expected amplifications in San Francisco, and recent more detailed examinations of liquefaction induced displacements by Hamada et al. (1987), and Youd & Bartlett (1988), warrant an additional study of this region.

In addition, more data are now available on sediment thickness as well as moduli and damping properties. Because the depth of liquefaction is very important in estimating earthquake induced displacements, effective as well as total stresss analyses will be carried out. Previous analyses carried out by Byrne (1978) indicated that significantly smaller depths of liquefaction were predicted from effective stress in comparison to total stress analyses.

SOIL CONDITIONS AND PROPERTIES

The Fraser Delta has a plan area of about 350 Km^2 as show in Fig. 1. The area has a mix of commercial industrial and residential development with a population of about 1/4 million people. The geology of the Delta is discussed in some detail by Wallis (1979). Basically the area is underlain by: (1) a surficial deposit comprised of a thin discontinuous veneer of clays, silts and peats up to 8 m in thickness, underlain by; (2) a sand and silty sand stratum generally 20 to 45 m in thickness, underlain by; (3) a silt-clay stratum in the North and more granular material in the South with a thickness generally in the range 100 to 300 m (Britton, 1990), underlain by; (4) a glacial till stratum with a thickness in the range 90 to 600 m (Britton, 1990), underlain by; (5) bedrock.

The water table is generally within a metre of the surface. A typical section in central Richmond is shown in Fig. 2.

Dynamic and liquefaction analyses require the following soils information: (1) G_{max} vs. Depth; (2) modulus reduction with level of shear strain; (3) damping as a function of shear strain; and (4) liquefaction resistance parameters

The maximum shear modulus versus depth relationship used in the analysis is shown in Fig. 2 and is based on Byrne and Anderson (1987) together with more recent in situ shear wave measurements reported by Finn et al. (1988) and Hunter (1990). Modulus reduction and damping values used were based on laboratory tests on similar materials as reported by Seed and Idriss (1970), Seed et al. (1986), Sun et al. (1988), and Idriss (1990). Recent test data on silt and clay material from the Fraser Delta reported by Zavoral (1990) were also examined and found to lie within the range of values considered. Liquefaction resistance of the sands is based on in situ penetration resistance values, both standard penetration and cone values (SPT and CPT). Average and lower bound values of the normalized standard penetration of the sands, $(N_1)_{60}$, are shown in Fig. 3 and are based on Byrne and Anderson (1987). The pore pressure rise parameters for the effective stress analysis are based on $(N_1)_{60}$ values as described by Byrne (1991).

ANALYSIS PROCEDURE

Total stress dynamic analyses were carried out using the computer code SHAKE. Effective stress dynamic analyses were carried out using the computer code 1D-LIQ (Byrne and Yan, 1990) based on the procedure outlined by Finn, Byrne and Martin (1976). The soil section analyzed and the soil properties used were as outlined in the previous section. Two different time histories of base acceleration were used. These were the CALTEC and GRIFFITH PARK records generated during the 1971 San Fernando earthquake. They were scaled to a peak base acceleration of 0.2 g, which corresponds approximately with the NBCC 1990 code value of 0.21 g for a probability of 10% in 50 years in Vancouver. They are both rock records, and from previous studies, have been found to be generally more severe than other rock records scaled to the same peak acceleration value.

438

The predicted amplifications of acceleration from base to surface are shown in Fig. 4. Also shown on this figure are results of recent analysis carried out by B. C. Hydro in connection with seismic assessment of transmission towers in the eastern Fraser Delta. The layer thicknesses varied widely in the Hydro analyses and the depths to firm ground were generally less than considered in Fig. 2. Also, significant peat layers were present in some of the Hydro profiles. The high surface accelerations computed by HYDRO were associated with a hard layer at a depth of 35 m. The low surface accelerations were associated with a weak surface layer.

The results of both analyses indicate that base accelerations of 0.2 g could generally be amplified to 0.3 g in the Fraser Delta. This is consistent with observations in San Francisco and Mexico City, and in agreement with the median recommendation proposed by Idriss (1990) also shown in Fig. 4.

The thickness of the till could range between 90 and 600 m. Additional analyses were carried out and indicated that the predicted surface accelerations were not sensitive to the till thickness. This is to be expected as the till under these high confining stresses would act much like a rock.

The thickness of the silt-clay layer could range between 100 and 300 m. Analyses carried out with a silt-clay layer of 270 m rather than 120 m indicated essentially no change in the predicted peak surface acceleration.

Analyses were also carried out to represent the condition of a sandy rather than a silt-clay layer in the depth range 30 to 150 m. This condition occurs in the southern portion of the Delta and resulted in essentially no amplification.

The predicted peak surface acceleration is approximately 0.30 g and represents an amplification factor of 1.5 for a base rock motion of 0.2 g. In a previous study Byrne and Anderson (1987) computed an amplification of 1.05. The amplification is essentially independent of the thicknesses of the till and silt-clay layers for the thicknesses likely to be encountered. Lower amplifications are predicted in the southern portion of the Delta where sands rather than silt-clays are present at depth.

The accelerations cause cyclic stresses in the ground, and it is these that may cause the soil to liquefy. Because the computed accelerations are higher than previously calculated by Byrne and Anderson (1987), the cyclic stresses will also be higher. The computed equivalent cyclic stress ratios (CSR) as a function of depth for the top 30 m are shown in Fig. 5. CSR is defined as 0.65 times the ratio of the computed peak shear stress to the effective overburden pressure. The values shown are for the soil conditions in Fig. 2 and vary significantly depending on the earthquake record and the choice of modulus reduction and damping values used. The higher CSR values correspond with surface accelerations that are in agreement with amplifications recorded in San Francisco. For this reason CSR's indicated by the dashed line shown on Fig. 5 are considered appropriate for design.

Liquefaction is triggered when the cyclic stresses from the design earthquake exceed the cyclic resistance of the soil (CSR > CRR). The CRR values were obtained from the $(N_1)_{60}$ values of Fig. 3 and the Seed et al. (1984) liquefaction chart, and are shown in Fig. 6. A correction value of 1.08 was included to modify the CRR for an M7 event rather than an M7.5 event. An M7 event is thought to be appropriate for a probability of 10% in 50 years. The design CSR from Fig. 5 is also shown on Fig. 6 and the results indicate that for the lower bound $(N_1)_{60}$ condition, liquefaction to the full 30 m depth is predicted to occur. For the mean $(N_1)_{60}$, liquefaction to a depth of 16 m is predicted.

Effective stress dynamic analysis were also carried out using the computer code 1D-LIQ. The pore pressuremeter parameters for the model were derived from $(N_1)_{60}$ values such that under constant amplitude cyclic load conditions, liquefaction would occur in 15 cycles in agreement with the Seed et al. (1984) liquefaction chart . The detailed procedure for doing this is outlined by Byrne (1991). The results are shown in Fig. 7a in terms of the excess pore pressure rise as a function of depth. It may be seen that in the depth range 3 to 23 m the pore pressure rises to equal the initial effective stress σ'_{vo} indicating liquefaction in this region. Below 23 m, while excess pore pressures are high, liquefaction is not predicted to occur.

The total stress 1D-LIQ analysis for the same lower bound $(N_1)_{60}$ condition is shown in Fig. 7b for comparison. It may be seen that the CSR exceeds the CRR for the complete depth of sand, indicating liquefaction to the full 30 m depth of the layer. Thus the total stress analysis is seen to give a conservative estimate of the depth of liquefaction, perhaps too conservative.

The reason for the reduced depth of liquefaction in the case of the effective stress analysis is that as pore pressure rise and liquefaction occurs in the layer most susceptible to liquefaction, this layer loses its ability to transfer shear load, and subsequent dynamic stresses are significantly reduced in all layers.

LIQUEFACTION INDUCED DISPLACEMENTS

Triggering of liquefaction may result in large horizontal and vertical displacements, and these can be estimated from the effective stress analysis carried out here. The computed maximum of vertical displacement at the surface was 0.22 m for the lower bound $(N_1)_{60}$ conditions. Liquefaction induced vertical displacements can also be estimated from empirical equations based on field experience during past earthquakes. Tokimatsu and Seed (1987) have presented a chart for predicting liquefaction induced strains based on in situ $(N_1)_{60}$ values and field experience. The predictions based on their chart for the various conditions are shown in Table I.

Table I
Liquefaction Induced Displacements

Type of Analysis	In Situ $(N_1)_{60}$ State	Thickness of Liquefied Layer, m	Vertical Displ., m	Horizontal Displacement, m	
				Byrne	Hamada
Effective stress	Lower bound	20	0.2	2.5	
Total stress	Lower bound	20	0.5*	2.7**	3.3
Total stress	Lower bound	27	0.7*	3.4**	3.9
Total stress	Mean	13	0.2*	0.6**	2.7

*Based on Tokimatsu and Seed, **Based on Byrne (1991a).

It may be seen that for the same $(N_1)_{60}$ conditions, and the same depth of liquefied layer, the vertical displacement based on total stress analysis and field experience is more than twice that computed from the effective stress analysis.

The computed horizontal displacements of the crust from the effective stress analyses are shown in Fig. 8. It may be seen that prior to triggering of liquefaction the displacements are small, but upon triggering, large displacements occur in one direction of the order of 2.5 m.

Byrne (1991a) presented a simple extension of the Newmark (1965) procedure for predicting liquefaction induced lateral displacements taking into account the post liquefaction stress-strain and strength behaviour and its dependency on $(N_1)_{60}$ value. He showed his procedure to be in good agreement with field and laboratory observations. The estimated displacements from this procedure are shown in Table I (Col. 5) and good agreement is obtained with the effective stress analysis (2.5 vs 2.7 m).

Lateral liquefaction induced displacements can also be estimated from empirical equations based on field experience. One such equation that is commonly used was proposed by Hamada et al. (1987) as follows:

$$D = 0.75 \, (H)^{1/2} \, (\theta)^{1/3} \tag{1}$$

in which D = the liquefaction induced displacement, m; H = the thickness of the liquefied layer, m; and θ = the slope of the ground surface in %.

The predicted liquefaction induced displacements based on this equation (θ = 1%) are also shown on Table I (Col. 6). They are significantly greater than the prediction from the effective stress analysis However, most of the Hamada data were associated with very loose sands, and this is discussed in more detail by Byrne (1991a).

It should be noted that all of the empirical methods of predicting liquefaction induced displacements are strongly dependent on the thickness of the liquefied layer. The total stress analysis procedure which predicts a greater thickness of liquefied layer will also predict greater liquefaction induced displacements as may be seen from Table I. Thus the commonly used total stress procedure may be unduly conservative when estimating both the extent of liquefaction and the magnitude of liquefaction induced displacements.

Extensive zones of liquefaction are predicted to occur in the Fraser Delta in the event of a major earthquake and are likely to result in severe damage. Experience at Niigata, Japan 1964, and San Francisco 1989 indicate that damage to buried services such as water, gas, sewer, electricity and telephone would be very severe due to the large differential movements of the surface crust. Damage to bridge and overpass structures, and the George Massey Tunnel could also be severe. The dyking system will likely suffer severe cracking, and flooding is a possibility. Light wood structures supported on the crust are likely to suffer light to moderate damage. However, older taller buildings supported on piles could suffer very severe damage due to loss of pile support.

SUMMARY

Dynamic analysis carried out in the Fraser Delta deposits indicate that in the event of a major earthquake, bedrock motions of 0.2 g may be amplified to 0.3 g at the surface. This is in accord with amplifications recorded in San Francisco in 1989 and Mexico City in 1985. The surface accelerations are not greatly influenced by the thickness of the till layer nor the thickness of the silt-clay layer. However, significantly lower surface accelerations are predicted if the silt-clay layer is not present as appears to be the case in the southern Delta.

These amplified accelerations would trigger liquefaction over most areas of the Delta for earthquakes with M7 or greater. The density and liquefaction resistance varies considerably in the various locations throughout the Delta and for the Lower bound condition liquefaction to the full depth of the sand layer could occur. This could result in vertical displacements of about 0.7 m and horizontal displacements of up to 3 or 4 m. Such movements could result in very severe damage to structures and lifeline facilities. Effective stress analyses indicate significantly smaller depths of liquefaction and lower displacements. Site specific evaluations are needed for major structures.

REFERENCES

Britton, J., Consultant, Vancouver, 1990. Personal Communication.

Byrne, P.M. 1991. "A Cyclic Shear Volume - Coupling and Pore Pressure Model for Sand", 2nd Int. Conference on Recent Advances in Geotechnical Earthquake Eng. and Soil Dynamics, St. Louis, Missouri, Paper 1.24, March 1991.

Byrne, P.M. 1991a. "A Model for Predicting Liquefaction Induced Displacements Due to Seismic Loading", accepted for publication in the 2nd Int. Conf. on Recent Advances in Geotechnical Earthquake Eng. and Soil Dynamics, St. Louis, Missouri, Paper 7.14, March 1991.

Byrne, P.M. and Yan, L. 1990. "ID-LIQ: A Computer Code for Predicting the Effective Stress 1-Dimensional Response of Soil Layers to Seismic Loading", Soil Mechanics Series No. 146, Dept. of Civil Engineering, University of British Columbia, September 1990.

Byrne, P.M. 1978. "An Evaluation of the Liquefaction Potential of the Fraser Delta", Canadian Geotechnical Journal, Vol. 15, No. 1, 1978.

Byrne, P.M. and Anderson, D.L. 1987. "Earthquake Design in Richmond: Version II", A report prepared for the Corporation of the Township of Richmond, Soil Mechanics Series No. 109, Dept. of Civil Eng., University of British Columbia, March 1987.

Finn, W.D. Liam, Byrne, P.M. and Martin, G.R. 1976. "Seismic Response and Liquefaction of Sands", Journal of the Geotechnical Eng. Division, ASCE, No. GT8, August 1976.

Finn, W.D., Woeller, D.J., and Robertson, P.K. 1988. "An Executive Summary of Liquefaction Studies in the Fraser Delta", Report prepared for Energy, Mines and Resources Canada, Geological Survey of Canada, December 1988.

Hamada, M., Towhata, I., Yasuda, S. and Isoyama, R. 1987. "Study on Permanent Ground Displacements Induced by Seismic Liquefaction", Computers and Geomechanics 4, pp. 197-220.

Hunter, J.A. 1990. "Preliminary Results of Surface Shear Wave Refraction Survey in the Southern Fraser Delta".

Idriss, I. "Response of Soft Soil Sites During Earthquakes", H. Bolton Seed Memorial Symposium Proceedings, Vol. 2, pp. 273-289.

Seed, H.B. and Idriss, I.M. 1970. "Soil Moduli and Damping Factors for Dynamic Response Analysis", Report No. UCB/EERC-70/10, University of California, Berkeley, December.

Seed, H.B., Wong, T.R., Idriss, I.M., and Tokimatsu, K. 1986. "Moduli and Damping Factors for Dynamic Analyses of Cohesionless Soils", Journal of the Geotechnical Engineering, Vol. 112, No. 11, November 1986.

Sun, J.I., Golesoskhi, R. and Seed, B.H. 1988. "Dynamic Moduli and Damping Factors for Cohesive Soils", Report No. UCB/EERC-88/15, University of California, Berkeley, August.

Tokimatsu, K.A.M. and Seed, H.B. "Evaluation of Settlement in Sands Due to Earthquake Shaking", Journal of Geot. Eng., ASCE, Vol. 113, No. 8, pp. 861-878.

Youd, T. and Bartlett, S. 1988. "U.S. Case Histories of Liquefaction-Induced Ground Displacements", Proceedings, First Japan-U.S. Workshop on Liquefaction, Large Ground Deformations and their Effects on Lifeline Facilities, Tokyo, Japan, November 1988.

Wallis, D.M. 1979. "Ground Surface Motions in the Fraser Delta Due to Earthquakes", M.A.Sc. Thesis, Dept. of Civil Engineering, University of British Columbia, April 1979.

Zavoral, D. 1990. "Dynamic Properties of an Undisturbed Clay from Resonant Column Tests", Dept. of Civil Engineering, University of British Columbia.

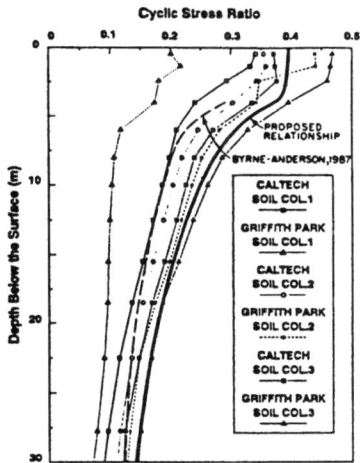

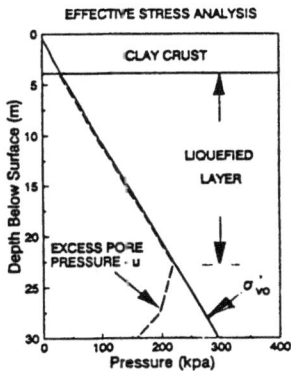

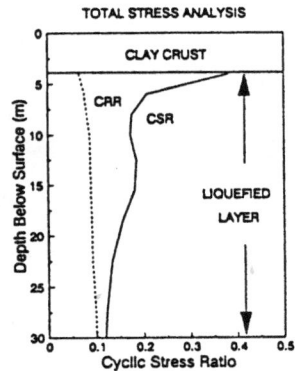

a

b

Fig. 5   Computed Cyclic Stress Ratios
for the Fraser Delta

Fig. 7   Liquefaction Assessment

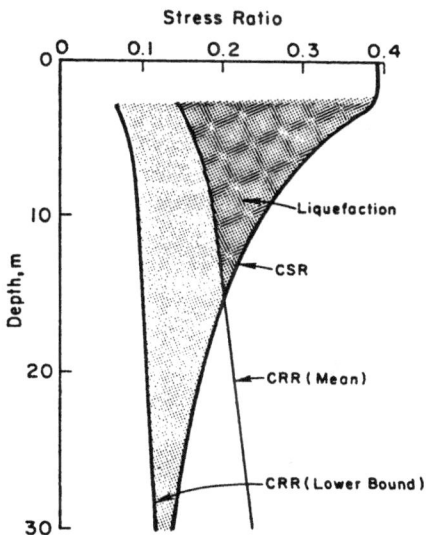

CSR = Cyclic Stress Ratio caused by
        Design Earthquake
CRR = Cyclic Resistance Ratio
CSR > CRR ⟹ LIQUEFACTION

Fig. 6   Liquefaction Assessment for Richmond,
0.2 g and M7

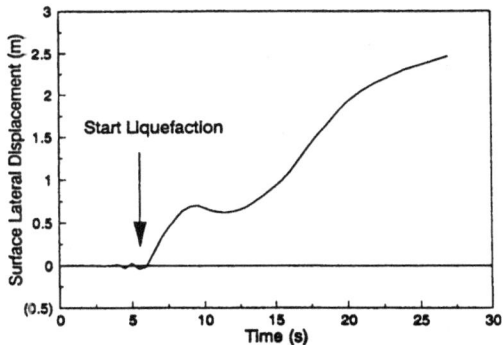

Fig. 8   Predicted Surface Displacements for
Lower Bound $(N_1)_{60}$ Values, 0.2 g and
M7

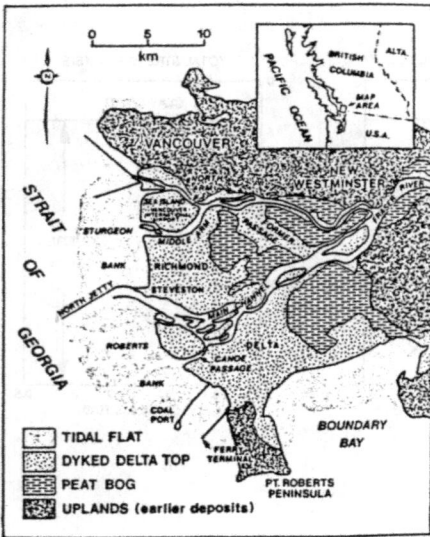

Fig. 1   Fraser Delta (adapted from Luternauer
and Finn, 1983)

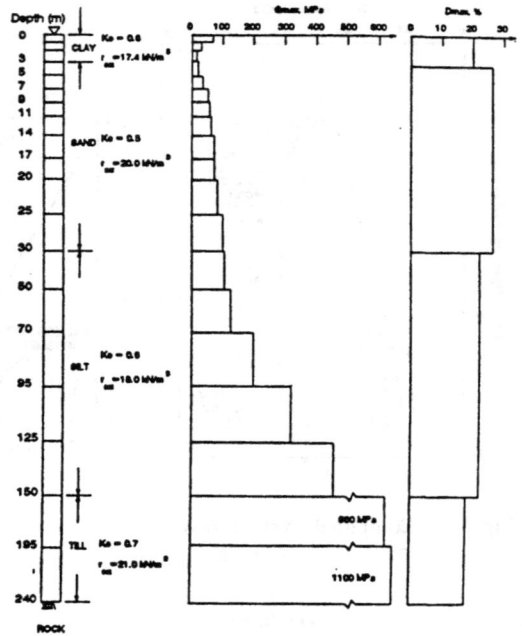

Fig. 2   Soil Profile and Soil Properties used in
Analysis

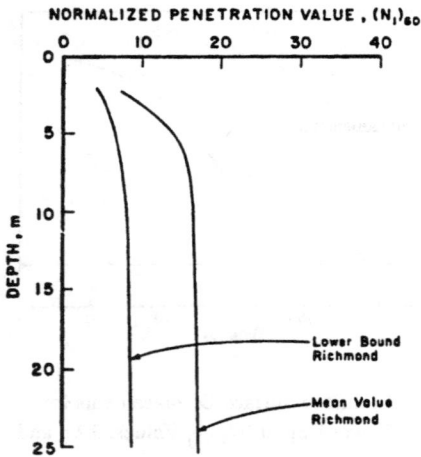

Fig. 3   Normalized Penetration Values in
Richmond

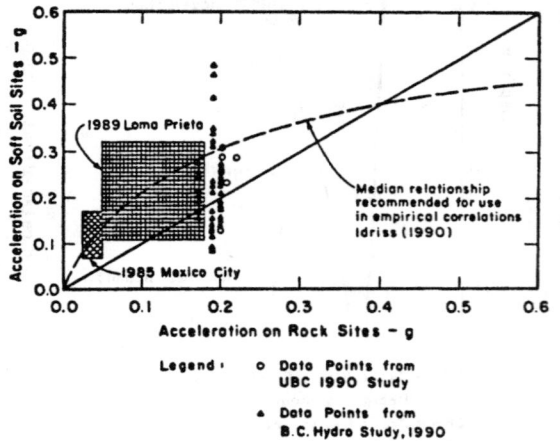

Fig. 4   Amplification of Acceleration

444

Response of reinforced soil slopes to earthquake loadings

M. Yogendrakumar[I], R.J. Bathurst[II], and W.D. Liam Finn[III]

ABSTRACT

Principal features of a direct nonlinear method for analyzing reinforced soil structures under earthquake loading is presented. The response of a soil slope reinforced with horizontal layers of polymeric reinforcement and subjected to a typical earthquake loading is presented. The seismic response of the unreinforced soil slope is also included to examine the influence of the reinforcement.

INTRODUCTION

The use of advanced polymeric materials as tensile soil reinforcement in the construction of embankments and slopes with side slopes steeper than the angle of internal friction of the soil fill is becoming more common. Conventional limit equilibrium–based methods are commonly used for analysis and design of these slopes (Bonaparte et al. 1986). These methods are primarily stress–based and do not consider deformations explicitly. Moreover, little attention has been given to the design of reinforced soil slopes subjected to seismic loading.

In recent years, however, the use of finite element analysis has been introduced into the study of the dynamic response of reinforced soil systems to seismic load. The study reported here presents a direct nonlinear method for analyzing reinforced soil structures under dynamic load conditions. The essential features of the method are implemented in the finite element program TARA-3 (Finn et al. 1986). This program has been used successfully to analyse the seismic response of a centrifuged model of a cantilever retaining wall (Finn et al. 1989) on the Cambridge geotechnical centrifuge. The program simulated satisfactorily not only the accelerations, dynamic displacements and dynamic moments but also the residual displacements and moments remaining after the earthquake.

I Research Associate, Royal Military College of Canada, Kingston, Ontario, Canada.

II Associate Professor, Royal Military College of Canada, Kingston, Ontario, Canada.

III Professor, University of British Columbia, Vancouver, B.C., Canada

TARA-3 has also been used to investigate key aspects of the seismic response of a reinforced soil retaining wall (Yogendrakumar et al. 1991). In this paper, the same method of analysis reported in these earlier studies is used to examine the influence of polymeric reinforcement on the seismic response of a soil slope.

METHOD OF ANALYSIS

This section presents the principal features of the direct nonlinear method that is implemented in the finite element program TARA-3 for total stress analysis of soil structures subject to dynamic loading. In this method an incremental approach has been adopted to model nonlinear behaviour of soil using tangent shear and tangent bulk moduli, G_t and B_t respectively. The incremental displacements during the base excitation are obtained by solving the incremental dynamic equilibrium equations given in Eq. 1 by a direct numerical integration method.

$$[M]\{\Delta \ddot{x}\} + [C]\{\Delta \dot{x}\} + [K]\{\Delta x\} = -[M]\{I\}\Delta \ddot{u}_g \qquad (1)$$

Here [M] is the mass matrix; [C] is the damping matrix; [K] is the stiffness matrix; $\{I\}$ is the unit vector; $\{\Delta \ddot{x}\}$, $\{\Delta \dot{x}\}$ and $\{\Delta x\}$ are incremental acceleration, velocity and displacement vectors of the nodes relative to the base and; $\Delta \ddot{u}_g$ is the increment in base input acceleration.

The stiffness matrix [K] is a function of the current tangent moduli during loading, unloading and reloading. The use of shear and bulk moduli allows the elasticity matrix [D] to be expressed as

$$[D] = B_t [Q_1] + G_t [Q_2] \qquad (2)$$

where $[Q_1]$ and $[Q_2]$ are constant matrices for the plane strain conditions usually considered in the analysis. This formulation reduces the computation time for updating [D] whenever G_t and B_t change in magnitude because of straining.

Soil model

The behaviour of soil in shear is assumed to be nonlinear and hysteretic and to exhibit Masing behaviour during unloading and reloading. The relationship between shear stress τ and shear strain γ for the initial loading phase under either drained or undrained loading conditions is assumed to be hyperbolic and given by

$$\tau = f(\gamma) = \frac{G_{max} \; \gamma}{\left[1 + (G_{max}/\tau_{max})|\gamma|\right]} \qquad (3)$$

where G_{max} is the maximum shear modulus and τ_{max} is the maximum shear strength. The initial loading curve is shown in Fig. 1a. The equation for the unloading curve from a point (γ_r, τ_r) at which the loading reverses direction is given by

$$\frac{\tau - \tau_r}{2} = f\left(\frac{\gamma - \gamma_r}{2}\right) \qquad (4)$$

446

or

$$\frac{\tau - \tau_r}{2} = \frac{G_{max}\,(\gamma - \gamma_r)/\,2}{\left[1 + (G_{max}/\,2\tau_{max})|\gamma - \gamma_r|\right]} \tag{5}$$

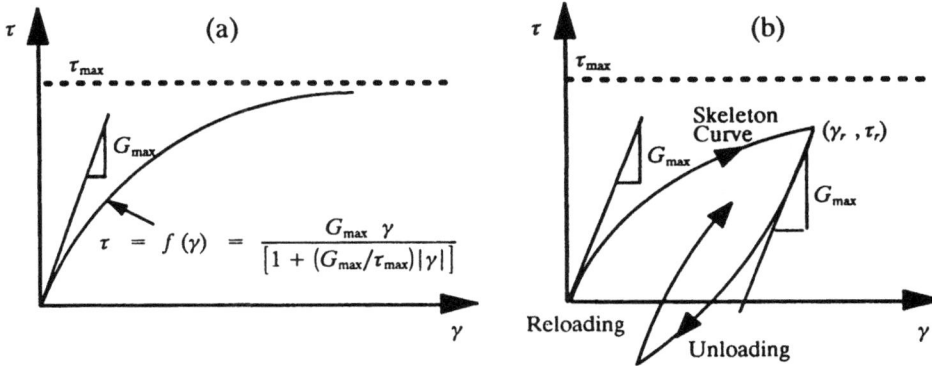

Figure 1. Nonlinear hysteretic loading paths

The shape of the unloading–reloading curve is shown in Fig. 1b. The tangent shear modulus, G_t, for a point on the skeleton curve is given by

$$G_t = \frac{G_{max}}{\left[1 + (G_{max}/\tau_{max})|\gamma|\right]^2} \tag{6}$$

and at a stress point on an unloading or reloading curve G_t is given by

$$G_t = \frac{G_{max}}{\left[1 + (G_{max}/\,2\tau_{max})|\gamma - \gamma_r|\right]^2} \tag{7}$$

The response of the soil to uniform all–round pressure is assumed to be nonlinearly elastic and dependent on the mean normal stress. Hysteretic behaviour, if any, is neglected in this mode. The tangent bulk modulus, B_t, is expressed in the form

$$B_t = K_b\,P_a\,\left(\frac{\sigma_m}{P_a}\right)^n \tag{8}$$

in which K_b is the bulk modulus constant, P_a is the atmospheric pressure in units consistent with mean normal effective stress σ_m and n is the bulk modulus exponent.

Reinforcement model

The reinforcement is modelled using one–dimensional beam elements with axial stiffness only. Slip elements of the type developed by Goodman et al. (1968) may be used to allow for the relative movements between the soil and reinforcement during earthquake excitations. Relatively inextensible type of reinforcement is assumed to be an elastic per-

447

fectly plastic material with the yield stress given by the elastic limit. The behaviour of relatively extensible reinforcement such as polymeric materials is assumed to be nonlinear. The relationship between axial load and axial strain for the initial loading is assumed to be of the nonlinear quadratic form (Chalaturnyk et al. 1988) given by

$$F = D_i \epsilon_a (1 - \frac{\epsilon_a}{2\epsilon_{af}})$$ (9)

in which F is the axial load per unit width (kN/m), D_i is the initial load modulus (kN/m), ϵ_a is the axial strain and ϵ_{af} is the axial strain at failure.

The details of the model parameters are shown in Fig. 2. The tangent load modulus D_t on the initial loading curve is calculated as

$$D_t = \frac{dF}{d\epsilon_a} = D_i (1 - \frac{\epsilon_a}{\epsilon_{af}})$$ (10)

During the analysis, compression is not allowed in the polymeric geosynthetic reinforcement and the hysteresis during unloading and reloading is not modelled. The unloading and reloading portions are approximated as straight lines and the unload–reload modulus is defined as

$$D_{ur} = K D_i$$ (11)

in which D_{ur} is the unload–reload modulus and K is a constant.

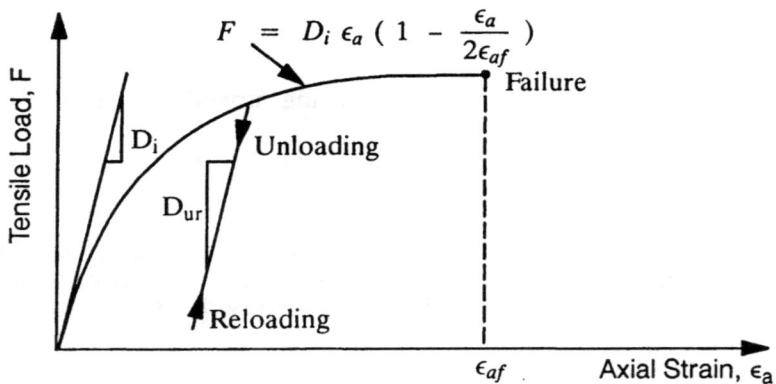

Figure 2. Nonlinear load–strain relationship

FINITE ELEMENT ANALYSIS

The dynamic responses of a reinforced and unreinforced soil slope resting on a rigid foundation were computed using the TARA–3 program. The slopes were assumed to be 12.0m high with a side slope of 1:1. It is lightly reinforced with polymeric reinforcements 12.0m in length. The reinforcement layers are placed horizontally with vertical spacing of

2.0m. The finite element representation of the reinforced soil slope shown in Fig. 3 consisted of 87 soil elements and 30 one–dimensional beam elements in the slope and 69 elements in the foundation. Slip elements which allow for relative movement to occur between the soil and the reinforcement have not been used in this analysis.

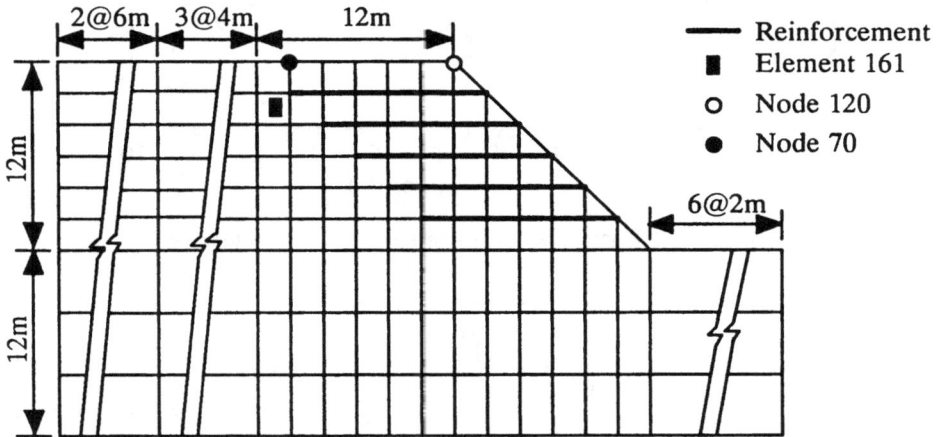

Figure 3. Finite element representation of the reinforced soil slope

The foundation soil was assumed to be very stiff and the shear modulus, Poisson's ratio and unit weight were taken as 3500 MPa, 0.49 and 20.0 kN/m^3 respectively. The following properties were selected for the soil in the slope; K_b = 2950.0, n = 0.5, Poisson's ratio = 0.40, cohesion = 35 kPa and angle of internal friction = 17°, unit weight = 20 kN/m^3. For the polymeric reinforcement, D_j, ϵ_{af} and K were taken as 778.0, 0.18 and 2.0 respectively. The response of the slope to the first 9.60 seconds of the N–S component of the 1940 El Centro earthquake scaled to 0.2g was computed using the program TARA–3. The input motion is shown in Fig. 4. The base was assumed to be rigid and the nodes on the left and right vertical boundary were supported on horizontal rollers for the dynamic analysis. A static analysis was first conducted to establish the stress–strain field prior to the earthquake excitation. The program simulated the incremental construction process of the slope.

NUMERICAL RESULTS

Fig. 5 shows the dynamic horizontal displacement time history of node 70 for the unreinforced and reinforced slopes. Node 70 is located at the top surface 10m from the crest of the slope. Because the nonlinear behaviour of both the soil and the reinforcement is modelled, the analyses show that there is residual dynamic displacement present after the earthquake in both the reinforced and unreinforced slopes. The comparison in Fig. 5 clearly shows the effect of the reinforcement in reducing slope deformations. The peak values and the size of oscillations of the displacements are reduced in the case of the reinforced slope. For this node, the maximum and the residual displacements are reduced to about 11% and 20% respectively of the values predicted for the unreinforced slope.

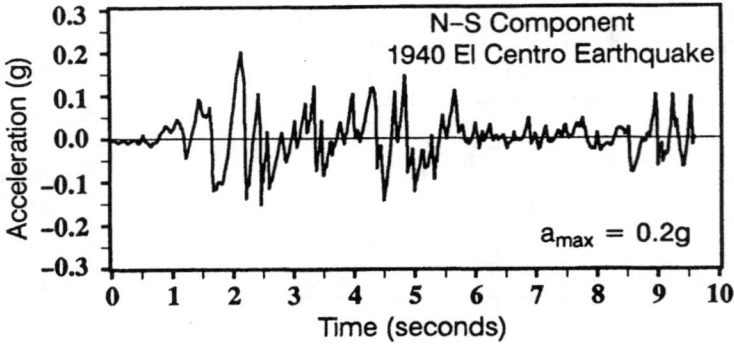

Figure 4. Base input motion for TARA-3 analysis

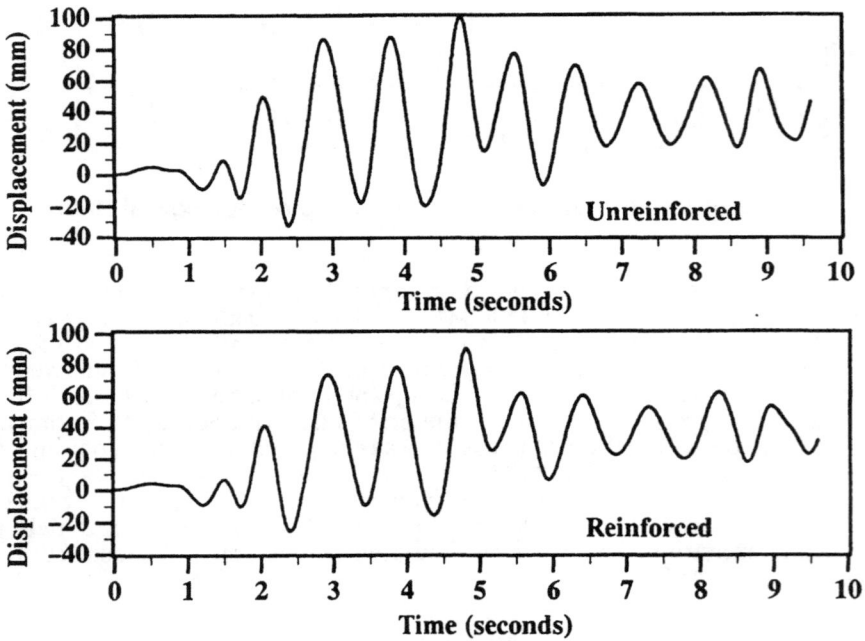

Figure 5. Displacement time history of node 70

The dynamic horizontal displacement time histories of node 120, the node at the top edge of the side slope, are shown in Fig. 6. Again as expected, analysis with the reinforced slope produces smaller displacements than the analysis of the unreinforced slope. The maximum displacement in the case of reinforced slope is 10% less than value with the unreinforced slope. A reduction of about 30% is achieved in the residual displacement. Similar reductions in maximum and residual deformations are observed at other locations within the soil slope indicating the beneficial effect of the reinforcement.

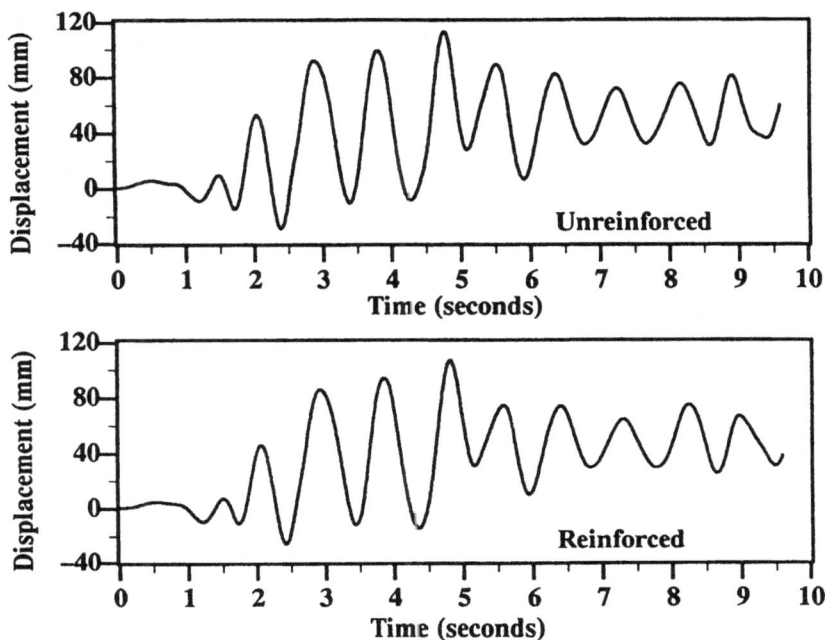

Figure 6. Displacement time history of node 120

The improved seismic performance of the reinforced slope is further illustrated in Fig. 7. This shows the shear stress–strain response of soil element 161 for the case of unreinforced and reinforced soil slope. Both show nonlinear response with stronger response occurring in the case of the unreinforced slope. The magnitude of the strain and the size of hysteresis loops are larger in the case of unreinforced slope than in the case of reinforced slope.

CONCLUSIONS

The TARA–3 analysis is capable of providing the key information needed to assess the seismic performance of a soil slope in terms of deformations. Because the nonlinear hysteretic behaviour of soil and the nonlinear behaviour of the polymeric reinforcement are modelled, the residual deformations after the earthquake can be calculated directly and the consequences of seismic shaking clearly seen.

An important concern is the contribution of the polymeric reinforcement in improving the seismic behaviour of the soil slope. This is readily determined by TARA–3 analysis as the program has the capability to model the nonlinear load–strain behaviour of the polymeric reinforcement. For the example considered it has been shown that the polymeric reinforcement has improved the seismic behaviour of the soil slope.

The results so far suggest that the program TARA–3 may be a very useful tool in assessing the seismic performance of reinforced soil slopes. Studies are planned to determine the influence of the different model parameters describing the nonlinear load–strain behaviour of the reinforcement.

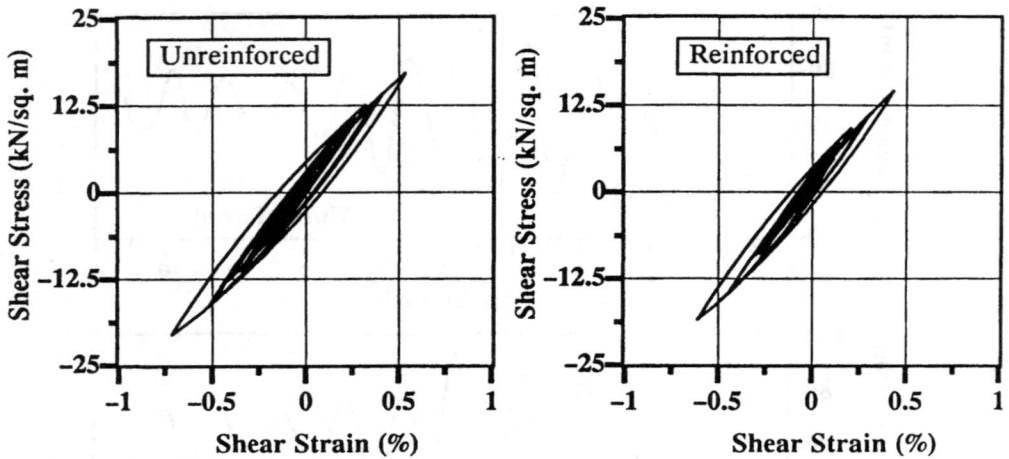

Figure 7. Stress–strain behaviour of element 161.

ACKNOWLEDGEMENTS

The funding for this study was provided by the Department of National Defence (Canada) as part of on–going research at RMC related to the design and analysis of blast and seismic resistant geosynthetic reinforced soil structures.

REFERENCES

Bonaparte, R., Schmertmann, G.R. and Williams, N.D. 1986 Seismic Design of Slopes Reinforced with Geogrids and Geotextlies. Technical Note, The Tensar Corporation, Morrow, GA, U.S.A.

Chalaturnyk, R.J., Scott, J.D., Chan, D.H. and Richards, E.A. 1988. Stresses and Deformations in a Reinforced Soil Slope. Third Canadian Symposium on Geosynthetics, Kitchener, Ontario, Canada, 79–89.

Finn, W.D. Liam, Yogendrakumar, M., Otsu, H. and Steedman, R.S. 1989. Seismic Response of a Cantilever Retaining Wall: Centrifuge Model Test and Dynamic Analysis. Proceedings of the 4th International Conference on Soil Dynamics and Earthquake Engineering. Mexico City, Mexico, October.

Finn, W.D. Liam, Yogendrakumar, M., Yoshida, N. and Yoshida, H. 1986. TARA–3: A Program to Compute the Response of 2–D Embankments and Soil–Structure Interaction Systems to Seismic Loading. Department of Civil Engineering, University of British Columbia, Vancouver, Canada.

Goodman, R.B., Taylor, R.E. and Brekke, T.L. 1968. A Model for the Mechanics of Jointed Rock. J. Soil Mech. and Foundation., ASCE, 94(SM3) 637–659.

Yogendrakumar, M., Bathurst, R.J. and Finn, W.D. Liam 1991. Response of Reinforced Soil Walls to Earthquake Loads. Proceedings of the IX Panamerican Conference, Vina del Mar, Chile, August (in press).

Dynamic behaviour of fly ash under seismic conditions of Eastern Canada

K.T. Law[I] and P.A. Chen[II]

ABSTRACT

The dynamic behaviour of a fly ash was studied with particular reference to seismic ground motions that prevail in eastern Canada. Resonant column tests, cyclic simple shear tests and cyclic triaxial tests at both low and high frequencies have been conducted. The resonant column test measures the initial shear modulus and damping ratio of the fly ash. The cyclic simple shear test shears the fly ash in a horizontal direction that is realistic with the transmission of seismic shear wave in the field. The cyclic triaxial test at high frequency (10Hz) is consistent with the high frequency content of the seismic vibrations observed in this region. The test results have been analyzed to study the deformational characteristics and liquefaction resistance of the fly ash. The practical implications from the study are also discussed.

INTRODUCTION

Large amount of fly ash is produced annually in Canada as a result of burning coal for power generation. Some are stockpiled in open yards while others are stored in ponds or hauled to fill pits from which the coal was mined. Some are utilized commercially in various ways to produce economical and environmental benefits. One way is to turn fly ash into structural fill material (Toth et al. 1988). When such fills and stockpiles of fly ash are located in earthquake zones, the potential of liquefaction and other problems associated with dynamic loading are of public concern.

During an earthquake, horizontal shear stresses are transmitted from the bedrock to the overlying soil strata. A good laboratory method for measuring soil behaviour under seismic conditions therefore should involve testing soil specimen under a horizontal shearing mode. In eastern Canada, seismic ground motions are found to contain a significant high frequency content in the range of 10

I,II: Associate Professor and graduate student, respectively,
    Department of Civil Engineering, Carleton University,
    Ottawa, Ontario, Canada K1S 5B6

to 15 Hz (Adams 1989, personal communication). Dynamic soil behaviour in this frequency range is little known because of the lack of equipment to test soil in this range.

This paper describes a laboratory study on the dynamic behaviour of fly ash. A resonant column test device was used to measure the initial shear modulus and damping ratio of the material. A cyclic simple shear apparatus was used to shear soil specimens in a horizontal direction. A unique cyclic triaxial equipment has been used to measure cyclic strength at a high frequency.

The test results show that the fly ash is stronger at the high frequency range than at the usual test frequency of 1 Hz and that it is weaker in the cyclic simple shear mode than in the triaxial mode. The practical implications from the results are also discussed.

TEST PROGRAM

A sample of fly ash was obtained from the ash lagoon at the Ontario Hydro Nanticoke power generating station. The ash has a specific gravity in the range of 2.20 to 2.26. It was oven dried for more than 24 hours before being passed through the #200 sieve. The resulting ash has a mean grain size of 0.012mm and a coefficient of uniformity of 13.7. The ash was then mixed with 25% of water to form specimens for the test program.

The cyclic triaxial equipment used herein is one its kind in the western world. It uses an electro-magnetic drive to provide cyclic loads sufficiently large to fail test specimens at frequencies as high as 20 Hz. The details of this equipment are described by Law et al (1991). Undrained cyclic test isotropically consolidated to 50 kPa at 1 Hz and 10 Hz have been carried out.

The Seiken cyclic simple shear device has been used to impose horizontal shear to fly ash specimens. Compared with other devices, this device has a number of unusual features. Firstly, it permits the application of both the cell pressure and the back pressure to ensure specimen saturation during the test. Secondly, it does not require a steel wire reinforced membrane to provide a zero horizontal strain to maintain the simple shear condition. Ordinary rubber membrane is acceptable. Thirdly, the cell pressure can be varied at will so that any consolidation stress system, either isotropic or anisotropic, can be applied. For the present test series, undrained cyclic tests isotropically consolidated to pressures of 50 kPa and 150 kPa, have been conducted.

The Stokoe resonant column test cell has been used to apply torsional vibrations to a cylindrical specimen configured to a fixed-free type column. The bottom end of the specimen is rigidly fixed to the base while the top end is connected to a drive system for applying the torsional vibrations. The initial shear modulus was determined from the measured shear wave velocity in the specimen at a strain amplitude around 0.001%. The damping ratio was obtained by means of the logarithmic decrement method.

Specimens used in all the different tests were reconstituted using the moist tamping procedure. The specimens for the cyclic triaxial tests and the resonant column tests were 3.91 cm in diameter and 8 cm high and were formed in 8 layers. The specimens for cyclic simple shear tests were 7 cm in diameter and 1.8 cm high and were formed in one layer. The specimens were saturated first by passing carbon dioxide through the specimen under a small confining pressure to displace the air in the void space. This was followed by passing distilled water to replace the carbon dioxide. Any trace of carbon dioxide that remained in the sample dissolved in the pore water when a back pressure of 200 kPa was applied. This saturation procedure produced a pore pressure parameter of $\bar{B}$ of 0.96 or higher.

During the undrained cyclic triaxial and simple shear tests, readings were taken with an IBM PC AT compatible data acquisition system equipped with an integrated circuit board which provided up to 8 analogue to digital channels for taking readings. The computer is programmed to take 100 readings per second per channel for 1 Hz vibrations and one thousand readings per second per channel for 10 Hz vibrations. For a cycle in the cyclic triaxial test, therefore, 100 readings were taken of each of the excess pore pressure, vertical cyclic load and vertical displacement. Similarly, in the cyclic shear test, 100 readings were taken of each of the excess pore pressure, horizontal cyclic load, horizontal displacement and vertical load. After the test, the data were processed and the results were displayed on the monitor and on the plotter. For the resonant column tests, a high speed strip chart recorder was used to trace the torque.

TEST RESULTS AND ANALYSIS

Deformational characteristics

A summary of the results of the resonant column tests is shown in Table 1. Within the strain amplitude tested, the average damping ratio is 1.3% and the modulus value drops slowly with strain amplitude for a given consolidation pressure. The modulus value at around 0.001% was taken as the initial modulus (G_0) to analyze the cyclic test data in the following.

Table 1. Summary of results from resonant column tests

Consolidation Pressure kPa	Peak Shear Strain %	Modulus MPa	Damping Ratio%
	0.0014	20.56	1.36
50	0.0038	20.16	1.24
	0.0065	19.96	1.90
	0.0123	19.37	1.94
	0.0010	37.40	1.20
	0.0026	37.40	0.67
150	0.0057	37.12	0.91
	0.0085	36.57	1.14
	0.0110	36.30	1.17

A hyperbolic strain model (Hardin and Drnevich, 1972) is used to analyze the modulus degradation with strain amplitude. From the cyclic loading and resonant column tests, the functional relationship can be written as:

$$G = 1/(a+b*\gamma) \tag{1}$$

where:

G = shear modulus
γ = shear strain
a,b = constants to be determined from the test results.

The values of "a" and "b" can be obtained by plotting 1/G vs γ as shown in Fig.1 for the cyclic simple shear tests at a consolidation pressure of 50 kPa. A straight line can be used to approximate the relationship with the intercept on the 1/G axis equal to "a" and the slope of the line equal to "b". By definition, the initial shear modulus, G_0, is the modulus value at zero shear strain. The value of "a" should therefore equal to $1/G_0$. This is the case in the present study as "a" from the cyclic simple shear tests agrees with $1/G_0$ from the resonant column tests.

Rewriting Eq.1, one obtains

$$G/G_0 = 1/(1+\gamma_1) \tag{2}$$

where:

$\gamma_1 = \gamma/\gamma_r$ (normalized shear strain)
$\gamma_r = a/b$ (reference strain).

Fig.2 presents the test data, on the basis of the hyperbolic strain model, from the resonant column tests, cyclic simple shear tests and cyclic triaxial tests. The same G_0 and γ_r have been used in all these test data. The result shows that the cyclic simple shear test and resonant column tests give more or less the same modulus degradation curve. The cyclic triaxial tests however yield a higher modulus than the cyclic simple shear tests at comparable shear strain amplitudes.

Liquefaction resistance

The liquefaction resistance (τ_l) of a granular soil is defined as the uniform cyclic shear stress applied to the soil until failure occurs. Failure corresponds to the condition when the excess pore pressure reaches the initial consolidation pressure (σ_c'). The liquefaction resistance normalized by σ_c' for a given soil is a function of shearing mode, void ratio, frequency of loading and number of cycles or energy dissipated to the point of failure. Fig.3 shows typical test results from a cyclic simple shear test for $\sigma_c' = 50$ kPa.

Fig.4 shows the test results for $\sigma_c' = 50$ kPa for different shear modes, frequencies, void ratios and numbers of cycles to failure. The results show that the liquefaction resistance increases with increase in frequency and decrease in void ratio and number of cycles to failure. As well, the cyclic triaxial test yields a higher strength than the cyclic simple shear test for the same void ratio. For example, at 10 cycles to failure and at the loose state of void ratio of 1.285, the cyclic triaxial strength at 1 Hz is 12% higher than the cyclic simple shear strength at the same frequency and the cyclic triaxial strength is 11% higher at 10 Hz than at 1 Hz.

A number of researchers (He 1981, Davis and Berrill 1982 and Law et al. 1990) have proposed another approach of assessing liquefaction resistance based on the total energy dissipated in the soil during dynamic loading. The total dissipated energy (ΣW) consists of two components: one from plastic deformation and the other from hysteretic damping. The value of ΣW can be obtained by summing the areas covered by the plastic deformation and the hysteresis loops on a stress-strain plot for each test. The stress and the strain at any point of a test were evaluated from the data acquired by the high speed computer system.

Fig.5 shows some of the typical test results based on the energy approach. The vertical axis represents the excess pore pressure, Δu, normalized by the initial consolidation pressure, σ_c'. The horizontal axis represents the normalized total dissipated energy, W_N, defined by $\Sigma W/\sigma_c{}'$. The test shown in the figure corresponds to $\sigma_c{}' = 50$ kPa and void ratio = 1.285 and includes both the cyclic simple shear and cyclic triaxial shear. For a given shear mode, the results suggest that there exists a single relationship between $\Delta u/\sigma_c{}'$ and W_N, similar to that of a clean sand studied by Law et al. (1990). The relationship can be written as

$$\Delta u/\sigma_c{}' = \alpha W_N{}^\beta \tag{3}$$

where α and β are constants to be obtained from the test results.

The results also show that for the same amount of dissipated energy, the triaxial test yields a lower excess pore pressure than the simple shear test. This is consistent with the earlier observation that soil sheared under the cyclic triaxial mode is stronger than under the cyclic simple shear mode.

Eq. 3 shows another aspect based on the energy approach. By definition at liquefaction failure, $\Delta u/\sigma_c{}'$ is equal to 1.0 and hence the normalized total dissipated energy at failure, W_{Nf}, is a constant. In other words, for a given shear mode, void ratio and consolidation pressure, liquefaction failure can be expressed by a single variable of W_{Nf}. In the conventional approach, however, two variables, the applied shear stress and the number of cycles of load application, have to be considered.

PRACTICAL IMPLICATIONS

The liquefaction resistance of the fly ash can be compared with that of similar soils reported in the literature. However it is recognized that the relative density of the materials being compared can not be ascertained as the grain size of the fly ash is too fine for determination of the maximum density based on the ASTM standard (D4253). The comparison presented here therefore can only be considered qualitative. By nature of formation, fly ash is similar to volcanic soil and by grain size distribution, it is similar to silt. Fig.6 compares the triaxial liquefaction resistance of the loose fly ash with a loose volcanic soil (Hatanaka et al. 1985) and a loose silt (Cao and Law 1991). The fly ash is stronger than the volcanic soil and silt. One possible reason for the relatively high strength is this fly ash has a friction angle of about 35°(Toth et al. 1988) even at a loose state.

The test results suggest that a conventional cyclic triaxial equipment that shears soil at a low frequency range around 1 Hz may yield reasonable liquefaction resistance for soils in eastern Canada provided that the field density can be maintained in the laboratory test. Here in this region, the seismic stress induced in the soil mass is horizontal and contains a high frequency content. The horizontal shear tends to lower the resistance while the high frequency loading tends to raise it. These two factors are compensating each other, hence giving support to the use of the conventional equipment which is far more commonly available than either the cyclic simple shear apparatus or the high frequency triaxial device.

The triaxial equipment, on the other hand, may overestimate the modulus as seen in Fig.4. This figure suggests that frequency has little effect on modulus as the same modulus degradation curve applies to both the resonant column test which was conducted at high frequency (about 25 Hz) and the cyclic simple test which was conducted at low frequency (1Hz). The modulus obtained from the triaxial will not be subject to a compensating effect similar to that for resistance.

SUMMARY AND CONCLUSIONS

A laboratory test study has been conducted on a fly ash from Nanticoke, Ontario. Tests have

457

been performed by means of a cyclic simple shear apparatus in order to be more realistic compared with the shear mode in the field under earthquake condition. A unique frequency triaxial device was also used to study the effects of high frequency vibration that exists in the ground motions of eastern Canada. The test results show the following:

1). The shear modulus degradation of the fly ash can be described by the hyperbolic strain model. The parameters for the model obtained from the cyclic simple shear tests agree well with those from the resonant column tests. Under cyclic triaxial loading state, however, the shear modulus degrades less rapidly than under the cyclic simple shear loading state.

2). The liquefaction resistance of the fly ash at loose state is stronger than a loose volcanic soil and a loose silt, possibly due to a high friction angle in the fly ash.

3). At 1 Hz loading, the cyclic triaxial test gives a higher liquefaction resistance than the cyclic simple shear test.

4). The cyclic triaxial test gives a higher strength at 10 Hz than at 1 Hz.

5). The energy approach is applicable for assessing liquefaction resistance of the fly ash.

ACKNOWLEDGEMENTS

The experimental phase of research described in this paper was conducted at the National Research Council of Canada (NRCC). The authors have benefited from the fruitful discussion of Y.L. Cao and the conscientious laboratory assistance of T. Hoogeveen, both of NRCC. The fly ash sample was supplied by C.B.H. Cragg of Ontario Hydro. Financial support was given by the Natural Science and Engineering Research Council through Operation Grant No.8741-01.

REFERENCES

Cao,Y.L. and Law,K.T.1991. Energy approach for liquefaction of sandy and clayey silts. Proc. 2nd International Conference on Recent Advances in Geotechnical Engineering and Soil Dynamics, Rollar Missouri, USA.

Davis,R.O. and Berrill, J.B. 1982. Energy dissipation and seismic liquefaction in sands. Earthquake Engineering and Structural Dynamics, Vol.10, 59-68.

Hardin,B.O. and Drnevich,V.P 1972.Shear modulus and damping in soils: Design equations and curves. J.SMFD.ASCE. Vol.198,No.SM7, 667-692.

Hatanaka, M., Sugimoto,M. and Suzuki, Y. 1985. Liquefaction resistance of two alluvial volcanic soils sampled by in situ freezing. Soils and foundations, Vol.25, 49-63.

He,G.N.1981. Energy analysis procedure for evaluating liquefaction potential (in Chinese) Chinese Journal of Geotechnical Engineering. Vol.3, 11-21.

Law,K.T. Cao,Y.L. and He,G.N.1990. An energy approach for assessing liquefaction potential. Canadian Geotechnical Journal. Vol.27,320-329.

Toth,P.S. Chan,H.T. and Cragg,C.B.H. 1988.Coal ash as structural fill, with special reference to Ontario experience. Canadian Geotechnical Journal. Vol.25, 694-709.

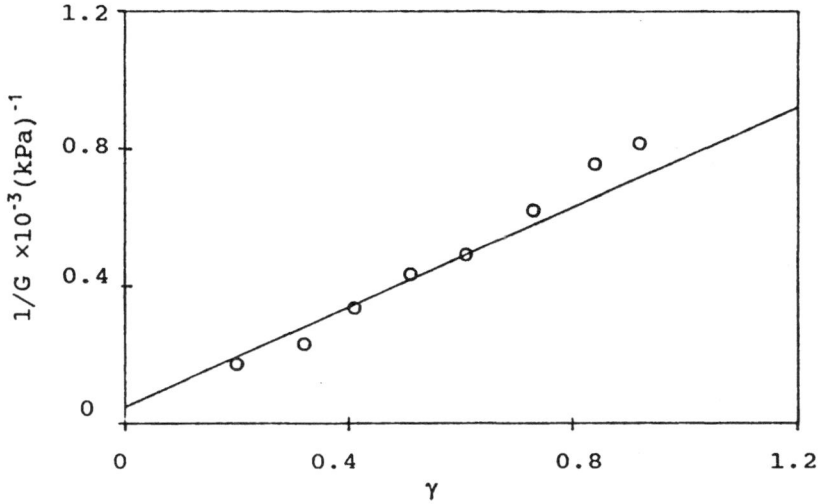

Figure 1. 1/G vs γ from cyclic simple shear test at a consolidation pressure of 50 kPa

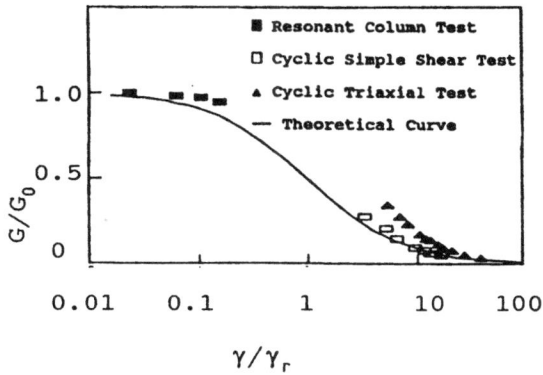

Figure 2. Comparison of modulus degradation from different tests

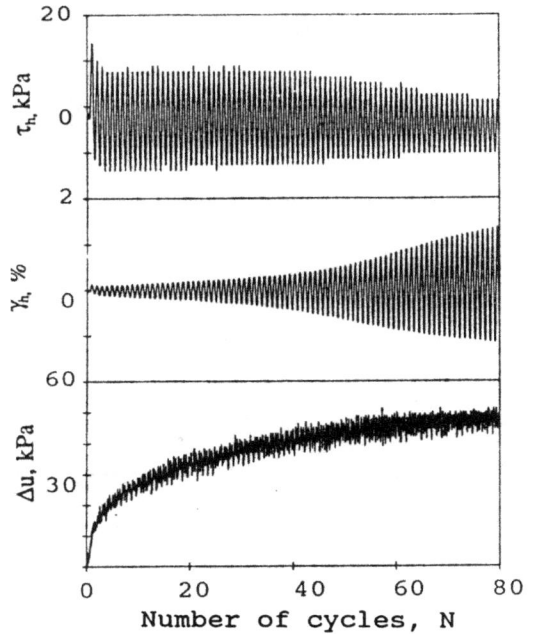

Figure 3. Typical horizontal stress (τ_h), horizontal strain strain (γ_h) and excess pore water pressure(Δu)

459

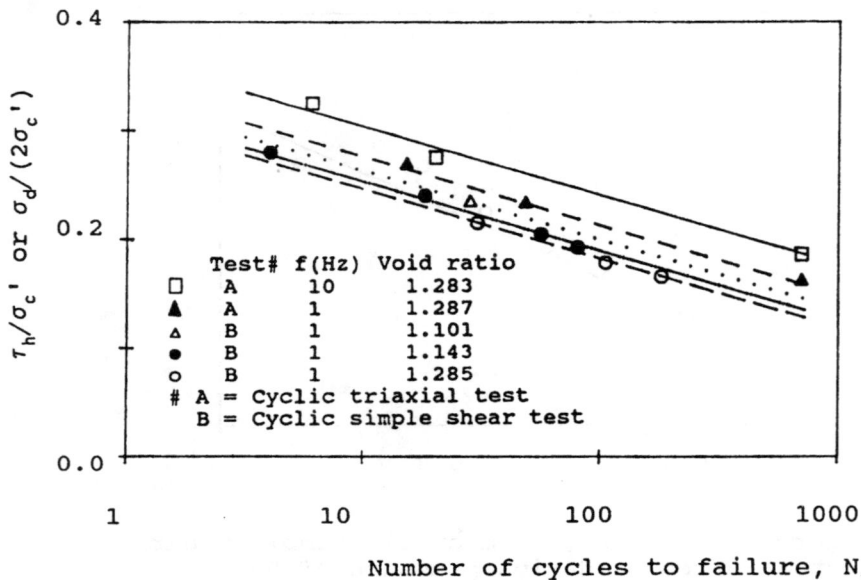

Figure 4. Normalized liquefaction resistance vs number of cycles to failure

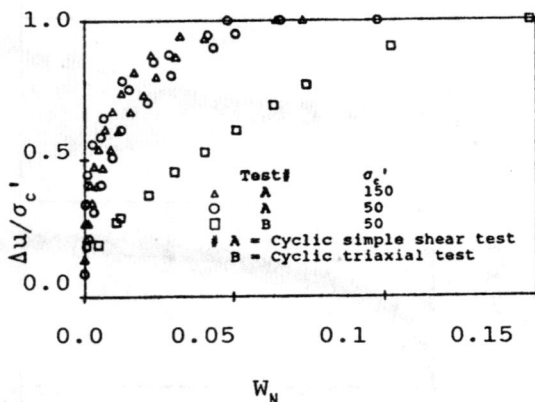

Figure 5. Normalized pore water pressure ($\Delta u/\sigma_c'$) vs normalized dissipated energy (W_N)

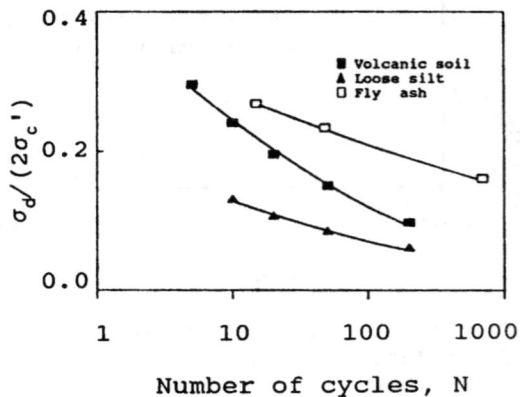

Figure 6. Comparison of liquefaction resistance of different loose soils

Empirical prediction of response spectra

W. Scott Dunbar[1]

ABSTRACT

Response spectra of earthquake accelerograms, including smoothed averages of such spectra, are descriptions of ground motion and may be used as a basis for defining a design spectrum. Two methods of deriving response spectra for use in sesimic design are reviewed. These are: scaling dynamic amplification factors by peak ground motions and computing response spectral ordinates by means of attenuation relationships. It is shown that, if comparisons are made in a consistent manner, application of each of these methods gives similar spectra. However, the standard procedure of scaling dynamic amplification factors by peak ground motions can lead to biased estimates of response spectral ordinates. Reasons for this are related to the fact that spectral shape is dependent on earthquake magnitude and to the manner in which a scaled spectrum is computed.

INTRODUCTION

The response spectrum of a particular accelerogram is a highly irregular function of frequency or period. For purposes of describing ground motion, it is desireable to have a smooth function of frequency or period which envelops the irregular spectra below a particular level. Such a smooth spectrum is also useful for design, in which case it is a specification of the performance requirements of a structure.

Two procedures for the determination of smooth response spectra are

1) scale dynamic amplification factors by peak ground motions, and

2) compute spectra by means of an attenuation relationship.

Each of these methods is based on empirical data but on quite different assumptions. However, in the following it will be shown that, if comparisons are made in a consistent manner, the above procedures result in similar response spectra with differences that can be explained.

SCALED RESPONSE SPECTRA

A response spectrum can be divided into three different period ranges within which amplification of peak ground motion is relatively constant: the low period or 'acceleration' range, the intermediate period or 'velocity' range, and the long period or 'displacement' range. In each region the corresponding peak ground motion is amplified the most.

Dynamic amplification factors for peak ground motions have been derived by numerous inves-

---

[1] Acres International Ltd., Niagara Falls, Ontario, Canada L2E 6W1

tigators using slightly different suites of accelerograms (Newmark and Hall, 1982 and references therein). The most recent are given in Newmark and Hall (1982) and reproduced in Table 1.

As originally suggested by Newmark *et al* (1973), these amplification factors are multiplied by peak ground motions resulting in acceleration, velocity and displacement spectral ordinates which can be plotted on four-way log paper to produce a smooth response spectrum. There are two variants of the original Newmark-Hall method of computing a scaled spectrum. These are given below.

Procedures for Computing a Scaled Spectrum
Four-way logarithmic construction

Procedure A

1) Obtain peak ground motions *pga*, *pgv* and *pgd*.

2) Plot *pga*, *pgv* and *pgd*, draw horizontal line through *pgv*, a line parallel to displacement axis through *pga* and a line parallel to acceleration axis through *pgd*. Connect all lines.

3) Obtain median or 84 percentile amplification factors, F_a, F_v and F_d from Table 1.

4) Multiply $pga \times F_a = A$, $pgv \times F_v = V$ and $pgd \times F_d = D$.

5) Plot A, V and D. Draw horizontal line through V, a line parallel to displacement axis through A and a line parallel to acceleration axis through D. Connect all lines.

6) Draw a line between the A ordinate at 0.125 sec period (8 Hz) and the *pga* line at 0.03 sec period (33 Hz).

Procedure B

1) Obtain peak ground acceleration *pga*.

2) Estimate *pgv* and *pgd* from published ratios pgv/pga, $pga \cdot pgd/pgv^2$ appropriate to site conditions.

3) Plot *pga*, *pgv* and *pgd*, draw horizontal line through *pgv*, a line parallel to displacement axis through *pga* and a line parallel to acceleration axis through *pgd*. Connect all lines.

4) Follow steps 3-6 of procedure A.

An important variable in each of these procedures is the percentile level of the amplification factors to be used. The use of 84 percentile amplification factors would result in a spectrum that represents uncertainty or variability in the ground motion. Accounting for ground motion variability may be desirable for purposes of ground motion description and is certainly necessary in some seismic design applications. However, care must be taken to ensure that the variability is not counted twice. For example, peak ground motions are often estimated by probabilistic or other procedures which account for ground motion variability. Using such estimates to scale 84 percentile amplification factors would be a case of 'compounding conservatisms' (Housner and Jennings, 1982; Donovan and Becker, 1986).

ATTENUATION RELATIONSHIPS FOR SPECTRAL ORDINATES

Attenuation relationships for spectral ordinates at a particular natural vibration period can be derived. This is a natural extension of the attenuation relationships used to compute peak ground acceleration given a magnitude and distance in that such relationships are for spectral response at low natural period. Several attenuation relationships for spectral ordinates exist and

TABLE 1

Spectral Amplification Factors

From Newmark and Hall (1982)

Damping ratio (%)	84 percentile			Median		
	F_a	F_v	F_d	F_a	F_v	F_d
0.5	5.10	3.84	3.04	3.68	2.59	2.01
1	4.38	3.38	2.73	3.21	2.31	1.82
2	3.66	2.92	2.42	2.74	2.03	1.63
3	3.24	2.64	2.24	2.46	1.86	1.52
5	2.71	2.30	2.01	2.12	1.65	1.39
7	2.36	2.08	1.85	1.89	1.51	1.29
10	1.99	1.84	1.69	1.64	1.37	1.20
20	1.26	1.37	1.38	1.17	1.08	1.01

TABLE 2

Spectral Attenuation Relationships

$$\log y = a + b(M - 6) + c(M - 6)^2 + d\log R + kR + s$$
y - randomly oriented horizontal component
After Joyner and Boore (1988)

T(sec)	a	b	c	d	h	k	s	σ
	Pseudo-velocity (cm/sec) 5% damping							
0.10	2.16	0.25	-0.06	-1.0	11.3	-0.0073	-0.02	0.28
0.15	2.40	0.30	-0.08	-1.0	10.8	-0.0067	-0.02	0.28
0.20	2.46	0.35	-0.09	-1.0	9.6	-0.0063	-0.01	0.28
0.30	2.47	0.42	-0.11	-1.0	6.9	-0.0058	0.04	0.28
0.40	2.44	0.47	-0.13	-1.0	5.7	-0.0054	0.10	0.31
0.50	2.41	0.52	-0.14	-1.0	5.1	-0.0051	0.14	0.33
0.75	2.34	0.60	-0 16	-1.0	4.8	-0.0045	0.23	0.33
1.0	2.28	0.67	-0.17	-1.0	4.7	-0.0039	0.27	0.33
1.5	2.19	0.74	-0.19	-1.0	4.7	-0.0026	0.31	0.33
2.0	2.12	0.79	-0.20	-1.0	4.7	-0.0015	0.32	0.33
3.0	2.02	0.85	-0.22	-0.98	4.7	0.0	0.32	0.33
4.0	1.96	0.88	-0.24	-0.95	4.7	0.0	0.29	0.33
	Peak acceleration (g)							
	0.43	0.23	0.0	-1.0	8.0	-0.0027	0.0	0.28
	Peak velocity (cm/sec)							
	2.09	0.49	0.0	-1.0	4.0	-0.0026	0.17	0.33

are reviewed in Joyner and Boore (1988).

As an example, Joyner and Boore (1988) assumed that a peak ground motion parameter or a response spectral ordinate, y, at a particular natural period could be predicted by a function of the form

$$\log y = a + b(M - 6) + c(M - 6)^2 + d \log R + kR + s \pm \epsilon. \qquad (1)$$

$$s = \begin{cases} \neq 0 & \text{soil site} \geq \text{5m thickness} \\ 0 & \text{rock site} \end{cases}$$

$$5 \leq M \leq 7.7$$

$$R = (r^2 + h^2)^{\frac{1}{2}}$$

where M is moment magnitude, r is the distance to the vertical projection on the earth's surface of the nearest point of rupture, and h is a constant. Base 10 logarithms are used. The constants a, b, c, d, h, k and s are determined empirically using regression methods.

The term ϵ is the random error in $\log y$ which has a Gaussian or normal probability distribution with zero mean and standard deviation σ. Thus $\log y$ has a normal distribution or, equivalently, y is lognormally distributed having a median value, y_{50}, given by Equation 1 with $\epsilon = 0$. The median plus one standard deviation or 84 percentile value is given by $y_{84} = y_{50}10^\sigma$. The standard deviation is believed to be mainly a reflection of travel path and local site condition variability. Note that the reason for the existence of the error term in Equation 1 is the same as the reason for the median and 84 percentile amplification factors in Table 1.

The constants in Equation 1 were estimated using the spectral ordinates of the horizontal components of a selected suite of accelerograms from western North America. Coefficients for the randomly oriented horizontal component were estimated by considering both horizontal components at a recording site as independent data. These coefficients are shown in Table 2.

A COMPARISON

A response spectrum computed using the JB attenuation relationships can be meaningfully compared with NH response spectra since the empirical basis of both types of spectra is horizontal component accelerograms from western North America. Assuming that a M6.0 earthquake occurs at a distance of $r = 20$ km, the Joyner and Boore attenuation relationships for median values on a rock site ($s = 0$) result in pseudo-velocity ordinates which are converted to pseudo-acceleration (PSA) ordinates to give the response spectrum labelled JB shown in Figure 1.

For the same earthquake, the relationships for pga and pgv in Table 2 give the median values

$$pga = 0.109g \qquad pgv = 5.34 \text{ cm/sec.}$$

These are used to compute the Newmark-Hall spectrum for 5% damping using median amplification factors in procedure A. This results in the spectrum labelled NH-A shown in Figure 1. This linear plot of an NH spectrum was produced by directly plotting the value $PSA = A = pga \times F_a$ in the short period 'acceleration' range and converting $V = pgv \times F_v$ to PSA in the intermediate period 'velocity' range. It is possible that the long period (> 1 sec) portion of the linear plot is in the 'displacement' range of the spectrum.

Assuming the same value of pga and that the value of pgv is not given, procedure B can also be used to compute a spectrum. An empirical ratio pgv/pga for rock sites is given in Newmark and Hall (1982)

$$pgv/pga = 91.4 \text{ cm/sec/g.}$$

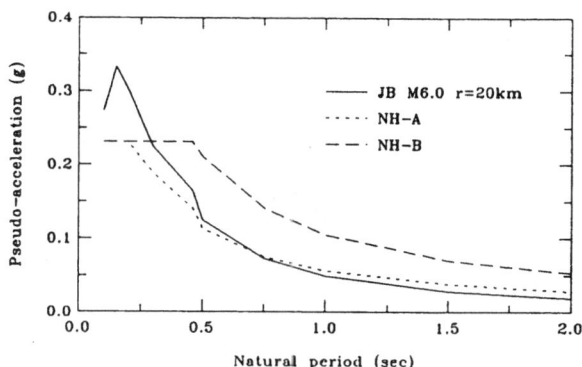

Figure 1. Comparison between Joyner and Boore spectrum and Newmark-Hall spectra computed by procedures A and B.

Given $pga = 0.109g$, the above ratio gives $pgv = 9.97$ cm/sec. This results in the spectrum labelled NH-B in Figure 1.

Despite the differences in their derivation, the JB and NH-A spectra are remarkably similar for periods greater than 0.5 sec. At periods less than about 0.3 sec, there is a significant difference between JB and both NH spectra. This is partly due to the assumption of constant pseudo-acceleration in this period range by procedures A and B. However, NH-B exhibits larger amplification than either JB or NH-A for periods greater than 0.3 sec. The likely reasons for these discrepancies are discussed in the next section.

DEPENDENCE OF SPECTRAL SHAPE ON MAGNITUDE

The amount of low frequency energy in ground motion generated by large magnitude earthquakes is relatively larger than that generated by small earthquakes. Figure 2a shows the accelerogram due to a M6.2 earthquake recorded by an accelerometer of the SMART array in Taiwan. Note the 1 Hz and lower frequencies in this ground motion. A M4.9 aftershock of this earthquake occurred and was recorded by the same accelerometer. The resulting accelerogram is shown in Figure 2b. The high frequency (> 1 Hz) in the ground motion is evident. Both these accelerograms were recorded at almost the same distance from the earthquake and at the same site, so that no complicating factors are present in this comparison.

Studies showing similar dependence of spectral shape on magnitude are described in Trifunac and Anderson (1978) and Iwasaki (1981).

There exists the possibility that the Newmark-Hall spectrum is biased toward that of a large magnitude earthquake. The 28 accelerograms used by Newmark *et al* (1973) to derive amplification factors were recorded during nine earthquakes, seven of which were greater than M6.0. The distribution of magnitudes of these events is shown in the bar diagram of Figure 3a. The preponderance of accelerograms due to earthquakes greater than M6.0 is evident. It should be noted that there were few accelerograms due to earthquakes of magnitude M< 6 at the time the amplification factors were derived.

Joyner and Boore used 64 horizontal component accelerograms recorded during 12 events in

465

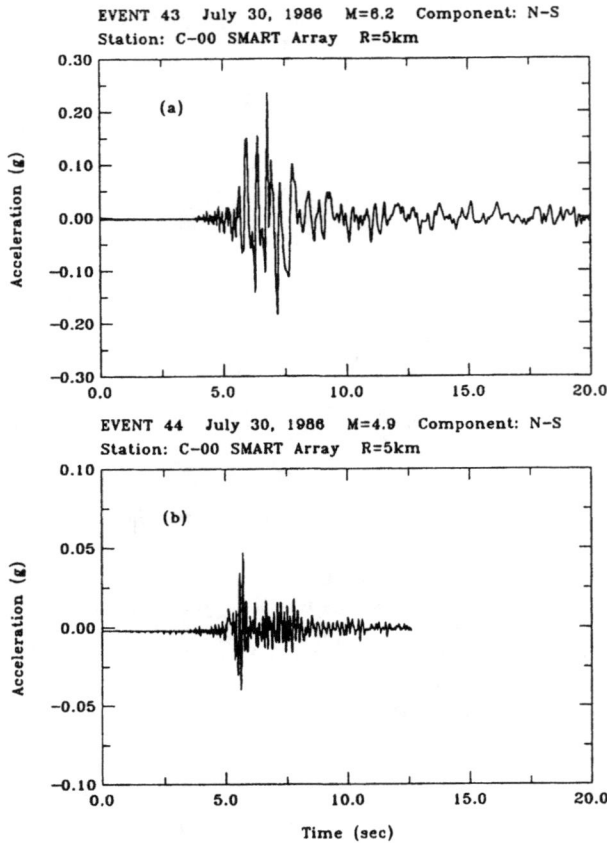

Figure 2. Accelerograms of M6.2 and M4.9 earthquakes recorded by SMART array in Taiwan.

their study. The distribution of magnitudes of these events is shown in the bar diagram of Figure 3b. It may be seen that the magnitudes of the events selected by Joyner and Boore are more evenly distributed in the range $5.0 \leq M \leq 8.0$.

Since the bias in the Newmark-Hall spectrum is evidently toward large magnitudes, it would explain the lower amplification of Newmark-Hall spectra in the low period range, as shown in Figure 1. However, at longer periods, the possible bias of the Newmark-Hall spectra toward larger values may not be significant.

Bias due to magnitude dependence can also be introduced by estimating values of pgv and pgd using published ratios between these ground motion parameters and pga. Figure 4 shows a plot of the ratio pgv/pga versus magnitude. This was also computed using the JB attenuation relationships for rock. From this plot it may be seen that the ratio pgv/pga is a strong function of magnitude. The dependence of pgv/pga on magnitude was also noticed in the seismic zoning maps of Canada by Heidebrecht *et al* (1983). Using the same data base as that used by Joyner and Boore (1988), Donovan and Becker (1986) empirically derived a relationship for the ratio pgv/pga which suggests the same strong magnitude dependence of this ratio. Thus, the practice

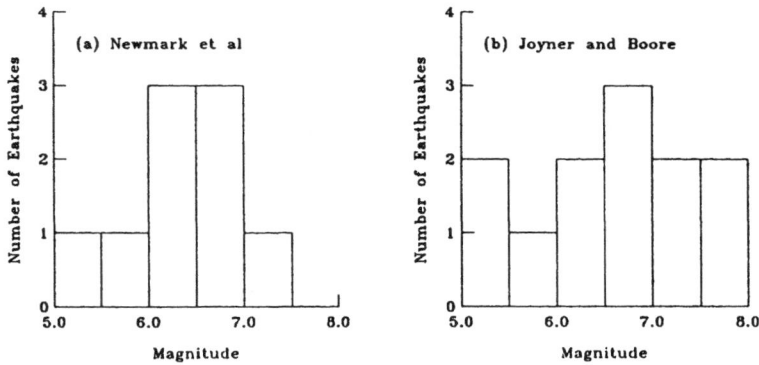

Figure 3. (a) Distribution of earthquake magnitudes used by Newmark *et al* (1973) to derive amplification factors. (b) Distribution of earthquake magnitudes used by Joyner and Boore (1988) to derive attenuation relationships for spectral ordinates.

of estimating *pgv* from published *pgv/pga* ratios can lead to significant magnitude bias in the resulting scaled spectrum. A similar conclusion likely applies to estimating *pgd* from the ratio $pga \cdot pgd/pgv^2$.

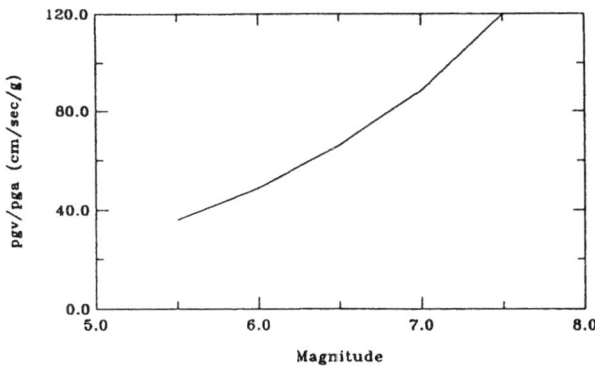

Figure 4. Plot of the ratio *pgv/pga* versus earthquake magnitude. This curve was computed using $r = 20$km in the Joyner and Boore attenuation relationships for rock given in Table 2.

CONCLUSIONS AND RECOMMENDATIONS

Since the shape of a response spectrum depends on earthquake magnitude, the procedure of scaling dynamic amplification factors by peak ground motions to obtain a smooth response spectrum can lead to biased response spectra. The bias can be reduced at natural periods greater

467

than about 0.3 seconds by using procedures which scale amplification factors by *independent* estimates of *pga*, *pgv* and *pgd*. However, at lower periods the Newmark-Hall amplification factors are unconservative owing to the relatively high number of large magnitude earthquakes used in the derivation of these amplification factors.

It is recommended that dynamic amplification factors be derived using a set of accelerograms due to earthquakes whose magnitudes are evenly distributed in the range $5.0 \le M \le 8.0$.

REFERENCES

Donovan, N. C. and Becker, A. M., 1986. Response spectra for building design, in *Proceedings of the Third U.S. National Conference on Earthquake Engineering*, pp. 1153-1163, Earthquake Engineering Research Institute.

Heidebrecht, A. C., Basham, P. W., Rainer, J. H. and Berry, M. J., 1983. Engineering applications of new probabilistic seismic ground motion maps of Canada, *Canadian Journal of Civil Engineering*, 10, 670-680.

Housner, G. W. and Jennings, P. C., 1982. *Earthquake Design Spectra*, Earthquake Engineering Research Institute, Berkeley, Ca.

Iwasaki, T., 1981. Free field design motions during earthquakes, in *International Conference on Recent Advances in Geotechnical Engineering Earthquake Engineering and Soil Dynamics*, S. Prakash, ed., University of Missouri, Rolla.

Joyner, W. B. and Boore, D. M., 1988. Measurement, characterization, and prediction of strong ground motion, in *Earthquake Engineering and Soil Dynamics, II*, J. L. von Thun, ed., Park City, Utah, ASCE Geotechnical Special Publication No. 20, pp. 43-102.

Newmark, N. M., Blume, J. A. and Kapur, K. K., 1973. Seismic design criteria for nuclear power plants, *Journal of the Power Division, ASCE*, 99, 287-303.

Newmark, N. M. and Hall, W. J., 1982. *Earthquake Spectra and Design*, Earthquake Engineering Research Institute, Berkeley, Ca.

Trifunac, M. D. and Anderson, J. G., 1978. Preliminary empirical models for scaling pseudo relative velocity spectra, in *Methods for Prediction of Strong Earthquake Ground Motion*, U. S. Nuclear Regulatory Commission, NUREG/CR-0689, A7-A90.

Use of the uniform hazard spectrum in characterizing expected levels of seismic ground shaking

Gail M. Atkinson[I]

ABSTRACT

This paper provides an overview of the uniform hazard spectrum in comparison to the traditional scaled spectrum approach, for applications in both eastern and western North America (ENA and WNA). Examples are used to show that a scaled spectrum overestimates linear response for intermediate frequencies for some types of earthquakes, by as much as 300%. The result is that scaled spectra for a constant probability level can be in significant error (100% or more) for some seismic environments. A new algorithm for constructing spectra for building code mapping purposes is proposed. The new algorithm replaces peak ground acceleration and velocity with two spectral parameters ('dynamic acceleration' and 'dynamic velocity'). Spectra constructed by the new algorithm are as simple as scaled spectra, but much more accurate.

INTRODUCTION

Traditional probabilistic seismic hazard studies, following the well-accepted Cornell-McGuire method (Cornell, 1968; McGuire, 1977), have often been used to estimate expected levels of peak ground acceleration (PGA) and velocity (PGV) for a specified probability level. Response spectra for engineering design purposes were then constructed by scaling a standard spectrum (eg. Newmark and Hall, 1982) to the site-specific PGA (and/or PGV) levels. This practice formed the basis for Canadian and U.S. building codes from 1970 through 1990 (eg. Basham et al., 1982).

It has been recognized since the mid-seventies that a more direct route of developing the response spectra for any desired probability (McGuire, 1977), known as the 'uniform hazard spectrum' (UHS) approach, is conceptually superior and less subject to error. The UHS is also based on the Cornell-McGuire approach, but the hazard computations are performed for response spectral ordinates for specified frequencies, rather than for peak ground motion parameters. This eliminates the need to scale

I 125 Dunbar Rd. S., Waterloo, Ontario, Canada N2L 2E8

standard spectral shapes to PGA and PGV. The two processes are
compared in Table 1.

Although the UHS method has been available for nearly 15
years, it has not been widely used in standard engineering
practice until recently. This is largely because reliable ground
motion relations for response spectral ordinates, which are
needed for the calculations, have only recently been developed,
based on data that did not exist when the concept was first
proposed. Within the last 5 years, the UHS method has been
applied to a range of facilities in Canada and the U.S.
(including dams, nuclear power plants, and offshore structures),
and has been recommended as the basis for the next revision of
seismic zoning maps for use in building code applications
(Whitman, 1989). However it is still unfamiliar to most
engineers.

The purpose of this paper is to provide an overview of the
UHS in comparison to the traditional scaled spectrum approach,
for applications in both eastern and western North America (ENA
and WNA).

GROUND MOTION RELATIONS FOR ELASTIC RESPONSE SPECTRA

Ground motion relations provide the mathematical link
between the occurrence of earthquakes and the resulting site
ground motions. Given a suitable database, ground motion
relations can be developed for any parameter of interest, such as
PGA, or the maximum response velocity for given frequency values
(PSRV). Response spectra convey more information regarding the
amplitude and frequency content of the earthquake than does PGA
and PGV, and are more directly applicable to dynamic analysis
methods. However PGA and PGV were more often used until
recently, partly due to the availability of applicable ground
motion relations. (Another is the widespread use of empirical
design checks which are based solely on PGA.)

In WNA, ground motion relations have been largely based on
regression of recorded strong ground motions. Since the late
1970's, the database for these relations has greatly improved, to
the point that empirical relations for PGA, PGV and PSRV are now
reliable except for large earthquakes (M > 7) at close distances
(R < 50 km) (see Joyner and Boore, 1988 for a review of these).
It must be understood, however, that such relations provide
median or average ground motion levels, and any specific
observation may deviate from the relations by a factor of two
(about one standard deviation) or more.

More recently, simple seismological models of the earthquake
source, in conjunction with random process methods, have been
used to derive ground motion relations (eg. Boore, 1983; Atkinson
and Boore, 1990) for both WNA and ENA. Comparisons of the
theoretical relations with empirically-based relations and actual

data have verified the model assumptions for WNA (Boore, 1983) and provided confidence in the applicability of the method. Relations derived for ENA by this method are compared to available data in a separate paper in these proceedings (Atkinson, 1991).

The recent improvement in ground motion relations allows systematic comparison of ENA and WNA earthquakes. Ground motion characteristics for the two regions differ due to the high-frequency enrichment of eastern earthquakes, and differences in attenuation and crustal properties. At frequencies less than 10 Hz, eastern and western PSRV values are comparable at near-source distances, but eastern motions decay more slowly with distance. For frequencies greater than 10 Hz, ENA ground motions are significantly larger than their western counterparts. These differences have been well-substantiated by data.

The concept of using a standard spectral shape for all earthquakes was developed in the 1960's and 1970's from a WNA database, dominated by records of M 6 to M 7.5 at distances of 20 to 40 km. The ground motion relations for elastic response spectra can be used to test the applicability of the scaled spectral shapes to earthquakes of various types. Results of such comparisons (Atkinson, 1989) show that the best agreement is achieved for spectra based on both the PGA and PGV, using the median amplification factors of Newmark and Hall (1982). Obviously, since scaled spectra are simplified versions of actual spectra (eg. two straight lines representing a smooth curve) the agreement between scaled and actual spectra is not expected to be perfect. However, the errors incurred by the use of PGA and PGV to construct the simple bilinear spectrum are surprisingly large for some types of earthquakes. In Table 2, these errors are listed for earthquakes of various types, for several frequencies. (Note: A detailed package of plots from which the tables were constructed is available to the interested reader; a sample is given in Figure 1 for illustration.) To estimate these errors, 'actual' response spectra (median observations, based on the relations of Joyner and Boore, 1982, for WNA, and Atkinson and Boore, 1990, for ENA) were compared to the corresponding 'standard' spectra (obtained by scaling PGA and PGV, also from the relations of Joyner and Boore, and Atkinson and Boore, by the median amplification factors of Newmark and Hall, 1982). In these comparisons, and for the remainder of this paper, all ground motion values are median horizontal component values for rock site conditions; spectral response parameters are for 5% damping.

The PGA-PGV scaled spectrum approach works well for ENA events of M5 to M6 at frequencies above 1 Hz. (Caution: spectra for ENA based on PGA alone - not shown - overpredict frequencies less than 10 Hz by as much as an order of magnitude, and should never be used for ENA.) However the scaled PGA-PGV spectra overpredict WNA motions for M5 earthquakes, and for large distances. ENA motions are overpredicted for M5 at frequencies less than 2 Hz, and M7 at intermediate frequencies. In many cases the error is greater than 100%. The implications for hazard analysis are discussed in the next section.

DEVELOPMENT OF SPECTRA FOR A SPECIFIED PROBABILITY LEVEL

Seismic hazard analyses to develop response spectra for use in engineering analyses are usually geared to some target probability level. The degree of agreement between scaled spectra and UHS will depend on the magnitude and distance ranges that contribute most strongly to the hazard. This in turn depends on the region, level of seismicity, and probability level. In order to characterize the resulting effects, response spectra were developed by the Cornell-McGuire method for probabilities of 0.002 per annum (or 10% chance in 50 years) and 0.0001 p.a. (1% in 100 years), for six example cases (Atkinson, 1989). The cases represent areas of low seismicity (eg. Toronto), moderate seismicity (eg. Cornwall) and high seismicity (eg. Charlevoix) in both the ENA and WNA tectonic settings. The difference in results between the two regions arises solely from the different ground motion relations. Table 3 lists the differences between the scaled spectrum (eg. hazard analysis performed for PGA and PGV, then spectrum constructed by scaling) and the UHS (eg. hazard analysis performed directly for PSRV at several frequencies), for the moderate probability level used in building code maps. The scaled spectrum appears to be a good approximation to the expected PSRV values for active areas of WNA (although note its applicability if large subduction earthquakes are considered has not been addressed), and for areas of low to moderate seismicity in ENA. In many cases of interest (eg. regions of low to moderate seismicity in WNA, and regions of high seismicity in ENA), however, the scaled spectrum significantly overestimates the UHS. This is because the most significant contributions to hazard are due to earthquakes that have spectral shapes very different from those assumed by the scaled spectrum approach.

DISCUSSION

Elastic response spectra obtained through scaling standard shapes to PGA and PGV contain significant errors for many types of earthquakes. Errors are even larger if spectra are scaled based on PGA alone. It is concluded that the use of scaled spectra is in general ill-advised, especially since there are now practical, more direct methods to obtain response spectra. The UHS is the recommended method of correctly depicting linear response parameters.

For the purpose of preparing building code hazard maps, it is desirable to depict the amplitude and frequency content of expected earthquake motions in as simple a form as possible. From examination of the test cased described in this paper, it

has been determined that the following two-parameter algorithm accurately depicts the elastic response spectra. For ease of reference the two parameters have been named 'dynamic acceleration', a_d, and 'dynamic velocity', v_d. The dynamic acceleration is defined as the maximum response acceleration (eg. PSRV * 2 * pi * freq), at a frequency of 10 Hz in ENA, or at 5 Hz in WNA. The dynamic velocity is simply the PSRV at a frequency of 1 Hz. A simple spectrum can be constructed on a log-log plot of pseudo-acceleration vs. frequency, as follows:
  - Plot v_d*2*pi at f = 1 Hz.
  - Plot a_d at f = 10 Hz (ENA) or f = 5 Hz (WNA).
  - Draw a horizontal line (eg. constant acceleration) for all
      frequencies above a_d.
  - Draw a straight line connecting a_d to v_d (extrapolate this
      line to obtain lower frequencies).

    This new algorithm is as simple as the PGA-PGV scaled spectrum in that it is also a bilinear shape based on two parameters. However because the two parameters are directly tied to the response spectrum, it is subject to dramatically less error. For WNA, errors for M 5 to M 7 earthquakes, at distances of 10 to 100 km, are generally less than 20% for all frequencies in the range 0.5 to 10 Hz. For ENA, errors are less than 20% for M 5 to M 7, at distances of 10 to 100 km, for the frequency range 1 to 10 Hz; at lower frequencies errors are more significant, but less than those for the scaled spectrum approach. When the new algorithm is applied to hazard computations for the six example cases, it is found to be much more accurate than the scaled PGA-PGV approach for moderate probabilities, with typical errors of less than 20%. Thus a_d and v_d would form a simple basis for national seismic hazard maps that could be used in building code applications to construct elastic response spectra. Note that PGA, where required, could be estimated by dividing a_d by the approximate dynamic amplification, which is about a factor of two (Newmark and Hall, 1982).

    The UHS has recently been endorsed by several bodies. In 1989 a workshop on future U.S. building codes, sponsored by the National Center for Earthquake Engineering, revealed a strong consensus that national hazard maps for building codes should now be based on spectral ordinates, rather than PGA and PGV (Whitman, 1989). The UHS has also been recognized by the EERI Committee on Seismic Risk (1990), and the U.S. National Research Council's Panel on Seismic Hazard Analysis (1988). The Canadian Council for Earthquake Engineering is currently considering the UHS as a basis for seismic parameter mapping for the 1995 edition of the National Building Code.

    Linear response spectra based on the UHS approach are a major improvement over scaled spectra and, by definition, very useful in dynamic analyses for structures which are expected to remain elastic. For extreme loading conditions, economic design of most structural systems requires that limited inelastic deformation be permitted. Recent work (Cornell and Sewell, 1988; Turkstra et al, 1989; Atkinson et al., 1990) has shown that design spectra can be constructed based on damage potential of motions, rather than linear response parameters. Current research efforts are aimed at developing appropriate ground motion relations for damage-potential parameters (such as ductility demand, defined as the ratio of system displacement to displacement at yield) as a function of magnitude and distance. Once these relations are developed, they can be readily used in seismic hazard analyses (eg. replace Step 3 in the right hand side of Figure 1 with the definition of a nonlinear response parameter for several frequency values).

The potential utility of ground motion relations for nonlinear response parameters is very significant. A hazard analysis for nonlinear spectra could be provided in a building code map, based on the UHS approach. This information, in conjunction with the estimated ductility and damping for the structural system, would provide the required yield force, to be used by a designer in proportioning members. Thus the nonlinear spectral approach has the potential to provide seismological information in a format which is of direct use to designers.

In conclusion, seismic hazard mapping should now be based on uniform-hazard linear response spectra. In the future, an increased emphasis on nonlinear response spectra is expected.

REFERENCES

Atkinson, G. (1989). Use of the uniform hazard spectrum in characterizing expected levels of seismic ground shaking. Rpt. to Ontario Hydro.

Atkinson, G. (1991). A comparison of Eastern North America ground motion observations with theoretical predictions. Proc. 6th Can. Conf. Earthq. Eng., this issue.

Atkinson, G., and D. Boore (1990). Recent trends in ground motion and spectral response relations for North America. Earthquake Spectra, 6, 15-36.

Atkinson, G., C. Turkstra and A. Tallin (1990). Damage potential of earthquake ground motion. Rpt. to Ont. Hydro.

Basham, P., D. Weichert, F. Anglin and M. Berry (1982). New probabilistic strong seismic ground motion maps of Canada - A compilation of earthquake source zones, methods and results. Earth Phys. Branch Open-file Rpt. 82-33, Ottawa.

Boore, D. (1983). Stochastic simulation of high-frequency ground motions based on seismological models of the radiated spectra. Bull. Seism. Soc. Am., 73, 1865-1894.

Cornell, C. (1968). Engineering seismic risk analysis. Bull Seism. Soc. Am., 58, 1583-1606.

Cornell, C. and R. Sewell (1988). Non-linear-behavior intensity measures in seismic hazard analysis. Proc. Intl. Seminar on Seismic Zonation, Guangzhou, China, Dec. 1987.

EERI Committee on Seismic Risk (1990). The basics of seismic risk
    analysis. Earthquake Spectra, 5, 675-702.
Joyner, W. and D. Boore (1982). Prediction of earthquake response
    spectra. Proc. 51st Annual Conv. Struct. Eng. Assoc. Calif.
Joyner, W. and D. Boore (1988). Measurement, characterization,
    and prediction of strong ground motion. Proc. ASCE Conf.
    Soil Dynamics, Park City, Utah, June.
McGuire, R. (1977). Seismic design spectra and mapping procedures
    using hazard analysis based directly on oscillator response.
    Intl. J. Earthq. Eng. Struct. Dyn., 5, 211-234.
Newmark, N. and W. Hall (1982). Earthquake spectra and design.
    EERI Monographs. Earthq. Eng. Res. Inst., El Cerrito, Calif.
Turkstra, C., A. Tallin, M. Brahimi and H. Kim (1988). The use of
    ARMA models to measure damage potential in seismic records.
    NCEER Rpt. 88-0032, SUNY-Buffalo.
U.S. National Research Council Panel on Seismic Hazard Analysis,
    Committee on Seismology (1988). Probabilistic Seismic Hazard
    Analysis. Natl. Academy Press, Washington, D.C.
Whitman, R. (ed.) (1989). Workshop on ground motion parameters
    for seismic hazard mapping. NCEER Technical Report.

TABLE 1 - Steps to Obtain Response Spectra for a Specified
          Probability

SCALED PGA-PGV SPECTRUM	UNIFORM HAZARD SPECTRUM
1. Use tectonic information to subdivide region into source areas or faults.	Same
2. Calculate magnitude-recurrence statistics for each source.	Same
3. Define ground motion relations for PGA and PGV based on empirical and theoretical earthquake database.	Define ground motion relations for PSRV at several frequency values based on empirical and theoretical earthquake data.
4. Compute PGA and PGV for selected probability.	Compute PSRV at several frequency values for selected probability. This is the UHS.
5. Estimate spectral values based on scaling algorithm using PGA and PGV. This is the scaled spectrum.	Not required
6. Modify spectrum for local site conditions, if required.	Same

TABLE 2 - Percentage errors for Scaled Spectrum Approach:
Specified magnitudes and distances

CASE	ENA 0.5 Hz	2 Hz	10 Hz	WNA 0.5 Hz	2 Hz	10 Hz
M5 R10	220			320	30	
R30	200			320	40	
R100	160			250	100	100
M6 R10	50			50		- 40
R30	50			40		
R100				30	40	100
M7 R10	40	70				- 25
R30	30	80			25	
R100		60			100	100

TABLE 3 - Percentage Errors for Scaled Spectrum Approach:
Expected values for probability of 0.002 per annum

CASE	ENA 0.5 Hz	2 Hz	10 Hz	WNA 0.5 Hz	2 Hz	10 Hz
low				70	120	180
moderate	50			200	120	
high	50	60		50	30	

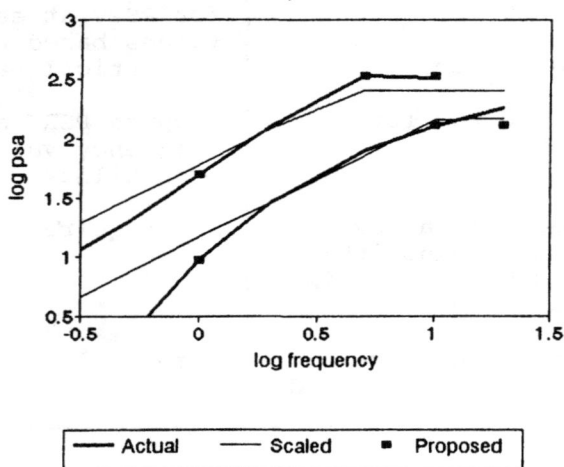

FIGURE 1 - Spectra at R = 20 km
M5 ENA, M6 WNA

— Actual   — Scaled   ■ Proposed

Results of a parametric study on site response effects

K. Elhmadi[I], A.C. Heidebrecht[II], and N. Naumoski[III]

ABSTRACT

This paper presents a brief overview and some results of a parametric study on site response effects due to strong earthquake ground motions. Influence of the soil type, site thickness as well as frequency content and intensity of the seismic motion are investigated. The results are presented in the form of dynamic foundation factors. Comparisons are made with the proposed foundation factors in the National Building Code of Canada (NBCC) 1990. These show that the computed foundation factors significantly exceed the code values when the predominant period of the earthquake motion is close to the site period.

INTRODUCTION

Seismic waves propagating through near-surface soil layers can amplify and produce free field ground motions much larger and with different characteristics than those at rock. The combined effects of earthquakes and local site conditions are commonly referred to as site response effects. There are many examples when extensive damage has been caused by site response effects. During the 1985 Mexico City earthquake, site amplification caused substantial damage and collapse of many buildings. Similarly, the site response effects during the 1989 Loma Prieta, California earthquake were the major cause of failure for many facilities.

In the seismic provisions of the NBCC 1990 (Associate Committee on the National Building Code 1990), the site effects are represented by foundation factor. The soil conditions are categorized into 4 types, and values are assigned to the foundation factor, depending on soil type and depth. In order to evaluate the code foundation factor, an extensive investigation of site response has been

I   Post-Doctoral Fellow, Department of Civil Engineering and Engineering Mechanics, McMaster University, Hamilton, Ontario, Canada, L8S 4L7.

II   Professor, Department of Civil Engineering and Engineering Mechanics, McMaster University, Hamilton, Ontario, Canada, L8S 4L7.

III Research Engineer, Department of Civil Engineering and Engineering Mechanics, McMaster University, Hamilton, Ontario, Canada, L8S 4L7.

conducted at McMaster University (Heidebrecht et al. 1990, Henderson et al. 1990). One phase of this investigation is the parametric study of site response effects (Elhmadi et al. 1990). The parameters include soil type, site thickness, frequency content and intensity of earthquake motion. This paper presents an overview and some results of this parametric study.

METHODOLOGY OF THE PARAMETRIC STUDY

Soil Categories

Four categories of soil deposits are considered in the parametric study. These categories are: (i) normally to lightly overconsolidated clay (NC), (ii) heavily overconsolidated clay (OC), (iii) alluvial sand and silt (AS), and (iv) dense sand (DS). For each category, site thicknesses of 5, 15, 40 and 100 m are considered in the analysis. These types of soil and thicknesses cover a wide range of sites expected to produce site response effects. The soil properties required for the site response analysis are based on measured data from actual sites (Elhmadi et al. 1990). Figure 1 shows the low strain shear modulus, G_0, versus depth for the four soil types. The low strain fundamental periods for each of the sites considered are given in Table 1.

Table 1. Low strain site periods (sec)

Site. Categ.	Thickness (m)			
	5	15	40	100
NC	0.37	0.65	1.10	1.70
OC	0.12	0.32	0.70	1.34
AS	0.18	0.40	0.83	1.65
DS	0.08	0.18	0.39	0.77

Input Motion

It is well recognized that the frequency content of strong seismic ground motion can be characterized by the a/v ratio, in which a is the peak horizontal ground acceleration (g) and v is the peak horizontal ground velocity (m/sec). High a/v ratios are characteristic of earthquake motions with high frequency (low period) content; low a/v ratios are characteristic of motions with low frequency (long period) content. It should be mentioned that the NBCC 1990 uses peak ground velocity, v, to scale the intensity of ground motion and three ranges of a/v ratio to define different force coefficients for structures of low period (i.e., <0.5 sec).

The strong motion database at McMaster was used to select three ensembles of actual time histories, recorded on rock or stiff soil. The three ensembles and their relationships to the zonal combinations in NBCC 1990 ($Z_a>Z_v$, $Z_a=Z_v$ and $Z_a<Z_a$) are defined as follows: H - high a/v ratios, mean a/v ≈ 2 ($Z_a>Z_v$); I - intermediate a/v ratios, mean a/v ≈ 1 ($Z_a=Z_v$); and V - very low a/v ratios, mean a/v ≈ 0.5 ($Z_a<Z_v$), where Z_a and Z_v represent respectively acceleration and velocity related seismic zones. Each ensemble contains 15 different time

histories. A detailed listing of all records is given in Elhmadi et al. (1990). Figure 2 shows the mean acceleration response spectra for each of the three ensembles of records, scaled to peak ground velocity of 1 m/sec. The relations between the a/v ratios and the frequency content of the ensembles are evident from this figure. The predominant periods for H, I and V ensembles are in the neighbourhood of 0.15, 0.3 and 1.0 sec respectively.

For the purposes of the parametric study, all the time histories are scaled to four levels of peak ground velocity, v. These levels correspond to v=0.05, 0.1, 0.2 and 0.4 m/sec. It should be noted that this range of v covers the full range of zonal velocities defined in the seismic zoning maps used with NBCC 1990.

Response Analysis

In the parametric study, the soil deposits are modelled as one-dimensional layered systems with propagation of shear waves only in the vertical direction. The computer program SIREN developed by J.W. Pappin at Ove Arup & Partners (Henderson et al. 1989) is used for the response analysis. The soil is represented as a series of lumped masses connected by nonlinear springs. The soil behaviour is modelled by a hysteretic stress/strain relationship satisfying the Massing principles (Pyke 1979).

For each site and value of v (0.05, 0.1, 0.2 and 0.4 m/sec), surface acceleration time histories were computed using each time history in each ensemble as input motion at the base of the soil profile (rock level). The response characteristics of the motions at both the rock and surface level were also computed and then analyzed statistically for each ensemble. The following response characteristics were computed and analyzed in the parametric study (all are for 5% damping): (i) elastic base shear coefficients for wall- and frame-type structures, (ii) base shear coefficient ratios (surface to rock), and (iii) dynamic foundation factors. Detailed discussion for all of these characteristics is given in Elhmadi et al. (1990). This paper presents some of the results for foundation factors.

FOUNDATION FACTORS

For structures of normal importance, the elastic base shear, V_e, in the NBCC 1990 is defined as

$$V_e = F(vSW) \tag{1}$$

in which F is the foundation factor, having a value of 1 for rock or stiff soil sites and a value of 1.3 to 2 for other site conditions (see NBCC 1990 for detailed description of soil categories); v is the zonal horizontal velocity coefficient, as specified in the seismic zoning map, and equivalent to peak velocity in m/sec; S is the seismic response factor or unit velocity base shear coefficient; and W is the dead load. Note that in terms of F, NBCC 1990 provides for low period "caps" on base shear by specifying upper limit on the product FS, namely: FS need not be greater than 4.2 when $Z_a>Z_v$, and FS need not be greater than 3.0 when $Z_a=Z_v$ or $Z_a<Z_v$.

In order to evaluate the NBCC 1990 foundation factor F, it is necessary to compare the code base shear (Eq. 1) with the surface base shear implied by the response analysis. The implied base shear at the surface level is designated by V_s and can be expressed in the following form:

$$V_s = C_{rs}vW \qquad (2)$$

The coefficient C_{rs} is the dynamic unit velocity elastic base shear coefficient for input motion at the rock level and structures located at the surface. In this equation, v represents peak velocity of the rock motion and it is identical to the code velocity coefficient in Eq. 1. For easier comparison, Eq. 2 can be rewritten as follows:

$$V_s = (C_{rs}/S)vSW = F^*(vSW) \qquad (3)$$

The quantity F^* is a "dynamic" foundation factor which can be compared with F in Eq. 1, including the NBCC 1990 provision that specifies an upper limit on the product FS, as described above.

In the parametric study, the base shear coefficients C_{rs} were computed for each of the surface level time histories for two types of structures, i.e. uniform frame and (shear) wall structures. Simple continuum models of frame and wall structures were used (Heidebrecht and Stafford Smith 1973). The computation was performed using five modes and 5% modal damping. Only frame structures results were analyzed since they exceeded those for wall structures in the low and intermediate period regions. For each site and each set of input motion (H, I and V ensemble; v=0.05, 0.1, 0.2 and 0.4 m/sec), the mean plus one standard deviation dynamic foundation factor spectra, F^*, versus fundamental structural period, T, were computed. Recognizing that the site period, T_s, is one of the most important factors in site response effects, the F^* spectra are converted into normalized spectra (i.e., F^* versus T/T_s).

This paper presents only results for two soil categories (normally to lightly overconsolidated clay, NC, and dense sand, DS) and two site thicknesses (5 m and 100 m). These represent limits in terms of the soil categories and thicknesses considered in the parametric study. The results for the other two soil categories and thicknesses follow the same trends and they are within the range of the results from these limit cases. In order to compare the dynamic foundation factors with the provisions of the NBCC 1990, the code foundation factor, F, for the sites whose results are presented here, is prescribed as follows: F=1 for NC - thickness of 5 m, and for DS - thickness of 5 m and 100 m; F=2 for NC - thickness of 100 m. Figure 3 shows the code foundation factor F=2 versus fundamental structural period, for the three zonal combinations; note that the effective value is reduced at low periods because of the limitation of the product FS as proposed in NBCC 1990. The code foundation factor F=1 has a value of 1.0 for all periods and zonal combinations.

Figures 4 to 7 show the dynamic foundation factors for the soil categories NC and DS, for the thicknesses of 5 m and 100 m. As can be seen, the peak values of F^* are between 1.0 and 6.5. These peak values occur in the range of T/T_s from 0.5 to 1.5. Among the parameters studied, the intensity level, v, is the most influential. The values of F^* are a decreasing function of v. This is due to the lower energy absorbtion and damping at relatively low strain levels. The influence of v, however, is less pronounced for the dense sand deposit. In terms of the influence of the soil category, the peak values of F^* for the soft soil deposit, NC, are in general larger than those for the stiff soil deposit, DS. An exception to this is the 5 m thick soil deposit subjected to H ensemble, which shows larger F^* values for DS site than those for NC site, for $v=0.2$ and 0.4 m/sec. Concerning the influence of site thickness and frequency content of the input motion, the results follow expected trend for DS site but not for NC site. As is shown in Fig. 6, the largest values for the 5 m DS site are associated with H ensemble ($F^*_{max}\approx3.5$), while the effect of V ensemble is very small ($F^*_{max}\approx1$). However, the V ensemble produces significant peak values at the 100 m DS site ($F^*_{max}\approx2$). In terms of NC site (Figs. 4 and 5), the effect of the frequency content on the peak values cannot be recognized when one compare the results for 5 m with those for 100 m soil deposit. For example, the 5 m and 100 m NC sites show similar peak values, when subjected to H ensemble. This is also true for I and V ensemble. An explanation for this is associated with the effect of the top layers. Based on the results presented in Figs. 4 and 5 and those for thicknesses of 15 m and 40 m given in Elhmadi et al. (1990), it seems that the response of the NC site is characterized by that of the top 5 to 10 m soil deposit.

Comparing the dynamic foundation factors, F^*, with the corresponding NBCC 1990 foundation factors, F (i.e., a value of 1.0 for all periods, for F=1; Fig.3 for F=2), one can see that the F^* values in the neighbourhood of the site period exceed the code values in almost all cases. The difference is especially large for low intensity levels of the input motion. The same conclusion is true for the results corresponding to thicknesses of 15 m and 40 m (Elhmadi et al. 1990).

CONCLUSIONS

Results of a parametric study on site response effects due to strong earthquake ground motion are summarized in this paper. The parameters include the soil type, site thickness, frequency content and intensity level of the seismic motion. The results are presented in terms of dynamic foundation factors, F^*, versus the ratio of the fundamental structural period to the site period, T/T_s. The main conclusions are:

- Among the parameters studied, the intensity level, v, is the most influential parameter. Values of F^* are a decreasing function of v. The influence of v is less pronounced for denser deposits.
- The site period, T_s, which is a function of almost all parameters (i.e. intensity level, soil type, site thickness, shear modulus and damping ratio), is a very important factor in studying site response effects.
- In the neighbourhood of the site period, the dynamic foundation factors exceed the NBCC 1990 foundation factors, in almost all cases.

REFERENCES

Associate Committee on the National Building Code 1990. National Building Code of Canada. National Research Council of Canada, Ottawa, Ontario, Canada.
Elhmadi, K., Heidebrecht, A.C. and Hosni, S. 1990. Parametric study of seismic site response effects. EERG Rept. 90-01, Earthq. Eng. Res. Group, McMaster University, Hamilton, Ontario.
Heidebrecht, A.C. and Stafford Smith, B. 1973. Approximate analysis of tall wall-frame structures. ASCE J. Str. Div., 99, 199-221.
Heidebrecht, A.C., Henderson, P., Naumoski, N. and Pappin, J.W. 1990. Seismic

Response and design for structures located on soft clay sites. Can. Geotech. J., 330-341.

Henderson, P., Heidebrecht, A.C., Naumoski, N. and Pappin, J.W. 1990. Site response effects for structures located on sand sites. Can. Geotech. J., 27, 342-354.

Henderson, P., Heidebrecht, A.C., Naumoski, N. and Pappin, J.W. 1989. Site response study - methodology, calibration and verification of computer programs. EERG Rept. 89-01, Earthq. Eng. Res. Group, McMaster University, Hamilton, Ontario.

Pyke, R. 1979. Non-linear soil models for irregular cyclic loadings. ASCE J. Geot. Eng. Div., 105, 715-726.

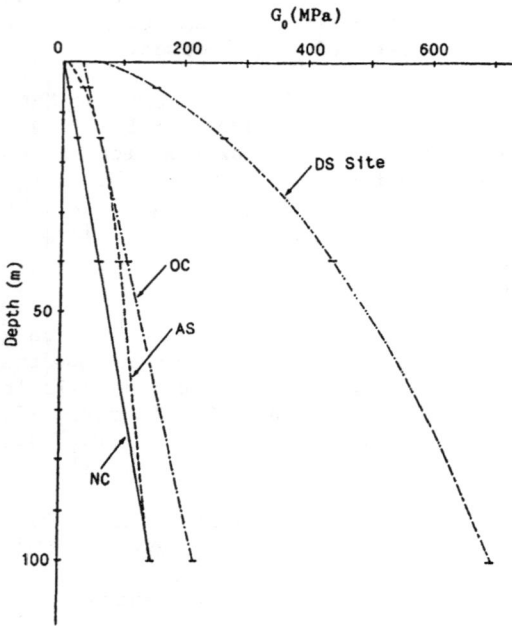

Figure 1. Low strain shear moduli profiles for all four soil categories

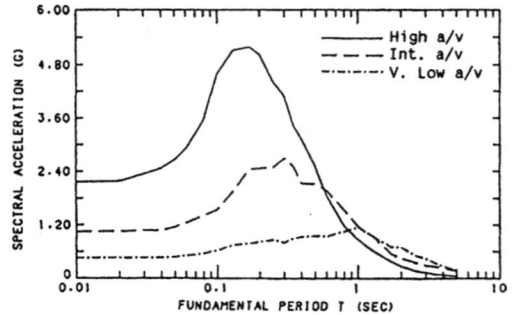

Figure 2. Mean acceleration response spectra for the three ensembles of input motions; 5% damping, v=1 m/sec

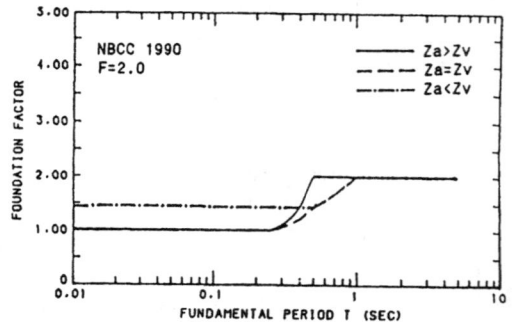

Figure 3. NBCC 1990 foundation factor F=2

Figure 4. Dynamic foundation factors for NC site, 5 m thickness

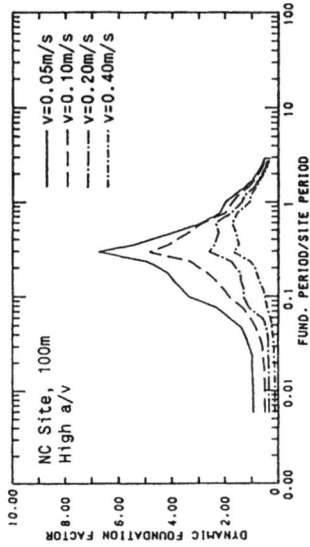

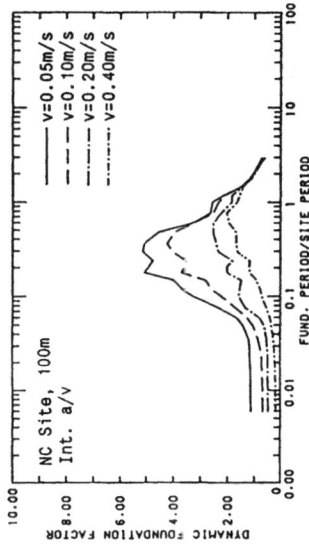

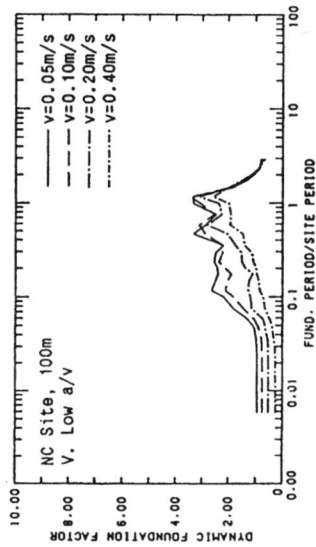

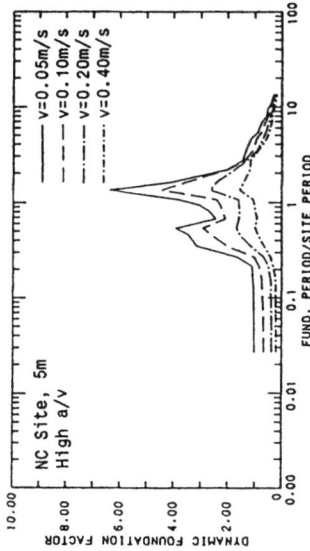

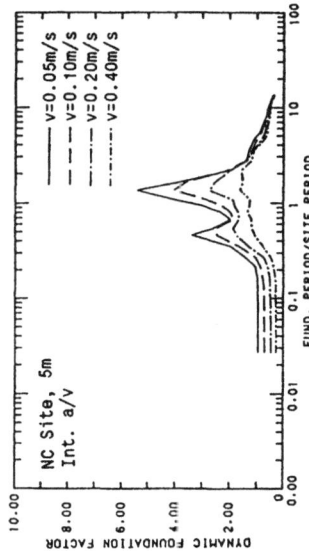

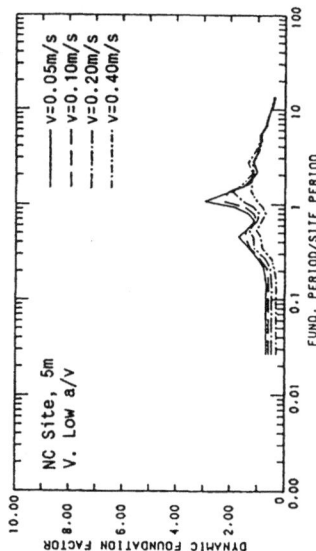

Figure 5. Dynamic foundation factors for NC site, 100 m thickness

483

Figure 6. Dynamic foundation factors for DS site, 5 m thickness

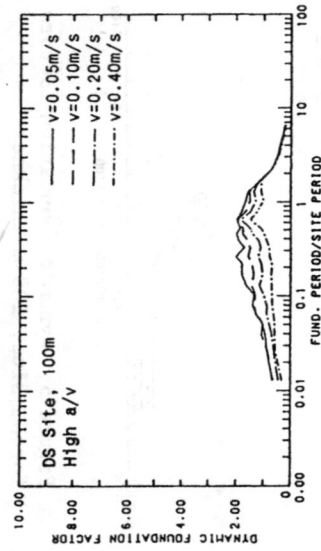

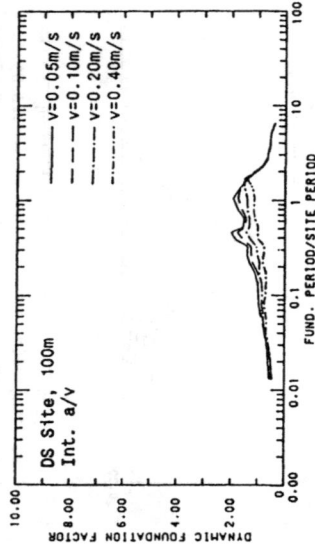

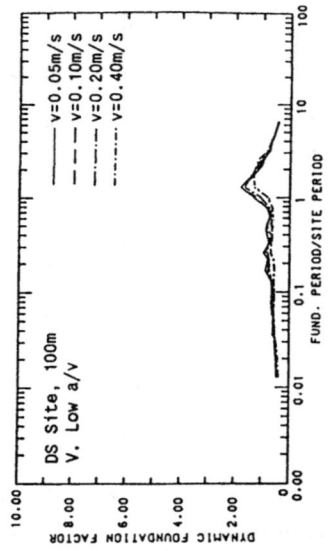

Figure 7. Dynamic foundation factors for DS site, 100 m thickness

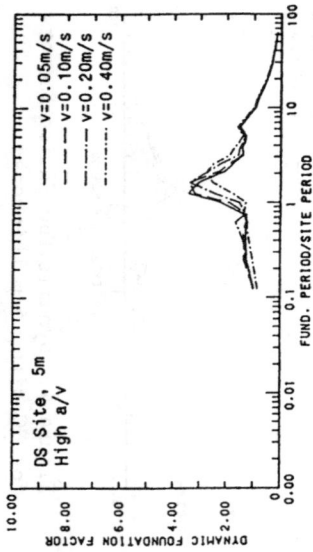

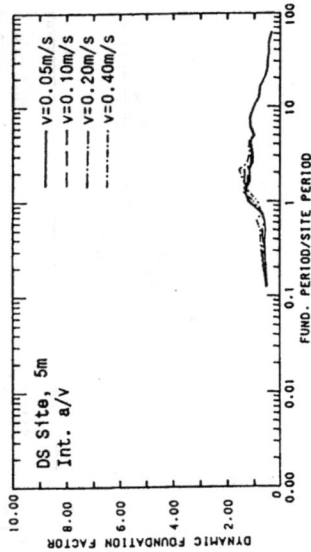

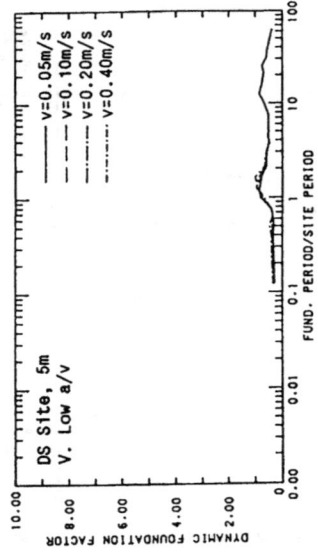

Seismograph records processed by Fourier transform

J. Hans Rainer[1]

ABSTRACT

The method presented of processing strong-motion accelerograph film records employs the Fourier transform and boundary conditions of zero mean and initial and final acceleration, velocity and displacement. Ways of achieving instrument correction, integration, baseline correction and filtering are described and the mathematical expressions summarized. Numerical results for the records of the Saguenay earthquake of 1988 are compared with those obtained by the Geological Survey of Canada using the time-domain method. Good agreement is achieved for acceleration and velocity, but substantial differences are encountered for displacements. This is due to differences in boundary conditions used for displacement. It is recommended that the time-domain method adopt the zero mean displacement instead of the current zero initial displacement.

INTRODUCTION

Recordings of motions of the ground and of structures during earthquakes have contributed immeasurably to the technical advancements in seismology and earthquake engineering. From such records, engineers and scientists are able to develop seismic motions for design and analysis. Before such recordings can be used, however, they need to be "processed" so that inaccuracies and distortions that invariably accompany an instrumental recording process are properly accounted for.

Instruments that have been most successful in recording strong seismic motions obtain the acceleration at the instrument location. Early versions, from the 1940's to the 60's, employed a lightbeam that was reflected off a mirror mounted on a damped single-degree-of-freedom oscillator and focused onto photo-sensitive paper. Then up to the 1980's, photographic film replaced the paper. There are now

[1]   Senior Research Officer, Institute for Research in Construction, National Research Council Canada, Ottawa, Ontario   K1A 0R6

thousands of these film accelerographs in use all over the world, including in Eastern and Western Canada (Weichert and Munro, 1987). A description of film instruments, the recording process, and the numerical treatment of the data is given by Hudson (1979). More recently, force-balance or servo accelerometers have been introduced which produce an electrical signal that is proportional to acceleration. This permits the use of magnetic tape or digital storage media and retention of "pre-trigger" information by using a continuously recirculating memory and saving this pre-trigger portion along with the subsequent signal from the seismic motion. Triggering occurs when the instrument is activated as the arriving seismic motion exceeds a pre-set trigger level. The rest of the paper deals specifically with records from film instruments. For other instruments, certain of the steps described need to be appropriately modified.

PROCESSING OF SEISMOGRAPH RECORDS

After the film is retrieved from the instrument, it is photographically developed and then converted to digital format by manual or automated digitizing equipment. This represents the "raw" digital record. Before it can be considered to represent the "true" seismic motion at the recording station, the record needs to be processed with instrument correction, baseline adjustment and filtering. This can be carried out by a step-by-step time-domain process, generally based on the methods developed at the California Institute of Technology in Pasadena, California (Trifunac and Lee, 1973) and further adapted by the U.S. Geological Survey (Converse 1984).

The Fourier transform technique described here provides an alternative means of seismograph processing and can therefore be used as an independent check on the time-domain results. The technique also permits easier introduction of numerical procedures that are inherently frequency dependent, such as filtering and instrument corrections. In the following, the Fourier transform procedure will be summarized and numerical comparisons presented for the results from the Saguenay Earthquake of November 25, 1988. Further details of the mathematical basis for this procedure can be found in Mulder and Rainer (1989).

Instrument Correction

Since the light beam that exposes the film is reflected off the single-degree-of-freedom oscillator in the instrument, the resulting trace is actually proportional to the displacement of the highly damped moving mass. This happens to be fairly close to the instrument base acceleration, but it is still "contaminated" by the dynamic response characteristics of the oscillator. In the frequency domain, this oscillator characteristic is given by the complex transfer function $H(\omega)$ of the transducer,

$$H(\omega) = \frac{\omega^2/\omega_0^2}{1 - \omega^2/\omega_0^2 - 2i\,\beta\,\omega/\omega_0} \tag{1}$$

where ω_0 = natural frequency of transducer, β = critical damping ratio of transducer, $i = \sqrt{-1}$, and ω = frequency variable.

If $\ddot{X}(\omega)$ is the Fourier transform of the raw time signal $\ddot{x}(t)$, then the instrument-corrected function $\ddot{Y}(\omega)$ in the frequency domain is obtained from

$$\ddot{Y}(\omega) = \ddot{X}(\omega) / H(\omega) \tag{2}$$

and the division is carried out for every frequency increment. The instrument-corrected time trace of the acceleration, $\ddot{y}(t)$, is then obtained by taking the inverse Fourier transform of $\ddot{Y}(\omega)$.

Integration and Baseline Adjustment:

Integration in the frequency domain is carried out simply by dividing the Fourier transform of the acceleration successively by $i\omega$ and $(i\omega)^2$ for every frequency increment to obtain the Fourier transform of the velocity, $\dot{Y}(\omega)$, and displacement, $Y(\omega)$, respectively. The inverse Fourier transform of $\dot{Y}(\omega)$ and $Y(\omega)$ then yields the time-domain traces for velocity, $\dot{y}(t)$, and for displacement, $y(t)$, respectively.

Unfortunately, this integration produces large discrepancies with what are the known or desired boundary conditions for the records, i.e. nominally zero start and finish of the motion. A realistic boundary condition for an elastic vibratory process is zero mean and zero initial and final values for acceleration, velocity and displacement. This would of course, not be valid if permanent displacements had occurred, but to the author's knowledge such a possibility has so far not been taken into account in actual seismograph processing.

It is now possible to derive correction terms to the acceleration trace so that after integration in the frequency domain, the velocity and the displacement traces satisfy the boundary conditions. This derivation is carried out in steps: step 1, the zero mean acceleration is obtained; step 2, zero mean velocity; and finally step 3, zero mean displacement along with zero start and finish. For purposes of observing the efficacy of the correction procedures and for improved behaviour of the filtering process, "tails" of zeros have been added to the start and finish of the record, from 0 to t_1, and t_2 to T, respectively. Thus $(t_2 - t_1) = \tau$ is the original record length. After processing, these tails are then truncated so that only the original record length is displayed in the resulting plots. In the following summary, the subscripts 1, 2 and 3 refer to the signals after the correction steps 1, 2 or 3.

Step 1. From the acceleration trace $y(t) = a$ compute the first-level corrected acceleration trace $a_1(t)$ by subtracting the mean $\bar{a}'$;

$$a_1 = a - \bar{a}' = a - \frac{1}{\tau} \int_{t_1}^{t_2} a(t) \, dt \tag{3}$$

Step 2. Integrate a_1 to compute velocity v_1 and its mean $\bar{v}'$; derive a correction term to the acceleration a_1 so that both the acceleration and velocity trace have zero mean and start and finish. This results in velocity-corrected acceleration a_2 and velocity v_2:

$$a_2 = a_1 + \frac{6\,\bar{v}'\,T}{\tau\,(T-\tau)}\,\left(\frac{2t-\sigma}{\tau}\right), \quad \text{where } \sigma = t_2 + t_1 \tag{4}$$

Step 3. Integrate velocity v_2 to obtain displacement, u, and its mean $\bar{u}'$; derive correction term to the acceleration trace a_2 so that the new velocity v_3 and displacement u_3 have zero mean and start and finish. This results in the fully corrected acceleration trace a_3:

$$a_3 = a_2 - \frac{30\,\bar{u}'\,T}{\tau^2\,(T-\tau)}\,\left\{3\left(\frac{2t-\sigma}{\tau}\right)^2 - 1\right\}. \tag{5}$$

Filtering

Filtering is needed to remove undesirable components of low frequency (high-pass filter) and high frequency (low-pass filter). A high-pass filter is needed to remove those low-frequency components whose presence would overshadow the higher frequency components in the displacement trace. Low-pass filtering is needed to avoid aliasing in the numerical process, and to remove high-frequency components that may not be realistic.

Filtering in the frequency domain is achieved by applying the filter function $F(\omega)$ to the Fourier transform $\ddot{Y}(\omega)$ of the signal $\ddot{y}(t)$:

$$\ddot{Y}_f(\omega) = F(\omega) \cdot \ddot{Y}(\omega) \tag{6}$$

The inverse Fourier transform of $\ddot{Y}_f(\omega)$ then yields the filtered time trace, $y_f(t)$. $F(\omega)$ can be any appropriate filter window; popular ones are the nth order Butterworth filters for low pass and high pass, respectively:

$$F(\omega)_l = 1/(1 + (\omega_0/\omega)^{2n}); \quad F(\omega)_h = 1/(1 + (\omega/\omega_0)^{2n}) \tag{7}$$

Another filter that can be used is the "cosine taper", which reduces the frequency amplitudes to zero via a quarter-cycle cosine curve between two strategically chosen frequencies.

RESULTS FOR SAGUENAY EARTHQUAKE

The Fourier transform method described has been applied to the processing of the ground motions from the Saguenay earthquake of November 25, 1988 as recorded by the Geological Survey of Canada. An example of a record is shown in Fig. 1 for Site 20, Les Éboulements, P.Q., for the transverse instrument direction. For comparison, the corresponding record processed by Weichert and Munro (1989) is shown in Fig. 2. Identical filter values of 0.50 Hz high pass and cosine taper from 50 to 100 Hz were used. The two methods are seen to give results that agree very

CORRECTED ACCELERATION, VELOCITY AND DISPLACEMENT
SAGUENAY EARTHQUAKE OF 1988 11 25 2346 UT
SITE 20, LES EBOULEMENTS, PQ; COMPONENT:- 270.
RECORDED BY: GEOLOGICAL SURVEY OF CANADA
ANALYSIS BY: IRC/NRC, OTTAWA, CANADA
FILTER: 4TH-ORDER BW.: 0.500 HZ, COS.: 50 - 100 HZ; 200 SPS
PEAK VALUES: ACCEL.--100.08 CM/S/S, VELOCITY- -2.64 CM/S, DISPL.- 0.187 CM

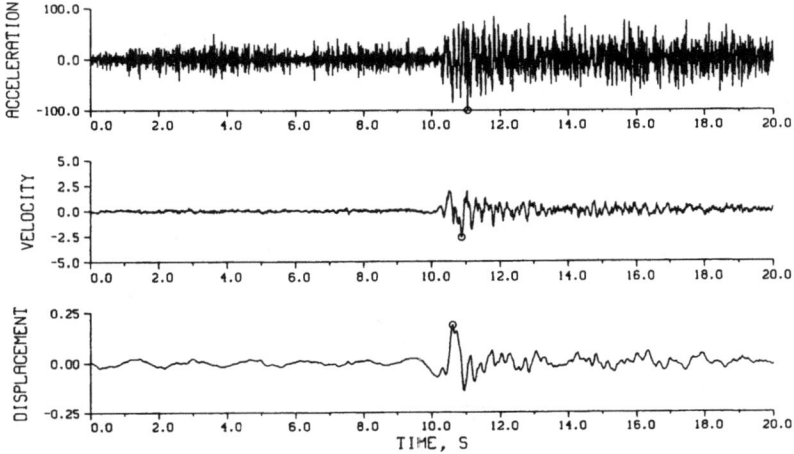

Figure 1.   Seismograph record processed by Fourier Transform Method, Filtered at 0.50 Hz

CORRECTED ACCELERATION. VELOCITY. AND DISPLACEMENT 200.00 SPS
GEOLOGICAL SURVEY OF CANADA
SAGUENAY EARTHQUAKE OF 1988 11 25 2346 UT
SITE 20: LES EBOULEMENTS. QUEBEC
+T = 270 DEGREES; AZ. = 13° DEG.; DIST.- 90 KM
4TH-ORDER BUTTERWORTH AT 0.500 HZ
PEAK VALUES: ACCEL=-100.27 CM/SEC/SEC. VELOCITY=-2.65 CM/SEC. DISPL=0.18 CM

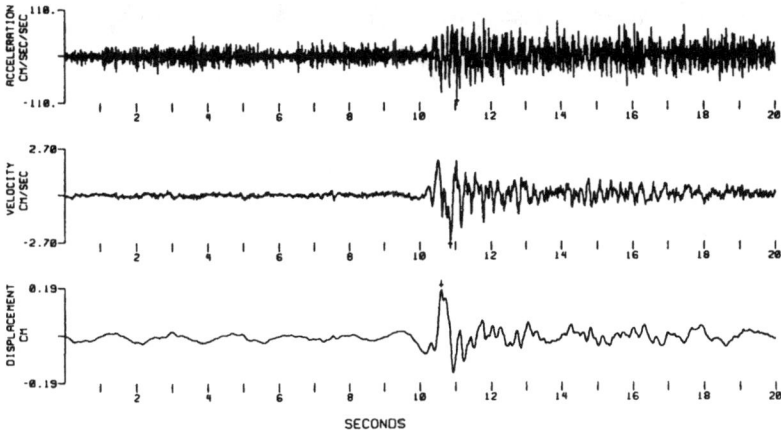

Figure 2.   Seismograph record processed by time-integration method (Weichert and Munro, 1989)

Table 1: Comparison of peak ground motions for Saguenay Earthquake 1988 between time domain (GSC) and Fourier transform (IRC) methods

SITE	STATION	FILTER (HZ)	COMP	ACCELERATION, CM/S/S			VELOCITY, CM/S			DISPLACEMENT, CM		
				GSC	IRC	% DIFF	GSC	IRC	% DIFF	GSC	IRC	% DIFF
1	ST-FERREOL	0.50	L	118.79	118.75	0	-2.71	-2.70	0	-0.11	-0.11	0
			V	-61.21	-62.05	1	1.71	1.71	0	0.13	0.14	7
			T	-95.44	-95.08	0	-2.45	-2.42	1	0.09	0.10	10
2	QUEBEC	0.33	L	-49.61	-49.38	0	1.50	1.49	1	0.21	0.18	17
			V	-19.49	-19.38	1	-0.96	-0.95	1	0.14	0.13	8
			T	-49.68	-50.02	1	2.16	2.15	0	-0.16	-0.18	11
5	TADOUSSAC	0.33	L	26.37	25.83	2	0.58	0.58	0	0.10	-0.07	43
			V	52.27	44.43	18	-1.05	-1.05	0	0.15	0.12	25
			T									
7	BAIE-ST-PAUL	0.33	L	122.84	-122.31	0	3.76	3.73	1	0.56	-0.51	10
			V	-121.31	114.20	6	2.43	2.46	1	-0.40	0.27	48
			T	170.62	170.31	0	-5.34	-5.34	0	0.87	0.52	67
8	LA MALBAIE	0.33	L	-121.82	-128.71	5	-4.65	-4.66	0	0.41	0.40	2
			V	66.51	67.17	1	1.72	1.71	0	0.11	0.12	8
			T	58.73	57.37	2	1.33	1.32	1	-0.12	-0.13	8
9	ST-PASCAL	0.40	L	-45.46	-45.59	0	-2.60	2.56	2	-0.34	-0.35	3
			V	35.91	36.13	1	-1.85	-1.85	0	-0.13	-0.14	7
			T	-54.70	-54.19	1	-2.62	-2.60	1	0.19	-0.20	5
10	RIVIERE-OUELLE	0.40	L	-39.59	-40.74	3	2.21	2.20	0	0.28	0.26	8
			V	22.85	22.98	1	-1.30	-1.30	0	-0.12	0.12	0
			T	-55.92	-55.96	0	-3.52	-3.50	1	0.42	0.40	5
14	STE-LUCIE	0.40	L	13.58	13.37	2	-0.64	-0.64	0	-0.04	-0.05	20
			V	-22.91	-23.63	3	-1.23	-1.34	8	-0.23	0.15	53
			T	22.84	20.98	9	1.03	1.11	7	0.19	-0.12	58
16	CHICOUTIMI-NORD	0.67	L	-104.48	107.69	3	1.51	1.57	4	0.08	0.07	14
			V	100.50	103.65	3	-1.85	-1.86	1	-0.51	-0.13	292
			T	128.66	128.39	0	2.52	2.56	2	-0.20	-0.20	0
17	ST-ANDRE	0.80	L	-152.92	-140.11	9	1.83	1.85	1	-0.07	0.08	12
			V	44.36	45.35	2	-0.88	-0.86	2	-0.05	0.03	67
			T	89.36	98.53	9	0.94	0.94	0	-0.04	-0.04	0
20	LES EBOULEMENTS	0.50	L	-123.07	-123.09	0	4.40	4.42	0	-0.32	-0.32	0
			V	-229.89	-260.70	12	-5.01	-5.04	1	-0.43	0.28	54
			T	-100.27	-100.08	0	-2.65	-2.64	0	0.18	0.19	4

NOTES: GSC = Geological Survey of Canada, Energy Mines and Resources Canada
IRC = Institute for Research in Construction, National Research Council of Canada
L = longitudinal, V = vertical, T = transverse - component directions of instrument

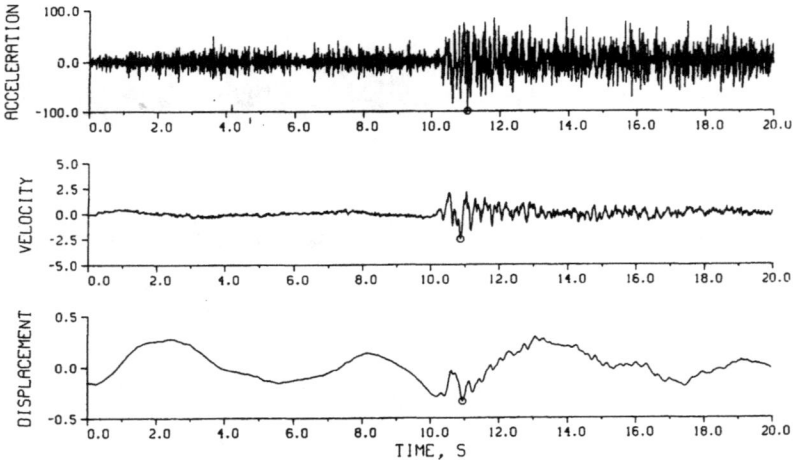

PEAK VALUES: ACCEL.- -99.87 CM/S/S, VELOCITY- -2.54 CM/S, DISPL.- -0.336 CM

Figure 3.   Seismograph record as in Fig. 1 processed by Fourier transform method filtered at 0.1 Hz

closely, as can be seen by comparing the peak values and the detailed amplitude variation with time.

Table 1 presents a comparison of peak values of accelerations, velocity and displacement for all Saguenay earthquake records processed by Weichert and Munro (1989) and by the Fourier transform method. The former results are designated by "GSC", the latter by "IRC". Percent differences of results for the two methods are computed by taking the difference in absolute values and dividing by the IRC value. It may be seen that with few exceptions the peak accelerations and peak velocities agree very closely. Large differences can be seen in the displacements, however. These differences occur because the GSC time-domain method employs an initial condition of zero displacement, after filtering, whereas the IRC method uses a zero mean and a zero start and finish before filtering. This difference is even further accentuated when the 0.10 Hz high-pass filter is employed, with results shown in Fig. 3. While there is still good agreement in velocity and acceleration peaks, large low-frequency components are evident in the displacement trace. The non-zero displacement at the start of the record is clearly discernible and arises from the filtering operation; the comparable GSC displacement trace would be shifted upward by the amount of the offset. After the record has been filtered, the zero mean displacement is considered to be a more reasonable initial boundary condition than the zero initial displacement; thus a change from the latter to the former is recommended for the standard time-domain processing.

Whether the low-frequency components in the displacement trace are real ground motions or an artifact of the recording or processing methods is not clear yet. But since these long-period components occur in the displacement traces filtered at 0.1 Hz for both methods (Munro, 1991), it can be concluded that these components are not the result of the processing methods for instrument correction, integration, baseline correction or filtering.

CONCLUSIONS

The Fourier transform method of processing strong-motion seismograph records retrieved from film instruments presents an independent alternative to the commonly used time-domain method. Comparisons between results from the time-domain method and the Fourier transform method from the Saguenay earthquake of 1988, shows good agreement in peak values of acceleration and velocity and in amplitude variations with time. The displacement traces, however, show large deviations which are mainly due to different boundary conditions used by the two methods. It is recommended that the time-domain method adopt a zero mean displacement boundary condition rather than the current zero initial displacement.

Low-frequency components predominate in the displacement record when high-pass filters of 0.10 Hz are used. The origin of these components is shown not to be due to the integration, instrument correction and filtering procedures.

491

ACKNOWLEDGEMENT

The assistance by L.J. Mulder and J.V. Marans in processing the Fourier transform results is gratefully acknowledged, as are the fruitful discussions with P.S Munro of the Geological Survey of Canada, Ottawa.

REFERENCES

Converse, A. 1984. AGRAM: a series of computer programs for processing digitized strong-motion accelerograms. Version 2.0. U.S. Geological Survey, Open-File Report 84-525, Denver Colorado, July 1984.

Hudson, D.E., 1979, Reading and Interpretating Strong Motion Accelerograms, Berkeley: Earthquake Engineering Research Institute, 112 p.

Mulder, L.J. and Rainer, J.H. 1989. Processing of Seismograph Records-Procedures and Results. Internal Report No. 587, Institute for Research in Construction, National Research Council of Canada, Ottawa, 74 p.

Munro, P.S. 1991. Private Communication.

Munro, P.S. and Weichert, D. 1989. The Saguenay Earthquake of November 25, 1988 Processed Strong Motion Records. Geological Survey of Canada Open File Report 1996, Ottawa, February 1989.

Trifunac, M.D. and Lee, V. 1973. Routine processing of strong-motion accelerograms. Report EERI-73-03. California Institute of Technology, Pasadena, Calif., 360 p.

Weichert, D.H. and Munro, P.S. 1987. Canadian Strong-Motion Seismograph Networks. Proc. Fifth Canadian Conference on Earthquake Engineering, 6-8 July, Ottawa.

Analytical evaluation of site-specific response spectrum

Howard H.M. Hwang[I] and Chen Sam Lee[II]

ABSTRACT

An approach for nonlinear response analysis of sites in the eastern United States is presented to evaluate the characteristics of earthquake ground motions. Since strong-motion data in the eastern United States are very scarce, we use a seismologic model to generate a synthetic acceleration time history at the base of the soil profile. The site response analysis is performed by using the MASH computer program, in which the hysteretic models for sand and clay proposed by Hwang and Lee (1990) is used. The effect of soft-rock on the site-specific response spectrum is discussed.

INTRODUCTION

Estimating the characteristics of ground motions induced by large earthquakes occurring in the eastern United States is quite challenging because only a few strong-motion data were recorded. In addition, soil conditions at a site have significant effects on the characteristics of earthquake ground motions and corresponding response spectrum. Earthquake motions at the bedrock level can be drastically modified in frequency contents and amplitude as seismic waves are transmitted through a soil deposit. Using the site of the Sheahan pumping station, Memphis, Tennessee, as an example, we propose an approach to evaluate the site-specific response spectrum for design of structures.

DYNAMIC SOIL MODEL

[I] Professor, Center for Earthquake Research and Information, Memphis State University, Memphis, TN 38152.

[II] Research Associate, Center for Earthquake Research and Information, Memphis State University, Memphis, TN 38152.

The existing boring log of the Sheahan pumping station terminates at 52 ft (16 m). The soil profile is extended from 52 ft (16 m) to 200 ft (61 m) using a water-well log in the vicinity of the Sheahan pumping station. The soil at this depth is very stiff and thus the base of the soil profile is chosen at this level. The bedrock in the Memphis area is about 3000 ft (909 m) below the ground surface. The material between 200 ft (61 m) and 3000 ft (909 m) is denoted as soft-rock and its effect on the ground motion is included in the input synthetic earthquake time history. The soil profile of the Sheahan pumping station is shown in Fig. 1.

The site response analysis is performed using the MASH computer program (Martin and Seed 1978). The dynamic soil model in the MASH program consists of a horizontally multi-layered soil profile with a fixed base. Static and dynamic properties of soil layers and location of water table are needed to define a dynamic soil model. In this study, static soil properties are either taken from available existing boring logs in the Memphis area or estimated based on empirical correlations. Soil exhibits pronounced nonlinear behavior under cyclic loadings. The secant shear modulus G is strain-dependent and decreases with increasing shear strain levels γ. In the MASH program, the secant shear modulus is expressed as

$$\frac{G}{G_0} = 1 - \left[\frac{[\gamma/\gamma_0]^{2B}}{1 + [\gamma/\gamma_0]^{2B}} \right]^A \tag{1}$$

where G_0 is the low-strain shear modulus; γ_0 is the reference strain; and the parameters A and B describe the shape of the normalized shear modulus reduction curve. In the MASH program, G_0 in psf for sand is estimated from the following empirical equation:

$$G_0 = 61000[1 + 0.01 (D_r - 75)] (\bar{\sigma})^{1/2} \tag{2}$$

where D_r is the relative density in percentage and $\bar{\sigma}$ is the average effective confining pressure in psf. The reference strain γ_0 is expressed as τ_{max} / G_0. τ_{max} is the maximum shear stress under dynamic loadings and is computed using the formula suggested by Hardin and Drnevich (1972). The parameters A and B were determined by Hwang and Lee (1990) as 0.941 and 0.441, respectively.

Several studies have demonstrated that the plasticity index PI is the most dominant factor affecting the shape of the shear modulus reduction curve for clay. In general, the shear modulus exhibits a smaller reduction with increasing plasticity index at the same shear strain level. The low-strain shear modulus G_0 for clay in the MASH program is computed as $G_0 = 2500 S_u$, where S_u is the undrained shear strength of clay and is taken as one half the unconfined compressive strength q_u. In this study, τ_{max} is taken as S_u and G_0 is taken as 2500 S_u; thus the reference strain γ_0 is equal to 0.0004. Sun et al. (1988) suggested the shear modulus reduction curves for clay corresponding to different ranges of plasticity indices. Using nonlinear regression analysis, Hwang and Lee (1990) determined the values of parameters A and B for these curves.

INPUT EARTHQUAKE MOTION

494

Seismic hazard in Memphis and Shelby County, Tennessee, is entirely dominated by the New Madrid seismic zone (NMSZ). In this study, a New Madrid earthquake of moment magnitude M = 7.5 is assumed to occur at Marked Tree, Arkansas, which is near the southern end of the NMSZ. The epicentral distance R from the source to the Sheahan pumping station is about 67 km. Since strong-motion data in the eastern United States are very scarce, a seismologic model is used to estimate the horizontal motions at the base of the soil profile. In this model, source mechanism, path attenuation and soft-rock effects are considered to establish ground motions primarily due to shear waves generated from a seismic source.

The Fourier amplitude spectrum $A(f)$ is formulated following the approach proposed by Boore and Atkinson (1987).

$$A(f) = C \times S(f) \times D(f) \times I(f) \times AF(f) \tag{3}$$

where C is a scaling factor; $S(f)$ is a source spectral function; $D(f)$ is a diminution function; $I(f)$ is a shape filter; and $AF(f)$ is an amplification factor. These factors except the amplification factor are explained in detail by Boore and Atkinson (1987). The values of the parameters that defines these factors are summarized in Table 1.

Table 1. Summary of seismic parameters

	SYMBOL	VALUE
Moment Magnitude	M	7.5
Epicentral Distance	R	67 km
Focal Depth	h	10 km
Radiation Pattern	$\langle R_{\theta\phi}\rangle$	0.55
Horizontal Component	V	0.71
Shear Wave Velocity	β	3.5 km/sec
Source Rock Density	ρ	2.7 gm/cm^3
Quality Factor	$Q(f)$	$1500 f^{0.4}$
Stress Parameter	$\Delta\sigma$	150 bars
Cut-Off Frequency	f_m	30 Hz
Strong-Motion Duration	T	16 sec

The amplification factor $AF(f)$ accounts for the soft-rock effects resulting from the decreasing shear-wave velocities in the soft-rock layers. $AF(f)$ can be calculated as (Boore 1987):

$$AF(f) = \sqrt{\frac{\beta}{\beta_r}} \tag{4}$$

where β is the shear-wave velocity at the source and β_r is the effective shear-wave velocity. The method to calculate β_r proposed by Boore (1987) requires that the shear-wave velocity and layer thickness of all soft-rock layers be known. Based on the shear-wave velocity and thickness of the soft-rock layers in the Memphis area as suggested by Jacob, the amplification factors are calculated and shown in Table 2.

Table 2. Calculation of the amplification factor

H_i (m)	ΣH_i (m)	β_i (m/s)	T_n (sec)	f_n (Hz)	β_r (m/s)	AF
9	9	400	0.02225	11.11	400.00	2.96
13	22	600	0.04392	5.69	500.95	2.64
17	39	1000	0.06092	4.10	640.22	2.34
100	139	950	0.16618	1.50	836.44	2.04
300	439	1100	0.43891	.57	1000.89	1.87
200	639	1400	0.58176	.43	1129.91	1.76
200	839	1700	0.69941	.36	1197.00	1.71
100	939	2000	0.74941	.33	1254.98	1.67
2000	2939	3000	1.41607	.18	2071.00	1.30
7061	10000	3500	3.43350	.07	2892.60	1.10

An earthquake accelerogram generally shows a build-up segment followed by a strong-motion segment and then a decay segment. The frequency content of earthquake accelerograms is found to be approximately constant during the strong-motion segment. Thus, the strong-motion segment of an acceleration time history is considered as a stationary random process and the one-sided power spectrum $S_a(f)$ can be derived from the Fourier amplitude spectrum.

$$S_a(f) = \frac{2}{T_e} |A(f)|^2 \tag{5}$$

where $A(f)$ is the Fourier amplitude spectrum in Eq. 3. T_e is the strong-motion duration and is equal to the source duration which is the reciprocal of the corner frequency f_0, that is, $T_e = 1 / f_0$. In this study, the synthetic time histories are generated using the method proposed by Shinozuka (1974). Given the power spectrum, the stationary acceleration time histories $a_s(t)$ can be generated as follows:

$$a_s(t) = \sqrt{2} \sum_{k=1}^{N} \sqrt{S_a(\omega_k)\Delta\omega} \ \cos(2\pi\omega_k t + \phi_k) \tag{6}$$

where $S_a(\omega_k)$ = one-sided earthquake power spectrum; N = number of frequency intervals; $\Delta\omega$ = frequency increment; $\omega_k = k \ \Delta\omega$; ϕ_k = random phase angles uniformly distributed between 0 and 2π. The nonstationary acceleration time histories $a(t)$ can then be obtained from the multiplication of an envelope function $w(t)$ to the stationary time history. The envelope function w(t) used in this study is composed of three segments: (1) a parabolically increased segment simulating the initial-rise part of the accelerogram and its duration is chosen as one fifth of T_e, (2) a constant segment representing the strong-motion portion of an earthquake excitation and has a duration equal to T_e, and (3) a linearly decayed segment extending four fifths of T_e. Thus, the total duration is $2T_e$. Real earthquake records are commonly observed with long coda durations; however, the coda durations are considered unimportant in most engineering applications.

SITE-SPECIFIC RESPONSE SPECTRA

Using the MASH program, the acceleration time history and corresponding response spectrum with 5% damping ratio at the ground surface are computed based on the earthquake-site model described above. The response spectra (denoted as Case 1) at the ground surface and at the base of the soil profile are shown in Fig. 2. The frequency contents of the base motions have been drastically modified. The spectral values of ground acceleration are significantly higher than those of base accelerations between the period of 0.2 to 1.1 seconds. On the other hand, the high frequency contents of base acceleration, say less than 0.2 second, are significantly reduced as shear waves transmit through the soil deposit.

SOFT-ROCK EFFECT

The approach described above requires that the properties of all the soil and soft-rock layers between the ground surface and the bedrock be known. For a site with a great depth to the bedrock such as the site of the Sheahan pumping station used in this study, the properties of soft-rock layers may be obtained using the reflection analysis or other techniques. However, the results cannot be determined precisely. Assuming we do not have soft-rock data at a site, we still can establish the site model using the upper soil layers as described in this paper. However, the input earthquake time history is generated based on the bedrock data and including the free-surface effects, i.e., neglecting the existance of soft-rock and soil layers. Using this approach, the response spectum at the ground surface and the input response spectrum, denoted as Case 2, are also shown in Fig. 2. It can be seen that even though the input response spectra have significant difference, the resulting ground response spectra are very similar. Thus, for engineering application, if we perform a nonlinear site response analysis with reasonable deep soil profile, we can obtain a fairly accurate response spectrum for the design of structures.

CONCLUSIONS

This paper presents an approach for the nonlinear response analysis of sites in the eastern United States to evaluate the characteristics of earthquake ground motions. The conclusions from this study are as follows:

1. The analytical approach described in this paper combines a seismologic model for input motion and a nonlinear site response analysis. This approach is a reasonable way to generate ground response spectrum for a site in the eastern United States, since strong-motion data are very scarce in this region.

2. The response spectrum at the ground surface is quite different from that at the base of the soil column because of the nonlinear behavior of soil under cyclic loadings. Thus, nonlinear site response analysis must be performed, when a site is subject to large earthquakes.

3. For a site with deep soil profile and several layers of soft rocks overlain the bedrock, the ground response spectrum can be approximately estimated by using the input base motions taken as the bedrock motion at the free surface.

ACKNOWLEDGMENTS

This paper is based on research supported by the National Center for Earthquake Engineering Research (NCEER) under contract No. NCEER-88-3016 and 89-3009 (NSF Grant No. ECE-86-07591). Any opinions, findings, and conclusions expressed in the paper are those of the writers and do not necessarily reflect the views of NCEER or NSF of the United States.

REFERENCES

Boore, D.M. 1987. The prediction of strong ground motion." In Erdik, M.O., and Toksoz, M.N., eds. Strong Ground Motion Seismology, D. Reidel Publishing Company, Boston, MA, pp. 109-141.

Boore, D.M., and Atkinson, G.M. 1987. Stochastic prediction of ground motion and spectral response parameters at hard-rock sites in eastern North America. Bull. Seismol. Soc. of Am., 77 (2), 440-467.

Hardin, B.O., and Drnevich, V.P. 1972. Shear modulus and damping in soils: Design equations and curves. Journal of the Soil Mechanics and Foundations Division, ASCE, 98 (SM7), 667-692.

Hwang, H., and Lee, C.S. 1990. Parametric study of site response analysis. Accepted for publication in International Journal of Soil Dynamics and Earthquake Engineering.

Jacob, K. Shear-wave velocities of soil and rock profile in the Memphis area. Personnal Communication.

Martin, P.P., and Seed, H.B. 1978. MASH, a computer program for the nonlinear analysis of vertically propagating shear waves in horizontally layered deposits. Rep. UCB/EERC-78/23, Earthquake Engineering Research Center, University of California, Berkeley, Calif.

Shinozuka, M. 1974. Digital simulation of random processes in engineering mechanics with the aid of FFT technique. In Ariaratnam, S.T., and Leipholz, H.H.E., eds. Stochastic problems in mechanics, University of Waterloo Press, Waterloo, Ontario, Canada, 277-286.

Sun, J.I., Golesorkhi, R., and Seed, H.B. 1988. Dynamic moduli and damping ratios for cohesive soils. Rep. UCB/EERC-88/15, Earthquake Engineering Research Center, University of California, Berkeley, Calif.

0'

STIFF CLAYEY SILT & SILTY CLAY (ML-CL)

γ_s = 120 pcf, PI = 10-20, S_u = 1500 psf, V_s = 1003 fps

12'

VERY STIFF CLAYEY SILT & SILTY CLAY (ML-CL) $\quad$ Water Table

γ_s = 125 pcf, PI = 10-20, S_u = 3000 psf, V_s = 1389 fps $\quad$ ▽ 18'

34'

DENSE CLAYEY SAND (SC)

γ_s = 130 pcf, K_o = 0.42, D_r = 0.75, ϕ' = 35°, V_s = 838 fps

44'

DENSE CLAYEY SAND TO SAND (SC-SP)

γ_s = 130 pcf, K_o = 0.41, D_r = 0.75, ϕ' = 36°, V_s = 872 fps

52'

VERY DENSE CLAYEY SAND

γ_s = 130 pcf, K_o = 0.41, D_r = 0.93, ϕ' = 36°, V_s = 972 fps

60'

VERY DENSE GRAVELLY SAND

γ_s = 130 pcf, K_o = 0.40, D_r = 0.93, ϕ' = 38°, V_s =1027 fps

99'

VERY STIFF CLAY

γ_s = 125 pcf, PI = 20-40, S_u = 3000 psf, V_s = 1389 fps

141'

HARD SANDY CLAY

γ_s = 130 pcf, PI = 20-40, S_u = 6000 psf, V_s = 1930 fps

172'

VERY DENSE SAND

γ_s = 130 pcf, K_o = 0.40, D_r = 0.93, ϕ' = 38°, V_s =1245 fps

182'

HARD CLAY

γ_s = 130 pcf, PI = 20-40, S_u = 6000 psf, V_s = 1930 fps

200'

/////////////////////////////////////

BEDROCK

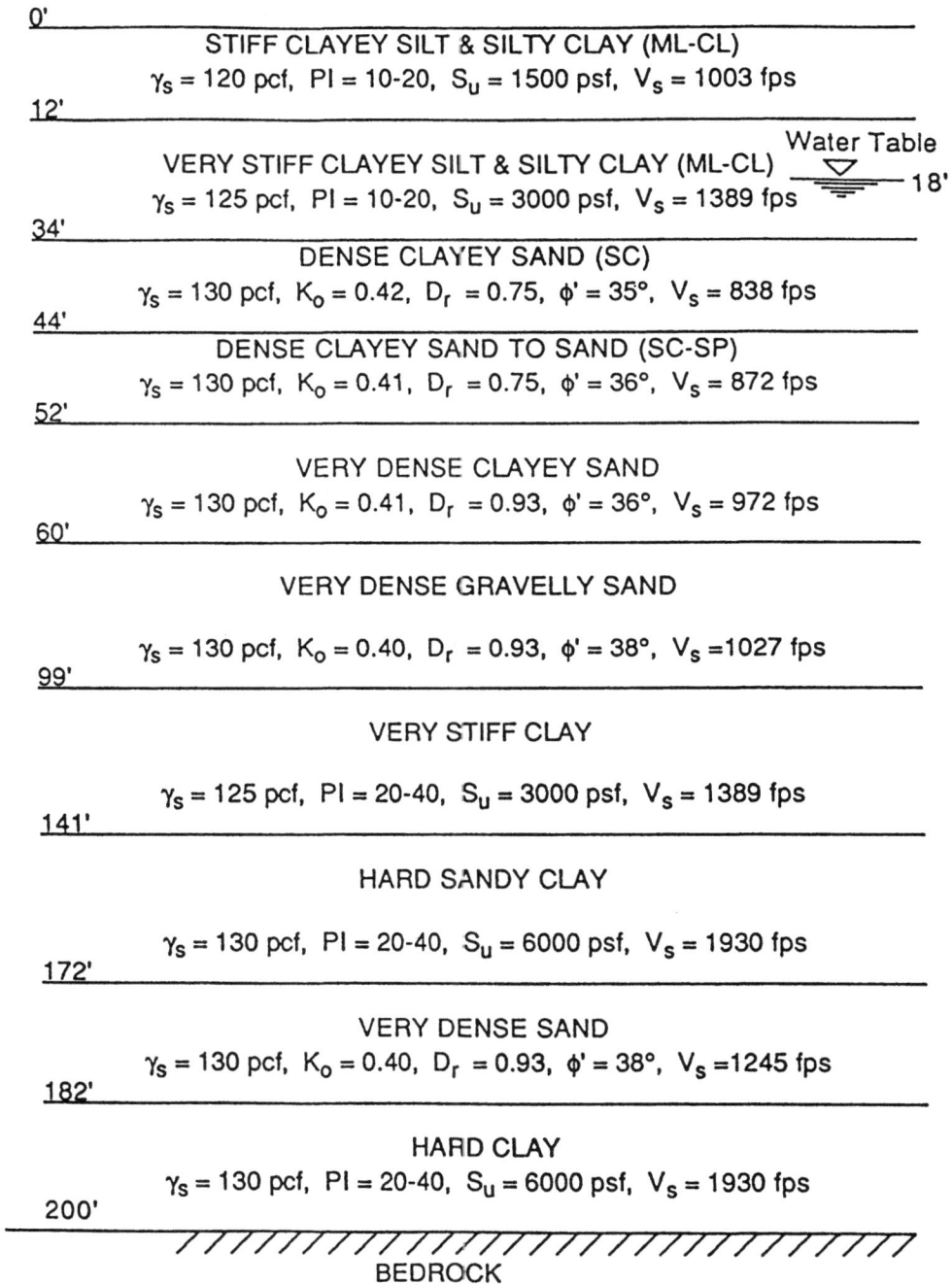

Figure 1. Soil profile for Sheahan pumping station

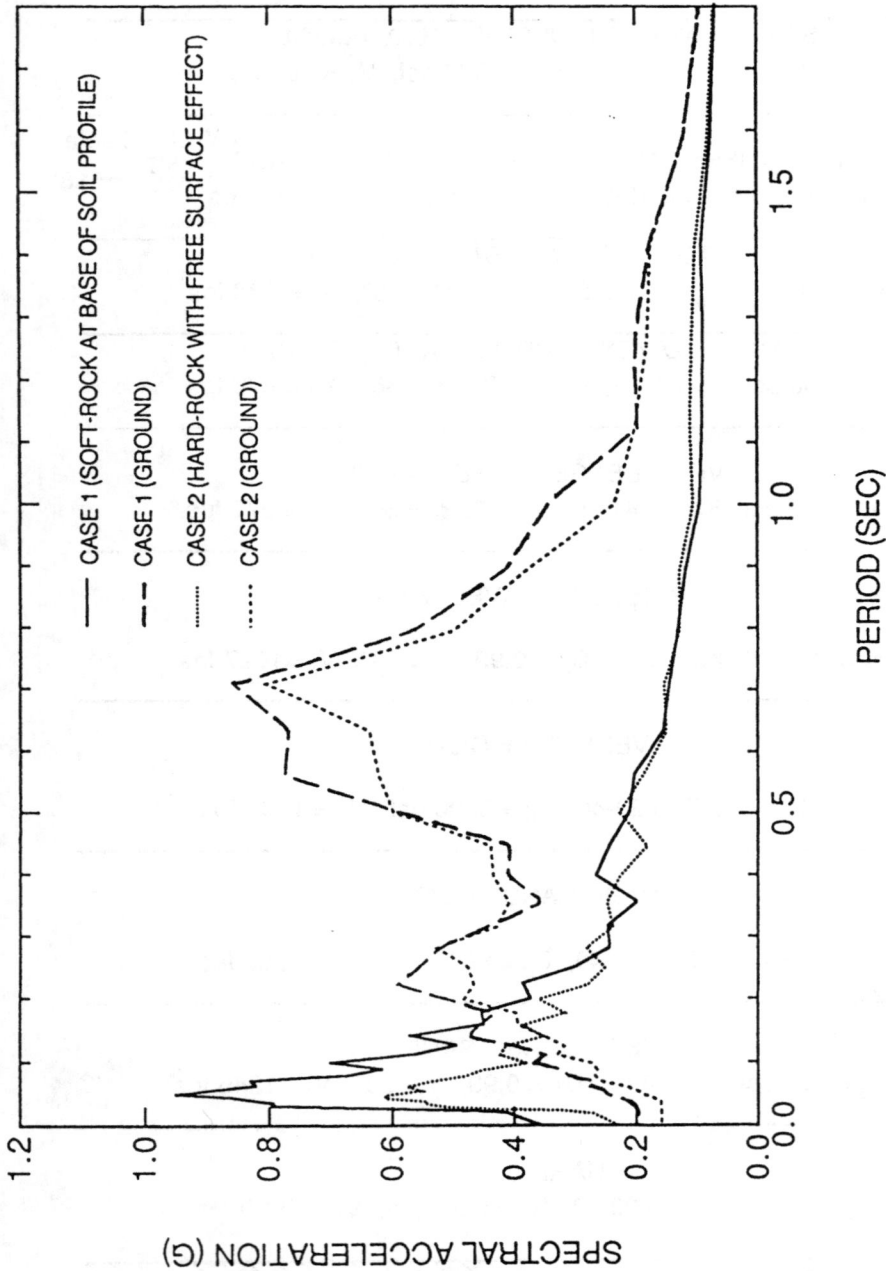

Figure 2. Comparison of response spectra

The implications of recent ground motion observations for theoretical predictions for Eastern North America

Gail M. Atkinson[1]

ABSTRACT

Theoretical predictions of ENA ground motion parameters based on a stochastic model (Boore and Atkinson, 1987; Atkinson and Boore, 1990) are evaluated in light of recent data, including data from the 1988 Saguenay, Quebec earthquake. Data are consistent with the theoretical model on average, although high-frequency ground motions from the Saguenay earthquake are underpredicted. A more general two-corner model of the source is proposed, which would define the source spectrum of any earthquake by both moment and Nuttli magnitude.

INTRODUCTION

The Nov. 25, 1988 Saguenay, Quebec earthquake produced high-frequency ground motions that were significantly greater than those predicted by recent ground motion relations for eastern North America (ENA)(Boore and Atkinson, 1987; Toro and McGuire, 1987; Atkinson and Boore, 1990). The ENA ground motion relations were based on a simple theoretical model. High-frequency radiation was treated as finite-duration bandlimited white noise, whose amplitude spectrum was given by a seismological source model; the source model was a simple bilinear shape, defined solely by the moment magnitude of the earthquake and stress parameter (Brune model with 100 bar stress drop).

The model had been validated, to some extent, by comparisons with small to moderate ENA earthquakes, but the Saguenay earthquake (moment magnitude 6) was seen as a test of its applicability to larger events. (Note: The 1985 Nahanni earthquakes were also considered a test for larger events (Wetmiller et al., 1988), but were not as widely recorded and extensively studied.) The fact that the stochastic model did a poor job of predicting the Saguenay ground motions has raised important questions concerning the validity of the underlying seismological model. Before jumping to conclusions based on a single earthquake, it is worthwhile to step back and consider the

[1] 125 Dunbar Rd. S., Waterloo, Ontario, Canada N2L 2E8

implications of the new data in the broader context of the data set as a whole.

The purpose of this paper is to review the theoretical predictions of the ENA ground motion model in light of all the currently available data. The predictions are represented by the equations of Atkinson and Boore (1990) for the random horizontal component of pseudo-relative velocity (PSRV), for 5% critical damping, on ENA rock sites. The data are ENA PSRV observations (horizontal component or equivalent), for frequencies of 1 to 10 Hz (see Atkinson, 1991a for detailed tables and plots). Data were included only if they were recorded at rock sites in ENA, and the seismic moment had been reliably determined.

The Nahanni earthquakes are considered a valid inclusion in this data set because they exhibit the dominant characteristics of ENA events (Wetmiller et al., 1988): the events occurred in a region of low seismicity and high horizontal compressive stress, thrust mechanisms are dominant, surface ruptures are lacking despite shallow focal depths, and the rocks in the focal region have high seismic velocities.

COMPARISON OF OBSERVATIONS WITH PREDICTIONS

The predictions are evaluated by examining residuals, defined as the log (to the base 10) of the ratio of observations to predictions. Figure 1 plots residuals calculated for the Atkinson and Boore equations, as a function of magnitude and distance, for frequencies of 1 and 5 Hz (points which plot above the zero line represent observations larger than predicted, while points below the zero line are observations that were smaller than predicted). Frequencies of 2 and 10 Hz were also examined, but are not plotted. The Atkinson and Boore equations were derived for magnitudes 4.5 to 7.5 at distances of 10 to 500 km, so these comparisons actually stretch the range of validity of the equations considerably.

At 1 Hz, the residuals show a marked increase with R, and a decrease with M. For frequencies of 2 Hz and greater, the equations predict the data quite well for magnitudes as small as 3.6, to a log distance of 2.8 (R = 630 km), with the exception of the Saguenay ground motions (all the M 6.0 data). The Saguenay motions are much larger than predicted (by almost a factor of 10) for frequencies of 5 to 10 Hz, and show an unusually large degree of scatter. Somerville et al. (1990) interpret the Saguenay observations in terms of wave propagation effects. They argue that strong postcritical reflections produce large ground motion amplitudes in the 60 to 200 km distance range, due to the focal depth and crustal structure. However the analyses of Atkinson and Boore (1991) suggest that the large amplitudes persist over a much larger distance range.

The mean residual at all frequencies is somewhat greater than 0 (equivalent to a factor of about 1.25). The standard deviation of residuals is larger than typically observed in the west (e.g. about 0.35, as compared to typical western values of about 0.25 (Joyner and Boore, 1982)). The positive mean residuals and large standard deviations are at least partly due to the influence of the Saguenay data, but may also be attributed to the fact that the predictive equations were not derived from the data set.

The latter observation suggests that it may be enlightening to derive regression equations from the data set, and compare these to the equations based on the theoretical model. Regression analyses based on these data are described by Atkinson (1991a). The data-based regression equations produce residuals which show no persistent trends with magnitude or distance, except for the Saguenay data, which systematically exceed the overall best-fit equations at high frequencies. In other words, the equations which best fit the data set as a whole cannot fit the Saguenay observations. The equations based entirely on the data do not differ greatly from the theoretical equations, as illustrated in Figure 2. The agreement is rather remarkable considering the very different nature of the derivation methods (empirical versus theoretical).

DISCUSSION

The Atkinson and Boore (1990) equations appear to be conservative with respect to empirical results, even though the Saguenay ground motions are an important part of the empirical data set. However the scatter of the empirical relations is large, with standard deviations of the residuals being about 0.35 log units. This is greater than typical western values of 0.25 (Joyner and Boore, 1982), and may have important implications for seismic hazard evaluations.

The fact that the high-frequency ground motions are overpredicted at Nahanni and underpredicted at Saguenay raises the question as to whether the Nahanni motions are perhaps lower than 'average' due to differences in tectonic setting and stress drop between the Nahanni region and ENA. To address this possibility, Figure 3 compares the amplitude levels of the Nahanni source acceleration spectra, at frequencies above the corner frequency, to the levels for other intraplate events. The data sources and their interpretation are described by Boore and Atkinson (1989).

To show the implications of the data points for source scaling, lines have been drawn corresponding to stress parameter values of 10 and 100 bars in the simple bilinear (Brune) source model. The data suggest that, with the exception of the Saguenay mainshock, stress parameters for intraplate events scatter over the range 25 to 150 bars and have little systematic dependence on moment for M 3.5 to 7. The Nahanni mainshock appears to be consistent with other intraplate events, whereas the Saguenay mainshock does not. This discrepancy cannot be explained solely by focal depth, since the Saguenay foreshock and aftershock do not appear to have anomalous high-frequency amplitudes (all three Saguenay events occurred at depths of 25 to 30 km).

The high stress parameter implied by Figure 3 for the Saguenay mainshock (about 800 bars) does not adequately describe the nature of the discrepancy. Recall that while the model underpredicts the Saguenay earthquake's high-frequency amplitudes, it overpredicts its low-frequency amplitudes (the cross-over point is around 1 Hz). It is the nature of the simple model which fails in this case, rather than the value of the stress parameter.

The discrepancies between theoretical and empirical ground motion predictions can be used to suggest possible refinements to the underlying seismological source model. As suggested by Atkinson (1991a), the shape of the source spectrum can be revised to produce qualitative agreement between theory and observations by the introduction of a second corner frequency. However, a simple two-corner model would still underpredict high-frequency amplitudes of the Saguenay mainshock (and overpredict Nahanni), unless additional parameters are used to more closely reflect specific earthquake characteristics. In order to significantly improve the fit of the theoretical model to any particular earthquake, it appears necessary to introduce modifications to account for earthquake-specific source complexities, or distinctive source characteristics of local tectonic settings.

A simple way to accomplish this is to describe the earthquake by both its moment magnitude, M, and Nuttli magnitude, MN. Atkinson (1991b) describes a scheme for defining the source spectrum, in which the low frequency portion of the spectrum, including the lower corner frequency, is specified by the moment magnitude. The location of the higher corner frequency, and the level of the high-frequency portion of the spectrum, is specified by MN. This new shape is still simple enough to be used in developing predictive equations, but better reflects inter-earthquake variability in the high-frequency level of the spectrum.

The new source spectral model has been tested against the observations of the Saguenay earthquake (Atkinson, 1991b). The fit to the data is much improved, but still not completely satisfactory at high frequencies. In particular, PSRV is underpredicted by a factor of 1.5 to 2 at frequencies of 5 to 10 Hz. (At frequencies of 2 Hz and less, the agreement is good.) Possibly, the refinement of using a high-frequency spectral level tied to MN does not entirely predict the actual high-frequency spectral level. Alternatively, the underprediction may be attributable to wave propagation effects within the relatively small distance range examined, as suggested by Somerville et al. (1990). To examine these possibilities, the model needs to be evaluated and refined based on systematic analysis of a larger data set. Research to accomplish this goal is currently in progress. With proper modeling of the underlying source spectrum and salient propagation effects, ground motion relations can be given a sounder theoretical underpinning. This should ultimately lead to better agreement with ground motion data.

ACKNOWLEDGMENTS

Ground motion data were provided by the Geophysics Division, Geological Survey of Canada. Research was funded by Ontario Hydro.

REFERENCES

Atkinson, G. (1989). Attenuation of the Lg phase and site response for the Eastern Canada Telemetred Network. Seism. Res. Let., 60, 2, 59-69.

Atkinson, G. (1991a). A comparison of ENA ground motion observations with theoretical predictions. Seism. Res. Let., in press.

Atkinson, G. (1991b). Simple ground motion prediction for the Saguenay earthquake. Proc. EPRI/USGS/Stanford Workshop on Ground Motion Prediction, in press.

Atkinson, G., and D. Boore (1990). Recent trends in ground motion and spectral response relations for North America. Earthquake Spectra, 6, 15-36.

Atkinson, G., and D. Boore (1991). Evaluation of ground motion data from the 1988 Saguenay Quebec earthquake. Bull. Seism. Soc. Am., submitted.

Boore, D. and G. Atkinson (1987). Stochastic prediction of ground motion and spectral response parameters at hard-rock sites in eastern North America. Bull. Seism. Soc. Am., 77, 440-467.

Boore, D. and G. Atkinson (1989). Spectral scaling of the 1985-1988 Nahanni, Northwest Territories, earthquakes. Bull. Seism. Soc. Am., 79, 1736-1761.

Joyner, W. and D. Boore (1982). Prediction of earthquake response spectra. Proc. 51st Annual Conv. Struct. Eng. Assoc. Calif.

Somerville, P., J. McLaren, C. Saikia and D. Helmberger (1990). The Nov. 25, 1988 Saguenay, Quebec earthquake: source parameters and the attenuation of strong ground motion. Bull. Seism. Soc. Am., 80, 1118-1143.

Toro, G. and R. McGuire (1987). An investigation into earthquake ground motion characteristics in eastern North America. Bull. Seism. Soc. Am., 77, 468-489.

Wetmiller, R., R. Horner, H. Hasegawa, R. North, M. Lamontagne, D. Weichert, and S. Evans (1988). An analysis of the 1985 Nahanni earthquakes. Bull. Seism. Soc. Am.,78, 590-616.

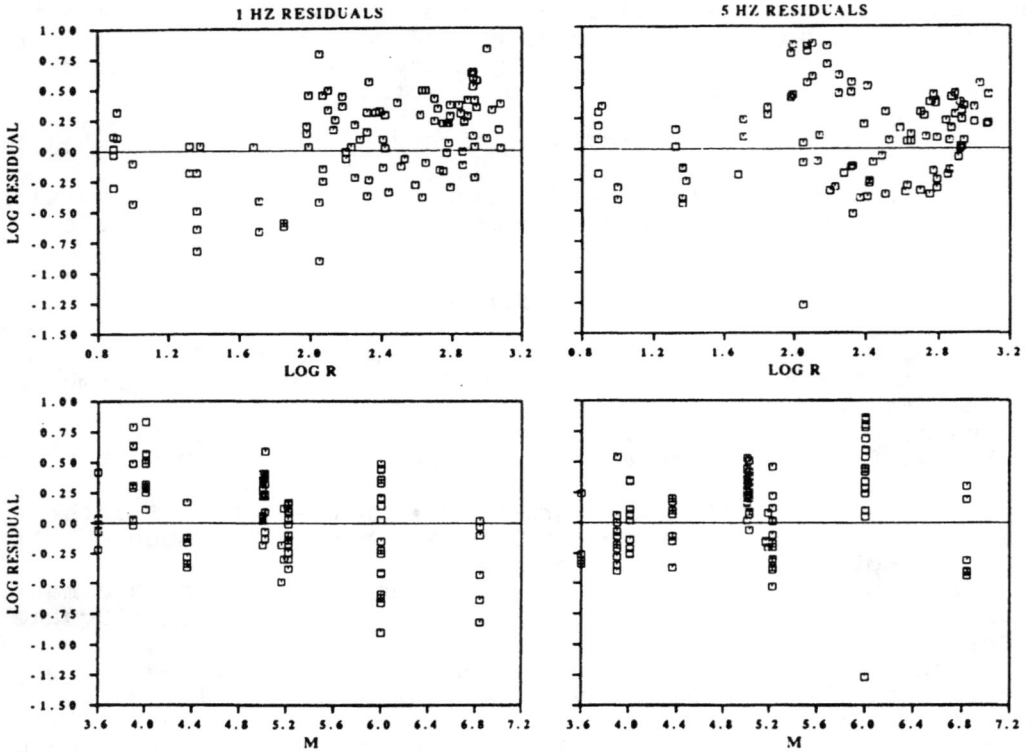

Figure 1. Residuals from comparison of PSRV data with theoretical predictions, as a function of distance and magnitude.

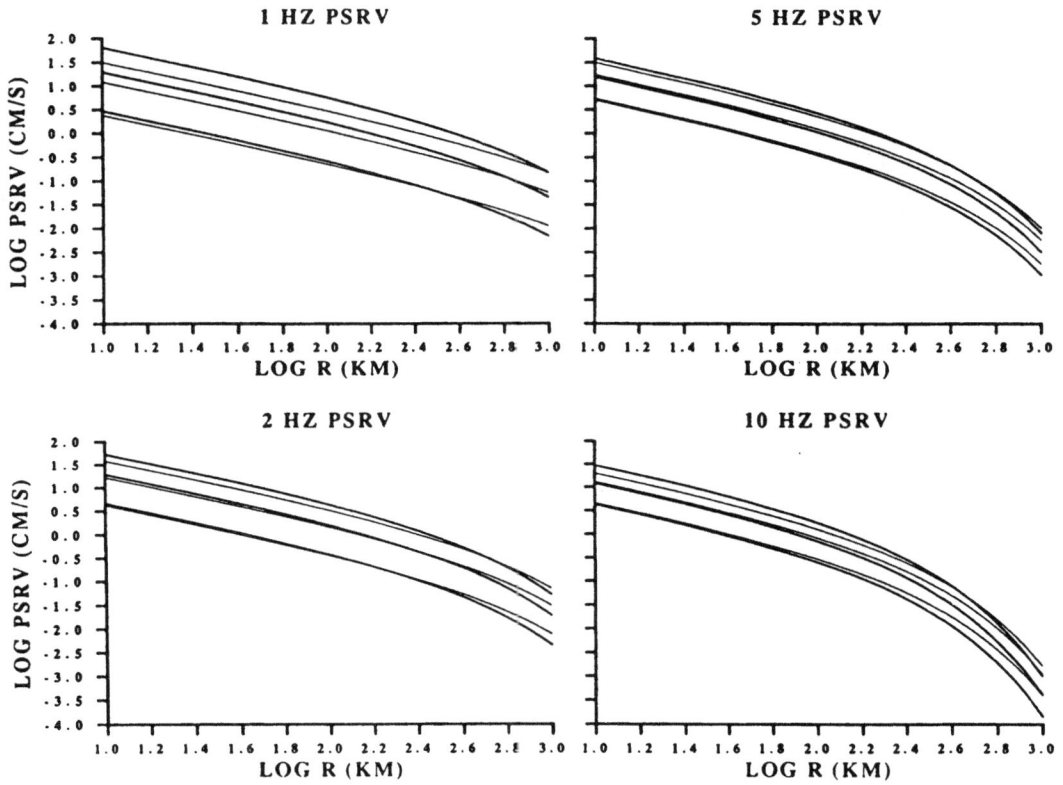

Figure 2. Comparison of theoretical ground motion equations of
Atkinson and Boore (1990) (heavy lines) with empirical
equations from regression of data.

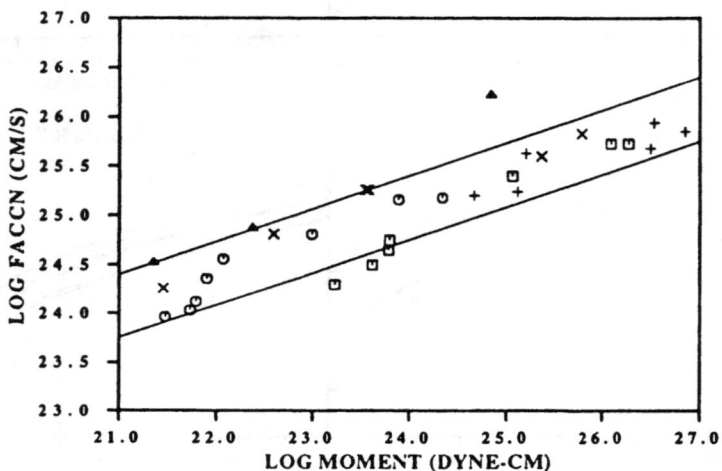

Figure 3. High-frequency acceleration spectral levels for
intraplate earthquakes. Mainshocks of all but the Nahanni,
Miramichi and Saguenay earthquakes are shown as x's (ENA data)
or crosses (intraplate data of Boatwright and Choy); the
Nahanni, Miramichi and Saguenay data are given by solid
squares, circles, and triangles, respectively.

Seismic site amplification study for Fraser Delta, British Columbia

Robert C. Lo[I], Alex Sy[I], Paul W. Henderson[I], David Y. Siu[I], W.D. Liam Finn[II], and Arthur C. Heidebrecht[III]

ABSTRACT

A site amplification study was carried out for the Fraser Delta of British Columbia. The study consisted of three major components: (1) definition of representative soil profiles and selection of dynamic soil properties to simulate typical conditions in the delta, (2) selection of input motions compatible with the local seismo-tectonic setting, including possible megathrust earthquakes offshore, and (3) dynamic response analyses for the selected sites using current state-of-the-practice procedures. The foundation factor computed from these analyses were compared to the provision in the 1990 National Building Code of Canada. The comparison appears to indicate that for typical sites involving deep soil deposits in the Fraser Delta, the current code value of 2 as the foundation factor is generally adequate, except for the period range of about 0.5 to 2 sec depending on: the type of building, the site conditions and the level of spectra considered.

INTRODUCTION

Severe structural damages on deep soil sites during recent major earthquakes around the world highlight the importance of amplification of ground motions at these sites. Notable examples are the 1985 Mexico earthquake, the 1988 Armenian earthquake, and the 1989 Loma Prieta earthquake. A site amplification study was, therefore, carried out to assess the level of protection offered by the seismic provisions of the current National Building Code of Canada (NBCC 1990) to buildings founded on deep soil deposits in the Fraser Delta. The delta is located in the southwest corner of the British Columbia mainland within the Greater Vancouver Regional District, and is in one of the highest seismic areas in Canada.

Ground response studies for sites in the Fraser Delta have been carried out by others (Wallis 1979; Byrne and Anderson 1987; Finn and Nichols 1988). The present study

---

[I]   Klohn Leonoff Ltd., Richmond, British Columbia
[II] University of British Columbia, Vancouver, British Columbia
[III] McMaster University, Hamilton, Ontario

extends the previous works and includes: (1) site specific field and laboratory dynamic soil data, (2) the latest understanding of the areal seismicity, and (3) dynamic soil and structural analyses to study possible performance of buildings of different types. The study focuses on the site amplification aspect, and does not include ground failure aspects such as liquefaction and lateral spreading, which require special foundation treatments. Possible surface wave phenomena involving two or three dimensional effects are also outside the scope of this study.

FRASER DELTA AND STUDY SITES

The Fraser River has been developing its delta since the retreat of the last glaciation approximately 11,000 years ago. Figure 1 shows the areal extent and a typical north-south section of the delta. The soil deposits can be broadly classified into Fraser

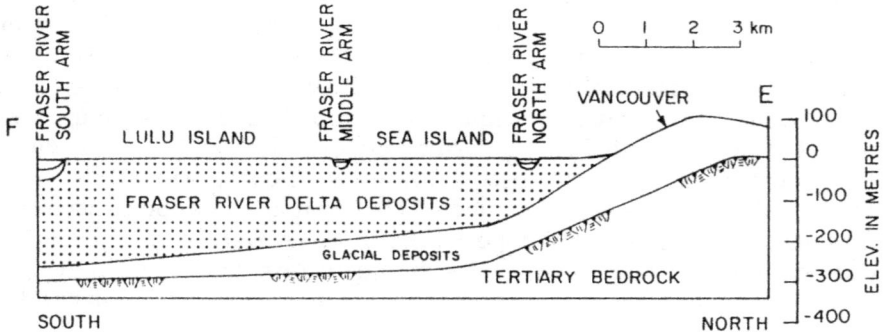

Fig. 1 Areal extent (left) and typical section (above, after Blunden 1975) of Fraser Delta

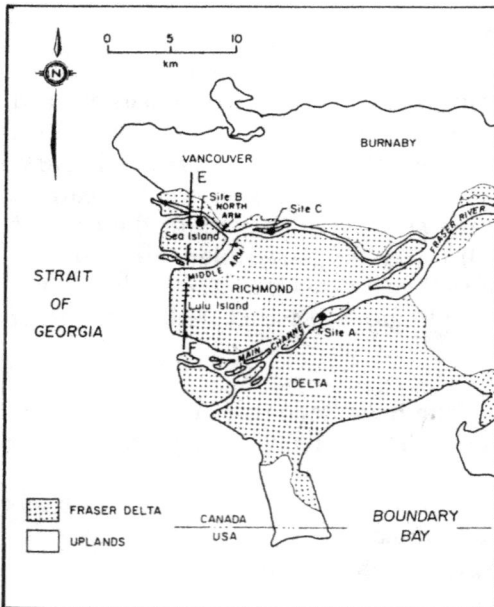

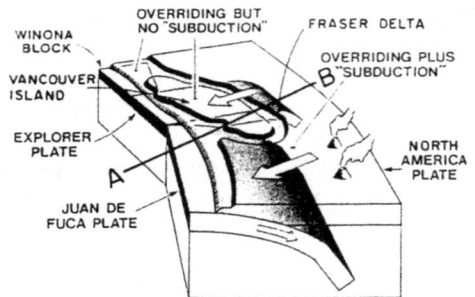

Fig. 3 Tectonic setting of Fraser Delta Area (above, after Rogers 1988)

River sediments of Holocene age and glacial deposits of the Late Wisconsin age or older. The river sediments generally consist of unconsolidated fine-grained marine sediments (silts and clays) overlain by sand-sized marine and tidal flat deposits which are in turn overlain by fine silty overbank deposits. These postglacial deposits are up to about 200 m thick. The underlying glacial deposits consist of ice-loaded tills, glaciomarine and glaciofluvial deposits which overlie Tertiary sedimentary bedrocks. The bedrock surface is estimated to occur at depths of 250 m or deeper under most of the delta (Blunden 1975).

Three representative deep soil sites in the Fraser Delta (Sites A, B and C) were selected for ground response analyses. The selection was based on site locations relative to urban and industrial development as well as available site specific data. The idealised soil and small-strain shear modulus profiles are shown in Fig. 2, based

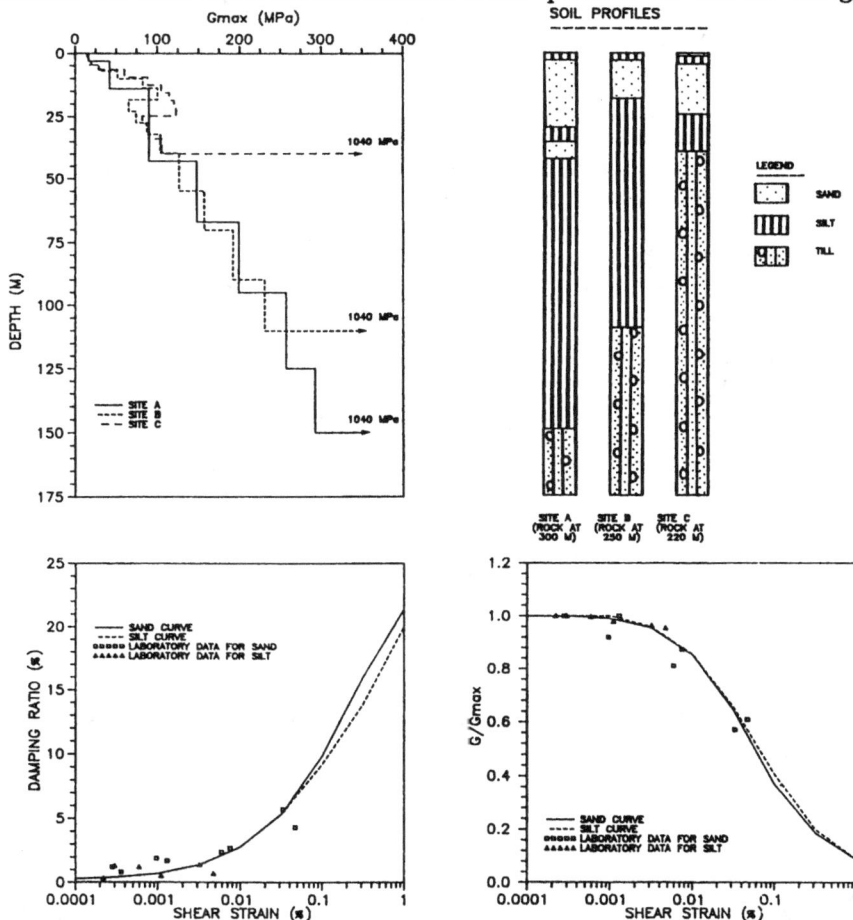

Fig.2 Subsoil and small-strain shear modulus profiles and modulus reduction and damping curves for study sites

on data from deep boreholes, insitu shear wave velocity measurements and other geotechnical and geophysical data. Shear modulus and damping ratio values as functions of shear strain were based on laboratory resonant column test data supplemented by other published data. Figure 2 also shows the modulus reduction and damping curves for sand and silt in the soil profiles. In the small-strain range mobilized in the glacial till, its dynamic properties are represented adequately by the curves indicated for both sand and silt in the figure.

SEISMO-TECTONIC SETTING AND EARTHQUAKE MOTIONS

Most of the seismic activities affecting the Fraser Delta are associated with the interactions among the following tectonic plates: the North America plate, the Juan de Fuca plate, the Explorer plate (see Fig. 3) and the Pacific plate further to the west. The Juan de Fuca plate is moving eastward and down as shown in Fig. 4. There has been no recorded interplate subduction earthquakes here of the type experienced at most other subducting plate boundaries around the world. Current NBCC provisions cover both crustal and intraplate subduction seismic events but not interplate subduction events. However, recent seismological development indicates possible occurrences of megathrust earthquakes (interplate subduction or hereinafter simply referred to as subduction events) off the coast of British Columbia resulting from the rupture at the contact zone of the Juan de Fuca (a segment of the Cascadia) subducting plate and the overlying North America plate. Such a rupture could lead to a magnitude 8.5 to 9.1 subduction event (Rogers 1988).

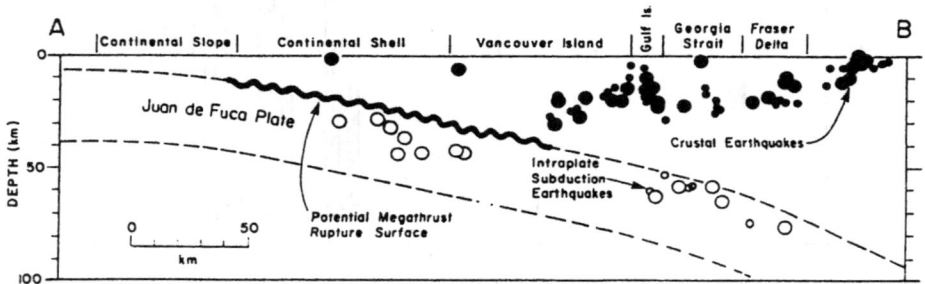

Fig. 4 Section of Juan de Fuca subducting plate and related seismic activities (after Rogers 1990)

A seismic risk analysis for the Fraser Delta, using the Cornell-McGuire method and the NBCC seismic model, yields a peak ground velocity ,v, of 0.21 m/s and a peak ground acceleration ,a, of 0.21 g, corresponding to a probability level of 0.0021 per annum as adopted in the current code. These ground motions give an a/v ratio of unity. The analysis also indicates that the dominant contributions to the NBCC ground motions come from earthquakes of magnitudes 6.3 to 7.3 at epicentral distances of about 30 km to 70 km and a depth of 20 km. A suite of 22 records with 0.8 < a/v < 1.2 (including 1971 San Fernando, two 1979 Yugoslavia, 1986 Taiwan and 1989 Loma Prieta earthquakes) were selected to meet the above criteria and to minimize undue influence by any specific earthquake. Records satisfying the above magnitude and epicentral distance ranges but with a/v ratios significantly higher and lower than unity were also included in another part of the study (Sy et al. 1991).

Rogers (1990) indicated that the potential subduction thrust earthquake could occur at a focal depth down to about 40 km and as close as 100 km west of the study area. The rupture surface could extend another 150 km further to the west. A search for earthquake records on rock at epicentral distances between 100 km and 250 km from subduction events with a magnitude greater than 7.5 resulted in 12 records (including 1985 Mexico and two Japan earthquakes). In addition, the study also examines the potential excitation of relatively moderate ground motions from distant subduction events (>250 km), comprising 10 records from the 1985 Mexico earthquake.

DYNAMIC RESPONSE ANALYSES

The computer program SHAKE was used to compute the response of the soil deposits to the selected input motions. For the NBCC model earthquakes, the selected records were all scaled to the same peak ground velocity for the area (0.21 m/s). For the subduction earthquakes, the closer suite of records (100 to 250 km) were scaled to 0.2 g, while the distant records (>250 km) were scaled to 0.05 g. Attenuation relations proposed by Kawashima et el. (1984) and Youngs et al. (1988) for subduction events were used to arrive at these levels of peak ground motions.

The mean and mean plus one standard deviation (mean + SD) pseudo-acceleration response spectra at the surface and at rock were computed by the SHAKE program for each suite of records. Heidebrecht and Stafford-Smith (1973) showed that the dynamic behaviour of most structural systems lies somewhere between uniform "frame" and "shear wall" type structures. Consequently, the base shear was computed for these two types of structures using simple continuum models (Heidebrecht and Lu 1988). The base shear was obtained by combining the first five modal values using the square root of the sum of the squares method. For each mode, the base shear is the product of the modal mass and the pseudo-acceleration response spectra. To facilitate later discussions, the computed base shears at the surface and at rock are designated as V_{cs} and V_{cr}, respectively.

The elastic base shear, V_e, in the 1990 NBCC is given by: $V_e = vSIFW$; where v = zonal velocity ratio, S = seismic response factor, I = importance factor, F = foundation factor, and W = dead load. For convenience, assuming a structure of normal importance (I = 1) and unit weight (W = 1), then $V_e = vSF$.

The 1990 NBCC specifies F = 2 for "very soft and soft fine-grained soils with depth greater than 15 m", which applies to most of the Fraser Delta. The S factor is specified in the code as a function of the structural period. The code further limits the product "FS" to a "caping" value of 3 in the short-period range.

For all NBCC model earthquakes, the computed foundation factor F_c is obtained as the ratio of V_{cs}/vS, where $v = 0.21$ m/s, and S corresponds to the curve designated as $Z_a = Z_v$ in the code. Figure 5 shows the comparison of computed versus code values of the foundation factor for this suite of events. For the subduction events, since the input motions were not scaled to $v = 0.21$ m/s, it is more appropriate to compare the computed base shears at rock ,V_{cr}, and at the surface, V_{cs}, with the corresponding code values, vS and vsF, as shown on Fig. 6.

DISCUSSIONS OF RESULTS

The fundamental site periods from the SHAKE analyses are 3 sec for Site A, 2.5 sec for Site B and 1.5 sec for Site C. The computed spectral ratio of surface to rock pseudo-acceleration indicates a maximum amplification at the site periods as well as significant amplification between 1-2 sec for Sites A and B and lesser amplification at about 0.5 sec for Site C. The amplification at these shorter periods is likely due to resonance of higher modes.

In Fig. 5, the 1990 NBCC foundation factor reduces from 2.0 to 1.0 at short periods because of the ceiling imposed on the product "FS". For the mean level, the foundation factor for "frame" buildings at Sites B and C exceeds the code factor at

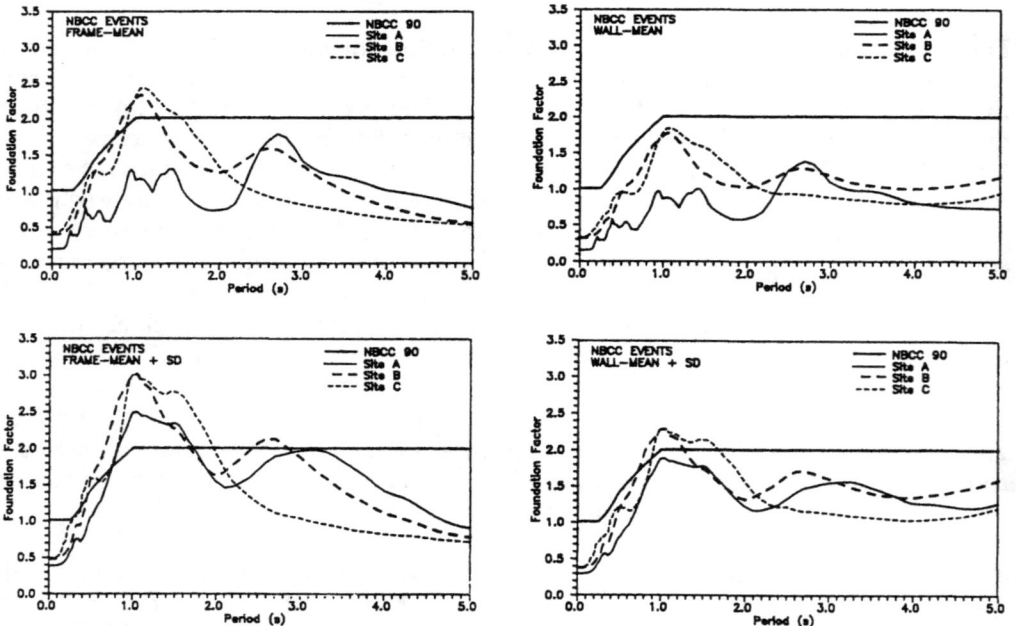

Fig. 5 Comparison of computed versus code values of foundation factor F - NBCC events

periods close to 1 sec, with maximum values of about 2.4. For the mean+SD level, the foundation factor for both types of structures exceeds the code factor for periods between 0.5 and 2 sec, except for "wall" structures on Site A. The peak F values vary between 2.5 and 3 for "frame" buildings and between 1.9 and 2.3 for "wall" buildings.

Only the structural response at Site A was calculated for the two suites of subduction earthquakes. In Fig. 6, the computed rock and surface level base shears are less than the corresponding code values for all periods except for the 'spike' at 0.15 sec for closer subduction events (100-250 km). These 'spikes' are caused by similar spikes present in the response spectra of the accelerograms of two Japanese events recorded at the Miyako Harbour station, and may reflect specific local conditions at that station.

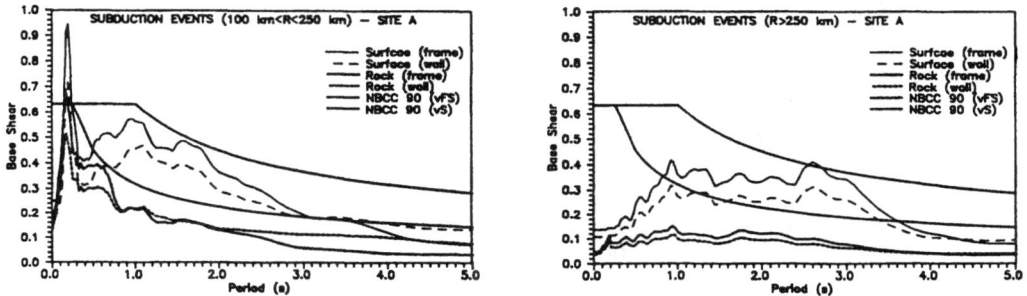

Fig. 6 Comparison of computed versus code values of base shear - subduction events

Currently, no local strong motion records are available to allow for calibration against the results presented herein. However, reasonable inferences can be drawn from these results. In general, for the NBCC model earthquakes, the code foundation factor appears to be adequate, except for the period range of about 0.5 to 2 sec depending on: the type of building, site conditions, and whether the mean or (mean + SD) level spectra are considered. Since this period range corresponds to the majority of buildings in the area, further studies are needed to investigate possible influences on the results from (1) the shortcomings of the equivalent-linear method embodied in the SHAKE analysis currently used in the practice, and (2) the peculiarities of some input motions. For the suites of subduction events, the computed base shears are overall less than the code specified values. This favourable situation, of course, requires further confirmation when more insight into the subduction events and more subduction records become available.

CONCLUSIONS

In summary, for typical deep soil sites in the Fraser Delta the increase of the code foundation factor, F, from 1.5 (NBCC 1985) to the present value of 2, as a result of lessons learned from the 1985 Mexico earthquake, appears to be justified. Further studies are needed to confirm the overall applicability of the present code value of 2 to typical buildings in the Fraser Delta.

ACKNOWLEDGEMENTS

Financial support for this study was provided by the National Research Council under an IRAP-M grant, which was administrated by Mr. Alan Toon. The assistance of Dr. Garry C. Rogers in describing the subduction earthquake scenarios, Dr. Nove Naumoski in providing earthquake records from the MUSE database, and Mr. Adrian Wightman in reviewing the study is also acknowledged.

REFERENCES

Blunden, R.H. 1975. Urban Geology of Richmond, British Columbia. Department of Geology Report No. 15, University of British Columbia.

Byrne, P.M. and Anderson, D.L. 1987. Earthquake Design in Richmond, British Columbia, Version II. Soil Mechanics Series No. 109, University of British Columbia.

Finn, W.D.L. and Nichols, A.M. 1988. Seismic Response of Long-Period Sites: Lessons from the September 19, 1985 Mexican Earthquake. Canadian Geotechnical Journal, Vol. 25, pp. 128-137.

Heidebrecht, A.C. and Stafford-Smith, B. 1973. Approximate Analysis of Tall Wall-Frame Structures, ASCE Vol. 99, No. ST2, pp. 199-221.

Heidebrecht, A.C. and Lu, C.Y. 1988. Evaluation of the Seismic Response Factor Introduced in the 1985 Edition of the National Building Code of Canada. Canadian Journal of Civil Engineering, Vol. 15, No. 3, pp. 283-288.

Kawashima, K., Aizawa, K. and Takahashi, K. 1984. Attenuation of Peak Ground Motion and Absolute Response Spectra. Proceedings of 8th WCEE, Vol. II, pp. 257-264.

Rogers, G.C. 1990. Personal Communications.

Rogers, G.C. 1988. An Assessment of the Megathrust Earthquake Potential of the Cascadia Subduction Zone. Canadian Journal Earth Sciences, Vol. 25, pp. 844-852.

Sy, A., Lo, R.C., Henderson, P.W., Siu, D.Y., Finn, W.D.L. and Heidebrecht, A.C. 1991. Ground Motion Response in Fraser Delta, British Columbia. Proceedings of 4th International Conference on Seismic Zonation, August, San Francisco.

Wallis, D.M. 1979. Ground Surface Motions in the Fraser Delta Due to Earthquakes. M.A.Sc. Thesis, University of British Columbia.

Youngs, R.R., Day, S.M. and Stevens, J.L. 1988. Near Field Ground Motions for Large Subduction Earthquakes. Proceedings of ASCE Specialty Conference - Earthquake Engineering and Soil Dynamics II, June, Utah, pp. 445-462.

Attenuation of strong ground motion from the Saguenay, Quebec earthquake of November 25, 1988

Paul G. Somerville[1]

ABSTRACT

The November 25, 1988 Saguenay earthquake, which occurred at the unusually deep focal depth of 29 km and had a moment magnitude of 5.8, produced by far the largest set of strong motion recordings of any earthquake in eastern North America. The attenuation of recorded strong ground motions is very gradual in the distance range of 50 to 120 km, and only becomes steep beyond 120 km. A profile of synthetic seismograms reproduces these features, and allows us to understand why the attenuation relation has this shape. In both the recorded and synthetic seismograms, the peak amplitudes inside 120 km are due to large postcritical reflections from the Conrad and Moho discontinuities. These observations support the model for the attenuation of strong ground motion proposed by Burger *et al.* (1987) in which the shape of the attenuation curve within 200 km of the source is controlled by focal depth and crustal structure. The distances over which ground motion amplitudes are elevated by postcritical reflections generally lie in the overall range of 50 to 200 km, with the specific distance range depending on the focal depth of the earthquake and on the crustal structure. Because of the deep focal depth of the Saguenay earthquake, the critical distances for these reflections were short, causing the ground motion amplitudes to be elevated in the distance range of 50 to 120 km. The recorded ground motions were significantly underpredicted by attenuation relations based on random process models which do not take account of these effects.

INTRODUCTION

The November 25, 1988 Saguenay earthquake produced by far the largest set of strong motion recordings of any earthquake in eastern North America (Munro and North., 1988). These recordings provide an opportunity to test methods, presently based on a limited data set, for estimating strong ground motions of eastern North American earthquakes. The Saguenay earthquake occurred within the Grenville Province, close to the southern margin of the Saguenay Graben in southern Quebec and about 100 km northwest of the St. Lawrence River. The earthquake occurred at 23:46:04. GMT on November 25, 1988 at latitude 48.117°N, longitude 71.184°W (North *et al.*, 1989). The mechanism of the Saguenay earthquake was nearly pure thrust with a P axis oriented east-northeast, consistent with that of the larger earthquakes in the northeastern United States and southeastern Canada (Ebel *et al.*, 1986, Somerville *et al.*, 1987). The Saguenay earthquake originated at a depth of 29 km, which is greater than the depth range of 5 to 15 km that is characteristic of the larger earthquakes in eastern North America (Ebel *et al.*, 1986; Somerville *et al.*, 1987). The overall source duration τ of the earthquake of 1.8 seconds, combined with a seismic moment M_o of 5 x 10^{24} dyne-cm, corresponds to a stress drop $\Delta\sigma$ of approximately 160 bars. This is within the uncertainty of the median value of 120 bars obtained from thirteen previous eastern North American events using the same methods (Somerville *et al.*, 1987).

---

[1] Associate, Woodward-Clyde Consultants, 566 El Dorado Street, Pasadena, CA 91101

The seismic moment of the Saguenay earthquake estimated from long-period regional (P_{nL}) and teleseismic body waves is 5.0×10^{24} dyne-cm, with an uncertainty of 25% (Somerville *et al.*, 1990a). This seismic moment value corresponds to a moment magnitude of 5.8 (+/- 0.07). The reported magnitudes of the Saguenay earthquake are an M_s of 5.8 (NEIS), an m_b of 5.9 (NEIS), and an m_{bLg} of 6.5 (North *et al.*, 1989). North *et al.* (1989) noted that this m_{bLg} value is substantially larger than the teleseismic m_b of 5.9. It exceeds the value of 6.0 estimated from the seismic moment of the event using an empirical relation between moment magnitude M and m_{bLg} for the larger historical eastern North American earthquakes. The empirical relation ($m_{bLg} = 0.89$ M + 0.90) is derived from the twelve eastern North American earthquakes listed in Table 1 of Somerville *et al.* (1987). Somerville *et al.* (1990a) concluded that the observed m_{bLg} of the Saguenay earthquake is consistent with their source and crustal structure models.

MODELING OF STRONG GROUND MOTION

Because of the small size of the strong motion data base for eastern North American earthquakes, synthetic seismogram techniques have played an important role in the development of ground motion attenuation relations for eastern North America in recent years. Burger *et al.* (1987) investigated the effect of wave propagation in the crustal waveguide on the shape of the ground motion attenuation curve. The principal seismic phases associated with wave propagation in a crustal waveguide are shown schematically in simplified form in Fig. 1a. At close distances, peak horizontal ground motions are controlled by direct upgoing shear waves. As distance increases, the reflections of the shear wave from interfaces in the lower crust (such as those at 30 and 40 km in Fig. 1a) reach the critical angle and undergo total reflection. The strong contrast in elastic moduli at these interfaces, especially at the Moho (the base of the crust) causes these reflected phases to have large amplitudes. Using synthetic seismograms (Fig. 1b), Burger *et al.* (1987) found that peak ground motion amplitudes at distances beyond about 50 km are controlled by these post-critical reflections from velocity gradients in the lower crust (Fig. 1c). They found evidence supporting this result in empirical strong motion data (Toro and McGuire, 1987, Figs. 6 through 9) that showed a flat trend in the distance range of 60 to 150 km.

Further studies of ground motion attenuation using digital network data and strong motion recordings in the northeastern United States and adjacent Canada (Barker *et al.*, 1989) lent support to the hypothesis that crustal structure and focal depth play an important role in determining the shape of the strong ground motion attenuation curve. Using procedures similar to that of Burger *et al.* (1987) and Barker *et al.* (1989), Gariel and Jacob (1989) and Ou and Herrmann (1990) analyzed the attenuation of strong motion from the 1988 Saguenay earthquake and obtained results similar to those of Somerville *et al.* (1990a) which are summarized below.

Our objective in analyzing the strong motion recordings of the Saguenay earthquake is to test the suggestion of Burger *et al.* (1987) that post-critical reflections from velocity gradients in the lower crust control the attenuation of strong ground motion in the distance range of about 50 to 200 km. Synthetic acceleration and velocity seismograms were computed using the seismic moment and focal mechanism derived from long-period body wave modeling. We used an empirical source function derived from the tangential component of motion recorded at station SM17 which allows us to simultaneously match the recorded amplitudes of strong motion acceleration, strong motion velocity, and teleseismic short-period and long-period body waves (Somerville *et al.*, 1990a).

The recorded and synthetic tangential velocity seismograms are compared in Fig. 2, with all seismograms shown at their absolute times and scaled to their peak values which are indicated by arrows. At a distance of 100 km, the arrivals are, in order of increasing arrival time, the direct S, the Conrad refraction S_c, the Conrad reflection S_cS, the Moho refraction S_n, and the Moho reflection S_mS. At the two closest stations (SM16 at 43 km and SM17 at 64 km), the direct upgoing wave produces large motions on both the recorded and synthetic seismograms. However, at 90 km, the direct arrival has very small amplitude on both the recorded and synthetic seismograms. The largest phase in the recorded data at this distance is the post-critical reflection from the Conrad discontinuity; this phase is also present in the tangential component of the synthetic seismograms. Between 110 and 120 km, the recorded seismograms show large post-critical reflections from both the Conrad and the Moho; these are also present in the synthetic seismograms, and both reflections contribute to the largest motions. Beyond 150 km, the Moho reflection is masked by the Conrad reflection, which controls peak amplitudes out to 200 km.

In summary, the recorded and synthetic profiles both demonstrate that at distances beyond about 70 km, the direct shear wave arrival ceases to control peak ground motion amplitudes; instead, peak amplitudes are controlled by post-critical reflections from the velocity gradients in the lower crust. The strength of these postcritical reflections, and the distance ranges over which they are dominant, are controlled by the focal depth and crustal structure. Thus crustal structure and focal depth control the attenuation of strong ground motion.

ATTENUATION OF STRONG GROUND MOTION

The peak velocities and accelerations of the profiles of recorded and synthetic seismograms are compared in the lower part of Fig. 2. The attenuation of both the recorded and simulated peak motions is very gradual in the distance range of 50 to 120 km, and only becomes steep beyond 120 km. In the top part of Fig. 2, we have already shown that the peak amplitudes in this flat portion of the attenuation curve inside 120 km in both the data and the synthetic seismograms are due to large postcritical reflections from the Conrad and Moho discontinuities. These reflections become postcritical at close distances because of the depth of the source.

There is evidence that postcritical reflections from the lower crust also control peak ground motion amplitudes in at least some regions of the western United States. Preliminary analysis of accelerograms having absolute times that were recorded during the October 17, 1989 Loma Prieta earthquake shows that in the distance range of 50 to 100 km (which includes San Francisco and Oakland), the largest motions at a given station were due to postcritical Moho reflections (Somerville and Yoshimura, 1990). These motions were further amplified, presumably by impedance contrast and resonance effects, at soft soil sites. For both the Saguenay and Loma Prieta earthquakes, the short critical distance and the consequent elevation of ground motion amplitudes on rock sites between about 50 and 120 km are due to deep focal depth, or more precisely, to the proximity of the source to the base of the lower crust.

From the above analysis, it is clear that the focal depth of the Saguenay earthquake and the structure of the crust in which it occurred influenced the shape of the ground motion attenuation curve of this earthquake. Earthquakes occurring at other depths and in other crustal structures are expected to have different attenuation curves. Attenuation curves for eastern North America that are based in part on empirical strong motion data represent the averaging of attenuation curves for a range of focal depths and a variety of different crustal structures. It is therefore to be expected that the attenuation curve for the unusually deep Saguenay earthquake might differ from these empirical attenuation curves.

Several recent attenuation curves are based on random process theory (Hanks and McGuire, 1981; Boore, 1983), in which strong ground motions are modeled as segments of band-limited noise. In the simplest of these models, wave propagation effects are modeled by anelastic attenuation and by a geometrical spreading term, assumed to be $1/R$ within 100 km (corresponding to body wave spreading in a whole-space), and $1/\sqrt{R}$ beyond 100 km (corresponding to surface wave spreading in a half space). These effects produce a smooth, monotonically decreasing function of ground motion amplitude with distance (Boore and Atkinson, 1987; Toro and McGuire, 1987). This simple ground motion model does not include focal depth and crustal structure among its input parameters, and so it does not account for the influence of these parameters on the ground motion attenuation curve. For this reason, it is expected that the attenuation curve for the unusually deep Saguenay earthquake might differ from curves derived using simple random process theory.

519

The random process models are based on seismic moment as a description of earthquake size. Comparison of the Boore and Atkinson (1987) attenuation relation, computed using the moment magnitude of 5.8, with the recorded peak acceleration is shown in Fig. 3. It underpredicts the data, and allowing for a natural logarithmic standard error of 0.5, it is significantly different from the data at the 99.9% confidence level. The peak accelerations from the synthetic seismograms are also shown, and provide a better fit to the recorded data, specifically in modeling the amplitude level, the flat trend of the attenuation curve in the distance range of 50 to 120 km, and the rapid decay beyond 120 km. The Boore and Atkinson (1987) relation still underpredicts the recorded motions (at the 99% confidence level) if the m_{bLg} magnitude of 6.5, converted to a moment magnitude of 6.35 by the relation given above, is used in place of the measured moment magnitude of 5.8, but the underprediction is somewhat reduced (Somerville *et al.*, 1990b).

CONCLUSIONS

The attenuation of recorded strong ground motions from the Saguenay earthquake is very gradual in the distance range of 50 to 120 km, and only becomes steep beyond 120 km. A profile of synthetic seismograms reproduces these features, and allows us to understand why the attenuation relation has this shape. In both the recorded and synthetic seismograms, the peak amplitudes inside 120 km are due to large postcritical reflections from the Conrad and Moho discontinuities. Because of the deep focal depth of the Saguenay earthquake, the critical distances for these reflections are short, causing the ground motion amplitudes to be elevated in the distance range of 50 to 120 km.

These results support the model for the attenuation of strong ground motion proposed by Burger *et al.* (1987). According to this model, the shape of the attenuation curve within 200 km of the source is controlled by focal depth and crustal structure. Inside the critical distance for reflections from the lower crustal velocity gradient, peak ground motion amplitudes are controlled by upgoing shear waves. However, beyond the critical distance, peak amplitudes are controlled by post-critical reflections from the velocity gradients in the lower crust, which typically causes an elevation of ground motion amplitudes. Generally, the amplitudes are at most two to three times higher than those that would be obtained by extrapolating the attenuation curve from pre-critical distances. The distances over which the amplitudes are elevated generally lie in the overall range of 50 to 200 km, with the specific distance range depending on the focal depth of the earthquake and on the crustal structure. Regional variations in crustal structure therefore cause regional variations in ground motion attenuation.

When strong motion data from different crustal structures or from earthquakes having different focal depths are combined into a single data set, the detailed characteristics of regional attenuation relations are smeared out. This leaves a data set having a broad scatter that is reasonably modeled using a smooth attenuation curve which is applicable to the estimation of ground motions across eastern North America. However, the smooth character of such a curve, whether derived empirically or from simplified theoretical models such as random process theory, does not have a rigorous physical basis in wave propagation theory. If it is desired to estimate ground motions at a specific site or within a given region, then estimates having a lower degree of uncertainty can be obtained by using an attenuation curve that reflects the wave propagation characteristics of that region.

ACKNOWLEDGMENTS

We are grateful to many individuals in the Geological Survey of Canada for speedily providing data used in this study, including Mr Robert Halliday, Mr Philip Munro, Mr William Shannon, and Ms. Anne Stevens. Don Helmberger, James McLaren, Chandan Saikia participated in the analysis of the source parameters and the modeling of the strong motion records. This work was sponsored by the Electric Power Research Institute under the direction of Dr John Schneider and Dr J. Carl Stepp.

REFERENCES

Barker, J.S., P.G. Somerville and J.P. McLaren 1989. Modeling ground motion attenuation in eastern North America, *Tectonophysics, 167*, 139-149.

Boore, D.M. 1983. Stochastic simulation of high-frequency ground motions based on seismological models of the radiated spectra, *Bull. Seism. Soc. Am.* 73, 1865-1984.

Boore, D.M. and G.M. Atkinson 1987. Stochastic prediction of ground motion and spectral response parameters at hard rock sites in Eastern North America, *Bull. Seism. Soc. Am.* 77, 440-467.

Burger, R.W., P.G. Somerville, J.S. Barker, R.B. Herrmann, and D.V. Helmberger 1987. The effect of crustal structure on strong ground motion attenuation relations in eastern North America. *Bull. Seism. Soc. Am.* 77, 420-439.

Ebel, J.E., P.G. Somerville, and J.D. McIver 1986. A study of the source parameters of some large earthquakes in eastern North America, *J. Geophys. Res.* 91, 8231-8247.

Gariel, J.-C. and K.H. Jacob 1989. The Saguenay earthquake of November 25, 1988: the effect of the hypocentral depth on the peak acceleration: a modeling approach, in *The Saguenay earthquake of November 25, 1988, Quebec, Canada: strong motion data, ground failure observations, and preliminary interpretations*, K. H. Jacob (ed.).

Hanks, T.C. and R.K. McGuire 1981. The character of high-frequency strong ground motion, *Bull. Seism. Soc. Am.* 71, 2071-2095.

McGuire, R.K., G.R. Toro and W.J. Silva 1988. Engineering model of earthquake ground motion, EPRI Report NP6074.

Munro, P.S. and R.G. North 1988. The Saguenay earthquake of November 25, 1988: strong motion data. Geological Survey of Canada Open File Report 1976.

North, R.G., R.J. Wetmiller, J. Adams, F.M. Anglin, H.S. Hasegawa, M. Lamontagne, R. Du Berger, L. Seeber, and J. Armbruster 1989. Preliminary results from the November 25, 1988 Saguenay (Quebec) earthquake, Seismological Research Letters 60, 89-93.

Ou, G.-B. and R.B. Herrmann 1990. A statistical model for peak ground motion produced by earthquakes at local and regional distances, *Bull. Seism. Soc. Am.* 80, 1397-1417.

Somerville, P.G., J.P. McLaren, L.V. LeFevre, R.W. Burger and D.V. Helmberger 1987. Comparison of source scaling relations of eastern and western North American earthquakes, *Bull. Seism. Soc. Am.* 77, 332-346.

Somerville. P.G. and J. Yoshimura 1990. The influence of critical Moho reflections on strong ground motions recorded in San Francisco and Oakland during the 1989 Loma Prieta earthquake, *Geophysical Research Letters* 17, 1203-1206.

Somerville, P.G., J.P. McLaren, C.K. Saikia and D.V. Helmberger 1990a. The November 25, 1988 Saguenay, Quebec earthquake: source parameters and the attenuation of strong ground motion. *Bull. Seism. Soc. Am.* 80, 1118-1143.

Somerville, P.G., N.A. Abrahamson, J.F. Schneider, and J.C. Stepp 1990b. Attenuation of ground motion from the Saguenay, Quebec earthquake: comparison with eastern North America models. *Proceedings of the Third Symposium on Current Issues Related to Nuclear Power Plant Structures, Equipment and Piping*, pp. I/2-1 to I/2-20.

Toro, G.R. and R.K. McGuire 1987. An investigation into earthquake ground motion characteristics in eastern North America, *Bull. Seism. Soc. Am.* 77, 468-489.

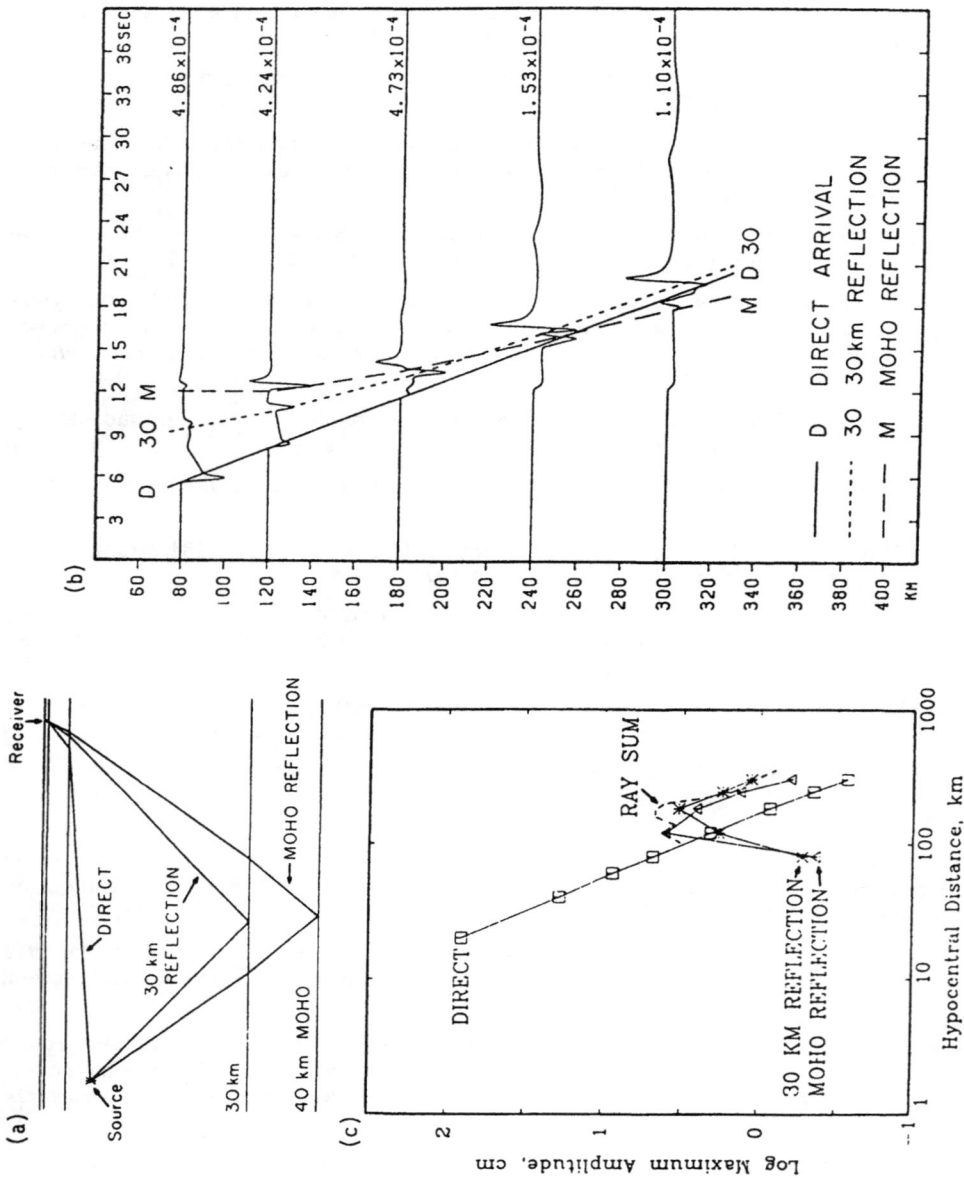

Figure 1. (a): Simplified model of wave propagation in a layered crust. (b): Synthetic displacement seismograms for the crustal model shown in (a). (c): Attenuation of direct and reflected phases shown in (b). Source: Burger et al., 1987.

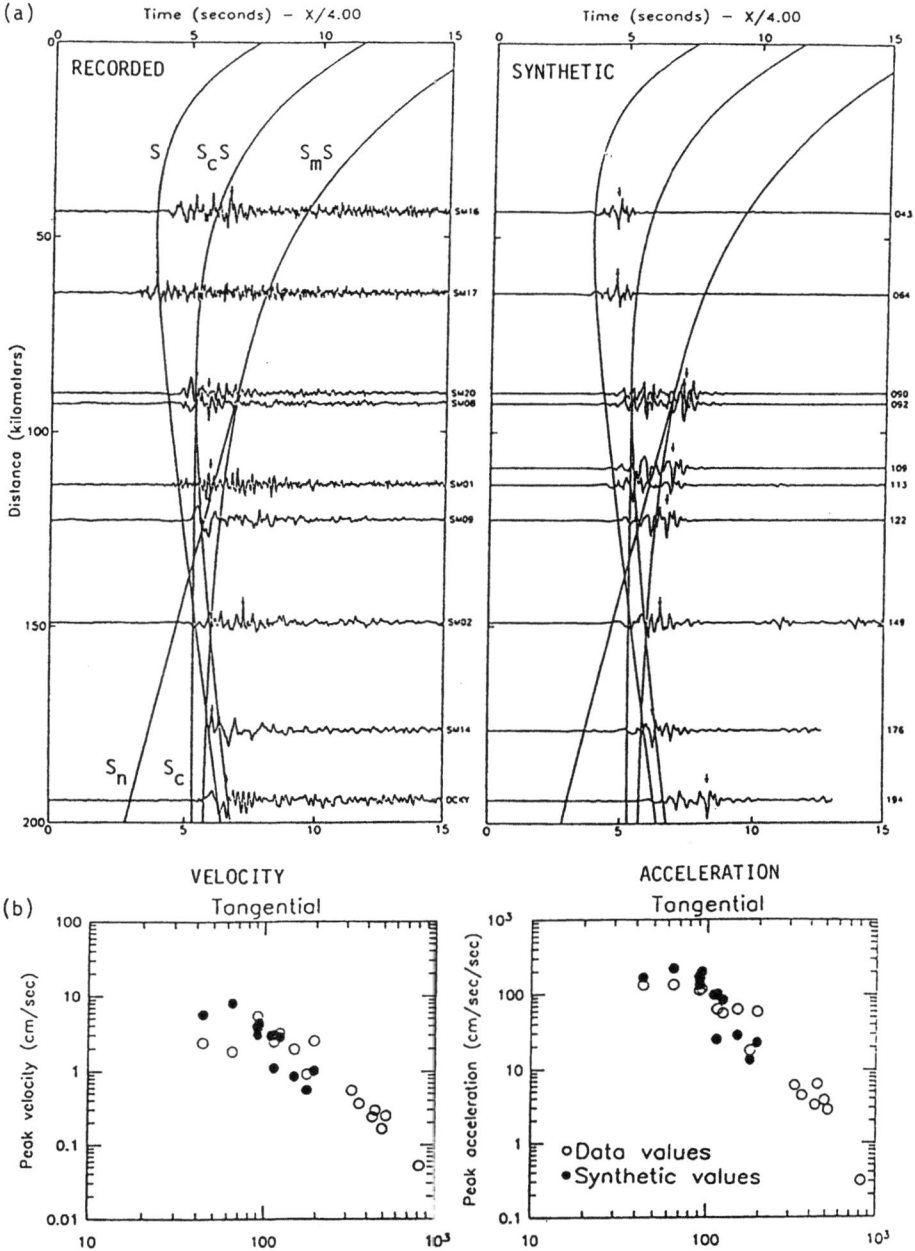

Figure 2. (a): Profiles of recorded (left) and synthetic (right) tangential velocity of the Saguenay earthquake, together with calculated travel times of principal arrivals. (b): Recorded (circles) and synthetic (dots) peak velocity (left) and peak acceleration (right) of the Saguenay earthquake as a function of epicentral distance. Source: Somerville *et al.*, 1990a.

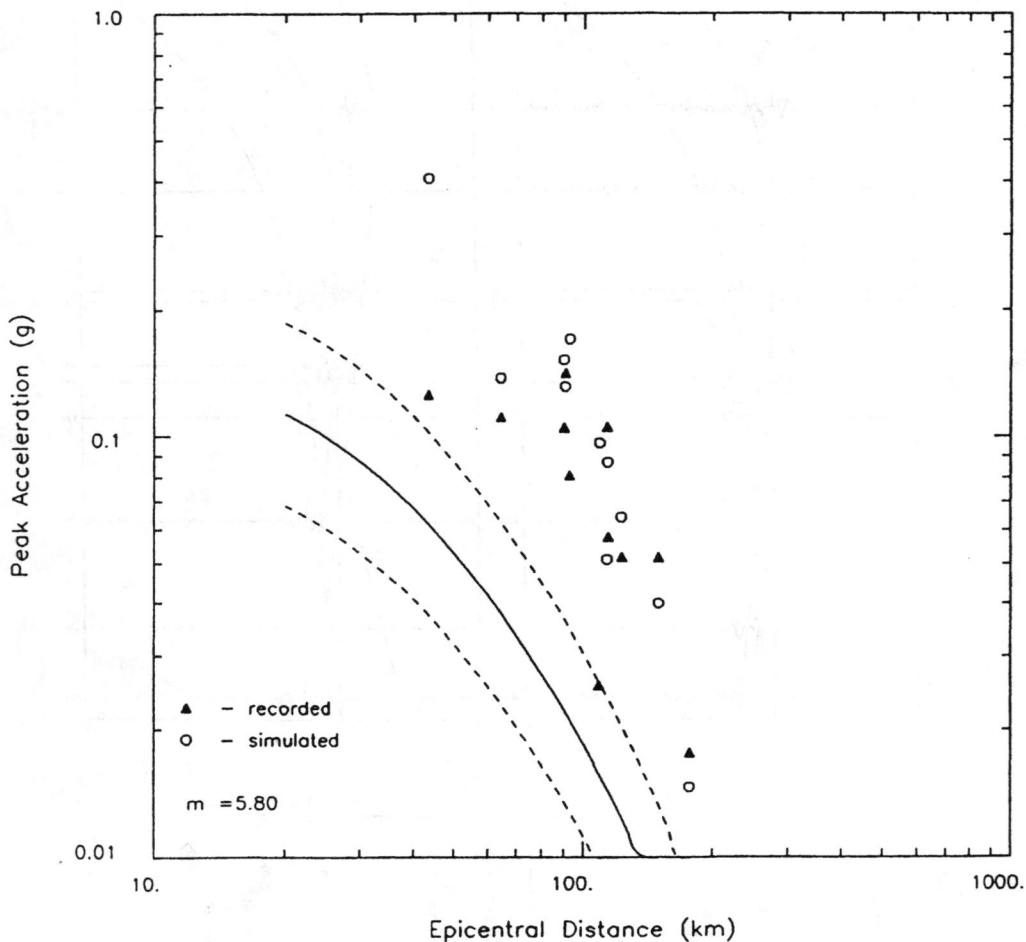

Figure 3. Comparison of recorded (triangles) and synthetic (circles) peak acceleration of the Saguenay earthquake with the Boore and Atkinson (1987) attenuation curve for epicentral distance using a moment magnitude of 5.8.

Ground motion characteristics and engineering implications of the 1988 Saguenay, Quebec earthquake

N. Naumoski[I], A.C. Heidebrecht[II], and J. Wang[III]

ABSTRACT

An analysis is performed on 31 horizontal components recorded at rock sites during the earthquake which occurred in the Saguenay region of the province of Quebec, Canada, on November 25, 1988. The peak ground motions, response spectra, and base shear coefficients are analyzed statistically. The results are compared with those from other studies and with equivalent parameters in the National Building Code of Canada. It is found that the Saguenay earthquake has produced substantially larger peak ground motions and response spectra than those predicted from other studies. The base shear coefficients for low period structures exceed those from the code provisions.

INTRODUCTION

On November 25, 1988, an earthquake of magnitude $M_s = 5.7$ occurred in the Saguenay region of the province of Quebec, Canada. The earthquake caused some damage in the sparsely populated epicentral region, caused power outages more regionally, and was felt over an extremely large area, as far south as New York City and as far west as Toronto, Ontario. This earthquake was the most significant earthquake in over fifty years in eastern North America (ENA). It was also the first earthquake of significant magnitude whose resulting ground motions were well recorded at large epicentral distances.

Due to lack of data from strong earthquakes in eastern Canada, the predictions of earthquake ground motions in this region have been based on data from western North America (WNA) (Hasegawa et al. 1981), or stochastic prediction techniques have been developed based on theoretical models (Atkinson and Boore 1990). Similarly, the response related parameters, such as the base shear coef-

I   Research Engineer, Department of Civil Engineering and Engineering Mechanics, McMaster University, Hamilton, Ontario, Canada, L8S 4L7.

II   Professor, Department of Civil Engineering and Engineering Mechanics, McMaster University, Hamilton, Ontario, Canada, L8S 4L7.

III   Graduate Student, Department of Civil Engineering and Engineering Mechanics, McMaster University, Hamilton, Ontario, Canada, L8S 4L7.

ficients incorporated in the National Building Code of Canada 1990 (NBCC 1990) are defined based on strong earthquake records from other regions in the world. The Saguenay earthquake records provide valuable information for verification and updating the present prediction techniques of earthquake ground motions in eastern Canada, and for evaluation of the base shear coefficients used in NBCC 1990.

In this study, the peak ground motions, response spectra, and base shear coefficients of the Saguenay earthquake are analyzed statistically. The peak ground motions and the response spectra are compared with the predictions for ENA and WNA. The base shear coefficients are compared with those specified in NBCC 1990.

GROUND MOTION DATA

The motions at rock sites during the Saguenay earthquake were recorded by ten accelerographs from the Eastern Canada Strong Motion Seismograph Network (Munro and Weichert 1989), at epicentral distances between 43 and 177 km. In addition, six rock records at distances between 200 and 525 km were obtained by the National Center for Earthquake Engineering Research (NCEER) (Friberg et al. 1988). In this study, the 31 horizontal components from these records are analyzed.

The peak values of the ground motion range from 0.13 to 15.6%g for acceleration, and from 0.1 to 4.65 cm/sec for velocity. The records are characterized with high peak acceleration, a, to peak velocity, v, (a/v) ratios. With exception of one component which has a/v-1.04, the a/v ratios of the other components range from 1.26 to 9.68. As is expected, the a/v ratios tend to decrease with increasing epicentral distance. However, it is surprising to get high a/v ratios (greater than 1.26) at very long epicentral distances (300 to 500 km), which is the case with the NCEER records. Because these are the first data recorded far from the epicentre of an earthquake in eastern North America, it is not clear whether the high a/v ratios at very long epicentral distances are due to the seismo-tectonic characteristics of this region, or some other factors.

ATTENUATION MODEL FOR PEAK GROUND MOTIONS AND RESPONSE SPECTRA

The attenuation functions of the horizontal peak ground acceleration, peak ground velocity and spectral velocity in terms of the epicentral distance were determined using least-square regression analysis. The following form of equation was used

$$\log y = b_1 + b_2 \log R + b_3 R \tag{1}$$

in which y is the peak ground motion variable being defined, R is epicentral distance, and b_1, b_2 and b_3 are coefficients determined by the regression analysis. Equation 1 represents a simplified version of the attenuation model proposed by Joyner and Boore (1988). In the original model, the coefficient b_1 is magnitude dependent and it controls the level of the attenuation curves for different magnitudes. In this study, the data set used is from a single earthquake and b_1 is assumed to be magnitude independent. Concerning the coefficients b_2 and b_3, the Joyner and Boore (1988) approach is followed for determining their values, e.g. only one of these two coefficients is determined by the regression analysis. Usually b_2 is set equal to -1, and the regression analysis is carried out to determine b_1 and b_3. In the case when a positive value is obtained for b_3, it is set to zero, and the coefficients b_1 and b_2 are determined by the regression analysis.

PEAK GROUND MOTIONS

The attenuation functions for mean horizontal peak ground acceleration, a, and peak ground velocity, v, determined by the regression analysis have the following form

$$\log a = 4.064 - \log R - 0.00215 \, R; \quad \sigma_{\log a} = 0.231 \tag{2}$$

$$\log v = 2.414 - \log R - 0.00101 \, R; \quad \sigma_{\log v} = 0.299 \tag{3}$$

where the acceleration is in cm/sec^2, the velocity is in cm/sec, and the epicentral distance, R, is in km.

Figures 1 and 2 show some of the features of the acceleration attenuation. Figure 1 shows the distribution of the acceleration data on which the attenuation curves (mean, and mean +/- 1 standard deviation) are superimposed. As can be seen, the dispersion of the data around the mean curve is relatively small. Figure 2 shows a comparison between the mean attenuation curve for peak ground acceleration for the Saguenay earthquake and those obtained by Hasegawa et al. (1981) for eastern Canada (EC), Atkinson and Boore (1990) for ENA, and Joyner and Boore (1988) for WNA, all of which are developed for rock. The curves from these studies are computed for the magnitude of the Saguenay earthquake of 5.7. The comparison with the Hasegawa et al. (1981) curve (referred to as Hasegawa curve in the further discussion) is of special interest because the same attenuation relation has been used for developing the probabilistic seismic zoning maps for eastern Canada. As is shown, at epicentral distances less than 300 km, the Saguenay earthquake produced larger accelerations than those predicted by Hasegawa equation. At short epicentral distances, below 100 km, the Saguenay earthquake curve is about 2 times higher than the Hasegawa curve. For long epicentral distances, beyond 300 km, the Saguenay earthquake peak ground accelerations are substantially lower than those predicted by the Hasegawa attenuation relation. Concerning the Atkinson and Boore (1990) curve and the Joyner and Boore (1988) curve, both are similar in shape with that of the Saguenay earthquake, but the predictions based on these curves are very low in comparison with the accelerations obtained during the Saguenay earthquake.

The results for the horizontal peak ground velocity are shown in Figs. 3 and 4. As can be seen from Fig. 3, the dispersion of the velocity data around the mean curve is somewhat larger than that for the acceleration data, which leads to a larger standard deviation (see Eq. 3). Figure 4 shows a comparison between the attenuation relation for the mean ground velocity from the Saguenay earthquake and those from the foregoing studies. In terms of the Hasegawa prediction, the Saguenay earthquake has produced larger velocities for a wide range of epicentral distances, below 500 km. At short epicentral distances, below 100 km, the mean velocities from the Saguenay earthquake are about 2.5 times larger than those predicted by Hasegawa. The Atkinson and Boore (1990) and the Joyner and Boore (1988) curves are substantially below the Saguenay earthquake curve for all epicentral distances.

RESPONSE SPECTRA

Attenuation of Response Spectra

To investigate the characteristics of the ground motions during the Saguenay earthquake in terms of the response of single-degree-of-freedom systems at different epicentral distances, the attenuation relations for 5% damped pseudovelocity response spectra were determined using regression analysis. The analysis was performed for 14 periods, between 0.01 and 4.0 sec, using the attenuation model given in Eq. 1. The coefficients and standard deviations obtained from the regression analysis are listed in Table 1. These enable one to determine the Saguenay earthquake spectra at any epicentral distance.

Table 1. Regression coefficients and standard deviations for pseudovelocity response spectra, 5% damping

$$(\log S_v = b_1 + b_2 \log R + b_3 R)^*$$

Period(s)	b_1	b_2	b_3	$\sigma_{\log Sv}$
0.01	1.300	-1	-0.213E-02	0.219
0.02	1.712	-1	-0.242E-02	0.217
0.03	1.951	-1	-0.251E-02	0.208
0.05	2.207	-1	-0.231E-02	0.205
0.08	2.477	-1	-0.228E-02	0.211
0.1	2.573	-1	-0.226E-02	0.251
0.2	2.780	-1	-0.175E-02	0.331
0.3	2.656	-1	-0.920E-03	0.301
0.5	2.529	-1	-0.466E-03	0.287
0.8	2.385	-1	-0.138E-03	0.334
1.0	2.250	-1	-0.557E-05	0.369
2.0	1.097	-0.620	0.	0.342
3.0	0.518	-0.452	0.	0.320
4.0	0.862	-0.696	0.	0.303

$^*S_v$ = pseudovelocity in cm/sec; R = epicentral distance in km.

Because of the limitation in the length of this paper, the shapes of the spectra in terms of the epicentral distance are not shown here. It should be mentioned, that the spectra are characterized with relatively low predominant periods which range from 0.2 sec at epicentral distance of 50 km to 0.6 sec at 400 km.

Figure 5 shows a comparison between the Saguenay earthquake spectrum at epicentral distance of 100 km and the spectra for rock sites predicted by Atkinson and Boore (1990) for ENA and by Joyner and Boore (1988) for WNA. As can be seen, the Saguenay earthquake spectrum is much higher than the predicted spectra. The predominant period of the ground motion is about 0.25 sec, and it is substantially lower than the predominant periods of the Atkinson and Boore (1990) spectrum, which range from 0.5 to 1.0 sec. Concerning the Joyner and Boore (1988) spectrum, it has a shape which is substantially different than that of the other spectra.

528

Spectral Amplifications

The codes for seismic design usually define the design spectra by spectral amplification factors in terms of the peak ground motions. These factors are commonly specified at mean + 1 standard deviation (M+1SD) level in order to ensure that there is a relatively small probability that the response will be above the specified design level.

In this study, the amplifications of the Saguenay earthquake spectra in terms of the peak ground acceleration and velocity are examined. To assess the influence of the epicentral distance on the spectral amplification, the spectra are analyzed in three groups, which correspond to the following ranges of epicentral distances: 40 to 100 km, 100 to 200 km and 325 to 525 km. The epicentral distances within these ranges will be conditionally referred to as short, intermediate and long epicentral distances. The first group comprises the spectra from 8 components, the second group from 13 components, and the third group from 10 components. The response spectra from each group are scaled to a peak ground acceleration of 1g, and to a peak ground velocity of 1m/sec. The M+1SD of the spectra from the acceleration scaling are shown in Fig. 6, and those from the velocity scaling in Fig. 7.

Figure 6 shows that the maximum amplification of the spectra in terms of the peak ground acceleration is about 3.3, in the period range between 0.1 and 0.3 sec. It is associated with the spectra from intermediate and long epicentral distance range. The spectrum from the short epicentral distance range shows maximum amplification of 2.5 at very short periods, around 0.04 sec. This is due to the significant presence of such short period components in the records at short epicentral distances. Concerning the velocity amplification of the spectra, Fig. 7 shows that the spectra from the three ranges of epicentral distances have the same maximum amplification of about 3. It should be noted that both the acceleration and velocity amplification of the Saguenay earthquake spectra of 3.3 and 3.0 respectively are larger than those suggested by Newmark and Hall (1982) (2.71 for acceleration amplification and 2.3 for velocity amplification), which with a slight modifications are accepted in various codes for aseismic design.

BASE SHEAR COEFFICIENTS

The unit velocity base shear coefficients were calculated for each component time history for uniform frame and wall-type structures. Simple continuum models of frame and wall structures were used. The response analysis was performed using time history superposition method, for five modes and 5% modal damping. The base shear coefficients were analyzed statistically for three groups of records obtained at different ranges of epicentral distances (short, intermediate and long epicentral distance range, as defined in the previous section). For each group, the M+1SD level of the base shear coefficients were determined. In this paper, the results are presented only for frame structures because they are larger than those for wall structures for fundamental periods below about 0.5 sec.

Figure 8 shows the M+1SD level of the unit velocity base shear coefficient spectra for the three groups of records, for frame structures with fundamental periods between 0.01 and 4.0 sec. Because the Saguenay records have high a/v ratios, the base shear coefficients are compared with that specified in NBCC 1990 for regions where the zonal acceleration, Z_a, is larger than the zonal velocity, Z_v ($Z_a > Z_v$). As shown in Fig. 8, the base shear coefficients for the records from the short and intermediate epicentral distance range exceed significantly the $Z_a > Z_v$ branch of the NBCC 1990 base shear coefficient, for periods below 0.3 sec. The base shear coefficients for the records obtained at long epicentral distances are slightly higher than that in NBCC 1990 for periods between 0.3 and 0.9 sec.

The features of the unit velocity base shear coefficient spectra for the three groups of records are related to the a/v ratios of the records from each group. For illustration, the mean a/v values of the records from the short, intermediate and long epicentral distance range are 5.5, 2.8 and 1.6 respectively. In terms of the Canadian seismic zoning maps, the largest values of a/v are in the neighbourhood of 2. Heidebrecht and Lu (1987) have shown that even for a/v=2 the base shear coefficient in the code underestimates those for short period frame structures. Taking into account that the a/v ratios of the Saguenay earthquake records from the short and intermediate epicentral distance range are larger than 2, it is obvious why the base shear coefficients for these records are higher than the comparable values in NBCC 1990.

SUMMARY

The analysis and results presented in this paper show that the ground motion during the Saguenay earthquake is characterized by very high peak ground motions and response spectra. It is found that the peak ground accelerations, peak ground velocities and response spectra are significantly larger, for a wide range of epicentral distances, than those from the prediction relations. The prediction relations for eastern Canada should be updated to account for the Saguenay earthquake characteristics.

The effects of the Saguenay earthquake on the response of multi-degree-of-freedom structures is presented in terms of base shear coefficients for frame structures. It is shown that the unit velocity base shear coefficients of the records at distances below 200 km substantially exceed that specified in NBCC 1990 for $Z_a > Z_v$. This is due to the very high a/v ratios of these records. Taking into account that the same observations have also been made in other studies for earthquake records with a/v>2, the base shear coefficient in NBCC 1990 for $Z_a > Z_v$ should be revised to recognize the effects of such earthquake motions.

REFERENCES

Associate Committee on the National Building Code 1990. National Building Code of Canada. National Research Council of Canada, Ottawa, Ontario, Canada.

Atkinson, G.M. and Boore, D.M. 1990. Recent trends in ground motion and spectral response relations for North America. Resp. Spectra, 6, 15-35.

Friberg, P., Rusby, R., Dentrichia, D., Johnson, D., Jacob, K. and Simpson, D. 1988. The M=6 Chicoutimi earthquake of November 25, 1988, in the province of Quebec, Canada. Preliminary NCEER strong motion data report, Lamont-Doherty Geological Observatory of Columbia University, Palisades, N.Y.

Hasegawa, H.S., Basham, P.W. and Berry, M.J. 1981. Attenuation relations for strong seismic ground motion in Canada. Bull. Seism. Soc. Am., 71(6) 1943-1962.

Heidebrecht, A.C. and Lu C.Y. 1987. Evaluation of the seismic response factor introduced in the 1985 edition of the National Building Code of Canada. Can.

J. Civ. Eng., 15, 382-388.

Joyner, W.B. and Boore, D.M. 1988. Measurements, characterization, and prediction of strong ground motion. Proc. ASCE Conf. Soil Dynamics, Geot. Spec. Publ., 20, 43-102, Park City, Utah.

Munro, P.S. and Weichert, D. 1989. The Saguenay earthquake of November 25, 1988 - Processed strong motion records. Earth Phys. Branch Open-file Rpt. 1966.

Newmark, N.M. and Hall, W.J. 1982. Earthquake spectra and design. Monograph series, EERI, Berkeley, California.

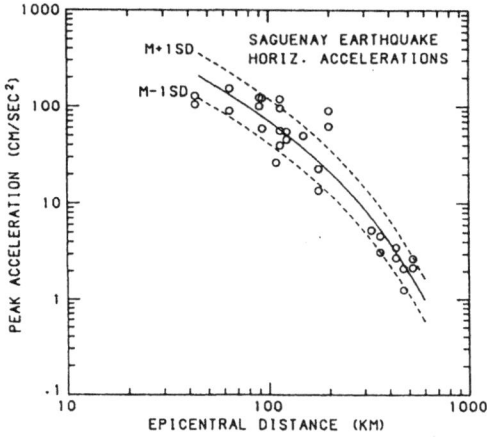

Figure 1. Peak acceleration from the Saguenay earthquake

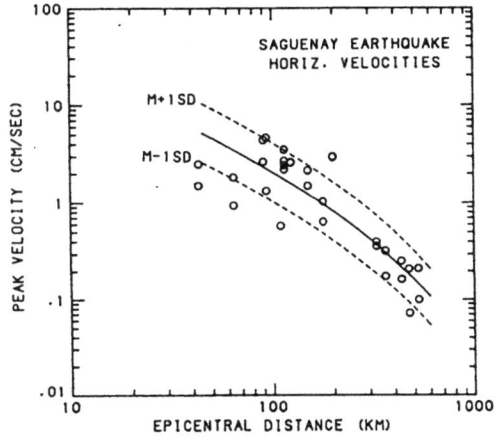

Figure 3. Peak velocities from the Saguenay earthquake

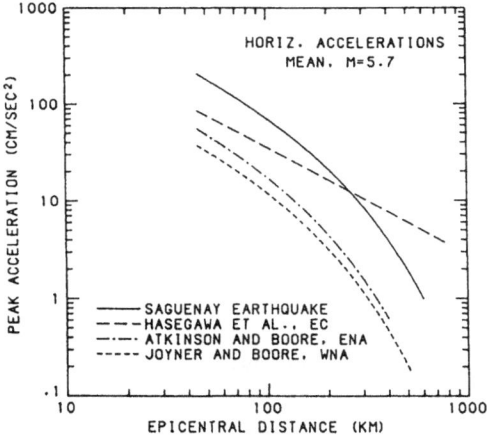

Figure 2. Attenuation curve for the Saguenay earthquake peak ground acceleration and predictions from other studies

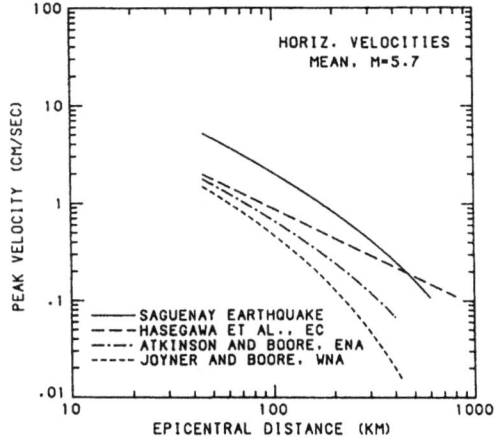

Figure 4. Attenuation curve for the Saguenay earthquake peak ground velocity and predictions from other studies

531

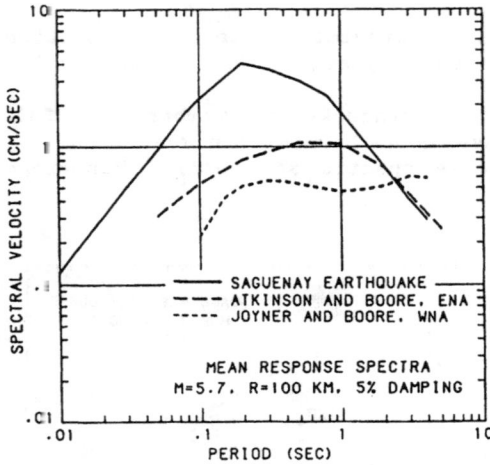

Figure 5. Saguenay earthquake response spectrum at distance of 100 km and predicted spectra from other studies

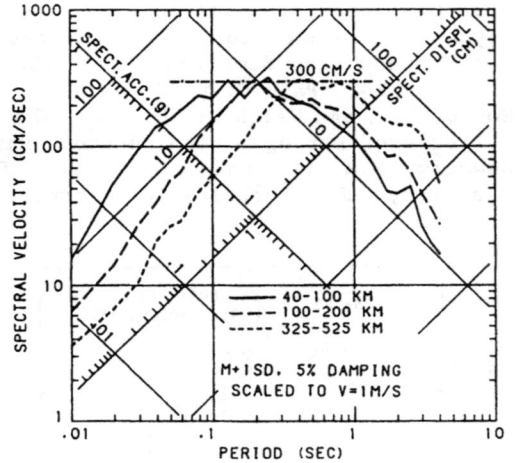

Figure 7. Saguenay earthquake response spectra scaled to v-1m/sec

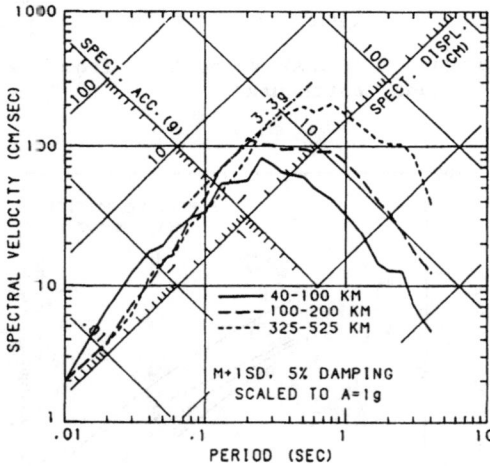

Figure 6. Saguenay earthquake response spectra scaled to a-1g

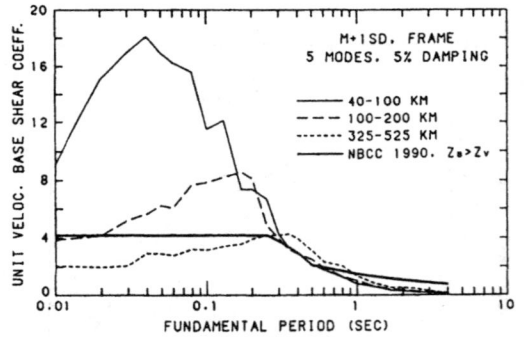

Figure 8. Unit velocity base shear coefficients, computed and NBCC 1990 for $Z_a > Z_v$

A digital approach to integration of accelerometer data

S.S. Law[1]

ABSTRACT

The direct integration of acceleration record introduces offset and slope in
the resulting displacement record. Several different balancing or baseline
correction procedures have been proposed over the years. All of these
approaches require the knowledge of the initial or final velocity and
displacement. A direct approach to the integration of accelerometer data is
presented which does not depend on the values of velocity and displacement
both at the beginning and end of the event. A digital filtering procedure is
developed and checked against simulated data and laboratory results. High
accuracy is obtained with the proposed method.

INTRODUCTION

Earthquake acceleration record represents a random physical phenomenon
which may be considered as a combination of multiple sine waves of different
magnitude at different frequencies. In the Fast Fourier Transform of the time
series, each sine wave is represented by a spectral peak in the frequency
spectrum of the signal. The offset to the mean and slope of a times series
are respectively the DC component and low frequency components in the power
spectrum.

Integration of acceleration data introduces offset and slope in the
resulting displacement record. Fig. 1 shows that for a signal with a given
slope, the slope is represented by low frequency components up to 1 Hz in the
frequency spectrum for a sampling rate of 512 samples/second. The magnitude
of the DC component is proportional to the value of slope. Several different
balancing or baseline correction procedures have been proposed over the years

---

I   Senior Lecturer, Civil & Structural Engg. Dept., Hong Kong Polytechnic

(Trifunac 1970; Pecknold and Riddell 1978). Dynamic programming approach (Trujillo and Carter 1982; Trujillo 1978) was also proposed tó solve the problem. All of these approaches require the knowledge of the initial or final velocity and displacement. The characteristics of the procedure which is of interest here is that it does not require any knowledge on the initial or final velocity and displacement of the event. The aim of the procedure described below is to minimize the DC and near DC components of the signal incurred during the process of integration.

FINITE IMPULSE RESPONSE (FIR) FILTER

The primary purpose of digital filtering is to alter the spectral information contained in an input signal. While this can be accomplished in either the time or frequency domain, FIR is used here in the time domain aspect of filtering. One of the characteristics of FIR is its linear phase response. An input signal goes through the filter would has its phase characteristic retained after filtering. This makes it most suitable for the present work.

THE PROCEDURE

Earthquake acceleration record usually lasts for several minutes and it is always possible to have a digitized time series to cover the whole event. It is assumed that a good knowledge of the zero datum of acceleration is available. Trapezoidal rule is used in the integration.

Step 1: Remove the offset in the acceleration record and perform first integration

2. The velocity record is fitted with a regressed straight line for the values of offset and slope which are then subtracted from the velocity record.

3. Perform second integration.

4. The power spectrum of acceleration record and displacement record obtained from Step 3 are compared to identify the frequency range at which error has been introduced.

5. The displacement record is then highpass filtered with a Finite Impulse Response (FIR) Filter at the cutoff frequency determined from Step 4.

SIMULATED DATA

A series of data is generated at 512 samples/second as a combination of sine wave at 10 Hz and an offset of 3.0 plus a slope of 1 percent. By following the procedure described above, the effect of offset and slope in the integration process is removed and Fig. 2 compares the original signal with the signal after integrating twice with appropriate scale factor included. Since the original signal is a sine wave, it is shown as the mirror image about the zero datum for comparison. The two signals are almost the same except for some distortion at the beginning and end of the filtered signal which are due to the FIR filtering.

EXPERIMENTAL DATA

The procedure is further tested by comparing the displacement obtained by integration and displacement obtained from displacement trandsucer. Two tests were carried out on a suspended structure in the laboratory. The experimental setup is shown in Fig. 3. Signal from accelerometer model B&K 8306 and linear voltage displacement tranducer (LVDT) model TML CDP-50 at a point of the structure were simultaneously lowpassed and digitized at 1024 samples per second. The analogue/digital conversion subroutine is such written that the time difference between adjacent channel so sampled is approximately 0.0001 second. There is virtually no phase difference between data in adjacent channel. 2048 weights Hanning's window was used in the subsequent highpass filtering of data. The transition bandwidth between passband and stopband is only 2 Hz for overshoot of less than 0.3% in the achieved filter response.

The input excitation are separately 20 Hz sinusoidal signal, and 31.6 Hz bandwidth narrow band random noise centred at 20 Hz. Figs. 4 to 7 show the comparison of the frequency responses from different sensors after integration. There are strong low frequency components introduced by integration for both types of input. Comparing with Figs. 8 and 9 shows that the magnitude of displacement spectral peaks at frequencies greater than 3 Hz are almost the same for the two sensors. This indicates that all the frequency components in the acceleration record are retained in the integration. The comparison of displacement time series in Figs. 10 and 11 show that in the case of sinusoidal input, the responses are very similar whereas they are not alike in the case of random input. There are also slight distortion at the beginning and end of the filtered signal.

ACCURACY

Table 1 shows details of the procedure on all the tests with the accuracy calculated basing on the root mean square (RMS) value of the data. It is known that FIR filter introduces distortion at the beginning and end of filtered data as seen in the experiment. Hence the first 300 samples and last 100 samples in the filtered signal are not considered in the calculation of RMS values.

The error in the simulated data is only +3.45 %. In the case of laboratory test, the RMS error is again low at +3.10 % for sinusoidal input and +8.29 % in the case for random input. This error is mainly due to the attenuation of the original signal near the cutoff frequency. Since the effect of slope in the spectrum is within DC to 0.1 Hz for 64 samples/second sampling rate as compared to DC to 1 Hz for 512 samples/second, this polluted region is isolated from the region containing significant information in the usual earthquake acceleration record. The exact cutoff frequency can be predicted by comparing the power spectrum of acceleration and integrated displacement record, and the undesirable low frequency components can be accurately eliminated with a more sharp cutoff transition by inclusion of a still larger number of filter weights in the highpass filter.

For earthquake record containing the major information within the frequency range from DC to 5 Hz, say, a digitizing rate of 64 samples/second would exhibit the relationship as shown in Table 2 between the transition bandwidth and the number of filter weights. This would still maintain an overshoot of less than 0.3% in the achieved filter response.

CONCLUSION

A simple digital approach has been tested on both sinusoidal and random acceleration records and accurate displacement history has been obtained after suitable digital filtering with a RMS error of 3% to 8%. The use of a larger number of weights in the highpass filter and appropriate sampling rate could reduce the upper limit of error to 5%.

REFERENCE

Pecknold D.A. and Riddell R. 1978. Effect of Initial Base Motion on Response Spectra. J. Engg. Mechanics, ASCE, 104(4).

Trifunac M.D. 1970. Low Frequency Digitization Errors and a New Method of Zero Baseline Correction of Strong-Motion Accelerograms. Earthquake Engg. Research Laboratory, EERL 70-07, California I. of Technology, Pasadena, Calif.

Trujillo D.M. 1978. Application of Dynamic Programming to the General Inverse Problem. Int. J. for Numerical Methods in Engg., Vol.12.

Trujillo D.M. and Carter A.L. 1982. A New Approach to the Integration of Accelerometer Data. Earthquake Engg. and Structural Dynamics, Vol.10.

Table 1

Data Type			No. of data	Highpass filter at	RMS value	% error in RMS
Simulated	Original	Sinusoidal plus noise	2048	-	2.8214	
	After integration		2048	6.0 Hz	2.9222	+3.45
Displacement	(LVDT)	Sinusoidal	4096	-	0.11699	
Acceleration	After integration	Sinusoidal	4096	10.0 Hz	0.12074	+3.10
Displacement	(LVDT)	Random	4096	-	0.22067	
Acceleration	After integration	Random	4096	2.0 HZ	0.24062	+8.29

Table 2

Transition Bandwidth	Filter weights
0.5 Hz	512
0.25 Hz	1024
0.125 Hz	2048

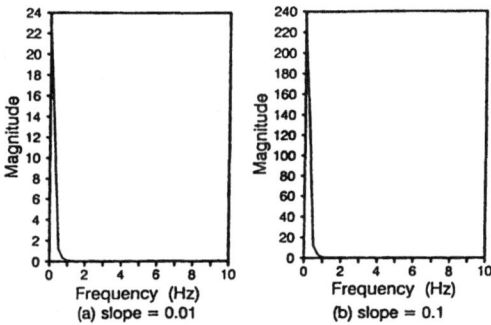

Figure 1: Representation of slope in data in power spectrum
(a) slope = 0.01
(b) slope = 0.1

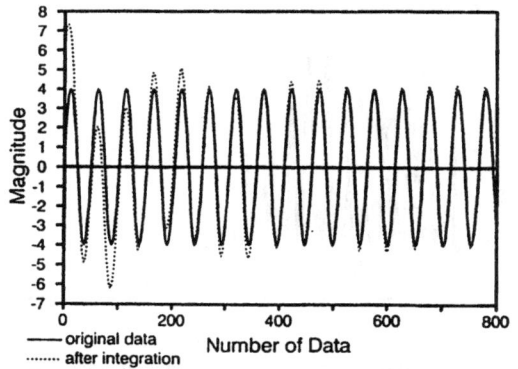

Figure 2: Comparison of simulated data in reverse and data after integration

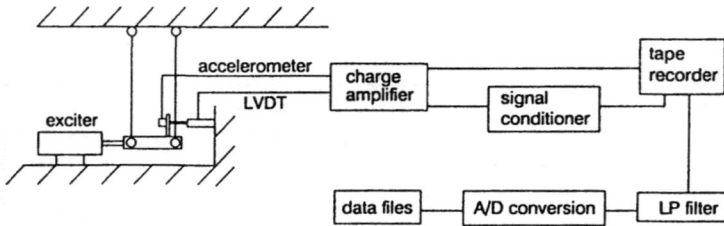

Figure 3: Arrangement of equipment in laboratory test

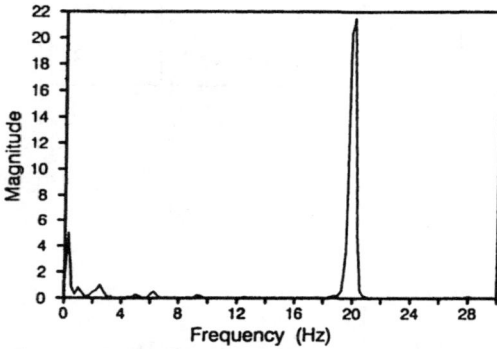

Figure 4: Power spectrum of acceleration response from accelerometer, 20 Hz sine wave input

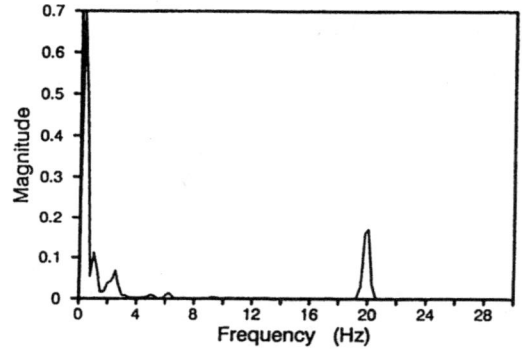

Figure 5: Power spectrum of displacement response from acceleration record after integration 20 Hz sine wave input

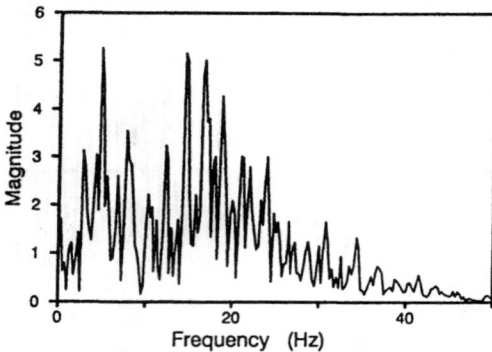

Figure 6: Power spectrum of acceleration response from accelerometer, random input

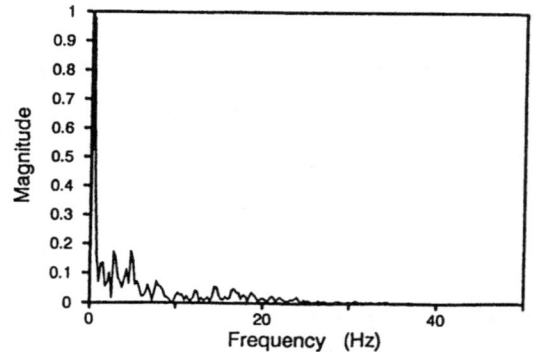

Figure 7: Power spectrum of displacement response from acceleration record after integration random input

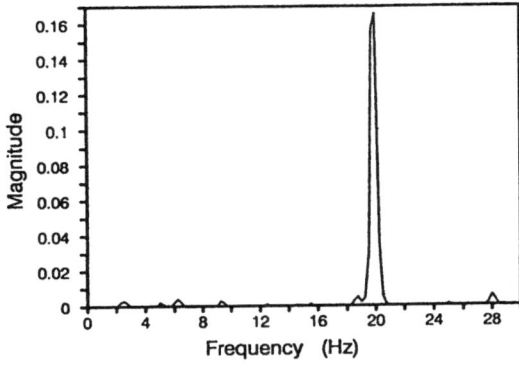

Figure 8: Power spectrum of displacement response from LVDT, 20 Hz sine wave input

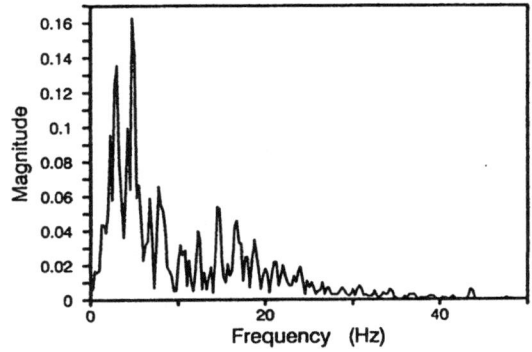

Figure 9: Power spectrum of displacement response from LVDT, random input

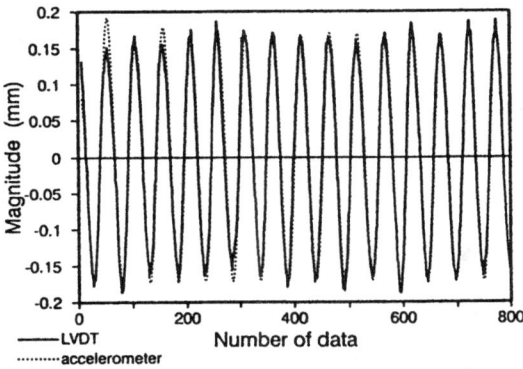

Figure 10: Comparison of displacement response from LVDT and accelerometer, 20 Hz sine wave input

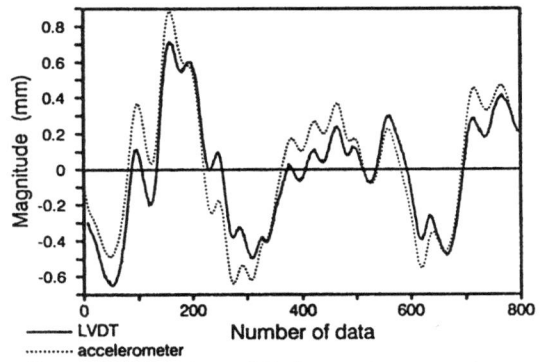

Figure 11: Comparison of displacement response from LVDT and accelerometer, random input

3 Seismic codes and structures
3 Codes sismiques et structures

Seismic design of concrete water retaining structures

M. Reza Kianoush[1]

ABSTRACT

This paper presents simplified approach for seismic design of concrete water retaining structures using the provisions of the National Building Code of Canada (NBCC). The proposed approach is based on serviceability criteria and the level of performance for concrete structures. Four levels of damage are identified and discussed. The application of the proposed design method is illustrated with the aid of an example. Results are then compared with two other widely acceptable design methods.

INTRODUCTION

Concrete water retaining structures are considered as essential facilities that require special care and accuracy in their design. In these structures, crack and leakage control are of prime design considerations. For this reason, serviceability becomes the major factor in the design.

The objectives of building codes such as CAN3-A23.3-M84 (1984) and ACI 318 (1989) are that reinforced concrete structures should be designed in a ductile manner to allow these structures absorb energy in the inelastic range during a major earthquake without major loss in strength. However, in the case of water retaining structures, these structures should be designed to remain safe during a major earthquake and maintain the required serviceability after a major earthquake. Therefore, the concept of serviceability in terms of performance criteria becomes an important parameter for the design of these types of structures.

I-Lecturer, Ryerson Polytechnical Institute, Toronto, Ontario, Canada

It is the objective of this paper to make recommendation for seismic design of concrete water retaining structures based on serviceability requirements. Four different levels of damage for concrete tanks are identified based on the importance and expected performance of the structure. In this fashion, substantial savings in construction cost can be achieved if some degree of damage can be tolerated in the structure without loss of life.

BEHAVIOUR OF LIQUID RETAINING STRUCTURES

The most traditional approach that considers hydrodynamic effects for water retaining structures has been developed by Housner (1957). This approach which applies to flat bottomed tanks of rectangular, circular or uniform sections separates the liquid inside the tank into two parts. As shown in figure 1, the lower portion of the liquid acts as a rigid mass of weight W_0 when accelerated by horizontal and/or vertical ground motion. This is defined as an impulsive force. The upper portion of the liquid is affected by the slosh motion which acts as a flexible mass of weight W_1. This is defined as a convective force. Although Housner's approach is based on simplifications and assumptions, it has proved to be sufficiently accurate for design purposes and is widely used by engineering profession. However, other researchers have made certain modifications to Housner's model. Some of these improvements have been made by Veletsos and Yang (1977), and Balendra (1982) for determining impulsive forces.

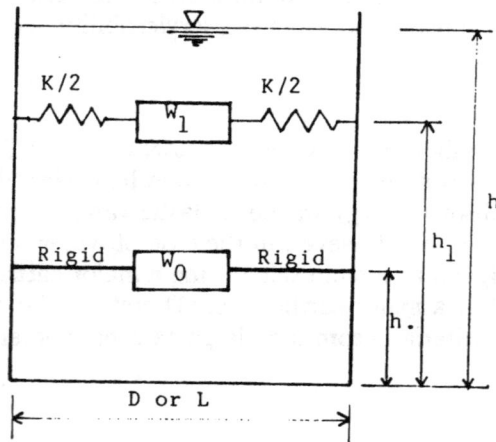

Figure 1.   Tank model

544

CURRENT DESIGN METHODS

Currently, there is a wide range of information available in the literature for the seismic analysis and design of water retaining structures. There are very few provisions in the North American building codes for their seismic design. The Uniform Building Code (1988) is the only recognized code that directly addresses seismic provisions for liquid holding tanks, and it does so only in a very general manner.

There are other standards that address seismic design for water retaining structures. These include the NSF report (1980), and the TID-7024 report (1963). These two methods are widely used in practice. The ANSI/AWWA standards (1986) and the Appendix E of the Petroleum Institute API-650 (1980) also address seismic design for tanks. In all of these standards, there are no specific statements related to concrete tanks. The performance criteria which is a crucial parameter for the design of water retaining structures is not clearly addressed in these standards.

The concept of performance criteria has been discussed by Priestley et al (1986) and recommendations have been made to the New Zealand National Society for Earthquake Engineering for inclusion in a code form for use by the design profession. A similar concept has also been discussed by Ikeda (1989) and has been adopted by the Committee on concrete in the Japan Society of Civil Engineering.

PROPOSED DESIGN CRITERIA

Based on the concept of performance criteria, an initial step in design of concrete water retaining structures is to categorize these structures in accordance with the type and importance of the structure. In categorizing water retaining structures, consequent disaster such as gas explosion and fires should be considered. In general, concrete water retaining structures can be divided into four different categories as follows:

Category 1 - Non-critical structures
For structures in this category, significant damage is not important. The entire system may be shut down after a major earthquake. These structures may undergo significant damage without collapse. Strengthening and/or repair of the structure will be necessary after a major earthquake.

Category 2 - Semi-critical structures
These structures are allowed to undergo limited damage without any required strengthening after a major earthquake. Major cracks and major leakage is acceptable. It should be possible to repair damages after a major earthquake.

Category 3 - Critical structures
In these structures, minor cracks and limited leakage in some local areas due to overstressing due to a major earthquake is acceptable.

Category 4 - Highly-critical structures
Structures in this category must remain fully functional and all components should retain their structural integrity. No cracks and no leakage is acceptable under the design loads.

Application of the proposed design method using NBCC

Based on the concept of the proposed design method using the performance criteria described above, it is possible to estimate the design forces for concrete water retaining structures using the NBCC (1990).
In the NBCC, the minimum lateral seismic force at the base of the structure is given by:

$$V = (V_e/R)U$$

where $U = 0.6$
V_e denotes the equivalent lateral seismic force
R denotes the force modification factor that reflects the capability of a structure to dissipate energy through inelastic behaviour.

In this formula, the R value can be assigned different values depending on the level of damage that can be tolerated. The suggested values for R are as follows:

Category 1 - Non-critical structures, $R = 2.5$
Category 2 - Semi-critical structures, $R = 1.8$
Category 3 - Critical structures, $R = 1.4$
Category 4 - Highly-critical structures, $R = 1.0$

Design example

The application of the proposed design method is illustrated here with the aid of an example. Results are then compared with two other well known design techniques that are commonly used by engineering profession. These are the TID-7024 (1963) method and the NSF (1980) method.

The TID-7024 approach is based on Housner's method. Using this approach, response spectrum data is necessary. In the NSF approach, information regarding the acceleration coefficients, the response modification factor and the period of vibration for the structure and sloshing water is necessary.

Details and dimensions of an above ground cylindrical tank used as an example are shown in figure 2. This example was taken from page VII-78 of the NSF report (1980). The seismic design forces are computed based on the following assumptions:

- The structure rests on a rigid foundation

- The structure is located in a high seismic risk zone. The acceleration coefficient is 0.40 and the zonal velocity ratio (v) is 0.4.

- The damping factor for structure and sloshing action is 5% and 0.5% of critical respectively.

- To determine the base moments, the point of application of impulsive force act at a distance of 3/8 h above the base of the tank. This is consistent with the assumption of parabolic distribution of lateral seismic hydrostatic loads.

- The working strength method is used in design.

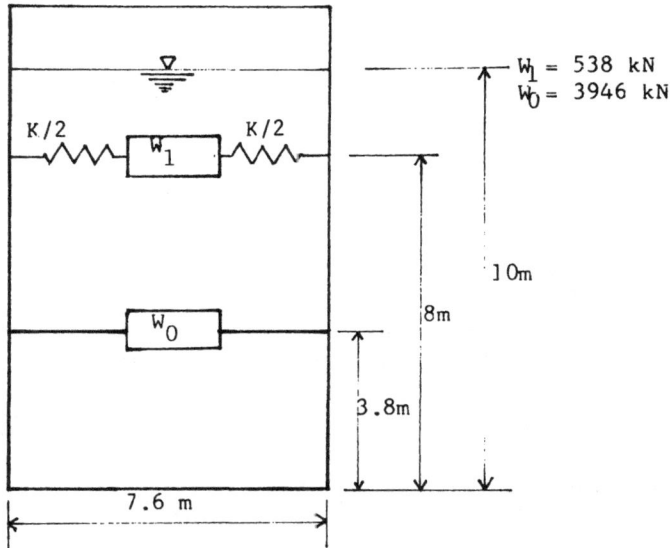

Figure 2. Details of cylindrical tank selected for design example

Table 1 shows a comparison in base shears and base moments using the three different approaches. These results show that the NBCC values for R=1.4 are very similar to the values obtained from the two other methods. This is because for values of R=1.4 or less, the behaviour of the tank approaches elastic response. The NSF approach and the TID-7024 approach are also based on elastic response. For values of R=1.8 and 2.5, the NBCC values are considerably less than the other two methods. This is clearly an indication that the response has been extended into the inelastic range.

Table 1 - Comparison of base shears and base moments for different design methods

Method		Base Shear (kN)	Base Moment (kNm)
TID-7024		1784	7137
NSF		1659	7945
Proposed NBCC	R = 1.0	2393	9309
	R = 1.4	1708	6649
	R = 1.8	1329	5172
	R = 2.5	956	3726

CONCLUSIONS

The design of concrete tanks for water retaining structures should ensure the required safety during a major earthquake and the required serviceability after a major earthquake. Four different levels of damage for concrete water retaining structures were identified based on their serviceability requirements and their importance. Based on such criteria, these structures are allowed to remain elastic; extend into elastic/inelastic range or behave completely inelastic. A comparison between different design methods showed that the National Building Code of Canada can be used to determine the design forces fairly accurately. Substantial savings in construction cost can be achieved if some degree of damage can be allowed without loss of life.

REFERENCES

ACI 318-77, 1989. Building code requirements for reinforced concrete. American Concrete Institute, Detroit, Michigan.

ANSI/AWWA D110-86, 1986. AWWA Standards for Wire Wound Circular Prestressed Concrete Water Tanks; American Water Works Association, Denver, Colorado.

API-650-E, 1980. Welded Steel Tanks for Oil Storage. Appendix E - Seismic Design of Storage Tanks; American Petroleum Institute.

Balendra, T., 1982. Seismic Design of Flexible Cylindrical Liquid Storage Tanks, Earthquake Engineering and Structural Dynamics, Vol. 10, pp 477-496.

CAN3-A23.3-M84, 1984. Code for the Design of Concrete Structures for Buildings. National Standard of Canada.

Housner, G.W., 1957. Dynamic Pressures on Accelerated Fluid Containers, Bulletin of the Seismological Society of America, Vol. 47, No. 1, pp. 15-35

Ikeda, S., 1989. Seismic Design of Concrete Structures based on Serviceability after Earthquakes. American Concrete Institute. Special Publication No. SP117-3, Detroit, pp 45-54.

National Building Code of Canada, 1990. Issued by the Associate Committee on the National Building Code, National Research Council of Canada, Ottawa.

NSF Report, 1980. Earthquake Design Criteria for Water Supply and Wastewater Systems. A National Science Foundation Report, Environmental Quality Systems Inc., Rockville, Maryland.

Priestley, M.J.N., Wood, J.H. and Davidson, B.J., 1988. Seismic Design of Storage Tanks. Bulletin of the New Zealand National Society for Earthquake Engineering, Vol. 19, No. 4, pp 272-284.

TID-7024, 1963. Nuclear Reactors and Earthquakes, Prepared by Lockhead Aircraft Corp. and Holmes and Nurver, Inc., U.S. Atomic Energy Commission.

Uniform Building Code, 1988. International Conference of Building Officials. Whittier, California.

Veletsos, A.S., and Yang, Y.Y., 1977. Earthquake Response of Liquid, Storage Tanks. Proc. of 2nd Annual ASCE Engineering Mechanics Specialty Conference, North Carolina State Univ., Raleigh, N.C., pp. 1-24.

Seismic performance of high-rise reinforced concrete frame buildings located in different seismic regions

T.J. Zhu[I], W.K. Tso[II], and A.C. Heidebrecht[III]

ABSTRACT

Seismic areas in Canada are classified into three categories for three combinations of acceleration and velocity seismic zones ($Z_a < Z_v$, $Z_a = Z_v$, and $Z_a > Z_v$). This paper evaluates the seismic performance of high-rise reinforced concrete frame buildings located in these three categories of seismic areas. Two frame buildings having 10 and 18 storeys are designed to the current Canadian seismic provisions, and their inelastic responses to three groups of ground motions are examined. The results indicate that the distribution of inelastic deformations is significantly different for high-rise frame buildings located in seismic regions with $Z_a < Z_v$, $Z_a = Z_v$, and $Z_a > Z_v$.

INTRODUCTION

Seismic regions in Canada are classified into three categories for three combinations of acceleration and velocity seismic zones ($Z_a < Z_v$, $Z_a = Z_v$, and $Z_a > Z_v$) (Heidebrecht et al. 1983). Seismic areas having $Z_a > Z_v$ are influenced mainly by small or moderate nearby earthquakes, and ground motions in these areas are expected to exhibit high frequency content and have high peak acceleration-to-velocity (A/V) ratios. Seismic regions with $Z_a < Z_v$ are affected mainly by large distant earthquakes, and ground motions in these regions are expected to have low frequency content and low A/V ratios. The specification of seismic design forces for buildings having fundamental periods longer than 0.5 sec is directly tied to zonal velocity, irrespective of seismic region category. For short-period buildings, three different levels of seismic design force are used for the three different categories of seismic areas having $Z_a < Z_v$, $Z_a = Z_v$, and $Z_a > Z_v$. The objective of this paper is to evaluate the seismic performance of high-rise reinforced concrete frame buildings located in the three different categories of seismic regions.

STRUCTURAL MODELS

Two reinforced concrete ductile moment-resisting frame (MRF) buildings having 10 and 18 storeys were considered. These two buildings are designated as 10S and 18S and have the same floor plan as shown in Fig. 1. The effect of seismic action was considered in the E-W direction. The elevations of the interior

---

[I]Post-Doctoral Fellow, Dept. of Civil Engineering, McMaster University, Hamilton, Ontario, Canada L8S 4L7

[II]Professor, Dept. of Civil Engineering, McMaster University, Hamilton, Ontario, Canada L8S 4L7

frames in this direction are shown in Fig. 1. The beam and column sizes are also depicted. The two buildings were designed for combined gravity and seismic effects in accord with the 1990 edition of the National Building Code of Canada (NBCC 1990) (Associate Committee on National Building Code 1990), and their structural members were proportioned and detailed according to the 1984 edition of the Canadian Reinforced Concrete Design Code (CAN3-A23.3-M84) (Canadian Standards Association 1984).

The design gravity loads are shown in Table 1. The seismic design base shear, V, for each frame was specified from the formula:

$$V = (V_e / R) U \tag{1}$$

in which V_e = the seismic design force representing elastic response, R = the force modification factor, and U = 0.6. For the ductile MRF buildings considered, R was taken as 4. V_e is given by

$$V_e = v \, S \, I \, F \, W \tag{2}$$

in which v = zonal velocity ratio, S = seismic response factor, I = importance factor, F = foundation factor, and W = dead weight. The two frames were assumed to be located in regions with high seismicity (Z_v=6). Accordingly, v was taken as 0.4. F and I were set to 1.0. To determine S, the fundamental periods of the two frames were estimated from the formula, T = 0.1 N, where N is the number of storeys.

The design base shear for each frame was distributed over its height based on the NBCC 1990 distribution formula. The frames were designed based on the following three load combinations:

$$
\begin{aligned}
&1.25 \, D + 1.5 \, L \\
&1.25 \, D + 1.0 \, Q \\
&1.25 \, D + 0.7 \, (\, 1.5 \, L + 1.0 \, Q \,)
\end{aligned}
\tag{3}
$$

in which D = dead load, L = live load due to use and occupancy, and Q = seismic load.

The effects of geometric nonlinearity (P-delta and slenderness effects) were considered according to CAN3-A23.3-M84. The factored beam design moments and column design axial forces were obtained from the elastic analyses of the frames under the three load combinations. The factored column design moments were related to the beam moment capacities by considering the equilibrium of each joint, coupled with a column overstrength factor. The column overstrength factor was set to 1.2 which is slightly higher than the minimum value of 1.1 required by CAN3-A23.3-M84. The final design results are given by Zhu (1989).

GROUND MOTION DATA

A total of 45 strong motion records recorded on rock or stiff soil sites were selected from the McMaster University Seismological Executive (MUSE) Database. The 45 records were obtained from 23 different events with magnitude ranging from 5.25 to 8.1, and they were recorded at epicentral distances ranging from 4 to 379 km. The 45 records were subdivided into three groups according to their A/V ratios, with 15 records in each group. The records having A/V < 0.8 g/m/s were classified into the low A/V group whereas those having A/V > 1.2 g/m/s were categorized into the high A/V group. The records with 0.8 g/m/s ≤ A/V ≤ 1.2 g/m/s were classified into the intermediate A/V group. These three groups of records were taken as representative ground motions in the three categories of seismic regions having $Z_a < Z_v$, $Z_a = Z_v$, and $Z_a > Z_v$. Fig. 2 shows the distribution of magnitude and epicentral distance for the three groups of records. It can be seen that the ground motions with high A/V ratios were obtained in the vicinity of small or moderate earthquakes whereas those having low or intermediate A/V ratios were recorded at large distances from large or moderate earthquakes.

To indicate their frequency content, 5% damped elastic response spectra were computed for the three groups of records scaled to a peak velocity of 0.4 m/s which is the design zonal velocity for the frames. Fig. 3 shows the mean response spectra for the three groups of records. The high A/V group of records has higher frequency content than the low A/V group. The elastic design spectrum (V_e/W) is superimposed in Fig. 3. Also shown in Fig. 3 is the inelastic design spectrum (V/W). The estimated and actual fundamental periods of the 10S and 18S frames are also shown in the figure. The duration of strong shaking was estimated for each of the 45 records based on the definition by McCann and Shah (1979). A statistical summary of the computed strong-motion durations for the three groups of records is presented in Table 2. The low A/V group of records has longer duration of strong shaking than the high A/V group.

ANALYSIS PROCEDURE AND RESPONSE PARAMETERS

Many hysteretic models have been proposed for reinforced concrete structural members (Otani 1980). In this study, it is assumed that sufficient transverse reinforcement has been provided for the structural members, and stiffness and/or strength deterioration due to shear or bond loss is not significant. Accordingly, the relatively simple dual-component element (Clough et al. 1967) was used to model the beams and columns. The effect of axial force on yield moment was considered for each column by a yield moment-axial force interaction curve. A simplified solution (Wilson & Habibullah 1987) was used to account for the second-order P-delta effect for each column. The general-purpose computer program DRAIN-2D (Kanaan & Powell 1973) was used to perform dynamic analysis for the frames. The DRAIN-2D was also modified to perform inelastic static analysis for the frames subjected to monotonically increased lateral loading (Zhu 1989).

The response parameters considered are interstorey drift for overall response and curvature ductility and cumulative plastic rotation for member response. The curvature ductility is representative of the response parameters for damage caused by large inelastic deformation excursions, and the cumulative plastic rotation represents the response parameters for damage due to sustained reversals of inelastic deformations.

STATIC ANALYSIS RESULTS

The inelastic behaviour of the frames subjected to monotonically increased lateral loading was studied. The lateral loading was distributed over frame height according to the NBCC 1990 distribution formula. The base shear versus roof displacement curves for the 10S and 18S frames are depicted in Fig. 4. The points for the first beam and column hinges are shown in the same figure. To indicate the overall lateral strengths of the frames relative to their seismic design forces, the design base shear levels are also shown in the figure.

The base shear-roof displacement response is nearly linear up to the formation of the first column hinge. Thereafter, the overall stiffness of the frames decreases drastically. The actual lateral strengths of the frames are higher than their design base shears. This is particularly true for the 10S frame. The strength for an overall drift of 0.5% is about 29% and 13% higher than the design base shear for the 10S and 18S frame, respectively. The lower overstrength for the 18S frame is mainly because the P-Delta effect for this frame is very significant. The seismic overstrength of the frames can be attributed to the following factors. First, the column strengths were keyed to the beam strengths based on the weak beam-strong column criterion. Second, strain-hardening effect was considered for the beams and columns in the analysis.

DYNAMIC ANALYSIS RESULTS

The inelastic responses of the frames to the three groups of records were analyzed statistically. All the records were scaled to a peak velocity of 0.4 m/s which is the design zonal velocity for the frames. The mean and mean plus one standard deviation (mean + σ) values of the response parameters were obtained for each group of records. The mean + σ level is appropriate for design purposes. A comparison between the mean and mean + σ values indicates the dispersion characteristics of the response parameters within each group of records.

Figs. 5 to 6 show the statistical results of the response parameters for the 10S and 18S frames. It can be seen that the distributions of inelastic deformations over frame height are significantly different for the three groups of ground motions. For the high A/V group of ground motions, the interstorey drifts in the upper storeys are higher than those in the lower storeys. The high A/V group of records also produces more inelastic deformations in the upper storey beams and columns. This "whiplash" effect can be ascribed to significant effect of higher mode participation. Since the high A/V ground motions have higher frequency content, they prompt the higher modal responses of the 10S and 18S frames.

For the low A/V group of ground motions, the interstorey drifts in the lower storeys are higher than those in the upper storeys. Beam inelastic deformations are concentrated in the lower storeys. This is particularly true for the cumulative plastic rotation. In the lower storeys of the 18S frame, the curvature ductility for the low A/V group of records is about twice that for the high A/V group whereas the difference increases up to about four times for the cumulative plastic rotation. The cumulative plastic rotation depends on both the peak inelastic response and the duration of strong shaking. The analysis of the ground motion data has indicated that the low A/V ground motions have longer duration of strong shaking than the high A/V ground motions. Since the beam peak inelastic response (indicated by the curvature ductility) for the low A/V group of records is already higher than that for the high A/V group in the lower storeys, the combined effect of strong-motion duration results in much higher cumulative plastic rotation for the low A/V group of records. The low A/V ground motions also produce higher inelastic deformations at the base of the first storey columns than the high A/V ground motions. The lower storeys of a high-rise frame are more vulnerable than the upper storeys due to the high axial forces carried by the lower storey columns. The concentration of inelastic deformation in the lower storeys would increase second-order P-delta effect. Therefore, low A/V ground motions are more damaging to high-rise frames than high A/V ground motions even when the frames are designed based on peak ground velocity.

To gain insight into the effect of ground motion frequency content on the inelastic response, the sequences of plastic hinge formation are shown in Fig. 7 for the 10S frame subjected to three example earthquake records, one from each of the three A/V groups. These three records are (a) the 1985 Mexico, Mesa Vibradora, N90W component; (b) the 1952 Kern County, Taft, S69E component; and (c) the 1935 Helena, Carroll College, N00E component, and they have an A/V ratio of 0.36, 1.01, and 2.03 g/m/s, respectively. It can be seen in Fig. 7 that for the Mesa Vibradora record which has a low frequency content, plastic hinges develop in the lower storey beams first and then migrate from the bottom to the top of the frame. Column plastic hinges are located only at the base of the first storey columns. For the Carroll College record which has a high frequency content, beam plastic hinges develop in the upper storey first, and column hinges are developed only in the upper storeys. The Taft record has a broad range of significant frequency content, and consequently, it produces column plastic hinges both at the base of the first storey columns and in the upper storeys.

An examination of the curvature ductility demands for the columns reveals that the use of a column overstrength factor of 1.2 does not prevent column yielding in the upper or middle storeys. This is particularly true for the high and intermediate A/V groups of records which tend to excite the higher modes of vibration. At a joint, the distribution of the beam moments to the columns above and below the joint during dynamic response can be considerably different from that assumed in the design. The effect of higher modal responses can alter the moment distribution at the joint to such an extent that the sum of the beam moments is resisted entirely by one of the two columns. Previous studies by Paulay (1981) have indicated that a dynamic magnification factor should be used in the design of columns to prevent upper or middle storey columns from yielding.

CONCLUSIONS

Because of their different dynamic characterist cs, ground motions in the three categories of seismic areas produce different distributions of inelastic deformation demands on high-rise frame buildings. For high A/V ground motions, the "whiplash" effect is very significant due to significant contribution of higher modal responses, and inelastic deformations in the upper storeys can be higher than those in the lower storeys. In particular, plastic hinges would be developed in the upper storey columns. Therefore, careful detailing should be given to the upper story columns and beams of high-rise buildings situated in high A/V seismic regions $(Z_a > Z_v)$.

For low A/V ground motions, inelastic deformations are concentrated in the lower storeys. This is particularly true for the cumulative plastic rotation because of the combined effect of the duration of strong shaking. Therefore, special attention should be given to the design and detailing of the columns and beams in the lower storeys of high-rise buildings located in low A/V seismic areas $(Z_a < Z_v)$. Because the lower storeys are more vulnerable than the upper storeys due to the high axial loads carried by the columns, low A/V ground motions are more damaging to high-rise buildings than high A/V ground motions even when the buildings are designed based on peak ground velocity.

ACKNOWLEDGEMENTS

The writers wish to acknowledge the support from the Natural Science and Engineering Research Council of Canada for the work presented herein.

REFERENCES

Associate Committee on National Building Code. 1990. National Building Code of Canada 1990. National Research Council, Ottawa, Ont.

Canadian Standards Association. 1984. Design of concrete structures for buildings, CAN3-A23.3-M84. Ottawa, Ont.

Clough, R.W. and Benuska, K.L. 1967. Nonlinear earthquake behaviour of tall buildings. J. Eng. Mech. Div., ASCE, 93, 129-146.

Heidebrecht, A.C., Basham, P.W., Rainer, J.H. and Berry, M.J. 1983. Engineering applications of new probabilistic seismic ground-motion maps of Canada. Can. J. Civ. Eng., 10, 670-680.

Kanaan, A. and Powell, G.H. 1973. General purpose computer program for inelastic dynamic response of plane structures. Report No. EERC 73-6, Univ. of Calif., Berkeley, Calif.

McCann, N.W. and Shah, H.C. 1979. Determining strong-motion duration of earthquakes. Bull. Seism. Soc. Am., 69, 1253-1265.

Otani, S. 1980. Nonlinear dynamic analysis of reinforced concrete building structures. Can. J. of Civ. Eng., 7, 334-344.

Paulay, T. 1981. Developments in the seismic design of reinforced concrete frames in New Zealand. Can. J. Civ. Eng., 8, 91-113.

Wilson, E.L. and Habibullah, A. 1987. Static and dynamic analysis of multi-story buildings, including P-delta effects. Earthquake Spectra, 3, 289-298.

Zhu, T.J. 1989. Inelastic response of reinforced concrete frames to seismic ground motions having different characteristics. Ph.D. thesis, McMaster University, Hamilton, Ont.

Table 1. Design gravity loads (kN/m²)

	Dead load	Live load
Roof	6.5	1.0
Floor	7.0	2.4

Table 2. Statistical summary of strong-motion durations (sec)

	Low A/V	Inter. A/V	High A/V
Mean	12.58	7.41	2.7
COV*	0.452	0.290	0.558

* Coefficient of variation

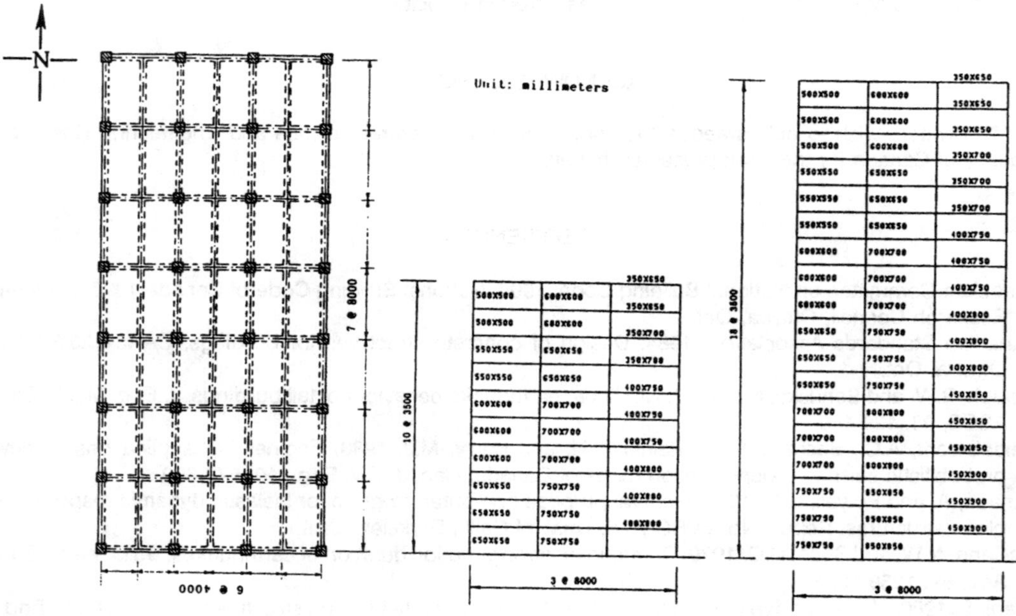

Figure 1. Floor plan and frame elevations

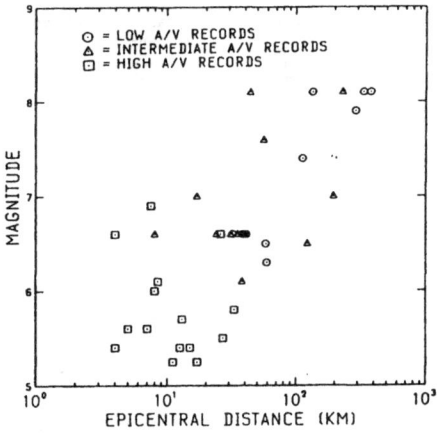

Figure 2. Distributions of magnitude and epicentral distance for three A/V groups of records

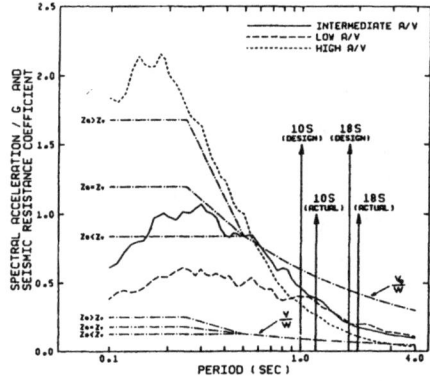

Figure 3. Mean elastic response spectra of three A/V groups of records and elastic and inelastic design spectra

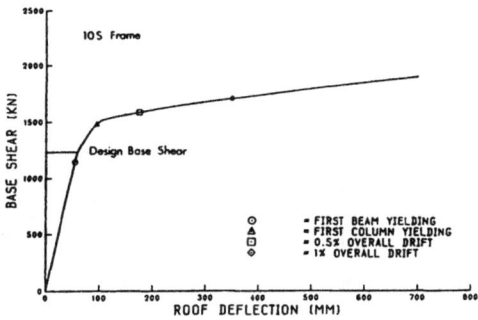

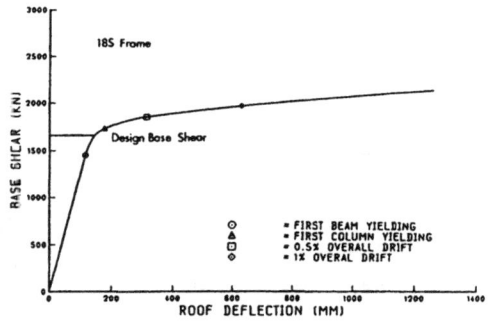

Figure 4. Base shear versus roof displacement responses for monotonically increased lateral loading

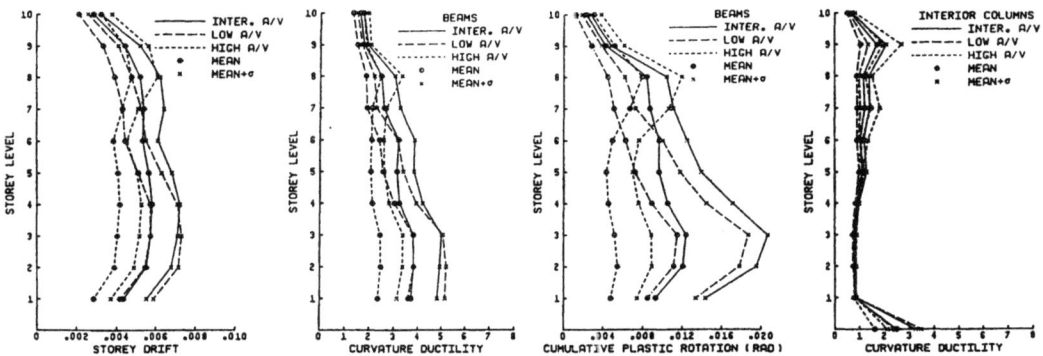

Figure 5. Statistical results of response parameters for 10S frame

557

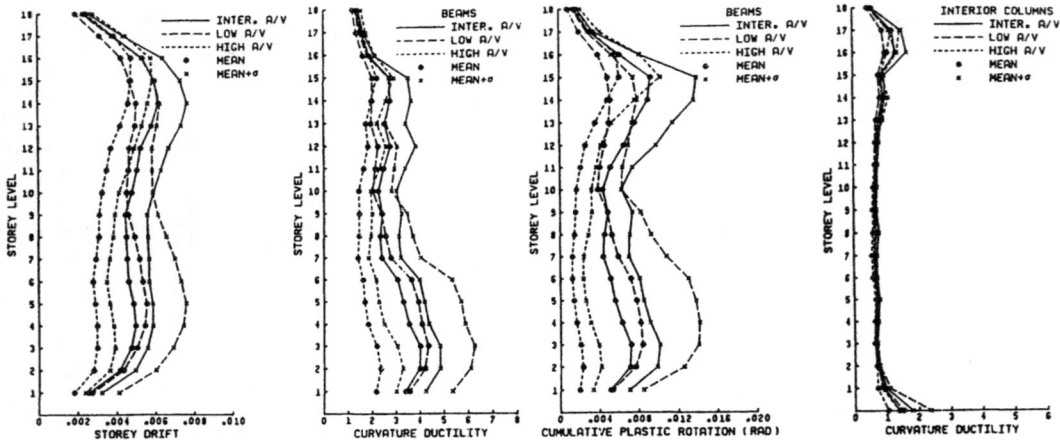

Figure 6. Statistical results of response parameters for 18S frame

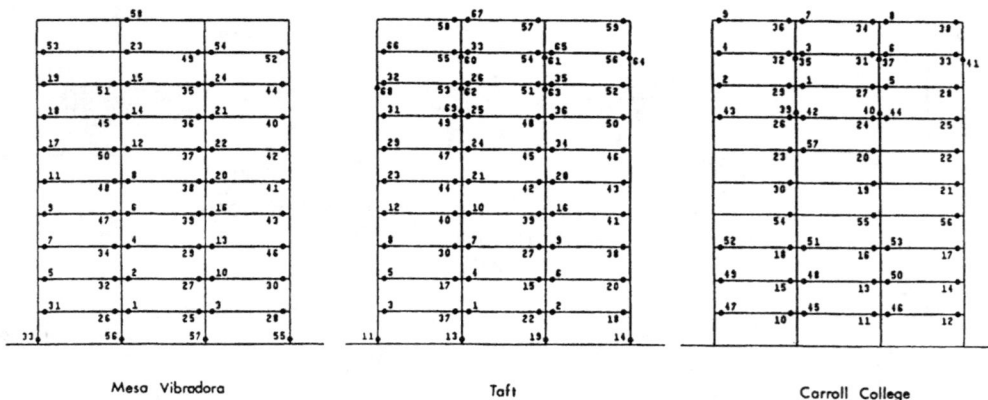

Mesa Vibradora   Taft   Carroll College

Figure 7. Sequence of plastic hinge formation for 10S frame
resulting from three example earthquakes

558

Two-level vs. one-level seismic design of buildings

Howard H.M. Hwang[I] and Ah Lim Ch'ng[II]

ABSTRACT

This paper evaluates the merits and shortcomings of the one-level design described in the NEHRP Provisions and the two-level design specified in the Tri-Services Guidelines. In this study, a four-story reinforced concrete hospital building is designed according to the NEHRP Provisions and the Tri-Services Guidelines, separately. Then, the capacities of these two structures at the stages of first-yielding and collapse are evaluated and compared. The advantages and shortcomings of these two design approaches are discussed.

INTRODUCTION

It is not economical to design structures to resist seismic forces induced by large earthquakes within the elastic limits, because large earthquakes have a low probability of occurrence during the expected life of the structure. To deal effectively with the combination of extreme loading and low probability, the philosophy of seismic-resistant design in model building codes is that a building designed according to the seismic provisions will (1) to resist a moderate earthquake without structural damage, and (2) to resist a large earthquake without collapse so that life safety can be maintained. This philosophy leads naturally to the concept of two-level seismic design as prescribed in the Seismic Design Guidelines for Essential Buildings (Tri-Services Guidelines 1986). However, as the current design practice in the United States, most structures are designed for one level of earthquake that usually has a 10% probability of being exceeded in 50 years as specified in the NEHRP Provisions (1988). For both design approaches, structure members are allowed to go into inelastic range during a large earthquake. The NEHRP Provisions uses the

---

[I] Professor, Center for Earthquake Research and Information, Memphis State University, Memphis, TN 38152

[II] Graduate Research Assistant, Center for Earthquake Research and Information, Memphis State University, Memphis, TN 38152

response modification factor to reduce the base shear from an elastic force level to an inelastic force level. On the other hand, the Tri-Services Guidelines uses the inelastic demand ratios applied to individual members. Thus, even though the design philosophy in the NEHRP provisions and the Tri-Services Guidelines is the same, the approaches used in these two documents are quite different. This paper evaluates the merits and limitations of the one-level seismic design as described in the NEHRP Provisions and the two-level seismic design as specified in the Tri-Services Guidelines. In this study, a four-story reinforced concrete hospital building is designed according to the 1988 NEHRP Provisions and the 1986 Tri-Services Guidelines, separately. Then, the capacities at the stages of first yielding and collapse are evaluated and compared.

DESIGN USING THE NEHRP PROVISIONS

A four-story hospital building is assumed to be located in a moderate seismic zone with the coefficient of effective peak-velocity related acceleration A_v equal to 0.2. According to the NEHRP Provisions, the Seismic Hazard Exposure Group for a hospital building is III. From these two conditions, the Seismic Performance Category is determined as E. Thus, the special moment-resisting (SMR) frame is required to be used to resist gravity loads and earthquake forces. A typical floor plan and an elevation of the four-story hospital building are shown in Figs. 1 and 2, respectively. The beam size is 12 inches by 20 inches and column size is 14 inches by 14 inches throughout the building. This study focuses on the design of a typical interior frame in the north-south direction. The seismic design base shear V is calculated using the following formula:

$$ V = C_S W = \frac{1.2\, A_v S}{R T^{2/3}}\, W \tag{1} $$

C_S is the seismic base shear coefficient and has 2.5 A_a/R as the upper bound. Since A_v is equal to 0.2, the seismic coefficient of peak acceleration A_a is also taken as 0.2. The soil condition of the site is assumed to be classified as S_2; thus the soil factor S is equal to 1.2. For an SMR frame, the response modification factor R is set equal to 8 in accordance with the NEHRP Provisions. The fundamental period of the building is determined as 0.547 second and the total dead load W of the building is 3394.9 kips (Ch'ng 1990). From Eq. 1, the design base shear V is calculated as 182.70 kips. Using the member forces caused by the base shear and gravity loads, the SMR frame is designed according to ACI code 318-89 (1989) as shown in Ch'ng (1990). The arrangement of flexural reinforcement is shown in Fig. 3.

DESIGN USING THE TRI-SERVICES GUIDELINES

According to the Tri-Services Guidelines, an essential building shall be designed to resist the following two levels of earthquakes EQ-I and EQ-II. EQ-I is the maximum probable earthquake, likely to occur during the life of a building, and is defined as an earthquake with a 50% probability of being exceeded in 50 years (a 72-year earthquake). On the other hand, EQ-II is the maximum theoretical earthquake that can occur at a site, but has a low probability of occurrence during the life of a building. EQ-II is defined as an earthquake with a 10% probability of being exceeded in 100 years (a 950-year earthquake). The four-story hospital building to be designed according to the Tri-Services Guidelines has the same lay-out as shown in Figs. 1 and 2. The size of beams is 15 inches by 24 inches and columns is 20 inches by 20 inches throughout the building. The design response spectrum is defined as

$$S_a = \frac{1.22 \ A_v \ S_i \ D_f}{T} \tag{2}$$

and S_a has the following upper bound:

$$S_a \leq 2.5 \ A_a \tag{3}$$

For the building located in a seismic zone with A_v and A_a equal to 0.2 as shown in the ATC3-06 (1984) contour map, A_v and A_a of EQ-I are equal to 0.08. For the S_2 soil condition, the soil factor S_i is equal to 1.2. According to the Tri-Services Guidelines, the damping adjustment factor D_f is 1.00 for the reinforced concrete structure with a damping ratio of 5%. Substituting A_v, S_i, D_f, and A_a into Eqs. 2 and 3, the design response spectrum of EQ-I is determined. In accordance with the Tri-Services Guidelines, an essential building such as a hospital needs to be analyzed dynamically even though the building is a regular building. In this study, the four-story hospital building is idealized as a multi-degree-of-freedom stick model with a fixed base. For low-rise buildings, say up to about 5 stories, the modal analysis can generally be performed by using only the fundamental mode. Thus, the story lateral forces of the fundamental mode are used to obtain the member forces. From the free vibration analysis, the fundamental period, the corresponding mode shapes and the modal participation factor PF_{j1} at level j can be determined. The story lateral forces at level j are computed as

$$F_j = PF_{j1} \ S_{a1} \ w_j \tag{4}$$

where S_{a1} is the spectral acceleration at the fundamental period (0.427 sec) and is equal to 0.2 g as determined from the EQ-I response spectrum. w_j is the weight assigned at the level j. From these story lateral forces, the member forces caused by EQ-I can be determined and combined with those from the gravity loads. The combined member forces are used to design structural members in accordance with the design procedure specified in the ACI code 318-89.

In level-two design, the structure is analyzed to determine its ability to resist the forces and deformations caused by EQ-II. For A_v and A_a equal to 0.2 in the ATC3-06 contour map, A_v and A_a of EQ-II are equal to 0.25. For a reinforced concrete structure, a 10% damping ratio is used for post-elastic analyses; thus, the damping adjustment factor D_f is equal to 0.80. Using Eqs. 2 and 3, the design response spectrum of EQ-II is determined. Structural member forces are calculated by means of modal analysis using EQ-II response spectrum. The member forces from gravity loads and EQ-II computed by means of the elastic analysis procedure are allowed to be larger than the design ultimate capacity. However, the inelastic demand ratio (IDR) is implemented to control the overstress within an acceptable limit for each member. Since the shear capacity is greater than the flexural capacity for a special moment-resisting frame, the flexural capacities of beams and columns are used to represent the capacities of structural members. Hence, IDR is defined as the ratio of the elastic demand moment M_D to the design ultimate moment capacity M_C. For essential buildings, the Tri-Services Guidelines specifies that the IDR of beams shall not be greater than 2, and the IDR of columns shall not be greater than 1.25. The smaller IDR of column is to ensure strong column-weak beam behavior, that is, to ensure that hinging will take place in beam rather than in column. The maximum IDRs of all the members are determined and shown in Fig. 4. It can be seen that IDRs of several columns and beams exceed

561

the allowable limit. Thus, the deficiencies must be corrected by redesign of these critical members. By a trial-and-error process, IDRs of all members are within the allowable limits set forth in the Tri-Services Guidelines. Thus, the frame structure has the capacity to resist both EQ-I and EQ-II. The flexural reinforcement arrangement of beams and columns is shown in Fig. 5.

LIMIT STATES AND STRUCTURAL CAPACITY

Two types of limit states are used in this study: first yielding and collapse of structure. For a frame structure, the first yielding is defined as the formation of first plastic hinge anywhere in the structure. If a structure subject to earthquakes does not reach the first yielding, then the structural response remains in the elastic range and the structure does not sustain any structural damage. The first-yielding limit state can be considered as a serviceability limit state. The collapse of structure is defined as the formation of a failure mechanism. Thus, the collapse limit state represents a strength limit state.

In this paper, the capacity spectrum method proposed by Freeman (1978) is used to construct the capacity curves for these two frame structures. The capacity curve displays seismic capacity in terms of the spectral acceleration versus the fundamental period. The capacity curves of the two frame structures designed according to the NEHRP Provisions and the Tri-Services Guidelines, are plotted in Fig. 6. It can be seen that these two curves are not in the same period range. To compare the capacities of these two structures, it is desirable to convert the spectral accelerations at the first yielding and collapse to the corresponding peak ground acceleration (PGA). To determine the PGA value corresponding to the first yielding of the frame, a response spectrum with a 5% damping ratio as specified in the NEHRP Provisions or the Tri-Services Guidelines is passing through the first-yielding point of the capacity curves. Then, the PGA values are obtained from the upper bound of the spectral accelerations divided by 2.5. The PGA values corresponding to the collapse of structures are determined in a similar manner. These PGA values are summarized in Table 1.

Table 1. Structural capacities in terms of PGA

Stage	NEHRP Provisions	Tri-Services Guidelines
First yielding	0.10 g	0.15 g
Collapse	0.31 g	0.47 g

CONCLUSIONS

From the design and evaluation of the two frame structures designed according to the 1988 NEHRP Provisions and the 1986 Tri-Services Guidelines, respectively, the advantages and short-comings of these two different approaches are discussed as follows:

1. The one-level seismic design, such as the approach used in the NEHRP Provisions, is easy to use by designers. However, the design earthquake as currently defined represents neither a

moderate earthquake nor a large earthquake. In addition, the designer is not required to evaluate the overall capacity of structure to resist a large earthquake. Therefore, there is some uncertainty regarding to the building performance in the event of a large earthquake.

2. The two-level seismic design, the approach employed in the Tri-Services Guidelines, follows naturally the current earthquake-resistant design philosophy. Two-level seismic design requires designers to evaluate the building performance and to discover possible weak spots that are susceptible to large earthquakes. Thus, the two-level seismic design can give designers confidence in the design of structures and can reduce the risk of catastrophic damage of buildings in the event of a large earthquake.

3. The building frame designed according to the Tri-Services Guidelines is much stronger than the one designed according to the NEHRP Provisions. The essential building designed according to the NEHRP Provisions seems to be unconservative based on the potential seismic hazard at the site.

ACKNOWLEDGMENTS

This paper is based on research supported by the National Center for Earthquake Engineering Research (NCEER) under contract number NCEER-89-1101 (NSF Grant No. ECE-86-07591). Any opinions, findings, and conclusions expressed in the paper are those of the authors and do not necessarily reflect the views of NCEER or NSF of the United States.

REFERENCES

Building Code Requirement for Reinforced Concrete 1989. ACI 318-89, American Concrete Institute, Detroit, MI

Ch'ng, A.L. 1990. One-Level versus Two-Level Seismic Design of Buildings. Thesis presented to the faculty of the graduate school of Memphis State University in partial fulfillment of the requirements of the degree of master of science in civil engineering.

Freeman, S.A. 1978. Prediction of Response of Concrete Buildings to Severe Earthquake Motion. Proceedings of the Douglas McHenry International Symposium on Concrete and Concrete Structures, ACI Publication SP-55, American Concrete Institute, Detroit, MI, 589-605.

NEHRP Recommended Provisions for the Development of Seismic Regulations for New Buildings 1988. Earthquake Hazard Reduction Series 17 and 18, Building Seismic Safety Council and Federal Emergency Management Agency, Washington, DC.

Seismic Design Guidelines for Essential Buildings (Tri-Services Guidelines) 1986. Departments of the Army, Navy, and Air Force, Washington, DC.

Tentative Provisions for the Development of Seismic Regulations for Building 1984. Applied Technology Council (ATC), Palo Alto, CA.

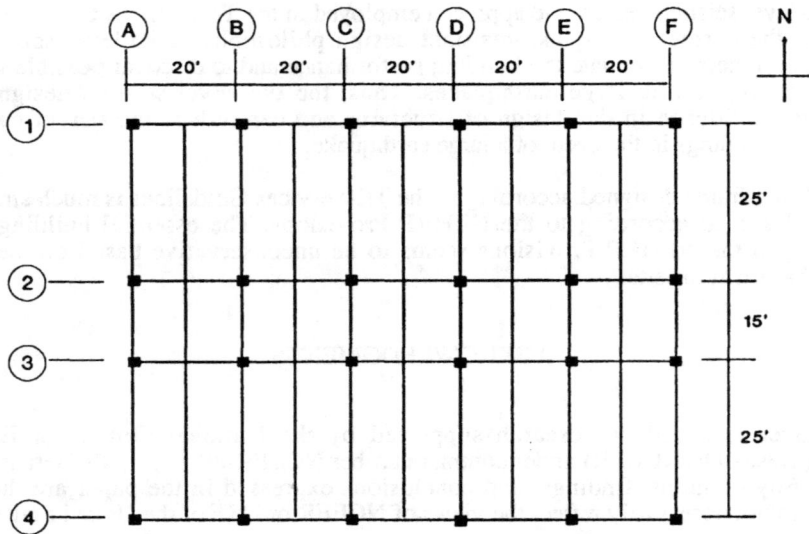

Figure 1. Typical floor plan of the four-story hospital building

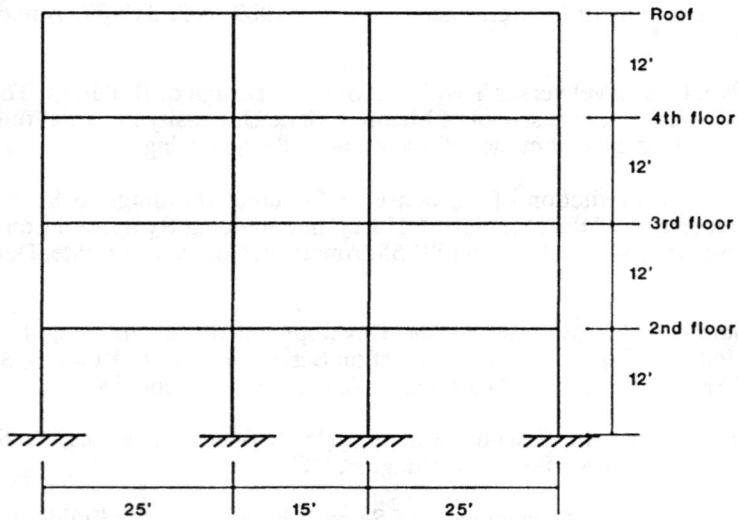

Figure 2. Elevation of an interior frame (N-S direction)

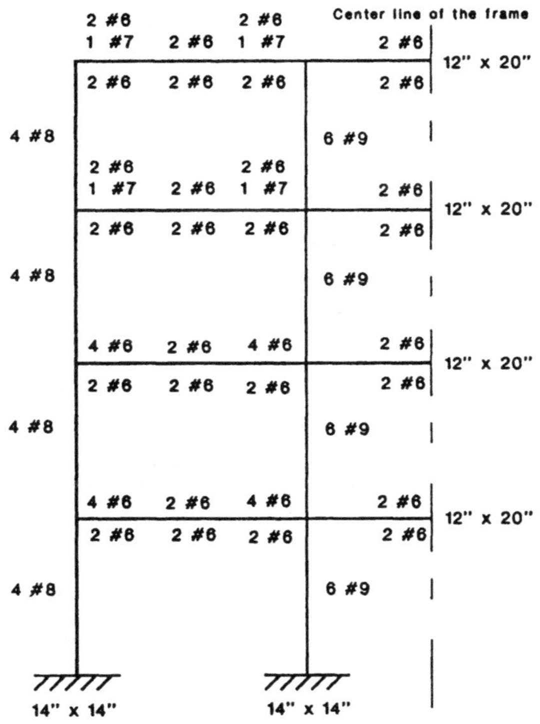
Center line of the frame
2 #6
1 #7
2 #6
2 #6
1 #7
2 #6
12" x 20"
4 #8
6 #9
14" x 14"
14" x 14"

Figure 3. Flexural reinforcement of the NEHRP frame

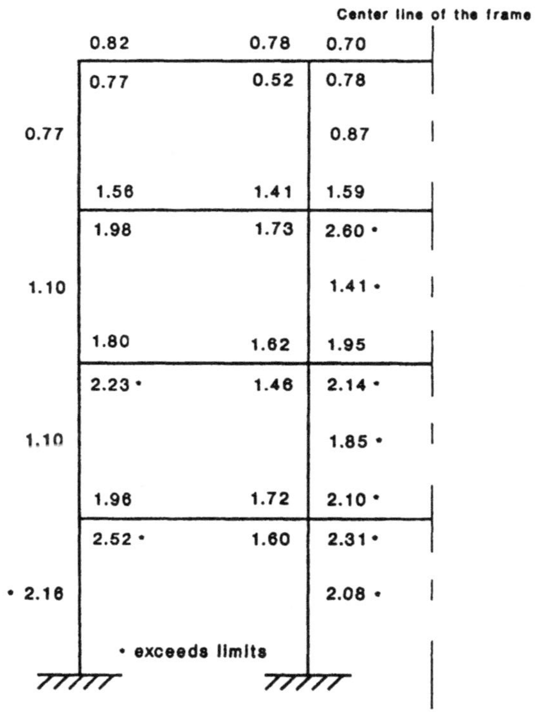
Center line of the frame
0.82
0.78
0.70
0.77
0.52
0.78
0.77
0.87
1.56
1.41
1.59
1.98
1.73
2.60 •
1.10
1.41 •
1.80
1.62
1.95
2.23 •
1.46
2.14 •
1.10
1.85 •
1.96
1.72
2.10 •
2.52 •
1.60
2.31 •
• 2.16
2.08 •
• exceeds limits

Figure 4. Inelastic demand ratios of the Tri-Services frame

Figure 6. Capacity curves of the NEHRP and Tri-Services frames

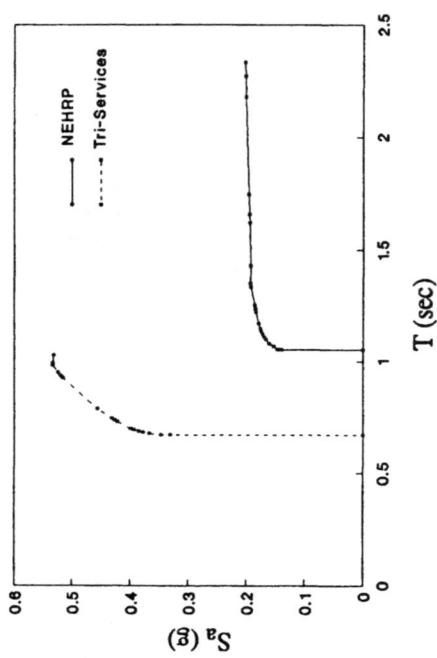

Figure 5. Flexural reinforcement of the Tri-Services frame

Design and construction features of a 37-story precast reinforced concrete moment frame building in Tokyo

Cloyd E. Warnes[1]

ABSTRACT

This paper describes design and construction features of a high-rise apartment building project in downtown Tokyo, Japan constructed during 1990. Not only is the building the tallest reinforced concrete building in this high seismic risk region, it was constructed using precast concrete frame "cruciform" elements fabricated at the jobsite. All connections use the "emulation design" approach for precast concrete to assure monolithic concrete performance of the structure.

Details of site fabrication of the cruciform elements are discussed, permitting lower ratio of labor required compared to that for formed cast-in-place structures.

Assembly of the erected elements is described. This includes column-to-column and column-to-beam connections, as well as the use of half-thickness factory-manufactured floor and balcony slabs, all of which are tied together with a cast-in-place concrete topping pour. One floor was structurally completed every six days, and other trades could begin work on the floors immediately below.

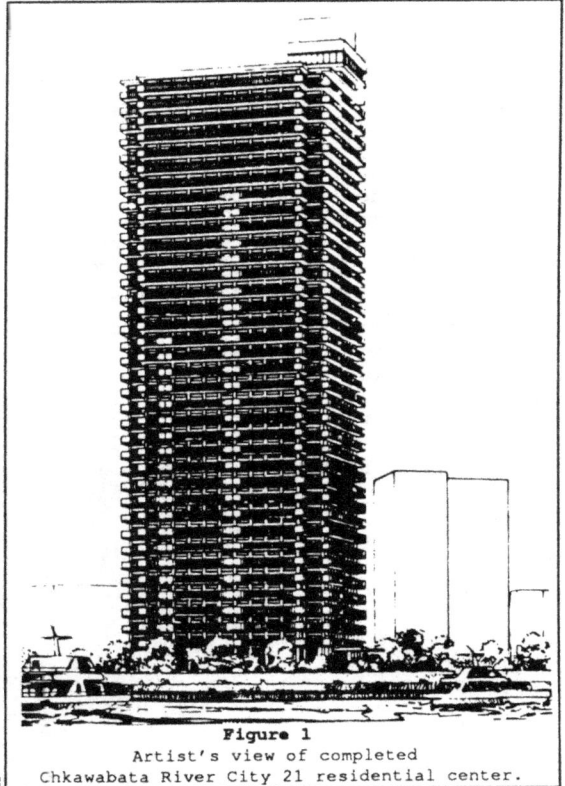

Figure 1
Artist's view of completed
Chkawabata River City 21 residential center.

GENERAL

The project, called Ohkawabata River City 21, is a residential center of mid- to high-rise buildings located on an attractive peninsula in the Sumida River near downtown Tokyo. (Fig. 1)

I. Principal, CPM Associates
    Sacramento, California, USA

It was as constructed by a joint venture of the Taisei, Shimizu and Ohbayashigumi corporations. Taisei prepared the detailed designs. The site area is 87,400 square feet (8,740 sq. m) and the building footprint is 15,272 sq. ft. (1,527 sq m). The building extends 32 feet (10m) underground to provide for parking and building support. The building extends 37 floors or 348 (106m) feet above grade. It was built in the form of a square doughnut with living spaces fronting both to the exterior and to the interior atrium, much like some hotels in North America.

At 37 stories, Ohkawabata Towers is the tallest reinforced concrete moment frame structure in Japan as of 1990. It is constructed predominantly of precast concrete structural elements joined together in such a manner as to create an equivalent monolithic cast-in-place concrete building.

Precast elements were manufactured both at the jobsite and in offsite precast concrete factories. Onsite precast manufacturing facilities produced columns and beams. Offsite factories produced half-thickness floor and balcony slabs as well as precast concrete non-bearing walls and partitions. Transit-mixed concrete from several different plants provided the concrete for the site precast elements.

Two precast concrete manufacturing facilities for the frame members were established directly at the base of the building, one on each side. (Figs. 2 and 3)

Conventional cast-in-place concrete topping was placed over the precast floor and balcony elements to provide a diaphragm. Simultaneously, open-

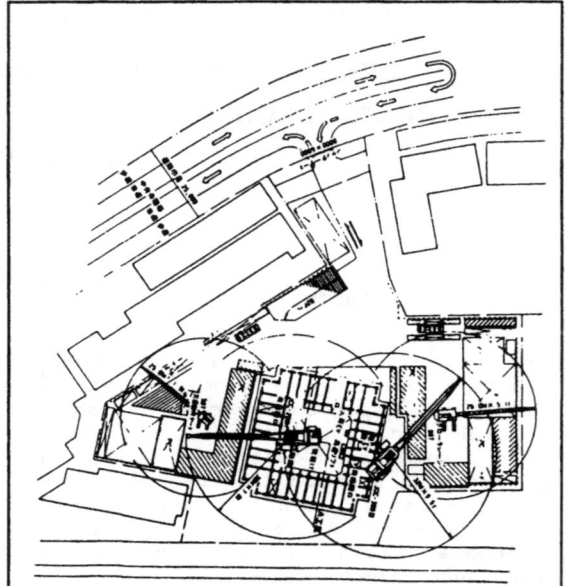

Figure 2
Two temporary precast concrete structural manufacturing plants were established at the base of the building, one on each side within the reach of a tower crane.

Figure 3
Aerial view of one of two precast concrete plants established on-site at the the base of the building.

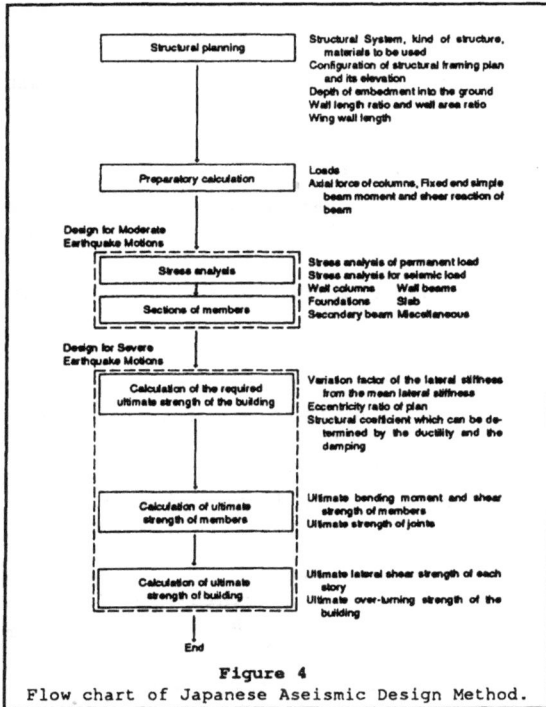

Figure 4
Flow chart of Japanese Aseismic Design Method.

ings between beam ends were filled with concrete of the same specification as that of the floor topping.

The basic element used in construction of the frame was a three-dimensional beam-column "cruciform tree" connected at locations away from the high moment region. During the early stages of erection, in the lower portion of the building, structural steel sections were embedded in the precast concrete cruciform elements.

Initially, column longitudinal reinforcing steel was joined by a special welding process and the closure area around the welds was completed with poured concrete. Subsequently, welding of the column longitudinal reinforcement was replaced with mechanical connections, providing a "blind connection", and eliminating the need for a closure pour.

Innovations included the use of a new type of high strength shear reinforcement in the form of prefabricated rectangular spirals for column confinement and a newly developed robot to fabricate reinforcement for columns and beams.

Items	Shape or thickness	Units	Maximum weight per unit (tons)	Cast at:
Column-Beams		24	11.9	PC factory in the job site
Columns		32	6.4	
Girder		8	5.4	
Small Beams		28	3.3	
Omunea Planks	Thickness 55 (2.25")	84	1.1	Outside precaster
Precast Balcony		56	3.4	Outside precaster
Exterior Precast Walls		30	2.7	

Figure 5
Types and numbers of precast concrete elements per floor

DESIGN

Seismic Considerations

The Japanese ASEISMIC DESIGN METHOD approach for the building described herein establishes the following base criteria (Ref. 1):

"(1) To prevent damage to the building from moderate earthquakes (about 100 gal) which will occur often during the service life of the building, and

(2) To prevent collapse of the building, although it may be partially damaged, and secure the safety of human life in the event of a severe earthquake (300-400 gal) which will occur rarely during the service life of the building.

On the basis of these basic criteria the aseismic safety is confirmed in definite terms by performing successively the following two stages of design:

(1)   First Stage (Primary design, elastic design, allowable stress calculation) Perform elastic design against seismic energy corresponding to 0.2G.

(2) Second Stage (Secondary design, plastic design, ultimate strength calculation) Secure the ultimate strength and ductility required to be able to absorb the seismic energy acting on the building in the event of an earthquake corresponding to 1.0G."

The Japanese New Aseismic Design Method procedure is diagramed in Fig. 4.

Emulation Design

Design of reinforced concrete structures in Japan is determined in accordance with the "AIJ Standard for Structural Calculation of Reinforced Concrete Structures" of the Architectural Institute of Japan. The design approach described below is called "Emulation Design."(Ref. 2) or "cast-in-place equivalent" design.

The Ohkawabata building was designed as if it was an ordinary moment-resisting frame which would be configured for aseismic response according to the AIJ criteria noted above. Had it been cast in-situ, the columns, the beams and floor slabs would have been formed and poured in the conventional manner. Using the concept of emulation design, the building was divided into discrete elements which could be fabricated separately and later joined to create an equivalent monolithic structure. These basic elements are shown in Fig. 5.

Connections

Figure 6
Close-up view of NMBSS mechanical connections. Each is accurately located by means of a special sleeve setting device installed in the base of the column form.

Connections of reinforcing bars in conventional reinforced concrete may be accomplished by one of three recognized methods: by lapping, by welding, or by use of mechanical connections. Any of these methods may also be used to join precast concrete structural members. All are recognized by the ACI building code, and by regional building code bodies.

Conventional lapped splices may be used quite effectively in emulation splicing of precast concrete beams when there is sufficient room for them, but they

570

Figure 7
Splicing of steel reinforcement by a special enclosed inert-gas welding process has been completed. The space between the beam ends was formed and filled with fresh concrete at the same time the floor topping is poured.

Figure 8
Addition of diaphragm steel over the half-thickness precast concrete "stay-in-place" forms. This was followed by a topping of fresh concrete which also included the "closure pours" completing the beam end connections.

are economically and practically of little advantage in splicing precast concrete columns.

Welded splices may also be employed for the same application as indicated above for the lapped splices, and within the same limitations. In the building described here, a special enclosed butt-welding process was used to insure ductility in the bar splices. This will be discussed later.

A third alternative for making rebars continuous is use of mechanical connections.

Design Concept

Initially, precast concrete column elements were connected by butt-welding the reinforcing bars in a relatively large space provided between the column ends which was located midway between floors. This space later was filled with poured concrete. As the building progressed, it was decided to eliminate the more expensive, difficult and time-consuming closure pour and to instead connect the column reinforcing bars with grout-filled mechanical connections called NMB Splice Sleeves (NMBSS). (Fig. 6).

These have been developed especially for making "blind" connections of precast concrete elements. The large closure pour was thus eliminated and it was replaced by a thin joint between the column ends which was filled at the same time and with the same high-strength, non-shrink grout as was used in the sleeves to develop the reinforcing bars.

The steel components of the beam ends were joined either with bolted steel plates or by butt-welding reinforcing bars, or by a combination of both methods.

571

(Fig. 7). The space between the beam ends was then formed and later poured at the same time as placing the topping concrete for the floor slab.

Half-thickness floor and balcony slabs called Omunea (similar to the Omni system) were set in place on the beams and upon temporary shoring. Additional slab reinforcement in the upper portion of the floor slab was installed as required. (Fig. 8). To tie the entire floor diaphragm together, ready-mixed topping concrete was then placed over the precast floor panels and at the same time, filled all of the spaces between the ends of the beams and all locations of negative reinforcing steel in the beam-column units.

The net result was that all of the openings between the precast concrete elements were joined by splicing the reinforcing bars and the openings were filled with grout or concrete. Accordingly, the reassembled precast elements were combined in such a manner as to create the equivalent of a monolithic reinforced concrete structure.

Figure 9
Completed precast concrete "cruciform" elements ready for erection.

MANUFACTURE OF PRECAST CONCRETE ELEMENTS

The contractor elected to use both plant and site precast concrete fabrication methods for the structural elements. Half-thickness floor and balcony slabs, as well as precast concrete in-fill non-structural partitions, were manufactured offsite and delivered to the site according to the schedule. Columns and beams were fabricated on site. (Fig. 9).

The entire operation was simple and straightforward. There was so much repetition of the precast elements that continuous modification of the forms and procedure was unnecessary. This is one of the key factors in improving the efficiency of field construction.

Ordinary ready-mixed

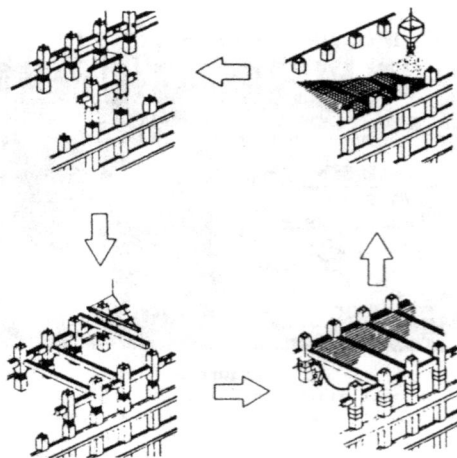

Figure 10
Sequence of erection.

concrete was delivered to the site plant by several different companies. The design strength of the concrete was 4200 psi (300 kg/cm2) at 28 days. A low slump of less than one inch was specified. Concrete was chuted from the truck directly into a bucket which was lifted by a small crane into position to pour the concrete.

ERECTION AND ASSEMBLY

Two climbing tower cranes were located at strategic positions, one outside the building and one in the atrium, both being able to reach the plant and material delivery sites at the base of the building. Fig. 10 shows a typical erection sequence for each floor.

Temporary shores were positioned and set to proper grade. The precast concrete cruciforms were lifted directly from the storage yard position to the erection floor. (Fig. 11). Fig. 12 shows the element being set into position on the temporary shoring.

The next operation was to install the beams. (Fig. 13) In some cases, these elements contained embedded steel plates which were simply bolted together. For those elements without plates, these too were set on temporary shores.

After installing the high strength spiral stirrups, the reinforcing bars in the beam ends were butt-welded using a special gas enclosed arc welding process.

At the same time, workmen installed a temporary grout dam around the interface opening between the column ends. When the dam was sufficiently strong (usually in about one day), using an electrically-driven grout pump, all of the mechanical connection sleeves as well as the bedding space were filled with a special high-strength non-shrink cementious grout in a single operation.

Half-thickness precast concrete floor and balcony slabs manufactured offsite were then set in place on the beams. Fig. 15 shows the structural frame and floor

Figure 11
Precast concrete beam/column cruciform element being raised by tower crane to its position on the structure on the 27th floor.

Figure 12
Cruciform element being lowered into position on shores which have been previously set to correction elevation. NMBSS sleeves in the column bases receive the rebar dowels extending upward from the lower column unit.

Figure 13
Precast concrete beam being installed between cruciforms.

Figure 14
Cruciform elements and half-thickness precast concrete
floor slabs in place. Outriggers extending to the
left support precast concrete balcony slabs.

system adjacent to the atrium. Reinforcing bars were installed in the deck, and a cast-in-place concrete floor was placed.

After sufficient curing time, the cycle was repeated. Each completed floor cycle took six days. The floors below were available to the follow-on trades shortly thereafter.

CONCLUSION

The paper describes a number of new and bold innovations for the construction of high-rise reinforced concrete frame structures in a highly active seismic regions. Not all of these techniques and procedures are universally applicable. Some are uniquely suited to Japanese labor and construction methods. However, some important lessons which can be learned might include the following:

High-rise reinforced concrete buildings can be and are being designed and constructed in a region of high seismic risk.

The use of precast concrete for portions of such structures is feasible and enhances the project schedule and reduces the number of skilled field laborers required in the field.

It would appear that construction technology in Japan may be out-distancing that of many of the other industrial countries. Innovation and concern for safety and quality seem to be major factors in the construction industry. The fact that a precast concrete manufacturing company, the Maeda Construction Company, Ltd., won the 1989 Deming Award for efficiency is an interesting commentary on their concern for excellence.

REFERENCES

1. Hamano, Y., Kameda, Y., Kobayaski, M., Koyama, M., Sasaki, T., (1985) "PASH SYSTEM, Development and Execution of the Panel System Prefabrication Method for High-Rise Housing," Technical Report 88-06, Kajima Corporation, paper given at Japan-Thai Civil Engineering Conference, Chulalongkorn University, The Engineering Institute of Thailand, March 1985, Bangkok, Thailand

2. Warnes, C. E., (1989) "Emulation of Cast-in-place Monolithic Design", Report to the Seismic Committee, Precast/Prestressed Concrete Institute, 1989

Earthquake type loading on R/C beam-column connections: special cases of wide beams and eccentric beams

T.R. Gentry[I], G.S. Raffaelle[I], and J.K. Wight[II]

ABSTRACT

Special reinforced concrete beam to column connections often found in framed structures have been tested under simulated earthquake loading at the University of Michigan Structural Engineering Laboratory. These connections include wide beam-column connections and eccentric beam-column connections.

Wide beam-columns, which are currently prohibited in high seismic zones (ACI-ASCE Committee 352 1985) will perform acceptably if design parameters are carefully controlled. The major parameters which must be controlled are the beam-width to column-width ratio, the fraction of the total longitudinal steel anchored in the column core, and the lateral stiffness of the structure.

The other special case considered is eccentric beam-column connections. In this type of connection the beam width is less than the column width, and the beam axes are eccentric to the column axis. The experimental program investigates the effect of varying the beam width, the beam depth, and the amount of longitudinal beam reinforcing on the behavior of the connection.

EXPERIMENTAL SETUP

Both the wide beam specimens and the eccentric beam specimens were tested by imposing a lateral displacement at the top of the column. The base of the column and the beam ends were pinned, consistent with the assumption that these locations were the inflection points for the lateral load moment diagram [see Fig. 1]. Increasing story drifts from 0.5% to 5.0% were applied to the specimens. An axial load of 20 kips was applied to the column to preclude tension in the column during testing.

WIDE BEAM-COLUMN CONNECTIONS

Wide beam-column connections, whose beams are wider than their supporting columns, are often found in one-way concrete joist systems and in other buildings where floor-to-ceiling heights are restricted. Testing of these connections stems from the recommendation by ACI-ASCE Committee 352 that these connections be experimentally evaluated for use in high seismic zones (1985).

The wide beam-column experimental program consisted of the testing of four exterior 3/4-scale specimens, including transverse beam with reinforcement. The effects of joint shear stress level, fraction of beam longitudinal reinforcement anchored in the column core, and beam-width to column-width ratios [b_w/c_w] were explored as part of this experimental research. In addition to the experiments, computer simulation of R/C frames using wide beams and an analytical model of the wide beam force transfer mechanism were formulated. Information from interior

---

[I] Ph.D. Candidate, Dept. of Civil Eng., Univ. of Michigan, Ann Arbor, MI 48109.
[II] Professor, Dept. of Civil Eng., Univ. of Michigan, Ann Arbor, MI 48109.

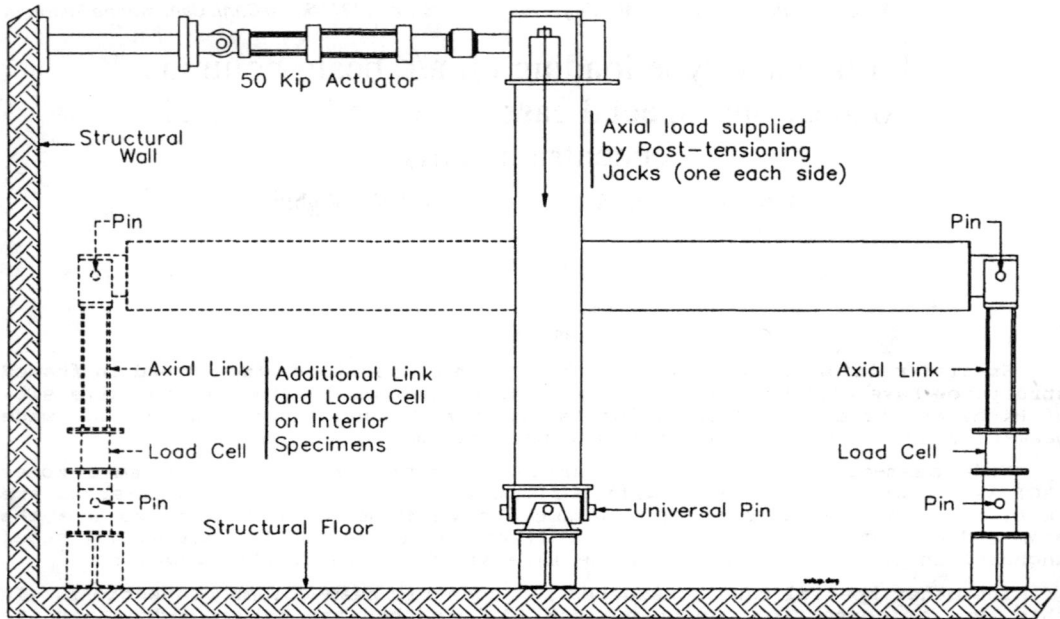

Figure 1. Experimental testing setup.

wide beam testing performed in Japan (Hatamoto 1990) supplemented the data from the testing performed at the University of Michigan. Table 1 and Fig. 2 outline the design parameters for the four specimens. All four specimens employed 14 in. square columns and 12 in. deep beams. At this date Specimens 1 and 2 have been tested; Specimens 3 and 4 will be tested during the Spring of 1991. Analytical and computer work will begin after testing is complete.

Specimen 1

Results from the test of Specimen 1 indicated that torsional distress in the transverse beam, along with anchorage loss for the wide beam flexural reinforcement, were the primary causes of wide beam connection failure. This specimen was too wide; its transverse beam did not have the capacity to transfer the torque applied by the wide beam to the column. The extensive torsional cracking in the transverse beam caused a loss of anchorage in the wide beam reinforcement.

Due to this anchorage loss, four of the six hooked top bars anchored in the

Table 1. Design parameters for wide beam specimens.

No.	b (in.)	b_w/c_w	Beam Longitudinal Reinforcement Top	Bot.	M_R	$\gamma = V_J/\sqrt{f'_c}$	% Steel in Core Top	Bot.
1	34.0	2.43	9#5	7#5	1.50	15.18	33%	43%
2	30.0	2.14	8#5	6#5	1.69	13.47	50%	33%
3	34.0	2.43	2#6, 2#5, 6#4	3#5, 6#4	1.49	14.79	56%	44%
4	34.0	2.43	10#5, 2#4	8#5, 2#4	1.18	19.24	35%	35%

576

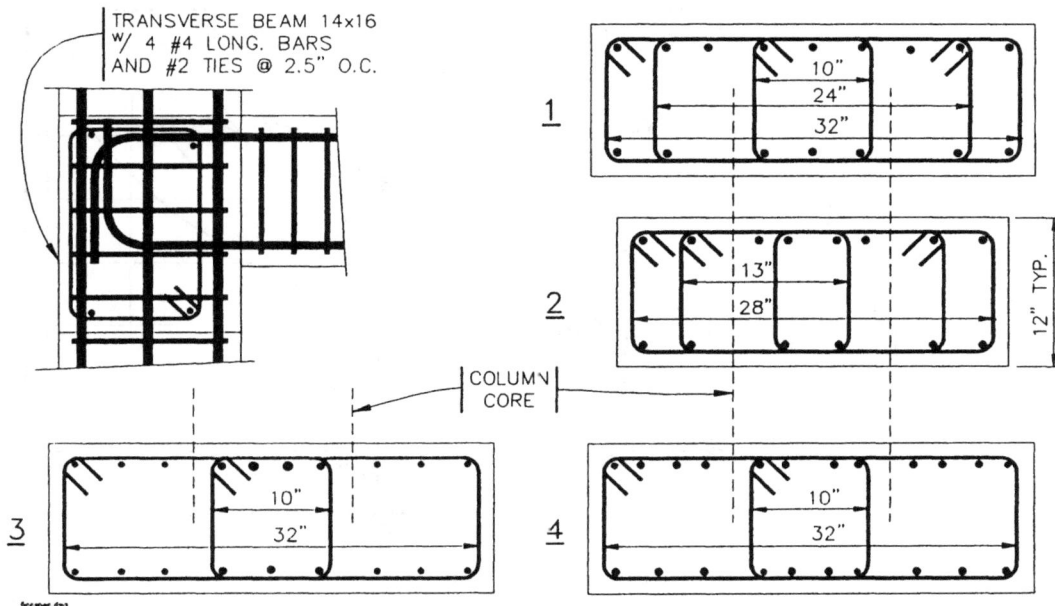

Figure 2. Wide beam specimens.

transverse beam never yielded during the testing of Specimen 1. The two bars that yielded did so only at a high lateral drift [3%, see Fig. 3]. The nominal story shear, that is, the story shear corresponding to full yielding of the beam plastic hinge, was not obtained from this specimen. At 2% lateral drift only 70% of the nominal story shear had been mobilized.

Specimen 2

Testing of Specimen 2, whose beam-width to column-width ratio $[b_w/c_w]$ was 2.14, demonstrated that wide beam connections can perform well if torsional distress and anchorage loss of bars anchored in the transverse beam are eliminated. For Specimen 2, as compared to Specimen 1, this transverse beam distress was controlled by reducing its b_w/c_w ratio and by reducing the amount of steel anchored in the transverse beam [see Fig. 3]. Full yielding of the wide beam reinforcement occurred early in the displacement history for Specimen 2, between 1.5% and 2% story drift. The yielding of the exterior-most bars, which were anchored in the transverse beam, showed only a slight lag behind those bars anchored in the column core.

Specimen 2 reached a higher story shear than Specimen 1, even though its predicted design strength was 12% less than that of Specimen 1. Specimen 2 also exhibited increasing story shear with increasing drift up to the maximum drift of 5% imposed on the specimens. At 2% lateral drift Specimen 2 had mobilized 95% of its nominal story shear.

Specimens 3 and 4

As this paper is written, Specimens 3 and 4 have not yet been tested. These specimens were designed to probe the limits established during the testing of Specimens 1 and 2. Specimen 3 has dimensions identical to Specimen 1, with modifications made to improve the behavior. A higher fraction of the total wide beam reinforcement is placed in the column-core in an attempt to reduce the demand on the transverse beam. Specimen 4 utilizes an increased beam steel ratio

577

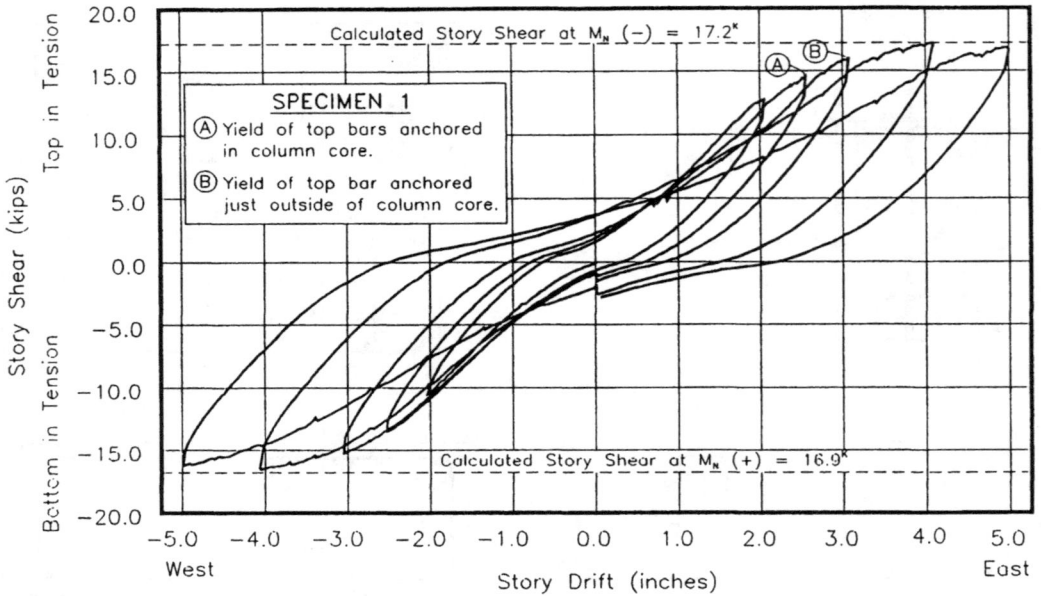

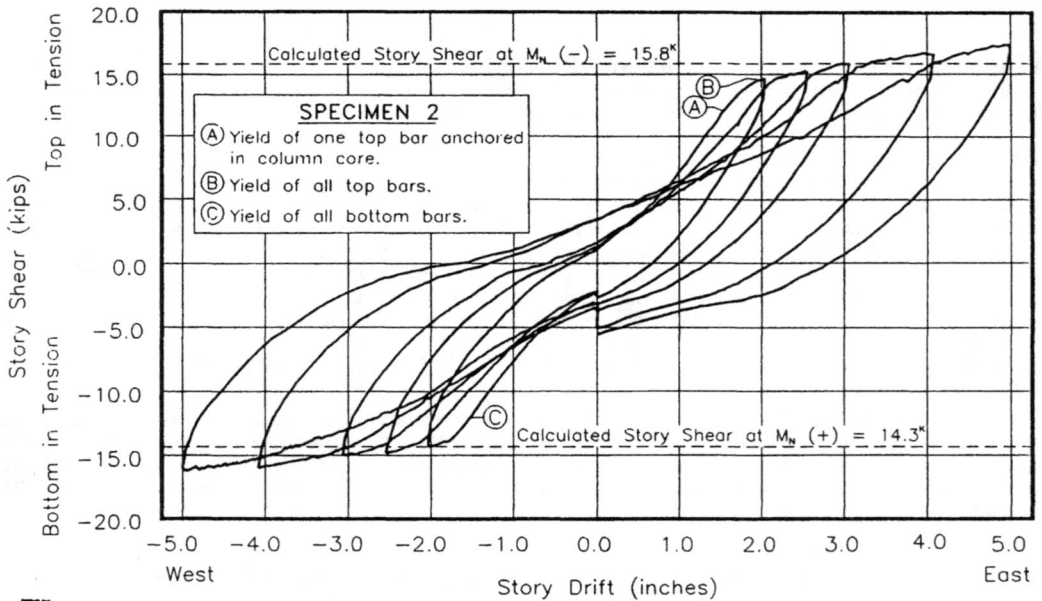

Figure 3. Story shear versus drift for Specimens 1 and 2. Drift cycles 2%, 2.5%, 3%, 4%, and 5% extracted from total curve.

[ρ] to anchor more wide beam reinforcement both in the column core and in the transverse beam. Both specimens benefit from increased torsional reinforcement in the transverse beam. Also, for both specimens, the anchorage of the wide beam reinforcement in the transverse beam is enhanced by increasing the cover over the hooked bars.

Analytical Analyses

Results from the experimental testing of Specimens 1 and 2 indicate that wide beam frames may have as much as 40% less lateral stiffness than comparable conventional R/C frames. Results from testing performed in Japan confirms this finding (Hatamoto 1990). The analytical component of this research, not yet complete, will probe the ramifications of this reduced lateral stiffness on wide beam frame behavior.

ECCENTRIC BEAM-COLUMN CONNECTIONS

A related program investigated the behavior of reinforced concrete eccentric beam-column connections subjected to earthquake-type loading. The specimens represent a beam-column subassemblage from an exterior moment-resisting frame in which, for architectural purposes, the beams were made flush with the exterior face of the column. This eccentric type connection is quite common yet is recognized by ACI-ASCE Committee 352 as an area of needed research (ACI-ASCE Committee 352 1985).

Experimental Program

All of the specimens have identical column sections and two beams framing into opposite sides of the column in such a way that one side of the beam is flush with the adjacent column face. The major design parameters varied were the beam width, the beam depth, and the amount of top and bottom longitudinal reinforcing steel in the beam. By varying the beam width, the eccentricity between the column centerline and the beam centerline and the effective joint shear area also varies. The amount of reinforcing steel in the beam influences the horizontal shear demand on the joint and changes the beam-to-column moment strength ratio. The design parameters for the four specimens are given in Table 2. The following items will be examined to determine the adequacy of these eccentric type connections: joint shear strength, deterioration of anchorage of longitudinal reinforcement, plastic hinge rotations, hysteretic behavior and energy dissipation, and loss of stiffness due to cyclic loading.

Test Results

At the time of this writing, only the results for the first two specimens are available. The spandrel beams on the first specimen had a cross-section of 10 in. x 15 in. and were longitudinally reinforced with 3-#6 and 3-#5 in the top and bottom of the beam, respectively. The column was 14 in. wide. Therefore, an eccentricity of 2 in. existed between the column centerline and the beam centerline. The load-displacement response of Specimen 1 to cyclic loading is shown in Fig. 4. The strength of the specimen continued to increase as the cycles reached story drifts of 4%. The stiffness, however, deteriorated rapidly as shown

Table 2. Design parameters for eccentric connections.

Specimen No.	Beam Size	Column Size	Eccen-tricity	Top Bars	Bottom Bars	Joint Shear Factor	Moment Strength Ratio
1	10"x15"	14"x14"	2"	3-#6	3-#5	14.0	1.42
2	7"x15"	14"x14"	3.5"	2-#6	2-#5	10.6	2.13
3	7.5"x15"	14"x14"	3.25"	3-#5	2-#5	10.7	2.03
4	7.5"x22"	14"x14"	3.25"	3-#5	2-#5	9.9	1.30

in Fig. 8. Torsional cracks on the sides of the column where the beams framed in were evident within the joint depth. Diagonal cracking in the joint was more extensive on the outside face of the column than on the inside face.

The second specimen had 7 in. x 15 in. spandrel beams with 2-#6 and 2-#5 in the top and bottom of the beam respectively. The same column section as in the first specimen was used. Since the beam was only 7 in. wide, an eccentricity of 3-1/2 in. existed between the beam and column centerlines. The load-displacement response of Specimen 2 is shown in Fig. 5. The hysteresis loops show excessive pinching, and consequently small amounts of energy were dissipated. The pinching is primarily caused by the closing of cracks and the slippage of the beam bars through the joint. The development of cracking in the joint was similar to that of the first specimen. Torsional cracks in the column within the beam depth developed early in the loading sequence, but did not grow larger as the cycles progressed.

The stiffness of Specimen 1 is compared to the stiffness of Specimen 2 in Fig. 8. The stiffness of both specimens decreased rapidly as the drift, which is given in percent of the story height, increased. The stiffness of Specimen 2, however, deteriorated at a rate faster than Specimen 1. A reason for this faster rate of deterioration was partially do to a quicker loss of anchorage of the beam bars in the joint.

The measured strains in the reinforcement indicate that slippage of the beam bars did occur. The strains recorded in one of the top longitudinal beam bars at the beam-column interface for Specimen 1 and Specimen 2 are shown in Fig. 6 and Fig. 7, respectively. From simple beam theory it would be expected that the bar would be in compression for positive story drifts and in tension for negative story drifts. The strains in the figures for both specimens show that the bar is in tension for negative story drifts, but for positive story drifts the strain changes from the expected compression to tension as the percent story drift increases. The change from compression to tension indicates the anchorage of the beam bar in the joint has been partially lost. For Specimen 1, the bar did not go into tension until 2% story drift was reached and at 3% story drift the tensile strain was still very small. For Specimen 2, the bar was in tension after 0.5% story drift and reached a tensile strain of approximately one-half the yield value at 1.5% story drift. This indicates that the beam bars experienced more slippage in Specimen 2 than in Specimen 1. The slippage of the beam bars may be the predominate cause of the pinching in the hysteresis loops and the rapid rate of stiffness deterioration in Specimen 2.

The premature loss of anchorage in Specimen 2 may be explained by examining the joint distortion on both sides of the connection. Displacement transducers were mounted on the faces of the column in an arrangement designed to measure the shear deformation. Fig. 9 shows, for Specimen 2, the joint deformation measured on the face of the joint flush with the spandrel beams (flush face) and on the opposite face where an offset exists between the column face and beam face (offset face). As may be expected, the figure shows the flush face of the joint experienced large deformations and significant diagonal cracking. The beam bars passed through the column close to the flush face of the joint where most of the damage was concentrated. This is believed to have been the cause of the anchorage loss of the beam bars in the joint.

The rotations over the plastic hinging zones in the beams were also measured. Figs. 10 and 11 show the beam end load versus the rotation in the west beam for Specimen 1 and 2, respectively. Similar behavior was also recorded for the east beams. Comparing the behavior of Specimen 1 to Specimen 2 it can be seen that the beams in Specimen 1 were stiffer and stronger. This would be expected because larger and more heavily reinforced spandrel beams were used in Specimen 1. Comparing the rotations for cycles with the same story drifts it can be seen that Specimen 2 underwent larger rotations. This would indicate larger fixed end rotations occurred in Specimen 2 as a result of bar slippage.

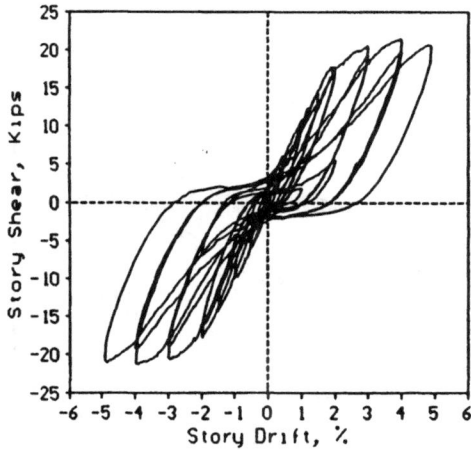

Figure 4. Load-displacement response of specimen 1.

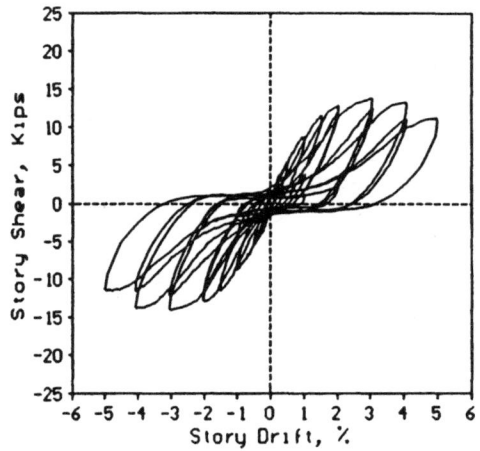

Figure 5. Load-displacement response of specimen 2.

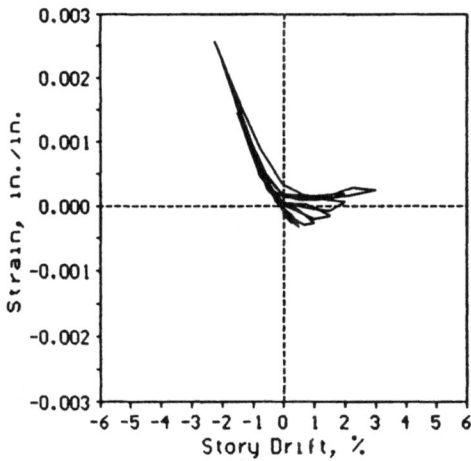

Figure 6. Strain in beam top longitudinal bar at column face in specimen 1.

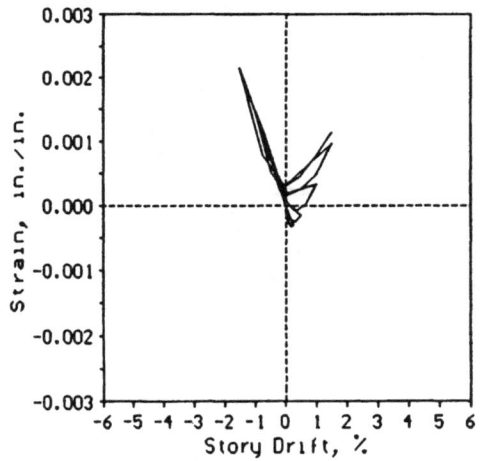

Figure 7. Strain in beam top longitudinal bar at column face in specimen 2.

Summary of Eccentric Beam-Column Tests

The poor performance of the eccentric beam-column connections are primarily a result of the longitudinal beam bars losing anchorage within the joint. The joints showed some torsional distress at the beginning of the reversed cyclic loading pattern, but this damage did not increase as the cycles progressed. The poor anchorage condition is a result of the beam bars passing through the side of the joint where most of the deformation and cracking are concentrated. Comparing the stiffness deterioration for the two specimens, the stiffness deteriorated more rapidly for the narrower spandrel beams.

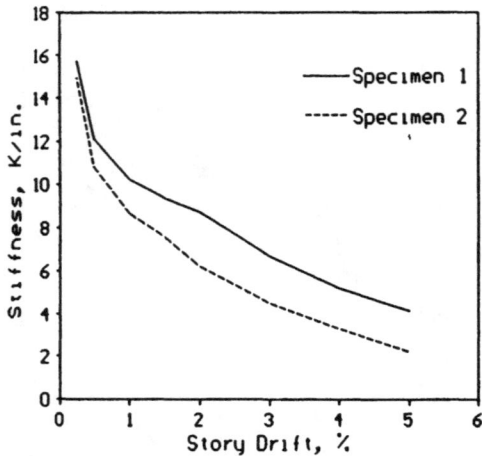

Figure 8. Stiffness deterioration of
specimen 1 and 2.

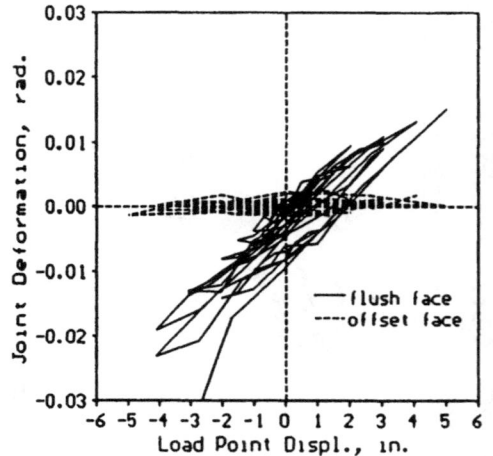

Figure 9. Joint shear deformation of
specimen 2.

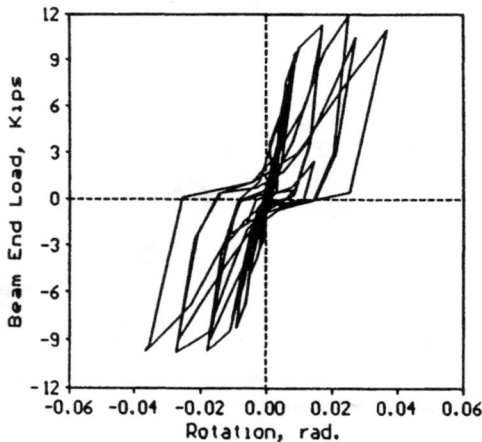

Figure 10. Plastic hinge rotation in
west beam of specimen 1.

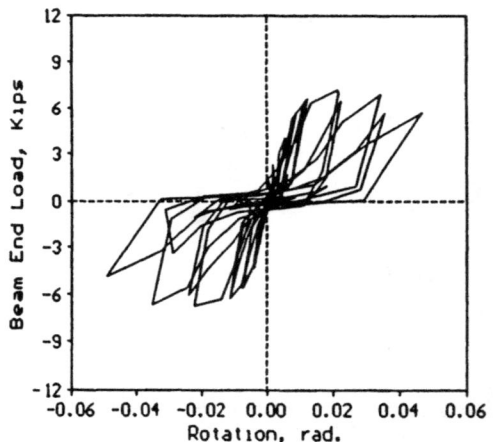

Figure 11. Plastic hinge rotation in
west beam of specimen 2.

ACKNOWLEDGEMENTS

Support for this research by the United States National Science Foundation
is gratefully acknowledged. The results and conclusions expressed here are those
of the authors and do not necessarily reflect the views of the sponsor.
Information from Hitoshi Hatamoto at the Kajima Construction Corporation in
Japan, along with his colleagues Satoshi Bessho at Kajima and Prof. Yasuhiro
Matsuzaki at the Tokyo Science University, is also gratefully acknowledged.

REFERENCES

ACI-ASCE Committee 352. May-June 1985. Recommendations for Design of Beam-Column
    Joints in Monolithic Reinforced Concrete Structures. ACI Journal, 82, 266-283.
Hatamoto, Hitoshi. 1990. Personal Communication including two Reports on Testing
    of Wide Beam Subassemblages (in Japanese) and experimental data.

582

Effects of beam width on the cyclic behavior
of reinforced concrete

David L. Hanks[1], Steven L. McCabe[2], and David Darwin[3]

ABSTRACT

Lateral force resisting concrete frames in regions of moderate seismic risk are generally designed to dissipate applied energy through the formation of plastic hinges in the beam elements. Previous research strongly suggests that, in addition to load history, applied shear stress, and nominal stirrup strength, member width also influence hinge performance. To facilitate a better understanding of the effects of beam width on the inelastic behavior of reinforced concrete members subjected to severe seismic loading, four lightly reinforced specimens were fabricated and tested. The performance of these specimens is compared to those of narrow beams fabricated and tested in a similar manner. This study indicates that, for specimens with the same flexural strength, nominal concrete strength, and stirrup spacing, an increase in beam width improves beam performance under cyclic load.

INTRODUCTION

Failures in reinforced concrete frames subjected to moderate earthquakes are minimized when energy is dissipated through the formation of plastic beam hinges. An understanding of the factors influencing beam performance is necessary if a structure's integrity is to be maintained throughout the duration of the seismic loading.

Numerous studies have been undertaken to determine the influence of various parameters on the hysteretic behavior of reinforced concrete beams. This research indicates that load history, applied shear stress, and nominal stirrup strength significantly affect hinge performance. Unfortunately, the development of a consistent measure of cyclic performance has been complicated due to variations in design and test parameters within and between studies.

Several measures of cyclic performance have been proposed for which the goal has been to characterize the net effect of variations in member properties and testing techniques. These measures include the Work Index and Modified Work Index, I_w and I_w', respectively (Gosain,

---

[1]Grad. Res. Asst., Dept. of Civil Engrg., Univ. of Kansas, Lawrence, KS 66045.
[2]Asst. Prof. of Civil Engrg., Univ. of Kansas, Lawrence, KS 66045.
[3]Deane E. Ackers Prof. of Civil Engrg., and Dir., Structural Engrg. and Materials Laboratory, Univ. of Kansas, Lawrence, KS 66045.

Brown and Jirsa 1977), the Energy Index, I_E (Hwang 1982), the Energy Dissipation Index, D_i (Nmai and Darwin 1984), and the Normalized Energy Index, I_{EN} (Ehsani and Wight 1990).

Previous research by Darwin and Nmai (1986) and Hanks and Darwin (1988) suggests that beam width may have a substantial influence on energy dissipation and cyclic performance. An increase in beam width reduces the maximum applied shear stress, which significantly affects member response and should increase the number of cycles to failure.

The purpose of this research is to investigate the influence of beam width on the cyclic behavior of lightly reinforced concrete beams. The results from the experimental portion of this study are compared to those of previous research (Nmai and Darwin 1984) using narrow beams. Beam hinge performance is based on an evaluation of the energy dissipated and the Energy Dissipation Index, D_i.

EXPERIMENTAL INVESTIGATION

Four cast-in-place reinforced concrete specimens were fabricated with a beam width of 15 in. and overall depth of 18 in.. Reinforcement ratios, ρ, of 0.34% (Beam H-1, H-3, and H-4) and 0.51% (Beam H-2) were used. Other details of the beam dimensions and properties are shown in Table 1 and Fig. 1. Nominal stirrup strength, $v_s = A_v f_{vy}/(bs)$, (A_v = total area of stirrup, f_{vy} = yield strength of stirrup, b = beam width, and s = stirrup spacing) was about 78 psi in all beams. The maximum applied shear stress, $v_m = V_m/(bd)$, (V_m = maximum shear force, and d = effective depth), varied from 64 to 105 psi depending on the area of flexural steel.

All specimens were fabricated with two layers of #4 bars as negative moment reinforcement, A_s (top steel). Beams H-1, H-2 and H-3 contained one layer of #4 bars as positive moment reinforcement, A_s' (bottom steel), while Beam H-4 was fabricated with two layers of #4 bars as A_s'. Transverse reinforcement was fabricated from 7/32 in. nominal diameter smooth rod and welded to form a closed hoop. The first stirrup was placed at 1 in. from the vertical face of the formed column, subsequent stirrups were spaced at 3 5/8 in. centers. The flexural reinforcement of the beam was welded to a 3/4 in. bearing plate to prevent anchorage failure within the column (Fig. 1).

The specimens were post-tensioned to a structural floor and the ends of the beams were loaded using a 110 kip capacity hydraulic actuator. Throughout testing, specimens were subjected to constant nominal displacement ductility factors, μ, ranging from 4.3 to 8.5 (Table 1) in both positive and negative bending. Strains in the longitudinal and transverse reinforcement were measured with foil gages. Beam tip displacement and displacements at various locations in the beam and column were measured with linear variable differential transformers, LVDT's.

TEST RESULTS

Studies relative to beam hinge behavior generally incorporate energy dissipated by the specimen into the measure of cyclic performance. Energy dissipated is the area bounded by the load-displacement plot for each hysteresis loop. Fig. 2 is a representative plot of the hysteresis loops for Beam H-3. A summary of the principle experimental results for the beams of this study is presented in Table 1.

EVALUATION OF TEST RESULTS

Energy dissipated, E, is a function of the number of cycles required to cause failure. To eliminate ambiguity in the definition of failure, several researchers have defined the energy dissipated by a member as the summation of energy for cycles in which the maximum load, P_n, is at least 75% of the yield load, P_y (i.e. $P_n \geq 0.75P_y$). The influence of beam width on the cumulative energy dissipated by a member can be obtained by comparing the results of specimens fabricated with different widths, yet similar flexural reinforcement, effective depth, stirrup spacing, concrete strength, and load history. Beams F-2 and F-3 tested by Nmai and Darwin (1984) and H-1 and H-2 of this study provide for direct comparison.

By comparing the energy dissipated for Beam F-2 (b = 7.5 in., ρ = 1.02%, A_s'/A_s = 0.5, μ = 5.1, E = 169 inch-kips) and H-2 (b = 15 in., ρ = 0.51%, A_s'/A_s = 0.5, μ = 5.3, E = 315 inch-kips) for all cycles in which $P_n \geq 0.75P_y$, it can be seen that the 86% increase in E for H-2 may be attributed to the increase in beam width. Similarly, a 22% increase in E for Beam H-1 (b = 15 in., ρ = 0.34%, A_s'/A_s = 0.5, μ = 4.3, E = 245 inch-kips) as compared to F-3 (b = 7.5 in., ρ = 0.69%, A_s'/A_s = 0.5, μ = 4.4, E = 201 inch-kips) also results from the larger beam width.

The increase in energy dissipated is primarily attributed to the reduction in the maximum applied shear stress. However, increased beam widths also improve confinement of the core concrete and thus delay buckling of the compression reinforcement which may increase both the number of cycles to failure and the energy dissipated.

Since the cyclic behavior of a member depends upon both strength and displacement ductility factor, energy dissipation alone is not a viable means in which to evaluate inelastic performance. One measure of cyclic performance which appears to provide a consistent evaluation of a wide range of design parameters is the Energy Dissipation Index, D_i, developed by Nmai and Darwin (1984). The Energy Dissipation Index is expressed as

$$D_i = \frac{\sum E}{0.5P_y\Delta_y[1+(\frac{A_s'}{A_s})^2]} \tag{1}$$

in which $\sum E$ is the summation of the energy dissipated for cycles where $P_n \geq 0.75P_y$, P_y and Δ_y = initial yield load and yield deflection in negative bending, A_s' = area of bottom steel, and A_s = area of top steel. The normalizing term $0.5P_y\Delta_y[1+(A_s'/A_s)^2]$ approximates the total elastic energy at yield for both negative and positive bending at the near and far ends, respectively, of a full span beam in a frame subjected to lateral displacement.

Nmai and Darwin (1984) show that the contribution of nominal stirrup strength, concrete strength and maximum applied shear stress correlate well with D_i. Their results, based on a linear regression analysis of selected data, show that these design parameters, expressed in the form $(v_s f_c')^{0.5}(v_m)^{-1.5}$, provide a reasonably good estimation, or prediction, of D_i. The dominance of the applied shear stress, and subsequently the influence of beam width, is evident, as seen by the relatively large magnitude of the v_m exponent.

The influence of increased beam width on D_i is seen by performing a linear regression analysis on the data for Beams F-2 & H-2 and F-3 & H-1. The regression of D_i on $(v_s f_c')^{0.5}(v_m)^{-1.5}$ results in a best fit equation of

$$D_i = 81.8[(v_s f_c')^{0.5}(v_m)^{-1.5}] + 13 \qquad (2)$$

with a correlation coefficient, $r = 0.957$ (Fig. 3). A comparison of D_i values in Fig. 3 for Beams F-2 and H-2 ($A_s = 6\#4$ and $A_s' = 3\#4$) and Beams F-3 and H-1 ($A_s = 4\#4$ and $A_s' = 2\#4$) suggests that an increase in beam width appears to be more effective at increasing D_i as the amount of flexural reinforcement increases.

The relationship represented in Eq. 2 appears to be independent of variations in μ and A_s'/A_s on D_i for specimens with similar stirrup spacing, effective depth, and f_c'. When Beam F-1 ($b = 7.5$ in., $\rho = 1.03\%$, $A_s'/A_s = 0.5$, $\mu = 3.9$, $E = 287$ inch-kips) from the research by Nmai and Darwin (1984) and Beams H-3 ($b = 15$ in., $\rho = 0.34\%$, $A_s'/A_s = 0.5$, $\mu = 8.5$, $E = 178$ inch-kips) and H-4 ($b = 15$ in., $\rho = 0.34\%$, $A_s'/A_s = 1.0$, $\mu = 4.7$, $E = 507$ inch-kips) of this study are included in the analysis of the specimens shown in Fig. 3, the best fit equation becomes

$$D_i = 84.9[(v_s f_c')^{0.5}(v_m)^{-1.5}] + 12 \qquad (3)$$

with $r = 0.972$. Fig. 4 shows the seven data points represented by Eq. 3. As this figure illustrates, there is a positive correlation of data represented by the best fit line for specimens fabricated with widths of 7.5 or 15 in., $3.9 \le \mu \le 8.5$, and A_s'/A_s of 0.5 or 1.0. For the range of data previously discussed, Eq. 3 provides a reasonably consistent prediction of D_i. Both Figs. 3 and 4 indicate that an increase in D_i may be obtained by increasing the nominal stirrup strength and concrete strength and decreasing the maximum applied shear stress. For the range of data analyzed in this study, the most effective means in which to improve D_i appears to be obtained through an increase in beam width.

CONCLUSIONS

For beams with similar flexural strength, stirrup spacing, and concrete strength, an increase in beam width increases the energy dissipation capacity. The increase in energy dissipated by the member appears to be primarily the result of a decrease in maximum applied shear stress. Increased beam widths improve concrete confinement and delay buckling of the compression reinforcement. As a result, the number of cycles to failure and the energy dissipation capacity of the member are increased.

The Energy Dissipation Index, D_i, appears to be a consistent measure of cyclic performance. Improvements in D_i may be readily obtained by increasing the width of reinforced concrete beams.

REFERENCES

Darwin, D. and Nmai, C.K. 1986. Energy Dissipation in RC Beams Under Cyclic Load. J. Struct. Eng., ASCE, 112(8), 1829-1846.

Ehsani, M.R. and Wight, J.K. 1990. Confinement Steel Requirements for Connections in Ductile Frames. J Struct. Eng., ASCE, 116(3), 751-767.

Gosain, N.K., Brown, R.H. and Jirsa, J.O. 1977. Shear Requirements for Load Reversals on RC Members. J. Struct. Div., ASCE, 103(ST7), 1461-1476.

Hanks, D.L. and Darwin, D. 1988. Cyclic Behavior of High Strength Concrete Beams. SM Report No. 21. University of Kansas Center for Research, Inc., Lawrence, KS, 120.

Hwang, T.H. 1982. Effects of Variation in Load History on Cyclic Response of Concrete Flexural Members. Thesis submitted to the University of Illinois at Urbana-Champaign in partial fulfillment of the requirements for the degree of Doctor of Philosophy, 232.

Nmai, C.K. and Darwin, D. 1984. Lightly Reinforced Concrete Beams Under Cyclic Load. J. American Concrete Institute, 83(5), 777-783.

Table 1. Beam properties and principle experimental results

Beam	H-1	H-2	H-3	H-4
Length, L (in.)	68	68	68	68
Height, h (in.)	18	18	18	18
Width, b (in.)	15	15	15	15
Effective depth, d (in.)	15.69	15.81	15.63	15.75
Effective depth, d_1 (in.)	16.81	16.88	16.69	15.75
Core width, b_c (in.)	13.0	13.0	13.0	13.0
Core depth, d_c (in.)	15.91	16.03	15.85	15.97
Shear span, a (in.)	60	60	60	60
a/d	3.8	3.8	3.8	3.8
Top reinforcement ratio, ρ (%)	0.34	0.51	0.34	0.34
Top reinforcement, A_s	4#4	6#4	4#4	4#4
Bottom reinforcement, A_s'	2#4	3#4	2#4	4#4
A_s'/A_s	0.5	0.5	0.5	1.0
Yield str. of flex. reinf. (ksi)	66.4	66.4	66.4	71.7
Stirrup diameter (in.)	0.222	0.222	0.222	0.222
Stirrup spacing, s (in.)	3.6	3.6	3.6	3.6
f_{vy} (ksi)	56.6	55.5	54.3	54.3
v_s (psi)	81	79	77	77
v_m (psi)	64	105	74	76
f_c' (psi)	4200	4400	4120	4060
Slump (in.)	3.25	4.0	2.0	4.0
Yield load (kips)	12.8	20.9	13.0	14.8
Maximum load (kips)	15.1	24.8	17.4	18.0
Yield deflection (in.)	0.30	0.34	0.24	0.37
Maximum deflection (in.)	1.29	1.80	2.04	1.73
Displacement ductility factor, μ	4.3	5.3	8.5	4.7
Number of cycles:				
$P_n \geq 0.75 P_y$	13	7	4	17
Total	21	13	5	21
Cumulative Energy Dissipated (inch-kips):				
Cycles where $P_n \geq 0.75 P_y$	245	315	178	507
Energy Dissipation Index, D_i	102	71	91	93

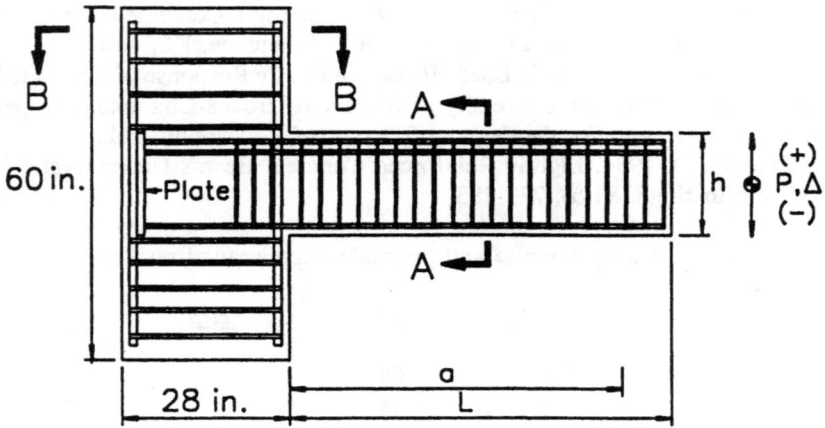

Elevation

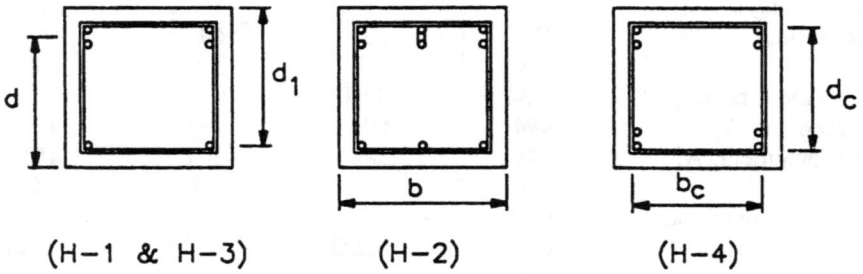

(H-1 & H-3)   (H-2)   (H-4)

Section A-A

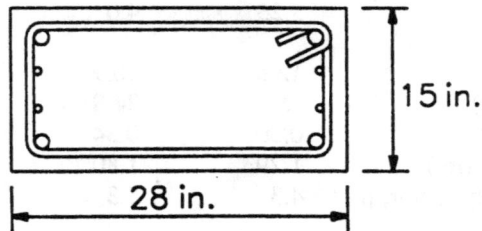

Section B-B

Figure 1.   Test specimen and reinforcing details

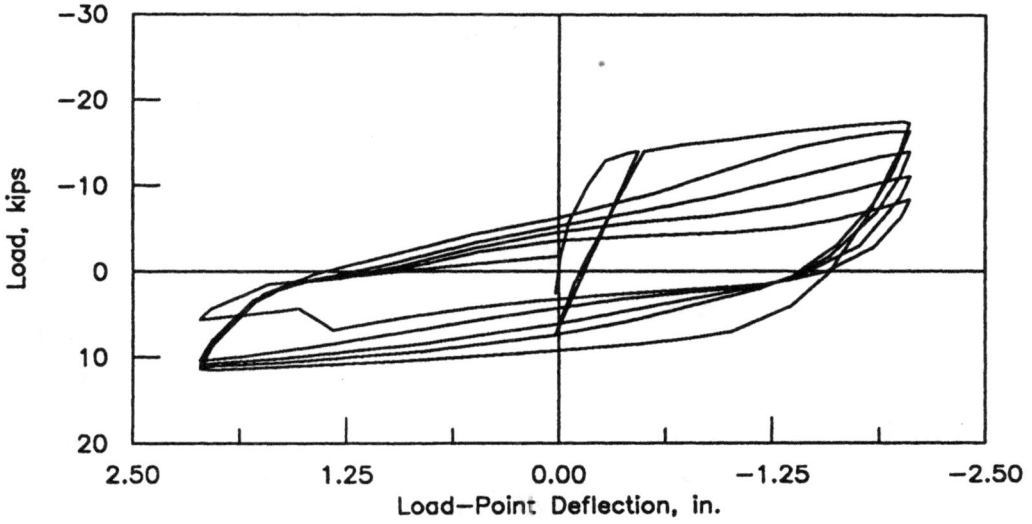

Figure 2. Load–deflection curve, Beam H-3

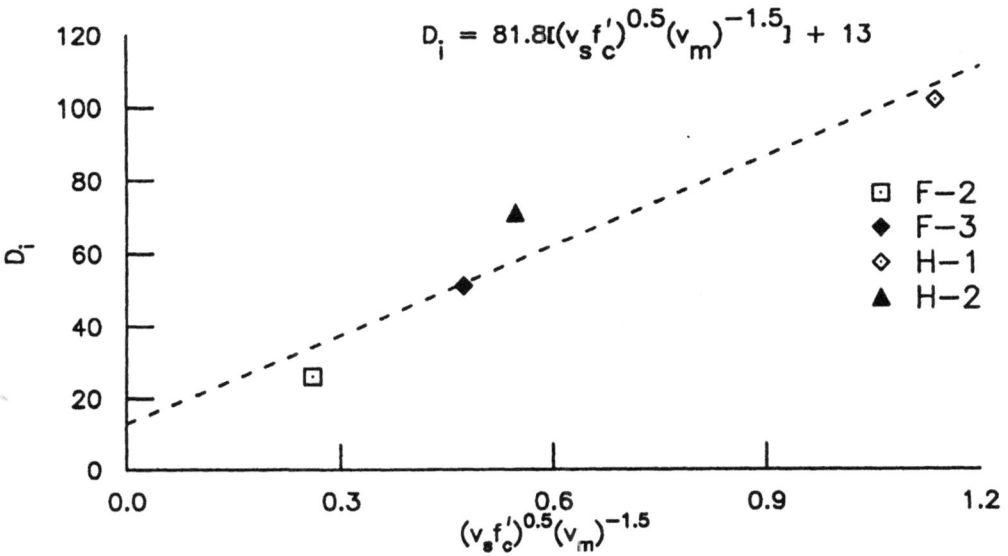

$$D_i = 81.8[(v_s f_c')^{0.5}(v_m)^{-1.5}] + 13$$

Figure 3. D_i versus $(v_s f_c')^{0.5}(v_m)^{-1.5}$

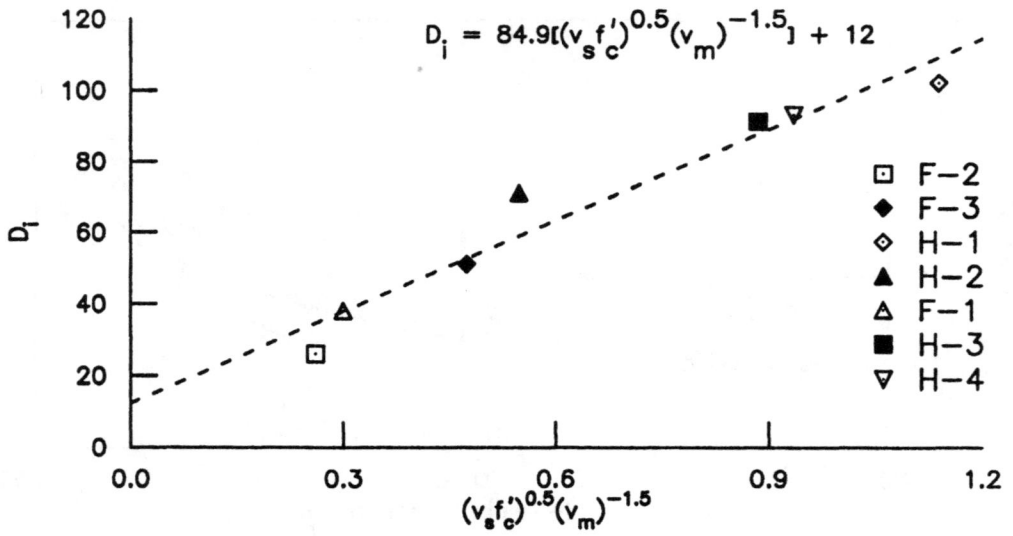

Figure 4. D_i versus $(v_s f_c')^{0.5}(v_m)^{-1.5}$

The effect of floor-slabs on the behaviour of beam-column connections under seismic loads

S.J. Pantazopoulou[1] and A. Bas[2]

ABSTRACT

The experimental behaviour of a full-scale reinforced concrete beam-column subassembly with floor-slabs is presented. The specimen, tested at the University of Toronto, modeled an interior frame connection and was subjected to bi-directional cyclic loads simulating earthquake effects. Effective slab width, joint behaviour, transverse beam torsion / weak axis bending, and interaction between the two orthogonal directions are investigated. The essential features of the mechanism controlling the flow of connection forces into the joint are discussed in light of the experimentally obtained strain fields.

INTRODUCTION

Behaviour of beam column connections with floor-slabs has been the focus of many research studies since the early 1980's, after reports from the U.S.-Japan cooperative experimental program indicated that monolithic floor-slabs may cause dramatic overstrength to the supporting beams in frame structures subjected to lateral loads (Durrani and Wight 1982; Otani et al. 1984). However, beam flexural overstrength if not accounted for in the design phase may be undesirable because it may increase the shear demand on the joint or defeat the weak-beam / strong-column philosophy of connection design (Paulay et al. 1978). Experimental studies conducted during the past decade (Zerbe and Durrani 1985 and 1988; French and Boroojerdi 1989; Kurose et al. 1988; Pantazopoulou and Qi 1990; Bas 1990) have helped to identify the cause of the so-called "slab contribution" effect which is now attributed to the kinematic constraints that exist between monolithically cast floor-slabs and their supporting beams along the interface of the two elements (Pantazopoulou et al. 1988). During earthquake action, lateral loads induce a linear moment variation along the beam spans, the moments reaching a maximum value at the faces of the columns within each bay. If the moments are negative (hogging), the upper beam fibres are in tension, and because of the limited tensile resistance of concrete, the neutral axis is well below the slab thickness. The deformation of the longitudinal beam fibres is approximately proportional to the distance from the neutral axis of the beam. Because of the kinematic constraints mentioned above, longitudinal deformations in the slab are identical to those of the beam along the common boundary of the two elements. However, the beam elongation is felt to a lesser extent at large transverse distances from the main beam, and this decay is manifested by a shear-lag effect in the slab diaphragm. Longitudinal elongations in the slab require that the slab reinforcement be in tension; the total force acting over the distance from the neutral axis generates an additional moment resistance which, when detected in experiments, is termed "slab contribution".

Because of the brittle tensile behaviour of concrete, surface elongations (caused by hogging moments) become more dramatic after cracking, and are more pronounced at higher levels of lateral displacements. For this reason, slab membrane actions increase with the extent of cracking and therefore with increasing lateral storey displacement (or lateral drift).

The initial results described above, were naturally followed by an array of unanswered questions (Ammerman and French 1989; Bas 1990). For example, the role of transverse beams in the mechanics

---

[1] Assistant Professor, Dept. of Civil Engrg., University of Toronto, M5S-1A4, CANADA

[2] M.A.Sc., Billur Sok. 70/4, K.Esat, Ankara, Turkey 06700

of the connection is not yet clearly understood. The mechanism by which forces are transferred from the slab to the joint must be explained; there is a need to establish whether joint shear demand increases as a result of slab participation, and, if so, to determine the degree of improvement of joint resistance which is also likely to result from the confining action of the slab. A rigorous explanation is necessary for the conflicting experimental and theoretical evidence of slab-beam behaviour under positive (sagging) moments, particularly in cases of interior slab-beam-column connections where the slab is nominally in tension on one face of the column, and in compression on the other. It is imperative that such an explanation be unbiased from the particular details of the simplified specimen tested in the laboratories; certain pronounced features of the response that have been observed in laboratory tests of single connection specimens are believed to be directly related to the details of the load setup and are not representative of the behaviour of connections in actual continuous structures.

This paper documents results of a single connection test carried out at the University of Toronto (Bas 1990). The specimen was a full-scale, statically determinate assembly of an interior slab-beam-column connection. Loading was applied along the two principal directions of the specimen geometry, simulating bi-directional earthquake effects. The aim of the paper is to provide experimental information as to such aspects of the behaviour that relate to the physical arguments outlined in the preceding.

EXPERIMENTAL PROGRAM

Specimen Geometry

The specimen represents an interior connection of a typical reinforced concrete multi-storey frame structure with the plan view shown in Fig. 1. The scale, geometry and load setup of the structure are similar to those of the larger number of specimen tested recently by other investigators, so as to enhance the available data base. The beam ends of the specimen are assumed to correspond to the approximate location of the inflection points at the midspans of the complete frame structure, and are loaded so that the force-boundary conditions of zero flexural moment are approximately satisfied. Although this convention is common amongst all single connection specimen that have been tested thus far, it must be emphasized that the remaining (displacement and shear) boundary conditions are insufficiently modeled at the edges of the specimen; this generally leads to concentrated load effects which must be taken into consideration in interpreting the results.

Specimen dimensions are illustrated in Fig. 2. Column ends were supported by spherical hinges; because of the physical dimensions of the hinges, the actual inflection points were somewhat shifted during the test outside the column height. (The resulting effective height of the column was 3.0 m which led to a relatively low level of shear stress in the column.)

Beam cross-sections were 380 mm high by 220 mm wide in both directions. The columns had a 400 mm square cross-section. Slab was 100 mm thick. Beam ends were extended beyond the slab edge (Fig. 2) to provide a loading region which would support the load actuators during the test, and to minimize the concentrated load effects which are induced to the slab by the loading system. To reduce the likelyhood of a localized failure, beam cross-sections were wider at these locations with dimensions of 380 mm by 320 mm. All four beams were reinforced with 4-#15 (15 mm diameter) top, and 3-#15 bottom bars (corresponding reinforcement percentages computed over the gross-area were 0.96% and 0.71%). Beams were confined by #10 (10 mm) hoops placed at 90 mm spacing on centres (o.c.) along their entire length. The slab was reinforced at top and bottom with 8 mm bars placed at 200 mm spacing o.c. in the longitudinal and transverse directions. Slab and beam longitudinal bars running in the North-South direction were placed between the respective longitudinal bars in the East-West direction since the two groups of reinforcement crossed each other, and consequently the flexural strength of the beams in the North-South direction was slightly lower than that of the beams in the East-West direction.

The connection was designed with the aim to minimize the possibility of yielding in the column under simultaneous development of the beam strengths in the two orthogonal directions. As a result, columns were heavily reinforced with 12-#25 bars (ρ= 3.75%) and were confined by 4-leg #10 hoops placed at a 90 mm spacing o.c. The transverse reinforcement used for column confinement was extended along the height of the beam-column joint. Joint shear stress under unidirectional loads (estimated assuming that all beam bars and two slab bars at each face of the beam had yielded, and that actual steel properties would be enhanced by 25% due to strain hardening) was 0.83 $\sqrt{f'_c}$ (MPa) which satisfies the limits established by the CSA (1984) Code. The ACI-ASCE 352 Recommendations (1985) were also satisfied in the joint design except for the limitation of the maximum column bar diameter passing through the joint.

To avoid localized shear failure at the loading points (beam ends), in addition to the hoops, #10 bent bars were placed as additional shear reinforcement. Because of the limited space in the laboratory, the anchorage length requirements at the beam ends were not satisfied and for this reason beam bars were welded to 1 in steel plates.

Average compressive strength of the concrete measured at the time of the test was 44.5 MPa. Yielding stresses of the reinforcement were 607 MPa, 454 MPa, 484 MPa, and 506 MPa for #8, #10, #15 and #25 bars respectively. The corresponding ultimate stresses were 685 MPa, 640 MPa, 646 MPa, and 687 MPa. Load and displacement records, and strain measurements were obtained at critical locations in all the components of the specimen. In addition, surface strains of the concrete were obtained using Zurich gauges at several stages during the test.

Test Setup and Loading Programme

In the beginning of the test, the column was loaded axially at the top by a 700 kN force (approximately 20% of the column balanced axial load, P_b) which was held constant throughout the test. This load was applied to simulate the self weight of the superstructure in real buildings, and also to confine the spherical bearings against possible lateral slippage due to shear.

Cyclic loads were applied on the beam ends, introducing moment transfer at the connection similar to that occurring under seismic attack. The loading program was carried out under displacement control, with the associated storey drift being the controlling parameter of the loading history. Displacement levels corresponding to 0.5%, 1.0%, 1.5%, 2.0% and 3% drift were subsequently applied to the specimen.

At each drift level, the pattern of the cyclic load history adopted was as shown in Fig. 3. The stronger beam (E-W direction) was the main direction of the test. During the first stage at each displacement level, only the E-W direction was loaded, introducing uni-directional loading to the connection. At the second stage, both directions were loaded simultaneously, modeling bi-directional earthquake loads occurring at a 45 degree angle with the E-W direction. At the end of the planned displacement history, the E-W direction was subjected to an additional cycle up to the limiting stroke of the actuators.

EVALUATION OF THE TEST RESULTS

In the following discussion, positive displacement or load is defined as upwards, and positive rotations or moments are associated with positive displacements and loads respectively. Tensile strains are considered positive for both steel and concrete.

Discussion of the Overall Response

The load-displacement histories of the East and South (strong and weak) beams of the specimen are given in Fig. 4. First yielding occurred in the strong beams at 1% storey drift under positive load, at which point the load-displacement response curve is characterized by a marked change of apparent stiffness. In contrast, for negative bending, there is a gradual increase of load with displacement without abrupt stiffness change up to 3% storey drift. The relative magnitudes of the loads in the positive and the negative bending cases were at a ratio of 1:2 for the stronger beams (East and West), and approximately 1:2.5 - 1:3 for the weaker beams (North and South). Due to uneven yielding under positive and negative bending, displacement ductilities in the former case were several times those observed in the latter case, although the overall displacements applied in the two directions were the same.

The ultimate flexural capacity of the strong beams reached 155 kN-m and 304 kN-m under positive and negative bending respectively. It is noteworthy that the corresponding estimates of beam flexural resistances using actual material properties were 106 kN-m and 135 kN-m, computed by ignoring the slab contribution. The discrepancy of values in the case of negative bending highlights the dramatic influence of the slab reinforcement in enhancing the negative flexural resistance of the supporting beam, and the significant increase of joint shear as compared with the estimate that results

593

from the ACI-ASCE 352 (1985) model.

Storey Shear Orbits

Although all beams maintained their strength throughout the test, in consecutive load cycles at a fixed displacement level, strength and stiffness degradation was observed and was more pronounced for negative bending (slab in tension) than for positive bending (slab in compression). Resistance reduction was manifested in the storey shear orbits shown in Fig. 5, along the two orthogonal directions of specimen geometry throughout the test. It was seen that during the bi-directional load cycles (2-nd and 3-rd load cycles at a given displacement level), a sharp decrease in resistance occurred in the direction of constant displacement, while load in the orthogonal direction was increased. The reduction exceeded 25% of the maximum load at any given cycle, and highlights the extent of interaction that exists between orthogonal directions in the connection. The reason why interaction is more pronounced under negative bending can be understood by reviewing the requirements of equilibrium of the slab panel; consider a quadrant of the specimen (including the slab), with the main beam subjected to a negative vertical displacement, while the transverse beam is restrained at its end against vertical movement. Due to the shear-lag effect, tensile stresses develop in the longitudinal slab reinforcement along the face of the connection between the slab and the transverse beam. Because of these stresses, equilibrium of moments in the slab about any vertical axis dictates development of tensile stresses in the transverse slab reinforcement along the face of the connection between the slab and the main beam. Therefore, under unidirectional loads, the entire slab reinforcement is placed under tension. Addition of loads in the orthogonal direction (bi-directional load case), produces the same pattern of stresses on the reinforcement. Equilibrium in the first direction at a given displacement level can be now attained at a lower level of externally applied load; the resulting load difference is detected as strength reduction in the storey shear orbits.

Transverse Beams

Torsional cracking was observed in the North and South beams while unidirectional loads were applied in the East West direction at 1.5% storey drift. The cracks were located near the column, and were at right angles in the opposite sides of the beam. The cracks on the side of the transverse beam facing the slab-in-compression quadrant were about twice as wide as those measured on the other side (facing the slab-in-tension quadrant).

Other than torsional cracks, some vertical cracks at the side of the transverse beam, where the main beam was under positive bending, were observed at high deformation levels (3% storey drift). This, together with the existence of wider torsional cracks on that side illustrates that the amount of weak axis bending was significant in the transverse beam. Torsional deformations were observed to increase under bi-directional loading, when transverse beams also carried out-of-plane flexural loading. This is likely to be an indication of stiffness reduction resulting from the interaction of weak and strong axis moments with torsion along the beam.

Behaviour of the Slab

Slab participation was evident from the crack patterns observed on the top and bottom faces of the slab, and from extensive strain measurements obtained either by strain gauges attached on the reinforcement or from Zurich targets attached on the surfaces of the concrete. The surface cracks depicted in Fig. 6 are characterized by two separate patterns which became interwoven as the level of lateral drift increased. The first pattern consists of extensive cracks parallel to the faces of the column; these cracks formed around the connection and spread transversely into the slab, in directions perpendicular to that of the applied load. The second pattern consists of inclined cracks fanning at 45° near the beam ends; these cracks were initiated around the loading points and spread with increasing level of load towards the middle of the beam length. They also extended at 45° downwards in the beam web and are believed to be an indication of concentrated load effects associated with the experimental setup used to enforce the condition of zero moment at the beam ends. (The second crack pattern described is not representative of the general behaviour of connections with floor slabs in indeterminate structures.)

Strains recorded in the slab reinforcement were tensile regardless the direction of loading, except for the top slab reinforcement adjacent to the column, which experienced compressive strains under positive bending and up to 1.5% storey drift. Figures 7(a)-(b) plot the strain distribution of top and bottom slab bars at the East face of the column under negative and positive bending respectively. Longitudinal slab strains recorded at the face of the transverse beam were typically higher near the column, decreasing with transverse distance from the support in the case of negative bending. This result concurs with the proposition of "shear-lag" in the slab diaphragm, and illustrates the degree of approximation implicit in the traditional assumption of plane-sections which is commonly used in analyzing monolithic R.C. tee-beams.

The mechanism of slab participation is better illustrated in the measured surface strains of concrete; these strains are averaged over a square grid of targets spaced at 20 cm and therefore are less sensitive to local slip disturbances than the reinforcement measurements. Table 1 summarizes the principal strains recorded during the first uni-directional excursion to 2% drift. Whereas the slab in tension side is under net elongation in both principal directions, in contrast, a diagonal compression field characterizes the slab in compression case, particularly around the column (at some distance from the area of influence of the concentrated loads which act at the beam ends). It is noteworthy that longitudinal steel strains at the same locations were tensile (Fig. 7(b)), suggesting than an amount of relative slip occurred between concrete and slab reinforcement.

To quantify the contribution of slab to the beam flexural resistance from the experimental results, beam properties were computed analytically assuming various effective slabs widths, equal to multiples of the beam depth, d. These computed moment-curvature relationships were compared with the experimentally obtained moment-curvature envelopes for the strong beam. From this comparison, it was found that for negative bending, at 2% lateral drift, the effective slab was 1.5 times the beam depth (measured on each side of the beam web); however, this quantity increased to 2-2.5 times the beam depth for 3% and 6% storey drift respectively. Under positive bending, the effective slab remained almost constant throughout the test, (approximately twice the beam depth on each side of the beam web); it is believed that the insensitivity of the computed moment-curvature relations to further increases in the assumed effective widths is probably a consequence of the limited amount of tension reinforcement at the bottom of the beam when the slab is in compression.

Joint Performance

Figure 8 plots hoop strains recorded in joint hoops in the E-W direction. Evidently, bottom hoops were overstressed when compared to the top ones; this indicates superior shear resistance at the upper part of the joint, likely to have resulted from the confining action of the floor-slabs. The total shear force introduced to the joint is estimated by considering equilibrium of forces in the column. (Column moments occurring at the top and bottom faces of the beam are computed from equilibrium of the entire specimen; joint shear is the slope of the column moment diagram within the beam depth.) Horizontal shear introduced to the joint directly by forces acting on the beam web was estimated using the actual yielding stress of the beam reinforcement. The resulting joint shear was 69%, 62% and 56% of the total joint shear measured under unidirectional loads at 2%, 3%, and 6% storey drifts respectively. This illustrates the increasing participation of other connection components (slab and transverse beams) in indirect loading of the joint, an effect that is not accounted for in the current design philosophy for joints.

CONCLUSIONS

Participation of slabs to the response of frame connections was investigated in light of experimental data obtained at the University of Toronto. Under uni-directional loads an effective width of slab approximately equal to 1.5-2 beam depths measured on each side of the beam web contributed to the flexural resistance of the supporting beams. However, a significant amount of interaction was detected between orthogonally placed beams when loads acted at a 45° angle relative to the principal directions of the connection geometry. The interaction was result of combined slab participation in the two orthogonal bending axes, but was also an indication of reduced joint shear resistance under bi-directional loads. Concrete strain records on the surfaces of the slab, in addition to reinforcement strain readings were considered in examining the mechanism of force transfer in the connection.

ACKNOWLEDGEMENTS

The experimental work presented in this paper was carried out at the University of Toronto, Canada. Financial support for the study was provided by NSERC grant No. OGP0042033.

REFERENCES

ACI Committee 318, Building Code Requirements for Reinforced Concrete, (ACI 318-83), American Concrete Institute, Detroit 1983.

ACI-ASCE Committee 352. 1985. Recommendations for Design of Beam-Column Joints in Monolithic Reinforced Concrete Structures. ACI Journal, Vol. 82(3), 266-283.

Ammerman, O. V., and French, C. W. 1989. R/C Beam-Column-Slab Subassemblages Subjected to Lateral Loads. J. Struct. Eng., ASCE, 115(6), 1289-1308.

Bas, A. 1990. Behaviour of Reinforced Concrete Beam-Column Connections With Floor Slabs Under Bi-directional Loads. Thesis submitted in conformity with the requirements for the degree of Masters of Applied Science in the Department of Civil Eng. of the Univ. of Toronto.

CSA Standard. 1984. Design of Concrete Structures for Buildings. CAN3-A23.3-M84, Canadian Standards Association, Rexdale, Ontario.

Durrani, A.J. and Wight, J.K. 1982. Experimental and Analytical Study of Internal Beam to Column Connections Subjected to Reversed Cyclic Loading. Report No. UMEE 82R3, Department of Civil Engineering, University of Michigan, Ann Arbor.

French, C. W. and Boroojerdi, A. 1989. Contribution of R.C. Floor Slabs in Resisting Lateral Loads. J. Struct. Eng., ASCE, 115(1), 1-18.

Kurose, Y., Guimaraes, G. N., Liu, Z., Krieger, M. E., and Jirsa, J. O. 1988. Study of R.C. Beam-Column Joints Under Uniaxial and Biaxial Loading. Report No. PMFSEL 88-2, Department of Civil Engineering, University of Texas at Austin, Austin, Texas.

Otani, S., Kabeyasawa, T., Shiohara, H., and Aoyama, H. 1984. Analysis of the Full-Scale Seven-Story Reinforced Concrete Test Structure. Earthquake Effects on Reinforced Concrete Structures, U.S.-Japan Research, ACI SP-84, American Concrete Institute, Detroit.

Pantazopoulou, S. J., Moehle, J. P., and Shahrooz, B. M. 1988. Simple Model for the Effect of Slabs on Beam Strength. J. of Struct. Eng., ASCE, 114(7), 2000-2016.

Pantazopoulou, S. J., and Qi, X. 1990. Proceedings of the 4-th U.S. National Conf. on Earthquake Eng., EERI, V. 2, 137-146, Palm Springs.

Paulay, T., Park, R., and Priestley, M. J. N. 1978. Reinforced Concrete Beam-Column Joints Under Seismic Actions. ACI Journal, 75(11), 583-593.

Zerbe, H. E., and Durrani, A. J. 1985. Effect of a Slab on the Behavior of Exterior Beam to Column Connections. Report No. 30, Department of Civil Eng., Rice University, Houston, Texas.

Zerbe, H. E., and Durrani, A. J. 1988. Seismic Behavior of Indeterminate R.C. Beam to Column Connection Subassemblies. Proceedings, 9-th World Conf. on Earthquake Eng., Tokyo, Japan.

TABLE 1
Principal Strains of the Slab - Unidirectional Loads at 2% Drift

Distance from column CL (mm)	Slab in tension			Slab in compression		
	ε1	ε2	θ (degrees)	ε1	ε2	θ (degrees)
1500	0.0025	0.00043	115.90936	8E-05	0.0011	51.623497
1300	0.0022	0.00015	115.04993	1E-05	0.0009	44.919891
1100	0.0019	0.00025	108.34632	-1E-05	0.0009	47.154427
900	0.0022	0.00056	86.115557	-4E-05	0.0008	57.353075
700	0.0028	0.00058	89.209529	-0.00023	0.001	63.541019
500	0.0035	0.00085	105.65342	-0.00061	0.0014	61.535667
400	0.0057	0.0008	112.52891	-0.00132	0.0022	59.415723

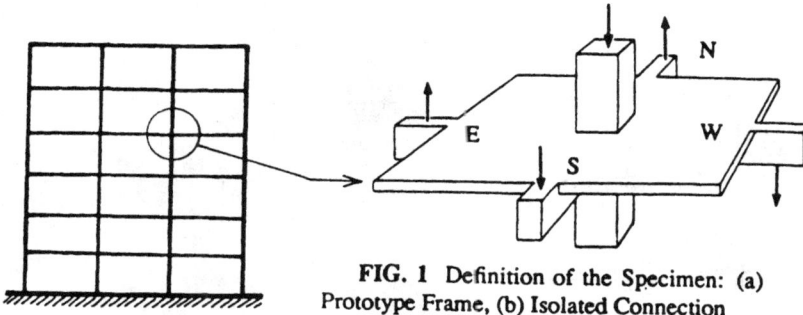

FIG. 1 Definition of the Specimen: (a) Prototype Frame, (b) Isolated Connection

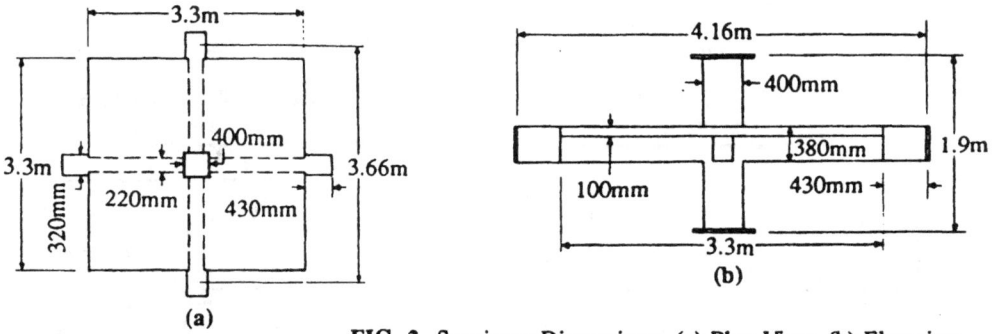

FIG. 2 Specimen Dimensions: (a) Plan View, (b) Elevation

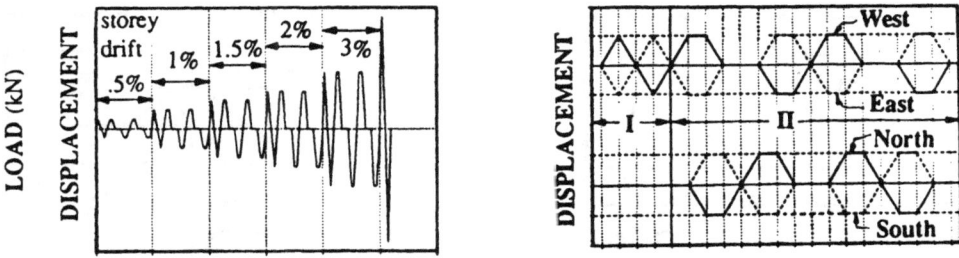

FIG. 3 Cyclic Load History: (a) Displacement History of the Strong Beams, (b) History of All Beams at Each Displacement Level

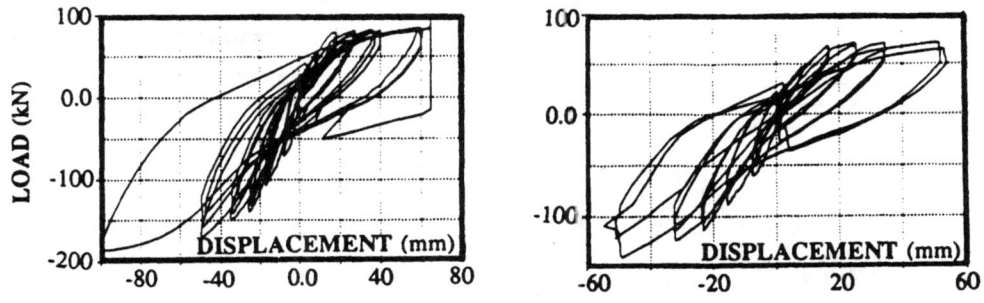

FIG. 4 Load-Displacement Response: (a) East Beam, (b) South Beam

STOREY SHEARS

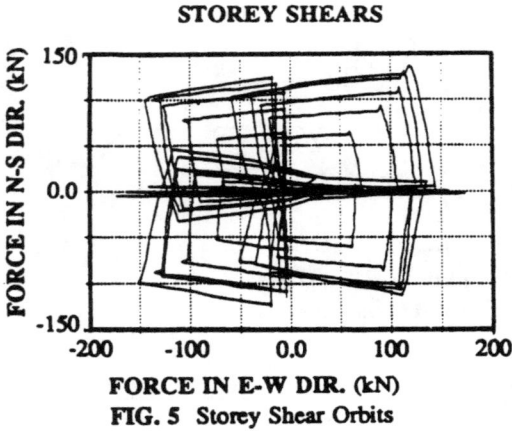

FIG. 5 Storey Shear Orbits

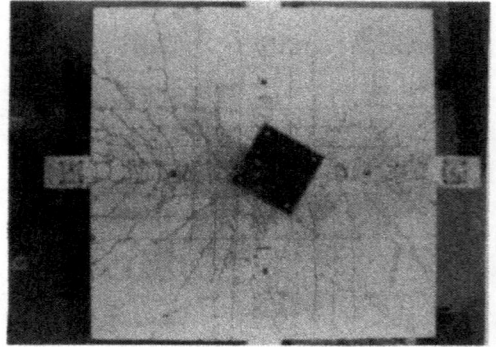

FIG. 6 Cracking Patterns on the Top of Slab

SLAB STRAINS: AT EAST FACE OF COLUMN
Negative Bending

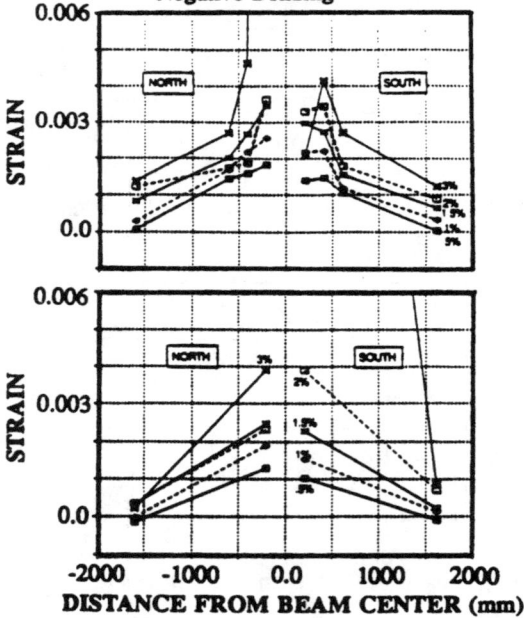

DISTANCE FROM BEAM CENTER (mm)

SLAB STRAINS: AT EAST FACE OF COLUMN
Positive Bending

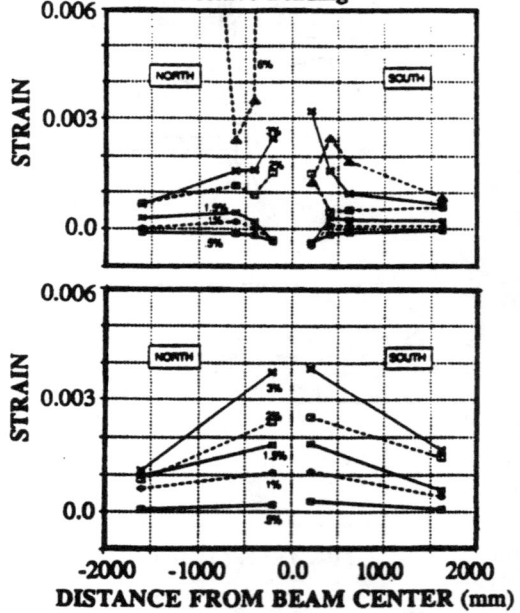

DISTANCE FROM BEAM CENTER (mm)

FIG. 7 Strains in Slab Reinforcement:

(a) Negative Bending,

(b) Positive Bending

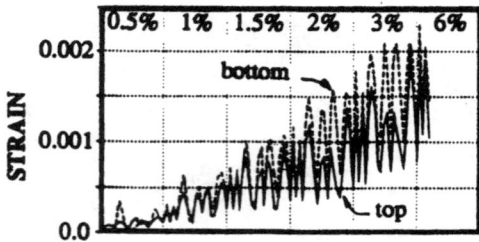

FIG. 8 Joint Hoop Strains

Load carrying capacity of top slab reinforcement in punched flat-plate floors

Robin T. Miller[I] and Ahmad J. Durrani[II]

ABSTRACT

Flat-plate buildings can experience punching failure at connections during strong earthquakes or when subjected to gravity overload. Such failures occur when the slab-column connections do not have adequate shear resistance. Once the punching failure has occurred, the integrity of the floor system depends entirely on the load carrying capacity of the slab reinforcement continuous through the column. As a defense against progressive collapse, the present seismic building codes require slab bottom reinforcement to be continuous through the columns, thus acting as hanger bars in the event of failure. Many existing flat-slab buildings do not have such reinforcement detail and thus rely entirely on the capacity of top slab reinforcement to carry the gravity load should a punching failure occur. This paper presents the results of an experimental investigation to determine the load carrying capacity of the top slab reinforcement in punched flat-plate floors. Results indicate that with a 3 in. spacing the top slab reinforcement was able to carry the load corresponding to one third of the total theoretical punching capacity of the slab-column connection while the top rebars spaced at 6 in. were able to support one half of the punching load. Increasing the length of embedment and the size of the bars increased the load carrying capacity by a small margin.

INTRODUCTION

The current design procedures for flat-plate buildings ensure sufficient margin of safety against punching of slab-column connections under normal service load conditions. Therefore, very little research has been done to evaluate the load carrying capacity of the top slab reinforcement for punched conditions. It is believed that following punching failure, the top slab reinforcement immediately rips through the slab cover and becomes ineffective in carrying the load (Mitchell and Cook 1984). Recent concerns over the safety of older, non-ductile flat-plate buildings under seismic loadings have motivated the need for evaluating the capacity of connections to sustain the gravity loads after punching has occurred.

For gravity load design the ACI 318 Building Code (1989) requires minimum extensions into adjacent spans for top slab reinforcement in slab-column frame systems. For a column strip without drop panels using straight

---

[I] Undergraduate Student, Department of Civil Engineering. Rice University, Houston, Texas 77251

[II] Associate Professor, Department of Civil Engineering, Rice University, Houston, Texas 77251

bars with a minimum of 50% of the bottom reinforcement continuous into the column, the suggested minimum bar length from the face of the column for the top reinforcement is 0.30 of the clear span. This minimum length is required to provide adequate development under design loads in order to carry the negative bending moment produced at the slab-column connection. For seismic loading the ACI Committee 352 (1989) recommends certain minimum continuous bottom reinforcement through the column in each principal direction in order to maintain the integrity of the floor system.

TESTING PROGRAM

The primary objective of this research was to investigate the load carrying capacity of the top slab reinforcement at slab-column connections following a punching failure. To achieve this objective, three variables were chosen on the basis of common design details found in flat-plate floors. Because the thickness of the slab and the amount of cover used are standard in most systems, the bar size, the spacing, and the length of the bar were chosen as variables. Eight half-scale slab-column concrete specimens were tested during this investigation. Table 1 lists the specimen designations and the respective reinforcement details for each specimen.

Table 1. Specimen details

Specimen	Bar Size	Bar Length (Fraction of Clear Span)	Center-to-Center Spacing*
B3L3S3	No. 3	0.30	3 in.
B3L3S6	No. 3	0.30	6 in.
B3L4S3	No. 3	0.40	3 in.
B3L4S6	No. 3	0.40	6 in.
B4L3S3	No. 4	0.30	3 in.
B4L3S6	No. 4	0.30	6 in.
B4L4S3	No. 4	0.40	3 in.
B4L4S6	No. 4	0.40	6 in.

* 1 in. = 25.4 mm

The cover and spacing were kept constant by tying each steel bar to four 0.75 in. plastic chairs. Two well greased metal plates 6.5 in. deep, placed

Figure 1. Reinforcement detail

back-to-back, were inserted vertically on each side of the column faces to simulate partially pre-cracked punching failure. Approximately 1.5 in. of concrete was left intact at each column face (see Fig. 1). Two hooks were embedded in each specimen about 2.5 in. in the concrete and about 2 ft. from each end to allow easy maneuverability in and out of the test machine.

All specimens were cast simultaneously using ready mixed concrete of 3000 psi (20.7 MPa) specified compressive strength at 28 days. Furthermore, high early strength cement was used to be able to test the specimens over the short duration of the project. Control cylinders (6 in. x 12 in.) were tested on each day a specimen was tested to determine the concrete compressive strength. In addition, two beams were tested to determine the modulus of rupture of the concrete.

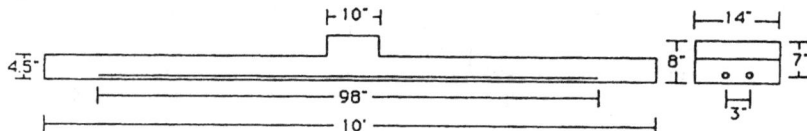

Figure 2. Configuration of Specimen B4L4S3

Instead of testing the specimens with full slab width, it was considered appropriate to model only the strip of slab in the column region since only the reinforcement passing through the column affected the load carrying capacity of the connection (Fig.2). Each specimen was tested in an inverted position to facilitate load application. A general test set up is shown in Fig. 3. The specimens were tested under displacement control to monitor the

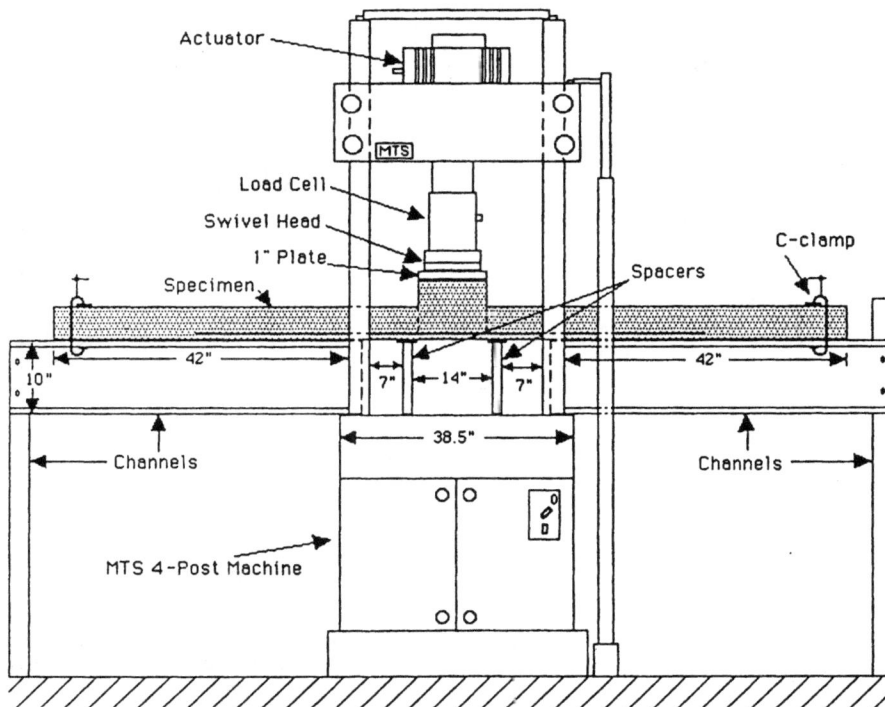

Figure 3. Test set up

601

progressive pull-out of reinforcement through the slab cover. The displacement was applied at a constant rate while a computer automatically recorded both the load and the deflection. Visual observations at appropriate stages of the ripping of the slab reinforcement were made during each test while the displacement was held steady.

TEST RESULTS

A typical load-deflection plot as recorded during the test is presented in Fig. 4. All of the load-deflection plots show a common trend: an initial peak during which the column punched through the remaining intact concrete followed by a drop and a gradual increase in load, peaking after several inches of displacement with a sudden loss of resistance after approximately 75% of the length of the reinforcement was pulled out. Some of the specimens, had an additional peak just before failure. This peak is extraneous and appears in specimens where friction developed between the column and the punched slab surface at large deflections. This friction occurred where the area through which the column was pushed was slightly smaller than the dimensions of the column. The load was thus carried by both the column and the slab instead of the reinforcement. Because of the great irregularities of its data due to excessive friction, the results for test B4L4S3 are not included in computations and comparisons.

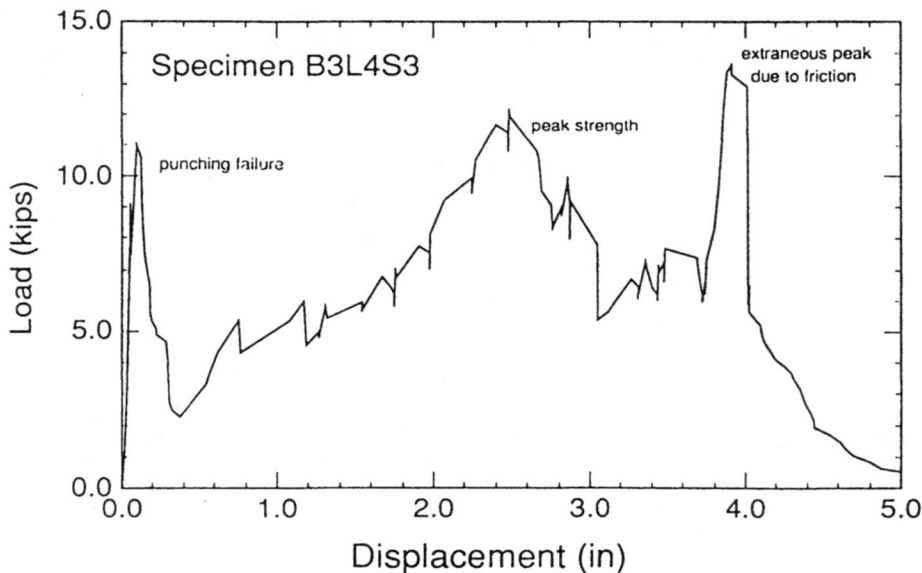

Figure 4. Typical load-displacement plot

The effect of each variable upon the load carrying capacity of the top slab reinforcement is presented in Fig. 5 . As Fig. 5a shows, an increase in the bar size resulted in a small increase in the carrying capacity of the top slab reinforcement. Increasing the bar length also increased the load carrying capacity as indicated in Fig. 5b. The load carrying capacity of the top slab reinforcement was most affected by the spacing of the rebars as indicated by the comparison of the results of specimens B4L3S6 and B4L3S3 shown in Fig. 5c.

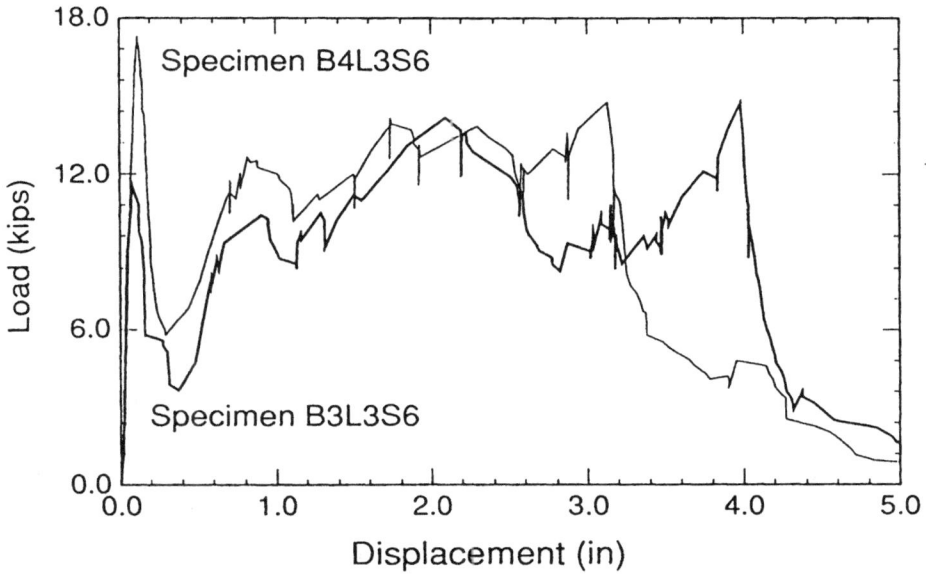

Figure 5a.   The effect of bar size

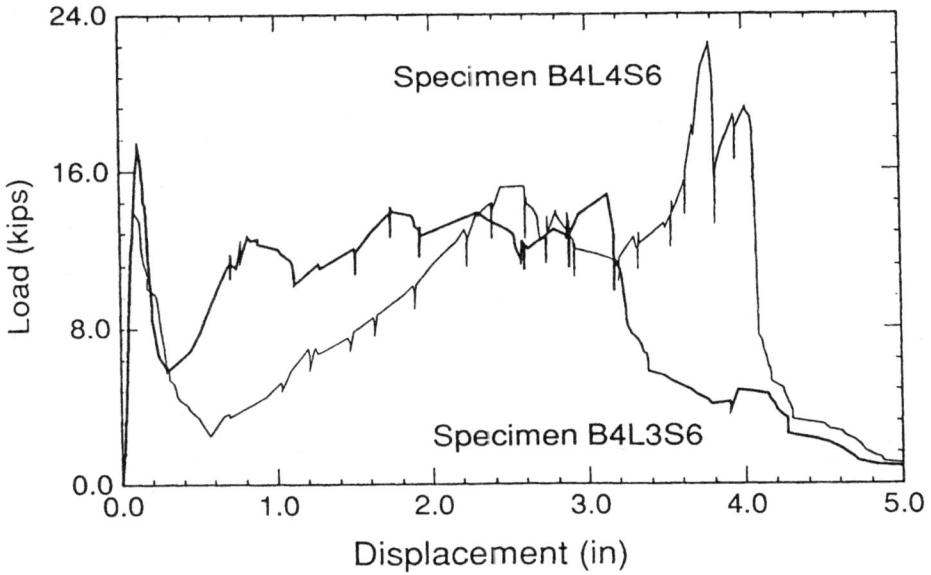

Figure 5b.   The effect of bar length

603

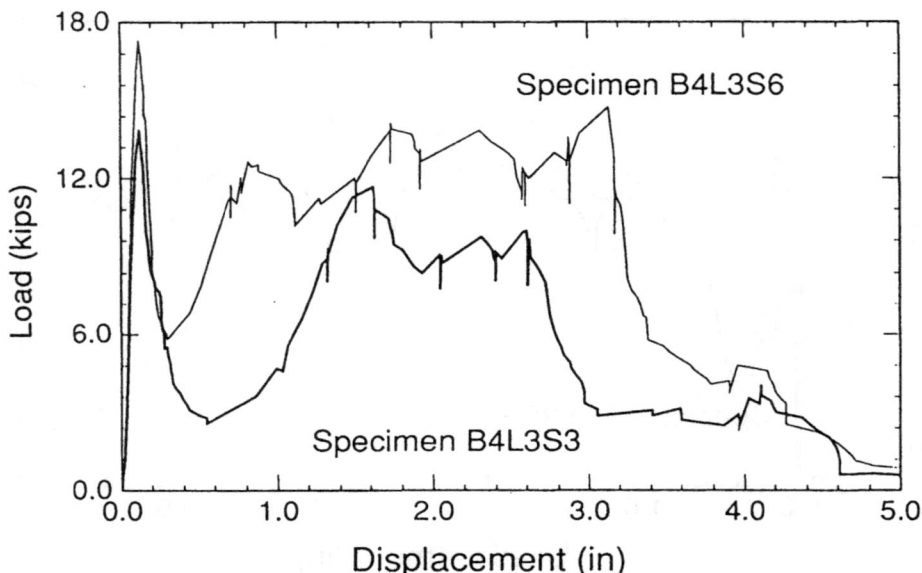

Figure 5c. The effect of bar spacing

The theoretical punching strength of each slab-column connection calculated using the ACI Code (1989) procedure and the actual capacity of the reinforcement as hanger bars represented by the second average peak load in the load-deflection plots are given in Table 2. The ratio of the theoretical punching load of the column to the observed load carrying capacity of the top slab reinforcement shown in Table 2 clearly suggests the significant effect of the bar spacing on the capacity of the top rebars. The average of the ratios for the specimens with a spacing of 3 in. is 0.38 while those with a spacing of 6 in. have and average ratio of 0.55. This depicts an increase in capacity of the top slab reinforcement by 43%. Looking at the specimens grouped by bar size and by bar length, the carrying capacity increased by an average of 13% when using the larger bar diameter and by 16% when using the longer bar length. The overall average ratio of all specimens excluding specimen B4L5S3 is 0.48, thus giving the average capacity of the top slab reinforcement almost half of the punching capacity for the connection. As indicated by the test results, the top slab reinforcement can support one third to one half of the total theoretical punching load of the connection depending upon the spacing of the reinforcement chosen.

Table 2. Observed load capacity of top reinforcement

Specimen	Average Peak Load P (kips)	Punching Capacity V_c (kips)	Ratio (P/V_c)
B3L3S3	8.0	24.0	0.33
B3L3S6	12.0	24.3	0.50
B3L4S3	9.0	22.2	0.41
B3L4S6	12.5	22.1	0.57
B4L3S3	10.0	24.2	0.41
B4L3S6	13.0	24.2	0.54
B4L4S3	20.0	24.4	0.82
B4L4S6	14.0	24.4	0.57

CONCLUSIONS

Based on the results of this experimental investigation, the following conclusions can be drawn:

1) both the bar size and the bar length have minimal effect on the load carrying capacity of the top slab reinforcement following punching;

2) the bar spacing has the most significant effect on the load carrying capacity of the top slab reinforcement; and

3) the top slab reinforcement can support one third to one half of the theoretical punching load of the interior slab-column connection, depending upon the spacing of the reinforcement chosen.

ACKNOWLEDGEMENTS

Great appreciation is extended to Dr. John Merwin and Mr. Hugh Hales for their valuable assistance throughout this investigation. The research was funded by the National Science Foundation under grant BCS-9003563 which is gratefully acknowledged.

REFERENCES

ACI-ASCE Committee 352. 1989. Recommendations for design of slab-column connections in monolithic reinforced concrete structures. American Concrete Institute.

ACI Committee 318. 1989. Building code requirements for reinforcements for reinforced concrete (ACI318-89) and commentary - ACI 318R-89. American Concrete Institute.

Mitchell, D. and Cook, W.D. 1984. Preventing progressive collapse of slab structures. J. Struct. Eng., ASCE, 110(7), 1513-1532.

Effects of loading rate on the behaviour of reinforced concrete coupling beams

H.C. Fu[I] and M.A. Erki[II]

ABSTRACT

This paper presents the preliminary test results of six identical rectangular concrete coupling beams, diagonally reinforced and subjected to monotonic and reversed cyclic loading, using two different loading rates. These rates were slow or quasi-static loading, and fast, simulating earthquake induced high-strain-rate loading. Compared to the test results for slow rate of loading, the fast rate of loading resulted in consistently, but moderately, higher strengths (less than 10%), small increases in initial stiffness and no significant change in ductility of the specimens. Failure of all specimens was precipitated by buckling of the secondary reinforcement and concrete crushing.

INTRODUCTION

In seismic design of coupled shear walls, the goal is to design the coupling beams such that the beams yield before the walls, thereby minimizing wall damage. Current seismic design provisions (ACI 1989; CSA 1984) have been developed primarily on the basis of results from quasi-static tests. The loading rates for such tests are substantially lower than those corresponding to the dynamic conditions expected in earthquake vibrations. Experiments to date, as reviewed by Fu et al. (unpublished), have shown that under dynamic conditions reinforced concrete members may exhibit brittle failure modes, with less hysteretic energy absorption. Also, the formation of plastic hinges may occur at higher than anticipated loads, thereby overloading some parts of the structure.

The behaviour of a diagonally reinforced coupling beam depends primarily on the behaviour of the diagonal steel. The concrete stabilizes the diagonal compression bars, but has little additional influence on the behaviour of the beam (Park and Paulay, 1975). Information from existing literature on the

[I] Research Associate, Department of Civil Engineering, Royal Military College of Canada, Kingston, Ontario, Canada K7K 5L0

[II] Assistant Professor, Department of Civil Engineering, Royal Military College of Canada, Kingston, Ontario, Canada K7K 5L0

effects of rate of loading on the tensile behaviour of steel has been reviewed by Fu et al. The yield and ultimate strengths, the modulus of elasticity, the yield strain, the strain at which strain hardening begins, as well as the length of the yield plateau in the stress-strain diagram of steel, all increase at higher rates of loading, but the modulus of elasticity and the percentage elongation at failure remain roughly unchanged.

Experimental studies of diagonally reinforced coupling beams has been limited to static loading (Park and Paulay, 1975; Barney et al, 1980). This paper reports the preliminary results of an experimental program investigating the behaviour of diagonally reinforced coupling beams, tested using slow and fast loading rates, in monotonic and reversed cyclic loading.

TEST PROGRAM

Six diagonally reinforced concrete beams, with identical geometric and material properties, were tested in this study. The beams were similar in size and detailing, to coupling beams used in buildings with coupled shear walls. The beams were designed for seismic loading in accordance with the seismic design provisions of the Canadian Standards Association - Design of Concrete Structures for Buildings, 1985 - Chapter 21: Special Provisions for Seismic Design.

All specimens had a 1100 mm prismatic test region, with a cross section 300 mm by 400 mm. Main reinforcement was placed diagonally and each diagonal consisted of four M20 bars. The M20 bars were enclosed by 6 mm diameter rectangular hoops, at 100 mm spacing. Outside hoop dimensions were 120 mm by 120 mm. Figure 1 shows the dimensions and reinforcement details of the specimens. For all specimens, the yield strength of the diagonal reinforcement was 405 MPa, and the concrete strength was 26.7 MPa. Secondary reinforcement of the beams was provided within the test region for confinement of the concrete. It consisted of six longitudinal M10 bars and 6 mm diameter stirrups, at 100 mm spacing. The yield strengths of the M10 and 6 mm diameter bars were 400 MPa and 430 MPa respectively.

The cross section at both ends of the specimens was enlarged to 300 mm by 700 mm, to ensure proper load transfer at the beam ends and to ensure that failure was confined to within designated test region. The test set-up is shown in Figure 2. Load was transferred to the specimens through two odd-shaped, thick steel plate loading beams, attached to the enlarged ends of the beams by means of eighteen 25 mm diameter steel rods (yield strength 490 MPa) at each end. The external load was applied to the specimen through the free end of one of the loading beams, which was connected to a 500 kN capacity actuator, while the free end of the other loading beam was connected to a pinned joint. With this loading arrangement, each beam specimen was subjected to equal moments at the ends of the test region, and zero moment and pure shear at the mid-span.

Each beam was subjected to a different combination of load history and loading rate, as summarized in Table 1. For monotonic loading, the load on the specimen was steadily increased up to the static strength or dynamic strength of the specimen, depending on whether the rate of the load application was slow or fast. Incrementally cyclic loading was applied according to the load history shown in Figure 3, and simulated earthquake induced forces. The ductility ratio,

608

referred to in Figure 3, is defined as the ratio of the displacement amplitude to the measured displacement at the static yield load (i.e., measured yield displacement for beam D1). The loading rate of the tests was measured in terms of the actuator displacement with respect to time (stroke control). For static tests, the external load was applied at a rate of 0.033 mm/sec. For dynamic tests, the rate of loading was 30 mm/sec, or approximately a thousand times faster than for the static tests. Beams D1, D2, D5 and D6 were tested monotonically. Beams D3 and D4 were subjected to reversed cyclic loadings. Beams D1, D4 and D6 were tested with slow or static loads, and beams D2, D3 and D5 were tested with fast or dynamic loads.

Table 1.   Summary of load history and results of test specimens.

Beam	D1	D2	D3	D4	D5	D6
Type of Loading	M	M	C	C	M	M
Rate of Loading	S	F	F	S	S	F
Failure Loads (kN)						
CW Moments	307	330	327	314	248	303
CCW Moments	–	–	–329	–309	–329	–359

Note : M – monotonic; C – cyclic; S – 0.033 mm/sec; F – 30 mm/sec;
CW – clockwise (actuator pulling away from beam);
CCW – counterclockwise (actuator pushing down on beam);

To begin each test, the loading frame lower pin, which had a 1 mm clearance with the pin hole, was brought into bearing contact. By suspending the specimens freely from the actuator, their dead weight, together with the connecting steel plates, was found to be 40 kN. The starting zero–load and zero–displacement position for all specimens was set at half of the dead weight or ±20 kN, depending on whether the first cycle of load was applied in a clockwise or counterclockwise direction. Consequently, before loading, specimens D1 to D4 were suspended freely from the actuator and were subjected to a preload of 20 kN. Similarly, specimens D5 and D6 were subjected to a preload of −20 kN. The equivalent load of ±20 kN resulted in a relatively small measured displacement of between 1 mm and 1.5 mm. Therefore, the failure loads for individual specimens, as given in Table 1, have been adjusted accordingly, unless otherwise specified.

During each test, continuous time record of the applied load, displacements and strains were monitored. The external load applied to the beam specimen was measured by a load cell attached to the actuator. The displacements at the beam ends, and those along the test region, as well as the deformations of the concrete core at midspan of the beam, were measured using a total of 23 linear variable differential transducers (LVDT's). Steel strains were measured using twelve electrical resistance strain gauges attached to the main diagonal reinforcement and to the secondary transverse stirrups.

A PDP11 series mini-computer was used to control the load application through the MTS actuator and to perform data acquisition for each test.

RESULTS AND DISCUSSION

According to the Special Provisions for Seismic Design as outlined in CSA Standard CAN-A23.2-M84, both shear and flexure shall be resisted by the diagonal reinforcement in both directions. The maximum allowable shear force that the beam section can carry is equal to $A_s f_y (2 \times \sin a)$, where f_y is the yield strength, A_s the area of the diagonal reinforcement, and a the angle between the axis of the diagonal reinforcement and the longitudinal axis of the beam (equal to 17 degrees for the specimens herein). Using the area of steel of 1200 mm^2 for each diagonal, and the yield strength of 405 MPa, the calculated theoretical failure load of the beam specimens is 284 kN. With the exception of D5, all observed failure loads were greater than this value, as seen in Table 1.

The measured load-displacement curves, for all six specimens, are given in Figures 4(a) to (e). Specimens D1 and D2 were tested to failure under clockwise end moments. Specimens D3 and D4 were subjected to incremental cyclic clockwise and counterclockwise end moments. Specimens D5 and D6 were subjected to counterclockwise end moments upon failure, and then clockwise end moments for one complete cycle of loading.

Figures 4(d) and (e) compare the load-displacement response between specimens under monotonic loading and cyclic loading and the same rates of loading. Very little difference can be seen in either maximum load and stiffness, for D1 (307 kN) and D4 (314 kN), which were tested at a slow rate of loading, and for D2 (330 kN) and D3 (327 kN), which were tested at a fast rate of loading. The following compares pairs of specimens tested with the same type of loading, but different rates of loading.

Specimens D1 and D2

For D1, first flexural and diagonal shear cracks were observed at loads of 98 kN and 133 kN, respectively. As load increased, the diagonal cracks widened. The yield load was 217 kN at a displacement of 18.5 mm. Once the diagonal cracks started opening rapidly, the maximum load of 307 kN was reached, at a displacement of 35 mm. Failure of the beam was initiated by the large diagonal cracks, progressive crushing of the concrete in the compression zones, and eventual buckling of the secondary reinforcement (M10 bars) at the end of the test region. At the end of the test (70 mm displacement), large rotations at the beam ends were observed, and the residual load was 278 kN.

For D2, the time for initial loading to failure was only a few seconds. To ensure that failure had occurred before stopping the loading at the end of the test, the target final displacement was increased from the 70 mm for D1 to 90 mm for D2. D2 was loaded at a rate of a thousand times faster than for D1, and failed at 330 kN, at a displacement of 40 mm. Failure mechanism, buckling of secondary reinforcement and crushing of concrete, was similar to that observed for D1.

It should be noted that fewer cracks and less severe concrete crushing were observed at the end of the test for D2. The initial stiffness of the load-displacement curve for D2 is higher than for D1. The strain readings for D1 and D2 indicated that the diagonal reinforcement yielded prior to failure, and that steel strains for D2 were slightly higher than for D1 at any particular load level. Because only the secondary reinforcement buckled, D1 and D2 continued to carry substantial load after the maximum load had been reached, as indicated in Figure 4(a).

Specimens D3 and D4

Specimens D3 and D4 were subjected to greater than twelve cycles of loading and unloading (Figure 3). D3, under fast rate of loading, and D4, under slow rate of loading, followed identical loading history, such that direct comparison between the two specimens could be conducted. Both tests were stopped midway through cycle 13, when the specimens were found to have been badly damaged.

For D4, the first flexural and shear cracks were recorded at a load of 104 kN at cycle 1 (98 kN for D1) and 136 kN at cycle 4 (133 kN for D1). Before the maximum loads were reached at cycle 7, no new flexural cracks were found and the shear cracks lengthened, but did not visibly open. Maximum loads for clockwise and counterclockwise moments were 314 kN and 309 kN respectively. At the end of cycle 7, concrete in the compression zone started to spall, and many more shear cracks appeared, with shear cracks opening to about 5 mm. By the end of cycle 11, much of the concrete in the compression zone, between two large diagonal cracks, had fallen off. At the end of the test, all reinforcement within the exposed area, including the diagonal reinforcement in both directions, and the longitudinal and transverse secondary reinforcement were found to have buckled.

The first large diagonal crack, extending from the top of one end of the test region to bottom of the other end, was observed during cycle 7 of the clockwise end moments for D3. It was not until cycle 9, counterclockwise end moments, that the first large diagonal shear crack was visible. Maximum loads of 327 kN (314 kN for D4) and 329 kN (309 kN for D4) were obtained for clockwise and counterclockwise end moments, respectively. Failure of D3 was similar to that for D4, but exhibited fewer flexural and shear cracks. Figure 4(c) compares cycle by cycle the load-displacement response of D3 and D4. As for D1 and D2, the initial stiffness of the curves is greater for D3, which was tested with fast dynamic loads, than for D4, which was tested with slow static loads.

Specimens D5 and D6

Specimens D5 and D6 were loaded to failure under counterclockwise end moments, then unloaded and reloaded under clockwise end moments until failure. First flexural and diagonal shear cracks for D5 occurred at loads of 93 kN (98 kN for D1 and 104 kN for D4) and 136 kN (133 kN for D1 and 136 kN for D4), respectively. As load increased, more flexural cracks and shear cracks developed. At a load of 300 kN, and a displacement of 42 mm, diagonal shear cracks were approximately 5 mm wide and extended across the length of the test region. A maximum load of 329 kN was obtained at a displacement of 34 mm. At failure, the diagonal crack had opened to approximately 10 mm, along the entire length of the test region. Concrete crushing at the ends of test region, and buckling of both main diagonal reinforcement and secondary longitudinal and transverse reinforcement, were observed.

A maximum load of 359 kN at a displacement of 49 mm (329 kN at 34 mm for D5) was recorded for D6 under dynamic loading. At failure, no crushing of concrete occurred and shear cracks were generally long but had not significantly opened. Both D5 and D6 failure under counterclockwise end moments and were subsequently reloaded under clockwise end moments. Maximum loads reached in clockwise loading were 248 kN for D5 and 303 kN for D6. Figure 4(b) presents the load-displacement behaviour of the two specimens.

CONCLUSIONS

The experimental results were presented for six diagonally reinforced concrete coupling beams, subjected to different types and rates of loading. Slow static, fast dynamic, monotonic and cyclic loading tests were performed. It was found that specimens subjected to dynamic loads tended to carry somewhat higher loads and showed higher initial stiffness, than specimens subjected to static loads. Specimens under fast rate of loading also developed fewer and less severe cracks than specimens under slow rate of loading. Ductility was not significantly affected by the different rates of loading. These conclusions are consistent with the effect of loading rate on the behaviour of reinforcing steel, indicating that the behaviour of coupling beams is primarily a function of the steel rather than the concrete behaviour.

Code provisions adequately predict the behaviour and the failure load of the beams tested. Also, whether the loading was monotonic or cyclic appeared to have minimum effects on the behaviour of specimens. Failure of all specimens was precipitated by buckling of the secondary reinforcement and concrete crushing. Therefore, additional secondary steel may be required for concrete confinement, at the ends of the test region for the beams, or, for coupled shear walls, adjacent to the vicinity of load transfer between the wall and the beam.

REFERENCES

ACI Committee 318, American Concrete Institute, 1989. Building Code Requirements for Reinforced Concrete (ACI 318-89). Detroit, Michigan.

Barney, G.B., Shiu, K.N., Rabbat, B.G., Fiorato, A.E., Russell, H.G. and Corley, W.G. 1980. Behaviour of Coupling Beams Under Load Reversals, Portland Cement Association, RD068.01B.

Canadian Standards Association 1984. Code for the Design of Concrete Structures for Buildings (CAN3-A23.3M84).

Fu, H.C., Seckin, M., and Erki, M.A. Effects of Loading Rates on Reinforced Concrete. Submitted to the Journal of Structural Engineering ASCE.

Park, R., and Paulay, T. 1975. Reinforced Concrete Structures, John Wiley and Sons, Chapter 12.

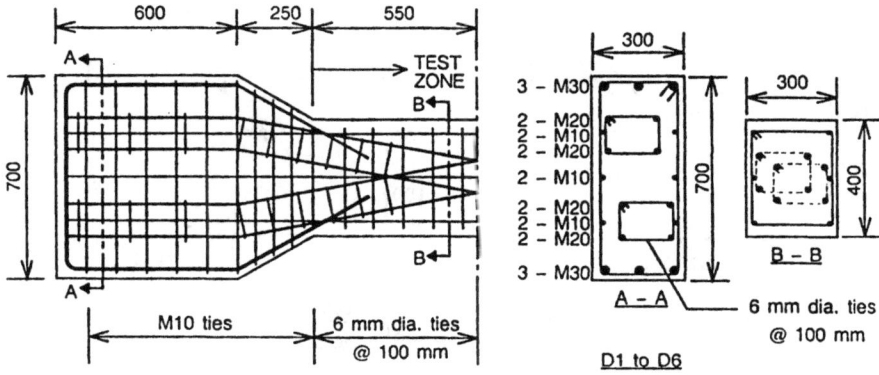

Figure 1.   Details of specimens

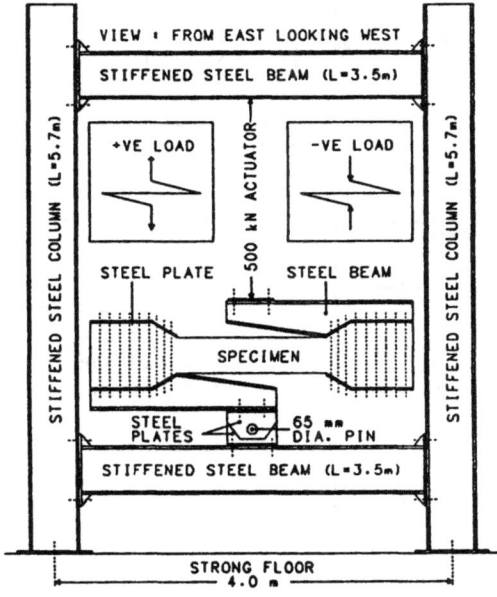

Figure 2.   Test set-up

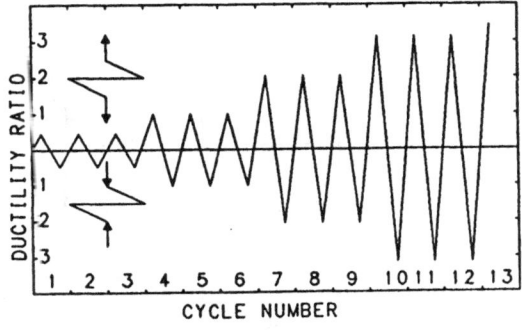

Figure 3.   Cyclic loading history

613

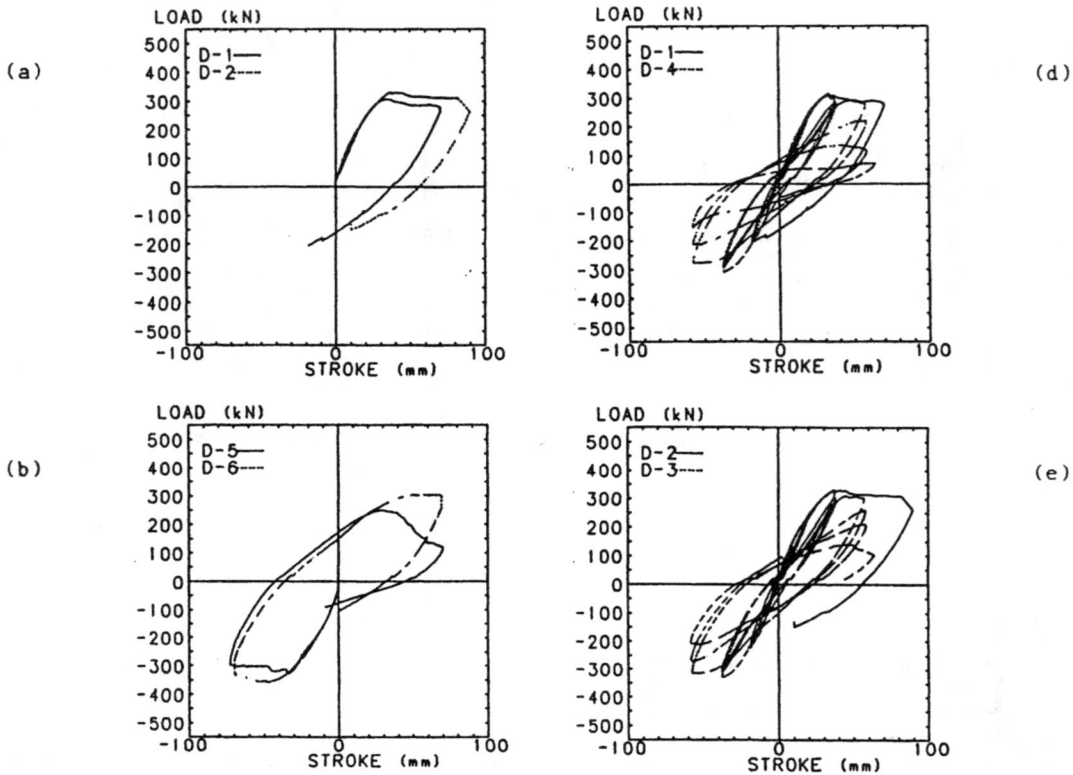

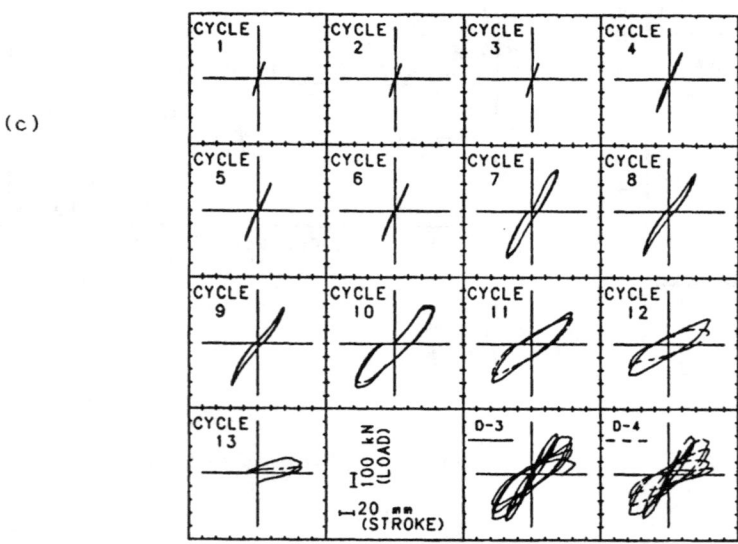

Figure 4. Load displacement curves

(a) D1 and D2
(b) D5 and D6
(c) D3 and D4
(d) D1 and D4
(e) D2 and D3

Reversed cyclic loading tests of precast concrete columns

Claude F. Pilette[1] and Denis Mitchell[2]

ABSTRACT

Eight full-scale precast concrete columns were tested to determine their reversed cyclic loading responses. The specimens represent typical one-storey precast concrete columns including the connections to the foundation pedestals. This experimental study illustrates the performance of columns designed using the current approach recommended by PCI and CPCI. Methods of improving the performance by increasing the strength and stiffness of the connections and improving the levels of ductility are presented. The resulting performance of the columns are assessed in terms of the expected R factors of the NBCC (1990).

INTRODUCTION

The National Building Code of Canada (NBCC 1990) gives force modification factors, R, for different structural systems exhibiting different levels of ductility and energy absorption. Precast concrete construction is not covered specifically in the NBCC and by default the designer might use the category for "other lateral-force-resisting" reinforced concrete systems and use an R value of 1.5. In addition the CSA Standard A23.3 (CSA 1984) does not provide special provisions for the seismic design of precast buildings. A little guidance for the seismic design of precast buildings is given in the design handbooks of the Prestressed Concrete Institute (PCI 1985) and the Canadian Prestressed Concrete Institute (CPCI 1987). There is a surprisingly small amount of experimental evidence (Dolan et al. 1987, PCI 1986) in the literature for the reversed cyclic loading response of typical precast columns. This paper summarizes the results of a series of tests of precast columns tested at McGill University as part of a larger research programme (Pilette and Mitchell 1991) investigating the seismic behaviour and design of precast concrete structures.

---

[1]Ph.D. Candidate, Department of Civil Engineering and Applied Mechanics, McGill University, Montreal, Canada, H3A 2K6.

[2]Professor, Department of Civil Engineering and Applied Mechanics, McGill University, Montreal, Canada, H3A 2K6.

Table 1: Details of test specimens with dimensions in mm

Specimen	Column details			Connection details	
	Size	Longitudinal bars	Transverse steel	Plate	Bolts
R1.5-1	475 × 475	8-No. 20	No. 10 @ 300	725 × 725 × 25.4	4-25.4
R2-1	400 × 400	8-No. 20	No. 10 @ 150	550 × 550 × 25.4	4-25.4
R4-1	300 × 300	8-No. 15	No. 10 @ 60	450 × 450 × 15.9	4-15.9
R4P-1	300 × 300	8-No. 15	No. 10 @ 60	550 × 550 × 19.0	4-19.0
R1.5-2	475 × 475	8-No. 20	No. 10 @ 300	675 × 475 × 31.7	4-28.6
R4-2	300 × 300	8-No. 15	No. 10 @ 60	500 × 300 × 44.5	4-25.4
R1.5-3	475 × 475	8-No. 20 dowels 4-No. 20 cage	No. 10 @ 300	475 × 475 × 25.4	4-34.9
R4-3	300 × 300	8-No. 15 dowels 4-No. 15 cage	No. 10 @ 60	300 × 300 × 25.4	4-34.9

TESTING PROGRAMME

Eight precast column-connection-foundation specimens were tested under reversed cyclic loading in the testing frame shown in Fig. 1. The lateral load was applied at a height of 2.75 m measured from the bottom of the base plate. Each column was subjected to a constant axial load of 184 kN to simulate the dead load acting on the column. Figure 2 illustrates three standard types of column connection details typically used (CPCI 1987). Type 1 has a base plate which is oversized on all sides of the column, Type 2 is oversized only in the direction of loading and Type 3 uses recessed anchorage pockets. For Types 1 and 2 the longitudinal column reinforcing bars pass through holes in the base plate and are welded to the underside of the base plate connection. Connection Type 3 uses dowel bars, welded to the steel anchor bolt pockets. The column reinforcing cage is slipped over these dowel bars and adequate lap slice length is provided between the dowels and the longitudinal column bars. For all specimens a nut was used above and below the base plate on all anchor bolts to facilitate erection. Grout, having a minimum compressive strength of 60 MPa was packed between the base plate and the top of the pedestal.

Table 1 gives the details of the specimens. The designation for each specimen contains the R value followed by a number indicating the connection type (e.g., Specimen R1.5-1 was designed with an R of 1.5 and has a connection Type 1). The compressive strengths of the concrete at the time of testing were 37, 33 and 42 MPa for connection Types 1, 2 and 3, respectively. The No. 10, No. 15 and No. 20 reinforcing bars had yield stresses varying from 423 to 516, 443 to 456, and 443 to 494 MPa, respectively. The anchor bolts had yield stresses varying from 263 to 315 MPa and the yield stress of the base plates varied from 252 to 365 MPa.

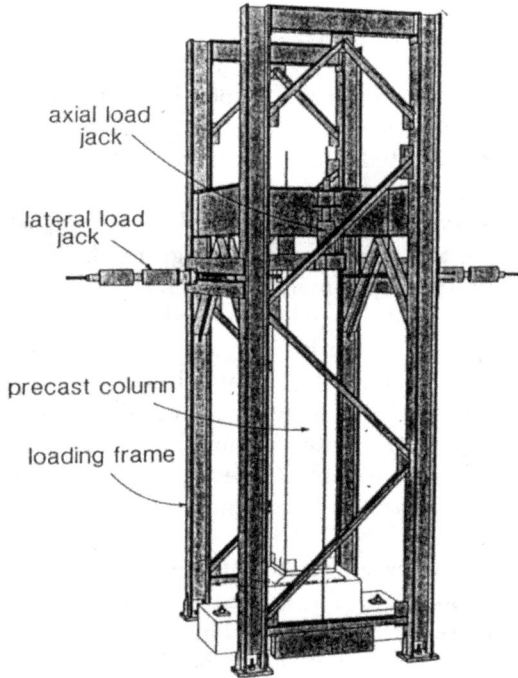

(a) Loading frame   (b) Specimen R4P-1 under load

Figure 1: Test setup for reversed cyclic loading of precast column-
connection-foundation specimens

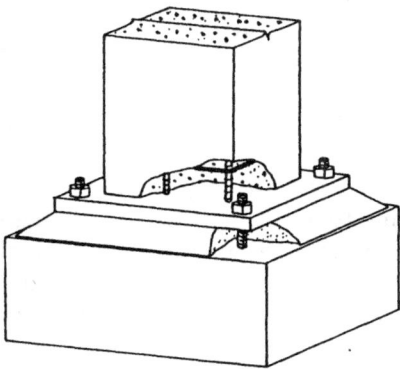
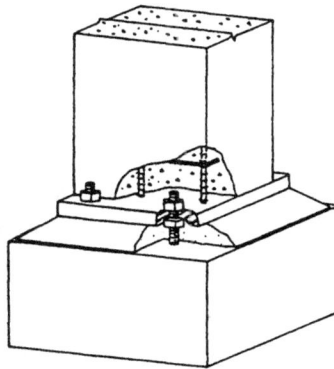
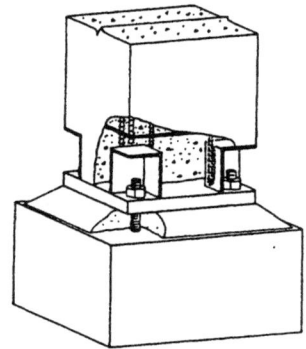

(a) Type 1   (b) Type 2   (c) Type 3

Figure 2: Types of connection details investigated

TEST RESULTS

Figure 3 compares the connection deformations of connection Types 1, 2 and 3 for specimens designed with R = 1.5. Although the standard design method (PCI 1985, CPCI 1987) was used for Specimen R1.5-1 the base plate exhibited extremely large deformations due to bending in the direction of loading as well as in the transverse direction (see Fig. 3a). Excessive base plate yielding took place in the transverse direction while the column just reached yielding at its base. It is interesting to note that the standard design methods do not consider the transverse bending of the base plates. Specimen R1.5-2, having a base plate oversized only in the direction of loading, exhibited a much improved performance over Specimen R1.5-1 because of the elimination of transverse bending of the base plate (see Fig. 3b). The improved strength and stiffness of the connection enabled inelasticity to spread into the column. The base plate of Specimen R1.5-3 was designed using yield line analysis to account for bending across the corner between the plates forming the anchor bolt pocket. The connection stiffness and strength was similar to that of Specimen R1.5-2 with localized yielding taking place in the column just above the top of the anchor bolt pockets.

Figure 4a shows the base moment versus the measured top displacement of Specimen R1.5-2, indicating a displacement ductility of about 2.0. The overall performance was controlled by the anchor bolt deformations as can be seen from Fig. 4b. The pinching of the hysteretic response is due mainly to the irrecoverable elongation of the anchor bolts between the nuts located above and below the base plate. Figure 4c shows the contribution of the base plate bending together with the longitudinal column bar elongation over the thickness of the base plate. This displacement contribution is about equal to the anchor bolt displacement contribution. Figure 4d indicates that the bottom portion of the column exhibited some yielding consistent with the design level of ductility.

Figure 1b shows the deformed shape of Specimen R4P-1 which had a column designed to exhibit significant ductility but no special consideration was given to the transverse bending of the base plate. The deformed shape of the column shows that the base plate yielded and contributed to most of the top displacement while only localized yielding of the column bars took place at the base. Specimen R4-2 was designed with an oversized base plate in the direction of lateral loading and a capacity design approach was used to ensure that the connection was stronger than the column. As can be seen from Fig. 5a, the stiffer and stronger connection details allowed significant plastic hinging to develop at the column base with yielding extending over a height of 500 mm. As expected, the lapping of the dowel and column bars at the base of Specimen R4-3 produced a region of overstrength which did not result in a significant zone of plastic hinging at the base, see Fig. 5b. This overstrength together with the termination of the dowels resulted in a plastic hinge starting at the free end of the dowels and extending over a height of 250 mm. Due to the formation of the plastic hinge at a height of 750 mm above the column base, the resulting ductility was significantly reduced compared to that observed in Specimen R4-2.

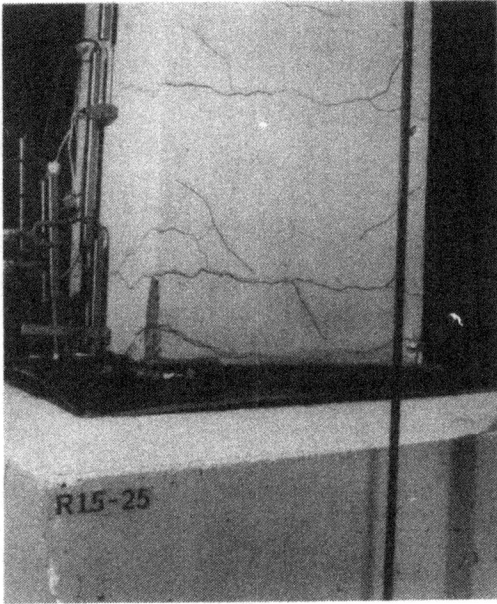

(a) Specimen R1.5-1

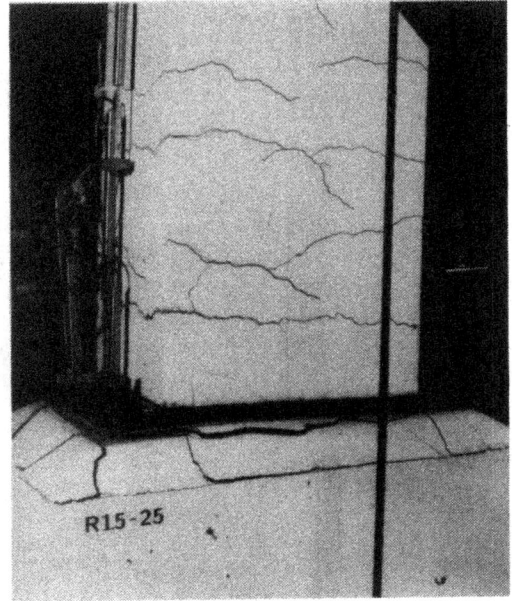

(b) Specimen R1.5-2

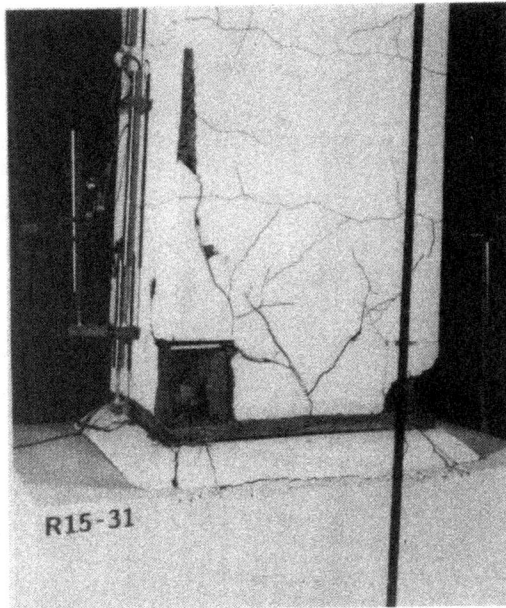

(c) Specimen R1.5-3

Figure 3: Deformations for connection Types 1, 2 and 3

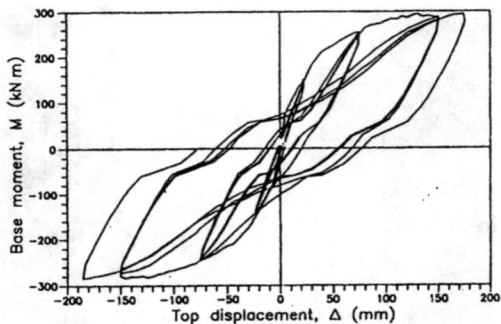

(a) Total deformations

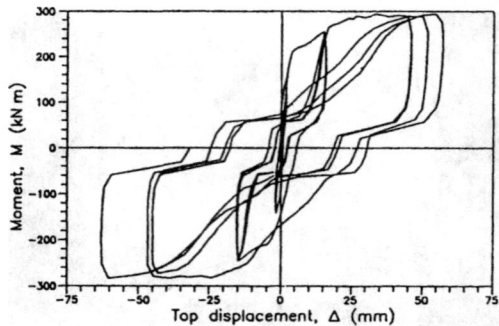

(b) Anchor bolt component

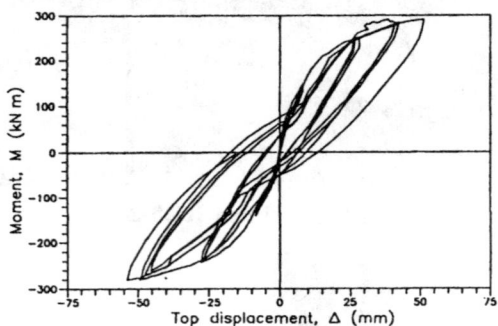

(c) Deformations in base plate
and joint

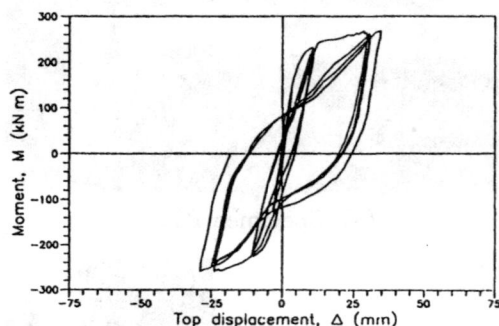

(d) Deformations in bottom 415 mm
of column

Figure 4: Moment versus top displacement components for Specimen R1.5-2

Specimens R4-1 and R4-3 exhibited displacement ductilities of about 4.0 with Specimen R4-1 showing poor hysteretic response. Specimen R4-2 had excellent hysteretic response and due to limitations of the loading system the testing had to be stopped at a displacement ductility of 4.0, however larger ductilities would have been possible.

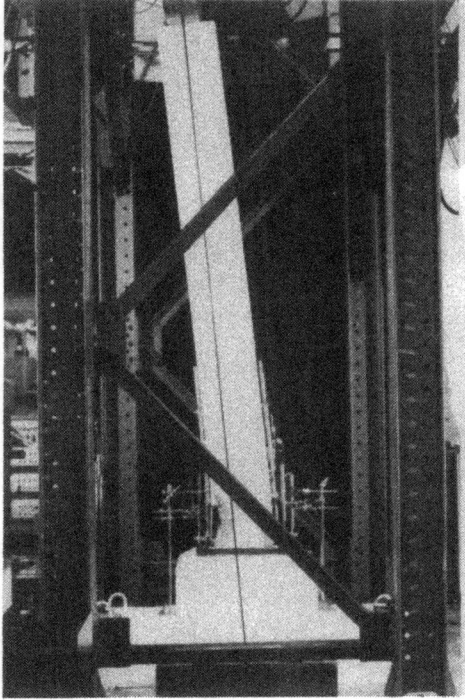

(a) Specimen R4-2   (b) Specimen R4-3

Figure 5: Influence of connections on the overall responses

CONCLUSIONS

The following conclusions can be drawn from this test series:

1. The current design procedure for precast column connections using oversized base plates (Type 1) results in poor seismic performance due to excessive base plate bending and yielding, without significant yielding developing in the column. The transverse bending of the base plate was a major factor contributing to the reduced stiffness and strength of the connection. This type of connection should not be used for situations requiring R greater than 1.5.

2. The use of an oversized base plate in the direction of loading (Type 2) improved the reversed cyclic loading response. If this is combined with a capacity design approach for the connection, excellent hysteretic response and significant displacement ductilities are attainable.

3. The use of recessed pockets (Type 3) gave acceptable reversed cyclic loading response for $R = 1.5$. However, if significant ductilities are to be attained then lap splices should not be used in regions where plastic hinges are expected.

ACKNOWLEDGEMENTS

The authors gratefully acknowledge the financial support from the Natural Sciences and Engineering Research Council of Canada. The research was carried out in the Jamieson Structures Laboratory in the Department of Civil Engineering and Applied Mechanics at McGill University.

REFERENCES

CPCI. 1987. Canadian Prestressed Concrete Institute. Metric design manual – Precast and prestressed concrete. Ottawa, Canada, second edition.

CSA. 1984. Canadian Standards Association. Design of concrete structures for buildings, CAN3-A23.3-M84. Rexdale, Canada.

Dolan, C.W., Stanton, J.F. and Anderson, R.G. 1987. Moment resistant connections and simple connections. Prestressed Concrete Institute Journal, 32(2):62–74.

NBCC. 1990. National Building Code of Canada. Associate Committee on the National Building Code. National Research Council of Canada, Ottawa, Canada.

PCI. 1986. Prestressed Concrete Institute. Moment resistant connections and simple connections. Chicago, U.S.A., R&D 1/4, 436 pp.

PCI. 1985. Prestressed Concrete Institute. PCI design handbook – Precast and prestressed concrete. Chicago, U.S.A., second edition.

Pilette, C. F. and Mitchell, D. 1991. Seismic response and design of one-storey precast concrete buildings. Structural Engineering Series No. 91-2, Department of Civil Engineering and Applied Mechanics, McGill University, Montreal, Canada.

Confined concrete columns with varied concrete strength

Shamim A. Sheikh[1] and Shafik Khoury[2]

ABSTRACT

Results from a select group of column specimens which are part of an extensive research program in the area of concrete confinement, are presented in this paper. The prismatic specimens were 305 mm square and 2.44 m or 2.74 m long. The non-prismatic specimens were 305 mm square and 1.45 m long with a 510 x 760 x 810 mm stub. The target values for concrete strength were 30 MPa and 60 MPa. A total of 24 specimens were tested under monotonic or cyclic flexure and shear while simultaneously subjected to large constant axial load. The objectives of this research are to evaluate the performance of confined concrete members as influenced by concrete strength, presence of stub, type of loading, amount of lateral steel and distribution of steel. Confinement provisions from the design codes are also critically evaluated in the light of the test results.

INTRODUCTION

In a framed structure during a severe earthquake, it is preferable to restrict the inelastic deformations to the beams in general to ensure structural stability and to maintain its vertical load carrying capacity. Accordingly, the concept of "strong column - weak beam" is suggested in most design codes to force the plastic hinges in beams rather than in columns. However, with uncertain earthquake demands, the occurrence of plastic hinges in columns cannot be avoided completely (Paulay 1986). Therefore, the potential plastic hinge regions of columns must be detailed for ductile behavior which can be achieved by confining the concrete effectively. Although the primary purpose of confining concrete is the improvement in ductility, confinement also enhances concrete strength significantly. The overall behavior of a member may not improve due to confinement if the length of the plastic hinge is underestimated or shear reinforcement is designed based on an underestimated flexural capacity. An accurate evaluation of the section behavior is therefore necessary for safe design.

CURRENT RESEARCH

The work reported here is part of an extensive research that started with a study of confinement mechanism under concentric compression (Sheikh and Uzumeri 1980). In the current phase of research, experimental work involved large-size specimens, the details of which are listed in Table 1. The first

[1] Associate Professor, Department of Civil Engineering, University of Toronto, Canada, M5S 1A4

[2] Ph.D. Candidate, Department of Civil Engineering, University of Houston, Houston, Texas, USA

Table 1 - Details of the Test Specimens

Spec.	Concrete strength (MPa)	Longitudinal steel No. & Size	ρ (%)	Transverse steel Size & Spacing (mm)	ρ_s (%)	f_{yh} (MPa)	$\dfrac{P_a}{f'_c A_g}$	$\dfrac{M_{max}}{M_{i\,ACI}}$
				Prismatic Specimens				
E-2	31.4	8-19 mm	2.44	12.7 @ 114	1.69	483	0.61	1.08
A-3	31.8	8-19 mm	2.44	9.5 @ 108	1.68	490	0.61	1.23
F-4	32.2	8-19 mm	2.44	9.5 @ 95	1.68	490	0.60	1.22
D-5	31.2	12-16 mm	2.58	9.5 @ 114	1.68	490	0.46	1.26
F-6	27.2	8-19 mm	2.44	12.7 @ 173	1.68	483	0.75	1.15
D-7	26.2	12-16 mm	2.58	6 @ 54	1.62	469	0.78	1.22
E-8	25.9	8-19 mm	2.44	9.5 @ 127	0.84	483	0.78	0.96
F-9	26.5	8-19 mm	2.44	9.5 @ 95	1.68	490	0.77	1.25
E-10	26.3	8-19 mm	2.44	9.5 @ 64	1.68	490	0.77	1.10
A-11	27.9	8-19 mm	2.44	6 @ 108	0.77	469	0.74	0.97
F-12	33.4	8-19 mm	2.44	6 @ 89	0.82	462	0.60	0.98
E-13	27.2	8-19 mm	2.44	12.7 @ 114	1.69	483	0.74	1.01
D-14	26.9	12-16 mm	2.58	6 @ 108	0.81	469	0.75	1.01
D-15	26.2	12-16 mm	2.58	9.5 @ 114	1.68	490	0.75	1.17
A-16	33.9	8-19 mm	2.44	6 @ 108	0.77	558	0.60	0.95
E-13H	57.6	8-19 mm	2.44	12.7 @ 114	1.69	464	0.63	1.43
F-9H	58.3	8-19 mm	2.44	9.5 @ 95	1.68	507	0.64	1.30
A-17H	59.1	8-19 mm	2.44	9.5 @ 108	1.68	507	0.65	1.30
				Non-Prismatic Specimens				
ES-13	32.5	8-19 mm	2.44	12.7 @ 114	1.69	464	0.76	1.40
FS-9	32.4	8-19 mm	2.44	9.5 @ 95	1.68	507	0.76	1.37
AS-3	33.2	8-19 mm	2.44	9.5 @ 108	1.68	507	0.60	1.37
AS-17	31.3	8-19 mm	2.44	9.5 @ 108	1.68	507	0.77	1.53
AS-18	32.8	8-19 mm	2.44	12.7 @ 108	3.37	464	0.77	1.70
AS-19	32.3	8-19 mm	2.44	9.5 @ 108	1.30	507	0.47	1.32
				6 @ 108		469		

letter in specimen designations refers to the steel configuration shown in the sketches included in Table 1. All the prismatic specimens were 305 mm square. The normal strength concrete specimens were 2.74 m long and the length of high-strength concrete specimens was 2.44 m. The prismatic specimens were tested under constant axial load and two lateral point loads such that the middle 0.91 m length of the specimens was free from shear due to lateral loads (Figure 1). The prismatic normal strength concrete specimens were tested under monotonically increasing flexural deformations until the lateral loads dropped to zero. The high-strength concrete specimens were subjected to cyclic lateral loads until the specimens could not maintain the axial load.

The non-prismatic specimens were 305 x 305 x 1450 mm columns with 510 x 760 x 810 mm stubs and were tested under constant axial load and cyclic point lateral load applied on the stub near the column-stub junction. The critical section adjacent to the stub was thus subjected to a constant axial load and cyclic shear and flexure. Standard cyclic loading consisted of one cycle of deflection to $0.75\Delta_o$ followed by two cycles each to a displacement of Δ_o, $2\Delta_o$, $3\Delta_o$, --- until the specimen could not maintain the axial load. The Δ_o is the estimated deflection at the critical section that causes curvature ϕ_o at that section. The ϕ_o is the curvature corresponding to the maximum unconfined section moment on a straight line joining the origin and a point corresponding to about 65% of the maximum moment. In the case of Specimens E-13H and F-9H, the upward displacement could not be effectively controlled due to malfunctioning of the loading system. A symmetrical cyclic loading was therefore not achieved. It is, however, believed that the envelope curves are relatively independent of the load excursions, and can be compared with the results from other specimens to study different variables.

For normal strength concrete ($f_c' \simeq 30$ MPa) specimens, the lateral reinforcement ratio (ρ_s) required according to the design codes' seismic provisions ("Building" 1983; "Code" 1984; "Recommended" 1980) is approximately 1.4% and for high-strength concrete ($f_c' \simeq 58$ MPa) specimens, this ratio is approximately 2.6%. Nominal yield strength of 400 MPa for steel was used to calculate ρ_s. With the actual yield strength of steel, the corresponding ρ_s values are approximately 1.2% and 2.1%, respectively, for normal and high strength concrete specimens.

RESULTS

The section moment capacities, non dimensionalized with respect to M_i, are listed in Table 1 for all the specimens. The M_i is the theoretical moment capacity of a section based on the concrete stress block suggested in the ACI and CSA design codes. Several points should be considered when the moment capacities of different sections are compared. The moment capacities reported are the ones experienced by the critical sections in the specimens. In the case of prismatic specimens failure occurred at the critical sections. In the case of non-prismatic specimens, the failure in the column occurred approximately 150 mm to 300 mm away from the critical section that was adjacent to the stub. Since the critical sections in the non-prismatic specimens did not fail, their capacities would be higher than the moment values shown in Table 1. In several specimens, the entire cover concrete was effective in compression when the section carried the maximum moment. This was especially the case for specimens which were not well-confined. The three high-strength concrete specimens fall in this category.

In several prismatic, normal-strength concrete specimens, the moment capacity was less than the theoretical capacity, M_i. All these specimens contained approximately 0.8% lateral reinforcement ratio and were subjected to large axial loads. It appears that the strength of concrete in these specimens that can be used to calculate section strength under axial load and flexure, is less than the concrete strength in a standard cylinder.

Results from a select group of specimens are shown in Figures 2 to 7 in the form of lateral load vs. deflection and moment vs. curvature curves. For ease of comparison between specimens made of normal strength and high-strength concretes, the lateral load is non-dimensionalized with respect to P_i, the load required to produce M_i at the critical section without considering P-Δ effect and the moment axis is non-dimensionalized with respect to M_i. From a comparison of specimens in Figures 2 to 5 it appears that the cover concrete in the case of high-strength concrete specimens is more effective initially and spalls off more rapidly than in normal-strength concrete specimens. Crushing strain in high strength concrete is larger than that in normal strength concrete and at that strain confinement of core becomes somewhat effective particularly because of internal cracking due to cyclic loading, possibly resulting in higher section capacity. Due to relatively larger lateral strain in lower strength unconfined concrete, the separation between the restrained core and the cover may result in weaker than normal strength of cover concrete. Higher section strength before cover spalling in the case of Specimen E-13H can perhaps be attributed to the fact that the separation of cover concrete from the core is less disruptive in Configuration 'E' than in other configurations.

A comparison of specimens of Configurations E and F in Figures 2 to 5 shows that a more efficient confinement is achieved if all the longitudinal bars are supported by tie bends. A rapid drop in the section capacity of Specimens F-9H and F-12 during the later part of the tests was caused by the opening of the 90° hooks (Sheikh and Yeh 1990). The combination of high axial load and small lateral reinforcement ratio results in this type of failure which can be prevented by the use of internal ties such as those used in Configuration A (Figure 6). The adverse effects of high axial load on ductility can also be observed by comparing the behavior of Specimens F-4 and F-9. The beneficial effects of increased amount of lateral reinforcement are obvious from a comparison of Specimens F-4 and F-12.

In the case of Specimens E-2, F-4 and F-12 the axial load was about 10-12% below the limit allowed by the ACI codes. For high-strength concrete specimens (E-13H and F-9H), however, applied axial load exceeded the code limits by about 10% although for all five specimens the index $P/f_c'A_g$ was approximately 0.62 and the ratio between the concrete stress caused by the axial load and f_c' was also constant at about 0.53. A limit on axial load that causes lower concrete stresses in high strength concrete columns is quite logical in view of the brittle nature of this material and the adverse effects of high axial load on ductility. For the same absolute amount of lateral steel, ductility in higher strength concrete specimens is lower indicating that the required amount of confining steel should be dependent on concrete strength; however, not for the purpose of compensating for the loss of strength due to cover spalling but to maintain the integrity of the core to provide ductile behavior. It appears that the required amount of steel is less than proportional to concrete strength.

For Specimens FS-9 and F-9, the moment-curvature relationships of the sections where failure occurred, are provided in Figure 7. The failure section was approximately 200 mm away from the critical section in Specimen FS-9. Moment at the critical section which did not fail was approximately 10% larger than the moment at the failed section indicating that the stub provided significant restraint to the adjacent critical section. The similarity in the moment-curvature behavior between the two specimens indicates that the effect of stub restraint on the failed section was minimal. In both the specimens, ultimate failure was caused by the opening of the 90° hooks. Behavior of Specimen FS-9 appears to provide healthy energy absorption properties and it is plausible to assume that Specimen F-9 has similar characteristics.

Amount of lateral reinforcement in Specimens F-4 and F-9 meets the code requirements for seismic design. Specimens F-12 and A-16 contain the amount of lateral reinforcement which is about 60% of that required for seismic design. A comparison of the behavior of these specimens in Figure 6 shows that column design according to code requirements may either be too conservative (Specimen F-4) or unsafe (Specimen F-9). Behavior of Specimens F-12 and A-16 that violate the code requirements should also be acceptable if the behavior of Specimens F-9 and FS-9 is acceptable. The expected performance of a section and a member, distribution of longitudinal and lateral steel and the level of axial load must therefore be considered, in addition to other parameters, in the design of confining steel.

CONCLUSIONS

The concept of providing confining steel to compensate for the loss of load carrying capacity of cover concrete does not provide a sound basis for design. Columns designed according to the code provisions display behavior which may range from unacceptably brittle to very ductile. The required structural performance, steel detailing and the load combinations must be considered, among other parameters, in the design of confining steel. Presence of heavy elements such as stub adjacent to the critical section must also be considered since their restraining effect enhances section strength significantly. For the same amount of confining steel, normal strength concrete specimens behave in a more ductile manner than the high strength concrete specimens. Currently the required amount of confining steel is directly proportional to the strength of concrete which may not remain practical for high strength concrete. However, it appears that in order to maintain ductile behavior of the core concrete, the required amount of lateral steel may be less than proportional to concrete strength.

ACKNOWLEDGEMENTS

The research reported here was supported by grants from the National Science Foundation and the Natural Sciences and Engineering Research Council of Canada.

REFERENCES

"Building code requirements for reinforced concrete" (1989). Amer. Concr. Inst., Detroit, Mich.

"Code for the design of concrete structures for buildings" (1984). CAN3-A23.3M84, Canadian Standards Assoc., Rexdale, Canada.

Paulay, T. (1986). "A critique of the special provisions for seismic design of the building code requirements for reinforced concrete (ACI 318-83)", ACI J., 83(2), 274-283.

"Recommended lateral force requirements and commentary" (1980). Seismology Committee, Structural Engineers Association of California (SEAOC), San Francisco.

Sheikh, S.A., and Uzumeri, S.M. (1980). "Strength and ductility of tied concrete columns". J. Struct. Div., ASCE, 106(5), 1079-1102.

Sheikh, S.A. and Yeh, C.C. (1990). "Tied Concrete Columns Under Axial Load and Flexure". J. Struct. Engrg., ASCE, 116(10), 2780-2800.

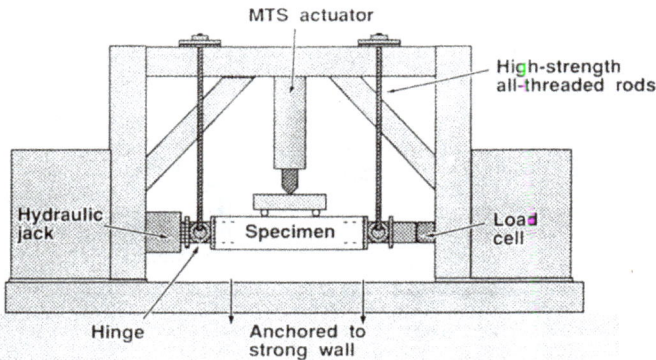

Figure 1 - Test Setup

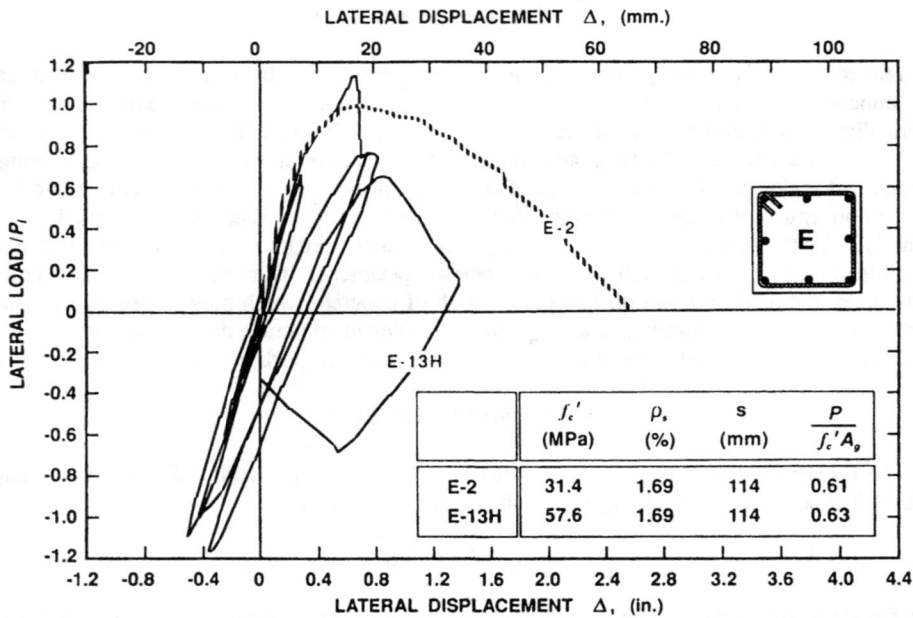

Figure 2 - Effect of Concrete Strength on Load-Deflection Behaviour
of Configuration "E" Specimens

	f_c' (MPa)	ρ_s (%)	s (mm)	$\dfrac{P}{f_c'A_g}$
E-2	31.4	1.69	114	0.61
E-13H	57.6	1.69	114	0.63

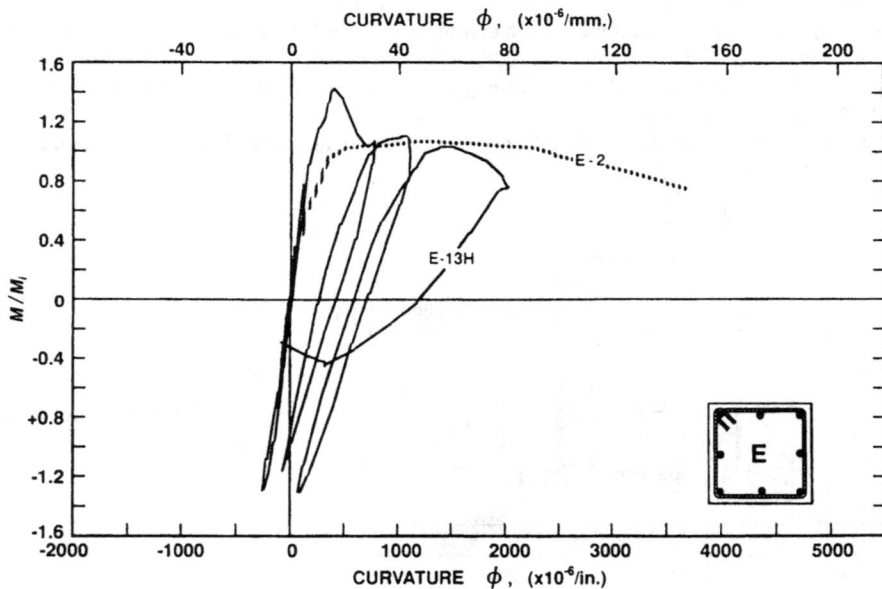

Figure 3 - Effect of Concrete Strength on Moment-Curvature Behaviour
of Configuration "E" Specimens

628

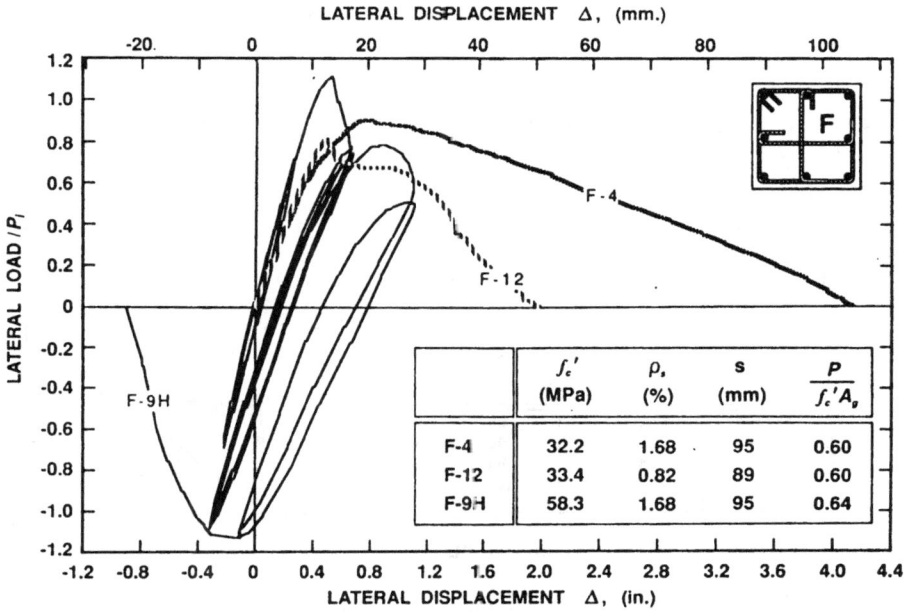

LATERAL DISPLACEMENT Δ , (mm.)

	f_c' (MPa)	ρ_s (%)	s (mm)	$\dfrac{P}{f_c'A_g}$
F-4	32.2	1.68	95	0.60
F-12	33.4	0.82	89	0.60
F-9H	58.3	1.68	95	0.64

LATERAL DISPLACEMENT Δ , (in.)

Figure 4 - Effect of Concrete Strength on Deflection Behaviour
of Configuration "F" Specimens

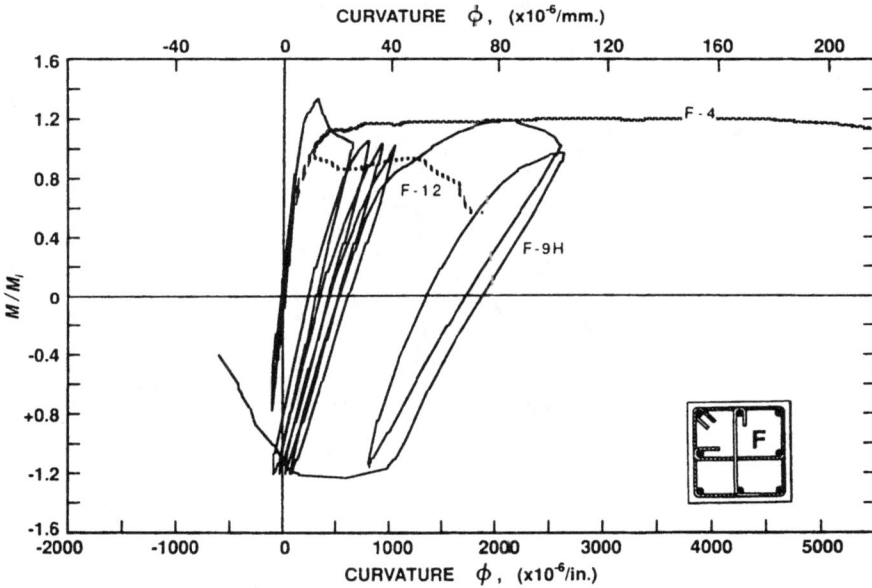

Figure 5 - Effect of Concrete Strength on Moment-Curvature Behaviour
of Configuration "F" Specimens

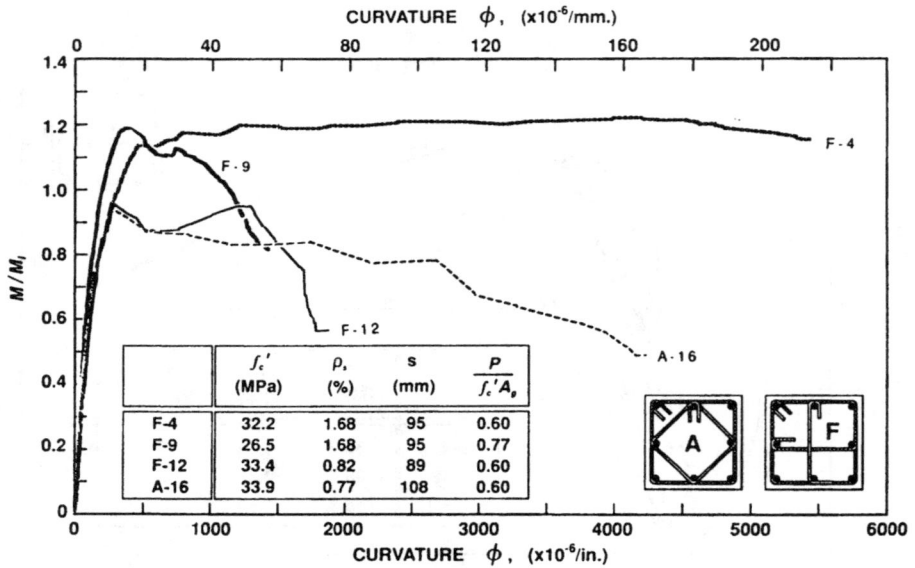

Figure 6 - Effect of Steel Configuration, Axial Load and Amount
of Lateral Reinforcement

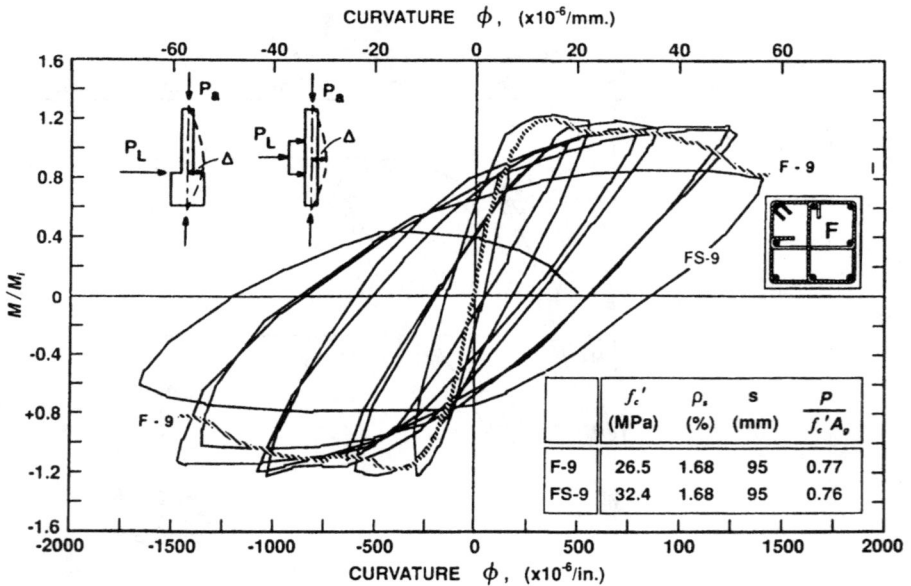

Figure 7 - Comparison of Specimens With and Without Stub

Jacketed columns subjected to combined axial load and reversed cyclic bending

U. Ersoy and T. Tankut[1]

ABSTRACT

Jacketed column behaviour is being investigated. Two of the four test series planned have been completed. The first series consisted of five specimens (two repaired, two strengthened and one monolithic reference) subjected to uniaxial loading. In the second series, five specimens (two repaired, one strenthened and two monolithic reference) were tested under combined constant axial load (the balanced axial load approximately) and either monotonically increased or reversed cyclic bending. In the third series in progress, effects of various bar (longitudinal reinforcement of the jacket) development techniques on the behaviour and strength are being studied. Tests indicated that the strengthening jackets generally performed satisfactorily, but the repair jackets were less successful, especially in the case of uniaxial loading.

INTRODUCTION

In the structural sense, rehabilitation can be defined as an operation to bring a structure (or a structural member) which does not meet the design requirements to the specified performance level. Depending on the state of the structure and the post-intervention performance level desired, rehabilitation is divided into two main categories. Repair is the rehabilitation of a damaged structure or a structural member with the aim of bringing the capacity back to the pre-damage level or higher. Strengthening is increasing the existing capacity of a non-damaged structure (or a structural member) to the specified level.

Load level at the time of intervention has a significant effect on the post-intervention performance. In practice repair/strengthening interventions in many cases are introduced while the structural member is still under load and sometimes after unloading the member by jacking. With these considerations, rehabilitation interventions can be classified in four main groups plus an auxiliary one.

a.   Repair under load (post-damage)
b.   Unloaded repair (post-damage)
c.   Strengthening under load (no damage)
d.   Unloaded strengthening (no damage)
e.   Soft interventions (non-structural measures and/or load limitations)

---

[1]Professors, Department of Civil Engineering, Middle East Technical University

A group of researchers led by the authors have been actively working on various aspects of the problem for the last four years. Current research projects being carried out by this team at Middle East Technical University and Gazi University in Ankara are concerned with post-intervention behaviour and strength of various structural members (columns, beams and slabs) and structural systems. Due to space limitations, only one major project (jacketed column behaviour) is reported in the present paper. A brief summary of the work related to the other projects is presented in another paper (Tankut and Ersoy 1991).

Jacketing is a technique widely used for repair or strengthening of reinforced concrete columns. Column cross section is enlarged by forming a jacket around the existing column, and additional longitudinal and transverse reinforcement is provided. The basic concept of jacketing is simple, but the actual behaviour involves many uncertainties. Questions such as,

* How is the load shared by the jacket and the column ?
* Does the jacket take a share from the present load ?
* When does it take a share from the new load ?
* How is the interaction with the neighbouring members ?
* Where and how should the reinforcement be developed ?
* How and to what extend can the bending capacity be improved ?

can not be easily answered, and they require further research. The present investigation was designed to obtain information in this respect.Only the experimental aspect of the work is discussed in the following paragraphs. The analytical work is not included due to space limitations.

SPECIMENS

In practice, jacketing is used mainly for increasing the axial load capacity. This relatively simple and basic case was investigated first (Aksan 1988, Tankut, Ersoy and Aksan 1989) and the first series consisted of four identical specimens representing four different jacket types and a monolithic specimen serving as reference,Table 1. Five specimens of the second series (Suleiman 1991, Tankut and Ersoy 1990) were designed to investigate the behaviour of repair and strengthening jackets under monotonic and reversed cyclic bending moment combined with axial load, Table 2. The dimensions and reinforcement of the specimens are given in Figure 1 and Table 3.

TESTS

Uniaxial loading tests (Series 1) were performed in a horizontal closed frame consisting of two reinforced concrete beams connected with steel tension bars. Axial strains were measured by four dial gauges along the two opposite faces of the specimen as shown in Figure 2. Dial gauge frames were attached to anchor bars embedded in the core. This kind of attachment enabled reliable readings beyond cover crushing and even after buckling. The test set-up for the combined bending and axial load series is shown in Figure 3. The axial load was applied by the universal testing machine while the bending moments were introduced by side loads applied by hydraulic jacks through prestressing cables. Instrumentation was basically the same as that of the uniaxially loaded series. However, measurement of the column midheight deflection was essential in this case, since second order moments could be very significant. A roller-to-roller stretched wire marking the deflection on a scale attached to the specimen was used as shown in Figure 3.

In Series 1 (uniaxial loading), following a few low level loading and unloading cycles (to check the alignment and minimize accidental eccentricity), each basic column was loaded upto a pre-determined load or deformation level at which jacketing was introduced. When the jacket concrete acquired a reasonable strength, jacketed column was loaded monotonically upto failure.

Mechanical bar development is a much preferred method of column repair/strengthening in practice. This method was used in all specimens Series 2 (combined bending and axial load). Brackets were formed by tightening two angles placed on top and bottom faces of the beam by a pair of high strength steel pull bars, and the longitudinal bars of the jacket were welded to these brackets.

In all tests of the combined bending and axial load series (basic columns and jacketed specimens alike) an axial load of 500 kN was applied first and kept constant throughout the test. This axial load corresponded approximately to 70% of the axial load capacity of the basic column or 35 % of that of the monolithic column. Bending moment was then introduced gradually. In monotonic loading cases, bending moment was increased upto failure. In cases of cyclic loading, moment was reversed with gradually increasing magnitudes. After testing the basic column to a predetermined level, the specimen was unloaded; the loose concrete was taken out and the jacket was cast. When the jacket concrete gained adequate strength, the jacketed specimen was tested in accordance with the predetermined load history. Tests results are given in Figures 4 and 5 in terms of moment-curvature and moment-strain diagrams.

In a third series (Yumak 1991), the effect various bar development techniques on the behaviour and strength are being investigated by performing similar combined bending and axial load tests on similar test specimens in which longitudinal reinforcement of the jacket is developed in different ways.

CONCLUSIONS

The following points seem to be valid within the limitations of the data obtained in the study.

Under uniaxial load,

a. Strengthening jackets proved to be very effective. Their axial load capacity, ductility, energy dissipation capacity and stiffness were quite close to those of the monolithic reference specimen. Both types (made unloaded and made under load) were equally good.
b. Repair jacket made unloaded was a little inferior to strengthening jackets; however, it was still quite successful.
c. Repair jacket made under load performed rather poorly displaying an axial load capacity around half of that of the monolithic reference specimen.

Under combined bending and axial load,

a. All the jacketed specimens performed satisfactorily under both monotonic and reversed cyclic loading.
b. Strengthening jacket made unloaded was very effective in improving the flexural capacity under constant axial load.
c. Repair jackets made unloaded were slightly less effective than the strengthening jacket.
d. All the test specimens displayed considerable ductility and satisfactory energy dissipation capacity.

REFERENCES

1. Tankut, A.T., Ersoy, U. 1991. Rehabilitation of Concrete Structures - Research in Progress, METU, Turkey. Proceedings, ACI International Conference on Evaluation and Rehabilitation of Concrete Structures and Innovations in Design, Hong Kong.
2. Aksan, B. 1988. Jacketed Column Behaviour Under Axial Load. M.Sc. Thesis, Middle East Technical University, Ankara, Turkey.
3. Tankut, A.T., Ersoy, U. and Aksan, B. 1989. Repair and Strengthening of Reinforced Concrete Columns. Seminar on Assessment and Redesign of Concrete Structures, Comité Euro-International du Béton, İzmir, Turkey.
4. Suleiman, R.E. 1991. Jacketed Column Behaviour Under Combined Bending and Axial Load. PhD Thesis, Middle East Technical University, Ankara, Turkey.
5. Tankut, T., Ersoy, U. 1990. Behaviour of Repaired/Strengthened Reinforced Concrete Columns. Proceedings, Ninth European Conference on Earthquake Engineering, Moscow, USSR.
6. Yumak, Y. 1991. Effects of Bar Development Methods on Jacketed Column Behaviour. MSc Thesis, Middle East Technical University, Ankara, Turkey.

Table 1 : Uniaxial loading specimens (Series 1)

Spec.	Jacket Type	Load at Jacketing
M	Monolithic	-
US	Strengthening	Unloaded
LS	Strengthening	Under Load
UR	Repair	Unloaded
LR	Repair	Underload

Table 2: Combined bending and axial load specimens (Series 2)

Spec.	Jacket Type	Test Loading		Type of Loading Simulated	
		Basic	Jacketed	Basic	Jacketed
MBM	Monolithic	--	Monotonic	--	Reference
MBR	Monolithic	--	Rev. Cyc.	--	Reference
RBM	Repair	Monotonic	Monotonic	Gravity	Gravity
RBR	Repair	Rev. Cyc.	Rev. Cyc.	Seismic	Seismic
SBR	Strengthening	Rev. Cyc.	Rev. Cyc.	Seismic	Seismic

Table 3: Dimensions and reinforcement of column jacketing specimens

Properties	Uniaxial Loading		Bending+Axial Load	
	Basic	Jacketed	Basic	Jacketed
Dimensions(mm)	130x130	180x180	160x160	230x230
A_{conf}/A_{gross}	0.82	0.87	0.83	0.95
Long. Steel	$4\phi 10mm$	$8\phi 10mm$	$4\phi 12mm$	$8\phi 12mm$
ρ_l	0.0186	0.0186	0.0177	0,0177
Ties	$\phi 4/40mm$	$\phi 4/40mm$	$\phi 4/100mm$	$\phi 8/100mm$
ρ_s	0.0085	0.0085	0.0029	0.0097

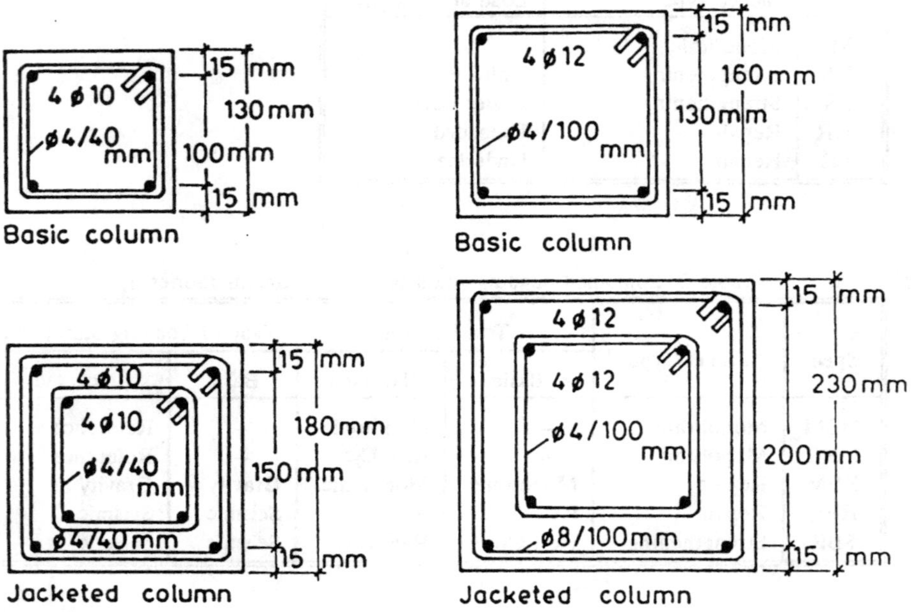

a. Uniaxial loading   b. Axial load & bending

Figure 1 : Sectional properties of test specimens

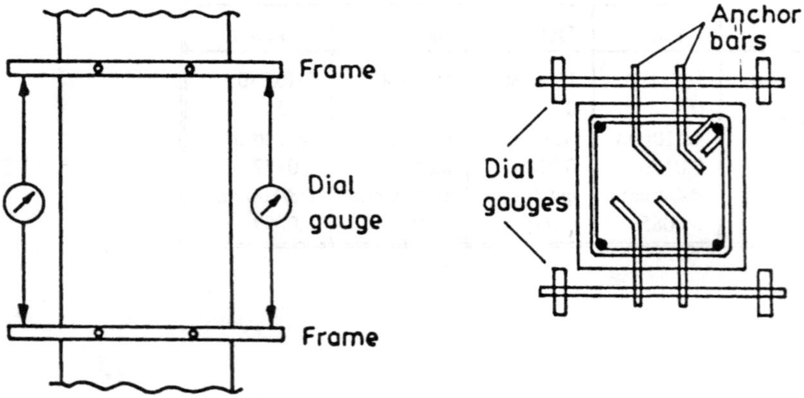

Figure 2 : Strain measurements

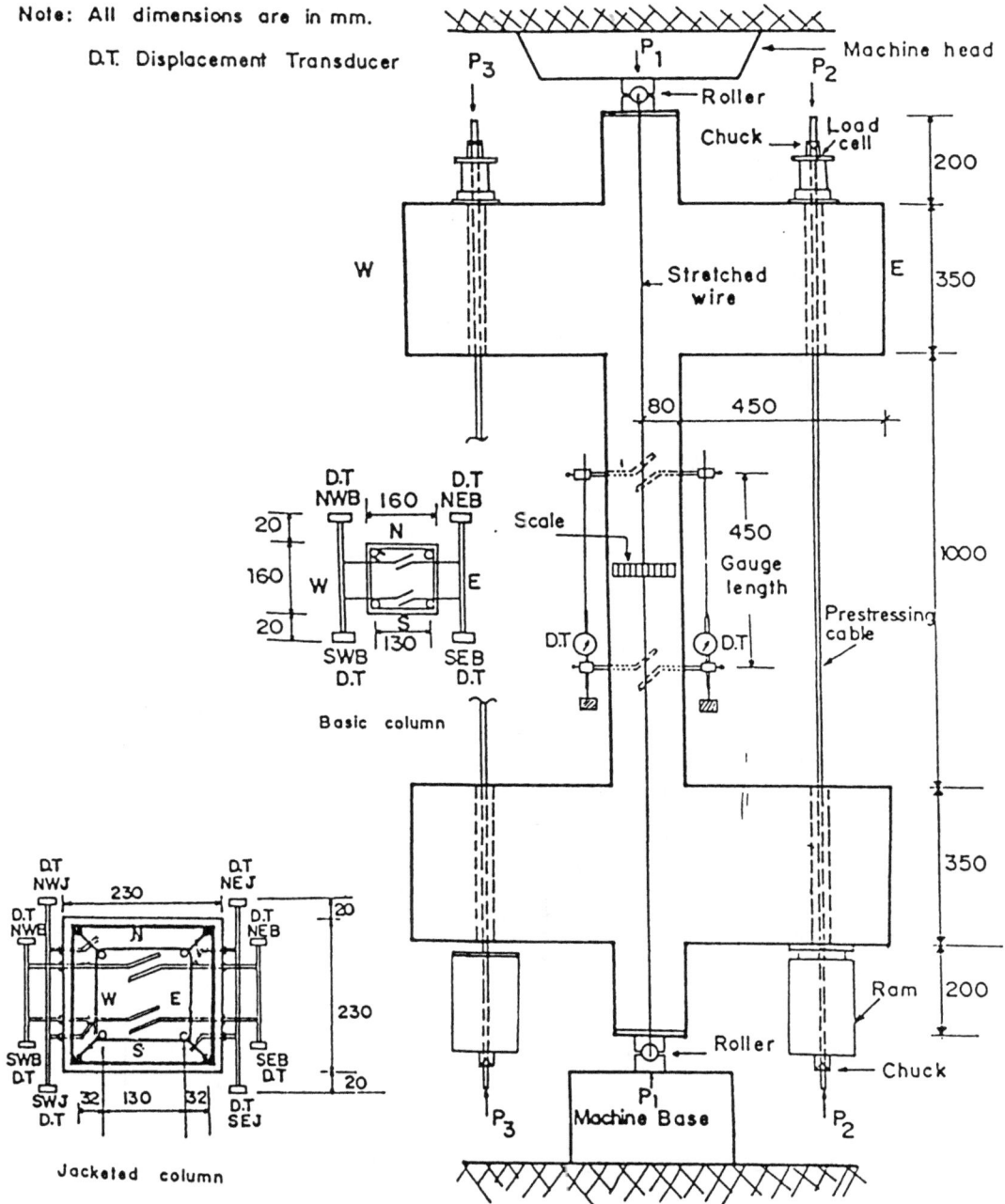

Figure 3: Combined loading test set-up

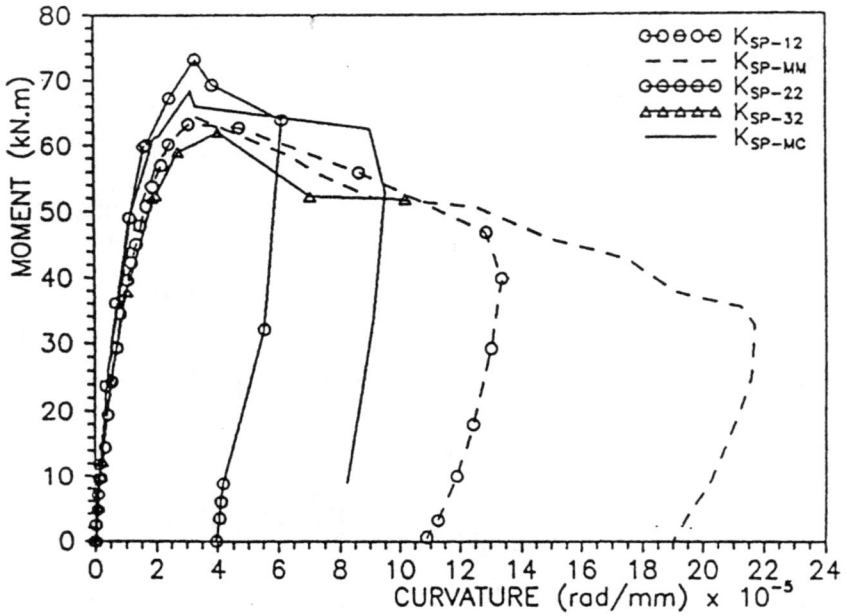

Figure 4 : Envelope (M-K) diagrams (Axial load & bending)

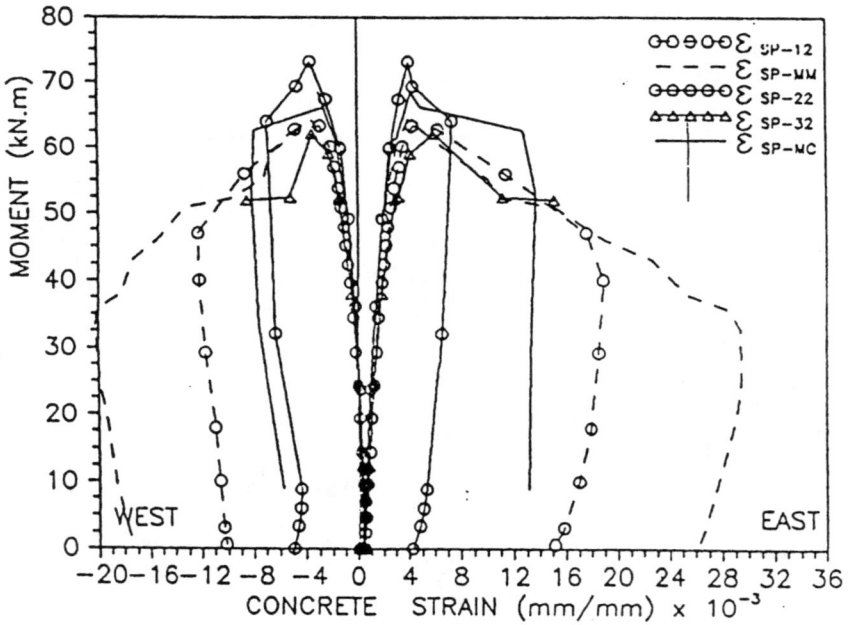

Figure 5 : Envelope (M-ε_c) diagrams (Axial load & bending)

638

Shake table test of a one-eighth scale three-story reinforced concrete frame building designed primarily for gravity loads

Adel G. El-Attar[I], Richard N. White[II], Peter Gergely[III], and Timothy K. Bond[IV]

ABSTRACT

Results of a 1/8 scale, 3-story, three-bay by one-bay lightly reinforced concrete (LRC) bare frame building tested on the Cornell University shake table are presented. The reinforcement details of the model were based on typical reinforced concrete frame structures constructed in the Central and Eastern United States since the early 1900's, in which the design was based primarily on gravity loads without regard to significant lateral forces. Special attention was paid to the duplication of the characteristic reinforcement details of this kind of building, especially at critical sections, such as joint regions and splices.

During the seismic tests, the model building showed a high degree of flexibility associated with a significant P-Δ effect and considerable stiffness degradation. The inadequate non-seismic reinforcement details did not form a serious problem to the model as most of the damage occurred in the columns, outside the joint regions. The building finally collapsed in a soft-story mechanism that took place in the first story columns. The seismic response of the model was compared with the predicted response obtained using program IDARC (Park et al 1987). The comparison indicated a very good agreement between predicted and measured global responses (top story displacement and base shear). Predicted specific member responses did not correlate well with the measured response.

I- Post Doctorate Associate, School of Civil and Environmental Engineering, Cornell University, Ithaca, N.Y. 14853.

II- James A. Friend Family Professor of Engineering, School of Civil and Environmental Engineering, Cornell University, Ithaca, N.Y. 14853.

III- Professor of Structural Engineering, School of Civil and Environmental Engineering, Cornell University, Ithaca, N.Y. 14853.

IV- Manager of Technical Services, George Winter Laboratory of Structural Engineering, School of Civil and Environmental Engineering, Cornell University, Ithaca, N.Y. 14853.

INTRODUCTION

The present investigation is part of a comprehensive research effort currently underway at Cornell University on the damage assessment and performance evaluation of LRC buildings subjected to seismic loads. The experimental work includes both full-scale and small-scale component tests in addition to small-scale complete building tests. In an early stage of this project, better microconcrete and model reinforcement were developed to enhance the simulation of reinforced concrete prototype specimen responses (hysteretic response, cracking patterns, and ultimate strength) (Kim et al 1988). The newly developed materials were used in a 1/6 scale 2-story office building model tested on the Cornell University shake table (El-Attar et al 1991 b). Test results of this model indicated the high flexibility and stiffness degradation associated with the discontinuous positive moment beam reinforcement pullout. The same model materials (with minor modifications) were used for the 3-story model. This model represented a more general case than the 2-story model, in that both exterior and interior joints were included in addition to the greater number of modes to be activated.

The main thrust of the experiment reported here is (a) to introduce some behavioral aspects of LRC structures during earthquakes, with special emphasis on the role of the non-seismic reinforcement details, and (b) to assess the reliability of one of the existing numerical modeling techniques (program IDARC) in predicting the response of this type of building.

DESCRIPTION OF THE TEST STRUCTURE

The test structure was a 1/8 scale true replica model of the prototype 3-story office building shown in Fig. 1. The model story height was 18", with a main frame span of 27". Members sizes were: column section 1.5" × 1.5", beam section 1.125" × 2.25", and a 0.75" thick slab. Reinforcement details of the model building are shown in Fig. 2; note the discontinuous positive beam reinforcement at the columns, the lack of confinement steel in the joints, and the column splice location.

The total weight of the structure was increased by adding lead blocks to meet the dead load similitude conditions. Special attention was paid to mounting the blocks to minimize the stiffening of the floor slab. In addition to the acceleration and displacement measurements at each floor level, the model structure was instrumented with internal force transducers at the mid-height of the first and the second story columns. Details of the similitude requirements, model design and fabrication, model materials, loading technique, and model instrumentation are provided in El-Attar et al (1991 a).

MODEL MATERIALS

The model microconcrete had mix proportions by weight of 0.95 : 1 : 3.6 : 2.4 (water : cement : model sand : model aggregate), where model sand was defined as particles passing #8 sieve and retained on #200 sieve, and model aggregates were defined as particles passing #6 sieve and retained on #8 sieve. At the time of model testing, the microconcrete had an axial compressive strength (f'_c) of 3.80 ksi and a splitting tensile strength (f'_t) of 0.34 ksi.

Threaded steel rods were used as longitudinal reinforcement in both beams and columns. The used sizes ranged from 0.099" dia. (3-48) to 0.164" dia. (8-32). All bars were heat treated to achieve the desired yield stress of about 40 ksi and to develop an adequate yield plateau. Shear reinforcement was provided by 0.05" dia. annealed steel wires.

TEST PROCEDURE

The model structure was subjected to four seismic tests using the time scaled Taft 1952 S69E earthquake at amplitudes of 0.05g, 0.18g, 0.35g, and 0.80g. Each seismic test was preceded and followed by a static test and a free-vibration test to determine the change in the structure properties (such as the fundamental period, damping ratio, etc.).

DISCUSSION OF TEST RESULTS

Global response

Story displacements and story shears recorded during run 0.35g are shown in Fig. 3 (a) and (b), respectively. It can be seen from both figures that all three stories were moving in phase, indicating the domination of the first mode of vibration. A brief summary of the seismic tests is provided in Table 1, where it can be seen that during run 0.18g, the model showed a large degree of flexibility associated with a high stiffness degradation (18% reduction of the fundamental frequency $\equiv [\sqrt{\frac{2.2}{1.8}} - 1] \times 100 \equiv 50\%$ stiffness reduction). The maximum base shear during this run (1.252 kips) represented 8.8% of the total load on the structure; the base shear was only 15% less than the model capacity. The damping ratio ζ expressed as a percentage of the critical damping ζ_{cr} (assuming a viscous damping model), increased significantly during this run due to the development of new cracks.

After the 0.35g run, the model fundamental frequency decreased to 1.65 Hz (25% reduction), indicating a 78% stiffness reduction. No significant change was detected in the damping ratio, indicating that few new cracks were developed during this run. The maximum base shear recorded during this run (1.384 kips) represented 97% of the model capacity.

641

Story shears and mode shapes recorded at the moment of maximum base shear are shown in Table 2. It can be seen that, despite the high non-linearity of the model, the mode shape remained essentially unchanged during all seismic tests. Also, except for the first low amplitude seismic test, the shear force distribution over the three stories remained the same for all subsequent runs. The model collapsed during run 0.80g in a soft story mechanism in the first story columns. Failure was initiated at one of the interior columns, followed by failure of the rest of the first story columns.

Local response

Cracks were detected at the top and bottom sections of the first and second story columns (especially at the structural hinge regions) after the 0.18g run. These cracks were localized at these areas and did not spread over the columns length even after the 0.35g run. No visual damage was detected in the beams, joint regions, or the splice areas, indicating that the non-seismic reinforcement details did not form a potential source of damage to this particular building.

The large degree of flexibility of the model resulted in a pronounced P-Δ effect. During the 0.18g run, the base shear obtained from the column load cell readings was 27% larger than that obtained from the story acceleration; the same effect was measured in all subsequent runs. It was also noticed that each column share of the total story shear was heavily dependent on its axial force. All columns showed a much larger flexural strength than that obtained using conventional flexural capacity calculations. The increase in strength was attributed to several factors such as the strain hardening of the model reinforcement and the strain rate effects.

Comparison with the analytically predicted response

The top story displacements and the base shears computed using program IDARC (Park et al 1987) for run 0.35g are plotted against the experimentally measured responses in Fig. 4 (a) and (b), respectively. It can be seen from both figures that the anaytical results were in good agreement with the experimental results. At the local level, column shears obtained using IDARC were less than the measured shears because the large P-Δ effect was ignored in the analysis. Individual column shares of the total story shears were also inaccurately predicted due to neglecting the effect of the change in the columns axial force on their yield moment.

SUMMARY AND CONCLUSIONS

A 1/8 scale 3-story, three-bay by one-bay bare frame LRC office building was tested on the Cornell University shake table under increasing versions of the Taft 1952 S69E earthquake. The building was designed to resist primarily gravity loads. The following conclusions may be drawn:

1. Lightly reinforced concrete buildings may be subjected to very large deformations associated with a significant reduction in stiffness during a moderate earthquake.

2. Although the non-seismic reinforcement details can form a potential source of damage to LRC buildings, they are probably not sufficient to develop a failure mechanism.

3. Due to their high flexibility, P-Δ effect is significant in LRC structures and should be considered in the analysis.

4. Low and medium rise LRC structures may be subjected to potential collapse in a soft story mechanism due to the higher flexural strength of the beams with respect to the columns.

ACKNOWLEDGEMENTS

This research was sponsored by the National Center For Earthquake Engineering Research, SUNY Buffalo, N.Y. The Center is funded by NSF Grant N. ECE 86-07591 and by a number of other sponsors including the State of New York.

REFERENCES

El-Attar, A.G., White R.N., Gergely, P., and Bond T.K., "Shake Table Test of a 1/8 Scale 3-Story Lightly Reinforced Concrete Building", NCEER Technical Report to be published in 1991.

El-Attar, A.G., White R.N., Gergely, P., and Conley, C., "Shake Table Test of a 1/6 Scale 2-Story Lightly Reinforced Concrete Building", NCEER Technical Report to be published in 1991.

Kim, W., El-Attar, A.G., and White, R.N., "Small Scale Modeling Techniques for Reinforced Concrete Structures Subjected To Seismic Loads", Technical Report NCEER-88-0041, November 22, 1988.

Park, Y.J., Reinhorn, A.M., and Kunnath, S.K., "IDARC: Inelastic Damage Analysis of REinforced Concrete Frame-Shear-Wall Structures", National Center for Earthquake Engineering Research. Technical Report NCEER-87-0008, July 20, 1987.

Table 1: Summary of Seismic Test Results

Test Amplitude (g)	Top Story Drift %	Max. Base Shear (kips)	Post-test Fundamental Frequency (Hz)	Post-test Damping Ratio % of ζ_{cr}
0.05	0.19%	0.338	2.20	1.30%
0.18	2.02%	1.252	1.80	2.74%
0.35	2.84%	1.384	1.65	2.76%
0.80	—	1.430	—	—

Table 2: Mode Shapes and Shear Distribution at The Maximum Base Shear

Run	Mode Shape	Shear Distribution
Taft 0.05-G	0.071" (100%) 0.055" (76%) 0.032" (45%)	0.110 kips (32%) 0.213 kips (63%) 0.338 kips (100%)
Taft 0.18-G	0.725" (100%) 0.571" (79%) 0.332" (46%)	0.580 kips (46%) 1.077 kips (86%) 1.252 kips (100%)
Taft 0.35-G	1.042" (100%) 0.824" (79%) 0.474" (46%)	0.646 kips (47%) 1.232 kips (89%) 1.384 kips (100%)

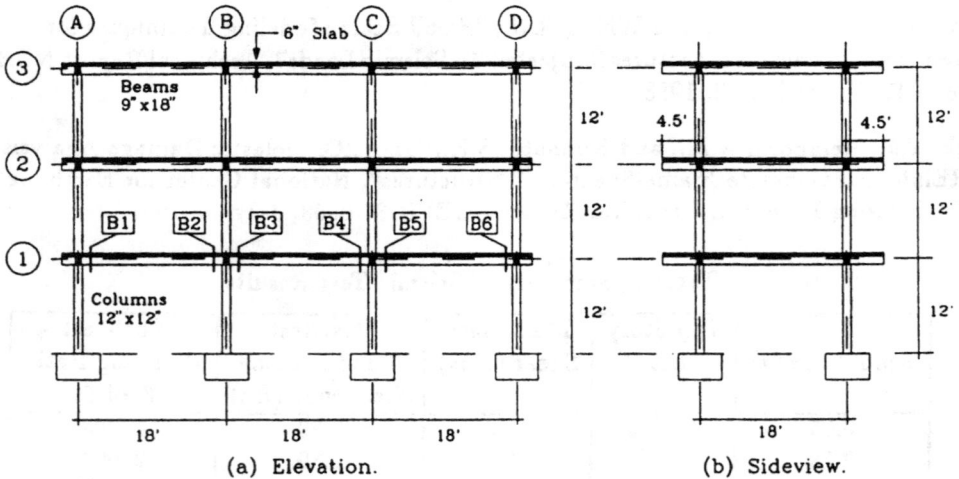

(a) Elevation.   (b) Sideview.

Figure 1: General Layout of The 3-Story Prototype Building.

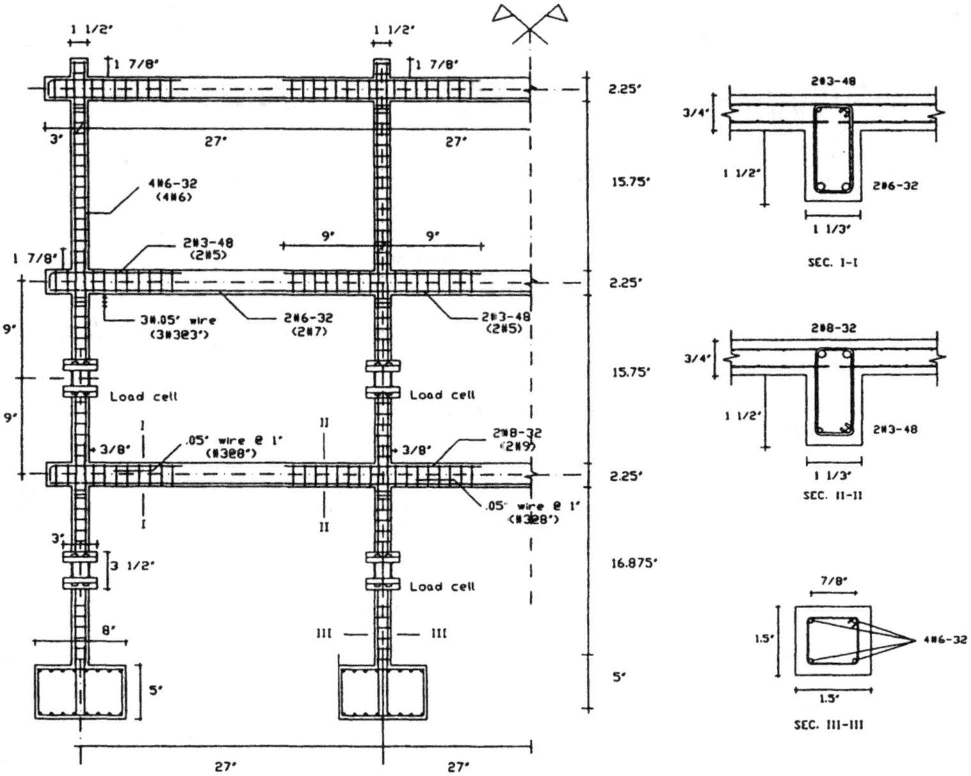

(a) Reinforcement Detail of The Model (Prototype) Main Frame.

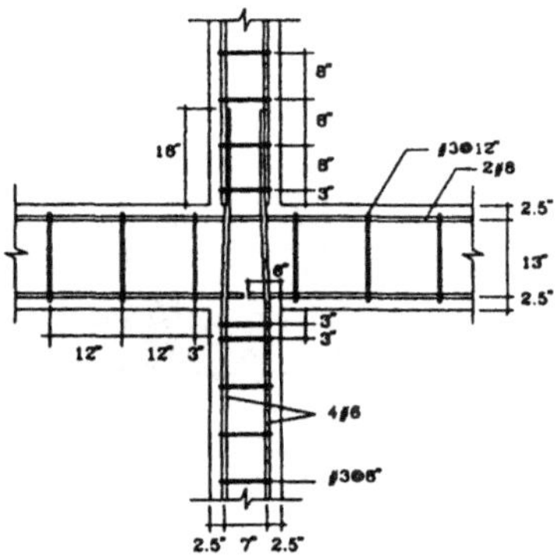

(b) Prototype Interior Joint

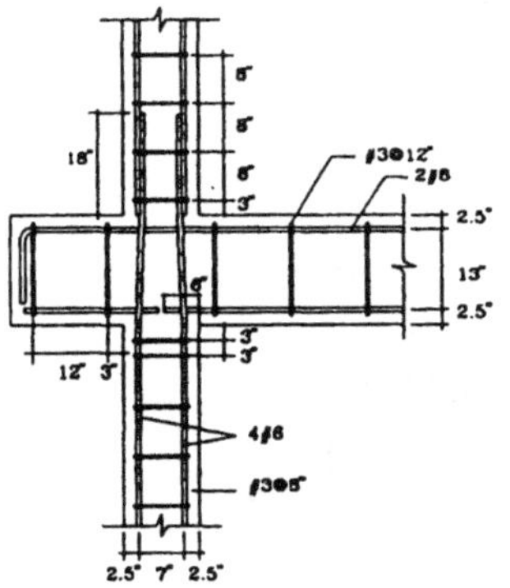

(c) Prototype Exterior Joint

Figure 2: Reinforcement Details of The 3-Story Office Building.

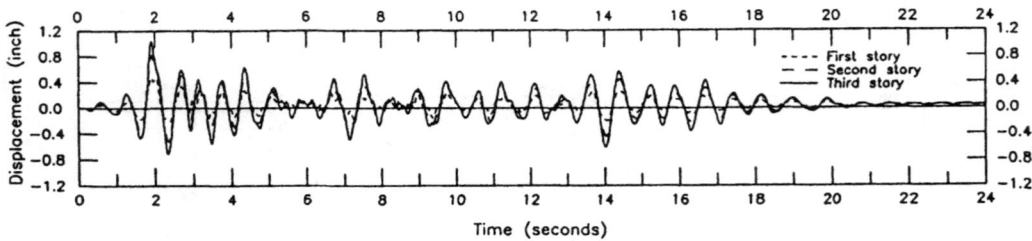

(a) Story Displacements.

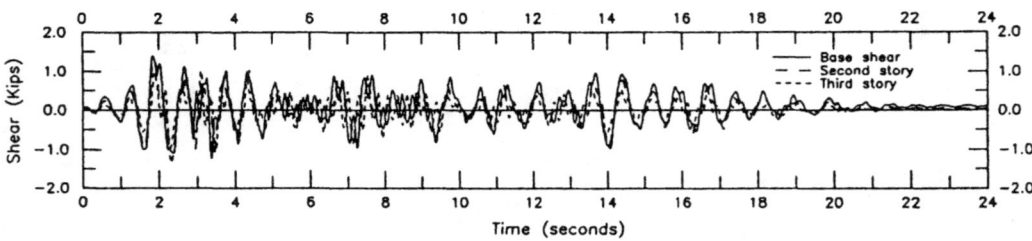

(b) Story Shears.

Figure 3: Model Response at Run Taft 0.35 g.

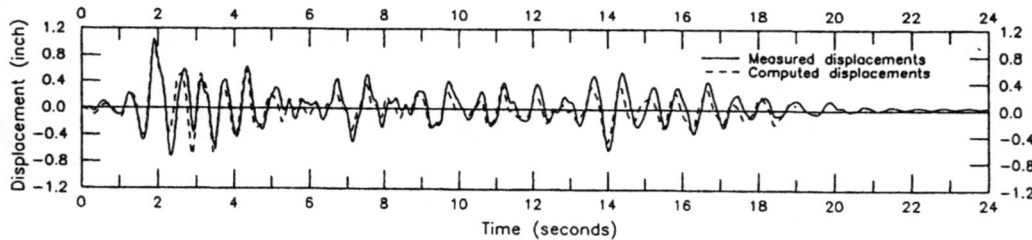

(a) Top Story Displacement.

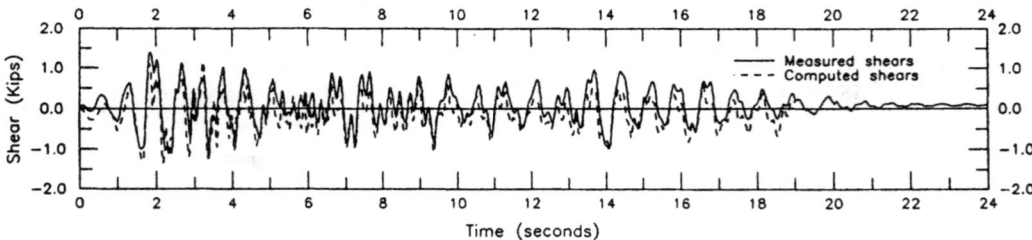

(a) Base Shear.

Figure 4: Computed Versus Measured Structure Response (Run Taft 0.35 g).

646

Experimental study of slab-wall connections

I. Imran[1] and S.J. Pantazopoulou[2]

ABSTRACT

This paper presents results of an experimental study on seismic behaviour of interior slab to wall connections. In the study, a half-scale model of a typical slab-wall connection was tested under the combination of cyclic lateral loads and distributed vertical loads (which simulated service gravity loads of the slabs in the full-scale prototype). Test results indicated that the connection between the slab and the wall experienced severe damage at early stages of the test, a result likely to influence the internal force distribution patterns occurring in indeterminate structures under lateral loads. Damage was accentuated by the presence of gravity load, and was manifested by the formation of shear sliding planes in the slabs.

INTRODUCTION

The study presented herein was motivated by the current design requirements for slab-wall and frame-wall structures which primarily consider the base of the walls as the critical sections for lateral load design. This practice reflects the assumption that the lateral loads are mainly resisted by the stiffer vertical elements of the structure, i.e. the shear-walls. In reality, the base-shear capacity of a structural wall can only be mobilized if the horizontal diaphragms of the structure are sufficiently stiff in their plane, so that lateral forces are distributed to the vertical elements (columns and walls) according to their lateral stiffness. In most reinforced concrete frame-wall structures, a large portion of the total mass is accumulated at the floor slabs of the structure; therefore, accelerations induced by earthquakes are likely to cause the development of large inertia forces in the slabs. If for the imposed lateral displacement levels the slab-wall connection possesses insufficient stiffness and strength, it is likely that horizontal forces will be diverted to the other vertical elements of the structure (e.g. columns); in such case, lateral forces will not be resisted by the wall, even if the wall possesses the desired strength, and therefore it is conceivable that in such circumstances it may not be possible to maintain the

[1]Grad. Research Assistant, Dept. of Civil Engrg., Univ. of Toronto, M5S-1A4, CANADA

[2]Assistant Professor, Dept. of Civil Engrg., Univ. of Toronto, M5S-1A4, CANADA

desirable hierarchy of failure in the structure.

Behaviour of the slab-wall connections is characterized by two features:
1) When horizontal diaphragms are flexible in their planes, the mechanism of in-plane shear transfer occurring between floor slabs and walls is governed by deep beam action developing in the slabs. Because of the deep beam action, a portion of the vertical interface between slab and walls is under normal tension, and is therefore expected to crack under moderate levels of lateral drift. Cracking is likely to cause redistribution of in-plane shear stresses in the flexural compression zone (Fig. 1).
2) The gravity loads which are inevitably present in the slab, induce negative moments at the supports. Associated with the normal tension which results from the flexural action of the moments are the following effects: a) the available preyielding range of the longitudinal slab reinforcement is reduced, and b) vertical flexural cracks which form at the connection, reduce the area of contact between slab and wall, over which shear transfer can take place.

The objective of this paper is to present the results of an experimental study of the lateral load behaviour of a slab-wall connection. The study focused on the characteristics of the deep beam action and gravity load effects summarized in the preceding. In the experimental program, a half scale model of a typical connection was tested under various levels of loading up to failure in order to obtain the load-deformation characteristics of the model and to identify its failure mechanisms.

EXPERIMENTAL PROGRAM

A series of experiments investigating the behaviour of slab-wall connections which had been previously carried out by Nakashima et al. (1981), (1982), had shown that in cases of strong walls and flexible slabs, failure occurring in the slab close to the connection can be the mechanism which controls the pattern of lateral load distribution through the structure. In these specimens longitudinal slab reinforcement was designed according to the standard practice, and was therefore cut-off at the location where moments associated with gravity loads vanished. However, during testing it was concluded that in-plane shear and bending moments which developed under the action of lateral loads were significant in that region. As a result, the section along the boundary where the longitudinal slab reinforcement was cut-off, was a weak region for in-plane shear transfer, and eventually controlled the failure of the specimen.

Because of the formation of a full-depth sliding plane at the location where reinforcement was cut-off in Nakashima's tests (1981), and contrary to common practice, top and bottom reinforcement was continuous over the point of zero gravity moment in the slabs of the specimen tested at the University of Toronto. The specimen was a half-scale model of a typical interior slab-wall connection (Fig. 2), and was subjected to the lateral load history shown in Fig. 3. The loads were introduced at the ends of the slabs, in order to simulate development and transfer of inertia forces from the slab to the wall. Displacement levels corresponding to 0.25, 0.5, 0.75, 1., 1.5, 2. and 3 times the theoretical yield displacement of the specimen were subsequently applied. (Theoretical yield displacement for the specimen, computed including both flexural and shear deformations was 7 mm, from the tip of the slab to the base of the wall).

648

The design of the specimen followed the CSA Standard CAN3-A23.3-M84 (1984). The floor slab was reinforced with 8 mm diameter bars at 175 mm and 240 mm o.c. in the longitudinal direction (top and bottom respectively). In addition, a minimum of $0.002*A_{gross}$ was provided as transverse reinforcement in the slab (6 mm diameter bars at 250 mm top and bottom). During the test, the self weight and the added distributed weights over the slabs amounted to 5.8 Kpa, which is approximately the load required to produce the cracking flexural moment at the face of the interior support. The span of the floor slab in the specimen corresponded to the approximate location of inflection points resulting under the action of gravity loads. The shear wall was designed to develop a maximum shear stress not exceeding approximately $0.5\sqrt{f'_c}$ Mpa (for ductile failure). The wall contained 0.37% and 0.54% vertical and horizontal web reinforcement respectively (2 curtains consisting of 8 mm horizontal bars at 150 mm o.c., and 8 mm vertical bars at 220 mm o.c.). The boundary elements contained 2% of the gross area of the wall in vertical reinforcement. Details of the slab and wall reinforcement are presented elsewhere (Imran 1990).

Concrete used for fabrication of the specimen had a cylinder compressive strength of 32.5 Mpa, and splitting strength of 3 Mpa. Reinforcing bars had a nominal yield strength of 400 Mpa, but actual yielding for the 8 mm diameter bars occurred at 607 Mpa, whereas for the 10 mm diameter bars, at 465 Mpa. The respective ultimate strengths were 680 and 630 Mpa. In addition to LVDT's, and reinforcement strain gauges, Zurich targets were placed on all the faces of the specimen on a 200 mm grid, and the configurations of these targets were monitored throughout the test in order to obtain surface strains of concrete.

During the test, the floor-slabs were propped at their free ends by means of roller supports. The specimen was post-tensioned to the Laboratory strong floor in order to simulate full fixity at its base.

RESULTS

The floor slabs of the specimen experienced a sharp stiffness reduction even at small-amplitude cycles corresponding to displacement levels of 0.5 Δ_y. The stiffness reduction was accompanied by formation of flexural-shear cracks in the slab, at the connection with wall; cracks forming in the upper part of the wall were continuation of the slab cracks in the vicinity of the connection (Fig. 4).

Full-depth major cracks, (resembling the sliding cracks which had been earlier reported by Nakashima (1982)), developed in the slabs parallel to the wall, but contrary to the reference tests these cracks did not control the failure mechanism of the specimen. The first crack was located at the face of the connection with the wall; the second was located at a distance of approximately a quarter of the span from the interior support. This point also corresponds to the approximate location of the inflection line, or point of zero gravity moments in the slab; it is therefore the location of a state of pure vertical shear, which acts in combination with the in-plane shears induced in the slab during the test. For this reason, the point of inflection in the slab is a potential weak link, which may control the in-plane shear resistance of the horizontal diaphragm if longitudinal reinforcement is cut-off (as seen in Nakashima's tests).

At large levels of lateral displacement, widening of flexural cracks was observed at the interface of the slab-wall connection. The growth of the slab in its plane in the direction of the load appeared to affect the pattern of cracks in the wall (Fig. 4), and it is believed that might have also affected the wall strength. The increasing elongation of the slab was also evident from the fact that the strain recorded on the transverse reinforcement remained in tension regardless the direction of loading (Fig. 5). Furthermore, because the specimen was statically determinate, expansion in the slab was unrestrained in both the longitudinal and the transverse directions. However, such an expansion is likely to be partially restrained by the adjacent slab panels in a continuous structure, therefore causing development of internal actions which would likely affect the internal force distribution throughout the structure. For this reason, further study to understand the effect of nonlinear slab deformations on the overall behavior of the structure is necessary.

The hysteretic load displacement relationship of the specimen is plotted in Fig. 6 (displacements were measured at the tip of the east slab panel relative to the support). Yielding of the slab and wall reinforcement occurred simultaneously at 0.36% total lateral drift. The respective relative contributions of slab and wall were, 0.13% of the slab length, (measured relative to the face of the connection) and 0.22% of the wall height, (measured relative to the base). At that displacement level, the average shear stresses developing at the vertical and horizontal faces of the connection were $0.19\sqrt{f_c}$ and $0.3\sqrt{f_c}$ respectively. The maximum total lateral displacement of the specimen attained during the test was 1.25% of the wall height; at that stage, the average in-plane shear stress in the slab was $0.25\sqrt{f_c}$, while shear stresses in the wall reached $0.39\sqrt{f_c}$. Previous tests (Nakashima 1981) also indicated similar or lower levels of average shear stress developing at the vertical face of the slab-wall connection. It is believed that this reduced level shear resistance of the slab side of the connection was primarily caused by the influence of gravity loads and by the lack of the beneficial action of self-weight which is present in vertically oriented elements.

Because the specimen was statically determinate, the only path of loads was through the wall. The capacity of the specimen was therefore limited by a ductile flexural failure at the base of the wall; this failure was manifested by widening of existing flexural cracks at that location. Displacement ductility factor of the specimen at failure was 7.

Fig. 7 depicts the deflection profiles in the slab and wall associated with shear deformations; it is evident that although both slab and wall were planar diaphragms subjected to in-plane shear and a bending moment linearly increasing from the tip of each element towards its respective support, the two demonstrated significantly different behaviours. This is believed to be primarily the result of the different boundary conditions of the two elements. Because of the restraining which was provided to the wall by the slab, the pattern of cracking in the wall was completely different from what is commonly observed in single wall tests, where cracking primarily is concentrated in the region of maximum flexural moment at the base (Figs. 8a and 8b). It is believed that this result is of particular importance in evaluating experimental data obtained from tests of single isolated walls loaded by direct actions rather than being subjected to the indirect diaphragm action which is provided by the slab in actual structures.

At low levels of lateral displacement, the in-plane shear forces were essentially transferred uniformly from the slab to the wall through the connection. This was evident from the formation of a uniform cracking pattern along the wall side of the connection (Fig. 8b). However, at higher levels of lateral displacement, as the horizontal flexural cracks along the connection became wider, it is believed that the transfer of shear forces occurred largely in the flexural compression zone of the connection. This is apparent from the distribution of tensile stresses in the longitudinal slab reinforcement plotted in Figs. 9a and 9b. (The stresses were computed from the measured strain histories of the reinforcement using the hysteretic stress-strain model for reinforcing steel proposed by Menegotto and Pinto (1977)). It can be seen from those figures that the tensile stress difference between two adjacent cross-sections in a single bar was approaching zero as the lateral displacement was increased. This suggests that the contribution of the flexural tension region of the connection in transferring in-plane forces from the slab to the wall was practically very small.

CONCLUSIONS

The mechanism of shear transfer at slab-wall connections subjected to cyclic lateral loads was investigated. The study included data from an experimental investigation of an interior slab-wall connection tested at the University of Toronto, under combined vertical and lateral loads simulating earthquake effects. Data from other related tests were also considered when applicable. Results of the study are as follows:

1. Although the pattern of in-plane actions was similar in the slab and the wall, the two elements exhibited different behaviours in response to the applied lateral loads; these differences are believed to have resulted from the action of gravity loads on the slab, and from the different displacement constraints present in the two elements.

2. The experiment revealed that vertical loads affect the in-plane stiffness and shear resistance of the floor-slabs particularly in the vicinity of the slab-wall connection. The available experimental evidence also indicates that vertical shear forces generated from gravity loads reduce the in-plane shear resistance of the slab at critical boundaries where top or bottom longitudinal slab reinforcement is cut-off (points of zero gravity moment).

3. Widening of flexural cracks at the slab side of the connection at high levels of displacement was observed to cause redistribution of shear forces along the connection. The in-plane shear resistance of slab-wall connections at the vertical face of the support is limited to approximately $0.25\sqrt{f_c}$ Mpa; in all test cases considered, experimentally obtained shear resistances were only 60% of the nominal values computed using the ACI Code equations (1983).

ACKNOWLEDGMENTS

The work presented in this paper was carried out at the University of Toronto, Canada. Financial support for the study was provided by NSERC grant No. OGP0042033.

REFERENCES

ACI Committee 318. 1983. Building Code Requirements for Reinforced Concrete (ACI 318-83). Detroit, 111.

CSA Committee A23.3. 1984. Design of Concrete Structures for Buildings (CAN3-A23.3-M84). Rexdale, 281.

Imran, I. 1990. Preliminary Study of the Lateral Load Behaviour of A Reinforced Concrete Slab-Wall Assembly. M.A.Sc. Thesis, Department of Civil Eng., University of Toronto, Toronto, 198.

Menegotto, M. and Pinto, P. 1977. Slender R/C Compressed Members in Biaxial Bending. Intl. Journ. of American Civil Engineers, 103(3), 587-605.

Nakashima, M. 1981. Seismic Resistance Characteristics of Reinforced Concrete Beam-Supported Floor Slabs in Building Structures. Ph.D. Thesis, Lehigh University, Bethlehem, March, 333.

Nakashima, M., Huang, T., and Lu, L. W. 1982. Experimental Study of Beam-Supported Slabs Under In-Plane Loading. ACI Journal, 79(1), 59-65.

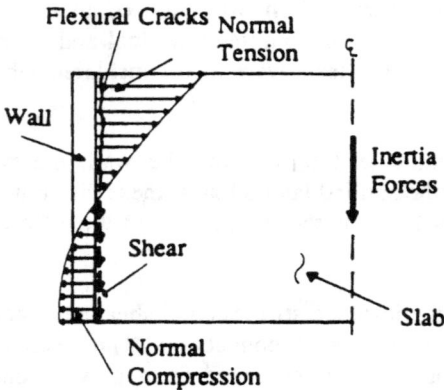

Figure 1. Deep beam action in the slab

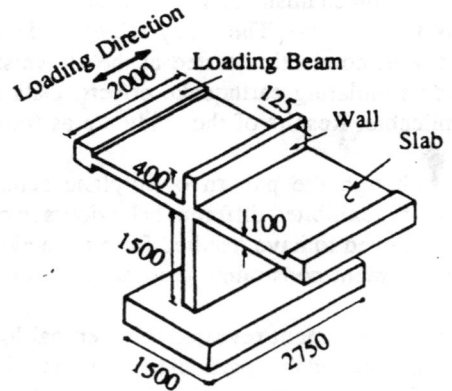

Figure 2. View of specimen

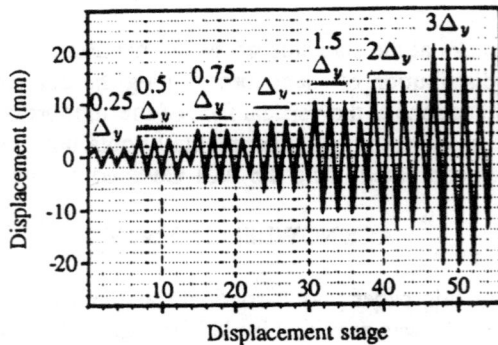

Figure 3. Displacement history

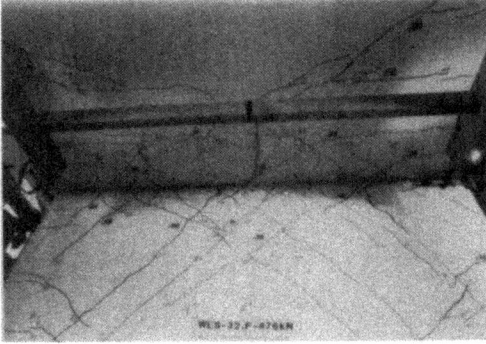

Figure 4. Cracking pattern at connection

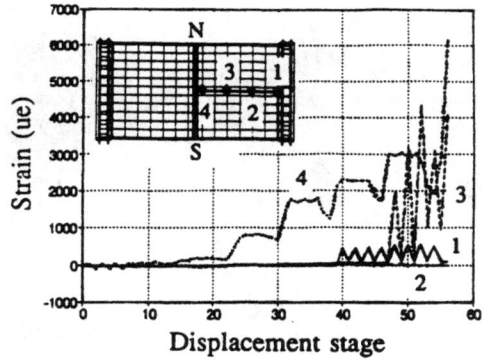

Figure 5. Strain history of transverse slab reinforcement

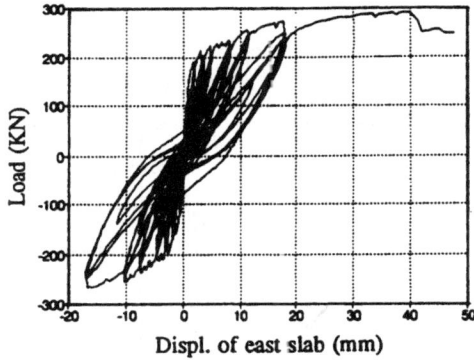

Figure 6. Total load-displacement curve

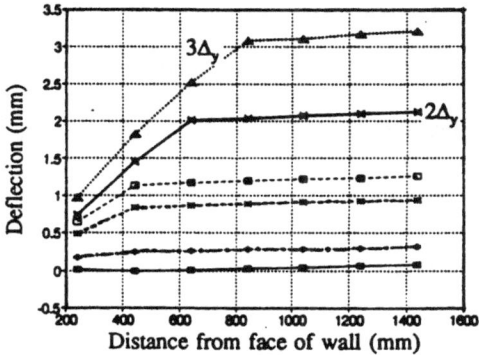

(a) Slab

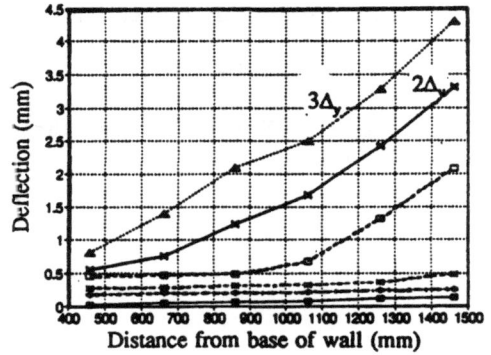

(b) Wall

Figure 7. Lateral deflections associated with shear

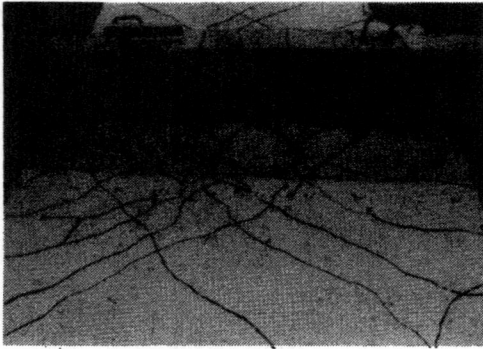

(a) Slab

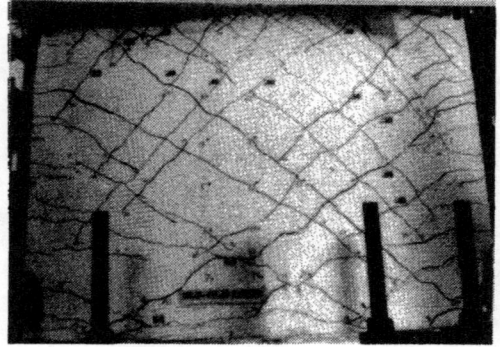

(b) Wall

Figure 8. Cracking patterns in specimen

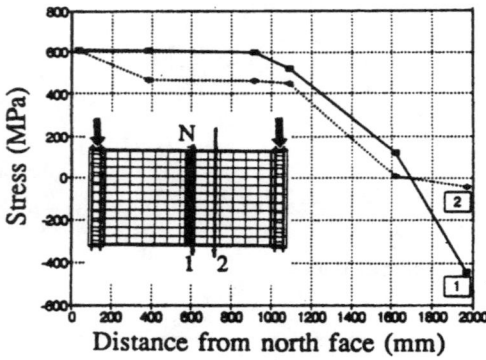

(a) at $\Delta = 2\Delta_y$

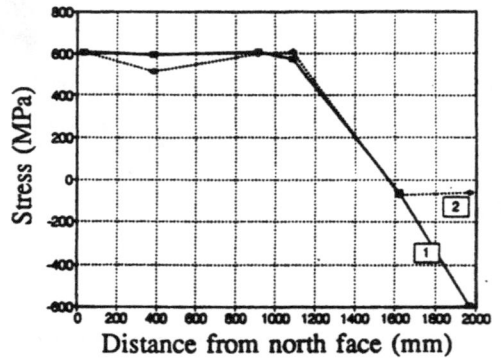

(b) at $\Delta = 3\Delta_y$

Figure 9. Stresses in longitudinal slab reinforcement along sections 1 and 2

Seismic performance of existing reinforced concrete frames designed primarily for gravity loads

Attila Béres[I], Stephen P. Pessiki[II], Richard N. White[III], and Peter Gergely[IV]

ABSTRACT

Full scale beam-column joint region specimens of lightly reinforced concrete frames were tested to study the seismic behavior of structures designed primarily for gravity loads. Details commonly used in the past were examined, including low percentages of longitudinal reinforcement, lap splices immediately above floor levels, widely spaced ties, and discontinuous positive beam reinforcement embedded in the beam-column joint. Fourteen interior and five exterior joints were loaded under combined constant gravity loads and reversing cyclic loads to and within simulate earthquake effects. Significant damage was concentrated close to the joint region. Interior specimens with discontinuous beam bars experienced a gradual decrease of load carrying capacity when this reinforcement pulled out. In exterior specimens, the pull-out was combined with joint shear failure and severe cracking along the splice exhibiting a less ductile behavior.

[I]Graduate Research Assistant, School of Civil and Environmental Engineering, Cornell University, Ithaca, NY, 14853.

[II]Assistant Professor, Department of Civil Engineering, Lehigh University, Bethlehem, PA, 18015.

[III]James A. Friend Family Professor of Engineering, School of Civil and Environmental Engineering, Cornell University, Ithaca, NY, 14853.

[IV]Professor of Structural Engineering, School of Civil and Environmental Engineering, Cornell University, Ithaca, NY, 14853

655

INTRODUCTION

There are many thousands of multi-story reinforced concrete frame structures in North America that were designed since the early 1900s without regard to significant lateral forces. The lateral load resistance of these structures is considered to be suspect for moderate to severe earthquakes because the details used are in sharp contrast to those now used in modern seismic design. In order to develop reliable seismic evaluation techniques for such frames, a research program is underway at Cornell University as part of a multi-university effort under the auspices of the National Center for Earthquake Engineering Research. The major goal is to provide a better understanding of the behavior of these kind of structures, by studying the effects of the critical parameters that influence (a) deterioration of load carrying capacity, (b) degradation of stiffness, (c) ductility, and (d) energy dissipation. Results will be used to improve inelastic dynamic analysis programs and to develop repair and retrofit schemes.

EXPERIMENTAL PROGRAM

Identification of critical details

Characteristic reinforcing details widely used in the past building construction were identified by review of ACI detailing manuals and design codes from the past five decades, along with input from practicing structural engineers. The following details were found to be typical and judged to be potentially critical to the safety of lightly reinforced concrete structures in an earthquake (Fig. 1).

1. Lapped splices of vertical column reinforcement located at the maximum moment region just above the construction joint at the floor level.

2. Widely spaced column ties which provide little confinement to the concrete.

3. Little or no transverse reinforcement within the joint.

4. Low percentage of longitudinal reinforcement in the columns and discontinuous positive beam reinforcement with a 6" embedment length into the column.

Test Plan and Variables Studied

A total of 19 specimens have been tested to date. The results from the tests of the first 10 specimens were reported previously by Pessiki, et al. (1990); a report on the remaining results will be published in 1991.

Six interior specimens (I1-I6) had continuous bottom beam reinforcement through the beam-column joint. These specimens were detailed to investigate the influence of the amount of the joint reinforcement and column bar arrangement on joints with spliced and unspliced vertical column rebars. Eight interior specimens (I7-I14) had discontinuous positive moment reinforcement extending into the columns. Variables studied included size of embedded reinforcement, column axial force level, and transverse confinement of the joint region by perpendicular stub beams.

Five experiments (E1-E5) have been conducted on exterior joints to study the effect of column axial force, transverse confinement, and ties within the joint.

Specimen Configuration and Loading Arrangement

Specimens were loaded in a computer-controlled testing facility constructed at Cornell (Pessiki, et al., 1988). Forces representing the gravity load (dead and service load) were applied to the column and at the ends of each beam, followed by reversed cyclic forces at the beam ends. The cyclic forces were controlled by a preset load history until the peak resistance was reached. After this point, the experiment was deformation-controlled, using measured rotations.

Specimen Geometry and Fabrication Details

Critical specimen dimensions were: 14"x24" beams with #3 (0.375" diameter) stirrups at 5' spacing, 16"x16" columns with 2% reinforcement and #3 ties at 16" spacing (with the first tie placed 8" above the joints as specified in earlier ACI Codes), extra #3 ties at the lower bending point of the offset vertical reinforcement, and 1.5" cover over ties and stirrups. Material strengths were $f_c' = 3500$ psi and $f_y = 60000$ psi. Some specimens had post-tensioned transverse beam stubs to simulate the presence of lateral confinement and 3-D frame effects in a real building.

EXPERIMENTAL RESULTS

Interior Joint Regions with Continuous Bottom Beam Reinforcement

Results from these 6 experiments have been summarized by Pessiki, et al. (1990a). A typical hysteresis and cracking pattern plot is shown in Figs. 2(a,b). In all specimens, damage to the column at the splice location was concentrated in a zone below the first column tie located 8 inches above the joint. In columns made with 8#7 bars, loss of cover in this zone contributed to the eventual failure by buckling of the offset reinforcement. Most of the energy dissipation and stiffness loss that occurred in the columns was also attributed to this region adjacent to the joint.

All specimens had extensive shear cracking in the joints at failure (Fig. 2(c)). The joint shear stresses at peak load (computed according to the guidelines of ACI352R) were in the range of 11.8 - 13.4 $\sqrt{f_c}$, with negligible influence of column bar size and arrangement.

Providing 2#3 ties within the joint distributed the cracks within the joint and shifted the failure zone from the joint to the splice region and decreased the rate of strength loss, but it did not increase the peak resistance significantly because of the weakness of the lightly confined splice zone.

Interior Joint Regions with Discontinuous Bottom Beam Reinforcement

Specimens I7-I14 were constructed with bottom beam reinforcement discontinuous 6 inches into the column. The cyclic load applied to each specimen was controlled by the values of the forces applied to the beams, with the "reference" position being 20 kips constant dead load on each beam. Three load cycles were applied to the beam ends at paired force levels of 30 and 10 kips, 40 and 0 kips, 50 and -10 kips (upward force), and 60 and -20 kips. Low-level cycles (30 and 10 kips) were applied after each set of 3 cycles. Loading beyond peak resistance was controlled by the values of positive beam rotation measured over a distance of 11 inches from the beam column joint.

Figs. 3(a,b) show plots of bending moment versus rotation measured close to the joint of a typical specimen (I-11). The individual hysteric loops are not symmetric since the reversing load cycles produce the superposition of the symmetric gravity loads and the antisymmetric loads simulating the lateral action.

Failure of each specimen with discontinuous beam reinforcement was by pullout of this reinforcement from the beam-column joint. At load cycles of 40 and 0 kips cracks appeared on the face of the joint in the vicinity of the embedded bars. These cracks further progressed at higher load levels to merge with diagonal cracks formed at lower load levels (Fig. 3(c)). Subsequent cycles gradually opened the cracks further causing loss in strength and stiffness. In a few cases, the cracks at the embedment zone did not progress towards the diagonal cracks, but proceeded vertically along the beam-joint interface. Spalling of concrete cover over a distance of 3-4 inches above and below the joint, and vertical cracking up to the first tie, occurred in the top column but the splices performed well. Joint shear stresses at the peak upward force were 20-40% less than in interior specimens with continuous reinforcement.

The column axial force was the most significant variable. Specimens loaded with larger axial force (350 kips) exhibited a significant increase of load bearing capacity at relatively low rotation levels, and had increased energy dissipation capacity and higher overall specimen stiffness at the beginning of the load history.

The size of the embedded reinforcement size (3/4" and 1" diameter) did not influence significantly the peak strength values, but rotations were larger in specimens with the smaller bars.

Three specimens had transverse beam stubs to simulate the possible lateral confinement by beams perpendicular to the plane of the main load bearing frame. Near the bottom of each stub, a 50 kips prestressing force was applied over an 8 by 14 inch area to simulate gravity load action in a 3-D framing system. The beam stubs produced no marked effect on the hysteresis envelopes nor on capacity.

Exterior Joint Regions with Embedded Bottom Beam Reinforcement

Five specimens (E1-E5) have been tested to study the behavior of exterior joints, using a load history that facilitated comparison with results from the interior joints.

A moment-beam rotation plot for a typical specimen without transverse beam stubs is given in Figs. 4(a,b) Initial cracks appeared on the face of the joint in the vicinity of the embedded bars during early load cycles. Under increasing loads, these cracks progressed diagonally across the joint into the splice region, and as cracking extended along the entire length of the splice, the load carrying capacity dropped suddenly. Additional load cycles induced a large opening of the construction joint above the beam. In addition, the cracks along the splice progressed vertically downwards toward the bottom column. The prying action of the bent-down beam reinforcement produced full separation of the 30-50 inch high concrete cover layer opposite the beam, as shown in Fig. 4(c). In contrast to the interior joints, downward loading on the beams had a major contribution to the failure of the exterior joints.

Specimens with transverse beam stubs showed a similar failure mechanism; however, cracking was less severe. Pullout of the bottom beam bars occurred at about the same load as intensive cracking occurred at the splices. Transverse confinement (either by beam stubs or by 2#3 ties) increased the peak load capacity by 25-40% and provided a more gradual strength degradation.

Only one specimen was tested at the higher level of column axial force (350 kips). It had no extra joint ties or transverse beam stubs, and experienced a very sudden failure at a relatively low rotation value.

The peak load capacity of exterior joints was nearly the same as obtained for interior joints. However, strength degradation of exterior joints was more rapid because of higher levels of damage to the splice region. Further analysis of data and the results of additional experiments will lead to firm conclusions about the behavior of joint regions with discontinuous embedded reinforcement.

SUMMARY

Experimental results have been presented on the seismic resistance of typical details of interior and exterior beam-column connection regions in existing lightly reinforced concrete frame structures designed primarily for gravity loadings. These experiments have provided new insight into damage mechanisms, failure mechanisms, and the influence of primary variables on connection region strength, stiffness, and ductility. The results show that these non-seismically detailed joint regions have significant initial strength and moderate ductility under simulated seismic loadings. Specimens with discontinuous positive beam reinforcement experienced considerable strength loss. At the same time these beam-column subassemblages became increasingly flexible reducing the demand for strength.

The measured bending moment-rotation hysteresis relations for these joint regions are complex and unsymmetrical, and point to the need for analytical models that properly incorporate strength deterioration, stiffness degradation, and pinching effects.

ACKNOWLEDGEMENTS

This research was sponsored by the National Center for Earthquake Engineering Research, SUNY Buffalo, NY. The Center is funded by NSF Grant No. ECE 86-07591 and by other sponsors including the State of New York. Advice on existing building details from Jacob Grossman, Raymond DiPasquale, and Glenn Bell is greatly appreciated. Opinions expressed herein are those of the authors only and not of the sponsors.

REFERENCES

ACI Committee 352, "Recommendations for Design of Beam-Column Joints in Monolithic Reinforced Concrete Structures", (ACI 352R-85), American Concrete Institute, Detroit, MI, 1985.

Pessiki, S.P., Conley, C., Bond, T., Gergely, P., and White, R.N., "Reinforced Concrete Frame Component Testing Facility - Design, Construction, Instrumentation, and Operation", Technical Report, NCEER-88-0047, National Center for Earthquake Engineering Research, State University of New York at Buffalo, New York, 1988.

Pessiki, S.P., Conley, C., White, R.N., and Gergely, P., "Seismic Behavior of the Beam-Column Connection Region in Lightly-Reinforced Concrete Frame Structures", Proceedings of the Fourth U.S. National Conference on Earthquake Engineering, EERI, Palms Springs, 1990a, Vol.2, pp.707-716.

Pessiki, S.P., Conley, C., Gergely, P., and White, R.N., "Seismic Behavior of Lightly-Reinforced Concrete Column and Beam Column Joint Details", Technical Report, NCEER-90-0014, National Center for Earthquake Engineering Research, State University of New York at Buffalo, New York, 1990b.

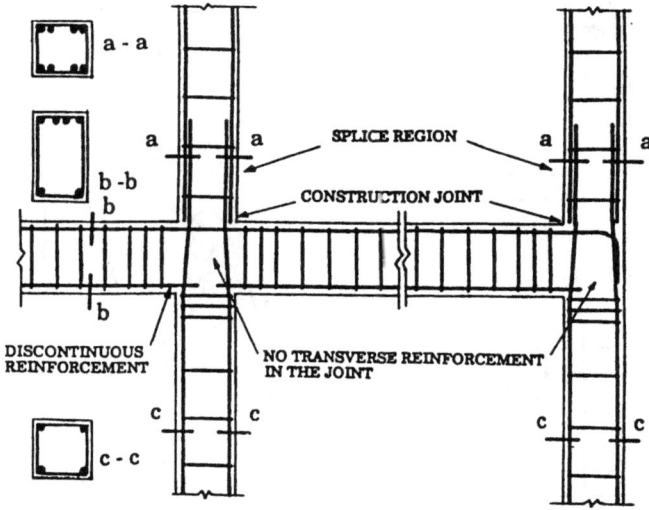

Figure 1. Elevation view of an interior and an exterior beam-column connection region.

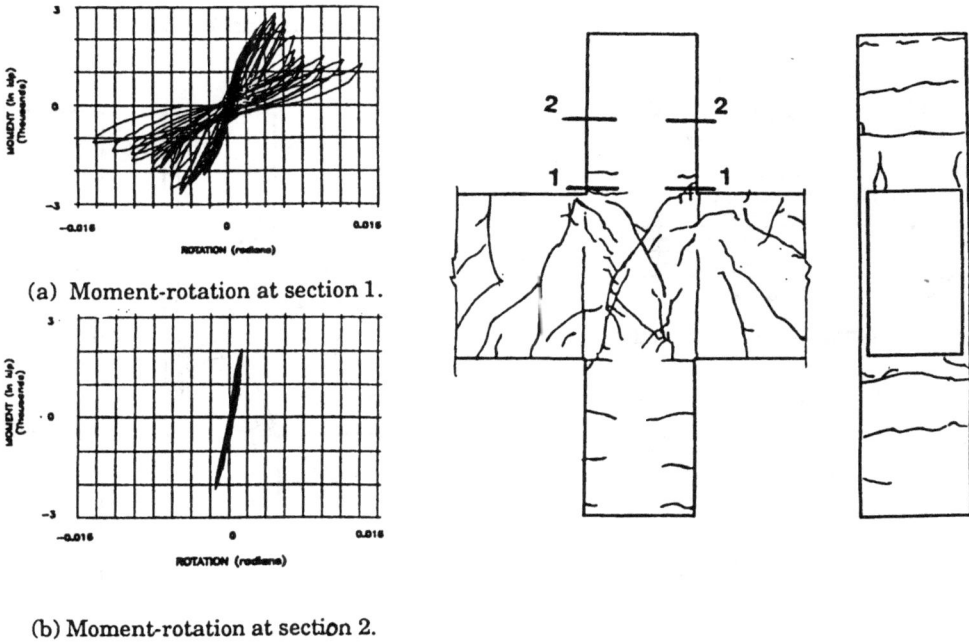

(a) Moment-rotation at section 1.

(b) Moment-rotation at section 2.

Figure 2. Interior joint (I2) with continuous reinforcement

661

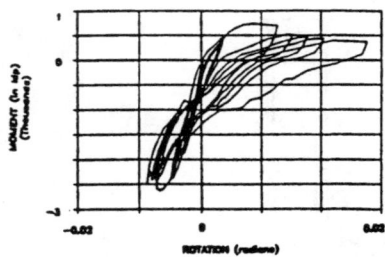

(a) Moment-rotation at section 1.

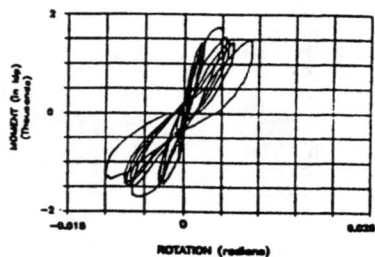

(b) Moment-rotation at section 2.

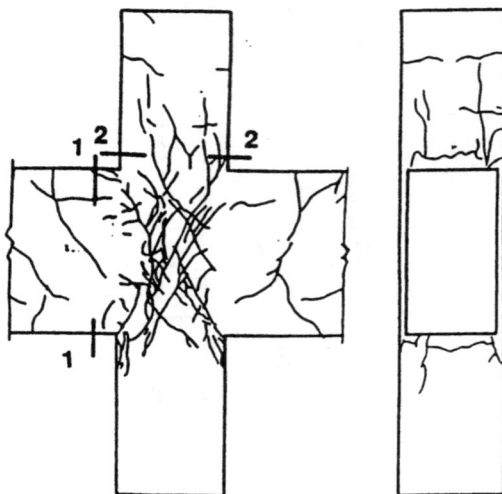

(c) Cracking pattern

- Figure 3. Interior joint (I11) with discontinuous reinforcement.

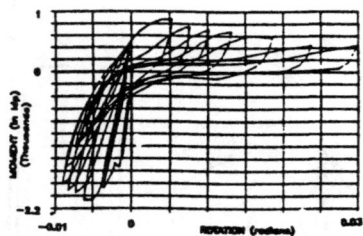

(a) Moment-rotation at section 1.

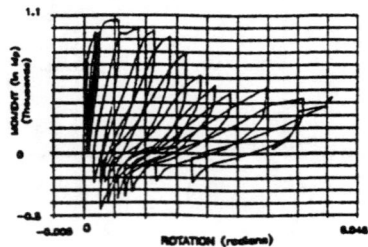

(b) Moment-rotation at section 2.

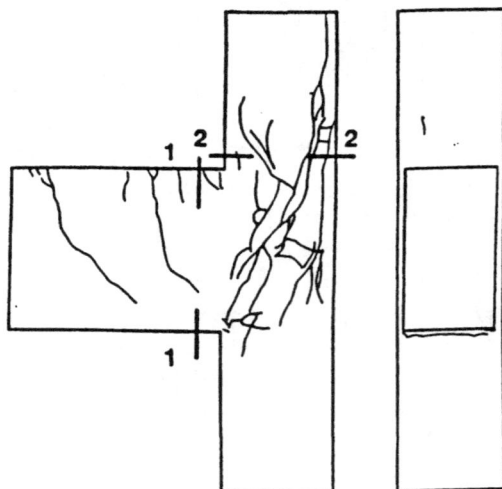

(c) Cracking pattern

Figure 4. Exterior joint (E1)

662

The seismic renovation and repair potential of ferrocement coatings applied to old brick masonry walls

H.H. Lee[I] and S.P. Prawel[II]

ABSTRACT

Older unreinforced brick masonry buildings that were originally designed with little or no provision for lateral loading occur in large numbers in most areas of the world. The risk represented by this old building stock has been recognized as one of the major problems facing the Earthquake Engineer today. The research described in this paper involves the development of an alternate type of upgrading procedure whereby a thin coating of a ferrocement like material is attached to both sides of the masonry. The method is shown to be very effective in the seismic upgrading of such walls for strength, displacement demand and energy dissipation. Enhancement in strength of up to three times is not uncommon. All of the other factors involved in the dynamic behavior of the walls are similarly improved. The results of the testing program shows quite clearly the value of the method for renovation.

INTRODUCTION

One of the major problems facing the earthquake engineers today is that represented by the large number of older masonry structures that were originally designed with little or no provision for lateral loading which can be found in most areas of the world. While some methods of upgrading these brick walls, such as shotcreting, grouting or providing external bracing have been found, in varying degrees, to be effective in actual earthquakes [1 to 3], they have either been quite costly or restricted in use to a certain type of structure.

The research described in this paper involves the development of an alternate type of upgrading procedure whereby a thin coating of a ferrocement like material is attached to both sides of the masonry. In a series of previous studies [4 to 7], it was found that ferrocement had a high potential as a material for increasing the earthquake resistance of a masonry structure. Three series of tests were carried out; the first was to determine the bonding and connector size and spacing by performing standard diagonal split tests, the other two were to study the hysteretic and dynamic behavior of the masonry walls upgraded with ferrocement, using both slow cyclic loading tests and the shake table.

I Post Doctorial Associate, Dept. of Civil Engineering, University at Buffalo
II Associate Professor, Dept. of Civil Engineering, University at Buffalo

In the first phase of study, ten pairs of thin ferrocement sheets, each 42 in. x 42 in. x 1/2 in. were subjected to a diagonal split test to determine the buckling behavior, and the required connector size and spacing. Four masonry walls made from old brick were also subjected to diagonal split. In the second group of tests, eight two wythe brick walls each 6 ft. wide 8 ft. high were tested. Half of the walls were coated with a 1/2 in. thick ferrocement overlay on both sides and were subjected to a series of pseudo-static cyclic loading tests. The third phase of the study involved earthquake simulation tests, in which eight additional walls were tested on the shake table under simulated earthquake ground motion. These tests were necessary to investigate the upper strength limits, ductility requirements, strength and stiffness degradation, and energy dissipation characteristics for the masonry walls.

EXPERIMENTAL STUDY I - DIAGONAL SPLIT TEST

The first series of standard diagonal split tests were to determine the connector size and spacing needed to prevent the delamination of the ferrocement from the brick masonry wall. It was determined from these tests that 1/4 in. bolts were required spaced at about 12 in.

The second part of this series of diagonal split tests were performed on four 42 in. x 8 in. wall specimens built from old reclaimed brick using ASTM type M mortar. Three of them were coated on both sides with the coated ferrocement overlay, each using a different size of mesh, as shown in Figure 1, while one bare wall was left as a control. All specimens were placed in a moist room to cure. Those to be coated were removed after 14 days and a 1/2 in. coating of ferrocement was applied. They were then put back into the moist room for an additional 28 days. After curing, each specimen was placed in a 300,000 lb. Universal testing machine, checked for alignment and loaded in diagonal split to destruction.

Fig. 2 shows the strength versus the vertical deflection for the test specimens. As can be seen, the uncoated specimen behaved almost linearly up to the cracking load at which point all load capacity was lost. The coated specimens all showed essentially, linear behavior. At nearly the same slope up to the point at which the brickwork split. This point was indicated by noises from within the specimen and a small drop in load carrying capacity. Visible cracks however did not appear in the ferrocement until after the load reached a value close to the brick cracking load. These small cracks formed in a band between the loaded corners and were more numerous with the smaller size mesh.

EXPERIMENTAL STUDY II - PSEUDOSTATIC TEST

Based on the preliminary diagonal split tests, a 1/2 in. x 1/2 in. x 19 gauge mesh was chosen and used in the ferrocement overlay attached to brick masonry walls. A total of 16 walls each 6 ft. wide, 8 ft. high and 8 in. thick were built from reclaimed old bricks, half of which were coated with a layer of ferrocement on each side. As was defined from earlier tests, 1/4 inch bolts spaced at 12 inches were used to prevent delamination between the coating and masonry wall.

A typical test set-up for a pseudostatic cyclic loading test is shown in Fig. 3. As shown, two wall specimens were mounted on the test base which was

664

anchored to the strong test floor. A ten-ton concrete ballast block was used to represent the overburden loads at the first floor level of a typical two-story masonry building. The block was attached to simulate a pin at the top and a fixed base at the lower end of the wall. A servo hydraulic actuator having a capacity of 55-kip and a 24-inch stroke was horizontally mounted between the heavily braced reaction frame and the ballast block which also acted as a rigid diaphragm. To simulate the usual nonrigid connection between the top of the wall and diaphragm or roof of an old masonry building, the end of the wall was allowed to rotate freely when subjected to out-of-plane loadings.

Sonic displacement transducers, calibrated statically and conditioned within the MTS 406 controller previous to the test, were mounted on the surface of the wall specimen at even intervals from the base as can be seen in Fig. 3. Tests for the in-plane and the out-of-plane behavior of the wall specimens were both carried out, for both coated and uncoated specimens.

Each pair of walls was selected from the results of free vibration tests and were subjected to a series of incremental lateral loading. The loads were controlled by specified displacements produced by the electro-hydraulic actuator. For each loading level, three cycles of saw-tooth type loading were applied at a very low frequency, i.e., 0.02 Hz, to the test specimens. the amplitude of the controlled displacement was gradually increased until the test specimens reached failure. Failure was defined as the load level which produced no loss in frequency with increasing amplitude.

All of the test wall specimens displayed a flexural mode of failure. During the early loading stages flexural cracks were initiated several bricks above the base for the plain wall specimens, while for the wall specimens coated with ferrocement, the damage usually took place somewhat above the connection at the base of the wall. These cracks later spread and allowed for a rocking and up-lifting type of motion about the interior crack in the brick. The difference in the cracking behavior for the coated wall and the plain wall specimens was that no major cracks developed in the coating.

Shown in Fig. 4(a-d) are the hystertic loops corresponding to either uncoated or coated walls in the cyclic loading tests. As shown for the out-of-plane test, both the coated and uncoated walls behaved in a ductile way, except that the uncoated walls had a much lower loading capacity and a very poor energy dissipation capacity. For the uncoated walls subjected to the in-plane loadings, the strength for one end dropped immediately after the first loading cycle while the strength for the other end of the wall specimen showed an increase. This nonsymmetric pattern was produced in subsequent loading cycles and reflected in the hysteretic loops because the loss of the resistance at one end of the wall, where sliding occurred when loading was applied. For the coated walls under in-plane loadings, elastic behavior was realized in the early loading stages. When the loading intensity was increased, the specimens started to yield at one end of the wall. Good energy dissipation was noted in the hysteretic loops.

EXPERIMENTAL STUDY III-SIMULATED EARTHQUAKE TEST

The basic test set-up was similar to the pseudostatic test except that the wall specimens were located on the shaking table. The shaking table constructed of a composite sandwich plate coated by ferrocement, has five degrees of freedom, of which three (vertical, lateral and roll) can be individually programmed.

Transducers, including accelerometers and sonic displacment transducers were both utilized in the shake table tests. The accelerometers used in the experiment were conditioned by PCB-conditioners/amplifiers with sensitivity of 5 V/g and with 0-50 Hz low-pass filters. These accelerometers were calibrated dynamically with respect to a reference whose calibration factor has been certified by the manufacturer, while the tempsonics used in this experiment were calibrated in the same manner as in the pseudostatic tests. The data acquisition and processing system employed was a digital PDP 11/34 minicomputer, a spectrum analyzer and several x-y recorders, that are a part of the shake table control system.

After pairs of walls showing the nearest frequency response were selected based on a vibration test, a banded white noise excitation having a frequency range of 0 to 20 Hz and time duration of 40 seconds was applied. From this preliminary white noise test, modal frequency, mode shape and modal damping factor could be estimated. To determine the failure mode, a series of incremental levels of simulated earthquake component with increasing intensity based on attenuated N-S component of the 1940 El Centro Earthquake, was applied to the wall specimens.

The physical behavior of both the coated and uncoated wall specimens was similar to the results observed in the pseudostatic loading test. A first crack was formed near the base in the early loading stages, and then the cracks propagated through the wall gradually corresponding to the intensity increase of the subsequent loadings. Typical response time history for the relative displacement corresponding to each temposonic transducer mounted on the wall specimens shown in Fig. 5.

DISCUSSION OF THE TEST RESULTS

The general characteristics discussed in this section consist of the hysteretic behavior including the ultimate strength and strength deterioration, stiffness degradation, energy dissipation, ductility, and general dynamic behavior such as frequency response and the variation of the damping factor for both the plain and coated walls. The only parameter considered was the effect of ferrocement overlay while other parameters such as overburden load, mortar strength, and the size of wire mesh were held constant.

The in-plane strength is improved by about 3 times when the brick wall specimen is coated by the ferrocement overlay. Considerable improvement in the flexural strengths of the coated walls was also indicated in the results of the pseudostatic tests. Fig. 6 (a,b) shows the hysteretic envelopes taken from the cyclic loading test results for the plain and the coated walls when subjected to both in-plane and out-of-plane loadings. Each point shown in the hysteretic envelopes was obtained from the hysteresis loops by averaging the absolute extreme values of the six peaks of three cyclic of loadings and the corresponding absolute values of the relative lateral displacement for each test stage. It can be seen that the ultimate strength of the masonry wall was upgraded by 3 to 4 times by the ferrocement overlay.

The hysteretic stiffness was defined as the slope of the hysteretic envelope. It is obvious that not only was the initial stiffness of the coated wall specimens improved but the degradation of stiffness was reduced during each incremental loading step. The improvement in the initial stiffness of the coated wall specimen loaded out-of-plane under simulated Earthquake loading was about two times.

Ductility, an important indicator of earthquake resistant ability, was obtained from the hysteretic envelopes and defined as the ratio of the maximum displacement at the failure point to the extrapolated displacement corresponding to the same loading point, which was noted in the envelopes. According to this definition, the ductility for the walls in the out-of-plane test was found to be 6.70 and 9.50 for the plain and coated walls respectively, while in the in-plane test, it was 3.50 and 4.00. It is evident that the ductility of the masonry walls was improved by the application of ferrocement coatings.

The energy dissipation capacity is taken as the average area contained in the hysteresis loops for three cycles of repeated loading for each stage of test. It can be seen in the hysteretic loops that in general, in the early loading stages, the energy dissipated in both the plain and the coated walls is very small since the response was still in the elastic range. In the later loading stages, significant dissipation of energy was found in the coated wall specimens, particularly for the wall specimen in the out-of-plane test. The energy dissipation capacity of masonry brick wall improved 3-6 times by the ferrocement overlay.

A banded white noise test with very small amplitude was performed after each level of simulated earthquake test loading. The frequency was then taken directly from a spectrum analyzer and from this data, a degradation curve of frequency with respect to the loading intensity was determined for both the coated and plain walls. The natural frequency was found to decrease with an increase of loading intensity. The degrading rate of frequency for the coated wall specimens appeared to be slower than for the plain wall specimens. With the ferrocement coatings, the wall specimens become more stiff and had a higher frequency.

The damping factor for both the coated and plain walls was calculated [7] and plotted against the corresponding loading intensity. Generally, the damping factor is increased corresponding to the increase of loading intensity. Higher damping was found in the wall specimens coated by ferrocement, especially during the later loading stages.

CONCLUSIONS

Brief conclusions that can be drawn from the results of these tests are as follows: (1) The mode of failure for both the coated and plain wall specimens subjected to either in-plane or out-of-plane loading was flexural. (2) The original stiffness of coated masonry walls was increased up to two times as much as that for the uncoated walls. (3) The shear strength is increased about 1.5-2 times, and the flexural strength of masonry walls in terms of moment capacity is increased about three times by the ferrocement reinforcement. (4) The energy dissipation capacity and ductility are improved when ferrocement coatings are applied. (5) With the ferrocement coatings, the wall specimens become more stiff and have a higher natural frequency. (6) The damping factor increased corresponding to the loading intensity. In the in-plane test, the coated wall

667

specimens appeared to have higher damping than the plain wall specimens at the same loading intensity. (7) It is evident that ferrocement is an appropriate material able to significantly improve the dynamic resistance of brick masonry.

REFERENCES

[1]Willie, L.A. and Dean, R.G. 1975. Seismic failures and subsequent
    performance after repair, Proc. ASCE National Structural
    Engineering Conference, New Orleans, April 1975.
[2]Sheppard, P. and Terceli, 1980. The effect of repair and strengthening
    method for masonry walls. Proc. 7th WCEE, Istanbul, Vol. 6, 255-264.
[3]Jabarov M. Kozharinov, S.V. and Lunyov, A.A. 1980. Strengthening of
    damaged masonry of reinforced mortar layers. Proc. 7th WCEE,
    Istanbul, Vol. 4, 73-82.
[4]Reinhorn, A.M. Prawel, S.P., and Jia, Z.H. 1985. Experimental Study on
    External Ferrocement coating for Masonry Walls. Journal of Ferrocement,
    Vol. 15, 247-260.
[5]Prawel, S.P., Reinhorn, A.M., and Quizi, S.A. 1988. Upgrading the Seismic
    Resistance of Unreinforced Brick Masonry Using Ferrocement Coatings.
    Proc. 8th International Brick/Block Masonry Conference, Dublin
    Ireland, September 1988.
[6]Prawel, S.P. and Lee, H.H. 1990. The Performance of Upgraded Brick Masonry
    Piers Subjected to In-Plane Motion. Fourth U.S. National Conference on
    Earthquake Engineering, Palm Spring, California, May, 1990.
[7]Prawel, S.P., and Lee, H.H. 1990. The Performance of Upgraded Brick Masonry
    Piers Subjected to Out-of-Plane Motion. Proc. of Fifth North American masonry
    Conference, University of Illinois at Urbana-Champaign, June, 1990.

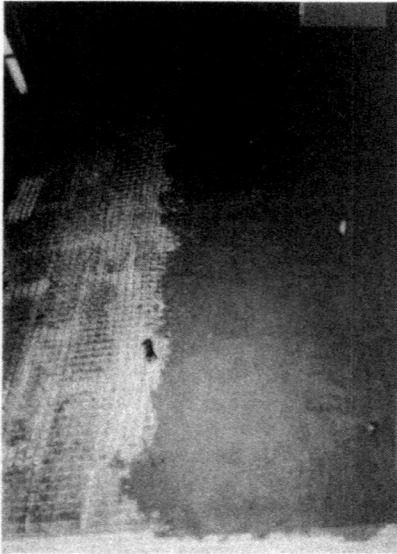

Fig. 1 The Ferrocement Overlay Attached to the Masonry Wall

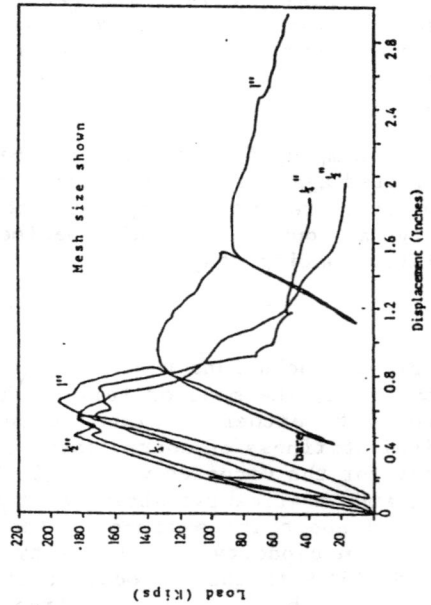

Fig. 2 Load vs Vertical Deformation for Diagonal Split Test

Fig. 3 Test Set-Up of Wall Specimens in Cyclic Loading Test

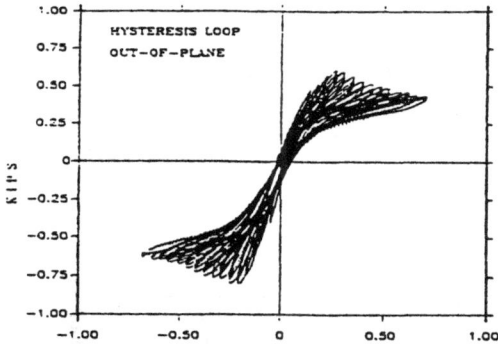

Fig. 4(a) Typical Hysteresis Loops for Plain Wall in
Out-of-plane Test

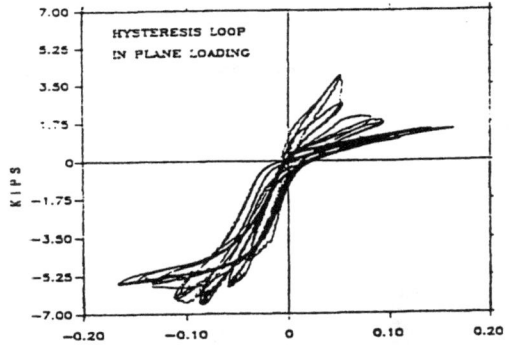

Fig. 4(c) Typical Hysteresis Loops for Plain Wall in
In-plane Test

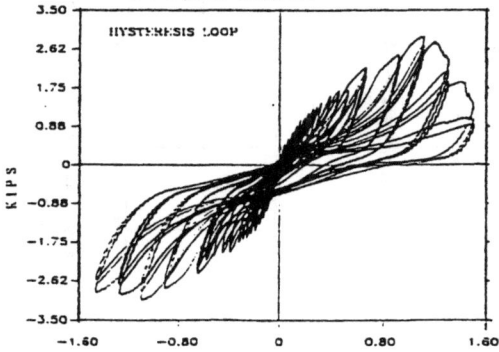

Fig. 4(b) Typical Hysteresis Loops for Coated Wall in
Out-of-plane Test

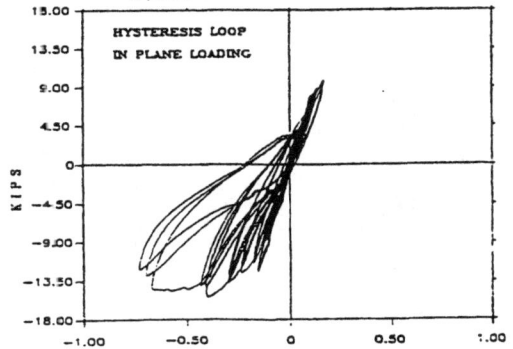

Fig. 4(d) Typical Hysteresis Loops for coated Wall in
In-plane Test

669

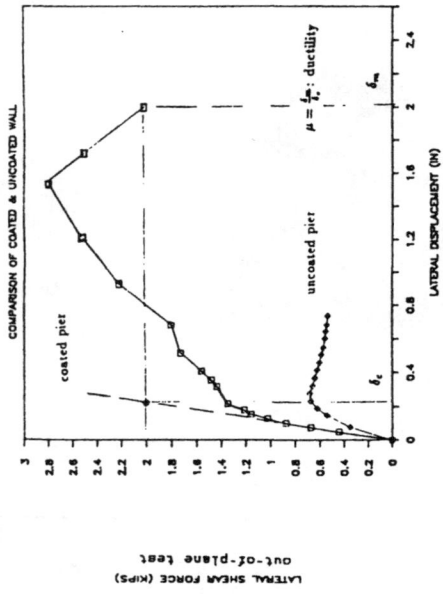

COMPARISON OF COATED & UNCOATED WALL

coated pier

uncoated pier

$\mu = \frac{\delta_m}{\delta_e}$: ductility

δ_m

δ_e

LATERAL DISPLACEMENT (IN)

LATERAL SHEAR FORCE (KIPS)
out-of-plane test

Fig. 6(a) Hysteresis Envelope for out-of plane test

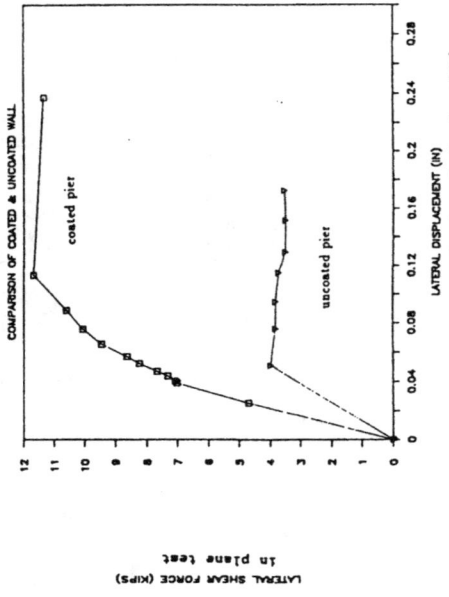

COMPARISON OF COATED & UNCOATED WALL

coated pier

uncoated pier

LATERAL DISPLACEMENT (IN)

LATERAL SHEAR FORCE (KIPS)
in plane test

Fig. 6(b) Hysteresis Envelope for in plane test

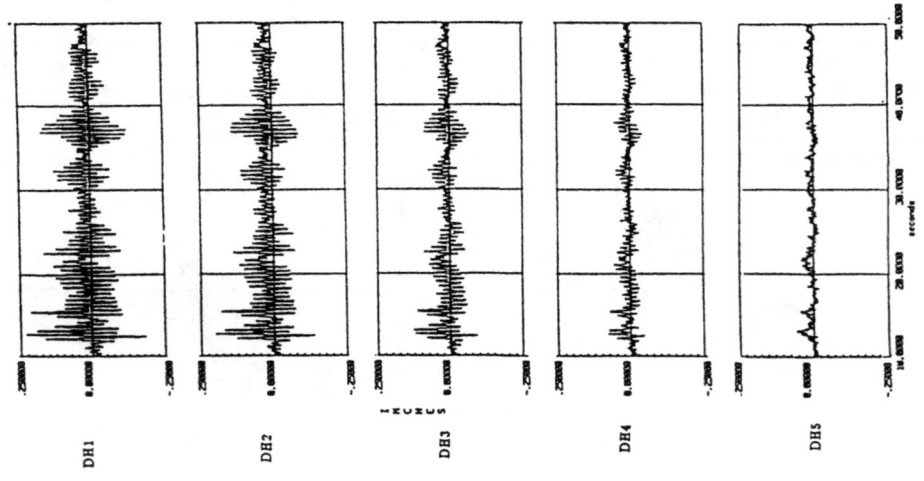

DH1

DH2

DH3

DH4

DH5

seconds

Fig. 5 Time Histories of Displacement Response for Plain Wall
in Shake Table Test

670

A study on elasto-plastic behavior of one-twentieth scale reinforced concrete frames

Hiroyasu Sakata[I] and Akira Wada[II]

ABSTRACT

This paper describes the structural behavior of the reinforced concrete frames based on the experiments and analyses on the one-twentieth scale models. Focusing on the influence of the beam elongation, the experimental program consists of monotonic loading tests on the four-story reinforced concrete frames with one, three or seven spans. The structural behavior of the frames was also analyzed by the finite segment method.

As conclusions obtained from the experiments and analyses, (1) the horizontal displacement vs. horizontal load relationship for each column in the frames was not identical due to the axial elongation of bending yield beams, (2) the lateral loads for all the frames was increased due to the axial restriction in beams caused by the lateral stiffness of columns, (3) the results of the analysis showed a good agreement with the experimental results.

INTRODUCTION

Reinforced concrete beams and columns forming up a structural frame exert influence on each other during loading and behave physically as one structural body.

Thus, whenever analysis of the mechanical behavior of the reinforced concrete frame is concerned, experiment carried out on the frame as a whole system makes more sense than just experimenting on the individual member.

A loading test of a statically indeterminate reinforced concrete beam-to-column connection subassemblies was carried out by Zerbe and Durrani(1988), who pointed out that compressive force was developed in a beam which subsequently increased its flexural strength. Focusing on the restraint of the beam elongation, the author and his colleagues studied the behavior of the reinforced concrete beam subjected to the restraint of axial deformation(Kokusho, Hayashi, Wada and Sakata 1988).

In order to clarify the influence of the axial elongation of beams, horizontal loading experiments and analyses of a beam sidesway mechanism type (with the number of spans used as a parameter), multi-story multi-span reinforced concrete plane frame were carried out. In this paper, the axial elongation of beams, the axial strain developed in the beams, the horizontal strength of the frame, the behavior of collapse mechanism, and etc. will be discussed.

I Research Associate, Tokyo Institute of Technology, Dr. Eng.
II Professor, Tokyo Institute of Technology, Dr. Eng.

Specimens

With reference to a certain reinforced concrete building with a pure rigid frame, a 1/20 scale plane frame was envisaged representing the lower portion of the building, and a total of three specimens; a four-story, one-span specimen (FR1), a four-story, three-span specimen (FR3) and a four-story, seven-span specimen (FR7) were manufactured. Each specimen represents the portion up to the intermediate height of the columns on the fifth floor erected on the fourth story of the specimen. The cross section of the members is shown in Fig.1. The dimension of the FR7 specimen is shown in Fig.2. For the purpose of later explanation, the columns are marked to show A, B, C, D, E, F, G and H in order from the left to the right.

In lieu of concrete, the mortar with cement and standard sand proportioned at mixing ratio 1:2.5, which was mixed with water at a water-cement ratio of 75%, was used. D3 deformed reinforcement(Murayama et al. 1982) used as main reinforcement was assembled with spiral hoops and stirrups manufactured from a 1.6mm diameter annealed wire.

Loading apparatus

The loading apparatus and the method of loading are shown in Photo1. and Fig.3.
<Axial force of column> : Fig.3 shows the method of introducing the axial force of column, (N). The plate attached to the top of the column of each specimen was connected to the spring located in the spring case underneath the reaction frame, by means of a round steel bar with pin-shaped ends. By tightening the nut underneath the spring, the vertical force of 13.7kN per column was caused to act as a fixed load.
<Horizontal force> : Fig.3 shows the method of applying a monotonic, horizontal force. A pin was fitted with each top of column, and a mini oil jack was installed between the pin and the

Table1. Physical Properties of Materials

Reinforcing Bar

	Sectional Area (cm^2)	Yield Strength (MPa)	Tensile Strength (MPa)	Elongation (%)
D3	0.0731	317.3	440.9	40.7
1.6ϕ	0.0201	253.7	383.8	

Mortar			Exp. Start	Exp. End
Material Age(Day)	7	28	40	56
Compressive Strength(MPa)	21.5	32.2	40.1	44.2
Strain at Compressive Strength($\times 10^{-6}$)	——	——	3526	3896
Young's Mondulus(MPa)	——	——	21379	22948
Tensile Strength(MPa)	1.8	2.4	3.0	3.4

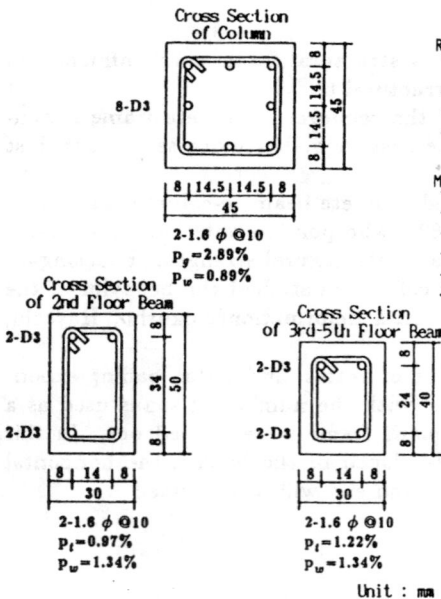

Figure1. Cross Section of Specimens

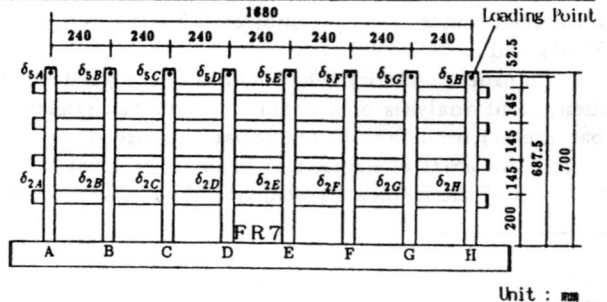

Figure2. Specimens(FR7)

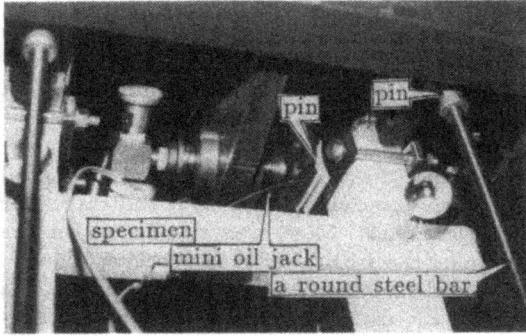

Photo1. Loading Arrangement

Table2. Horizontal Resistance Force Calculated
by Simple Analysis Assuming Mechanism of Collapse

FR1	FR3	FR7
3.53kN	9.12kN	20.40kN

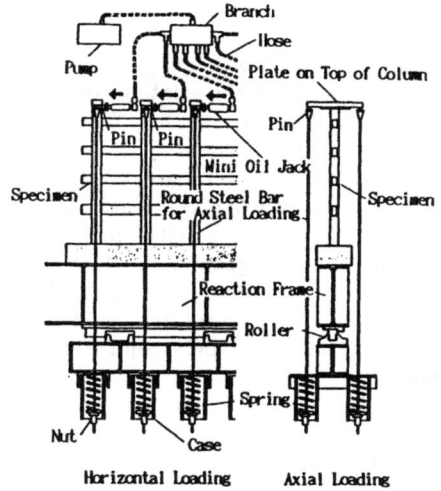

Figure3. Loading Arrangement

loading frame. These mini oil jacks were connected with one unit of hydraulic pump, and are capable of applying force uniformly to each top of column. However, in order for the columns at both ends to be subject to half the force in other columns, a mini oil jack was used with a cross section being half that of another mini oil jack.

Mechanical properties of materials

Table1 shows the mechanical properties of the D3 deformed reinforcement, the 1.6mm diameter annealed wire, and the mortar which were used.

Horizontal resistance force calculated by simple analysis assuming mechanism of collapse

Table2 shows the estimated horizontal resistance force of the specimens, which were obtained on the assumption that no axial force is developed in the beams of a collapse mechanism with a yield hinge assumed at the beam ends and column bases. For this model, consideration was given to the rigid zone in the beam ends.

RESULTS OF EXPERIMENTS AND DISCUSSIONS

Relationship between horizontal force and horizontal displacement

Fig.4 shows the relationship between horizontal force and horizontal displacement. In the figure, symbols such as δ_{5A}, δ_{5B}, ..., δ_{5H} are used corresponding with those shown in Fig.2. The value in parentheses denotes the angle of rotation. In the figure, the simple analysis represented by broken line considers the N-δ effect of the horizontal resistance force in Table2.

In all specimens, the horizontal displacements of all the columns on each floor were identical in an initial stage, which began to vary after initial cracking occurred, and the horizontal displacements of the columns on the left became more noticeable than the columns on the right when the whole average angle of rotation exceeded 1/200. This was due to the cumulative axial elongation of the flexuously yielded beams. Each specimen indicates a greater strength than simple analysis. The reason for such increase in strength would be that the rigidity of the columns restrained the axial elongation of the flexuously yielded beams which consequently led

673

to the increased flexural strength of the beams.

Axial elongation of beam

Fig.5 shows the relationship between horizontal force and the axial elongation of beams on the second floor. AB ...... following Δ represents the difference of horizontal displacement between columns A and B ....

Immediately after the occurrence of flexural cracking, the axial elongation was small, and began growing when the whole average angle of rotation exceeded some 1/200, and thereafter continued increasing with increasing horizontal displacement. Regarding the beams on the second floor, the greater the number of spans, the smaller the amount of elongation per span.

Fig.6 shows the axial elongation of the beams when the whole average angle of rotation is 1/20. From the figure, it is evident that the beams on the fifth floor indicate irregular elongation, but those on the second floor indicate elongation, much greater in the spans closer to the left.

Axial strain of beams

Fig.7 shows the distribution of strains in the beams when the whole average angle of rotation were 1/100, 1/50 and 1/20 for the FR3. Compressive strains were developed in the beams on the second floor; a maximum 50μ, maximum 40μ in FR1 and maximum 350μ in FR7, indicating that the greater the number of spans, the much greater the compressive strains in the beams on the second floor.

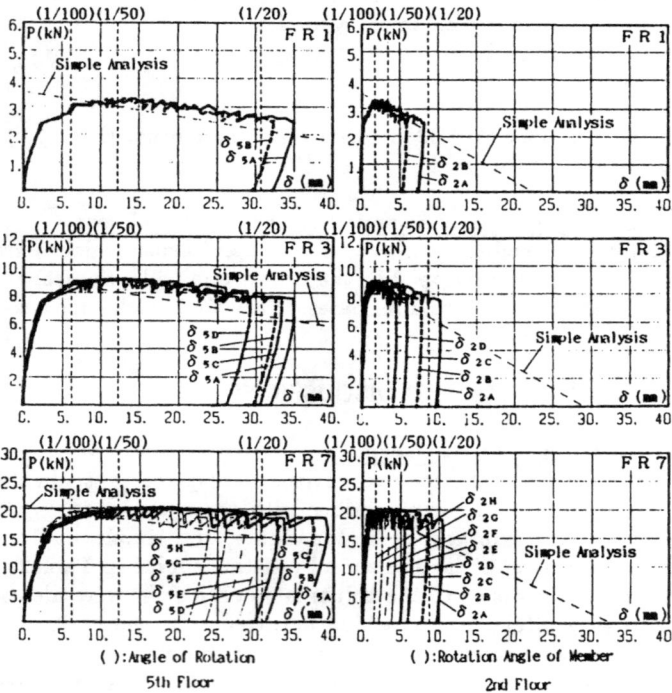

Figure4. Horizontal Force - Horizontal Displacement Relationships

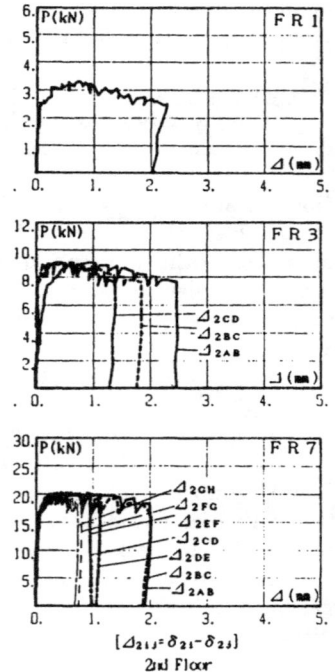

Figure5. Horizontal Force - Axial Elongation of Beam Relationships

State of deformation

Fig.8 shows the state of ultimate deformation of the FR3. In FR1, both columns A and B indicate the mode of deformation. Columns closer to the left indicate nearly straight deformation at the positions of the beams on the second floor. From the figure, it is evident that the columns on the right (column D) is bent in the form of > at the position of the beams on the second floor, indicating the effect of axial elongation of the beams on the second floor also. This phenomenon is shown in FR1 and FR7.

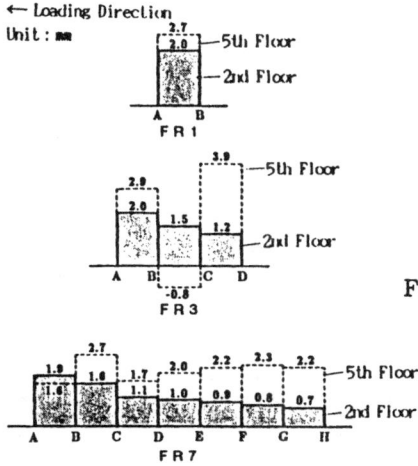

Figure7. Axial Strain of Beam

Figure6. Axial Elongation of Beam at
Whole Average Angle of Rotation 1/20

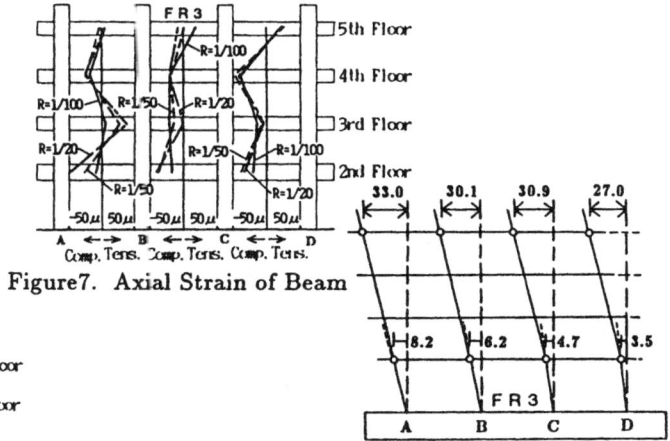

Figure8. Deformation of FR3 at
Whole Average Angle of Rotation 1/20

ANALYSIS METHOD AND ANALYSIS MODEL

By the implementation of the elasto-plastic analysis method of reinforced concrete frame taking into consideration shear deformation of beam-column joints and bond-slip using finite segment method, analyses of the present experiments were carried out. As the details of the elasto-plastic analysis method are given in Reference (Kokusho, Wada and Sakata 1988), explanation of the analysis method is omitted.

Analysis model

Fig.9 shows the model for the FR3 used in analyses. Each of the columns and beams between the panels was divided into 20 elements in the direction of its axis, the columns

Figure9. Model Used in Analysis(FR3)

on the fifth floor were divided into 10 elements, and the cantilever beams projecting on the left and the right were divided into 5 elements. The overall degree of freedom of the analysis models was 3722 in FR1, 7724 in FR3, and 15700 in FR7.

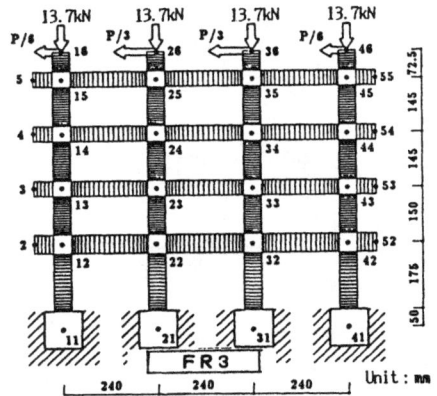

675

Mechanical properties of materials used in the analyses

Table3 shows the constants used in the analyses in correspondence to the relationship between bond stress and slip (Fig.10), between stress and strain of concrete (Fig.11), between stress and strain of reinforcing bar (Fig.12), and between stress and strain of concrete panel (Fig.13). The values obtained through material tests were applied as the values of the compressive strength of concrete, the yield stress of the reinforcement, etc., and the values involved in the bond slip of the reinforcement were modeled (Fig.10) with reference to the results of bond tests conducted by Murayama, et al.(1982).

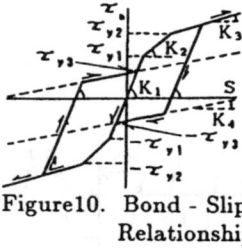

Figure10. Bond - Slip Relationship

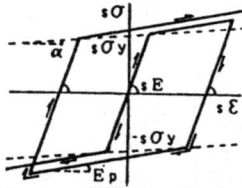

Figure12. Stress - Strain Relationship of Rebar

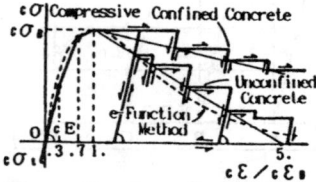

Figure11. Stress - Strain Relationships of Concrete

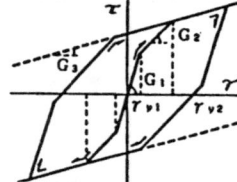

Figure13. Stress - Strain Relationship of Concrete Panel

Table3. Mechanical Properties of Materials Used in The Analysis

	FR1			
$_cE$ =21379.	MPa	$_c\varepsilon_B$=0.0035		
$_c\sigma_B$=40.1	MPa			
	FR3			
$_cE$ =22163.	MPa	$_c\varepsilon_B$=0.0037		
$_c\sigma_B$=42.2	MPa			
	FR7			
$_cE$ =22948.	MPa	$_c\varepsilon_B$=0.0039		
$_c\sigma_B$=44.2	MPa			
	Common to All			
$_sE$ =196133.	MPa	α=49.	MPa	
E_p=98.1	MPa	$_s\sigma_y$=317.3	MPa	
K_1=343.(1118.)	MPa/cm			
K_2=58.8(215.7)	MPa/cm	G_1 =5884.	MPa	
K_3 =0.98(15.7)	MPa/cm	G_2 =4707.	MPa	
K_4 =0.98(4.9)	MPa/cm	G_3 =1177.	MPa	
τ_{y1}=4.9 (5.9)	MPa	γ_{y1}=0.0005		
τ_{y2}=7.4 (9.1)	MPa	γ_{y2}=0.003		
τ_{v3}=0.5 (0.5)	MPa	( ):for Joint Panel		

RESULTS OF ANALYSIS AND DISCUSSIONS

Relationship between horizontal force and horizontal displacement

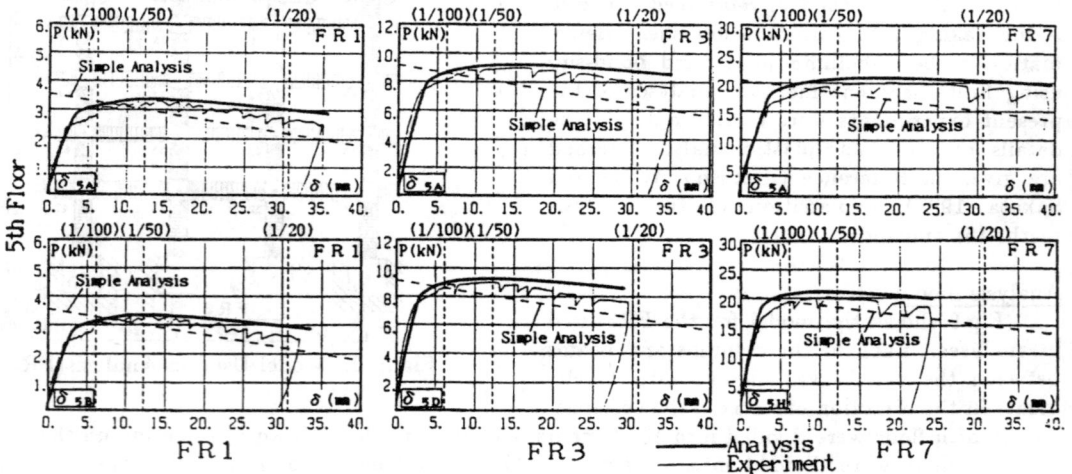

FR1   FR3   FR7

——— Analysis
——— Experiment

Figure14. Horizontal Force - Horizontal Displacement Relationships

Fig.14 shows the relationship between horizontal force and horizontal displacement. The bold solid line represents the analysis value and the narrow solid line represents the experimental value. In all specimens, the analysis value is slightly greater than the experimental value, indicating a good correspondence between the two. As in the case of the experimental value, the analysis value indicates a greater strength than simple analysis, in all specimens.

Axial elongation of beam

Fig.15 shows the axial elongation of the beams on the second floor when the whole average angle of rotation is 1/20. The experimental values and the analysis values concur well.

Relationship between bending moment at beam end and axial force

Fig.16 traces the bending moments generated in the ends of the A-B beams on the second floor and also the axial forces generated in the same beams. The solid line indicates the left end and the broken line indicates the right end. When a compressive force is developed in a beam, the axial force is shown in the positive, and the tensile side of the beam bottom is the positive bending moment. Due to the development of an axial force, the bending moment in the beam end varied with the compressive axial force which also changed along the interaction curve. This bending moment was greater than one with a zero axial force, and from this, it is feasible to explain the fact that increasing resistance strength was seen as in Fig.4.

Shearing force of first floor columns

Fig.17 shows the shearing force of the first floor columns with moment diagram for the FR3 when the whole average angle of rotation is 1/200. The shearing force shared by the first floor columns is not uniform because of the effect of axial elongation of the beam on the second floor, particularly a large difference of shear values at both ends.

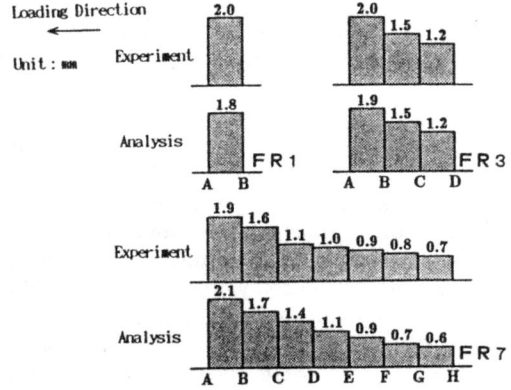

Figure15. Axial Elongation of 2nd Floor Beam at Whole Average Angle of Rotation 1/20

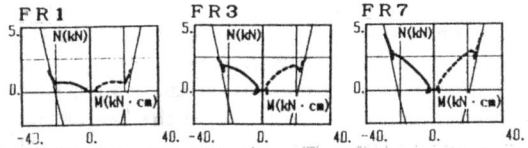

Figure16. Bending Moment at Beam End - Axial Force Relationships on 2nd Floor A-B Beam

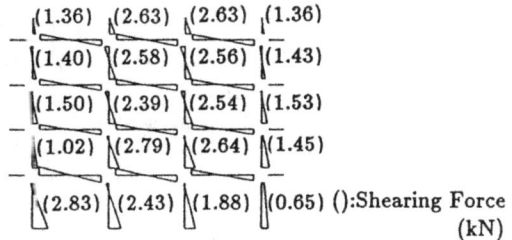

Figure17. Bending Moment Diagram and Shearing Force (FR3)

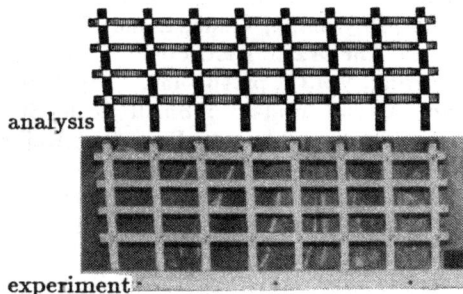

Figure18. State of Ultimate Deformation (FR7)

State of ultimate deformation

Fig.18 depicts how deformation occurs when the whole average angle of rotation is 1/20. Both the analysis results and the experimental photograph indicate a very close deformation mode.

CONCLUSIONS

As a result, the following conclusions were achieved.
1) When a beam sidesway mechanism type reinforced concrete plane frame is deformed due to a horizontal force, all columns are identically deformed at an early stage. Thereafter, the beams are flexuously yielded, and then elongated in the direction of axis, but the footing beams were deformed only a little, resulting in the varying angles of the columns on the first floor. When the frame is receiving leftward monotonic loading, the angles of the columns on the first floor become greater as they are located closer to the left, resulting in a large difference between those at the left end and the right end. This phenomenon is more noticeable with the increasing number of spans since all the yielding beams tend to elongate.
2) All specimens demonstrated that the horizontal strength is greater than the horizontal strength obtained using the ultimate bending moment on the assumption that no axial force is generated in a beam. As the results of the analysis indicated, this was because the bending strength of the beams increased due to a compressive axial force.
3) In the reinforced concrete plane frame, the columns are subject to extra forced deformation due to the axial elongation of the beams that have been flexuously yielded, hence the sharing of the shearing forces in the columns at both ends of the frame is greatly varied. This phenomenon is noticeable in the columns on the first and second floors.
4) When the reinforced concrete plane frame is monotonically loaded from the right to the left, the elongation of the beams on the left spans on the second floor is greater than that of the beams on the right spans.
5) The compressive strain on the second floor beams which tend to axially elongate after being flexuously yielded within the reinforced concrete plane frame is greater in the specimen with a greater number of spans.
6) The results of the analysis presented here cover most details of the experimental results, and represent the behavior of the multi-story multi-span reinforced concrete plane frame, taking into consideration the restraining effect of axial deformation of the flexuously yielded beams.

REFERENCES

Kokusho,S., Wada,A. and Sakata,H. : Computer Analysis of Inelastic Behavior of Reinforced Concrete Frame, 13th IABSE Congress in Helsinki, pp.665-670, June 1988

Kokusho,S., Hayashi,S., Wada,A. and Sakata,H. : Behavior of Reinforced Concrete Beam Subjected to The Axial Restriction of Deformation, Ninth World Conference on Earthquake Engineering, Vol.IV, pp.463-468, Aug., 1988

Murayama,Y. and Noda,S. : Small Scale Structural Model Tests by Using 3mm Diameter Deformed Re-bars, Annual Report of Kajima Institute of Construction Technology Vol.16, pp.31-40, July. 1982 (in Japanese)

Zerbe,H.E. and Durrani,A.J. : Seismic Behavior of Indeterminate R/C Beam-to-column Connection Subassemblies, Ninth World Conference on Earthquake Engineering, Vol.IV, pp.663-668, Aug., 1988

Seismic resistance of a moment resisting steel portal frame with composite concrete slab deck and simulated gravity loading

A.G. Gillies[I], L.M. Megget[II], and R.C. Fenwick[III]

ABSTRACT

The seismic resistance of structural steel moment resisting frames with composite concrete floor slabs which are subjected to appreciable gravity loading has received little attention from researchers. The gravity load can cause unidirectional plastic hinges to form in the beam with a high rotational ductility demand.

The majority of experimental tests reported to date for frame systems have concentrated on statically determinate sub-assemblages of beam/column joints. Relatively little testing has focussed on indeterminate test specimens which have more complex behaviour characteristics. This bias is reflected in code documents, for example, which discuss performance goals in terms of member ductility capabilities. Apparently obvious definitions such as "first yield" need to be reviewed when applied to indeterminate systems.

Reported in this paper is the observed response of a portal frame which was subjected to simultaneous vertical (gravity) and lateral loading. The frame comprised steel universal beam sections with a poured slab (105mm) atop a proprietary steel tray deck. The dimensions were selected to approximate at 3/4 scale a realistic floor system. Composite action was achieved with standard welded steel studs.

Introduction

Prior to 1989, composite slab construction was not addressed in the New Zealand steel design code. In 1989 NZS 3404:1989 was published and for the first time detailed provisions were included for composite construction. To date little experimental research has been undertaken within New Zealand to investigate the behaviour of composite beam structures. Consequently when drafting NZS 3404:1989 extensive reliance was placed on the results of overseas studies. The new New Zealand rules draw heavily on Canadian design practice in particular, since Canada has adopted limit states design methods which are consistent, at least in the non-seismic part, with New Zealand design philosophies (refer CAN/CSA-S16.1-M89, 1989, for example).

NZS 3404:1989 contains some limited guidance for special detailing of the composite member for ductile action. The provisions discuss the positive or negative moment region at a support and the positive moment region within the span. In the region near the support the more simple (and predictable) solution is to curtail the composite action at some distance away from the support

---

I Associate Professor, Lakehead University, Thunder bay, Ontario

II Senior Lecturer, University of Auckland, Auckland, New Zealand

III Associate Professor, University of Auckland, Auckland, New Zealand

face and design based on the capacity of the steel beam alone. For the
midspan region there is a suggested rule based on the plastic strain in the
steel section compared to the ultimate compressive strain in the slab (for a
category 1 member the Code requires that the steel beam shall achieve a maximum tensile strain of 24 times yield strain prior to the ultimate compressive
strength of the concrete slab being attained and a maximum concrete strain of
0.003 being exceeded). It is not common to expect a midspan hinge as a result
of seismic loading but the possibility cannot be ignored in frames which have
high gravity loads.

The two goals of the test program reported here were therefore to design
a steel portal frame with a composite deck slab and to study (i) the region of
the beam adjacent to the beam/column joint and (ii) the midspan region. For
reasons discussed later in this report there was limited success in forming
and maintaining the in-span hinge.

Experimental Program

Portal Geometry

Figure 1 shows in elevation one half of the span geometry (apart from the
loading details - the lateral load was applied as either tensile or compressive at one end only, and the central loads were slightly eccentric from the
beam span centreline - the portal was symmetrical about its mid-span
centreline).

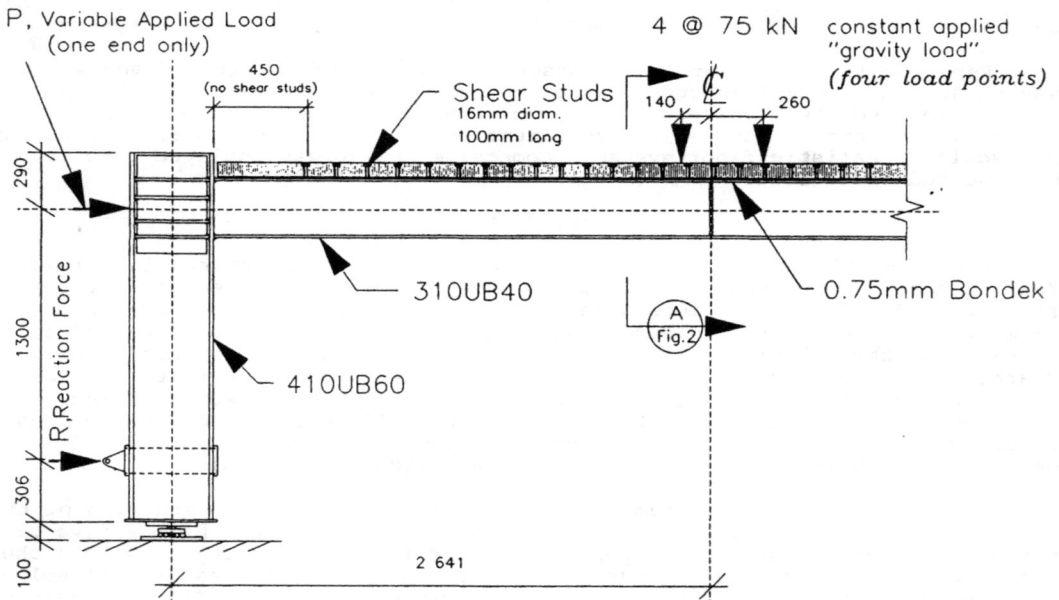

Figure 1. Elevation of Portal Frame

The two inner pairs of web stiffeners were eliminated in the beam-column joint
remote from the horizontal load application end. The portal was detailed as
pinned base. To determine the bending moments in each portal leg during the
testing, the horizontal reaction was measured by means of a load cell near
each column base. For this reason the columns were restrained by a horizontal

link approximately 300mm above their base, which provided translation restraint but rotation release. To maintain the simple pinned support it was necessary to detail a roller bearing detail under the column baseplates.

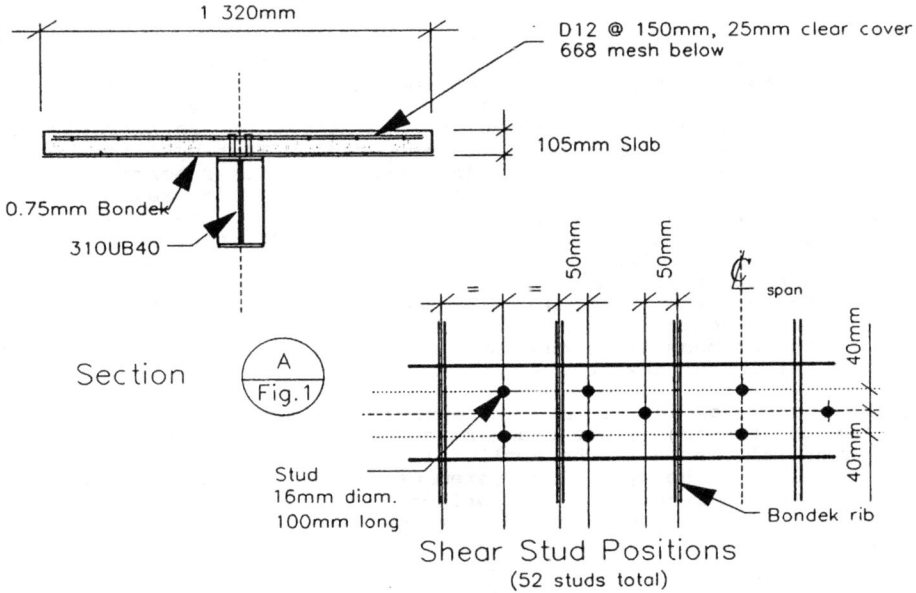

Figure 2.   Section Details of the Full Composite Beam and Slab

The typical section details are given in Fig. 2.   The shear studs were 16mm diameter and 100mm long and were welded through the deck pan to the top flange of the Universal Beam section.   Composite action was curtailed at the ends of the beam span by eliminating any studs within 450mm (1.5 x beam depth) from the column face.

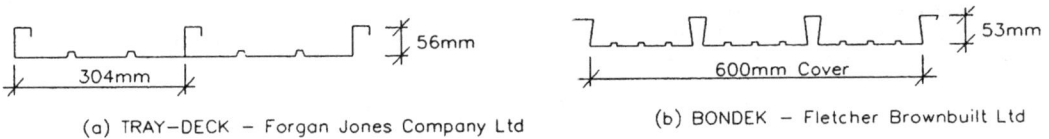

Figure 3.   Comparison of Steel Deck Profiles

The original design was based on the TRAYDEK profile which produces a solid composite slab.   The fabricator chose to substitute BONDEK on the assumption that it was an equivalent product.   BONDEK produces a hollow composite slab, however, and a simple substitution was not adequate.   As a result of this substitution the original 75mm long studs installed were too short and did not satisfy the Code minimum requirements.   The studs were ground off and the beams refabricated with 100mm long studs.   The depth of the slab had to be increased from 80mm to 105mm to permit full embedment of the longer studs. This increase in slab depth resulted in an increased midspan moment capacity without a corresponding in the capacity near the support (since there was no composite action in this region).   As discussed later this modified the intended behaviour of the portal during the test.

681

Standard concrete cylinder tests established a concrete compressive strength of 29MPa. Coupons tests from steel samples determined a steel yield strength of 310MPa and an ultimate strength of 465MPa.

Loading Arrangement

The fixed "gravity" load was applied vertically, in a downward direction, at midspan through four high tensile strands coupled via a manifold to a common hydraulic source. The cables passed through 50mm diameter holes bored in the deck slab and spanned between spreader beams at the top and the underside of the reaction floor at the bottom. A flexible loading system was necessary in order to permit the lateral displacement of the portal. Each spreader beam applied a concentrated point load to the slab through a rocker bearing at the beam web centreline. The original design assumed that the load was applied as a single concentrated load at midspan, but the limitation of the strand capacity resulted in the final load geometry of two point loads instead of the single point load. The substitution of the two loads for the single one did reduce the midspan peak moment and this did affect the proposed test program. The load was monitored through a load cell inserted between the strand anchorage and the spreader beam for each strand. The load system provided no torsional restraint to the beam since the four jacks remained coupled to a common hydraulic source throughout the test sequence.

During the cycling of the portal the vertical deflection at midspan of the beam and the cable geometry was changing constantly. This could have caused a consequential change in the applied load. In an attempt to maintain a constant vertical load the hydraulic system was linked to an accumulator and the pressure adjusted as required. The horizontal load was applied through a horizontal hydraulic jack mounted between the reaction wall and the portal frame. The applied load was measured with a load cell in series with the jack.

Instrumentation

One side of the beam was instrumented with portal-type displacement gauges set in a cross-braced truss pattern. The top and bottom chords of the truss measured the extension/compression of the bottom and top flanges of the beam section. Addition portal transducers were mounted at each end of the beam to measure total elongation/shortening of the beam and to measure deformations within the joint region. Portal displacement gauges spanned between the top flange of the steel beam and the slab soffit to measure any slip at the interface. A line of twelve portals down the centreline of the top of the concrete slab recorded strains in the slab. The gauges in this array were scanned semi-automatically via a Phillips Multichannel Scanning Datalogger to Personal Computer file storage.

On the upper surface of the concrete deck a grillage of DEMAC gauge points were used to measure the strains in the concrete in the midspan region and near the columns.

Load Sequence

Before the lateral load test, the portal was subjected to several loading and unloading cycles of vertical load only to verify the linearity of the system under elastic load levels.

The subsequent load tests were displacement rather than load controlled. A typical sequence beginning with cycles to 75% of "yield", then "first yield", and subsequent cycles to displacement ductilities of 2, 4 and 6 was envisaged. A definition problem arose however. Most tests undertaken by others in the past, and which have established the cycles of loading behind Code ductility rules, have been undertaken on statically determinate subassemblages. The important point being that there is no distinction between first "yield" ("yield" in this context implies full plastic moment at a section rather than first yield of the extreme outside fibre) and the formation of a collapse mechanism for such systems. There is a difference between first yield (i.e. the first hinge) and the formation of sufficient hinges to form a collapse mechanism for an indeterminate frame (such as the pinned based portal selected for this study).

The strategy adopted for this project was to first calculate the horizontal load intensity necessary to form the plastic collapse mechanism for the portal (250kN). This value was calculated theoretically, using the best estimates available for the material properties. The initial elastic lateral load-lateral deflection curve was plotted and the straight line curve extrapolated to find the displacement required to reach the theoretical collapse load (15mm). This can be considered a conservative displacement prediction since the portal stiffness was expected to soften with increasing load - particularly in the interval between the formation of the first hinge and the formation of the collapse mechanism. The displacement thus defined was treated as the "yield displacement" and subsequent cycles took the frame to 2, 4, and 6 times this displacement.

Theoretical Prediction of Beam behaviour

Prior to the load test, the anticipated performance of the portal frame was determined from plastic analyses. The results of the theoretical analyses are summarised in Steps 1-4 in Fig. 4. The horizontal load required to form the collapse mechanism was 250kN and the corresponding horizontal displacement 21.7mm. The equivalent figures at formation of the first hinge are a load of 92kN and a displacement of 3.8mm.

The flexural strength of the composite section of the beam member, 371kN-m, in the portal was based on the simple plastic section analysis recommended in NZS3404:1989. This value was supported by the results of a vertical load test of a simple span composite beam reported separately (HERA Report, 1990).

Figure 4 follows the predicted response of the portal frame through several cycles of lateral loading and contains several points of interest. Step 1 gives the bending moments which result from the self-weight of the portal and from the vertical "gravity load". During the first cycle of loading a hinge forms first at the beam face of the column remote form the applied lateral load (Step 2). During this increment of load the sway moments have little effect on the gravity moments at midspan. Beyond Step 2, however, the moments increase in proportion to the distribution identified as "Sway 2" in Step 2a. It is this step which introduces the possibility of a mid-span hinge. According to the idealised model, the second hinge (and mechanism) forms near the column face at the point of termination of the shear connectors, Step 3. (The mathematical model assumed a step change in member plastic moment capacity at this point.) The moments at midspan are approaching 95% of the section capacity.

Steps 3a & 4 trace the reversal of lateral load back to zero load. There is a substantial redistribution of the gravity moments in this new equilibrium position compared to the initial distribution in Step 1. Steps 5 through 7 trace subsequent cycles and demonstrate that little further redistribution of gravity moments is expected after the first mechanism forms. The lateral load required to form the first hinge increases fourfold as a consequence of the "gravity" moment redistribution, however the load required to form the collapse mechanism is little changed.

683

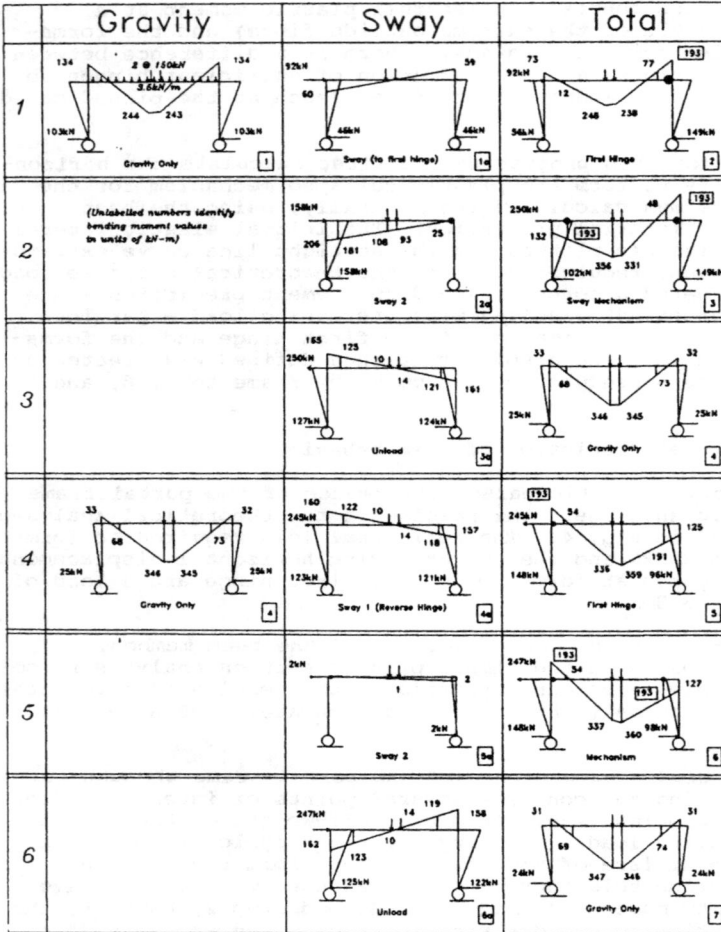

Figure 4. History of Sway Mechanisms and Moment Redistribution

Observed Response

In Figure 5 the cyclic variation of lateral load versus portal lateral deflection is plotted. The oval shape of the resulting plots suggests a robust seismic performance. Up to the end of the "ductility 4" cycle there was little strength loss, but beyond this level there was a softening of the structure. Large cracks began to open in the deck surface, radiating out from the shear stud locations in the deck at this displacement level.

Of interest for this paper was the collapse mechanism formed by the structure, in particular evidence of substantial gravity moment redistribution as predicted by the theoretical model.

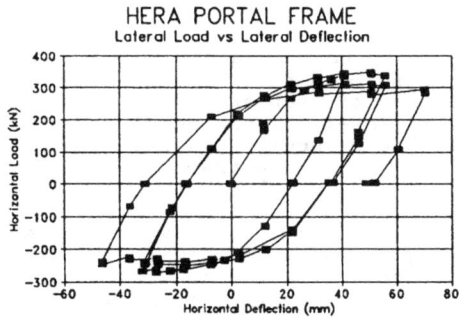

HERA PORTAL FRAME
Lateral Load vs Lateral Deflection

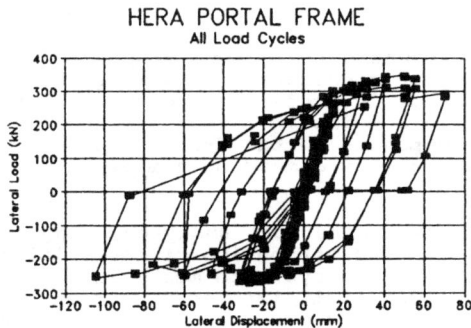

HERA PORTAL FRAME
All Load Cycles

(a) Ductility cycles 2 & 4 (b) Composite of all Cycles

Figure 5. Summary of Applied Horizontal Load versus Measured Horizontal
Deflection during Testing Cycles

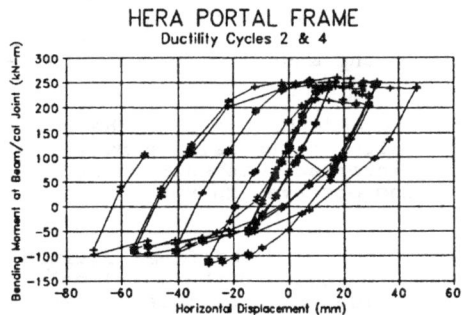

HERA PORTAL FRAME
Ductility Cycles 2 & 4

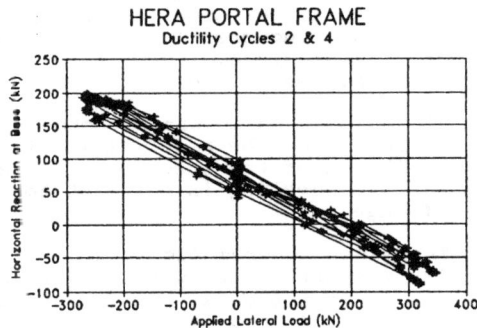

HERA PORTAL FRAME
Ductility Cycles 2 & 4

(a) Beam/Column Joint Moment vs
Applied Lateral Displacement

(b) Column Horizontal Reaction vs
Applied Lateral Load

Figure 6. Summary of Bending Moment, Applied Lateral Load and Applied
Lateral Displacement during Ductility 2 & 4 Testing Cycles

The redistribution of gravity bending moment is confirmed by the change
in horizontal reaction as measured at the base of the portal (the portal
column remote from the applied lateral load is plotted) which is plotted in
Figure 6(b). The initial reaction, under gravity only, of approximately 100kN
relaxes to less than half this value during successive cycles. The peak
bending moment at the joint (Figure 6(a)) approaches 275kN-m. This value
exceeds predicted value of Mp=193kN-m because of strain hardening effects and
because the plotted value is at the joint centreline whereas yielding takes
place at the column face.

685

Summary and Conclusions

It was difficult to proportion a portal frame to have a strength hierararchy which predicted that a midspan hinge would form as part of the collapse mechanism in preference to the more common hinges at each end of the beam. This is a point in favour of, rather than detrimental to, the performance of the composite portal frame. Because of their one direction plastic rotations the ductility demand on a midspan hinge can be more severe than a column face hinge.

The theoretical analyses, supported by measured results, suggested a substantial redistribution of the initial "gravity" moments particularly after just one load cycle to mechanism level. This redistribution is well in excess of the 30% limit imposed by current codes. Since such a redistribution is expected there would seem to be no reason to impose any limit on the amount of moment redistribution within a span for design load combinations which include seismic loads. The limit should be retained for normal gravity design to ensure a minimum strength distribution and acceptable service state performance.

The portal demonstrated a very stable resistance to the applied lateral loads with little pinching in the horizontal load vs lateral deflection hystertic loops. There was some torsion instability in the deck of the portal but this was a result of modelling a "plane frame" slice rather than a two-way deck spanning concurrently in an orthogonal direction to adjacent beams in a realistic floor system. Some simple propping at the ends of the span stabilised the torsion problem.

Although the slab was non-composite near the ends of the span there was clear evidence of the beneficial effect of the slab in restraining buckling in the top flange of the steel beam. The contact interface was sufficient to restrain the buckling in contrast to the buckling observed in the unrestrained lower flange.

There was no dramatic failure of the portal system. The final failure of the slab was associated with the propping added to restrain the torsional instability in the deck. During the last few cycles at large lateral displacement (+/- 100mm) there was shear slip in the beam part of the beam/column joint.

Acknowledgement

The research reported in this paper was funded by the Heavy Engineering Research Association (HERA), New Zealand. The authors are grateful for the financial support and for the cooperation provided by the HERA Structural Engineer, Mr Charles Clifton. Any opinions, findings and conclusions or recommendations are those of the authors and do not necessarily reflect those of HERA.

The Department of Civil Engineering, University of Auckland, provided laboratory space and resources, and the capable assistance from Senior Technician, Mr Hank Mooy, during the experimental program is recognised.

References

Canadian Standards Association. "Steel Structures for Buildings (Limits States Design)", CAN/CSA-S16.1-M89.

Gillies, A.G., Megget, L.M. and R.C. Fenwick. "Experimental Load Tests of a Full Composite and of a Partial Composite Simple Span Beam", Preliminary Report for Heavy Engineering Research Association, New Zealand, 1990.

Standards Association of New Zealand. "Code for Design of Steel Structures: Sections 12, 13, 14", NZS 3404:1989.

Bolted beam-to-column subassemblages under repeated loading

A. Osman[I], R.M. Korol[II], and A. Ghobarah[II]

ABSTRACT

Four beam-to-column subassemblages, representing parts of a typical moment resisting frame, were built and tested under a controlled cyclic displacement program. In these subassemblages, the beams were connected to the column flanges using extended end-plate connections. The tests were conducted to investigate the stiffness, strength, ductility, and the energy dissipation capacity of such a joint type and that of its individual components. Special emphasis was placed on the behaviour of the joint's elements, i.e the connection and the panel zone, and on their contribution to the overall response of the subassemblage.

INTRODUCTION

Current design specifications for steel structures in seismically active areas (UBC.1988 ;CAN3-S16.1-M89.1989) recommended that, the joints' panel zones in ductile moment resisting frames (MRFs) participate efficiently with the beams in dissipating the earthquake input energy. As a result, a design criterion that allows the columns' panels to yield and undergo sufficient inelastic deformation (about 2 to 4 γ_y , where γ_y is the panel average shear strain at yield) prior to the yielding of the beams was developed. In establishing such a criterion, the experimental results obtained by Krawinkler and Popov (1982) on beam-to-column subassemblages utilizing fully welded connections or connections with beam flange welded and beam web bolted to column flanges, were taken as the basis. The implication of adopting the same approach when other connecting media such as extended end-plate connections, employed to join beams to columns, was not investigated. The seriousness of such a problem can be understood from examining Fig.1. This figure shows the deformed shapes of the panel zones when both fully welded connections and extended end-plate connections are employed. As can be observed, the end-plate must deform in order to allow for the panel deformation, i.e the end-plate flexural stiffness contributes to the panel shear resistance. This contribution may be significant in the case of thick end-plates. Detailing the panel zones according to current design criteria without taking the end-plate contribution into account would generally result in

---

[I]Graduate Research Assistant, Department of Civil Engineering and Engineering Mechanics, McMaster University, Hamilton, Ontario L8S 4L7.

[II] Professors, Department of Civil Engineering and Engineering Mechanics, McMaster University, Hamilton, Ontario L8S 4L7.

relatively strong panels. As such, limited inelastic action would take place in the panel and dedicated most of the energy absorbtion to the beams. This contradicts the basis of earthquake design philosophy of requiring the input energy to be dissipated extensively throughout the structure.

Another important aspect that should be considered as a result of adopting such a joint type is the fact that the assumption of infinitely rigid connections can no longer be maintained. Recent research on end-plated connections showed that, relative rotation between the beam and the column can be observed at high loading levels (Ghobarah et al.1990). Consequently, the connection itself can contribute significantly to the overall drift of frames.

As a result, this study was conducted to investigate the cyclic behaviour of beam-to-column subassemblages utilizing extended end-plate connections. The main objectives are to gain information that can be helpful in detailing such a joint type in order to achieve good seismic performance.

EXPERIMENTAL PROGRAM

Description of subassemblages:

Four exterior joint subassemblages denoted as, CB-1, CC-1, CC-2, and CC-3, were tested. Each subassemblage consisted of a 2800 mm. long column connected to a beam at its mid height by an extended end-plate attachment (Fig.2). The connections were designed according to the design criteria proposed by Korol et al. (1991). The details of the joints are shown in Fig.3.

In designing each specimen, the strengths of the panel zones and the beams were deliberately changed relative to each other by altering the panel zone thicknesses. As a result, the effect of changing the panel zone strength on the overall behaviour of the subassembalge could be investigated. Table 1 shows the theoretical relative yield strength of the subassemblage components in terms of the beam tip load. It should be noted that the columns were designed to remain essentially elastic throughout the tests.

Test setup and procedure:

A special setup which allows application of the axial load to the columns while subjecting the beams to a cyclically controlled displacement was constructed. Fig.4 shows an illustration of the test setup. During the tests a constant axial load was initially applied to the columns, after which the beams subjected to a displacement program as in Fig.5.

The applied loads, the beam-tip deflection, the panel zone deformation, the connection rotation, and the internal stresses at various locations in the specimens were recorded. A more detailed description of the instrumentation used is reported elsewhere (Ghobarah et al.).

EXPERIMENTAL RESULTS

A summary of experimental results for the four tests is given in Table 2. The table shows for each test the maximum beam load recorded, P_{max}, the corresponding beam tip displacement, Δ_{max}, the maximum panel zone average shear strain, γ_{max}, and the maximum plastic beam rotation, θ_p. Also, the overall hysteretic behaviour for two of the tested specimens is presented in Fig.6.

Examining Table 2 and Fig.6 shows that with the exception of specimen CC-2, all the specimens exhibited stable hysteretic behaviour. The strength deterioration observed in specimen CC-2 was attributed

688

to severe local buckling of its beam. This buckling was triggered by the high demand imposed on the beam as a result of stiffening the panel zone. The latter responded elastically with only localized plastification. In specimen CB-1, the panel underwent several inelastic strain reversals and suffered shear buckling. However, this did not result in significant strength deterioration as there evolved the formation of an alternative system for transmitting the forces through the panel (diagonal tension field). In specimens CC-1 and CC-3, the panel participated jointly with the beam in dissipating the input energy. This imposed relatively low demands on each component resulting in good subassemblage performance.

Behaviour of connections:

Fig.7 shows a typical moment-rotation hysteretic behaviour of a connection in one of the tested specimens (CC-3). As can be seen, the connection suffered degradation in its stiffness with the progression of loading. Similar response was observed previously in cyclic tests conducted on extended end-plate connections. (Osman et al. 1990).

Behaviour of panel zones:

Figs.8 and 9 show the responses of the panels in specimens CB-1 and CC-3. In these figures, the theoretical applied moments required to yield the panels and the theoretical moments corresponding to the panel zones shear strengths are shown by a solid and dotted lines, respectively. It should be noted that, in calculating these moments the effect of axial force was neglected and the equation recommended by the codes to calculate the panel shear strength without taking the resistance provided by the end-plate into account was used. This equation is given by:

$$V_u - 0.55 d_c t_{cw} F_y [1 + \frac{3 b_c t_{cf}^2}{d_c d_b t_{cw}}]$$ (1)

where
  t_{cw} = the total thickness of the joint panel including doubler plates.
  d_b = the depth of the beam.
  d_c = the column depth.
  b_c = the width of the column flange.
  t_{cf} = the thickness of the column flange.

It should be noted that Eq.1 gives the shear resistance corresponding to a distortion of approximately 4 γ_y in the joint. Comparing these theoretical results with the experimental results revealed the following:
1) The predicted moments required to yield the panels in specimens CB-1 and CC-3 were 1.26 and 1.10 that of the actual recorded moments, respectively.
2) The predicted moments corresponding to the shear strengths of the panels in specimens CB-1 and CC-3 were 0.79 and 0.71 that of the actual recorded strengths at distortion of 4 γ_y, respectively.

Early yielding of the panels was attributed to the presence of the axial forces. However, the recorded high shear resistance of the panels in the post-elastic range was attributed to the stiffening effect of the end-plate. As can be seen, this effect counted for 21 to 29% of the total resistance.

Analysis of the panels' responses in other specimens (CC-1 and CC-2) supports the previous conclusions. Also, it was observed that the doubler plate welded to the panel participated efficiently in resisting the panel shear deformation.

689

Contribution of joints to overall behaviour:

The contribution of the individual components in the subassemblage to its overall behaviour can be expressed either in terms of the cumulative energy dissipated by each component separately, or in terms of the components' participation to subassemblage deflection. The first approach is considered important when assessing the ductility and the capacities of the components to undergo several strain reversals without failure. The second approach is desirable when drift is the issue. Both approaches were applied to the tested specimens. The results for specimen CC-3 are shown in Figs.10 and 11. As can be seen, the panel, the connection and the beam dissipated 28% ,13% ,and 59% of the total input energy ,respectively. In terms of the displacement, the panel, the connection, the column, and the beam contributed 21.5%, 16%, 17.5% and 45% to the total deflection. Similar results were obtained for other specimens. This shows the importance of incorporating joint behaviour in the analysis of MRFs.

CONCLUSIONS

Based on the previous investigation, the following conclusions may be drawn:

1. The panel zone is a very ductile element that can undergo several inelastic reversals without signs of distress.

2. The extended end-plate joints contributes significantly to the frame's interstorey drift and neglecting such an effect in the analysis will lead to serious errors.

3. Excellent seismic performance can be achieved by allowing both the panel zone and the beam to participate jointly in dissipating the input energy.

4. The end-plate as an adjoining element to the panel zone contributes significantly to the post-elastic panel shear strength.

5. Adopting current design criteria for detailing the panel zones in the case of extended end-plate joints will lead to strong panels that can impose higher demands than expected on the beams.

ACKNOWLEDGEMENTS

The authors wish to acknowledge the support of the Natural Science and Engineering Research Council of Canada. This work is carried out under NSERC grants to McMaster University.

REFERENCES

Canadian Standards Association. 1989. Limit state design of steel structures. CSA-Standard, CAN3-S16.1-M89, Canada.

Ghobarah, A., Osman, A.,and Korol, R.M.1990. Behaviour of extended end-plate connections under cyclic loading. Engineering structures,Vol.12, No.1,15-27.

Ghobarah, A., Korol, R.M.,and Osman, A. Cyclic behaviour of extended end-plate joints. J. Struct. Div., ASCE, to be published.

Korol, R.M., Ghobarah, A.,and Osman, A.1991. Extended end-plate connections under cyclic loading: behaviour and design. J. Const.steel Research, in print.

Krawinkler, H.,and Popov, E.P. 1982. Seismic behaviour of moment connections and joints. J. Struct. Div.,

ASCE, 108, ST2.

Osman, A., Korol, R.M.,and Ghobarah, A. 1990. Seismic performance of extended end-plate connections.Proceeding of 4th U.S National Conference on Earthquake Engineering, Palm Springs, CA, Vol.2.

Uniform Building Code. 1988. International Conference of Building Officials, Whittier, CA.

Table 1. Yield strength of specimen components

specimen	P_{ybeam}	P_{ypanel}	P_{yconn}
CB-1	159	105	103
CC-1	172	163	218
CC-2	172	275	218
CC-3	181	163	414

All loads are in kN.

Table 2. Summary of experimental results

specimen	P_{max} (kN.)	Δ_{max} (mm.)	γ_{max} (rad.)	θ_{max} (rad.)
CB-1	175	145	0.041	-
CC-1	250	135	0.015	-
CC-2	250	115	0.004	0.057
CC-3	260	140	0.012	0.032

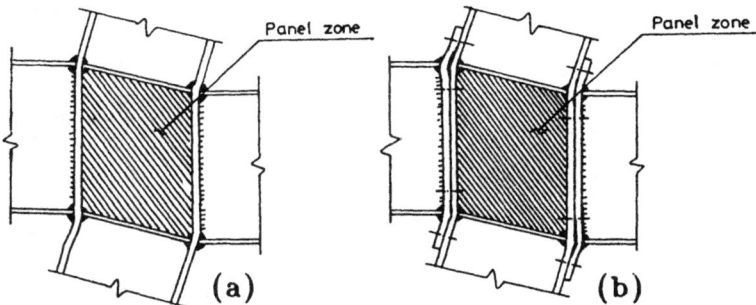

Fig.1 Deformed panel zones

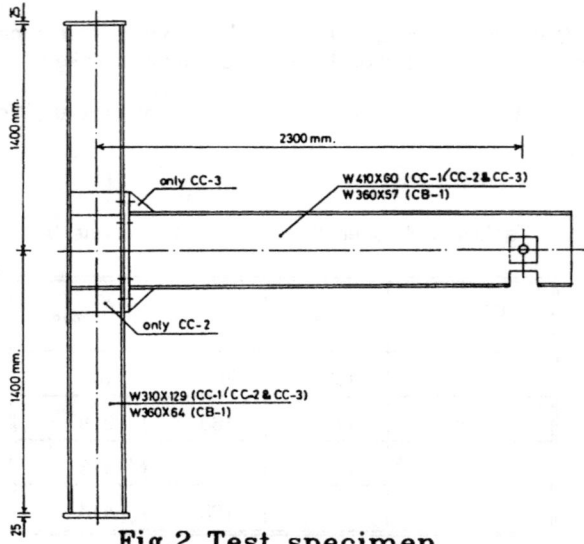

Fig.2 Test specimen

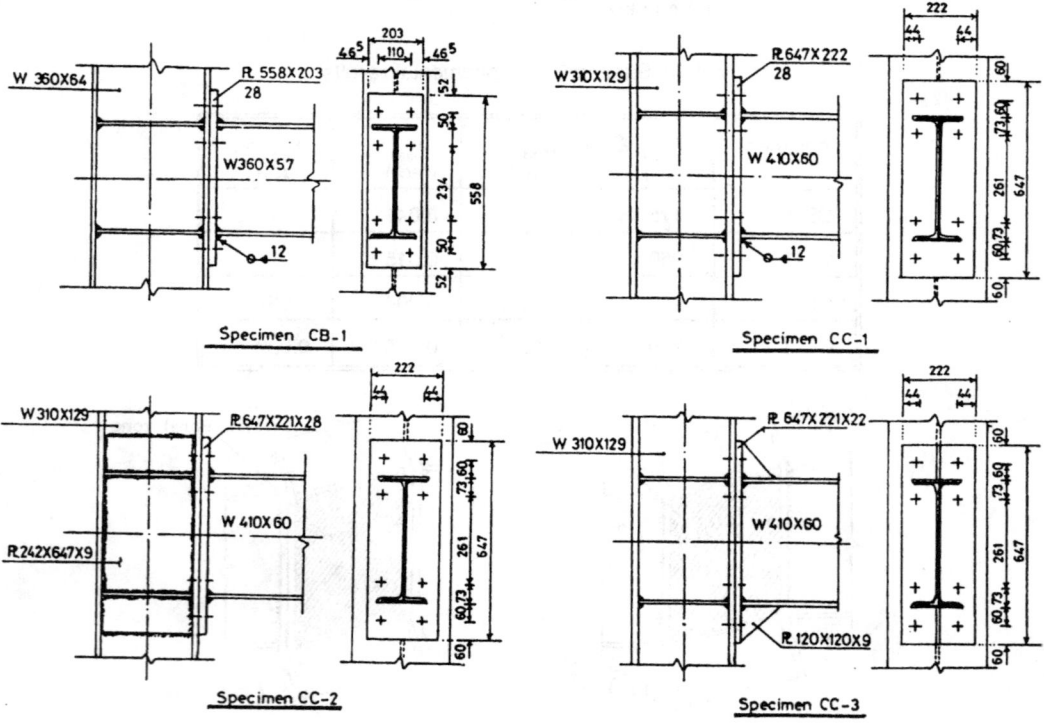

Specimen CB-1

Specimen CC-1

Specimen CC-2

Specimen CC-3

Fig.3 Joints details.

692

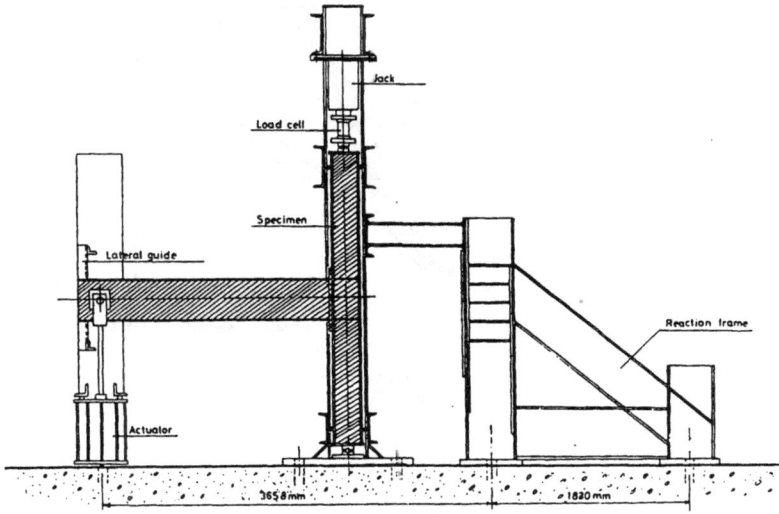

Fig.4 Test setup.

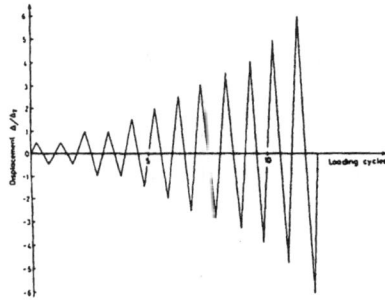

Fig.5 Loading routine.

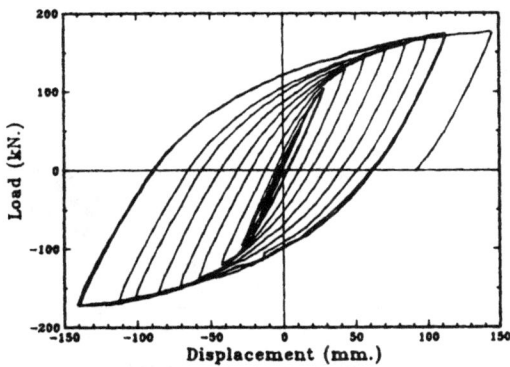

Fig.6(a) Beam tip load–deflection relationship
for specimen CB–1.

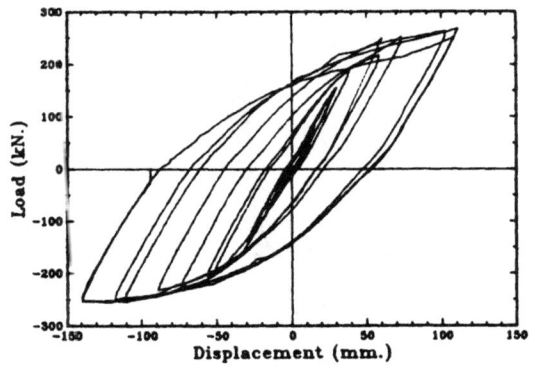

Fig.6(b) Beam tip load–deflection relationship
for specimen CC–3.

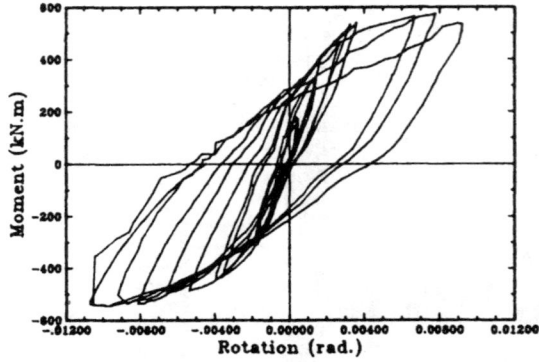

Fig.7 Connection moment–rotation relationship
(specimen CC–3).

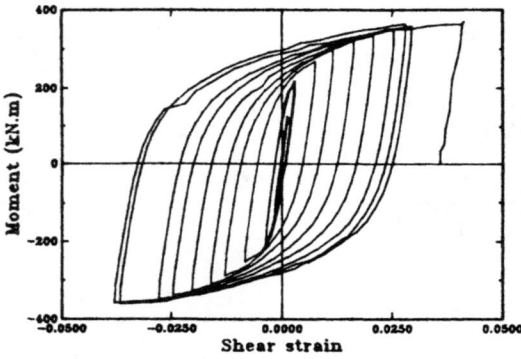

Fig.8 Applied moment–shear strain relationsip
for the panel zone (specimen CB–1).

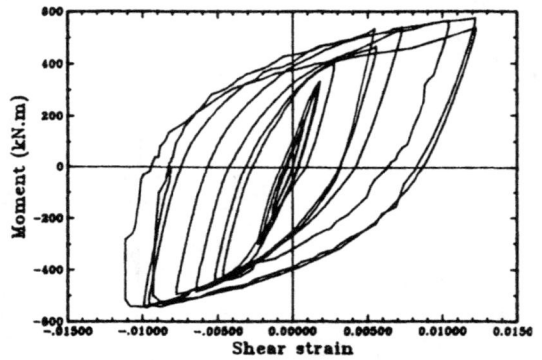

Fig.9 Applied moment–shear strain relationship
for the panel zone (specimen CC–3).

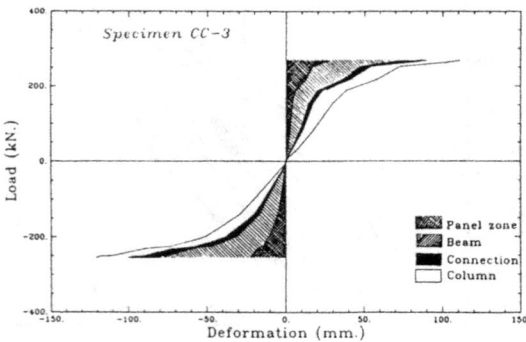

Fig.10 Contribution of specimen components
to beam–tip deflection (specimen CC–3).

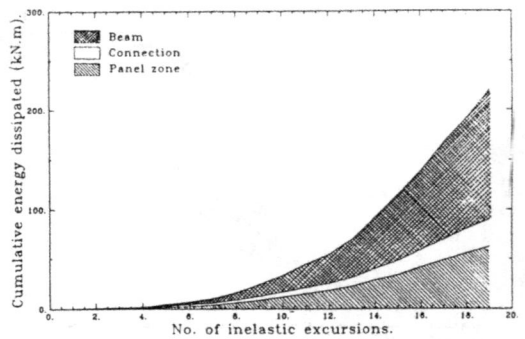

Fig.11 Cumulative energy dissipated by
each component in specimen CC–3.

694

Behavior of steel frame buildings with infill brick

Sigmund A. Freeman[1]

ABSTRACT

The City of Oakland, California, has a substantial inventory of mid- to high-rise steel frame buildings with unreinforced brick infill forming the exterior walls. Many of these buildings suffered significant damage during the October 17, 1989, Loma Prieta Earthquake. These buildings were built about the 1920s and their designs are similar to buildings built in cities east of the Rocky Mountains.

The ground motion felt in downtown Oakland was of fairly short duration with maximum peak horizontal ground accelerations estimated to range from 10% to 20% of gravity. The observations made on the performance of the infill steel frame buildings during the earthquake can be of great value in addressing earthquake hazard mitigation issues in the eastern half of North America.

Data obtained from observations and evaluations of brick infill steel frame buildings are summarized. A generalized case study is given to illustrate damage assessment techniques and methods of repair. The City of Oakland criteria for assessing and repairing earthquake damaged buildings and criteria for evaluating and upgrading existing undamaged infilled steel frame buildings is reviewed.

INTRODUCTION

North America has a substantial inventory of mid- to high-rise steel frame buildings with unreinforced brick infill forming the exterior walls. These buildings were generally not designed with earthquake forces in mind; wind being the governing factor for lateral forces. Not much is known on how these buildings will perform in a major earthquake. They are difficult to evaluate by conventional analytical procedures because of the uncertainty of how the brick walls interact with the structural steel frames.

This type of construction was very common for residential and commercial buildings in the downtown areas of San Francisco and Oakland, California,

---

[1]Branch Manager and Senior Consultant, Wiss, Janney, Elstner Associates, Inc., 2200 Powell Street, Suite 925, Emeryville, CA 94608.

during the early part of this century. Many were designed by Chicago and New York architectural/engineering firms. During the October 17, 1989 Loma Prieta Earthquake the San Francisco Bay Area experienced ground motion of fairly short duration with horizontal peak ground accelerations in the range of 10% to 20% of gravity. Reported damage to brick-infilled steel-framed buildings ranged from very minor to significant. The types of damage included cracked interior partitions and ceilings, spalled ornamental exterior terra-cotta and stone, and cracked and dislodged exterior brick. Although most of the buildings remained open, many were closed because of potentially hazardous conditions.

The post-earthquake damage assessments, structural evaluations, and rehabilitation proposals give us an opportunity to gain knowledge on how these types of buildings work.

BUILDING DESCRIPTION

The type of building discussed here is a steel frame structure with brick infill exterior walls.

The typical building has a complete three-dimensional steel frame that carries the gravity loads. The beams are generally rolled sections and the columns are generally riveted built-up sections. The exterior perimeter beams typically are offset from the columns' center lines so that the web of the beam lines up with the outer flange of the columns. An idealized representation of the brick-infilled steel frame is shown in Fig. 1. The riveted beam-column connections are primarily designed for gravity loads with some nominal moment resistant capacity for lateral forces. The moment resistant capacity is generally provided by two or four rivets attaching the top flange clip angle and the beam seat.

The infill masonry generally consists of three wythes of brick. Two interior wythes are supported by the top flange of the beam. The exterior wythe often includes decorative terra-cotta elements and its method of support varies for different buildings. In some cases a steel ledger plate cantilevers out from the steel beams. In other cases, it is supported by the keying action of header bricks interlocked with the interior wythes. Whereas the brick is generally tightly fitted above and below by the steel beams, the details of the brick interfacing with columns on each side are not usually well known.

The floor framing systems for these buildings may be wood or concrete. When wood is used, the floors generally have two layers of wood sheathing. When concrete is used, some of the structural steel framing may be encased in concrete fireproofing. Interior partitions may be plastered stud walls or hollow clay tile. Ceilings are lath and plaster hung from the floor framing above.

The buildings generally range in height from 6 to 7 stories for hotels to over 15 stories for office buildings. Floor areas range from 5,000 to 10,000 square feet. Most of the buildings have some shape irregularities such as flat-iron, L- or U-shaped in plan, light wells on the sides, light courts in the center, tall first stories, and open storefronts on one or two sides.

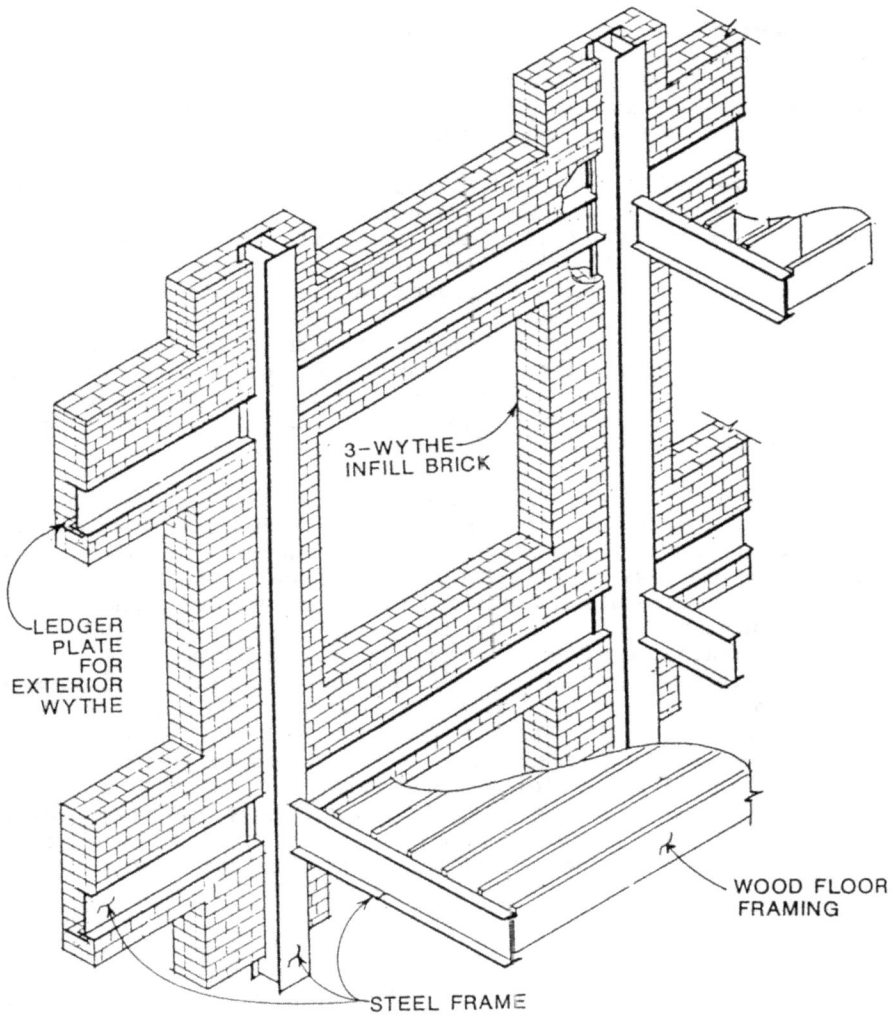

Figure 1. Schematic representation of brick infilled steel frame

STRUCTURAL EVALUATION

Lateral Force Resisting System

The lateral force resisting system of an existing brick infill steel frame building does not conform to structural systems in current codes (e.g., UBC 1988). If the steel frame alone is considered to be the structural system, it has deficiencies in connection details, strength, drift, and compatibility with the brick. If the brick is considered to act as a shear wall it is deficient because of lack of continuity and because it is unreinforced.

However, if we consider the interaction between the brick infill and the steel frame we can begin to develop a mathematical model that will define the lateral force resisting system. There are still some obstacles to fully understanding how the system works. For example, how does the brick interface with the steel frame? Is there horizontal shear transfer between beam and brick by bond or friction, or is the column confined enough by the brick to be fully engaged? What are the vertical reactions at the beam-column connections? How do the window openings affect the load paths through the brick? What is the difference between a brick pier that encases a steel column and an intermediate pier that is not on a column line? Fig. 2 shows an elevation of a typical exterior wall with some representative load paths.

Until better data is available, it appears reasonable to assume that the wider piers act as compression struts that fail in diagonal tension as shown in Fig. 2. For evaluation purposes, we have taken the area of a horizontal plane through the piers as an effective shear area. It is then assumed that the loads transfer to the steel frame and that the steel columns work axially to resist overturning. In addition to the exterior infill frame system, consideration must also be given to the horizontal diaphragms and to possible participation of interior partitions. If hollow clay tile partitions exist, they can play a major part in resisting lateral forces because of their rigidity. Lath and plaster partitions, to a lesser extent, can also participate in the lateral force resisting system; especially in cases where the diaphragms are flexible (e.g., wood flooring instead of concrete). Although these procedures may appear crude, they provide reasonable correlations between the performance observed for these types of buildings during the Loma Prieta earthquake and the data obtained from ground motion records.

Loma Prieta Earthquake Experience

During the Loma Prieta earthquake a substantial number of brick infilled steel frame buildings were damaged in San Francisco and Oakland. The results of the studies covered by this paper are primarily based on data from buildings in the downtown area of the City of Oakland.

The strong motion instrument located closest to the downtown area recorded peak ground accelerations of roughly 10% of gravity in a north-south direction and 20% of gravity in an east-west direction. These values were consistent with other recordings in the Bay Area. Response spectra developed from these

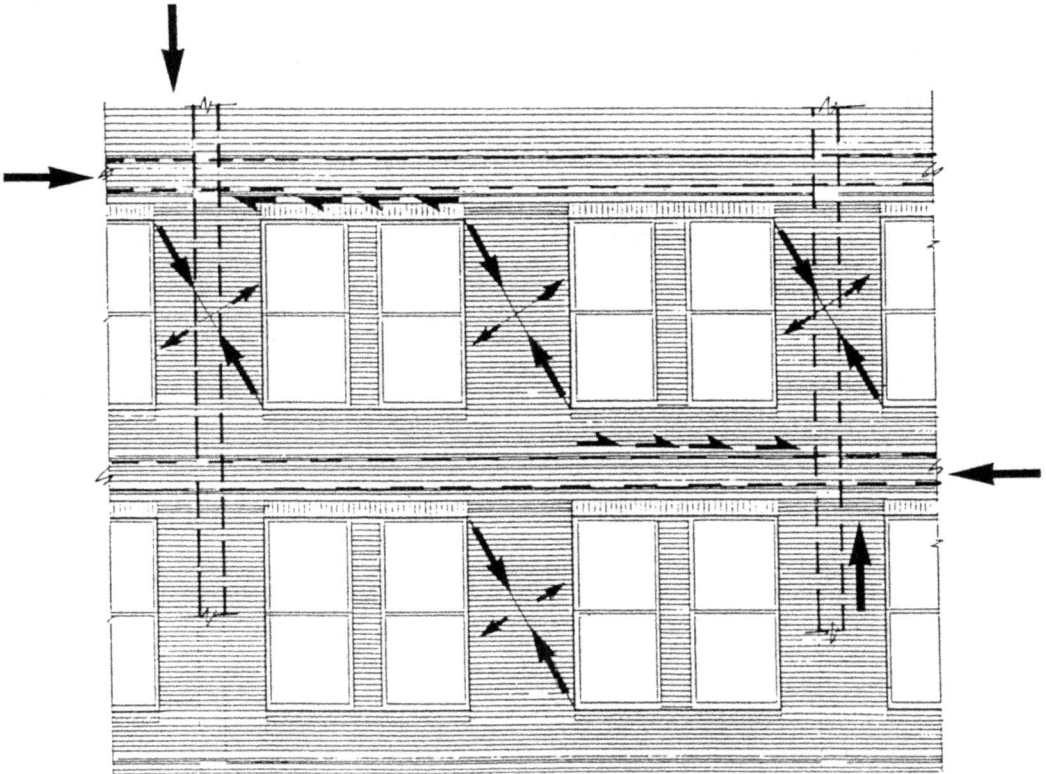

Figure 2. Partial wall elevation with representative load paths

recordings were fairly broad in shape indicating a significant ground motion amplification in the 0.5 to 1.0 period range (Fig. 3).

Soon after the earthquake the City of Oakland adopted an Emergency Order for Abatement of Hazardous Structures Associated with Earthquake Damage. This emergency order requires that a Damage Assessment Report (DAR) be prepared for earthquake damaged buildings. The DAR is to include an estimate of the loss of capacity to resist lateral forces that the building sustained during the earthquake. If the loss of capacity is more than 10%, the order requires a total seismic upgrade.

This process requires the engineer to estimate the lateral force resisting capacity of the structure prior to the earthquake. This capacity is stated in terms of an equivalent base shear coefficient calculated as the ratio of the horizontal seismic force to the weight of the building. If the post-earthquake capacity is at least 90% of the pre-earthquake capacity, the building need only be repaired to its former condition. However, if the capacity is less than 90% (i.e., the loss is greater than 10%) a full seismic upgrade is required.

The capacity of the building is determined in a rational, consistent manner, taking into account relative strengths and rigidities of the various materials. This includes materials not conforming to current codes if it is apparent that their participation was significant in resisting the forces caused by the earthquake. The results of the evaluations of both the pre-earthquake and post-earthquake structure should be consistent with the observed earthquake damage. In order to be rational, the values of the strengths of the elements have to reasonably represent the in-place materials. Sample guidelines were established to aid engineers. For example, a tentative recommendation of 75 psi was suggested for the shear capacity of infilled brick.

Sample Evaluation

As a result of studies of several damaged buildings some general observations can be made. Each of the studied buildings had some shape irregularities of the types described earlier. In each case, the initial observable damage did not appear severe. It was not always obvious that the loss of capacity exceeded 10%.

However, upon further evaluation of the load paths and the overall response characteristics, a pattern of the sequence of damage could usually be established. In this sample all buildings were about seven stories in height. Periods of vibration were estimated to be about 0.5 to 0.6 seconds in their pre-damaged state.

When buildings were evaluated on the basis of a code type allowable shear stress of 12 psi for brick infill, the resulting pre-earthquake equivalent base shear capacities ranged from about 2% to 4% of the weight of the building. For the steel frames, with their nominal moment resisting capacities, the base shears coefficients generally ranged from 1% to 3%. However, the brick infill was much stiffer than the frames and major loss of brick would have to occur before the very flexible steel frames would participate. If ultimate strengths, such as 75 psi for the brick infill, were used in the evaluation, the base shear equivalent would be about 15% to 25% of the building weight. In its cracked (or damaged) state the periods of vibration were estimated to be about one (1) second. Damping values were assumed to be 5% of critical damping for the uncracked state and 20% for the cracked state. Using this data with the smoothed response spectra in Fig. 3, approximations of the building capacities in terms of peak ground accelerations can be made as shown below.

	C_B	T	B	S_A	A_G
Uncracked	.02	0.5	5%	.025	.01g
Uncracked	.04	0.6	5%	.05	.02g
Cracked	0.15	1.0	20%	.20	0.13g
Cracked	0.25	1.0	20%	.30	0.20g

C_B = base shear coefficient
T = fundamental period of vibration
B = percent of critical damping
S_A = spectral acceleration = $C_B/0.8$ (approximate)
A_G = peak ground acceleration (acceleration at zero period of response spectra) obtained from ratio of S_A value above to S_A value in Fig. 3

This indicates that a peak horizontal ground acceleration of 1% or 2% of gravity could result in brick stresses of 12 psi; however, it would take a ground acceleration of 13% to 20% to reach an ultimate cracking strength of 75 psi. On the assumption of 10% peak ground acceleration in the north-south direction and 20% in the east-west direction during the Loma Prieta earthquake, it appears that one would expect minor or no damage in the north-south direction and significant cracking in the east-west direction of the building. This agrees with the observations. If the results did not seem reasonable, one would take a closer look at the assumptions and try again.

REPAIR AND UPGRADE

If a building loses more than 10% of its pre-earthquake capacity, Oakland requires that the entire structure be made to substantially comply with the structural requirements of the current code (i.e., 1988 UBC). However, some allowances are made because of the difficulties of bringing an existing building up to current standards. This includes the application of the equivalent of the 1991 UCBC for unreinforced masonry bearing wall buildings and the provision that the total design base shear coefficient need not exceed 0.133. The repair ordinance also states that the building official may approve an alternative procedure if it can be demonstrated by rational analysis that the modified structure provides the intended level of safety.

This last provision appears to allow for innovative solutions and the use of performance criteria or limit state procedures for seismic upgrading of existing structures.

CONCLUDING REMARKS

There is still a lot to learn on how steel frames with brick infill perform when subjected to major earthquakes.

If we take the opportunity to learn from data obtained from past earthquakes, we can improve our design procedures and criteria for upgrading existing buildings.

REFERENCES

UBC 1988. Uniform Building Code, International Conference of Building Officials, Whittier, California.

UCBC 1991. Uniform Code on Building Conservation, International Conference of Building Officials, Whittier, California.

ACCELERATION RESPONSE SPECTRUM
OAKLAND – 2STY STRUCTURE 5%, 20% DAMPING

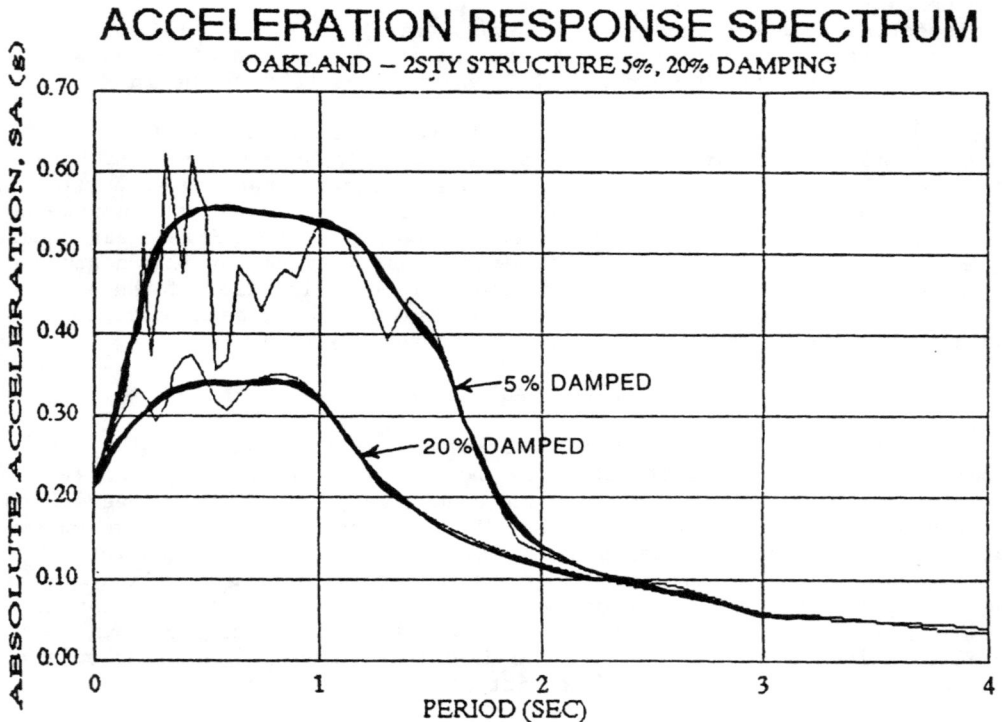

Figure 3.   Smoothed response spectra, Oakland, October 17, 1989

Ambient vibration testing and analysis of the Bosporus Suspension Bridge

Mustafa Erdik[I], Cetin Yilmaz[II], and Eren Uckan[I]

ABSTRACT

The paper presents the results of the ambient vibration survey of the Bosporus Suspension Bridge in İstanbul. In each of the experiment series several measurements of the ambient vibration of the bridge deck were obtained at different locations and orientations and at different traffic conditions. Between 0.07 and 1.0 Hz. most of the most of vibration of the bridge can be assessed from the experimental data. The substantial amount of the vibration energy on the bridge deck is contained between 2 and 3 Hz. The results of these experiments have been analysed in terms of the modal shapes, frequencies and damping ratios. These experimentally obtained mode shapes were compared with the theoretical ones obtained on the basis of two- and three-dimensional finite element analysis. Good correlation of experimental results with the theoretical ones are obtained.

INTRODUCTION

Bosporus Suspension Bridge, commissioned in 1973, joins the European and Asian Continents in İstanbul. Figure 1 shows the general structural sections and elevations of the bridge. Based on the project design sheets of Freeman, Fox and Partners, the design engineers of the bridge, the overall structural and physical parameters of the bridge are as follows: The main span is 1074 m long. The bridge deck is of hollow steel box girder type construction, has a total width of 33.4m and carries six lanes of traffic. The deck is supported by four steel towers of 165 m height. The axial distribution of the dead load mass is 14970 kg/m (deck: 11000, cables: 3850 and suspenders: 120 kg/m). Area moment of inertia of the deck about the two principal axis are 1.3 and 63.6 m^4. Torsional constant of the deck is 3.4 m^4. Nominal radius of the cables is 0.28 m.

Vibration characteristics and the dynamic properties of the suspension bridges are important design parameters controlling their wind and the earthquake safety. The determination of the dynamic parameters of the existing structures are, thus, very important studies that assist in the calibration of

----

I   Boğaziçi University, İstanbul, Turkey
II Middle East Technical University, Ankara, Turkey

the relevant analytical techniques (Bleich, et al.,1950). One of the best methods utilized to assess the dynamic characteristics of such massive structures is the measurement of the structural response to the ambient excitations, such as: wind and traffic noise. The structural vibrations caused by such excitations are termed the "Ambient Vibrations". Such measurements and experiments have been carried out on several suspension bridges (Abdel-Ghaffar and Housner, 1978). The ambient vibrations of the Bosporus Bridge has also been the subject of similar investigations conducted by Petrovski et al. (1974) and Tezcan et al.(1975), and Brownjohn et al.(1988).

EXPERIMENTAL SET-UP AND DATA ANALYSIS

The ambient vibrations of the Bosporus Bridge have been measured by a joint team of Boğaziçi and Middle East Technical Universities in two campaigns: in March, 1987 and November, 1988. In the measurements, depending on the amplitude of the signal, a four-channel data acquisition system consisting of either accelerometers or seismometers were employed. Figure 2 portrays the instrumental set-up.

The following equipment were used:
Analog-to-Digital Interface Board: A Metrabyte DAS-16 interface board installed in a PC to digitize analog data.
Tape Recorder: Hewlett-Packard Model 3964 A instrumentation tape recorder used to record 4 channel analog data.
Fourier Analyzer: Hewlett-Packard Model 3582 Spectrum Analyzer used to provide Fourier amplitude and phase transform, auto and cross correlation and transfer function of two channel analog data.
Acceleration Transducers: Four Shinkoh UA Series strain-gage type acceleration transducers with a range of ±1g and a flat frequency response up to 200 Hz.
Dynamic Strain Amplifier: Shinkoh DS-6002-F 6-Channel Dynamic Strain Amplifier used for the conditioning of accelerometer signals.
Seismometers: Four Kinemetrics SS-1 Ranger Seismometers with 1 Hz natural frequency.
Signal Conditioner: Kinemetrics CS-1 Signal Conditioner utilized to filtering and amplification of seismometer signals.
Analog Recorder: San-Ei Sokki Model 5M21 6 Channel Portable Direct Recording Oscillograph used to obtain direct record of the wave forms.

Measurements were taken along and across the main span and both inside and outside of the deck. Different set-ups were used to detect the vertical, lateral and torsinal vibrations. In each case, a reference station was established and simultaneous successive measurements were taken at the moving station. For the detection of torsional vibrations the vertically oriented transducers are placed at the outer edges of the deck and the difference of their signals were used. Measurements were also taken under light and heavy traffic conditions to assess the difference. The vibrations were recorded at each experiment for a duration of about 10 minutes to allow for satisfactory resolution and averaging in spectral analysis. A typical power spectrum of the acceleration obtained at the mid-span in lateral direction is provided in Figure 3. The high energy vibrations at 2.5 Hz is indicative of local vibrations of the deck.

A preliminary processing of data was carried out in-situ through the Fourier analyzer. The final processing was done in the office through playback of the recorded data via an analog-to-digital converter into a PC. For the identification of the modal vibration frequencies and shapes, frequency domain analysis involving Fourier amplitude and phase spectra, cross spectra and coherence functions were used. The half-power bandwidth method was used to estimate the modal damping ratios. The estimations vary upto ±50% from one measurement to another.

EXPERIMENTAL RESULTS

The results of these experiments have been analyses in terms of the peak frequencies. It was possible to associate some of these frequencies with the modal vibrations. In Table 1 a general assessment of the experimental modal frequencies of vibration and the identified mode shapes are provided. The first four of the experimentally obtained mode shapes are compared in Figure 4 with the theoretical mode shapes obtained on the basis of three dimensional finite element analysis. The experimental damping values are also indicated on thiese figures.

TWO DIMENSIONAL FINITE ELEMENT ANALYSIS

A finite element solution of the continuum approach developed by Abdel-Ghaffar (1978) for the lateral vibration of suspension bridges is utilized to obtain the theoretical mode shapes for the lateral vibration of Bosporus Bridge. The approach is based on the derivation of the equations of motion through the use of the Hamilton's principle under the assumptions that: (1) vibration amplitudes are sufficiently small to remain in linear ranges, (2) coupling between lateral and torsional modes are ignored and (3) the cable ends are immovable. Mathematical model of a suspension bridge should depend on the construction details and on the support conditions. In three span-two hinged bridges, the interaction between the main and side spans may not be significant. The analytically derived energy expressions are then used to express the stiffness and consistent mass matrices for the finite element application. As many equations are obtained as the number of active degrees of freedom used to express the mathematical model. The normalized interpolation functions with respect to the horizontal axis are used to obtain the displacement vector in terms of the nodal displacements. For the Bosporus Bridge a subdivision of elements is used resulting in 23 degrees-of-freedom.

The first five lateral modes of vibration of the Bosporus Suspension Bridge, computed on the basis of this simplified finite element analysis scheme are plotted in Figure 5. Even though the bridge is divided into only eight elements, the theoretical results are in good agreement with the experimental values. It is observed that in the first symmetric and anti-symmetric modes, the cable and the deck motion are in phase, whereas, in the fourth mode the two systems are in out of phase motion. The third, sixth, seventh and the eight modes can be considered as cable modes since the cable movements are dominant. In these modes the cable displacements are about 10 times that of the deck.

TABLE 1

EXPERIMENTAL AND ANALYTICAL MODAL PARAMETERS

MODE SHAPE	EXPERIMENTAL FREQUENCY	DAMPING	ANALYTICAL FREQUENCY 2-D.O.F.	3-D.O.F.
First Lateral Symmetric	0.072Hz	7%	0.077Hz	0.072Hz
First Vertical Asymmetric	0.150Hz	4%	-	0.141Hz
First Lateral Asymmetric	0.215Hz	4%	0.192Hz	0.223Hz
Second Vertical Symmetric	0.230Hz	4%	-	0.233Hz
First Torsional Symmetric	0.320Hz	-	-	0.325Hz
Second Vertical Asymmetric	0.330Hz	-	-	0.340Hz
Third Vertical Symmetric	0.370Hz	-	-	0.367Hz
Second Lateral Symmetric	0.380Hz	-	0.412Hz	0.387Hz

THREE-DIMENSIONAL FINITE ELEMENT ANALYSIS

The modal vibration frequencies and the shapes of the Bosporus Suspension Bridge is determined on the basis of SAP-90 finite element analysis program (Wilson and Habibullah, 1989). The bridge deck is modeled using equivalent shell elements. The effects of axial forces on the stiffness of the cables are considered through the use of equivalent frame elements. The model used is intended to be a preliminary one and does not explicitly include the towers and side spans. However, the boundary conditions are chosen to simulate the effect of the deck and cable supports at the towers. Figure 6 provides the isometric views of, respectively, the first lateral symmetric, first vertical asymmetric, first lateral asymmetric, second vertical symmetric, first torsional symmetric and second vertical asymmetric mode shapes. Results are also listed in Table 1. The experimental and the 3-D finite element analytical values of the modal vibration frequencies are given under each inset shape.

CONCLUSIONS

The study presented involves the experimental assessment and the theoretical verification of the modal shapes and the frequencies of the Bosporus Suspension Bridge. Following is a summary of the findings and conclusions:

1. As it has also been shown by previous investigators, the ambient vibration survey techniques can be satisfactorily utilized to obtain the vibration mode shapes and the frequencies of the suspension bridges.

2. A relatively large number of modes may be necessary to obtain a reasonable representation of the response.

3. Between 0.07 and 1.0 Hz. most of the vibration of the bridge can be assessed from the experimental data. However more refined techniques and denser instrumentation may be required for the reliable differentiation of the closely spaced modes.

4. The local spurious vibrations of the bridge deck contaminate the ambient vibration of the total bridge structure and makes it impossible to recover any modal data in the frequency ranges higher than about 1.5 Hz.

5. It was not always possible to determine reliable damping values by the use of the half-power method, due to closely spaced peaks.

6. Theoretical analysis of the free vibration of suspension bridges indicate that in certain modes the displacemnts of the deck are important (Deck Modes), whereas in others the displacements of the cables (Cable Modes). The deck modes can be classified as symmetric or asymmetric and lateral or vertical or torsional.

7. Theoretical 2D and 3D finite element analysis have produced results that correlate surprisingly well with the experimental data considering the assumptions made regarding the boundary conditions, equivalent shell element modeling the deck and the equivalent frame element modeling the cables.

8. The substantial amount of the vibration energy on the bridge deck is contained between 2 and 3 Hz. These vibrations associate with vertical accelerations peaks up to 0.3g. There may be reasons to believe that these high energy vibrations constitute one of the causes of the low-cycle fatique effects observed on the welds in the deck. Future investigations should concentrate on the source, characteristics and the remedies regarding these high amplitude local vibrations of the deck.

ACKNOWLEDGEMENT

This research has been supported by the Research Funds of Boğaziçi and Middle East Technical Universities. The permission and assistance of the Turkish General Directorate of Highways are gratefully acknowledged.

REFERENCES

Abdel-Ghaffar, A.M. and G.W.Housner (1978), Ambient Vibration Tests of Suspension Bridges, Journal of Engrg. Mech. Div., ASCE, v.104, No:EM5, pp. 983-999.

Abdel-Ghaffar, A.M. (1978), Free Lateral Vibrations of Suspension Bridges, Journal of Structural Division, ASCE, V.104, ST3, 503-525.

Bleich, F., et al. (1950), The Mathematical Theory of Vibration of Suspension Bridges, Bureau of Public Roads, Department of Commerce, U.S. Government Printing Office, Washington D.C.

Brownjohn, J.M.W., R.T. Severn, A.A. Dumanoğlu and A. Blakeborough (1988), Ambient Vibration Survey of the Bosporus Suspension Bridge, Report No. UBCE-EE-88-1, Department of Civil Engineering, University of Bristol, U.K.

Petrovski, J., T. Paskalov, A. Stojkovich, D. Jurokovski (1974), Vibration Studies of Istanbul Boğaziçi Suspension Bridge, Report OIK 74-7, Institute of Earthquake Engrg. and Engrg. Seismology, IZIIS, Skopje, Yugoslavia.

Tezcan, S., M. İpek, J. Petrovski and T. Paskalov (1975), Forced Vibration Survey of Istanbul Boğaziçi Bridge, Proc. 5th ECEE, Vol.2, Istanbul, Turkey.

Tezcan, S.S. and S.Cherry (1969), Earthquake Analysis of Suspension Bridges, Proc. 4th. World Conf. on Earthq. Engrg., V.2, Chile, 1969.

Wilson, L.E. and A. Habibullah (1989), SAP90: A Series of Computer Programs for the Static and Dynamic Analysis of Structures, Computers and Structures, Inc., Berkeley, CA., USA.

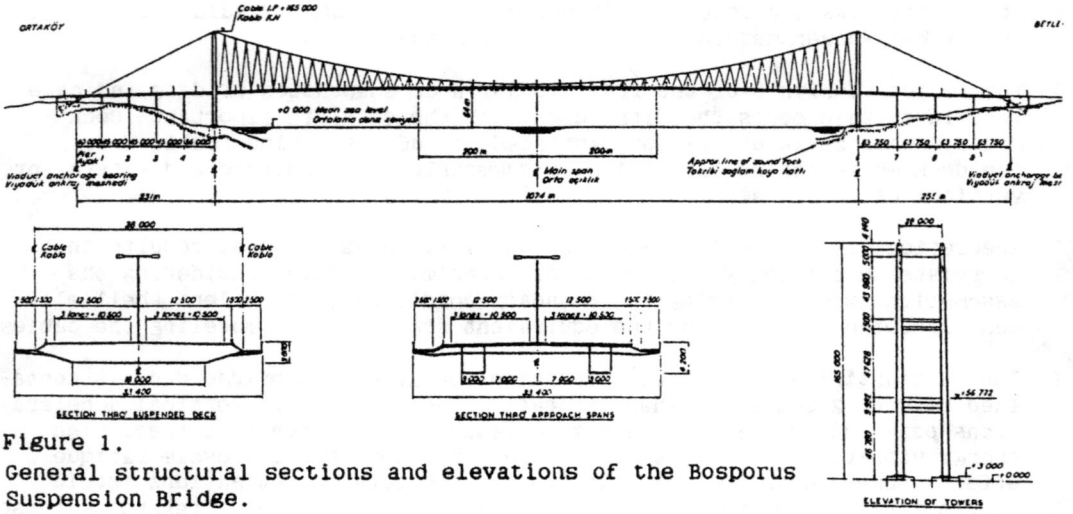

Figure 1.
General structural sections and elevations of the Bosporus
Suspension Bridge.

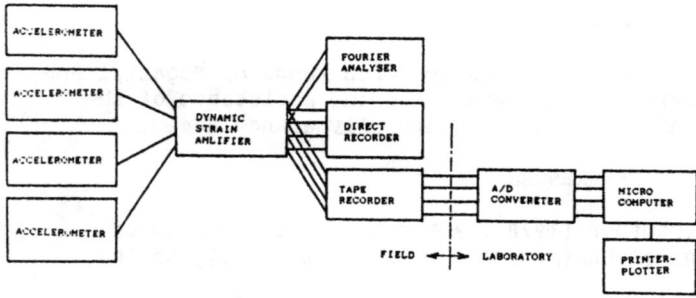

Figure 2.
Instrumental set-up
using accelerometers

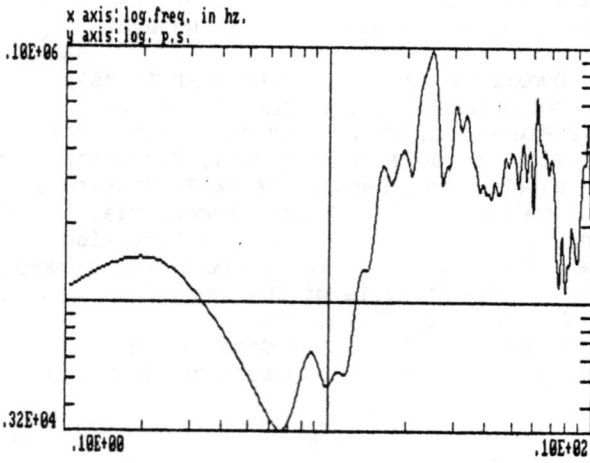

Figure 3.
A typical power spectrum of the
acceleration obtained at the midspan in lateral direction. Note
the high energy peaks at 2.5Hz.

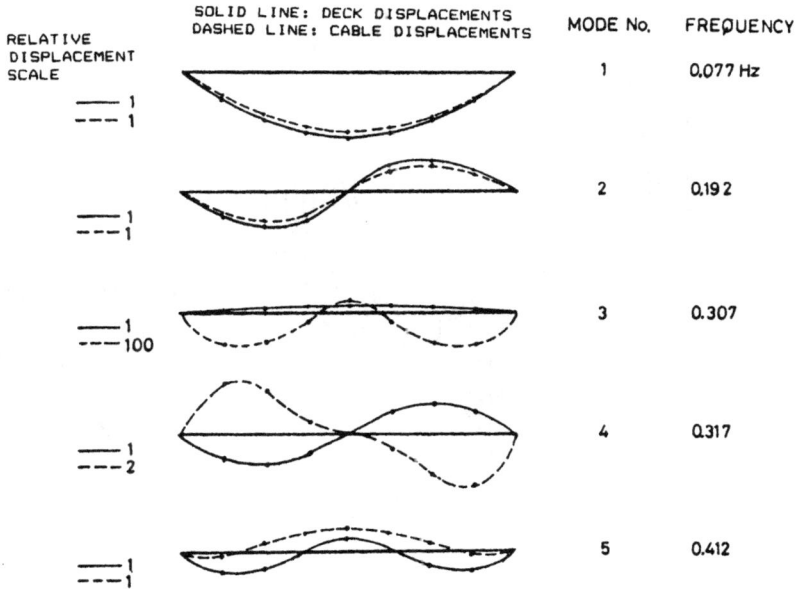

Figure 5. The first five lateral modes of vibration of the Bosporus Suspension Bridge, computed on the basis of 2 D.O.F. finite element analysis.

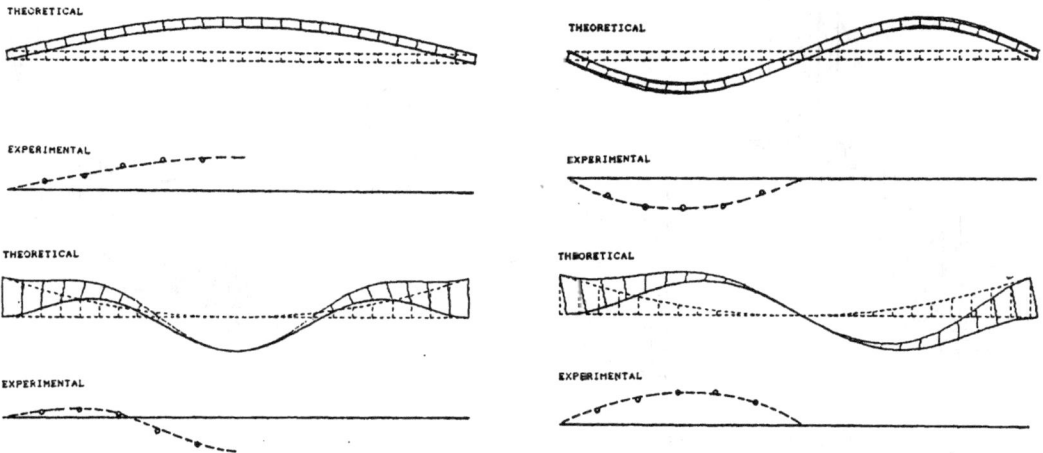

Figure 4. Comparison of the first four experimentally obtained mode shapes with the theoretical ones obtained on the basis of three dimensional finite element analysis. The experimental damping values are respectively 7%, 4%, 4% and 4% at each mode. (Also see Table 1)

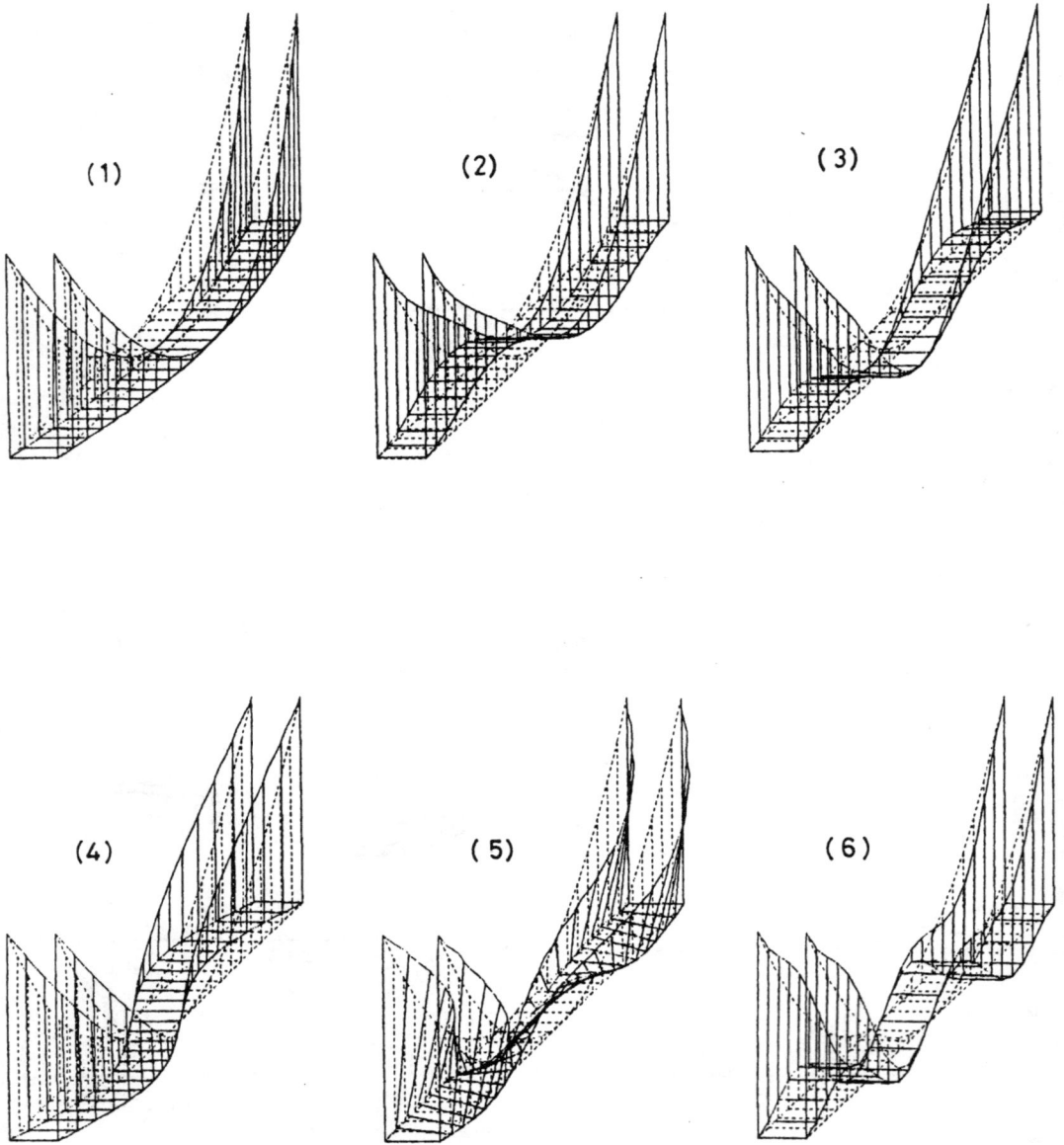

Figure 6. The isometric views of, respectively:
(1) The first lateral symmetric, (2) First vertical asymmetric,
(3) First lateral asymmetric, (4) Second vertical symmetric,
(5) First torsional symmetric and (6) Second vertical asymmetric
mode shapes. (Please refer to Table 1 for the experimental and the
3-D.O.F. finite element analytical values of the modal vibration
frequencies.)

Impact of the 1989 Loma Prieta earthquake from a Canadian perspective

D.T. Lau[I], A.M. Jablonski[II], K.T. Law[I], J.-R. Pierre[III], and J.H.K. Tang[IV]

ABSTRACT

Shortly after the earthquake, the Institute for Research in Construction of the National Research Council of Canada (IRC/NRCC) organized a team of engineers to visit the San Francisco Bay area. This paper gives an overview of the findings by the reconnaissance team. It covers important observations regarding the geotechnical aspects of the ground motion, the performance of building structures, liquid storage tank and other industrial facilities, electric power generating and distributing facilities, and transportation structures. A summary of the lessons learned from the Canadian perspective is also presented.

INTRODUCTION

On October 17, 1989, 17:04 Pacific Daylight Time, a strong earthquake, called the Loma Prieta Earthquake, of Richter magnitude $M_L = 7.0$ originated from the San Andreas fault and shook the entire San Francisco Bay area. The average surface wave magnitude (M_s) was estimated by the U.S. Geological Survey as 7.1. The epicenter was located approximately 16 km northeast of the city of Santa Cruz. The felt area stretched from Las Vegas, Nevada to Los Angeles, California. The hypocentral depth was about 18.5 km, which is relatively deep for most events associated with the San Andreas fault. The fault movement includes not only the typical horizontal component of slip, but also a significant thrusting of the southwest side up and over the northeast side. The rupture zone was 40 km long. The overall felt duration of the strong shaking in some areas reached 20 seconds.

The closest accelerographs were located in Santa Cruz and in Watsonville, and they recorded the highest horizontal and vertical peak accelerations: Santa Cruz – 0.64g H, 0.47g V; Watsonville – 0.39g H, 0.66g V. The records from San Francisco stations showed differences depending on the type of soil deposits and other geologic features. Figure 1 presents the measured peak horizontal acceleration vs. epicentral distance according to the strong motion data collected from the California Strong Motion Instrumentation Program network (Shakal et al. 1989) and the U.S. Geological Survey network (Maley et al. 1989). The attenuation tendency is also shown.

[I] Dept. of Civil Engineering, Carleton University, Ottawa, Ont. K1S 5B6
[II] Inst. for Research in Construction, National Research Council, Ottawa, Ont. K1A 0R6
[III] Services Etudes et Normalisation, Hydro-Québec, Montreal, Québec H2L 4P5
[IV] Civil Eng. & Architecture Dept., Ontario Hydro, Toronto, Ontario M5G 1X6

This earthquake was one of the most costly natural disasters in U.S. history. The estimated cost of damage was in the range of $6 – 10 billion. At least 62 people were confirmed dead and 3,757 people were injured. More than 12,000 people were displaced from their homes by the earthquake.

GEOTECHNICAL ASPECTS

Ground motion amplification and soil liquefaction were two major problems due to soil conditions during this earthquake. The amplification of earthquake motions is a complex process and is dependent on soil properties and thickness, frequency contents of motion and local geological settings. For a given earthquake and geological setting, the amplification increases with the increase of soil compressibility and thickness. In the San Francisco Bay region, various thicknesses of a soft deposit known as Bay mud are found. This deposit is normally underlain by a layer of alluvium. Measurements showed that these deposits amplified the maximum peak horizontal accelerations by a factor of two to three. Structures founded on these deposits therefore have been subjected to high horizontal excitation during the earthquake, hence resulting in many major damages in the region. Examples are the collapsed 2 km section of the Nimitz Freeway, the failure of one section of the Bay Bridge and many severely damaged buildings in the Marina District.

This earthquake caused extensive soil liquefaction damage in the form of sand boils, ground cracks, ground heave, building collapse and differential settlement. Many of these damages occurred in areas known to experience similar failures in the 1906 San Francisco earthquake. Figure 2 shows the locations of the damages as well as the corrected standard penetration resistance (N_1) of the soil, below which liquefaction will occur for this earthquake. The N_1 values were obtained from an energy approach for assessing liquefaction potential (Law et al. 1990). As an example, this method suggests that the Marina District will suffer liquefaction failure when N_1 is less than 10. Such a value is consistent with that of the hydraulic fill which liquefied there.

PERFORMANCE OF BUILDINGS

In general, the performance of buildings during an earthquake depends on a large number of factors such as: the type of structure, year of construction, lateral resistance system and local ground effects (soil liquefaction, densification of soil, differential settlement, lateral spread, uplift or rupture of actual fault line under the buildings).

Wood Frame Housing

The performance of wood frame housing in this earthquake has been reported by Brueneau (1989), and Jablonski et al. (1990) and Rainer et al. (1990). Two types of wood-frame houses sustained heavy damage. The first type is different variations of single-family dwelling in the epicentral area. The second type is multi-storey wood frame construction, typically 3 - 4 storey wood frame apartment houses or 2 - 3 storey townhouses in the Marina District in San Francisco (about 100 km from the epicenter). In the epicentral area, most of the damage was caused by failure of "crippled" stud foundation walls (also called "pony" walls). Cripple stud walls are common in older buildings in California. They are short (less that 14"

(35.6 cm) high), but may reach a height of one full storey in modern wood villas. Improper bracing and inadequate connections to the foundation and the upper-floor framing system caused many older buildings displaced from their foundations. In general, majority of wood houses performed well except those situated on large ground fissures. Outside the epicentral region, extensive structural damage occurred in the Marina District in San Francisco. The highest recorded acceleration nearby, 0.21 g, was recorded in Presidio on rock, a few blocks northwest from the Marina District. The garage floor of the typical townhouses and apartment buildings in the area behaved like a "soft storey" usually without bracing, or in some cases with only limited lateral resistance provided by sheathed walls (horizontal boards nailed to posts) (Fig.3).

Unreinforced Masonry Buildings

Many unreinforced masonry buildings (especially older ones) suffered severe damage and in some instances complete collapse. Examples of damage include out-of-plane collapses and large in-plane distress of the masonry walls, and brick veneer, bearing walls, parapets, and in-fill masonry walls. Severe damage to masonry buildings were reported in Santa Cruz (the Pacific Garden Mall), Los Gatos and Watsonville (Fig.4). Upgraded old stone masonry buildings at Stanford University campus in Palo Alto performed well.

Engineered Buildings

Case studies from the Loma Prieta earthquake related to engineered concrete structure has been reported by Mitchell et al. (1990). Most of the high-rise buildings in the San Francisco Bay area were not seriously damaged. There were some exceptions in Oakland, such as the steel frame building at Franklin and 18th Street, built in the early 1960's, which experienced significant shear cracks on the shear walls and some damage to the first floor columns.

PERFORMANCE OF TRANSPORTATION STRUCTURES

The Loma Prieta earthquake had a major effect on transportation routes and bridges in the San Francisco Bay Area. Two examples of the damage to transportation structures are presented here: the collapses of the Nimitz viaduct of I-880 and the failure of the Struve Slough Bridge on Highway 1 near Watsonville. The damage to the Oakland Bay Bridge has been described by Astaneh et al. (1989).

Collapse of Nimitz Viaduct of I-880

The collapsed freeway was located a short distance south of the east side entrance intersection of the San Francisco-Oakland Bay Bridge, approximately 100 km from the epicenter. The double deck freeway was designed in 1954 and constructed according to the design standards applicable at the time. There were two common types of design layouts, different mainly in the placements of pin connections in the upper frame structure. For most of the collapsed bents along the 1.2 km failed sections shown in Fig.5, the columns failed at the hinge joint regions due to shear force from the lateral load induced by the ground motion. Other details are reported by Jablonski et al. (1990) and others.

Struve Slough Bridge

The Struve Slough Bridge is actually two bridges for the north and south bound traffic. Due to large horizontal displacement associated with laterally shifted old river bed, some columns were sheared off and displaced from the external bridge girders as shown in Fig.6. The southern deck dropped about 30 cm down and many columns punched through it. The northern deck columns barely withstood the earthquake.

PERFORMANCE OF INDUSTRIAL FACILITIES

The San Francisco Bay Area is a major commercial and industrial center on the west cost of North America. Although many industrial facilities throughout the regions experienced strong ground motions from the main shock and the many aftershocks, in general most facilities survived the earthquake with only minor or no property damage.

Electric Power Generating Plants

The Loma Prieta earthquake disrupted the electric power systems to about 1.4 million customers throughout the regions. Within 48 hours, full service to most areas was restored except for about 26000 customers. Located about 30 miles (50 km) south of the epicenter is the Moss Landing Power Plant, which was damaged by the earthquake. Only the 750 megawatts natural gas fired unit 6 built in 1967–68 of this seven-unit plant, was operating at the time of the earthquake. The superstructure of unit 6, consisting of a 200 ft (61 m) high steel braced frame supported on the ground floor, generally performed very well. Only two bracing members buckled. There was no crack in the foundation slab, although some ground settlement was observed outside.

No damage to piping was observed. A vertical strut, supporting the horizontal snubbers for the super heater by-pass between the primary heater and secondary heater, yielded in bending as a result of the large displacement experienced during the earthquake.

The two 500 ft high steel lined, reinforced concrete stacks supported on pile foundations, did not suffer any serious damage, except for the cracking of the valve leading to the water tank located at the bottom of the stack.

In spite of the damage mentioned above, the plant was on-line again within 24 hours.

Substation Damage

The substations at Moss Landing, Metcalf and San Mateo, with epicentral distances of 45, 30, and 75 km respectively, suffered considerable damage. From the acceleration records of the surrounding areas, the intensity of the ground motions at those locations are estimated to be between 0.12 g to 0.45 g.

At the Moss Landing Substation, many seismic strengthened gas circuit breakers, designed to withstand about 0.30 g acceleration, failed and were overturned. The anchoring devices were inadequate, and the steel supports were dragged down by the failing circuit breakers. Other damage included the failures of several current transformers, disconnectors, wire-traps, and an oil leak in a power transformer. Similar circuit breakers failed at the 230-kV San Mateo Substation and the 500-kV Metcalf Substation because of rupture at the base of the porcelain column.

The total load loss due to this 7.1-magnitude Loma Prieta earthquake was estimated at about 4300 MW on the PG&E network with an overall system capacity of about 18000 MW. As a comparison, the Saguenay earthquake (M=6) caused a load loss of 3000 MW on the Hydro-Québec system, which has a capacity of 30226 MW.

Liquid Storage Tank Damage

Many thin-walled, stainless steel wine storage tanks in a winery facility, located approximately 20 km east of the epicenter, suffered earthquake damage. The severity of the damage depended on the liquid content level at the time of the earthquake. In general, empty tanks suffered little or no damage. In other cases, the anchored bolts of many tanks were pulled off the concrete pad of the foundation (Fig.7) as the tanks rocked in response to the sloshing liquid during the strong ground motions. Following the failure of the anchorage system, the high compressive axial stress from the rocking motion of the tanks buckled the tank shells.

At least two 25,000-gallon anchored, stainless steel fermentation tanks suffered the commonly known "elephant foot" type of buckling, of which the characteristic is that the buckling is generally located just above the tank base and spreads around the entire or most of the circumference Nearby, several 19,000-gallon tanks suffered the diamond-shaped type of buckling shown in Fig.8.

The liquid storage tanks in a food processing plant, located a few city blocks from downtown Watsonville, suffered similar damage as described above due to the earthquake.

LESSONS LEARNED FROM A CANADIAN PERSPECTIVE

One of the important characteristics of the Loma Prieta earthquake is the significant effect of soil conditions on the intensity of ground motions, leading to a very variable pattern of ground motions, and the many examples of liquefaction and liquefaction-induced damage on fills around the San Francisco Bay Area.

The experience of this earthquake indicates that the most serious deficiency in Part 9 of the NBC (1985) is the lack of any requirements for wall bracing in wood-frame construction, as well as the need for tying ends of beams over supports and anchorage of masonry chimney to the roof and floors. Lateral collapse of foundation cripple-stud walls was also a serious failure mode. This is covered by Part 4 in the NBC lateral force requirement via Subsection 9.15.15 and Article 9.4.1.1. Many designers or builders, however, may not be fully aware of this.

It was again shown that conventional power transmission equipments would be susceptible to earthquake induced failures. The major tremor predicted for the Charlevoix-

Kamouraska area in Quebec could reach a magnitude of 7 to 7.5, which is similar to that of the Loma Prieta earthquake. The maximum rock acceleration predicted for the Charlevoix-Kamouraska area is 0.70 g, as compared to the maximum acceleration of 0.64 g recorded near the Lome Prieta earthquake epicenter. Thus from the experience of this earthquake, the estimated acceleration values for the Charlevoix-Kamouraska area are proved to be very realistic.

Finally, this earthquake strongly demonstrated that three conditions occurring together can create a major seismic disaster: a high population density, a significant earthquake and a thick deposit of compressible soil. These conditions can be found in some regions of Canada. The most prominent examples are the Fraser River Delta in British Columbia and Quebec City in the Province of Quebec. Both are highly populated and located in areas of significant seismicity and with thick layers of soft alluvial deposits. Seismic hazards are therefore real both in the western and eastern regions of Canada.

ACKNOWLEDGEMENTS

The authors wish to express their appreciation to numerous individuals and organizations in Canada and California, EERI, El Cerrito, Calif., EERC, Richmond, Calif., PG&E, San Francisco, EPRI, Palo Alto, and EQE Engineering, San Francisco, who assisted the team during the reconnaissance visit. The above list is by no means complete. The invaluable help and information provided by many others, too numerous to be listed here, are also gratefully acknowledged. The access to many damaged areas was made possible by the help of the Earthquake Engineering Research Institute.

REFERENCES

Asteneh,A., Bertero,V.V., Bolt,B.A., Mahin,S.A., Moehle,J.P., Seed,R.B. 1989. Preliminary report on the seismological and engineering aspects of the October 17,1989 Santa Cruz (Loma Prieta) earthquake. Report No. UCB/EERC-89/14,Earthquake Engineering Research Center, Univ. of Calif., Berkeley, CA.

Bruneau,M. 1990. Preliminary report of structural damage from the Loma Prieta (San Frnacisco) earthquake of 1989 and pertinence to Canadian structural engineering practice. Canadian Journal of Civil Engineering, 17(2): 198-208.

Jablonski,A.M., Law,K.T., Lau,D.T., Pierre,J.-R., Tang,J.H.K. 1990. The 1989 Loma Prieta (San Francisco area) earthquake: site visit report. Internal Report No. 594, IRC/National Research Council of Canada, Ottawa, Ont.

Law,K.T., Cao,Y.T., He,G.N. 1990. An energy approach for assessing seismic liquefaction potential. Canadian Geotechnical Journal 26(3): 320-329.

Maley,R. et al. 1989. U.S. Geological Survey strong-motion records from the northern California (Loma Prieta) earthquake of October 17, 1989. Open-File Report 89-568.

Mitchell,D., Tinawi,R. 1990. Damage to buildings due to the 1989 Loma Prieta earthquake – a Canadian code perspective. Canadian Journal of Civil Engineering, 17(2): 813-834.

NBCC. 1985. National Building Code of Canada. National Research Council of Canada, Ottawa, Ont.

Rainer,J.H., Jablonski,A.M., Law,K.T., Allen,D.E. 1990. The San Francisco area earthquake of 1989 and implications for the Greater Vancouver area. Canadian Journal of Civil Engineering, 17(2): 798-812.

Shakal,A. et al. 1989. CSMIP strong-motion records from the Santa Cruz mountains (Loma Prieta), California earthquake of 17 October 1989. Report No. OSMS 89-06, Division of Mines and Geology, Sacramento, CA.

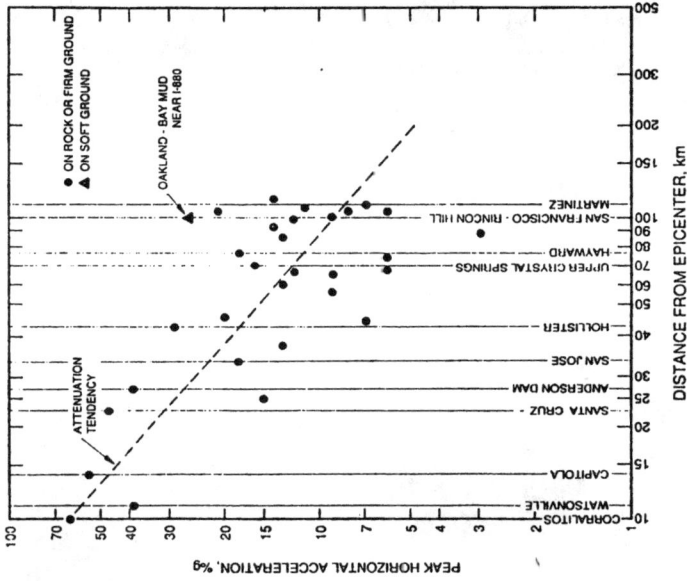

Fig.1 Measured peak horiz. accel. vs epicentral distance

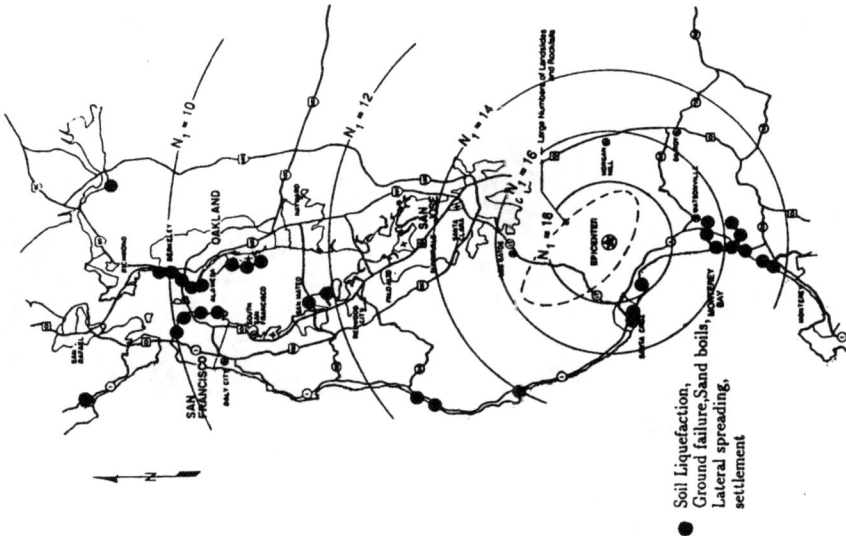

Fig.2 Location of soil failures and contours of standard penetration resistance (N_1) below which liquefaction will occur

717

Fig.3 Soft storey effect in Marina District

Fig.4 Collapsed unreinforced masonry
building in Watsonville

Fig.5 Collapsed Nimitz viaduct of I-880

Fig.6 Damaged Struve Slough Bridge
near Watsonville

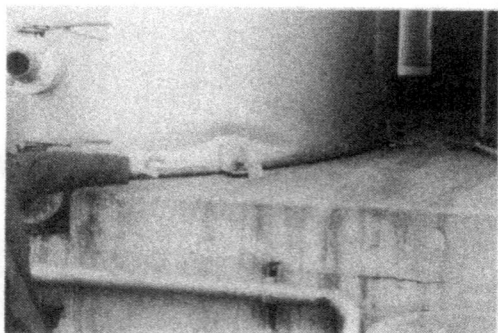

Fig.7 Pull out of anchor bolt

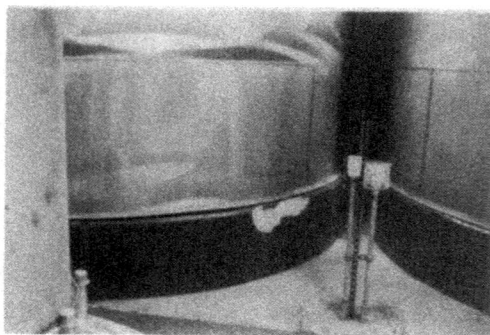

Fig.8 Buckling failure of tank

Distribution of damages and site effects during the 1988 Saguenay earthquake

Guy Lefebvre[I], Patrick Paultre[II], Jean-Philippe Devic[III], and Gaétan Coté[IV]

ABSTRACT

The 1988 Saguenay earthquake caused a large variety of geotechnical and structural damages. More than 1920 claim files, most of which submitted as part of a compensation program sponsored by the Ministry of Public Safety of the Government of Québec, have been studied. Results of this study are presented in terms of correlations between damages and earthquake motion charateristics, soil conditions, and type of structures. Site effects are pointed out as the main cause of damages to buildings even at large distances from the epicentre.

INTRODUCTION

The Saguenay earthquake of November 25, 1988 constitutes for Eastern North America a unique and considerable source of data on the characteristics of the ground motions and the related damages. This earthquake, whose epicentre was located 36 km south of Chicoutimi, was felt as far as Toronto and New York City. It caused a large variety of geotechnical damages (Tuttle et al., 1989, Lefebvre et al., 1990) but minor structural damages to large buildings (Mitchell et al., 1989).

This paper presents the results of the analysis of more than 1920 files of reported damages to buildings. Some of these files were submitted as part of a compensation program sponsored by the Ministry of Public Safety of the Government of Québec. Other files, concerning public buildings, were supplied by the government. Statistical studies show that the geographical distribution of damages are related to site effects. This study also reveals relationships between ground motion, type of soil and type of structural damage. The total repair cost of damage to buildings, inventoried in the present study, reached 44 millions of dollars.

[I]Professor, Département de génie civil, Faculté des sciences appliquées, Université de Sherbrooke

[II]Associate Professor, Département de génie civil, Faculté des sciences appliquées, Université de Sherbrooke

[III]Graduate Student, Département de génie civil, Faculté des sciences appliquées, Université de Sherbrooke

[IV]Ministère de la sécurité publique du Québec, Bureau de la protection civile

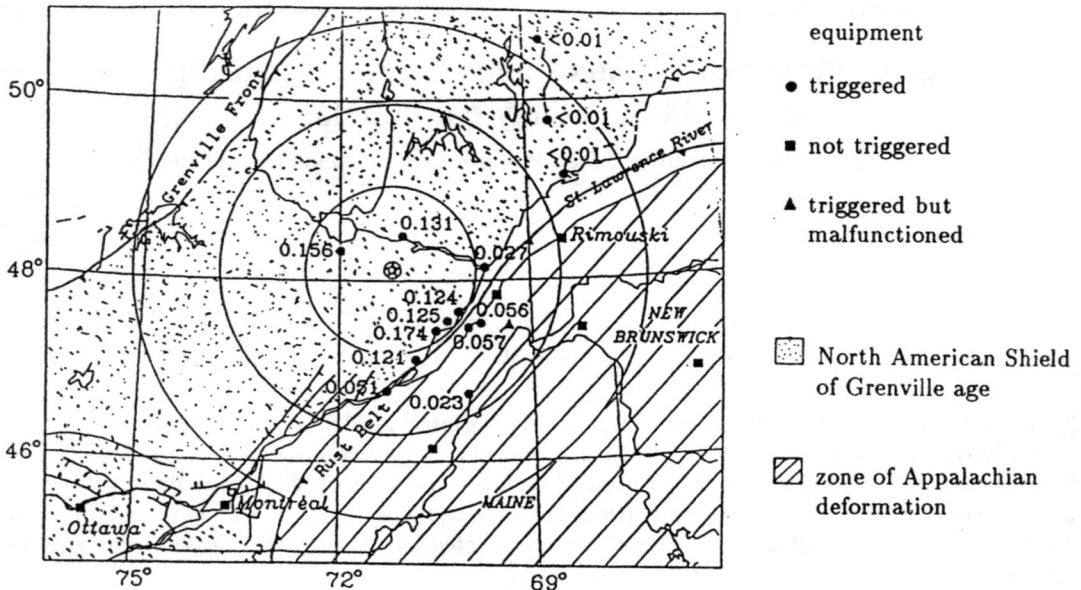

Figure 1. Maximum peak horizontal ground accelerations, in g, recorded during the Saguenay earthquake; circles: 100, 200, 300 km (after Munro and Weichert, 1989).

EARTHQUAKE MOTION CHARACTERISTICS

Figure 1 shows the largest of the two peak horizontal ground accelerations recorded for each intrumented site (Munro and Weichert, 1989). These accelerations were recorded at rock level, except at Baie-St-Paul, which was on a thick layer of alluvial deposits. This figure clearly illustrates the amplification of ground motion due to local soil effects. There is an amplification of 1.5 to 2 between the Baie-St-Paul station (on alluvial deposits) and the La Malbaie and Les Eboulements stations (on rock), at the same distance from the epicentre. The map shows a considerable decrease in the peak ground acceleration beyond a distance of 110 km from the epicentre. According to Sommerville et al. (1990), this is due to the large focal depth of the earthquake (29 km), and to the strong reflexions on discontinuities of the earth's crust along the St. Lawrence River, between the Appalachian deformation zone and the North American Shield (see Fig 1). Leboeuf and Lefebvre (1990) show that for epicentral distances of more than 110 km, the amplification for the spectral acceleration is less than 3.4 and the predominant periods are low (less than 0.075 s); whereas for epicentral distances less than 110 km, the amplification can be as high as 5, and the predominant periods vary from 0.13 s to 0.2 s.

THE DATA BASE

The 1927 files compiled made it possible to establish an identification sheet for each case indicating locality, description of the damaged building (age, number of storeys, type of foundation and structure, normalized municipal evaluation), and damages (cost of repair, damaged structural

720

Table 1. Distribution of claims and their amount according to usage of facilities

Usage of facilities	Number of claims	Amount (in thousands $)
Houses of less than 2 storeys	1155	5,617
Apartment buildings of more than 2 storeys	28	585
Commercial and industrial	55	474
Churches	51	2,066
Schools	226	25,000
Hospitals	36	6,678
Public services	73	2,050
Wells and aqueducts	228	1,013
Total	1852	43,928

elements). Some reserves must however be made on the objectivity of the collected data: (i) all damages due to the earthquake were not always reported; (ii) the determination of the cause of the damages was not always compiled – 25% of the files corresponding to important damages above compensation threshold were thoroughly investigated and are well documented, 5% (i.e., 75) of the claims were not attributed to the earthquake itself; (iii) the evaluation of the cost of damages caused by the earthquake has generally been made by the owner, except for public buildings and private houses with large damages where an independent assessment was made, and (iv) the normalized municipal evaluations used in this study to define the level of damage do not always reflect reality, particularly for public services buildings.

DISTRIBUTION OF DAMAGES

The geographical distribution of damages presented in Fig. 2 shows that the damages are concentrated along the St. Lawrence River where the population is denser. Some regions located more than 300 km from the epicentre, like the Boucherville region, have several damaged buildings. This map does not specify building type or the degree of damages. Table 1 gives the number of claims and their amount according to usage of the facilities. A total of 1200 claims was received for low rise buildings (detached or semi-detached houses, apartment buildings with less than two storeys), totaling 6.1 M$; 414 claims were received for medium rise buildings (appartment buildings with more than two storeys, commercial buildings, churches, hospitals and public services buildings), totaling 36.8 M$; and 228 claims were received for wells and aqueducts, totaling 1 M$.

Figures 3 a, b and c give the number of claims with respect to the epicentral distance for the three classes of facilities mentionned above. The number of claims decreases with distance except for a peak in the Quebec City region (100 to 150 km), and another peak for small buildings in the Montreal region (300 to 350 km). Figure 3 c shows three regions for damaged surface wells (drying up): the Ferland-Boileau region, where soil liquefaction occurred, the Shannon region and the south shore of the St. Lawrence River from Rimouski to the Beauce region. These hydrologic phenomena did not all occur immediately after the earthquake, some wells progressively dried up. Figure 3 d shows that the average damage ratio, DR_{av}, for single houses (ratio of damage repair cost to estimated building value), does not change much with distance. At large distances, few claims were reported but the amounts involved were important.

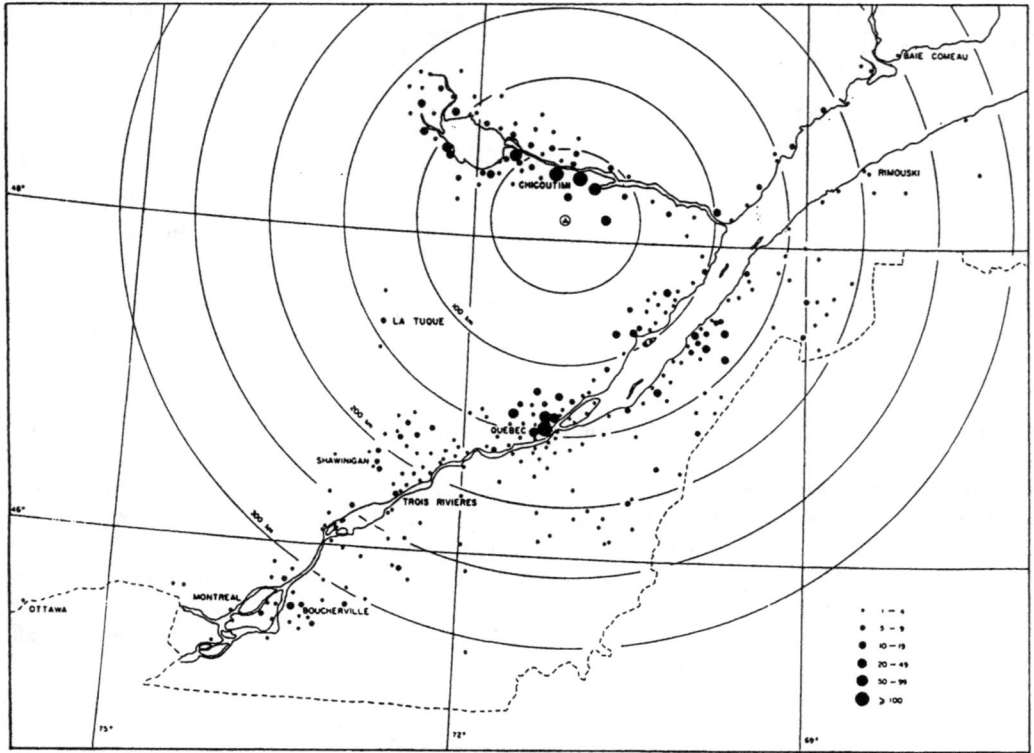

Figure 2. Geographical distribution of the total number of claims submitted to the compensation program of the Government of Quebec.

The distribution of damage to buildings less than two storeys high was examined in more details because it represents a more homogeneous type of building, constitutes a large sample (62% of all claims) and allows to consider the population density. The parameters used to evaluate the geographic distribution of damages are:

1. The house density $d_h = n_{dh}/n$ where n_{dh} is the number of damaged houses in a municipality and n is the total number of houses in that municipality obtained from the 1986 census (Statistics Canada, 1987).

2. The average damage ratio DR_{av} per municipality.

3. The average damage intensity, $DI_{av} = d_h \times DR_{av}$. This parameter is more representative than the two previous ones and can be compared to seismic intensity. The geographic distribution of this parameter is given in Figure 4.

Figure 4 shows the Ferland-Boileau, Jonquière and La Baie region which were particularly touched due to their proximity to the epicentre. Also shown are more distant regions where numerous claims were filed: Charlevoix County (Baie-Saint-Paul, Rivière du Gouffre, Saint-Siméon,

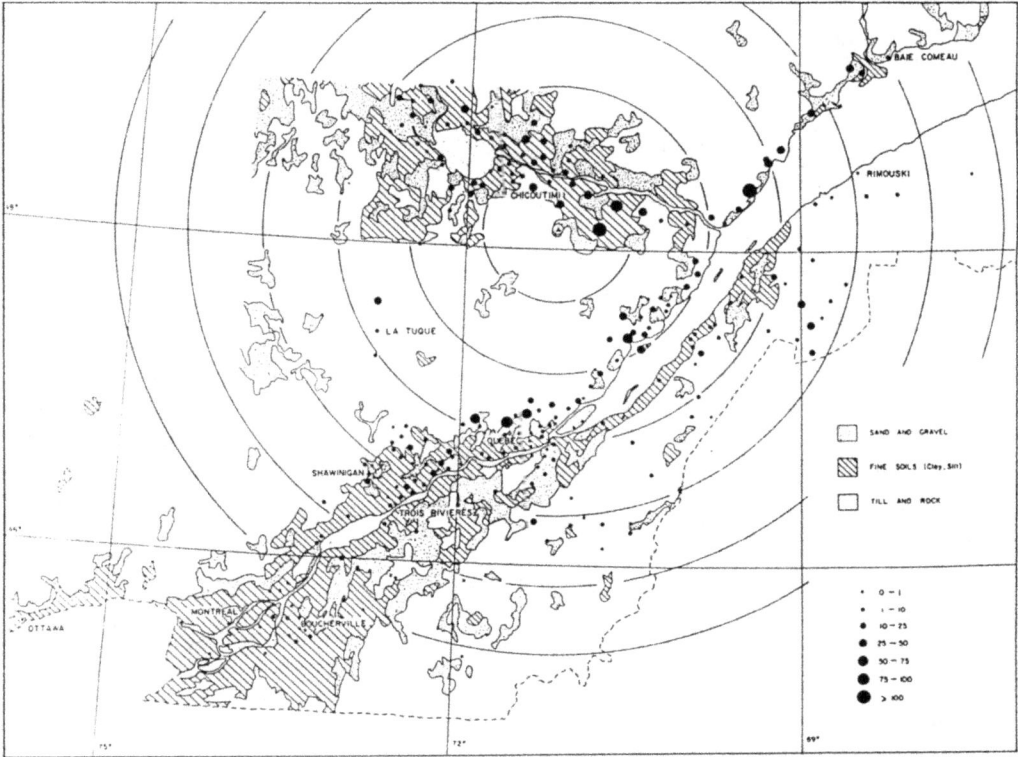

Figure 4. Distribution of the average damage intensity, DI_{av}, for houses superimposed on a surface deposits map.

profiles, 14% on sand). Damaged buildings involving clay foundations can be found at very large distances whereas comparable damages for sandy foundations are limited to an epicentral distance of 130 km.

Amplification due to the overburden is evident when comparing the damages in Chicoutimi, mostly on bedrock, close to the epicentre and the Boucherville area, mostly on clay deposits, at 320 km from the epicentre. Many damaged buildings were also associated with other local effects, such as sloping ground, proximity of a slope crest (generally clay) and foundation on fill material.

TYPE OF DAMAGES

Very few of the 1857 reported cases involved important damages. Essentially, the damages were cracks and fissures to walls and foundations. The analysis was carried out by dividing the buildings in two classes: (1) small buildings (less than 2 storeys high), and (2) large buildings (more than 2 storeys high).

723

Table 2. Proportion of number of claims according to damage and building types

Type of damage	Component	Small buildings (%)	Large buildings (%)
Structural	Structure	4	7
	Foundation	59	12
	Roof	3	10
	Chimney	38	4
Non-structural	Exterior walls	23	67
	Interior finish	18	45
	Openings, panes	14	19
	Staircase, steps	2	11
	Plumbing	4	4

Typically, the small buildings have a concrete or concrete block foundation walls, a wood frame structure and stone or brick exterior walls. The damages are concentrated in the stiff sections of the buildings. Table 2 summarizes the type of damages most commonly reported. No reported cases involved collapse or required demolition. Only 4.7% (54 houses) have a damage ratio DR_i greater than 50% due to the necessity of rebuilding the foundations or, in some cases, driving piles. For more than half the requests involving a damage ratio, DR_i, greater than 50%, the cause of the damages could be related to excessive settlements of fill. Soil liquefaction in the Ferland-Boileau area, close to the epicentre, also caused damage to foundations (Law, 1990).

Large buildings presented a higher level of damage in the unreinforced masonry exterior walls and the interior walls (see Table 2). Fewer damages to chimneys and foundations were reported. A striking difference in building behaviour can be observed: several unreinforced concrete block walls were displaced inside the building and they have to be demolished. Failures also occurred on the top of the walls, chimneys or parapets. Vibrations also worn out the cement bonds of the exterior masonry walls; in some cases, a separation of several inches between one non-anchored exterior wall and the building structure was noticed. Cracks occurred mostly around openings (windows, doors) or corners (walls and ceilings). It is important to note that symmetrical buildings suffered less damage than nonsymmetrical ones.

It appears that earthquake damage is controlled by the type of foundation (masonry, concrete, concrete blocks), the quality of construction, the wall construction (stucco, stones, bricks) and its connection with the structure. The age of the buildings did not seem to have played a significant role.

CONCLUSION

The comprehensive survey of building damages clearly pointed out that soil and site factors played a significant role in the damage observed and its geographical distribution. A possible explanation lies in the amplification of the seismic motions by the soil layers. Some towns, far from the epicentre, were strongly affected by the earthquake (for example, in Boucherville – 320 km from the epicentre, Shawinigan – 210 km, St. Alban – 170 km, Shannon and St-Brigitte-de-Laval – 140 km). Damages were widespread in thick overburden zones or in alluvial valleys and there are some evidences that local topography contributed to damages or failures. Damage distribution was very uneven, with a higher concentration on the North Shore of the St. Lawrence. It is also

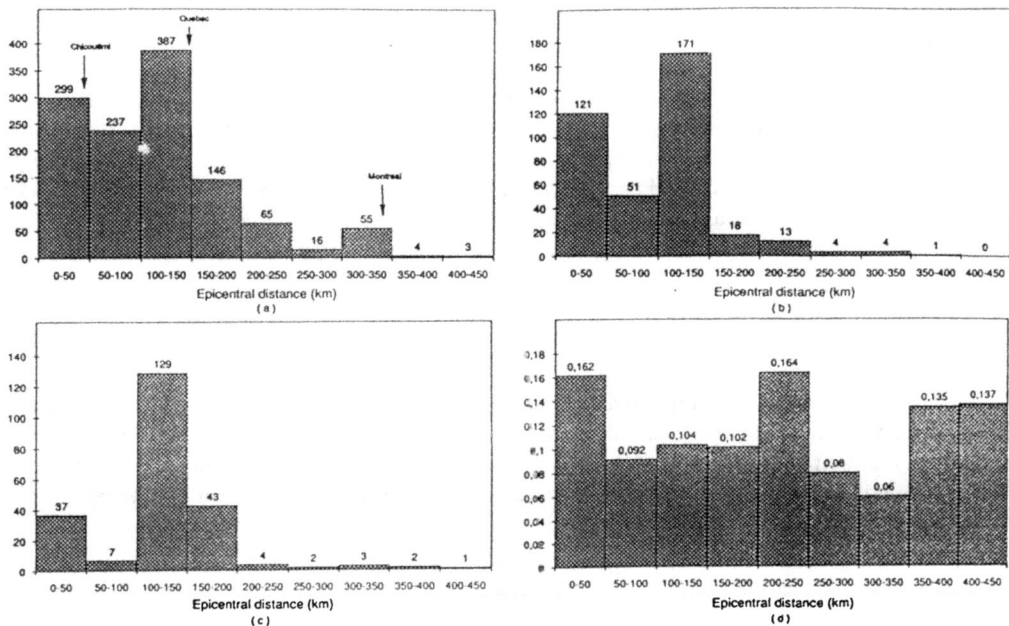

Figure 3. Distribution of the number of claims with respect to epicentral distance (a) for damaged houses and apartment buildings of less than two storeys, (b) for damaged churches, public services buildings, apartment buildings of more than two storeys, (c) for damaged wells and aqueducts, and (d) average damage ratio (DR_{av}) for detached houses vs. epicentral distance.

at 100 km from the epicentre), north of Québec (Shannon, Sainte-Brigitte de Laval, Stoneham, Clermont, at 130 km), the Portneuf region (Saint-Alban, La Pérade, at 170 km), Champlain (Batiscan, at 200 km), Saint-Maurice (Shawinigan, at 210 km), and the Montreal region (Boucherville, at 320 km).

Although with some irregularities, there is a clear trend of the average damage intensity, DI_{av}, decreasing with the distance to the epicentre. There are very few damages on the south shore of the St-Lawrence River, beyond the contact North American Shield–Appalachian zone (see Fig. 1). This is well correlated to the discontinuity observed in the recorded peak ground accelerations shown in Fig. 1.

DAMAGES AND TYPE OF SOIL

Any site effects should be examined individually. Figure 4 shows a map of the overburden deposits on which the average damage intensity DI_{av} is superimposed. It can be seen that most damages are associated with clay deposits. It should be noted that at this scale small extents of clay or mud deposit are neglected and could exist elsewhere locally. Very few damages were reported for buildings on bedrock or till foundation. For churches, the data showed that 96% of the reported cases are located on soil deposits (56% on clay deposits thicker than 20 m, 26% on multilayer soil

725

possible that local mechanisms (deep hypocentre) and "tectonic anisotropy" along the propagation path played a significant role.

The horizontal and vertical seismic vibrations induced several cracks and fissures in the foundations and weakened the peripheral walls but no major failures to structures were noticed. Most damages to detached houses involved the foundation walls and can be explained by the poor quality of the foundation soil or thickness variations. Soil-structure resonance did not seem to have played a role as the predominant frequencies of the ground motions were very high. Another cause of damages is the use of unreinforced masonry for wall construction and its poor connection to the structure.

ACKNOWLEDGMENTS

This research was partly supported by the Direction générale de la sécurité civile. The authors want to thank the officers of the local offices of the Protection civile du Québec, in Sherbrooke and Québec City, for their collaboration. Additional support from the Natural Sciences and Engineering Research Council and the Fonds pour la Formation de Chercheurs et l'Aide à la Recherche of the Government of Québec is also acknowledged.

REFERENCES

Law, K. T. 1990. Analysis of Soil Liquefaction During the 1988 Saguenay Earthquake. 43rd Canadian Geotechnical Conference. Prediction and Performance in Geotechnique, Québec, 189–196.

Leboeuf, D. and Lefebvre, G. 1990. Sécurité sismique de digues et ouvrages en terre: zonation et évaluation. Rapport SB8701 présenté à la Division sécurité des barrages Hydro-Québec.

Lefebvre, G., Leboeuf, D., Hornych, P. et Tanguay, L. 1990. Slope Failures During the 1988 Saguenay Earthquake – Québec, Canada. 43rd Canadian Geotechnical Conference, Québec. Prediction and Performance in Geotechnique, 197–207.

Mitchell, D., Tinawi, R. et Law, K.T. 1989. The 1988 Saguenay Earthquake: A Site Visit Report. Geological Survey of Canada Open File Report No. 1999.

Munro, P.S. et Weichert, D.H. 1989. The Saguenay Earthquake of November 25, 1988: Processed Strong Motion Records. Geological Survey of Canada Open File Report No. 1996.

Sommerville, P.G., McLaren, J.P., Saikia, C.K. and Helmberger, D.V. 1990. The 25 November 1988 Saguenay, Quebec, Earthquake : Source Parameters and the Attenuation of Strong Ground Motion. Bulletin of Seismological Society of America, 80(5), 1118–1141.

Statistiques Canada. 1987. Recensement Canada 1986. Caractéristiques de la population et des logements.

Tuttle, M.P., Such, R. et Seeker, L. 1989. Ground Failure Associated With the November 25, 1988 Saguenay Earthquake in Quebec Province, Canada, in The Saguenay Earthquake of November 25, 1988, Québec, Canada: Strong Motion Data, Ground Failure Observations and Preliminary Observations, Jacob K.H., Ed., National Center for Earthquake Engineering Research, Buffalo, New York.

Seismic response of low-rise buildings subjected to the 1988 Saguenay earthquake

P. Léger[I] and A. Romano[II]

ABSTRACT

This paper presents elastic and inelastic response spectra of strong motion accelerograms recorded during the 1988 Saguenay earthquake. Comparisons are made with the NBCC 1990 lateral forces requirements for the seismic resistant design of short-period structures. The seismic response of a typical low-rise steel building designed according to the NBCC 1990 and CAN3-S16.1-M89 is then investigated in the elastic and inelastic range. The use of a period-dependent force modification factor is proposed to take advantage of the energy dissipation capacity of short-period structures on a more rational basis. It is also shown that to obtain a realistic picture of the ductility demand of low-rise buildings, the structural overstrength, that is the supplied strength in excess of the seismic design base shear, should be explicitly considered in the design process.

INTRODUCTION

The 1988 Saguenay earthquake, which registered peak ground acceleration in the order of 10% g near the epicentre was found to contain high energy in the frequency range of relatively stiff structural systems (with periods smaller than 0.3 sec) such as low-rise buildings, concrete dams or nuclear containment structures. After the earthquake a site visit team reported that the earthquake did not cause significant structural damages. However, the poor serviceability performance of low-rise "tension-only" cross-braced buildings, structures containing "soft-storeys" as well as unreinforced masonry walls and nonstructural elements were observed (Mitchell et al. 1990). Although the Saguenay earthquake did not cause extensive damages for the level of acceleration recorded, it has raised serious concerns for the seismic performance of existing or new structures during future events with larger return periods.

[I]P. Léger, Associate Prof., Dept. Civil Eng. McGill University, Montreal, Qc.

[II]A. Romano, P. Eng., Tecsult Inc., 85 Ste-Catherine W., Montreal, QC.

This paper presents comparisons between the engineering characteristics of the 1988 Saguenay earthquake and the 1990 edition of the National Building Code of Canada (NBCC). The comparisons are first made in terms of elastic and inelastic response spectra for the lateral forces to be considered for the seismic analysis of short-period structures. The seismic response of a low-rise "tension-only" cross-braced steel building designed according to the NBCC 1990 and CAN3-S16.1-M89 for two sites with different levels of seismicity is then investigated. The acceleration records obtained at Chicoutimi, the nearest station to the epicentre, Baie St-Paul, corresponding to the largest PGA recorded, and Quebec City, a major city further away from the epicentre were scaled to the acceleration level specified by the NBCC to study the seismic behaviour of the structural systems. The peak ground motion parameters obtained from the selected sites are summarized in Table 1.

THE 1990 EDITION OF NBCC

In the 1990 edition of NBCC, the minimum lateral seismic force at the base of the structure is given by

$$V = (V_e/R) \; U \tag{1}$$

where U is a calibration factor equal to 0.6 to maintain a level of protection based on engineering experience and consistent with the previous code requirements, and R is the force modification factor that reflects the capability of a structure to dissipate energy through inelastic behaviour. The equivalent lateral force at the base of the structure representing the elastic response, V_e, is given by

$$V_e = v \; S \; I \; F \; W \tag{2}$$

where v is the zonal velocity ratio, S is the seismic response factor corresponding to an idealized 5% damped elastic response spectrum for unit velocity, "v" and weight "W". I is the importance factor, and F is the foundation factor to represent the local amplification of the ground motions due to soft surficial soil layers. In this formulation only the zonal velocity ratio, v, is specified explicitly whereas the zonal acceleration ratio, a, is used implicity in the seismic response factor, S.

RESPONSE SPECTRA

Many parameters such as the periods of vibration, ductility, overstrength, torsional coupling, and duration of ground motion will affect the value of the structural response modification factors. The NBCC 1990 defines the inelastic design response spectra by applying a force modification factor, R, to the elastic response spectra. The value of R is independent of the ground motion and vibration characteristics of the system. It ranges from 1, for nonductile structural systems expected to remain elastic under the design earthquake ground motion, to 4 for ductile moment resisting frames with good seismic detailing. In the case of a SDOF, the value of R is directly related to the displacement ductility demand of the system.

A more accurate procedure to obtain design loads is to develop inelastic design spectra from rigorous nonlinear dynamic analyses of structural models of specified initial stiffnesses (periods) and possessing realistic strengths, hysteretic behaviour, and damping properties, subjected to accelerograms that can be expected for the site under consideration. It is obvious that this procedure is highly impractical for preliminary design. However, the ground motion records obtain from the Saguenay earthquake provide an opportunity to investigate the adequacy of the NBCC 1990 in defining design loads for short-period structures located in Eastern North America.

The strength of the SDOF system is defined in terms of a non-dimensional strength parameter, n, expressed as the ratio of the design base shear at yield, V, to the maximum effective force applied during the earthquake:

$$n = V \ / \ M \ \ddot{V}_{gmax} = (vSIF)U \ / \ Ra \qquad (3)$$

where M is the mass of the system and $\ddot{V}_{gmax}$ is the PGA expressed in consistent units with the mass.

Constant strength and constant ductility elastic (R=1) and inelastic design spectra were constructed for initial periods ranging from 0.05 sec to 0.75 sec. The constant strength spectra represents the ductility demand for systems with different yield levels. The constant ductility spectra reflects the strength demand imposed on the structure by the earthquake for a specified ductility level allowing for a direct comparison with the code formulation. An elastic-perfectly plastic (EPP) hysteresis model assuming 5% viscous damping was used. The EPP model was selected because it is representative of the interstorey hysteresis of low-rise steel structures with tension-only cross-bracing that were affected most significantly by the earthquake. Moreover, previous studies that have considered the effect of EPP, bilinear and stiffness degrading hysteresis models (SDM) on the ductility demand of SDOF systems have concluded that using an EPP model for inelastic design is generally on the conservative side (Mahin and Bertero 1981).

Strength Demand -vs- Capacity

A typical constant strength spectra is shown in Fig. 1 for Baie St-Paul. For systems with initial periods greater than 0.3 sec, there is generally a reduction in the ductility demand as the period increases. For systems designed with n > 1 the ductility demand will remain bounded as the period is reduced and the system converges towards the static response. For systems designed with n ≤ 1 the ductility demand will grow unbounded if the EPP model converges towards the static response. The ductility demand is thus very high for short-period structures designed with a substantial strength reduction factor.

729

The smoothing effect of nonlinear behaviour on the strength demand can be observed by comparing the elastic ($\mu = 1$) and inelastic (constant ductility) response spectra as shown in Fig. 2. Note that a value of F=1.4 has been used to compute the NBCC values according to Mitchell et al. (1990). It is obvious that the design base shear will tend to be the same whether the structure is elastic or inelastic for very short-period structures since all the spectra curves converge to a strength coefficient of 1 for an infinitely rigid system with a zero period. The inelastic response spectra indicate that the required design strengths do not present significant variations to achieve prescribed ductility levels above $\mu = 3.0$. In this case, for a given ductility, μ, the required strength of the structure, n, generally decreases with an increase in the period. For a prescribed low ductility level, say, $\mu = 1.5$, the required strength exceeds the ultimate value supplied by the code over a small period interval for all the records analyzed. As the ductility level is increased the required strength to maintain the design ductility exceeds the code supplied values over a wider period range which approximately correspond to the portion of the short-period (acceleration controlled) range of the code design spectra which exhibit constant amplification values.

This observation, that code supplied strength is sometimes below the demand computed from dynamic analyses, has also been found by other researchers. It is sometimes believed that actually constructed structures will possess larger damping and ductility than specified in the design, which may compensate for some of the potential deficit. However, nonlinear dynamic analyses indicate that the damping effect is uncertain because of the impulsive nature of earthquake ground motion and that increased ductility will have a very small impact on the strength demand of short-period systems with ductility levels greater than 2, as is shown by the close agreement of the required strength coefficient for these cases in Fig. 2. It must therefore be concluded that some buildings will be able to resist the seismic excitation because they are constructed with real strengths that are far in excess of the code required value. Buildings with poor construction will fail.

It should be noted that in NBCC 1990 the force modification factor can be interpreted as the product of a global ductility factor varying from 1 to 4 and a constant overstrength factor, R_s, with $U = 1/R_s$ (Fischinger and Fajfar 1990). However, post-earthquake investigations of structural damages have indicated that the overstrength depends strongly on the type of structural systems and the number of stories (periods) of the structures. An improved overstrength (calibration) factor should therefore be considered in the next editions of NBCC.

Period-Dependent Force Modification Factors (R)

For short-period structures designed to remain elastic, the strength demand exceeds the code supplied capacity over a small-period interval. In this case, it might be appropriate to increase the elastic equivalent lateral force, V_e. One approach would be to increase the seismic response factor, S, in the short period range to account for the observed high a/v ratios. To satisfy the strength demand in the short-period range for systems designed with substantial ductility levels, a period-dependent strength reduction factor, $\bar{R}$, could be used by the NBCC code. The following bi-linear variation has been adopted by many codes around the world:

$$\bar{R} = 1 + (\mu_G - 1)\, T/T_1 \quad ; \; T < T_1 \qquad (4)$$
$$\bar{R} = \mu_G \qquad\qquad\qquad ; \; T \geq T_1 \qquad (5)$$

where μ_G is the global ductility of the structure, T is its fundamental period of vibration and T_1 is a parameter given as a function of the seismic zone and the type of soil. T_1 should also be related to the transition period between the short-period or acceleration controlled range and the medium-period or the velocity-controlled range. Using the dynamic procedure of NBCC with proper zonal amplification factor the following values can be identified for

$$Za/Zv > 1, \quad T_1 = 0.30 \; sec \qquad (6)$$
$$Za/Zv = 1, \quad T_1 = 0.42 \; sec \qquad (7)$$
$$Za/Zv < 1, \quad T_1 = 0.61 \; sec \qquad (8)$$

In the case of inelastic systems increasingly smaller values of T_1 are generally observed in Fig. 2 with an increase in ductility which will correspond to a decrease in code supplied strength. The hysteresis model might also influence the value of T_1. A preliminary investigation of a SDM inelastic spectra of the transverse acceleration record at Baie St-Paul has indicated a slight shift of T_1 toward shorter periods as compared to EPP model.

In the context of NBCC, the value of μ_G in Eqs. 4,5 will correspond to R. Figure 3 shows a comparison of the strength coefficients computed using the usual R values and new coefficients using $\bar{R}$ as the strength reduction factor in the short-period range. If a period-dependent strength reduction factor is used, there is a significant increase in the strength to be supplied by the code for all originally selected R values which are greater than one. To control the ductility demand for systems designed with $n \leq 1$ a new value of $\bar{R}$ can be defined to obtain a better balance between the earthquake strength demand and the code supplied capacity. This value of $\bar{R}$ could be selected such that the value of n converges to 1 as the period is reduced from T_1 to zero.

The following procedure is suggested for the design of short-period structures with $T < T_1$: (i) select R from NBCC, (ii) compute V from Eq. 1, (iii) compute n from Eq. 3, (iv) compute $\bar{R}$ from

$$\bar{R} = 1 + (R-1)(T/T_1) \qquad ; \; T < T_1, \, n > 1 \qquad (9)$$

$$\bar{R} = \frac{n\,R}{1 + (n-1)(T/T_1)} \quad ; \; T < T_1, \, n \leq 1 \qquad (10)$$

$$\bar{R} = R \qquad\qquad\qquad ; \; T > T_1 \qquad (11)$$

(v) obtain the adjusted equivalent lateral force at the base of the structure from

$$V = V \cdot (\bar{R}/R) \qquad (12)$$

731

This procedure has also been implemented in Fig. 3 for comparison with the actual code. Figure 4 shows the new strength coefficients obtained by the application of Eqs. 9-11. A much better balance between the actual strength demand from the earthquake and the code supplied ultimate capacity is now observed for all values of R selected from NBCC to take advantage of nonlinear behaviour.

BUILDING ANALYZED

The two-storey steel office building shown in Fig. 5 has been designed according to CAN3-S16.1-M89 and NBCC 1990 for Baie St-Paul, Chicoutimi and Quebec City. The lateral load resisting system is made up of orthogonal "tension-only" X-braced frames. For this structural system which possesses very little lateral redundancy and for which no special detailing of connections should be required, an R value of 1.5 has been selected. The fundamental periods of vibration in the X-dir. have been computed has 0.25 sec for the building located in Baie St-Paul and 0.34 sec for the other sites.

The seismic response of short period structures positioned in the acceleration-bound region of the spectra have been shown to be strongly dominated by the first mode of vibration. Building structures which are fairly regular can thus be modelled as SDOF systems according to the procedure described by Fajfar and Fishinger (1988). The strength and the stiffness of the MDOF system are determined by applying monotonically increasing lateral loads proportional to the first mode of the structure. For example, the initial yielding of the Baie St-Paul lateral load resisting system occurred at a base shear value of 1664 kN. The frame has been designed for a maximum base shear of 1320 kN according to NBCC. The 25% overstrength can be attributed to the use of a material factor (ϕ=0.9) and the member selection process using sections available from the CISC database. The actual building overstrength is likely to be larger if non-structural components are to be considered in the design process.

To demonstrate the effect of the earthquake on the structures, the PGA of each record was increased to reach the design ductility level of 1.5. The nonlinear response indicators are summarized in Table 2. The value of μ_{acc} is computed as the ratio of the sum of the absolute displacement values of all yield excursions that occurred during the record, to the yield displacement. This quantity is significant for structures that are susceptible to low-cycle fatigue. The maximum roof displacements, δ_{max}, are also given.

A PGA of 0.76g was required to reach the design ductility level at Baie St-Paul. This value represents a "safety margin" of approximately 1.9 that can be mainly attributed to the difference between the actual supplied strength and the design value. The Chicoutimi record, on the other hand, demonstrates a very low energy content at the fundamental period of the structure resulting in a low excitation level. The response from the Quebec City record demonstrates characteristics which are similar to the Baie St-Paul response.

732

CONCLUSIONS

The 1988 Saguenay earthquake provided many accelerogram records of very good quality characterizing the seismic behaviour of a significant event typical of the conditions found in Eastern North America (ENA). It was found that for short-period systems designed to exhibit significant nonlinear behaviour, the NBCC 1990 does not provide a rational control of structural damage that can be expected from seismic excitation with high a/v ratio typical of ENA. The use of a period-dependent strength reduction factor applied to the elastic strength demand of short-period structures has been proposed to take advantage of the inelastic energy dissipation capacity this type of system on a more rational basis.

The performance evaluation of the low-rise buildings has indicated that the ductility demand on the structure depends on the real strength of the lateral load resisting system including overstrength. To obtain a realistic picture of the ductility demand and a rational control of the collapse threshold, explicit consideration should thus be given to overstrength during the design process.

REFERENCES

Fischinger, M. and Fajfar, P. 1990, On the response modification factors for reinforced concrete buildings, Proc. Fourth U.S. Nat. Conf. EE, (2), 249-258.

Fajfar, P. and Fischinger M. 1988, N2 - A method for non-linear analysis of regular buildings, Proc. Ninth WCEE, Tokyo, Japan, V, 111-116.

Mithchell, D., Tinawi, R. and Law, T. 1990, Damage caused by the November 25, 1988 Saguenay Earthquake, Can. J. Civ. Engng., 17(3), 338-365.

Mahin, S.A. and Bertero, V., 1981, An evaluation of inelastic seismic design spectra, J. Struct. Eng., ASCE, 107, 1777-1795.

Table 1. Peak ground motion parameters

Site	PGA (g)	PGV (m/s)	NBCC a	NBCC v
BSP.	0.174	0.053	0.40	0.40
Chc.	0.131	0.025	0.20	0.15
Qué.	0.051	0.022	0.19	0.15

Table 2. Nonlinear indicators for steel building

Site	a (g)	n	μmax	μacc	δmax (mm)
BSP.	0.40	3.48	0.83	-	21.8
	0.48	2.90	1.00	1.00	26.2
	0.76	1.83	1.50	1.85	39.3
Chc.	0.20	3.97	0.22	-	6.4
	0.91	0.87	1.00	1.00	29.0
	1.60	0.50	1.50	2.25	43.6
Qué.	0.19	4.18	0.41	-	11.7
	0.47	1.69	1.00	1.00	28.9
	0.73	1.09	1.50	1.69	43.2

733

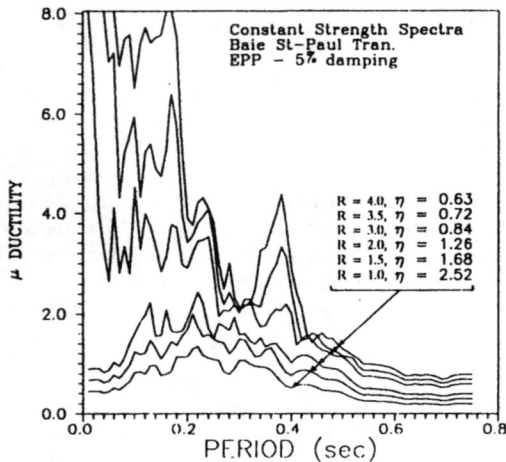

Figure 1. Constant strength spectra

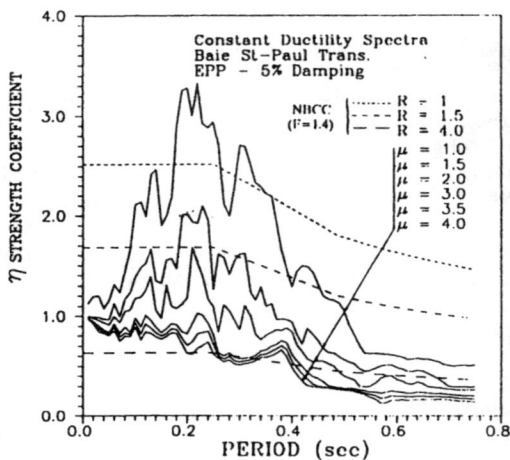

Figure 2. Constant ductilty spectra

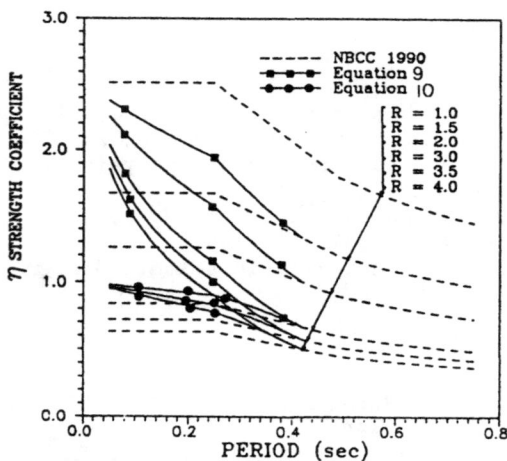

Figure 3. Ultimate strength capacity using R and $\bar{R}$.

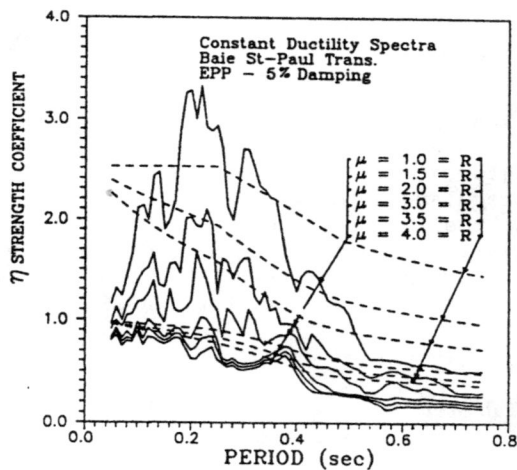

Figure 4. Strength demand -vs- capacity using $\bar{R}$.

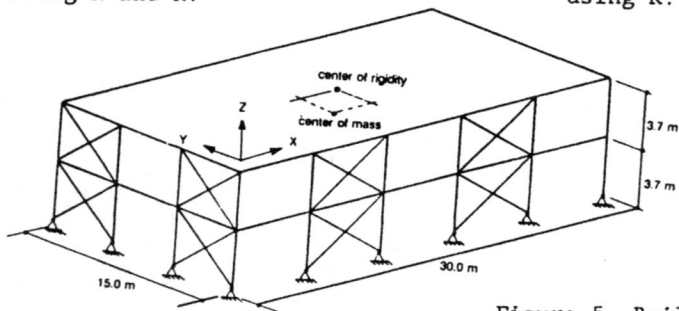

Figure 5. Building Analyzed

Observation of structural pounding damage from 1989 Loma Prieta earthquake

K. Kasai[I] and B.F. Maison[II]

ABSTRACT

The survey results on pounding incidents during 1989 Loma Prieta earthquake are summarized. They include the distribution of pounding damage in the specific areas, types of pounding damage, and examples for pounding damage involving major multi-story buildings.

INTRODUCTION

Pounding Incidents. - Structural pounding refers to the lateral collisions of adjacent buildings during earthquakes. Pounding occurs when building separations are insufficient to accommodate the relative motions of adjacent buildings. Pounding damage has been reported from most major earthquakes affecting metropolitan areas of the world. Pounding of adjacent buildings has made damage worse, and/or caused total collapse of the buildings. The earthquake that struck Mexico City in 1985 has revealed the fact that pounding was present in over 40% of 330 collapsed or severely damaged buildings surveyed, and in 15% of all cases it led to collapse (Rosenblueth and Meli 1986). This earthquake illustrated the significant seismic hazard of pounding by having the largest number of buildings damaged by its effect during a single earthquake (Bertero 1986).

Loma Prieta Survey. - The writers have surveyed the damage due to pounding in the San Francisco Bay area during the recent 1989 Loma Prieta Earthquake. The earthquake caused pounding between buildings over a wide geographical area including the cities of San Francisco (e.g., Fig. 1), Oakland (Fig. 2), Santa Cruz, and Watsonville. Significant pounding was observed at sites over 90 km from the epicenter thus indicating the possible catastrophic damage that may occur during future earthquakes having closer epicenters. This paper summarizes the writers' survey results on pounding (Kasai and Maison 1991).

GENERAL SURVEY FINDINGS

Survey Methodology. - This survey is compiled from data provided by: engineers, government officials and engineers, building owners, and block-by-block inspections performed by the writers. The database contains the input of about 90 interested parties and records more than 200 pounding occurrences involving more than 500 structures.

Classification of Pounding Damage. - Pounding damage patterns are classified as follows: Type-1, major structural damage; Type-2, failure and falling of building appurtenances creating a life-safety hazard; Type-3, loss of building function due to failure of key mechanical, electrical or fire protection systems; and Type-4, architectural and/or minor structural damage.

[I] Assist. Prof., Dept. Civil. Eng., Illin. Instit. of Tech., Chicago, IL 60616.
[II] Structural Engineer, SSD Engineering Consultants Inc., Berkeley, CA 94704.

Survey Findings and Comments. - The following are some of the general survey findings and comments:

(1) The majority of reported cases are in urban areas including San Francisco (e.g., Fig.1), Oakland (Fig. 2), Santa Cruz and Watsonville.

(2) Pounding typically involved multi-story buildings constructed prior to about 1930. They are typically of masonry construction with or without steel skeletal vertical load resisting systems. Very little consideration was given for separation between such buildings to preclude pounding. In many cases, they are in contact with each other.

(3) Fewer modern buildings suffered pounding. In such buildings, relatively larger separations exist. However, it is noted that many modern buildings having expansion joints suffered pounding due to small separations.

(4) There is evidence of correlation between occurrences of pounding and soft foundation soil conditions. This may be attributed to the more intense shaking typically reported for such soil conditions and/or from the possible settlement and rocking of the structures located on soft soils.

(5) Special pounding cases were also observed. They include; severe pounding at unsupported part (e.g., midheight) of columns or walls; pounding promoted by torsional behavior of building; and pounding between the buildings sharing a common wall.

(6) Older buildings that suffered Type-1 damage typically also had Type-2 damage (i.e., falling bricks). Modern buildings that pounded usually had Type-4 damage, and several of them also suffered Type-3 damage. The survey has relative distributions for damage Types 1 and 4 of 21% and 79%, respectively. Many of the present Type-4 damage cases will become damage Types 1, 2, and/or 3 when a future more severe earthquake affects the region. The Type-4 damage cases may be thought of as precursors for the major pounding damage yet to occur.

DISTRIBUTION OF POUNDING DAMAGE IN SELECTED AREAS

The following summarizes the occurrences of pounding for several specific areas. Discussion includes the distribution of pounding damage, earthquake intensity, maximum ground acceleration, soil condition, building density, and building type as well as height.

San Francisco Financial District and South-of-Market Area. - The writers conducted a block-by-block survey of this area. Fig. 1 shows the concentrated study area as specified by the dashed lines, and incidences of pounding are indicated by solid dots. Multi-story office buildings are the typical structures in this area. Heights typically range from 2 to over 20 stories. The area has a mix of older masonry buildings and modern high-rise buildings of concrete and steel construction.

The area experienced significant earthquake motions as indicated by the Modified Mercalli intensity of seven (MM VII) assigned by the United States Geological Survey (USGS). In the north-east corner of the area on a fill area, a peak horizontal ground acceleration of 0.2g was recorded by the California Strong Motion Instrumentation Program (CSMIP). In the region near the east corner of the study area, 0.09g was recorded, on Franciscan sandstone and shale. At about the center of the study area, 0.17g was recorded, on fill over bay mud.

North of Market street, pounding damage is concentrated near the eastern edge of the area associated with loose soil fill where liquefaction was reported. This area was filled in the late 19th century and recorded large ground deformations and soil liquefaction in the 1906 San Francisco earthquake. South of Market, pounding damage is widespread and somewhat concentrated at: (1) south of Mission street between 4th and 7th Streets including the area that was filled in the late 19th century which recorded large ground deformations and soil liquefaction during the 1906 San Francisco earthquake; and, (2) south of Market between 1st and 3rd Streets, which is near the same shore line.

Oakland City Center: The writers also conducted a block-by-block survey of this area. The survey area is shown by the dashed line in Fig. 2, and occurrences of pounding are shown by solid dots. The area has a variety of

736

building sizes that range from small low-rise buildings to large multi-story buildings. In general, the sizes and heights of the buildings are smaller than those in the San Francisco Financial District. The heights typically range from 1 to over 10 stories. The area contains a mix of wooden, older masonry, and modern high-rise buildings of concrete and steel construction.

The area experienced significant earthquake motions as indicated by the Modified Mercalli intensity of eight (MM VIII). At about the center of the concentrated survey area, a peak horizontal ground acceleration of 0.26g was recorded, on alluvium. The entire survey area rests on alluvium soil.

This survey area, which is smaller and having less building density than the San Francisco Financial District, still showed a high occurrence of pounding damage probably because of high ground accelerations in this area. Fig. 2 shows that the pounding damage is concentrated in a densely built area on Broadway and San Pablo Avenues. Significant pounding was present in older masonry as well as wooden buildings.

Other Areas. - The writers also performed a block-by-block review of other areas. The areas are: (1) San Francisco Marina District (MM IX), where 3 and 4 story residential buildings of wood construction are most common (e.g., Fig. 3); (2) The Pacific Garden Mall area (MM VIII) in downtown of Santa Cruz, where most buildings were constructed around the turn of the century. The heights range from 1 to 7 stories but most are typically up to 3 stories. Most of the severely damaged structures are of unreinforced masonry; (3) Watsonville downtown area (MM VIII), where the building heights are typically 2 to 3 stories. Most of these buildings were of masonry construction.

Pounding was present in many buildings in these areas (Kasai and Maison 1991). Further investigation is being conducted by the writers.

EXAMPLE CASES

This section presents actual examples of pounding involving major multi-story buildings. The buildings selected offer information on a variety of pounding mechanisms and consequences. The purpose is to illustrate the various types of major pounding damage caused by the Loma Prieta earthquake in order to clearly inform the engineering community of the significant seismic hazard that pounding poses.

Mission Street, South-of-Market, San Francisco: The building is 10 stories and constructed of masonry combined with a steel skeleton (Fig. 4). It was built in 1904. This building experienced pounding with an adjacent massive 5 story building which occupies most of the city block (Fig. 4(a)). Pounding was located at the 7th level in the 10 story building and at the roof level in the 5 story building. About 1 to 1.5 inch building separation is present. The 10 story building suffered structural damage above the pounding elevation as evidenced by the large diagonal shear cracks in the masonry piers (Fig. 4(b)). This case is classified as Type-1 damage.

Broadway Street, Oakland City Center: The 12 story building consists of an 11 story reinforced concrete structure constructed in 1924 with a penthouse steel structure added in the 1950's (Fig. 5(a)). The adjacent building is a modern 6 story reinforced concrete parking garage. A bridge spans from the parking garage top level to the building at the 8th floor level. The bridge has a sliding connection at the building that provides for about 4 inches of relative building movement. Severe pounding occurred resulting in extensive cracking of the building at the bridge location (Fig. 5(b)). Large interior diagonal shear cracks in the columns above and below the bridge elevation indicated the large magnitude of the local collision forces. After inspection, it was found that at least 4.5 inches of relative displacement must have occurred at least twice during the earthquake. The building also experienced extensive window glass cracking and falling, but whether pounding was the primary cause is uncertain. This building is classified as having Type-1 damage and possibly Type-2 damage as well.

Franklin Street, Oakland City Center: The building is 10 stories and constructed of reinforced concrete with a post-tensioned concrete floor system

(Fig. 6(a), building at right). It was built in 1965. This building pounded with an older 7 story building whose lower 4 stories are composed of reinforced concrete, and its upper 3 stories are of steel construction (Fig. 6(a), building at left). The buildings have about 2 inch separation.

The 10 story building suffered significant damage. Its lateral load capacity deteriorated due to severe damage at the slab-to-column connections. The horizontally placed reinforcing bars for the spandrel masonry walls were pulled out from the adjacent columns. Further study is needed to determine whether this damage is attributed to pounding. However, pounding caused the seismic hazard of falling building debris. Fig. 6(b) shows the debris that fell on and through the canopy located at the 2nd floor level. The brick veneer at the boundary of the two buildings was damaged due to impact, and a large amount of falling debris destroyed the canopy. This is classified as Type-2 damage and possibly Type-1 damage as well.

13th Street, Oakland City Center: The building is 11 stories plus a 2 story penthouse constructed of masonry combined with a steel skeleton. It was built in 1904. The building pounded between its 9th and 10th floors with the roof of an adjacent reinforced concrete parking structure (Fig. 7(a)) built in 1957. The buildings have about 2 inches wall-to-wall separation, but wooden formwork remained in this gap thus reducing the separation between the two buildings.

Pounding damage was observed along the entire contact length. Fig. 7(b) shows the location of severe pounding damage. The stairway is enclosed by the wall which pounded at the roof level of the parking structure. Since the stairway lacks a floor diaphragm at the pounding region, the wall was impacted and buckled out-of-plane (Fig. 7(b)). The stairway then collapsed at the pounding level and fell onto the lower levels, resulting in complete loss of the stairway. The pounding damage is classified Type-1 with incipient Type-2 damage.

11th Street, Oakland City Center. - The building is a large 6 story steel structure occupying an entire city block (Fig. 8(a)). It was built in 1981. The building in plan, consists of three segments separated by 4 inch expansion joints. All building segments have steel moment resisting frame systems. One segment was originally designed as a 20 story high-rise building, but only the bottom 6 story portion is presently constructed (this is the portion shown in the lower right corner of Fig. 8(a)). Accordingly, this segment is laterally much stiffer than the other two segments.

Due to the earthquake, these segments pounded at their floor slabs which are at common elevations. The slabs at the upper floors were locally damaged due to the contact forces, but there was no structural damage to the building. In general, pounding produces sharp irregular motions which results in large high frequency lateral accelerations (Kasai et al. 1990). It is possible that such accelerations contributed to other damage observed in the building. The windows facing the atrium fell down, and the computer equipments shifted and/or turned over. Heavy building equipments in the penthouse shifted significantly (Fig. 8(b)). Further, the plumbing for the fire protection system lacked seismic joints, and consequently it was destroyed at the building separations due to building relative motions. A considerable amount of water was released. The damage was quickly repaired and the separation distance between the slabs were increased. The pounding damage to this building is classified Types-2, 3 and 4.

Bluxome Street, San Francisco: - The building is a 4 story masonry structure having a large floor area. It was built in 1922. The building is located adjacent to a relatively modern 3 story reinforced concrete storage building having nearly the same floor area (Fig. 9(a)). The exterior walls of these buildings have virtually no separation. The building pounded at the roof level of the adjacent building, and the wall above the pounding level collapsed (Fig. 9(a)). The clear pattern of damage above the pounding level suggests that pounding may have contributed to this catastrophic event. As Fig. 9(b) shows, the cars parking on the street were completely crushed by the large amount of falling bricks and 5 people were killed. This case is classified as Type-1 and 2 damage.

CURRENT RESEARCH

The writers are conducting analytical research into pounding, sample references of which are listed below. The objectives are: to study the dynamics of pounding via correlative computer analyses of actual pounding incidents; to develop and provide engineers with practical analytical tools for predicting pounding response and damage; and to investigate possible methods of mitigating the damage due to pounding.

CONCLUSION

The Loma Prieta earthquake resulted in pounding between many buildings over a wide geographical area. The reported cases of pounding are concentrated in high density population areas of cities where maximum land usage is sought thereby promoting minimal building separations. Most incidents of poundings involved older multi-story buildings of masonry construction. Such buildings often have virtually no building separations.

It should be noted that the predominant architectural and/or structural damage of a minor nature (Type-4) is likely to be precursors to the more serious damage types (Types-1, 2, 3) that will occur during future more severe earthquakes affecting the area. It is imperative that rational methods be developed to mitigate the pounding hazard. The solution must involve major public policy regulations as well as research into the physical aspects of the problem such that practical engineering remedies can be formulated.

ACKNOWLEDGEMENTS

This survey is a part of the United States National Science Foundation (NSF) sponsored research project (Grant No. BCS-9003579). The cognizant NSF program official for this support is Dr. Shih-Chi Liu. The support is gratefully acknowledged. The writers also acknowledge the input of the many interested parties in providing the information regarding actual pounding incidents. D.J. Patel, P.C. Patel, and A. Jagiasi assisted in preparing this paper.

REFERENCES

Bertero, V.V. 1986. "Observation of Structural Pounding," Proc. Internat. Conf. : The Mexico Earthquake - 1985, ASCE.

Kasai, K., Maison, B.F., and Patel, D.J. 1990. "An Earthquake Analysis For Buildings Subjected to a Type of Pounding," Proc. 4th U.S. Nat. Conf. Earthq. Eng.

Kasai, K., Jeng, V., and Maison, B.F. 1990. "The Significant Effects of Pounding Induced Accelerations on Building Appurtenances," Proc. Seism. Design and Perform. of Equipmt. and Nonstruct. Compts. in Bldgs. and Indust. Structs. Appl. Tech. Council Seminar ATC-29.

Kasai, K., and Patel, D. 1990. "A Proposed Method of Evaluating Response For a Type of Collision Between Buildings", IIT-CE-91-02, Illin. Inst. of Tech.

Kasai, K., and Maison, B.F. 1991. "Structural Pounding Damage," Loma Prieta Earthquake Reconnaissance Report, Chapter 6, Struct. Engnrs. Assoc. of Calif.

Kasai, K. and Maison, B.F. 1991. "A Study on Earthquake Pounding between Adjacent Structures", Proc. 6th Canad. Conf. on Earthq. Eng.

Maison, B.F., and Kasai, K. 1990. "Analysis for Type of Structural Pounding," J. of Struct. Eng., ASCE, 116(4).

Maison, B.F., and Kasai, K. 1988. "SLAM: A Computer Program for the Analysis of Structural Pounding". Nat. Inf. Serv. for Earthq. Eng., U. of Calif.

Maison, B.F., and Kasai, K. 1990. "SLAM-2: A Computer Program for the Analysis of Structural Pounding". Nat. Inf. Serv. for Earthq. Eng., U. of Calif.

Rosenblueth, E., and Meli, R. 1986. "The 1985 Earthquake: Causes and Effects in Mexico City," Concrete International. 8(5).

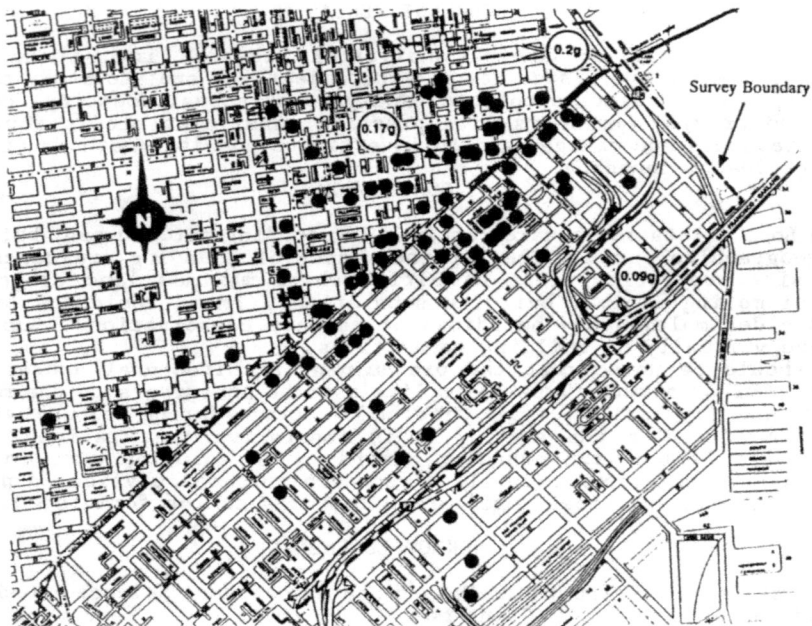

Fig. 1   Distribution of pounding damage in San Francisco Financial District and South-of-Market area.

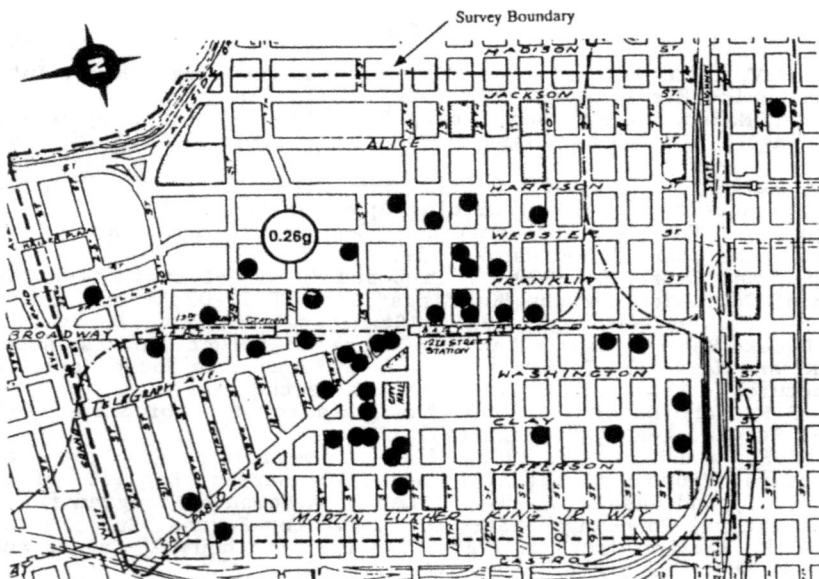

Fig. 2   Distribution of pounding damage in Oakland City Center.

Fig. 3 Damaged corner buildings in San Francisco Marina district.

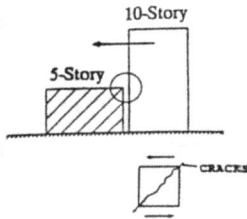

Fig. 4 Example pounding damage (Mission Street, San Francisco)

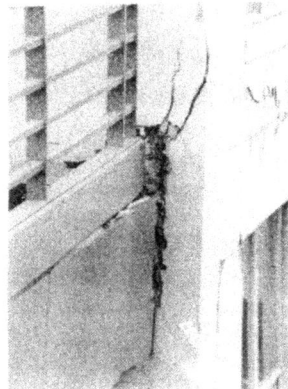

Fig. 5 Example pounding damage (Broadway Street, Oakland)

Fig. 6 Example pounding damage (Franklin Street, Oakland)

Fig. 7 Example pounding damage (13th Street, Oakland City Center)

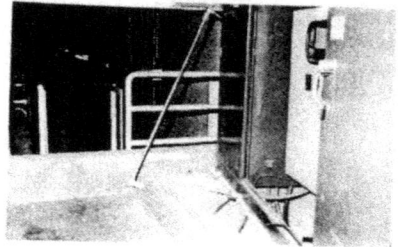

Fig. 8 Example pounding damage (11th Street, Oakland City Center)

Fig. 9 Example pounding damage (Bluxome Street, San Francisco)

Overview of seismic provision changes in national building code of Canada, 1990

W.K. Tso[1]

ABSTRACT

This paper highlights the major and minor changes of the seismic provisions in NBCC-1990. The major changes include (1) Load factor for earthquake load, (2) Base shear formula, (3) Force modification factor R in place of structural coefficient K, (4) Story drift estimation, and (5) Anchorage force for parts of portions of building. The minor changes from NBCC 1985 include: (1) Importance factor (I), (2) Foundation factor (F) and (3) Top concentrated force on structure (F_t).

Finally, a comparison of the base shear calculated according to NBCC-85 and NBCC-90 is given to show that the base shears for most structural systems are similar, except unreinforced masonry (URM). The necessity to increase the base shear for URM appears to be justify in view of the experience of the Saguenay Earthquake in 1988.

INTRODUCTION

The seismic provisions in the National Building Code of Canada, 1990 (NBCC 90) represent the second instalment of a two stage changes of seismic provisions in Canada since 1980. The first stage changes include the use of peak ground velocity instead of peak ground acceleration as a site seismic intensity index; the recognition of different ground motion characteristic effect in the design of short period structures; and the adoption of a new set of seismic zoning maps with contours presented based on the risk probability level of 10% exceedance in 50 years. The reasons for these changes have been summarized (Heidebrecht and Tso, 1985); and these changes have been incorporated in the National Building Code of Canada, 1985 (NBCC 85).

In this paper, the second stage changes since 1985 that appear in NBCC 90 will be discussed. The changes are classified into major and minor changes and specific changes in each class will be presented below.

MAJOR CHANGES

There are major changes in three broad areas, namely (a) the specification of building strengths, (b) the estimation of story drifts, and (c) the specification of anchorage forces for parts or portions of buildings.

---

[1]   Professor, Department of Civil Engineering and Engineering Mechanics, McMaster University, Hamilton, Ontario L8S 4L7

(a)   Building Strength Specification

Changes in building strength specification is more in form than in actual design values. These changes were made so that the information is presented in a more rational form to be implemented, and more reflective of the current seismic design philosophy. They will be discussed under three subheadings:

(a.1)   Load Factor for Earthquake Loading

Unlike NBCC 85, the base shear value given the NBCC 90, V_{90}, is already calibrated to incorporate the load factor a_Q, in the context of Limit State Design load combination. Any seismic load effects obtained based on V_{90} are factored seismic load already and a_Q for earthquake load should be taken as unity. The change involved from NBCC 85 can be summarized in equations (1a) and (1b).

$$(V_f)_{85} = a_Q V_{85} = 1.5 V_{85} \tag{1a}$$

$$(V_f)_{90} = a_Q V_{90} = V_{90} \tag{1b}$$

where V_f denotes the factored seismic base shear.

Therefore, one should compare V_{90} with $1.5V_{85}$ to assess the net change in design base shear between the two editions of the Code.

The reason for this change is the desire to make an important distinction between the wind and earthquake design philosophies. Both loads appear as lateral loads in the design process and are often compared in order to choose the "critical" lateral loading source. However, most building designed according to the codified factored seismic base shears can be expected to have exceeded some forms of ultimate limit state (inelastic deformation, severe cracking, etc.) when they are exposed to the "design" seismic event. Unlike wind design, seismic design deemphasizes the importance of strength in favour of good post-elastic performance, commonly referred to as ductile behaviour. It is more appropriate therefore to specify the codified seismic base shear directly in values compatible with the ultimate limit state rather than with the serviceability state as is done in wind loading.

(a.2)   Base Shear Formula

The base shear formula of NBCC 90 can be written in the form

$$V_{90} \left(\frac{1}{U} \right) = \frac{V_e}{R} \tag{2}$$

where V_e is the elastic base shear and R is the force modification factor dependent on the type of structural system used. U is called a factor representing level of protection based on experience and is taken to be equal to 0.6 for all structural systems. Alternatively, the factor $(1/U)$ can be considered as an overstrength factor. It has been observed that buildings designed using a base shear value of V actually have a lateral strength that is substantially higher than V (Osteraas and Krawinkler 1989; Miranda and Bertero 1989; Fishinger and Fajfar 1990). Osteraas and Krawinkler (1989) have studied the behaviour of steel frame structures in Mexico city during the 1985 Michanocan earthquake and found an overstrength which increases from 2 for long period structures to about 13 for very short period structures. Similar investigation for low-rise reinforced concrete frame structures in Mexico City has been carried out by Miranda and Bertero (1989) and they found an overstrength factor greater than 2.5 for four-story building and greater than 5 for two-story buildings. Fishinger and Fajfar (1990) have summarized the results from various investigations of a specially designed and experimentally tested seven-story reinforced concrete frame-wall structure and showed overstrength factor of 3-4.

The factors that contribute to the overstrength include the higher material strength realised than the nominal values specified in design, the many nominal or minimum design requirements in material codes irrespective of strength demand, the contributions to the lateral strength of elements such as stair and floor slabs, and the force redistribution effect due to the redundancy of most structural systems. The product of the base shear V_{90} and the overstrength factor $(1/U)$ leads to an estimate of the actual lateral yield strength of the building. In NBCC 1990, the overstrength factor is taken to be equal to 1.67 and is applicable to all structural systems with all periods. Equation (2) simply states that the actual lateral strength of the building should be equal to the elastic strength demand V_e, reduced by a factor R which is a function of the ductility capacity of the structural system concerned.

(a.3)   Force Modification Factor R

In all previous editions of NBCC, the influence of structural system choice is characterized by the "K" factor in the base shear expressions. The replacement of the K factor by the force modification factor is more than a mathematical exercise. For the mathematical oriented, it can be recognised that

$$K R \approx 3 \tag{3}$$

A comparison of the K and R values for different structural systems can be found in Table (1). There are two significant benefits using the R factor in NBCC 90. First, by relating it explicitly with the elastic force demand V_e, it draws attention to the designers the implication of choosing any R value in the seismic design load. For example using an R value of 4 implies that the design load is only a quarter of the design load needed if the building is to remain undamaged when exposed to the design seismic event. The building can therefore be excited well into the inelastic range and the survival (non-collapse) of the building relies heavily on its ductility. This line of reasoning will encourage designers to combine loading and detailing requirements when choosing the structural systems.

The second significant benefit is the direct linking of the R values assigned to different structural systems in NBCC 90 on one hand, to the design and detail requirements of these structural systems, as specified by the different material codes in Canada on the other. This direct linking is consistent with the limit state design approach. The material code of design specifications serve as the guidelines to provide appropriate capacities to satisfy the seismic demands set out by the seismic provisions of NBCC 90. Since careful design and detail requirements for ductile behaviour in the post elastic range have been stated in many Canadian material codes (Chapter 21 in CSA.A2.3.3-M-84 and Clause 27 and Appendix D of CAN/CSA-S16.1-M89), the designers are guided by NBCC 90 and these material codes to end up with a proper balance between strength and ductility in their design.

(a.4)   Comparison of Factored Base Shears Between NBCC 85 and NBCC 90

In concluding this discussion of change in strength specification, it is illustrative to compare the factored base shear specified according to NBCC 85 and NBCC 90 for different structural systems. For this comparison, it is assumed that both the foundation factor F and importance factor I are equal to unity and the buildings are located either in Vancouver, B.C., ($v = 0.2$, $Z_a = Z_v$) or Montreal, Quebec ($v = 0.1$, $Z_a > Z_v$). As shown in Fig. (1), the difference in factored base shear for ductile concrete or steel moment resisting frame is minimal. The same observation applies to ductile steel braced frame ($K = 1.0$, $R = 3.0$); and concrete moment resisting frame or steel braced frames with nominal ductility ($K = 1.3$, $R = 2.0$); as shown in Figs. (2) and (3) respectively. For the more ductile systems having $R = 3.0$ or 4.0, the NBCC 90 factored base shears are slightly less than those specified in NBCC 85, while the reverse is true for the systems with nominal ductility having $R = 2.0$. This bias in favour of more ductile structural systems is more dramatically shown in the comparison of base shear for ductile flexural walls on one hand, and the unreinforced masonry system on the other. Through research in New Zealand and United States, it has been shown that concrete walls properly designed and detailed, can behave in a ductile manner. These design rules have been incorporated in the Canadian concrete code for seismic design. In recognition of such advances, the design base shear for ductile flexural wall specified by NBCC 90 is substantially lower than those specified by NBCC 85, as shown in Fig. (4).

For design of unreinforced masonry structures, NBCC 90 prescribes a higher base shear than NBCC 85, as shown in Fig. (5). The justification of this increase in base shear is due to the poor seismic performance of such structures in many earthquakes. An evaluation of the base shear specification on this class of structure in view of their relatively poor performance in the 1988 Saguenay earthquake (M = 5.7) in Quebec province has been made (Tso and Zhu 1991). Damage survey in the epicentral area showed that a number of short period unreinforced masonry buildings have suffered damage (Mitchell et al. 1990).

The design base shear for short period unreinforced masonry buildings in the Saguenay region has bèen increased from NBCC 80 to NBCC 85, recognising the vulnerability of this type of structure to the high frequency content of ground motions likely to occur associated with earthquakes in this region. The level of design base shear is raised again in NBCC 90. The seismic resistance coefficient (design base shear to seismic weight) for short period unreinforced masonry buildings in the Saguenay region are shown in Fig. (6). Plotted in the same figure is the spectral accelerations computed on the two horizontal components of ground motions measured at Chicoutimi, a town 36 km from the epicentre of the Saguenay earthquake. The spectral acceleration plot can be interpreted as the seismic strength demand while the three codified seismic resistance coefficient curves can be treated as the strength supply curves. Comparisons of the supply and demand curves show that even the increase of base shear of NBCC 85 from NBCC 80 is not sufficient to ensure short period unreinforced masonry structures to withstand the ground shaking at Chicoutimi. The further increase of base shear for this type of structural system in NBCC 90 appears to be a step in the right direction.

(b)   Estimate of Story Drifts

Reliable estimates of the maximum story drift during the earthquake is essential to limit the damage of nonstructural elements. Many of the nonstructural elements need to be protected from damage as they are essential for the functioning of the buildings. This is particularly crucial in the cases of designing post-disaster buildings such as hospitals and fire stations. Another reason for building drift estimation is to allow adequate separation between adjacent buildings to prevent them from pounding each other. Observation of the many building damages caused by pounding in Mexico City in 1985 illustrates the importance of such consideration in design.

Unlike NBCC 85 where the story drift is taken as three times the elastic drift based on code load, NBCC 90 specifies that the story drift is estimated as the product of force modification R and the elastic drift based on the code load. This modification recognized that the inelastic deflection and the elastic deflection for buildings are approximately the same. R times the code load is essentially the elastic load. Drift provisions in NBCC 90 implies that the inelastic drift can be estimated by calculating the elastic deformation based on the elastic load demand on the structure.

To emphasize the importance of drift on post-disaster buildings, NBCC 90 limits the drift to 1% of story height for these type of buildings but consider drift up to 2% of story height to be acceptable for other buildings.

(c)   Anchorage Forces for Part and Portion of Buildings

Unlike NBCC 85, NBCC 90 has different clauses to specify design anchorage force for architectural components, and for mechanical and electrical components. The anchorage force for architectural components follows the same format as NBCC 85, namely, $(V_p)_{90} = v(S_p)(W_p)$. The S_p values are tabulated and the values presented in NBCC 90 is approximately 1.5 times the corresponding values in NBCC 85 to reflect that $(V_p)_{90}$ is the factored load already and no further load factor is needed.

Mechanical equipment is generally mounted on isolation dampers to reduce the vibrational effect under normal operation. As a result, it forms a dynamic system when attached to the building and further dynamic amplification can result depending on the condition of mounting.

Also a building filters and amplifies the ground motion and the resulting floor motion depends to a degree on the height of the building. Since the anchorage force depends on the severity of the floor motion, it must also be location dependent. Based on these considerations, NBCC 90 defines the S_p factor for mechanical/electrical equipment by

$$S_p = C_p A_r A_x \tag{4}$$

where C_p values are tabulated for different types of equipment, A_r is the connection factor and A_x is the location factor. A_r ranges from unity for rigid items rigidly mounted to 4.5 for flexible items flexibly mounted. A_x ranges from 1.0 at the ground floor to 2.0 at the top of the building. As a result of such subclassifications, the S_p values for equipment can be as large as 9 times those for the equipment that were located and mounted in the most favourable manner. It is felt that this large factor in design anchorage force represents a small premium to ensure that equipment will not lose its function because it dislodges from its normal position during the seismic shaking.

MINOR CHANGES

It should be noted that the minor changes described herein do not imply that they are insignificant changes. They are grouped under this category because the changes can be implemented in a relatively straight forward manner, assuming one is familiar with the corresponding NBCC 85 provision clauses.

(a)   Importance Factor I

While NBCC 85 assigned the same importance factor $I = 1.3$ to both post disaster buildings and schools, a distinction is made in NBCC 90 between these two types of buildings with a higher importance factor ($I = 1.5$) applicable to post disaster buildings. The poor performance in many of the hospitals in Mexico City during the 1985 earthquake re-emphasized the necessity of the availability of post-disaster buildings after an earthquake. This leads to a more stringent strength requirement as well as drift limitation for such buildings. The changes of the importance factor I can be summarized in Table (2).

(b)   Foundation Factor F

The amplification of ground motion by thick layers of soft soil was observed in Mexico City during the 1985 Michanocan earthquake and again in the San Francisco bay area during the 1989 Loma Prieta earthquake. In Mexico City, the amplification of peak ground surface acceleration can reach as high at 4 (Romo and Seed 1987), depending on the depth of soil. Typical amplification is in the range of 2 to 3. To accommodate the large amplification effect of deep soil deposit, NBCC 90 increases the soil factor F for such soil conditions from 1.5 to 2.0.

(c)   Top Force F_t

To allow for the higher modal contribution on the story shear distribution along the height of a building, NBCC adopted the procedure of adding a top force F_t in its formula for lateral seismic force distribution. Since the influence of higher modal contribution is much more related to the fundamental period T of the building, as opposed to its height to width ratio, NBCC 90 decides to relate F_t directly to the fundamental period T instead of the height to width aspect ratio of the building.

Many of the damages in buildings in Mexico City during the 1985 earthquake occurred around the mid-height level. This could be caused by inadequate allowance of the higher modal contribution. Based on the Mexican experience, NBCC 90 increases the upper limit of F_t from 15% to 25% of base shear. This upper limit of F_t would be reached when $T = 3.6$ sec. For buildings with a period beyond 3.6 second, it is likely that the actual earthquake force distribution will be determined by a more refined analysis than the formula given by the code.

CONCLUSIONS

Substantial upgrading and improvements have been made to the seismic provisions in the National Building Code of Canada since 1980. The present Canadian seismic loading code (NBCC 90) has incorporated a number of lessons learned from many major earthquakes around the world in the 1980's. It contains a number of innovative features among other seismic codes. First, it explicitly coupled the seismic design strength, through the force modification factor R, to specific material base design codes. This coupling provides direct guidance to Canadian engineers to design and detail structures such that these structures will have the expected seismic behaviour in the event of an earthquake. Second, the current Canadian code is one of the few seismic codes that explicitly recognizes the overstrength of buildings from their design values. This recognition is expressed in the form of (1/U) = 1.67 in NBCC 90, independent of structural systems used. As more research results become available, this factor may be refined to reflect its dependence on both structural period and structural systems used in future editions of the Code. Third, the Canadian Code recognises different seismic regions in Canada are likely to experience ground motions having substantially different frequency content, and makes allowance for the effect on these different types of ground motions on short period structures. The 1988 Saguenay earthquake occurred in a seismic region where the ground motions are expected to contain energy in the high frequency range and all measured records in the epicentral area confirmed this expectation. This demonstrated the necessity to differentiate the different types of ground motions as is currently done.

In summary, the current Canadian seismic code (NBCC 90) is at the forefront of seismic code for buildings and provides guidelines for the safe designs of buildings in seismic regions in Canada.

ACKNOWLEDGEMENT

The author would like to acknowledge the work of the current and past members of the Canadian National Committee on Earthquake Engineering (CANCEE) in arriving at the seismic provision changes described in this paper. Discussions within and between the various CANCEE task groups have contributed significantly to these developments.

REFERENCES

Associate Committee on the National Building Code 1985. National Building Code of Canada. National Research Council of Canada, Ottawa, Ontario.

Associate Committee on the National Building Code 1990. National Building Code of Canada. National Research Council of Canada, Ottawa, Ontario.

Fishinger, M. and Fajfar, P., 1990. The role of overstrength in seismic resistance of buildings. Proceedings, 9th European Conference on Earthquake Engineering, Moscow, Vol. 2, pp. 329-338.

Heidebrecht, A.C. and W.K. Tso, 1985. Seismic Loading Provision Changes in National Building Code of Canada, 1985, Canadian J. of Civil Engineering, vol. 12, No. 3, pp. 653-660.

Mitchell, D., Tinawi, R., and Law, T., 1990. Damage Caused by the November 25, 1988 Saguenay Earthquake, Canadian J. of Civil Engineering, vol. 17, No. 3, pp. 338-365.

Miranda, E. and Bertero, V.V., 1989. The Mexico Earthquake of September 19, 1985 -- Performance of low-rise buildings in Mexico City. Earthquake Spectra, Vol. 5, No. 1, pp. 121-143.

Osteraas, J. and Krawinkler, H., 1989. The Mexico earthquake of September 19, 1985 -- Behaviour of Steel Buildings. Earthquake Spectra, Vol. 5, No. 1, pp. 51-88.

Romo, M.P. and Seed, H.B., 1987. Analytical Modelling of Dynamic Soil Responses in the Mexico earthquake of September 19, 1985. The Mexico Earthquake 1985 - Factors involved and lessons learned, Ed. by M.A. Cassaro and E.M. Romero, ASCE, pp. 148-162.

Tso, W.K. and Zhu, T.J., 1991, Implication of the 1988 Saguenay earthquake on Canadian seismic strength specification. Canadian J. of Civil Engineering, vol. 18.

Table (1): K and R Factors for Structural Systems

Structural Systems	K	R
Ductile concrete or steel moment resistant frame	0.7	4.0
Ductile flexural wall	1.0	3.5
Ductile steel braced frame	1.0	3.0
Concrete moment resisting frame with nominal ductility	1.3	2.0
Steel braced frame with nominal ductility	1.3	2.0
Reinforced masonry	2	1.0

Table (2): Importance Factor I

Occupancy	NBC-85	NBC-90
Post disaster buildings	1.3	1.5
Schools	1.3	1.3
Other occupancy	1.0	1.0

Table (3): Foundation Factor F

Category	Type and Depth		F	
			NBC-85	NBC-90
1	Rock, stiff soil	< 15m deep	1.0	1.0
2	Stiff soil Loose coarse grained soil Soft fine grained soil	>15m deep < 15m deep	1.3	1.3
3	Loose coarse-grain soil	>15m deep	1.5	1.5
4	Soft fine-grained soil	>15m deep	1.5	2.0

DUCTILE CONCRETE & STEEL MRF
(K = 0.7, R = 4.0)

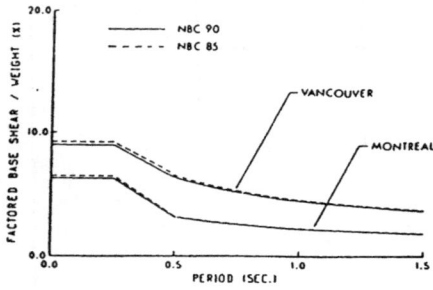

Fig.(1) Factored Base Shear for Ductile MRF

DUCTILE STEEL BRACED FRAME
(K = 1.0, R = 3.0)

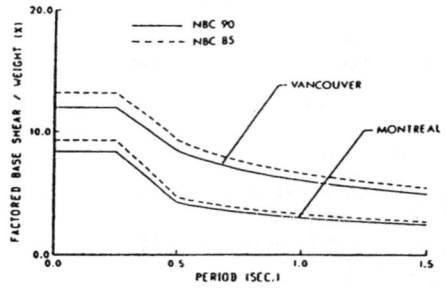

Fig. (2) Factored Base Shear for Ductile Braced Frame

CONCRETE MRF WITH NOMINAL DUCTILITY
STEEL BF WITH NOMINAL DUCTILITY
(K = 1.3, R = 2.0)

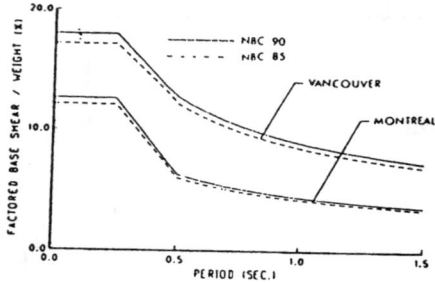

Fig. (3) Factored Base Shear of Frames of Nominal Ductility

DUCTILE FLEXURAL WALL
(K = 1.0, R = 3.5)

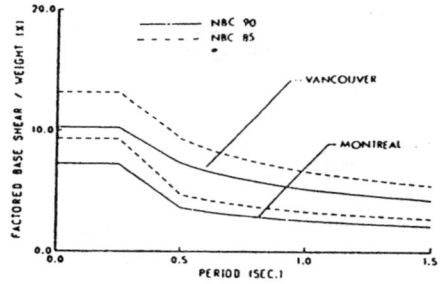

Fig.(4) Factored Base Shear of Ductile Flexural Walls

UNREINFORCED MASONRY
(K = 2, R = 1.0)

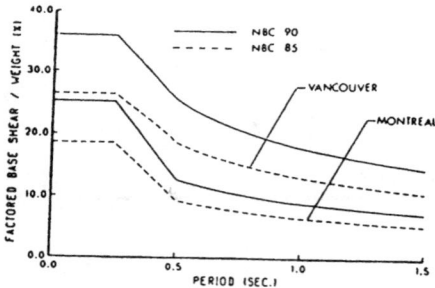

Fig.(5) Factored Base Shear of Unreinforced Masonry

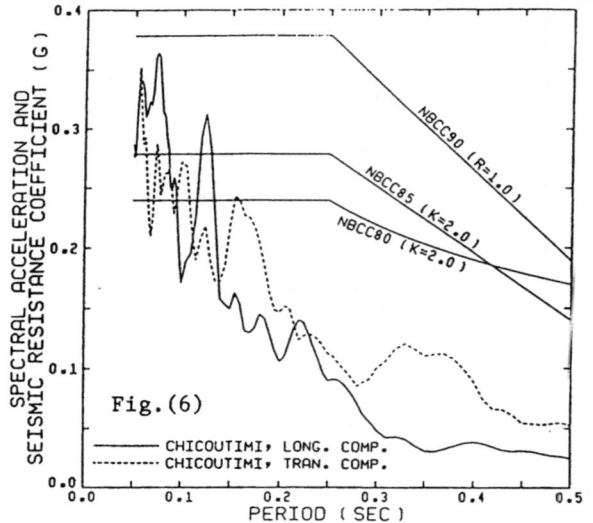

Fig.(6)

750

Evaluation of NBCC 1990 seismic force reduction factors

W.K. Tso[I] and N. Naumoski[II]

ABSTRACT

The seismic force reduction factors proposed in the seismic provisions of the National Building Code of Canada 1990 (NBCC 1990) are examined using ground motion records from two recent Canadian earthquakes. The displacement ductility demands are analyzed for structural systems with different ductility capacity. It is found that the NBCC 1990 force reduction factors, which are period independent, lead to a very high ductility demand for short period structural systems. To avoid this, two types of period dependent force reduction factor for short period structures are investigated. The results show that the linearly varying period dependent reduction factor represents a viable means to resolve the high ductility problems associated with short period structural systems.

INTRODUCTION

For economic reasons, the design strength specified in building codes to allow for the effect of earthquake motions is considerably smaller than the strength demand of the structure if it remains elastic. Therefore, structures so designed are expected to be deformed into the inelastic range when subjected to strong ground shaking. The permissible level of the strength reduction from elastic strength in codes is based on, among other considerations, observation of the seismic performance of structures during major earthquakes. In general, a larger strength reduction is permitted for structural systems capable of sustaining larger inelastic deformation without failure. One important task in seismic code specification is to ensure that the minimum specified strength is not reduced far in excess from the elastic strength in the sense that the resulting ductility demand of the structure when subjected to the design ground motions does not exceed its ductility capacity. This implies that evaluation should be carried out on the base shear specification of the building codes.

I   Professor, Department of Civil Engineering and Engineering Mechanics, McMaster University, Hamilton, Ontario, Canada, L8S 4L7

II   Research Engineer, Department of Civil Engineering and Engineering Mechanics, McMaster University, Hamilton, Ontario, Canada, L8S 4L7

In this paper, an attempt is made to evaluate the base shear expression of the Canadian seismic code provision (Section 4.1.9, NBCC 1990) (Associate Committee on the National Building Code 1990). There are three reasons that an evaluation is appropriate at this time. First, unlike the previous edition, the seismic design strength in NBCC 1990 is explicitly expressed in terms of the elastic strength. In other words, the reduction factor R^*, defined as the ratio of elastic strength to actual strength, can be obtained from the Code without ambiguity. Second, ground motion records are now available from two strong earthquakes recently occurred in Canada, which can be used as representative input for this evaluation. Third, the design and detailing requirements for the different structural systems covered in NBCC 1990 are much more specific than in the previous edition of NBCC. Each structural system mentioned in the code has to be designed and detailed according to the requirements of the current Canadian material design codes and standards. In other words, this is a direct linking between the seismic loading on one hand, and detailing requirements on the other, for each of the structural systems covered.

STRUCTURAL MODEL AND DESIGN STRENGTH SPECIFICATION

For buildings in the short to moderate period range, the higher modal contributions to the base shear are not significant and the major seismic response is from the fundamental mode. Single-degree-of-freedom (SDOF) systems are convenient structural model representations for these buildings, and they are used as structural models in this study.

The yield strengths of the SDOF systems are specified based on the base shear provision of the NBCC 1990. The minimum base shear, V, for a building structure is given as

$$V = (V_e/R)U \tag{1}$$

where V_e is the equivalent lateral seismic force representing elastic response. R is a force modification factor with assigned values between 1 for nonductile structural systems, to 4 for ductile structural systems. U is a factor representing a level of protection based on experience, with an assigned value of 0.6. To understand the implication of Eq.1 it is useful to rearrange it in the following form:

$$V(1/U) = V_e/R \tag{2}$$

The factor (1/U) can be considered as an overstrength factor. It has been observed that buildings designed using a base shear value of V have a lateral strength that is substantially higher than V. The factors that contribute to the overstrength include the higher material strength realised than the nominal values specified in design, the many nominal or minimum design requirements in material codes irrespective of strength demand, the contributions to the lateral strength of elements such as stair and floor slabs, and the force redistribution effect due to the redundancy of most structural systems. The product of the base shear V and the overstrength factor (1/U) leads to an estimate of the actual lateral strength of the building. Equation 2 simply states that the actual lateral strength of the building should be equal to the elastic strength demand V_e, reduced by a factor R which is a function of the ductility capacity of the structural system concerned.

Since the reduction factor R^* is defined as the ratio of elastic strength to actual strength, according to NBCC 1990,

$$R^* = R^*_{90} = R \qquad (3)$$

The subscript 90 denotes the reduction factor based on NBCC 1990. Since R^*_{90} depends on the structural system only, and is not a function of the period of the building, the reduction factor adopted by NBCC 1990 is period independent.

The yield strength F_y of the SDOF systems is taken equal to the actual strength, namely

$$F_y = V/U = V_e/R \qquad (4)$$

The equivalent elastic lateral seismic force, V_e, is specified as

$$V_e = vSIFW \qquad (5)$$

where, v is the zonal velocity ratio, S is the seismic response factor, I is the importance factor, F is the foundation factor, and W is the dead load. The zonal velocity ratio, "v", is defined as the ratio of the horizontal ground velocity to a reference velocity of 1 m/sec. In this study, "v" is taken as 0.2 in the strength calculation and all input ground motion records in the calculation are scaled to a peak ground velocity of 0.2 m/sec. For a rock site (F=1) and building of normal importance (I=1), the yield strength of the SDOF system model becomes

$$F_y = 0.2 \ SW/R \qquad (6)$$

The responses of three classes of structural systems are studied, having an R value equal to 4, 3, and 2. The more ductile structural systems with specified R values of 4 and 3, are modelled by SDOF systems having bilinear hysteretic force displacement relationship. The strain hardening stiffness of the system is taken to be 3% of the initial stiffness. The less ductile systems with an R value of 2, are represented by SDOF systems having a stiffness degrading hysteretic behaviour. Among a number of available stiffness degrading models, the Q-hysteretic model is adopted in this study. For all structural models, a 5% critical viscous damping is assigned to each SDOF system to represent other non-hysteretic form of energy dissipation during the earthquake shaking.

GROUND MOTION INPUT

The two sets of strong ground motion records available from two recent Canadian earthquakes are used as input. The first set was obtained in 1985 from the earthquakes which occurred in the North Nahanni river area of the North Western Territories of Canada. Six records (12 horizontal components) were recorded on rock sites from four earthquakes, with the strongest earthquake of magnitude $M_S = 6.9$. The records are characterized by extremely high peak acceleration, A, to peak velocity ,V, (A/V) ratios, which range from 1.2 to 10 (Heidebrecht and Naumoski 1988).

The second set of records was obtained from the 1988 Saguenay earthquake of magnitude $M_S = 5.7$ which occurred in the province of Quebec. Ten records (19 horizontal components) were recorded on rock sites during this earthquake at epicentral distances between 36 and 177 km. The records also have high A/V ratios which range from 1.62 to 9.68 (Tso and Naumoski 1990).

Figure 1 shows comparison of the 5% damped mean acceleration response spectra of the Nahanni and Saguenay earthquake records with the $Z_a > Z_v$ branch of the seismic response factor S used in NBCC 1990, all scaled to peak ground velocity of 0.2m/sec. This comparison is meaningful because the $Z_a > Z_v$ branch of the seismic response factor is recommended for regions where seismic ground motions are expected to contain major energy in the short period range. The mean spectra of both sets of records far exceed the $Z_a > Z_v$ branch of the seismic response factor in the short period range.

EVALUATION OF THE FORCE REDUCTION FACTORS

Since all input motions have high A/V ratio, and have their energy in the short period range, they are considered to be representative of ground motions that may occur in regions where $Z_a > Z_v$ in the Canadian seismic zoning maps (NBCC 1990). To be consistent, the strength of the SDOF systems used are calculated using the $Z_a > Z_v$ branch of the S curve. The mean displacement ductility demands for the three classes of structural systems having R=2, 3 and 4, subjected to the Nahanni and the Saguenay set of records are shown in Fig. 2. For all three classes of structural systems, the ductility demand decreases with increase of structural period. In view of the mean elastic response spectra of these two earthquakes as shown in Fig. 1, such a trend can be expected. Of more concern is the very high ductility demand of short period (say < 0.3 sec) structures, exhibited in all these plots.

Using mostly earthquake records from California to evaluate the inelastic responses of SDOF systems, the studies by Blume (1970) and Newmark and Hall (1973, 1982) showed that the overall ductility demand bears a simple relationship to the reduction factor, as can be expressed by

$$\mu \approx R^* \tag{7}$$

In other words, the ductility demand of the system is similar in value to the reduction factor used to specify the strength of the system. This relationship provides an important link between the design strength on one hand, and the maximum inelastic deformation demand on the other. In NBCC 1990, $R^*_{90} = R$. Therefore, this simple relationship can be represented by the line $\mu = R$ shown in each of the ductility demand plots. It can be seen that the ductility demand of short period buildings designed based on NBCC 1990 is much higher than that implied by Eq.7. Such high ductility demand may be beyond the ductility capacity of the structural systems. One alternative to reduce this very high ductility demand for short period structures is to cut down the reduction in strength from elastic strength for short period structures by adopting a period dependent reduction factor.

PERIOD DEPENDENT FORCE REDUCTION FACTOR

To reduce the high ductility demand for short period structures, two forms of period dependent reduction factor will be considered herein. The first form is that proposed by Newmark and Hall (1973, 1982). If R is the value of the reduction factor for medium and long period structures, their concept of period dependent reduction factor, denoted by R^*_{N-H} in this paper, can be written as

$$R^*_{N-H} = R \qquad \text{for } T > 0.5 \text{ sec} \qquad (8)$$
$$= \sqrt{2R-1} \qquad \text{for } 0.125 < T < 0.5 \text{ sec}$$

and R^*_{N-H} varies linearly from unity to a value of $\sqrt{2R-1}$ when T varies between 0.03 and 0.125 sec.

The second form of reduction factor is the linearly varying period dependent reduction factor, R^*_L. It takes the value of R for period longer than 0.5 sec. For period below 0.5 sec, it decreases linearly as the period decreases from 0.5 sec and has a value of unity at T=0 (rigid structure). Mathematically, it can be written as

$$R^*_L = R \qquad \text{for } T > 0.5 \text{ sec}, \qquad (9)$$
$$= 1+(R-1)T/0.5 \qquad \text{for } 0 < T \leq 0.5 \text{ sec}.$$

This form of reduction - period variation is similar to that proposed by Berrill et al. (1980). The period dependency of these two forms of reduction factor together with the period independent reduction factor R^*_{90}, is graphically illustrated in Fig. 3.

The mean ductility demands of SDOF systems designed based on the three forms of reduction factor R^*, subjected to the set of scaled ground motion records from the Saguenay earthquake are shown in Fig. 4. The plots represent ductility demands of structural systems designed corresponding to R=2, 3, and 4. The horizontal line representing the simple relationship $\mu = R$ is included in each of the plots. The Newmark-Hall type of period dependent reduction factor leads to some reduction of ductility demand from that associated with the period independent reduction factor. However, only the linearly varying period dependent reduction factor R^*_L leads to ductility demand which is below R for ductile structural systems (R=3 or 4), and slightly over R for systems with nominal ductility (R=2) in the very short period range.

The ductility demand of SDOF systems designed based on the linearly varying reduction factor when subjected to the set of ground records from the Nahanni earthquake follows a similar trend, as shown in Fig. 5. Therefore, the linear reduction factor R^*_L represents a viable means to resolve the high ductility demand problems associated with short period structural systems, when their strength is designed based on NBCC 1990.

CONCLUSIONS

For many short period structures which are designed with some ductility reserve, and hence are allowed to have lower strengths than the specified elastic strengths, it is shown in the present study that they may be exposed to a very high level of ductility demand if the reduction factor used is period independent, as exemplified by R^*_{90}. It is shown that the period dependent reduction factor, and in particular, the linearly varying period dependent reduction factor R^*_L represents a viable means to resolve the high ductility problems associated with short period structural systems. Using available Canadian earthquake records and realistic modelling of ductile systems, it is shown herein that R^*_L reduces the ductility demand of short period structures to such a level that the maximum ductility demand μ is approximately equal to the R factor in NBCC 1990. With this reduction factor, the ductility demand for buildings of all periods can be estimated using the relation $\mu = R$.

This paper is focused on the evaluation of the reduction factors of NBCC 1990 for regions where $Z_a > Z_v$. Presently, it is not possible to carry out similar evaluation for other seismic regions where $Z_a = Z_v$, or $Z_a < Z_v$ because there are no strong motion records from Canadian earthquakes to provide appropriate excitation for such evaluation.

REFERENCES

Associate Committee on the National Building Code 1990. National Building Code of Canada. National Research Council of Canada, Ottawa, Ontario.

Berrill, J.B., Priestley, M.J.N. and Chapman, H.E., 1980. Design earthquake loading and ductility demand. Bulletin of New Zealand National Society for Earthquake Engineering, Vol. 13, pp. 232-241.

Blume, J.A., 1970. Design of earthquake-resistant poured-in-place concrete structures. Earthquake Engineering (Wiegel, R.L., Coord. editor), Prentice-Hall Inc., Englewood Cliffs, N.J., pp. 449-474.

Heidebrecht, A.C. and Naumoski, N., 1988. Engineering implications of the 1985 Nahanni earthquakes. Journal of Earthquake Engineering and Structural Dynamics, Vol. 16, pp. 675-690.

Newmark, N.M. and Hall, W.J., 1973. Procedures and criteria for earthquake resistant design. Building Practice for Disaster Mitigation, Building Science Series 45, National Bureau of Standards, D.C, pp. 209-236.

Newmark, N.M. and Hall, W.J., 1982. Earthquake spectra and design. Monograph Series, Earthquake Research Institute, Berkeley, California.

Tso, W.K. and Naumoski, N., 1990. Period dependent seismic force reduction factors for short period structures. Accepted for publication in Canadian Journal of Civil Engineering.

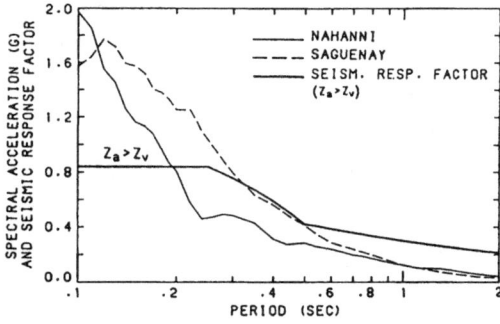

Figure 1. Comparison of the mean spectra of the Nahanni and Saguenay earthquake records with the $Z_a > Z_v$ branch of the seismic response factor; $V = 0.2$ m/s, 5% damping

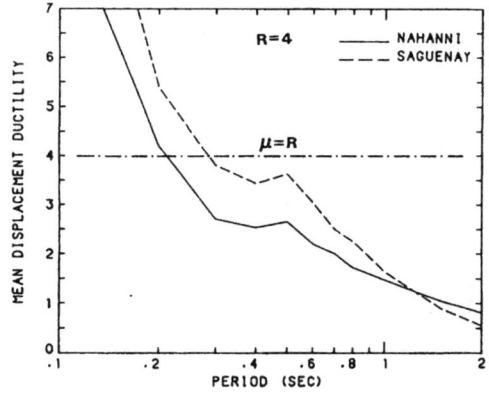

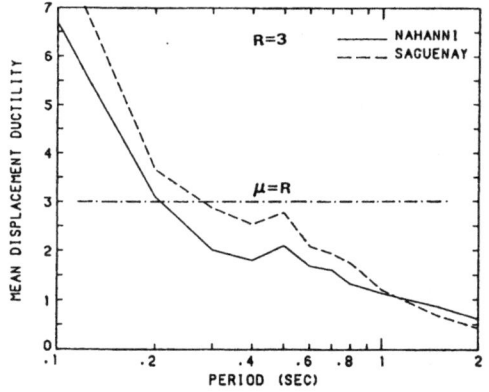

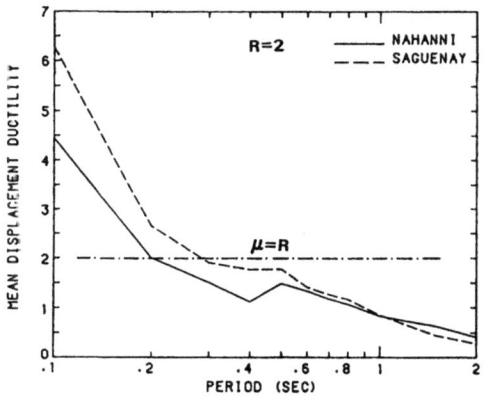

Figure 3. Period dependency of different types of reduction factor

Figure 2. Ductility demands for Nahanni and Saguenay earthquake records for period independent reduction factors as specified in NBCC 1990

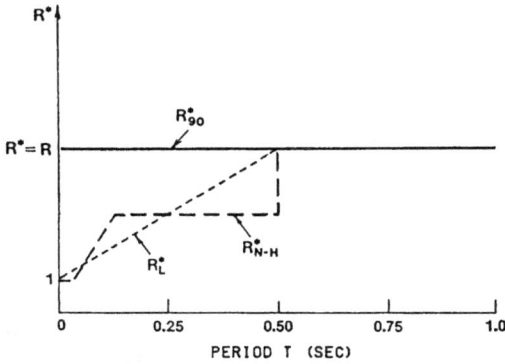

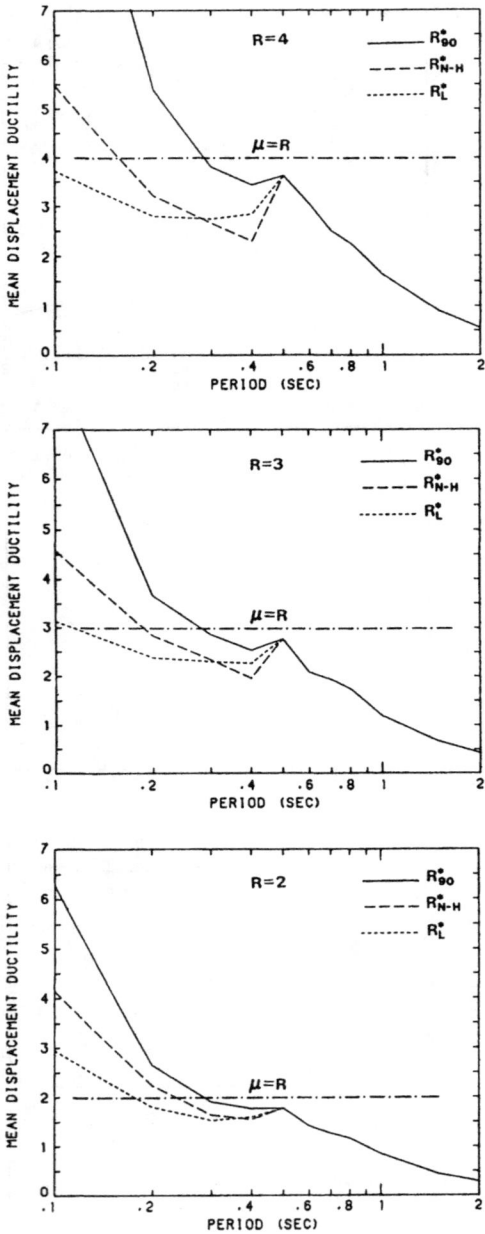

Figure 4. Ductility demands for Saguenay earthquake records for different types of reduction factor

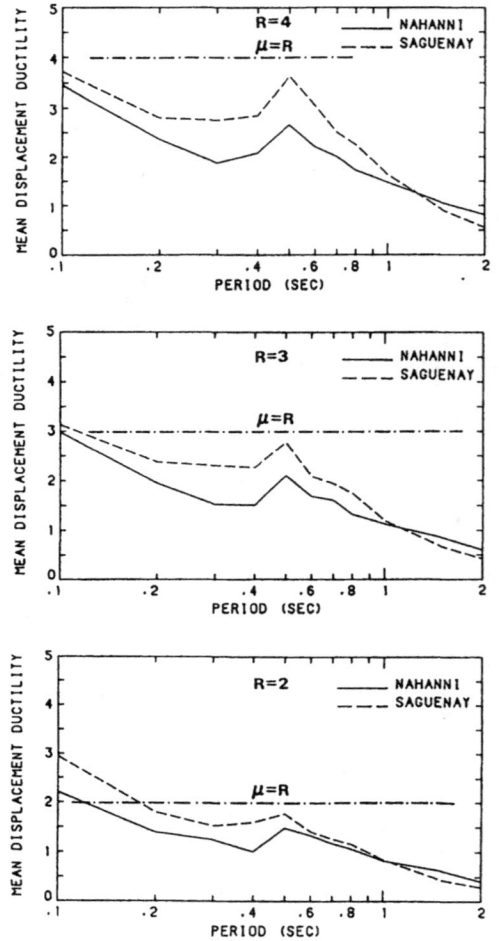

Figure 5. Ductility demands for Nahanni and Saguenay earthquake records for linearly varying period dependent reduction factor

Seismic resistant design of building: multinational codes and programs

Mario Paz[1]

ABSTRACT

The objective of this work is to implement the provisions of building codes current in the U.S. as well as those of other countries for the seismic resistant design of buildings. The work includes the implementation of a total of 28 selected building codes. The objective is to assemble in one volume the main provisions of each code followed by the corresponding computer program and illustrative examples for the seismic design of typical buildings. This paper presents a section in this volume corresponding to the implementation of the seismic provisions of the building code used in Spain, La Norma Basica de la Edificacion y Acciones en la Edificacion, NBE-AE-88.

INTRODUCTION

The design of buildings and other structures to resist the effects of earthquakes are generally based on building design codes promulgated for a specific country or region of a country. Building codes are intended to provide guidelines and formulas which constitute minimum legal requirements for design and construction within a particular country or region. These requirements are intended to achieve satisfactory performance of the structure when subjected to seismic excitation.

The objective of the present work is the preparation of a volume containing a description of the seismic codes for countries located geographically in regions of high seismic activity. This work also includes the development of computer programs for the implementation and application of the seismic codes of these countries. To illustrate this undertaken, the main provisions for seismic design of buildings as required by the building code currently in use in Spain is presented and applied to the seismic design of a six-story building.

PROVISIONS OF THE SEISMIC BUILDING CODE OF SPAIN

The Spanish Basic Norm for Building Construction (Ministerio de Obras Publicas y Urbanismo, Madrid, Spain, 1988) establishes that buildings should

---

[1]Professor, Civil Engineering Department, University of Louisville, Louisville, KY 40292

be analyzed under the action of equivalent seismic lateral forces F_i applied statically at discrete mass points located at the floor levels of the building. These equivalent seismic forces are calculated by

$$F_i = s_i \, Q_i \tag{1}$$

in which Q_i are the concentrated weights at the level of the floors and s_i is the seismic coefficient given by the following formula:

$$s_i = \alpha . \beta . \eta_i . \delta \leq 0.20 \tag{2}$$

The factors in this equation are designated as follows:

α — intensity factor
β — response factor
η_i — distribution factor
δ — foundation factor

The seismic coefficient s_i, through the factors α, β, η_i, and δ, depends on the fundamental period of the structure which may be determined using exact theoretical procedures or using approximate or empirical formulas provided by the code:

For buildings with structural walls:

$$T = 0.06 \; \frac{H}{\sqrt{L}} \sqrt{\frac{H}{2L + H}} \geq 0.50 \text{ sec} \tag{3}$$

For reinforced concrete buildings:

$$T = 0.09 \; \frac{H}{\sqrt{L}} \geq 0.50 \text{ sec} \tag{4}$$

For steel buildings:

$$T = 0.10 \; \frac{H}{\sqrt{L}} \geq 0.50 \text{ sec} \tag{5}$$

For reinforced concrete buildings with structural walls or with steel bracing, the values for the fundamental period T calculated by the above formulas should be multiplied by the factor f given by

$$f = 0.85 \sqrt{\frac{1}{1 + L/H}} \tag{6}$$

In these formulas H is the height of the building in meters and L the plant dimension (also in meters) in the direction of the seismic forces.

In lieu of a more precise determination for the period of the second mode T_2 and for the third mode T_3, the code provides the following empirical formulas:

$$T_2 = \frac{1}{3} T \geq 0.25 \qquad (7)$$

and

$$T_3 = \frac{1}{5} T \geq 0.25 \qquad (8)$$

where T is the fundamental period.

a) Intensity factor α

The intensity factor α is established by

$$\alpha = C R \qquad (9)$$

where C is the <u>basic seismic coefficient</u> which is equal to the spectral acceleration evaluated for different values of the seismic intensity G as indicated in Table 1 and where R is the <u>seismic risk coefficient</u> given in Table 2.

Table 1. Basic Seismic Coefficient C (NBE-NE-88)

Seismic Intensity G (Grades)	Basic Seismic coefficient C
V	0.02
VI	0.04
VII	0.08
VIII	0.15
IX	0.30

The seismic intensity Grade G is obtained from the seismic zone map provided by the code. In this map, the national territory of Spain has been divided into seismic zones with intensity degrees defined by the International Macroseismic Intensity Scale. The code also provides the correspondence between the seismic intensity Grade G and the spectral velocity S_v. This correspondence is given by

$$S_v = 1.5 \ (2)^{G-5} \ (cm/sec) \qquad (10)$$

761

Table 2. Seismic Risk Coefficient (NBE-NE-88)

Seismic Intensity G (Grades)	Seismic Risk R (Period in years)			
	50	100	200	500
VII	1	1	1	1
VIII	0.90	0.99	1	1
IX	0.72	0.92	0.99	1
X	0.53	0.78	0.95	1

b) Response factor β

The response factor β depends on the period of the structure and on its damping characteristics as given by

$$\beta = \frac{B}{\sqrt{T}} \geq 0.5 \qquad (11)$$

in which B = 0.6 for buildings with many internal partition walls and B = 0.8 otherwise. The values for the response factor β for the second and third modes are also calculated with Eq.(11) after replacing the fundamental period T respectively, for the period corresponding to the second and third modes.

c) Distribution factor η_i

For structures modeled with discrete concentrated weights Q_i, the distribution factor η_i corresponding to the level i is given by

$$\eta_i = \frac{\sum\limits_{k=1}^{N} Q_k X_k}{\sum\limits_{k=1}^{N} Q_k X_k^2} \qquad (12)$$

where
    N = Number of levels with concentrated weights
    X_k = Maximum displacement at level k
    Q_k = Concentrated weight at level k

The code provides also simplified formulas to calculate the values of η_i corresponding to the fundamental mode and to second and third modes. These simplified formulas are based on the assumption of linear lateral displacements of the building.

d) Foundation factor δ

The values of the foundation factor δ are given in Table 3 according to the nature of the soil and type of foundation.

Table 3   Foundation factor δ (NBE-NE-88)

			Type of Soil		
					Very
	Swamps	Loose sands and gravels	Consolidated sands and gravel	Compact rocks	comp rocks
Type of foundation 4000	C ≤ 500	500 < c ≤ 1000	1000 < c ≤ 2000	2000 < c ≤ 4000	c > 4000
Piles:					
Friction type	2.0	1.0	0.7	--	--
Bearing type	1.8	0.9	0.6	--	--
Spread Footing					
Isolated........	1.6	1.1	0.8	0.5	0.5
Continuous....	1.5	1.0	0.7	0.4	0.3
Slabs...........	1.4	0.7	0.5	0.3	0.2

c = velocity of elastic compression wave (meter per second)

e) Overturning moments

The lateral seismic forces produce overturning moments and axial forces in the columns, particularly at the external columns of the building. These overturning moments increase alternatively the gravitational forces in the external columns at one side of the building during the vibration of the structure. The overturning moment at a level of the building is determined as the moment produce at that level by the lateral seismic forces applied at the above levels. Therefore, the overturning moment M_i at the level i of height Z_i is given by

$$M_i = \sum_{k=1}^{N} F_k (Z_k - Z_i) \quad for\ i = 0,1,2 \ldots (N\text{-}1) \quad (13)$$

where the lateral forces F_k for each mode are calculated by Eq.(1).

f) Story Shear Forces

The shear force V_i at level i, on the resisting elements of story i of the building, is given by the sum of the lateral seismic forces above that level:

$$V_i = \sum_{k=1}^{N} F_k \quad (14)$$

763

g) Torsional Moment

Torsional moment M_{ti} at each story is equal to the story shear force V_i multiplied by the eccentricity e_i, the distance between the center of the above mass and the stiffness center of the story. This distance is measured normal to the direction of the seismic forces.

h) Lateral displacement

The lateral displacements at the various levels of the building may be determined by static analysis of the building subjected to the equivalent lateral forces F_i. When the structure is modeled as a shear building with rigid horizontal diaphragms at the floor levels, the relative displacement Δ_i in story i may be evaluated as

$$\Delta_i = \frac{V_i}{K_i} \tag{15}$$

in which V_i is the story shear force and K_i is the stiffness of the story.

The lateral displacement at level i with respect to the base of the building is then given by the sum of the relative displacements of the lower stories. Therefore, the displacement X_i at the level i is given by

$$X_i = \sum_{j=1}^{i} \Delta_j \tag{16}$$

Numerical Example

Use is made of a computer program developed to implement the seismic provisions of the Building Code of Spain. The plane steel frame of Fig. 1 serves to model a six-story building. Loads of 20,000 Kp (kilograms weight) are attributed at each level of the building except at the roof where the load is 10,000 Kp. The total flexural rigidity of the columns in any story is EI - 1.3 x 10^{11} Kp-cm^2. The risk period equal to 50 years and seismic intensity G - IX. The building is projected with isolated foundations for the columns in a soil of consolidated gravel and sand.

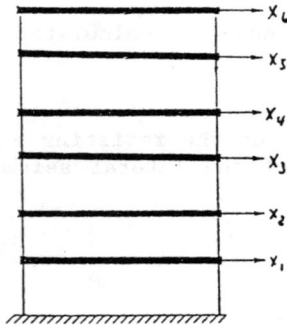

Figure 1 Modeled building for numerical example

INPUT DATA AND OUTPUT RESULTS FOR EXAMPLE

INPUT DATA:

INTENSITY FACTOR	NDI = 9
RISK PERIOD	RI = 50
BASE DIMENSION (EARTHQUAKE DIRECTION) (M)	L = 6
TYPE OF SOIL	NT = 3
FOUNDATION TYPE	NC = 3
NUMBER OF STORIES	ND = 6

FACTORS OF THE SEISMIC COEFFICIENT

FACTOR	MODE 1	MODE 2	MODE 3
INTENSITY FACTOR (ALPHA)	0.080	0.240	0.401
RESPONSE FACTOR (BETA)	0.689	1.194	1.541
FOUNDATION FACTOR (DELTA)	0.800	0.800	0.800

DISTRIBUTION FACTOR, ETA(I,J)

LEVEL	MODE 1	MODE 2	MODE 3
6	1.456	-0.103	0.331
5	1.235	-0.056	0.080
4	1.014	-0.009	-0.170
3	0.794	0.038	-0.240
2	0.573	0.085	0.010
1	0.353	0.075	0.260

SEISMIC FACTOR, S(I,J)

LEVEL	MODE 1	MODE 2	MODE 3
6	0.064	-0.024	0.163
5	0.055	-0.013	0.040
4	0.045	-0.002	-0.084
3	0.035	0.009	-0.119
2	0.025	0.019	0.005
1	0.016	0.017	0.129

RESULTS:

SEISMIC INTENSITY	GI = .3
FUNDAMENTAL PERIOD	T1 = 1.35
SECOND PERIOD	T2 = .45
THIRD PERIOD	T3 = .27
SEISMIC RISK	RS = .72

EQUIVALENT LATERAL FORCES (Kp):

LEVEL	MODE 1	MODE 2	MODE 3	EFFECTIVE FORCE
6	643	-238	1634	1790
5	1092	-259	792	1393
4	897	-43	-1683	1938
3	702	173	-2377	2502
2	507	389	99	621
1	312	346	2575	2616

SHEAR FORCE (Kp):

LEVEL	MODE 1	MODE 2	MODE 3	EFFECTIVE FORCE
6	643	-238	1634	1756
5	1735	-497	2426	2983
4	2632	-540	743	2735
3	3334	-367	-1634	3713
2	3841	22	-1535	4136
1	4153	367	1040	4281

OVERTURNING MOMENT (Kp-m):

LEVEL	MODE 1	MODE 2	MODE 3	EFFECTIVE MOMENT
5	3217	-1188	8169	8780
4	11893	-3672	20300	23527
3	25054	-6372	24013	34703
2	41724	-8208	15844	44631
1	60928	-8100	8169	61473
0	94151	-5162	16487	95584

LATERAL DISPLACEMENTS (cm):

LEVEL	MODE 1	MODE 2	MODE 3	EFFECTIVE DISPL.
6	2.91	-0.12	0.30	2.93
5	2.47	-0.07	0.07	2.47
4	2.03	-0.01	-0.15	2.03
3	1.59	0.04	-0.22	1.60
2	1.15	0.10	0.01	1.15
1	0.71	0.09	0.23	0.74

TORSIONAL MOMENT (Kp-m):

LEVEL	MODE 1	MODE 2	MODE 3	EFFECTIVE MOMENT
6	643	-238	1634	1756
5	1735	-497	2426	2983
4	2632	-540	743	2735
3	3334	-367	-1634	3713
2	3841	22	-1535	4136
1	4153	367	1040	4281

Reference

Ministerio de Obras Publicas y Urbanismo (1988),La Norma Basica de la Edificacion Acciones en la Edificacion, NBE-AE-88, Madrid, Spain.

Code provisions for structures on deep soft sites

S. Hosni[I] and A.C. Heidebrecht[II]

ABSTRACT

It is generally recognized that soil deposits can amplify seismic ground motions as these propagate from the underlying rock strata to the surface. During the 1985 Mexican earthquake, immense amplifications in the Lake Zone of Mexico City resulted in unprecedented loss of life and destruction.

In an effort to realize the engineering significance of the amplification effects observed in Mexico City, the foundation factor (F) in NBCC was increased to 2.0 for very soft and soft fine-grained soils with depth greater than 15 m. Emergency changes were also introduced to the design code for structures in the Federal District of Mexico City, increasing design forces for structures on deep soft deposits by as much as 67%. In this paper, the dynamic response of bilinear sdof systems is used to investigate the adequacy of these protection measures and the significant differences between both codes in accounting for amplification effects of deep soft deposits.

INTRODUCTION

On 19 September 1985, a large earthquake ($M_s = 8.1$) occurred near the south coast of Mexico. Mexico City, situated at 300-400 km from the epicentre, sustained severe building damage and heavy casualties. Buildings that sustained moderate or higher levels of damage were restricted exclusively to the Lake Zone (EEFIT 1986). Dobry and Vucetic (1987) demonstrated that the immense amplifications recorded in Mexico City were primarily due to the almost linear behaviour of Mexico City clays within the range of shear strains ($\leq .3\%$) induced by the earthquake. Taking into consideration that at such long epicentral distances, a substantial portion of the seismic energy at bedrock is associated with low frequencies, the levels of site amplification recorded in Mexico City

I Graduate Student, Department of Civil Engineering and Engineering Mechanics, McMaster University, Hamilton, Ontario, Canada, L8S 4L7.

II Professor, Department of Civil Engineering and Engineering Mechanics, McMaster University, Hamilton, Ontario, Canada, L8S 4L7.

should not be surprising. In fact, similar damage patterns were observed during the 1957 earthquake ($M_s = 7.5$). During that event, 96% of the damage sustained in Mexico City was restricted to the Lake Zone (Mitchell et al. 1986). Since soil deposits tend to focus the energy of the ground motions at frequencies in the neighbourhood of the natural frequency of the soil deposit, higher damage potential is expected to structures of fundamental periods close to the site period. This argument is substantiated by the damage pattern observed during the 1985 earthquake for which the most affected structures were over 6 storeys in height (Mitchell et al. 1986). Soil deposits in the Lake Zone (Zone III in Mexico City code) are characterized by natural periods in excess of 1.0 sec and depths exceeding 20 m of compressible soft clay (Gomez et al. 1988).

Following the 1985 earthquake, a presidential decree was published in Mexico giving emergency code changes for construction in the Federal District. The emergency changes involved increasing design forces by as much as 67% for buildings in the Lake Zone (Mitchell et al. 1986). Based on observations during the 1985 event, a foundation factor (F) of 2.0 was specified in the NBCC (National Building Code of Canada) for very soft and soft fine-grained soils with depth greater than 15 m (NBCC 1990). This implies a 33% increase in design forces as compared to the previous value of F=1.5.

Heidebrecht et al. (1990) had shown that the amplification levels associated with deep soft clay sites and subjected to low intensity, very low a/v (ratio of peak ground acceleration to peak ground velocity) motions are underpredicted by the factor F=2.0 provided in NBCC 90. Their results were based on computations of the response of simple elastic systems. In the current study, evaluation of code provisions for site amplification is extended to computations of the nonlinear response of simple bilinear sdof (single degree of freedom) systems. Ground motions included in the study are exclusively restricted to those recorded in Mexico City during the 1985 main shock. The immense amplifications recorded in Mexico City are a valid test of NBCC 90 provisions because as stated by Mitchell et al. (1986) :
1. The potential for large earthquakes near Vancouver could give rise to amplification effects by the thick recent sediments in the Fraser River delta similar to those recorded in Mexico City.
2. Sensitive clays are present in the seismically active parts of the St. Lawrence River valley.

As the Mexico City code is based on past experience with substantial site amplification effects, it is instructive to compare it to the NBCC 90 in this study. Mexico City code, henceforth, shall refer to the code after the 1985 emergency changes (Gomez et al. 1988).

ANALYTICAL MODEL

Bilinear sdof systems are used to model the nonlinear response of structures having fundamental periods in the range 0.1 to 4.0 sec. For longer structural periods, sdof systems are not a reliable model for actual structures due to lack of representation of higher mode effects. Strain hardening in actual structures is accounted for by a post-yield stiffness taken as 3% of the initial elastic stiffness. A viscous damping of 5% is allowed for in the model. The computer program used for computations of the dynamic response was provided by Zhu (1985).

The program has been modified to specify the yield strength (P_y) of the sdof systems according to either NBCC 90 or Mexico City code rather than NBCC 85. In this paper, the measure of the dynamic response is the peak ductility (μ). Peak ductility is the ratio of the peak relative displacement of the sdof system to its yield displacement.

GROUND MOTION RECORDS

Of the strong motion records obtained during the 1985 main shock, three records from the Hill Zone and five records from the Lake Zone are used as ground motion excitations in this study. Information on these records is presented in Tables 1 and 2. Locations of the recording stations are shown on Fig. 1. The Hill Zone (Zone I in Mexico City code) is characterized by shallow (less than 3 m) deposits of very competent soils and is generally referred to as firm ground.

For the purpose of this study, the 1985 main shock is considered to be the design earthquake for Mexico City due to the following reasons :
1. Design spectra in Mexico City code are based on motions recorded since December 1959 (Rosenblueth 1979) and the 1985 main shock is the largest event recorded since then.
2. Peak ground accelerations recorded in the Hill Zone during the 1985 main shock (≈ 70 year return period) are more consistent with the design spectra in Mexico City code than the .05 g value (g is the acceleration due to gravity) based on the longer return period of 100 years.

Consequently, the zonal velocity ratio used to specify design base shear in NBCC 90 is based on peak ground velocities recorded in the Hill Zone during this event and which are .1 m/s on average. Moreover, the records obtained during the 1985 event classify Mexico City as a region of low a/v motions. This is of consequence in specifying the seismic response factor, S, in NBCC 90.

DESIGN BASE SHEAR

NBCC 90

The design base shear, V, is given by :
$$V = (V_e/R)U \tag{1}$$

Where :
$$V_e = vSIFW \tag{2}$$
    = base shear for elastic response.
R   = force modification factor.
U   = calibration factor based on experience = 0.6
v   = zonal velocity ratio = .1 for Mexico City.
S   = seismic response factor (Z_a/Z_v less than 1.0 for Mexico City).
I   = importance factor = 1 for structures of normal importance.
F   = foundation factor = 1 for Hill Zone.
                    = 2 for Lake Zone (FS not to exceed 3.0).
W = dead load.

Eq. 1 may also be rewritten as :
$$V(1/U) = V_e/R \tag{3}$$

As discussed by Tso and Naumoski (1991), (1/U) in the left hand side of Eq. 3 is basically an overstrength factor reflecting the fact that actual buildings designed to the code base shear, V, will usually sustain larger loads prior to yielding. Since the analytical model in this study does not inherently model this overstrength effect, it should be explicitly incorporated in specification of the yield strength (P_y). Thus, for the purpose of achieving realistic estimates of the ductility demands in actual structures, P_y is specified as :

$$P_y = V(1/U) \tag{4}$$

Mexico City code

For buildings of normal importance, the design base shear, V, is given by:
$$V = C_s W \tag{5}$$

For a building having a fundamental period, T :
$$C_s = [(1+3T/T_a)c/4]/Q' \qquad T<T_a \tag{6}$$
$$Q' = [1+(Q-1)T/T_a] \tag{7}$$

$$C_s = c/Q \qquad T_a \leq T \leq T_b \tag{8}$$

$$C_s = [q(1-r(1-q))+1.5rq(1-q)]c/Q \qquad T>T_b \tag{9}$$
$$q = (T_b/T)^r \tag{10}$$

Q is a ductility factor that corresponds closely to the force modification factor, R, in NBCC 90. For the purpose of the current study, both factors are considered equivalent. Other factors in Eqs. 6 to 10 are given in Table 3. As stated earlier, overstrength should be allowed for in specifying P_y. For consistency and for comparisons to be made between both codes, the same overstrength factor (1/U) used for NBCC 90 shall be applied to the Mexico City code design base shear, V, to derive P_y as given by Eq. 4.

The design base shear, V, according to NBCC 90 and Mexico City code for the Hill Zone and Lake Zone is presented in Figs. 2,3 in terms of the base shear coefficient. This dimensionless coefficient is the ratio of the design base shear, V, to the dead load, W. It is obvious that V based on Mexico City code are higher than those based on NBCC 90 with larger discrepancies for the Lake Zone. An important feature of Mexico City code lacking in NBCC 90 is adoption of period dependent reduction factors for buildings having a fundamental period less than .2 sec (Hill Zone) and .6 sec (Lake Zone). This is in recognition of the fact that use of period independent reduction factors results in short period structures being subjected to excessive ductility demands. Tso and Naumoski (1991) propose adopting a period dependent reduction factor for structures of fundamental periods less than .5 sec. However, results based on the current study, and presented in this paper, tend to support the approach adopted in the Mexico City code, whereby the range of structural periods over which the reduction factor becomes period dependent is wider for structures on soil deposits than for structures situated directly on rock.

RESULTS

In this paper, results are presented for R(or Q)=2.0 and 4.0 to model structures with nominal ductility and ductile structures respectively.

Hill Zone (firm ground)

Mean plus one standard deviation (M+SD) peak ductility (μ) results are presented in Figs. 4 and 5 based on the six components recorded in the Hill Zone. On each figure is superimposed, as a solid line, the ductility level commensurate with the reduction factor used. This level signifies the design (tolerable) ductility limit associated with the specified reduction factor. Figs. 4 and 5 show that both codes provide an excellent level of protection for structures in the Hill Zone. Mexico City code appears to be more conservative by limiting μ to values less than 1.2 corresponding to a tolerable limit of 2.0 (Fig. 4) and to values less than 2.0 corresponding to a tolerable limit of 4.0 (Fig. 5). NBCC 90, on the other hand, maintains the ductility demands closer to, yet lower than, the specified tolerable limits. Since the design base shear based on NBCC 90 is lower than that based on Mexico City code as indicated by Fig. 2, it appears that NBCC 90 offers a more economical design for structures in the Hill Zone without violating the ductility potential for these structures. Based on the above, any violation of the ductility potential for structures in the Lake Zone, and designed to the base shear provided in NBCC 90, would imply deficiencies in the foundation factor, F, in accounting for the actual site amplification effects.

Lake Zone (compressible soil)

M+SD peak ductility results are presented in Figs. 6 and 7 based on the ten components recorded in the Lake Zone. For results based on NBCC 90, a foundation factor (F) of 2.0 is incorporated in the design base shear. Again, the design ductility limit is superimposed, as a solid line, on each figure. In Fig. 6, it is observed that for design based on NBCC 90, the tolerable ductility limit is exceeded with the exception of the structural period range of .2 to .8 sec. For structural periods close to the natural periods of the sites included in the study (1.9-3.9 sec), μ is highest at almost twice the tolerable ductility limit. For design based on Mexico City code, μ does not exceed 1.5, indicating adequate protection to structures of nominal ductility in the Lake Zone.

In Fig. 7, it is again observed that while peak ductilities up to 4.0 can be tolerated, design according to Mexico City code limits μ to values no higher than 2.2. As for design based on NBCC 90, ductility demands exceed the tolerable limits within the whole range of structural periods included in the study. The excessive ductility demand for structures of fundamental periods less than 1.0 sec is a direct consequence of using a period independent force modification factor. However, it can be observed that if a period dependent reduction factor is applied only to structural periods less than .5 sec, ductility demand would still be excessive in the .5 to 1.0 sec period range.

FOUNDATION FACTOR

Results for the Lake Zone indicate that the foundation factor of 2.0 provided in NBCC 90 does not adequately account for the site amplification effects recorded in Mexico City. To investigate this point further, the values of F required to limit M+SD ductility demands, for structures in the Lake Zone, and designed to the NBCC 90 base shear, to the tolerable ductility limits shown in Figs. 6,7 are computed through a process of iterations. In Fig. 8, results are directly compared to the F values provided in NBCC 90. It is observed that for structural periods close to the natural periods of the sites included in the study, actual requirements for F range from 3.0 to 3.5 compared to 2.0 in NBCC 90. This observation holds for both R=2.0 and 4.0. For the case of R=4.0, the demand for design forces higher than those provided in NBCC 90 for structural periods less than 1.0 sec is merely a consequence of using period independent reduction factors. This argument is substantiated by the fact that for the case of R=2.0, the foundation factor provided in the code is quite adequate for that range of structural periods.

Heidebrecht et al. (1990) had studied a deep soft clay site in Arnprior, Ontario and another in U.K. Their results indicate that for structural periods in the range of the site period, actual requirements for the foundation factor reach 3.0, higher than the value of 2.0 provided in NBCC 90.

CONCLUSIONS

Following are the main conclusions based on the current study :
1. F=2.0 provided in NBCC 90 for deep soft clay deposits does not adequately account for site amplification effects similar to those recorded in Mexico City during the 1985 earthquake.
2. There is a need for adoption of period dependent force modification factors in NBCC. The range of structural periods for which the force modification factor is period dependent should be a function of the geological conditions at the site being considered.
3. Mexico City code provides adequate protection to structures of nominal ductility and ductile structures in both the Hill Zone and Lake Zone.
4. Mexico City code is an example of how microzonation offers the better alternative for future aseismic codes.

REFERENCES

Dobry, R. and Vucetic, M. 1987. Dynamic properties and seismic response of soft clay deposits. Proc. Intl. Symposium on Geotech. Eng. of Soft Soils, Mexico City, Vol. 2, 51-87.
EEFIT 1986. The Mexican earthquake of 19 September 1985. A field report by EEFIT. The Institution of Civil Engineers, London, England.
Gomez, R. and Garcia-Ranz, F. 1988. The Mexico earthquake of September 19, 1985-Complementary Technical Norms for Earthquake Resistant Design, 1987 edition. Earthquake Spectra, Vol. 4, No. 3, 441-459.
Heidebrecht, A.C. et al. 1990. Seismic response and design for structures located on soft clay sites. Can Geotech. J., Vol. 27, 330-341.
Mitchell, D. et al. 1986. The 1985 Mexican earthquake - A site-visit report. Earth Physics Branch Open File Report No. 86-2. Energy, Mines and Resources Canada, Ottawa, Canada.
National Building Code of Canada 1990. Associate Committee on the National Building Code, National Research Council of Canada, Ottawa, Canada.
Rosenblueth, E. 1979. Seismic design requirements in a Mexican 1976 code. Earthquake Eng. and Struct. Dyn., Vol. 7, 49-61.
Tso, W.K. and Naumoski, N. 1991. Period dependent seismic force reduction factors for short period structures. Can. J. Civ. Eng. (accepted for publication).
Zhu, T. 1985. Effect of peak ground acceleration/peak ground velocity ratio on seismic response of simple nonlinear systems. M. Eng. Thesis. McMaster U.

Table 1. Ground motion records from the Hill Zone, Mexico City.

Station	Comp	PGA*	PGV**
CU01	S00E	.0206	.1020
	N90W	.0341	.0938
CUIP	N00E	.0323	.1025
	N90W	.0353	.0937
CUMV	S00E	.0381	.0919
	N90W	.0396	.1101

* Peak ground acceleration in g.
** Peak ground velocity in m/s.

Table 2. Ground motion records from the Lake Zone, Mexico City.

Station	Comp	PGA	PGV
CAF	S00E	.0821	.2485
	N90W	.0965	.3757
CAO	N00E	.0705	.3498
	N90E	.0820	.4186
SCT	S00E	.0999	.3874
	N90W	.1712	.6050
TLB	N00E	.1385	.6410
	N90W	.1087	.4461
TLD	N00E	.1199	.3490
	N90W	.1137	.3606

Table 3. Factors used to derive design base shear according to Mexico City code.

Zone	c	T_a(sec)	T_b(sec)	r
Hill Zone	.16	0.2	0.6	0.5
Lake Zone	.40	0.6	3.9	1.0

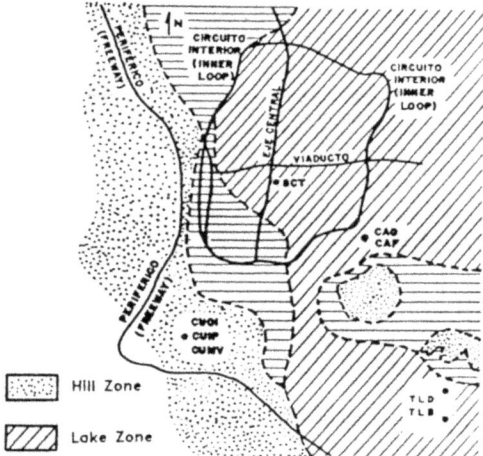

Figure 1. Locations of recording stations for the earthquake of 19 September, 1985.

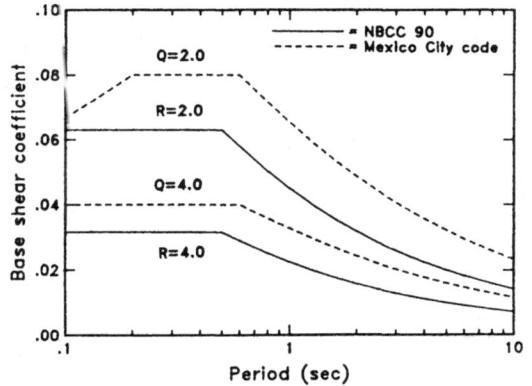

Figure 2. Design base shear for structures in the Hill Zone.

773

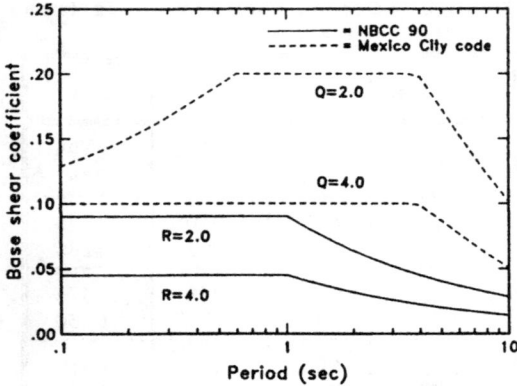

Figure 3. Design base shear for structures in the Lake Zone.

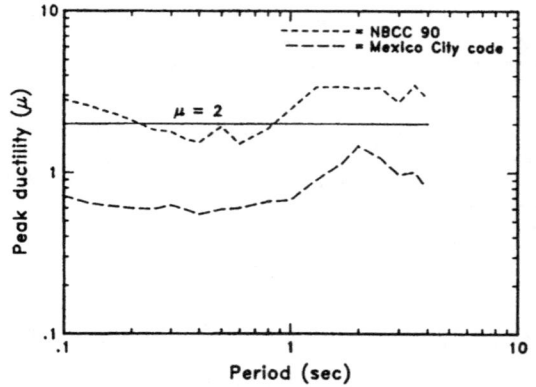

Figure 6. Peak ductility for structures of nominal ductility in the Lake Zone.

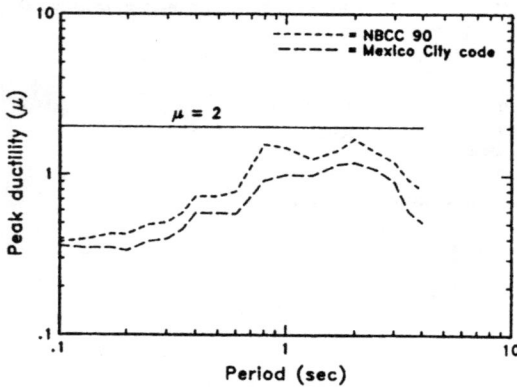

Figure 4. Peak ductility for structures of nominal ductility in the Hill Zone.

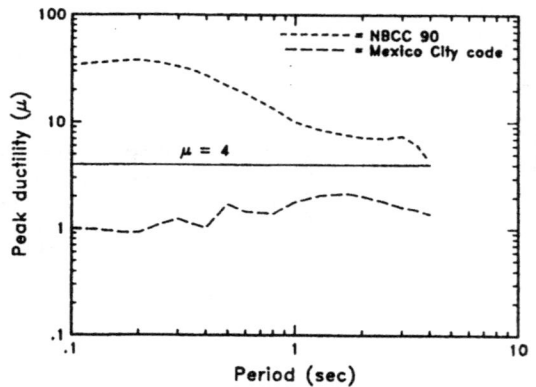

Figure 7. Peak ductility for ductile structures in the Lake Zone.

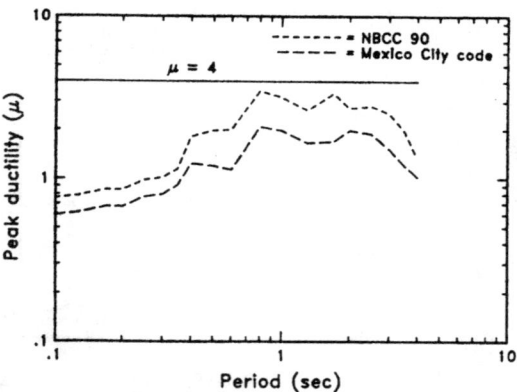

Figure 5. Peak ductility for ductile structures in the Hill Zone.

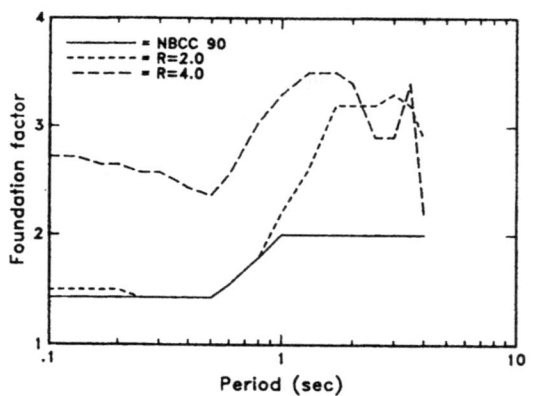

Figure 8. Required foundation factor for structures in the Lake Zone.

774

A comparison of the seismic provisions of the building codes of Canada and China

Jiajun Liu[I]

ABSTRACT

This paper compares the earthquake provisions of the National Building Code of Canada (NBCC 1990) with those of the Aseismic Design Building Code of China (ADBCC 1989). Items dealt with in this examination include: seismic zoning and the selection of ground motion parameters; general design approaches followed; analysis of earthquake action; and control of drift.

The presentation should be of interest to designers and code formulation bodies. The NBCC notation is used whenever possible.

1. INTRODUCTION

Over the past few decades significant advances in seismic resistant design have been incorporated in the earthquake codes of many countries. These developments result from research in earthquake engineering, and from the lessons derived from field observations of structures which have experienced major earthquake events. The earthquake design philosophy of Canada and the People's Republic of China (PRC) is the same and both countries have recently revised their earthquake codes. Various provisions in these regulations are compared and discussed in this paper.

2. SEISMIC ZONING AND GROUND MOTION PARAMETERS

Canada and China use their own, but analogous, methods to assess seismic risk and develop seismic zoning maps. These methods incorporate historical records and geological and tectonic information; they are based on a statistical analysis of the earthquakes that have been experienced by the different regions within and adjacent to these countries. Both countries use a probability of exceedence of 10% in 50 years to define zonal ground motion

(I) President & Senior Structural Engineer, Yunnan Design Institute, and Professor of Yunnan Institute of Technology, Xinwen Road, Kunming, People's Republic of China.

parameters and assign this probability level to identify a moderate earthquake.

To consider the influence of earthquake magnitude and epicentral distance on the ground motion spectrum at a given site, the Canadian Code uses both peak horizontal acceleration (PHA) and peak horizontal velocity (PHV) as zoning parameters for design. By using both parameters, the NBCC takes into account information on the frequency content of potential earthquakes.

China has used historical intensity information, together with modern instrument measurements and geological/tectonic conditions, as an indirect parameter for zoning. China's new intensity scale is similar to the modified Mercalli (MM) intensity scale; an approximate relationship between PHA and intensity established for China is:

$$A=10^{(I \lg 2-0.01)} \tag{1}$$

where A= peak horizontal ground acceleration and I=earthquake intensity.

As summarized in Table 1, China's seismic design code, the ADBCC, relates the PHA to its intensity scale for design purposes. This parameter represents only a single ground motion measurement. In order to account for the influence of earthquake magnitude and epicentral distance on the frequency content of the earthquake ground motion, the Engineering Mechanics Institute of the Chinese Academy has statistically analysed the spectra of the main shock and numerous after shocks of the 1976 Tangshang earthquake. The results, which are presented in Fig. 1, show that: (i) the short period oscillations are predominant for nearby, moderate earthquakes and the period corresponding to the peak point of the response spectra is around 0.1 seconds; (ii) the long period oscillations are predominant for large, distant earthquakes and the period corresponding to the peak point of the response spectra is around 1.0 seconds.

The ADBCC refers to a set of characteristic periods on the design spectrum; these are conceptually similar to "predominant periods." The codes stipulates the influence that earthquake magnitude and epicentral distance has on this important parameter (see Fig. 2 and Table 2). For the majority of seismic zones in China, only the influence of nearby earthquakes need be considered. Those zones in which the influence of distant earthquakes must be considered are clearly specified on the Chinese seismic zoning map. The ADBCC also recommends the use of microzoning results (if they are available) and the selection of ground motion records for the design of important structures.

3. GENERAL DESIGN APPROACH OF NBCC AND ADBCC

Various methods for evaluating the seismic response of buildings are suggested in the NBCC and ADBCC. The ADBCC permits an analysis using the base shear method, the elastic modal superposition method, or the nonlinear time-step dynamic method, depending on the type and importance of the structure.

The 1990 NBCC specifies that the minimum lateral seismic force, V, shall be calculated in accordance with the following formula:

$$V=(V_e/R)U \tag{2}$$

where V_e is an equivalent lateral seismic force representing elastic response. R is a modification factor reflecting the capability of the structural system to dissipate energy through inelastic behaviour (see NBCC 1990). U is a calibration factor (equal to 0.6); it is applied to maintain the design base shear at the same level stipulated in the 1985 edition of the NBCC for buildings with good to excellent capability of resisting seismic loads consistent with the R factors used.

The Chinese code expresses the base shear as

$$V=V_{em} \tag{3}$$

V_{em} is the equivalent lateral force corresponding to the elastic response produced by a "reduced earthquake", which is defined as a minor earthquake by the ADBCC. The minor earthquake and its ground motion parameter are established from the moderate earthquake by the following relationship:

$$\Delta I = I_b - I_c = \lg(1/c)/\lg 2 \tag{4}$$

In this equation, $c = A_c/A$ is defined as a "structural behaviour factor" (c has the same physical meaning as R); A represents the PHA corresponding to the basic intensity I_b of a moderate earthquake, and A_c the PHA corresponding to the reduced intensity I_c of a minor earthquake. The code specified values of c corresponding to different types of lateral load resisting systems, are listed in Table 3. The decrement of earthquake intensity ΔI can be derived from Eqn (1). Values of ΔI are listed in Table 3 for different structural systems. From the Table we note that the reduced intensity is between 1-2 scales lower than the corresponding basic intensity.

The basic intensity is defined as a moderate earthquake having a probability of exceedance of 10% in 50 years. On a probability density plot of the statistics of Chinese earthquakes, the mean intensity value, I_m, corresponding to frequent events, is about 1.55 scale units lower than the basic intensity I_b. The probability of exceedence in 50 years corresponding to the mean value intensity is approximately 63.2%. The ADBCC defines this I_m as the intensity corresponding to a "minor earthquake" and assumes the PHA of a minor earthquake at about 1/3 of the PHA of a moderate (basic) earthquake; that is $A_m = A/3$. This decreased intensity, or PHA, corresponds to c=0.34 in the Chinese code.

The c factor is introduced to reflect the capability of a structure to dissipate energy during inelastic behaviour. The ductility of various structures (or their elements) can be quite different, so that the factor c can differ from 0.34. To account for this variation, and to consider ductility more carefully, the ADBCC introduces a regulation factor in the design process.

For events which are stronger (and rarer) than the moderate earthquake, the ADBCC defines an earthquake intensity with a probability of exceedence of 2-3% in 50 years as a "major or violent earthquake". Using the ground motion parameters corresponding to this major earthquake (see Table 1), the ADBCC requires an elasto-plastic analysis be undertaken to ensure that the deformations of structures (especially ductile buildings or buildings with weak partitions) are controlled and that collapse does not occur. Based on the probability of exceedence of various intensities, and the economic conditions of China, the ADBCC specifies that PHA of major earthquakes in VII, VIII, and IX intensity zones as 6, 5, and 4 times the PHA corresponding to minor earthquakes for these zones (see Table 1).

In the ADBCC the ground motion parameters corresponding to a minor earthquake are used for cross-section design or checking strengths, and those for a major earthquake are used for checking building drift. The moderate earthquake is used for the detailing of structures.

4. ANALYSIS OF EARTHQUAKE ACTION

Equivalent Lateral Seismic Force and Response Spectra

The main earthquake analysis approaches of both the NBCC and the ADBCC are based upon response spectrum theory. In Eqn (2), which denotes the NBCC provisions for establishing the minimum lateral seismic force V, the equivalent lateral seismic force respresenting elastic response, V_e, is calculated in accordance with the following formula:

$$V_e = v. \text{ SIFW} \tag{5}$$

v is the zonal velocity ratio, i.e. the specified zonal PHV expressed as a ratio to 1m/s, and S is the seismic response factor for unit value of zonal PHV ratio. Some general comments relating to the foundation factor, F, are offered below. The importance factor of the structure, I, and the gravitational load, W, will not be discussed in this paper. Note that only the PHV is specified explicitly by the NBCC; the PHA is introduced implicitly by means of the seismic response factor S, which depends on the fundamental period of vibration of the building.

In the ADBCC, the horizontal earthquake force corresponding to a single degree of freedom structure is calculated in accordance with the formulae:

$$V_e = KSG \tag{6}$$
$$\text{or } V_e = \alpha\,G \tag{7}$$

In these formulae, G is defined as the "representative gravitational load". K is the ratio of zonal PHA to gravity acceleration. Because these formulae are used to check the strength of a structure in the first stages of the design, K is the value corresponding to a "minor earthquake". The coefficient α is directly used in the ADBCC instead of K and S. α is defined as an earthquake influence coefficient.

$$\alpha = K.S \tag{8}$$

Although the acceleration ratio, K, varies with zonal intensity, it remains constant for every given intensity. Therefore, the curves of the influence coefficient α are similar to the curves of the response factor S.

The ADBCC stipulates that the earthquake influence coefficient α is determined from Fig. 2 and Table 1, with consideration given to the geology of the site, epicentral distance and the periods of vibration of the structure. The maximum values of α listed in Table 1 are obtained from

$$\alpha_{max} = S_{max}.K \approx 2.25a \tag{9}$$

The code requires that the minimum values of α shall not be less than $0.2\alpha_{max}$.

The Influence of Sub-soil

The NBCC accounts for soil amplification potential by classifying soil conditions into four types and assigning a foundation factor F to each type (see Eqn (5)). The classification takes into account both the material and depth of the surficial layers. On the other hand, althougth the ADBCC also classifies the subsoil into four types (and also the site), it does not introduce the concept of a foundation factor. Instead, the ADBCC changes the characteristic period of the spectrum to account for the predominant period of the site soil.

Soil-structure interaction is not considered explicitly in the the NBCC and this phenomenon is only dealt with in a perfunctory manner. In the ADBCC, although the analysis of earthquake action is based on the assumption that structures are founded on rigid bases, for the design of high-rise reinforced concrete buildings constructed on type III and IV sites and having concrete box foundations or raft foundations of good rigidity, soil-structure interaction can be accounted for by reducing the resulting earthquake forces by 80-90%, depending on the type of structure and site. The drift calculation of a rigid-based building can also be reduced in this same manner.

Apart from offering some appropriate references outlining the assessment of liquefaction potential of foundation soils, the NBCC does not provide any code provision to handle this important material behaviour. On the other hand, the ADBCC deals with this matter by: (1) specifying a formula to calculate a liquefaction index which is used to assess the potential severity of liquefaction, i.e. to assess depth and scope of liquefaction and the thickness of the liquified soil; (2) suggesting countermeasures depending on the category of importance of buildings and the liquefaction of their subsoils.

5. CONTROL OF DRIFT OF STRUCTURES

Drift and Separation

Deformation limits, which are imposed to minimize non-structural damage and to avoid collapse, often control the design of multi-storey buildings. Many codes require certain deformation criteria to be satisfied which may be more stringent than the traditional strength approach and may therefore govern the earthquake design. The NBCC's stated principles for checking deformation are: (i) The drift obtained from an elastic analysis (using loads given by Eqn 2) shall be multiplied by 3 to give realistic values of anticipated deflections i.e., to account for some plastic deformation in a structural system. (ii) The separation of adjacent structures is required to be 2 times the combined deflection of these structures. (iii) Drift limitations should be established in consideration of acceptable damage to the non-structural components. An inter-storey drift limitation of 0.005 times the storey height is recommended. (iv) The effect of the drift on the vertical load carrying capacity of the lateral force resisting system should also be considered.

The ADBCC requires that the drift be checked not only in consideration of the acceptable damage to the non-structural components, but also to ensure that collapse of the structures is prevented. The check on drift involves two separate calculations:
(1)   Evaluation of the elastic drift under the "minor earthquake": The elastic incremental drift or the inter-storey drift of frame-shear wall structures resulting from a minor earthquake shall satisfy

$$\Delta u_e \leq (\theta_e)h \qquad (13)$$

where Δu_e is equal the elastic incremental drift calculated from the α_{max} for minor earthquakes; in this calculation all load factors are taken as 1.00. θ_e is the elastic incremental drift limitation (see Table 4) and h is the storey height.
(2)   Evaluation of elasto-plastic drift under major earthquakes: The elasto-plastic deformations of soft stories or weak portions of structures shall be checked with respect to "major earthquakes". To distinguish the weak storey, a yielding strength coefficient is used. The yielding strength coefficient of a storey is the ratio between the storey shear force carrying capacity V_y and the storey shear force V_e imposed by a major earthquake:

$$\xi_y = V_y/V_e \qquad (14)$$

V_y is evaluated from the specific shear strength and the actual reinforcement of the structural members involved and V_e is determined by elastic analysis under the action of a major earthquake.

The elasto-plastic incremental drift of a soft storey or a weak portion of a structure resulting from a major earthquake should satisfy

$$\Delta u_p \leq (\theta_p)h \qquad (15)$$

where Δu_p is the elasto-plastic incremental drift. The ADBCC has suggested a practical method for calculating Δu_p. θ_p is the elasto-plastic incremental drift limitation (see Table 5). The ADBCC also allows some conditions under which this limitation may be increased.

TABLE 1   DESIGN PHA AND α_{max} IN ADBCC

Level of Earthquake	Probability of Exceedance in 50 Years	K and α_{max} For Various Intensities					
		Ratio of PHA to gravity acceleration K			Earthquake Influence Coefficient α_{max}		
		VII	VIII	IX	VII	VIII	IX
minor, frequent event	63.2 %	0.04	0.08	0.16	0.08	0.16	0.32
moderate (zonal) event	10.0 %	0.125	0.25	0.50	0.23	0.45	0.90
major, rare event	2 ~ 3 %	0.25	0.40	0.65	0.50	0.90	1.40

* The earthquake influence coefficient is the product of the ratio of PHA to gravity acceleration K by the maximum magnification factor of the response spectrum; see section 3 of this paper.

*The Chinese earthquake intensity scale is similar to the Modified Mercalli scale. In this Table VII, VIII, IX are Chinese intensity scales.

Table 2   Characteristic Period of Site Soil

Distance from Epicenter	Category of Site Soil			
	I	II	III	IV
nearby	0.20	0.30	0.40	0.65
distant	0.25	0.40	0.55	0.85

Table 3   Structure Influence Factor C and Intensity Decrement ΔI

Type of Lateral Load Resisting System		C	ΔI
Ductile moment-resisting frame	steel	0.25	2.0*
	R.C.	0.30	1.74*
R.C. moment-resisting frame with R.C. wall		0.30–0.35	1.74*–1.51*
R.C. Wall		0.35–0.40	1.51*–1.32*
Unreinforced masonry		0.45	1.15*
Hinged Bents	steel column	0.30	1.74*
	R.C. column	0.35	1.51*
	masonry column	0.45	1.32*
Chimneys, water-tank towers, tall but flexible structures	steel	0.35	1.51*
	R.C.	0.40	1.32*
	masonry	0.50	1.00*
Timber structures		0.25	2.00*

Table 5   Elasto-Plastic Incremental Drift Limitation

Type of Structure	θ_p
Bent of single storey workshop with reinforced concrete columns	1/30
Frame and infilled frame	1/50
Frame of ground storey under upper masonry storeys	1/70

Table 4   Elastic Incremental Drift Limitation

Type of Structure	Condition	θ_{e}
Frame	incorporation of infilled walls	1/550
	interaction not considered	1/450
Frame-Shear Wall	higher standard decorated public buildings	1/800
	other common buildings	1/650

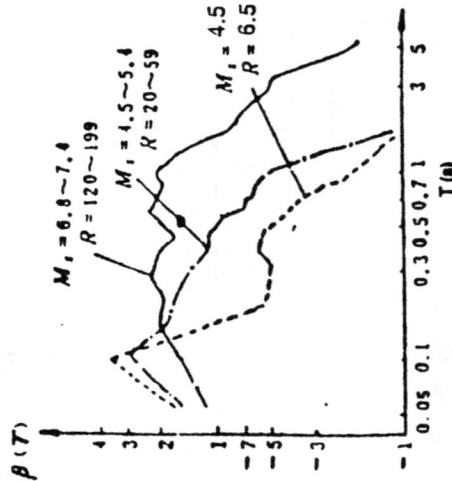

Fig. 1 CONFIGURATION OF RESPONSE SPECTRA FOR VARIOUS EARTHQUAKE MAGNITUDES AND EPICENTRAL DISTANCES

M_s—Magnitude; R—Distance from epicenter; β—PHA magnification factor

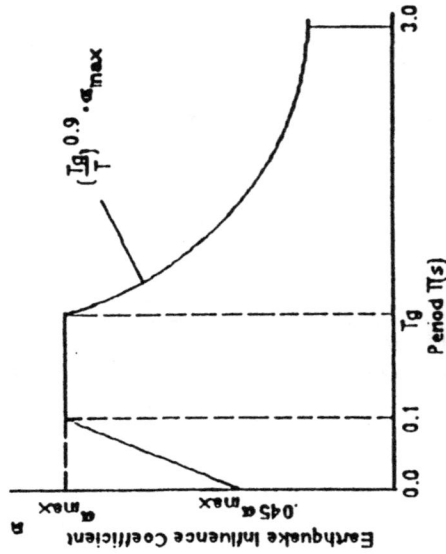

Fig. 2 SEISMIC RESPONSE INFLUENCE COEFFICIENT (ADBCC)

T_g—Characteristic Period (see Table 2)

U.S. national program on seismic repair and retrofit of structures

S.C. Liu[I] and H.J. Lagorio[II]

ABSTRACT

A five-year national research program in the U.S.A. on seismic repair and retrofit of structures initiated in 1990 is described in this paper. The objective is to promote closer ties between research and practice by facilitating the transfer of knowledge and rehabilitation techniques through a concentrated effort leading to the development of design guidelines and performance standards applicable to the needs of U.S. practice. The first year's projects implemented are reported and future priorities indicated. Procedures and mechanisms for coordinating this multidisciplinary and multi-investigator program are described.

INTRODUCTION

Recent damaging earthquakes that have occurred in the United States (Whittier-Narrows 1987, Loma Prieta 1989) and around the world (Mexico 1985, Chile 1985, Armenia 1988, Iran 1990, and the Philippines 1990) once again attest to the importance of earthquake-resistant design in the mitigation of potential property damage and life loss. The acute economic impact ($25 billion) and large number of casualties (20,000 deaths) which resulted from building damage and collapse in Mexico City during the 1985 Michaocan earthquake (Magnitude 8.1) constitute documentary evidence of the intense impact that a major seismic event could have on urban environments. In the U.S., the smaller October 1989 Loma Prieta earthquake (Magnitude 7.1) in the Santa Cruz Mountains of California gave further notice of the vulnerability of urban communities to seismic events with 72 deaths and a cost of about $10 billion in direct and indirect damages.

Over the last decade considerable research has been done in the United States and abroad on various aspects of seismic repair, strengthening, retrofit, and rehabilitation of existing, hazardous structures. During post-earthquake reconstruction periods, structures are typically repaired or strengthened to improve their seismic load capacity and performance. On other

---

I   Program Director for Structural Systems, Earthquake Hazard Mitigation Program, National Science Foundation, Washington, D.C. 20550

II Program Director for Architectural and Mechanical Systems, Earthquake Hazard Mitigation Program, National Science Foundation, Washington, D.C. 20550

occasions, in anticipation of future earthquakes, the seismic rehabilitation of an existing building may occur on a voluntary basis. Knowledge and technology have been advanced as a result of all these efforts. However it has also been apparent that the pace of application and implementation of research results in reducing the vulnerability of existing structures is falling somewhat short of expectations in the U.S. In reality, it is admittedly clear that the problem of designing repair and retrofit schemes for existing buildings is more complex than the design process for new construction. Problems involved in rehabilitation of structures can be approached through a 3-step process as indicated in Figure 1: (1) vulnerability estimate, (2) behavior analysis, and (3) options and decision making.

In late 1989 the Earthquake Hazards Mitigation Program of the National Science Foundation decided that it was timely to address this critical situation by initiating a concentrated research program on a national scale to: (1) integrate the existing body of knowledge, and (2) carry out additional research as required to develop practical engineering measures for immediate implementation to safeguard seismically vulnerable existing structures throughout the entire U.S.

This paper describes the purpose and objectives of the NSF coordinated research program specifically initiated for the seismic repair and rehabilitation of existing structures. The program was publicly announced through a formal program initiative that was distributed nationally to researchers, design professionals, and academicians in January 1990 soliciting research proposals. The formal NSF announcement of this program initiative was issued as NSF Document 90-23, entitled "Repair and Rehabilitation Research for Seismic Resistance of Structures".

TECHNICAL BACKGROUND

As a background to the program initiative, a brief discussion of the technical problems involved is made in this section. Along with these discussions, an attempt is also made to assess the current state of knowledge and practice. These assessments will provide certain clues to new research and development efforts needed to deal with those problems.

Reliable Evaluation of Seismic Vulnerability

The current state-of-the-practice for seismic vulnerability evaluation of buildings is represented by the ATC-14 methodology (1987) which was developed over a 5-year study supported by NSF. This methodology was subsequently adopted by FEMA as reports FEMA-175 and 178. The essential features of the ATC methodology are: (1) it is based on life-safety hazard consideration, (2) buildings are classified into fifteen model types and typical weak-links of the building in question are identified from records of past seismic performance of similar buildings, and (3) standard code-consistent procedures for new design are used to guide the evaluation of the building's capacity vs. demand on the level of structural elements.

Limitations and Remaining Problems

It is apparent that the current knowledge base and practice suffer from several limitations and a number of studies to resolves these problems should be made to improve the situation.

These limitations are:

(1) Fifteen buildings classification types may not be sufficient. A more broadened classification approach, as suggested by Gavlin, et al (1989) which considers the structural framing system types in terms of lateral force resistant sources and their functions might be more appropriate;

(2) Linear elastic procedures, via use of elastic spectra, for assessing local element demands do not adequately address situations involving inelastic load distributions and interaction effects. Limited state analysis, using more refined analytical models which take into account all critical failure mechanisms, systems interaction, and uncertainties in modeling, etc. will improve the evaluation (Aktan, 1990);

(3) Economic consequences as a result of various degrees of damage, i.e., the damageability patterns due to loss under moderate seismic actions are also important factors;

(4) Field methods for evaluation of the current state of materials and buildings i.e., the capacity of the building whether or not it had undergone previous earthquake motions, such as nondestructive tests (NDT) might provide valuable tools in practice as compared with methods based purely on mathematical modeling and analysis. Such NDT methods might include ultrasonic tests, radars with image processing techniques, etc. for detection of fractures, cracks, yielding, and non-structural damages.

The above highlight but a few deficiencies in the current knowledge base for seismic redesign or upgrade as reflected in the state-of-the-art practice. To improve our ability to deal with such urgent problems, particularly for seismic regions in the Central and Eastern U.S. where seismicity as well as design/construction types are vastly different from the West Coast, a broad based effort covering building inventory, behavior (vulnerability) evaluation, and upgrade techniques based on sound engineering measures and cost benefit analysis is clearly needed. The NSF program initiative was developed behind such a technical background. To focus its effectiveness, efforts are concentrated on development of generally applicable engineering measures only, including design techniques and engineering details for typical or common situations.

NSF PROGRAM ON SEISMIC REPAIR AND RETROFIT

Program Description

The purpose of the NSF program initiative is to promote closer ties between research and practice by facilitating the transfer of knowledge and techniques through the development of engineering design guidelines or manuals applicable to U.S. practice. To foster the transfer of information, collaborative efforts combining researchers and design professionals are encouraged. Although not mandatory for participation in the program, this clearly stated support of collaborative and coordinated efforts was aimed toward practical engineering solutions based upon the integration of existing techniques and new research results.

The fundamental objectives of this new NSF program initiative were established to:

- Provide technical information for the realistic evaluation of the vulnerability of existing structures for various levels of seismic excitation, and
- Develop and document cost-effective construction techniques for repairing or strengthening structures identified as hazardous in seismic events.

To achieve these objectives, research proposals were encouraged to study key problems identified for immediate study as high priority items in the following topic areas:

- Performance evaluation of existing structures and foundations.
- Load-transfer mechanisms.
- Retrofitting criteria and techniques.
- Problems and solutions applicable to seismic zones on a nation-wide basis.
- New materials, methods, systems, and devices for seismic retro-fitting, and their design, manufacture, fabrication, and field installation.

Levels of Financial Support

The program is established as a long-term effort in which approximately $2.5 million is made available to the earthquake hazard mitigation research community over the period of five years. The approximate distribution of funds over the program's duration will average $0.5 million in each of the fiscal years 1990 through 1994 contingent on the quality of the proposals submitted in response to the program announcement.

Research Parameters

The research to be supported under this program announcement is intended to generate innovative, practical approaches and techniques which can be used to increase the seismic safety of existing structures. Proposals which involve only the repackaging of existing methods, manipulation of existing data, or investigation of approaches that are limited to incremental improvements to the existing state of practice are being discouraged. Other NSF resources, under the standard process established for submittal of unsolicited research proposals, are still available for the study of the safety and design of new structures.

Structures designed without seismic provisions or with older provisions utilizing the following materials appear to represent the greatest hazard: (a) unreinforced masonry infill walls or bearing wall systems, (b) ordinary concrete structures (including structures using tilt-up systems), (c) ordinary steel structures with non-seismic connection design, and (d) precast concrete structures. In addition, special attention is to be given to buildings of high occupancy and structures housing critical emergency services and/or toxic materials. Research in hazard reduction techniques is being encouraged to address the problems associated with these existing building types. One of the first steps in the existing building rehabilitation process is to determine the actual strength of the existing structure in order to establish the anticipated seismic performance level of the rehabilitated building system.

The coordinated research program is intended to utilize input from practicing engineers to assure that the research data acquired are applicable to current problems and that the technologies developed may be transferred rapidly and efficiently. It is expected that this coordinated seismic study program will provide design professionals with the necessary guidance in selecting analytical design procedures for establishing the seismic resistance of older existing structures that, in turn, will lead to economic and cost-effective upgrade measures. The fundamental goal is to facilitate the development of nationally applicable design recommendations for the reduction of earthquake hazards posed by existing structures.

The following areas of interest, within the topic areas listed above, were provided to prospective researchers as prototype examples noteworthy of study. These topics, however, were not intended to limit the scope of proposals from potential investigators:

- Performance evaluation of existing structural or nonstructural systems at various levels of seismic excitation.
- Performance characteristics of retrofitted or strengthened members and/or structural systems.
- Interaction of new and existing systems.
- Techniques and procedures involving materials, devices, or systems which provide significantly improved earthquake response performance.
- Foundation retrofitting techniques.
- Application of intelligent systems such as high-tech sensors and/or computerized performance monitoring/control systems.
- Cost effectiveness of proposed schemes.
- Program coordination, information exchange and technology transfer.

Research Coordination

One important feature of this program is that all projects underway are fully coordinated in terms their progress, information exchange, etc., so that collectively the final research products can be delivered as intended. All projects are expected to contribute directly to the development of a comprehensive technical summary document including design requirements and details for engineering practice. To achieve this purpose a program coordinator was selected by NSF based on technical depth and breadth in the overall subject areas of research. The program's development process will involve this coordinator in the organization of joint meetings and other effective means of communication to discuss research progress and technical exchange of information among all investigators.

Through specific exercises the coordinator will monitor the necessary information flow resulting from the research work so as to insure that needed technical materials are obtained for the final summary document. The first coordination meeting of all U.S. researchers involved in seismic repair and rehabilitation studies is scheduled for February 15, 1991, in Salt Lake City, Utah, during the 1991 Annual Meeting of the Earthquake Engineering Research Institute (EERI).

The coordinator for the seismic repair and rehabilitation program is Professor James O. Jirsa of the Ferguson Structural Engineering Laboratory, The University of Texas at Austin.

<u>Project Selection</u>

In response to the repair and rehabilitation program announcement, NSF received during the initial year (1990) a total of 56 proposals, of which ten proposals were collaborative research projects developed jointly between respective universities and professional engineering design offices. All proposals were evaluated by a panel of seven peer reviewers appointed by NSF under the standard review process with each proposal receiving a minimum of three reviews. Of the 56 proposals received, a total of nine were recommended for multiple (2-3) year funding during the first fiscal year 1990 at a total budget of $580 thousand. Five of these nine proposals funded represented collaborative research projects in joint association with respective universities and professional engineering design offices.

Repair and rehabilitation research proposed for study by these nine projects include the following general topic areas:

1. Strengthening and repair of non-ductile reinforced concrete frames using external steel jackets and plates.
2. Seismic retrofit with energy dissipators.
3. Evaluation of the strength, stiffness, and ductility of older steel frame structures.
4. Experimental evaluation of slab-column frame buildings and the use of ductile steel.
5. Behavior of clay brick and concrete masonry infill walls.
6. Repair and rehabilitation research coordination mechanisms.

OTHER REPAIR AND REHABILITATION PROGRAMS

<u>NCEER Program</u>

In addition to the NSF seismic repair and rehabilitation program, on a national basis, the National Center of Earthquake Engineering Research (NCEER) of the State University of New York at Buffalo also has a separate program in this research topic area. The two NSF and NCEER programs are being coordinated respectively with one another in order to focus work and leverage research studies in the respective topic areas to their maximum capacities. While duplication of work is being avoided, some overlap in some specific topic areas with critical long-term objectives and high priority study, such as old unreinforced masonry structures for example, are being jointly pursued. Seismic repair and rehabilitation areas of study among others in the NCEER program include the following:

1. Retrofit of lightly-reinforced concrete frames.
2. Retrofit of non-ductile flat plate connections.
3. Retrofit of old unreinforced brick masonry walls.
4. Behavior of infill walls in frame structures.

Between the two organizations, plus other seismic repair and rehabilitation work by other individual principal investigators funded by NSF after the 1985 Mexico, 1987 Whittier-Narrows, 1987 Chile, 1989 Loma Prieta, and 1990 Philippines earthquakes, research study in this area is being carefully orchestrated on a national basis. By the end of the five-year cycle of funding, 1994, it is anticipated that a comprehensive set of design guidelines for the seismic repair and rehabilitation of existing structures in the United States will be substantially finalized for distribution to professional design offices for immediate application and implementation in the field.

FEMA Program

The Federal Emergency Management Agency (FEMA) has among its responsibilities to develop design practices and manuals though translation of research and technology into seismic hazard reduction approaches. Policies, products, and programs over the years, through its affiliated Building Seismic Safety Council (BSSC) and subcommittees under BSSC, a number of design practices and manuals have been published. In connection with the seismic design of existing buildings, efforts were made to identify nationally applicable strengthening techniques and to prioritize seismic rehabilitation in light of complex public policy and societal implications. A series of FEMA publications (1989) indicating the results of these efforts have been issued and are available through FEMA's Earthquake Program and BSSC in Washington, D.C.

SUMMARY AND CONCLUDING REMARKS

A national research program on seismic repair and retrofit of existing structures by NSF and its companion support programs conducted by NCEER and FEMA are described. These programs represent a concentrated, strategical attack to the complex problems the country is facing in its seismically vulnerable regions. While many technical problems remain that could be approached and improved on a piecemeal basis, it is considered important to integrate the current knowledge and develop generic applicable techniques for broad engineering applications. It is expected that the program described in this paper will contribute to the nation's pressing problems in the seismic safety of existing structures.

REFERENCES

Aktan, A.E., and Hu, I.K. 1990. Seismic Vulnerability Evaluation of Existing Buildings. Earthquake Spectra, Vol. 6, No. 3.

Applied Technology Council Report 14. 1987. Evaluating the Seismic Resistance of Existing Buildings.

Federal Emergency Management Agency Report 174/May 1989. Establishing Programs and Priorities for the Seismic Rehabilitation of Buildings, Earthquake Hazards Reduction Series 45, Washington, D.C.

Gavlin, N.L., Sozen, M.A., Wight, J.K., and Wood, S.L. 1990. A Research and Development Plan for Understanding and Reducing Earthquake Hazard to Construction in Regions of Low to Moderate Seismicity. Civil Engineering Studies, SRE No. 551, Department of Civil Engineering, University of Illinois, Urbana, Illinois.

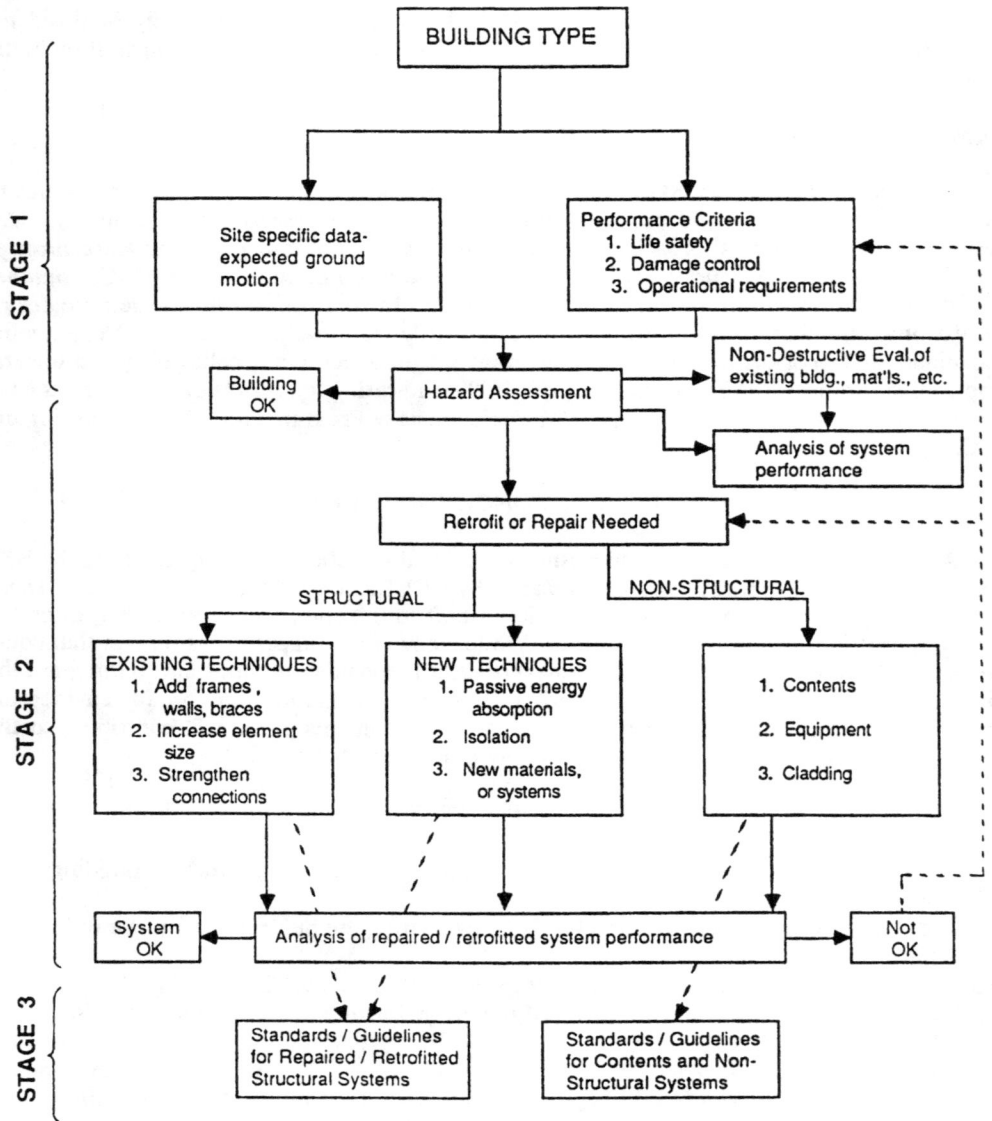

Figure 1. Stages in Hazard Reduction of Existing Buildings (from Los Angeles Workshop)

Earthquake education curricula: a necessary step in mitigating loss of life in the schools

Katharyn E.K. Ross[1]

ABSTRACT

Knowing what to expect, how to prepare and respond to an earthquake is a proven method of mitigating loss of life and property. Teaching this knowledge is the responsibility of today's educators.

Effective earthquake education teaches students both earthquake science and safety, clarifies misperceptions, empowers the school community to realize they can survive a major earthquake, contributes to building a generation of earthquake-cognizant adults, and reduces future loss of life.

In order to begin an earthquake education program, educators need to have information about available earthquake education materials.

In May, 1988, an education project was initiated at the National Center for Earthquake Engineering Research. A primary task of this project was determining who offered earthquake education and identifying the materials being used by students in grades K-12.

This presentation will provide a general overview of earthquake education curricula highlighting similarities and differences and noting areas in need of development.

INTRODUCTION

Earthquake education has an important role in both school and society. All 50 states are vulnerable to earthquakes and at least 39 of these are subject to moderate or major seismic risk, as are the most heavily populated parts of Canada. Millions of people are exposed to significant earthquake hazards. When the mobility of society is taken into consideration, the number of individuals who may one day experience a damaging earthquake is even greater. In this century, earthquakes in North America have resulted in more than a thousand deaths and billions of dollars in property damage. Earthquake education that provides an understanding of the causes of earthquakes, their

---

[1] Education Specialist, National Center for Earthquake Engineering Research

impact, and the necessary steps to reduce loss of life and property is essential to our physical and emotional well-being.

Children spend a significant portion of their day in schools. The school community needs to be well-prepared to meet school earthquake emergencies in order to protect the welfare of students and staff both during and after the ground shaking. The development of an effective disaster preparedness program requires an understanding of the natural processes involved and the type of dangers they pose to the school community. Psychological issues of anticipatory anxiety, emotional trauma, response and recovery must also be considered. Appropriate countermeasures to reduce earthquake damage and personal harm can then be included in school and home emergency response plans.

Children and their teachers do not automatically understand earthquake-generating mechanisms or know appropriate responses during an earthquake. A preliminary study by Ross and Shuell (1989), in which 35 students in kindergarten through sixth grade were interviewed, found that elementary school children had misconceptions about earthquakes, what occurs during an earthquake, what causes an earthquake, and what one should do during an earthquake. Subsequent interviews of students in fourth to sixth grades have continued to support this observation (Ross & Shuell, 1990a, b).

The cause of an earthquake and knowing what to do in response to such an event were especially difficult for the students in these studies. No student in the K-6 group (Ross & Shuell, 1989) mentioned plate movement as a cause of earthquakes. Causes indicated by students included the earth's core getting too hot, faults, lots of wind, loud noises, thunderstorms, drilling in the sidewalks, heat from the sun on the earth, and lava. Students showed confusion between earthquakes and volcanoes, and earthquakes and tornadoes.

After the October, 1989 Loma Prieta earthquake, some fourth graders in a school outside of California wrote the following: "If there were an earthquake I would go on a plane and I would go to Mexico and I would stay there until the earthquake was over and then I would go back." "If an earthquake came I would hurry up and go to the store and get some food and go back home and go in the basement and stay there." "If there was an earthquake I would go to the airport and get on the airplane and go to New York City and stay. I would go buy me a gun."

In the K-6 group (Ross & Shuell, 1989) only 9% of the students gave a clearly correct answer for what to do in an earthquake. Some fourth graders in New York (Ross & Shuell, 1990a, b) specifically mentioned metal when stating that one would be safe in an inside doorway. Probing of this response indicated that some children felt metal would protect them rather than the structure of the building. As one fourth grader stated, "An earthquake doesn't do metal. It does concrete."

Subsequently, a question was added to the "appropriate action" section of an Earthquake Information Test (Ross & Shuell, 1990a), that stated persons should hold on to something metallic. Thirty-three percent of 194 fourth, fifth, and sixth graders from New York and Utah marked this question "true."

Research shows that students have both difficulties in understanding the science of tectonic processes and applying scientific principles to mitigation strategies. Taking action appropriate to the particular environment seems to require understanding of the concept behind that action.

In order to meet the needs of our children in this important area, it is imperative that educators are provided with curricular information and background support materials so that valuable time and resources are not spent redesigning what is already available. Time can then be devoted to regionalizing existing materials, deciding what concepts are most crucial to teach at each age, developing teaching strategies that maximize conceptual understanding, designing materials for those groups of students that are currently not being reached, and intervening with at-risk students to encourage interest in science.

THE ROLE OF ENGINEERS

The role of engineers in earthquake hazards mitigation is to evaluate and understand the seismic performance of buildings and other structures, and to develop methods to enhance that performance. In terms of earthquake education, the engineering community is encouraged to play a dual role. First, as a visible, accessible presence to the school community advising them on structural weaknesses, and the expansion of the earth science and earthquake education curricula to include seismic design concepts; and second, as educators to our future engineers. As earthquake awareness and education grow, more and more students will take an interest in earthquake mitigation measures, including structural design. There is a need for engineers to develop some simple activities that can be done in classrooms to encourage those students.

EARTHQUAKE EDUCATION CURRICULA

Some curricula provide only scientific or safety information. Because a well-balanced earthquake education curriculum is an integration of the two, it is important that the missing half is simultaneously provided. Failure to do so may inadvertently perpetuate misperceptions, cause students to view science as something separate from daily life, and as a result, fail to encourage them to select science and engineering careers.

The following is a brief overview of some existing curricula:

CALEEP Curriculum   For grades 4-8
Lawrence Hall of Science

"Mini-Kit" consists of 14 hands-on earthquake education activities:

a.   Teacher's Guide - including blackline masters
b.   Computer Disk - (Apple II+ and/or IIe with disk drive)
     Quake: A Computer Simulation and
     Survival: A Computer Simulation Game
c.   Filmstrip
d.   Audio Cassette Tape - disc jockey, Mr. Pate,
     experiencing 1964 Alaska Earthquake

e.   AAA map California

Quake BINGO, Await the Quake game, and Simulator Kit may be purchased separately. The Complete CALEEP Kit contains 22 activities.

I Can Make **X** the Difference   For Primary Grades,
Utah State PTA   written at 4th grade
                                                        reading level.

This contains a series of units on a number of areas involving emergency preparedness: fire, earthquake, flood, nuclear war, and weather problems. Each unit is organized according to the same format and includes: a picture of a house in the student's community which becomes a home when each child imagines he lives there; an introductory poem; "What Would I Do" exercises; "Things I Should Know;" and games and puzzles. The earthquake section includes a map showing Utah earthquakes, an earthquake work hunt, and safety rules crossword puzzle.

Crustal Evolution Education Project   Designed primarily for
                                                        grades 7-12

This consists of 33 individual activity modules designed to provide students with an understanding "of the concepts behind plate tectonics and the physical Earth." Each module is individual, self-contained and designed for the Earth Science classroom. Modules include: "Locating Active Plate Boundaries by Earthquake Data," "Earthquakes and Plate Boundaries," "Plate Boundaries and Earthquake Prediction," "Hot Spots in the Earth's Crust," "Volcanoes: Where and Why?" and "Quake Estate," a board game to be played by two to four students, whose goal is, "to achieve success in net income based on accuracy of assessing earthquake risks" (copyright, 1979).
The CEEP is not intended to be a complete curriculum but designed to supplement any teacher's curriculum.

Earthquake Awareness and   For grades Pre-K-6;
Preparedness Curriculum   has been used with
Junior League of Oakland-East Bay   students up to 8th grade

This is a 1 hour curriculum that anyone can pick up and do. It is aimed at elementary students. There is a curriculum guide that provides lessons for each grade level, an Instructor's Guide from Environmental Volunteers, Inc., and role playing situations from CALEEP. There are also supporting videotapes that show each level of the curriculum that were prepared by JLOEB, the Albany Unified School District, and The Audubon Nature Training Society: preschool level, middle school, high school-adult (not included in the curriculum), and "School Facilitation." These can be borrowed from BAREPP.

Earthquakes: A Teacher's Package   For grades: K-6
for K-6/FEMA 159
Federal Emergency Management Agency

This 280 page curriculum was developed by the National Science Teachers Association and includes background material; sets of lessons and classroom activities on earthquake science and safety topics for each of three grade levels (K-2, 3-4, 5-6); scope and sequence charts depicting multidisciplinary connections; masters for reproduction; references; and resources. This package is designed for teachers who have little or no science background.

** Currently, there are plans to also develop "earthquake" theme hands-on classroom activities for grades 7-12 in supplements to Physics, Earth Science, Geometry, Trigonometry, and Calculus.

Earthquakes (Module)   For grades: 8-10
"Minorities in Engineering" Project

This is a module designed to interest students in earthquakes through activities, modeling, engineering applications, and simulation strategies. Has 12 lessons: 1-5 introduce students to earthquakes; 6-9 talk about observed precursors of earthquakes and introduces seismograms; and 10-12 try to make earthquake investigation relevant to students. Includes directions for making related items and doing experiments, i.e. making your own tiltmeter, creepmeter, shoebox model of a fault simulator, liquefaction simulation, resonating building demonstration, and earthquake simulation. Includes reproducible charts and maps. Can be used in part or total in an earth science or general science course.

** NCEER has been given permission to reproduce copies of this module on request.

Guidebook for Developing a   Designed to assist
School Earthquake Safety Program/   school community
FEMA 88   to develop and tailor
Federal Emergency Management Agency   an earthquake safety
                                                    program for the school.

This is a 50-page guide plus appendices that include reprints of FEMA 46, 48, and 113.

The Guidebook includes:
"The Planning Process"
"Hazard Identification"
"Earthquake Drills"
"Immediate Response and Care Requirements"
"Communication"
"Post-Earthquake Shelter Planning"

Appendices include: "Teacher's Package on Earthquake Drills," and example of an earthquake safety program plan; sections on "Children and Disasters" and "Non-Structural Earthquake Damage."

This is designed mainly as a guidebook, not a curriculum. It allows the school to be its own planner. It is included in this listing because many districts noted that it was the curriculum they were using.

** FEMA 88a Earthquake Safety for Children. This 4-part booklet contains excerpts from Units 5 and 6 of <u>Earthquakes: A Teacher's Package for Grades K-6</u> (FEMA 151).

<u>Hands-On Earthquake Learning Package</u>
Environmental Volunteers

1. Instructor's Guide
    a. 17 illustrated, plastic-protected Activity Folders
    b. 16 information/activity inserts (including quake myths, games, puzzles, math activity, "tremor tales").
    c. Illustrated text on basic earthquake geology: <u>The Story of the Earth</u>
    d. <u>Red Cross' Safety and Survival in an Earthquake</u>
    e. "Getting Ready for a Big Quake" - <u>Sunset</u> magazine
    f. Complete guide to school earthquake planning
    g. Neighborhood Preparedness Guide
    h. "Plans for the Teaching Materials"

2. Hands-On Teaching Materials
    a. Plate Tectonics Globe (removable plates)
    b. Earth Hemisphere Model
    c. Plate Puzzle map (ocean floor features)
    d. Wood Plate/Fault Blocks
    e. 9 ft. sq. plate tectonics rug (pattern also available)
    f. Sea Floor Basalt rock sample
    g. Sea Floor spreading box
    h. Time cards, markers and time-tape
    i. Continental Drift film (computer-generated)
    j. Fault Zone Model
    k. Magni-tube Model
    l. Motor driven shaking table and accessories

<u>I-Science Mate Program</u>   For grades: K-6
<u>(Integrating Math, Science and Technology)</u>
Math Science Nucleus

<u>Plate Tectonic Cycle - The Earth on the Move</u> (part of a master science curriculum consisting of six master themes and 24 subthemes).
1. Lab manuals for grades 2-6
2. Shaker tables (made of cardboard, marbles, wood, etc.)
3. Lessons/with experiments and worksheets for grades K-6. Plate Tectonics Cycle includes: Volcanoes, Earthquakes, Plate Tectonics, and Hazards. NCEER has copies of the lessons, experiments, and worksheets from K-6 and some books used in the lessons.
4. Also available from Math Science Nucleus:
    a. Historical Earthquake Slides
    b. Recent Earthquake Slides
    c. Inflatable globe
    d. Glue Balls - to illustrate faults have memory
    e. Physiographic Relief Globe

Plan to Live   For Secondary grades;
Utah State PTA   written at 11th grade
                                                reading level

This includes a series of lessons on various natural and manmade hazards,
including earthquakes. Earthquake related lessons include: "What to Do in
Case of an Earthquake," "How to Prepare for an Earthquake," and "Information
You Should Know About Earthquakes." Test questions are included at the end
of each lesson.

Teaching Earthquake Safety   For grades: K-3
in the Elementary Classroom
Utah Museum of Natural History

A 1/2 hour session gives children basic earthquake information utilizing simple
activities, myths and factual information. It includes the Kamchatka Myth
poster (originally obtained from CALEEP), Wasatch Fault poster and five follow-
up activities (adapted from CALEEP to reflect the Utah scene). A Fault
Blockset available from NASCO science is recommended. This curriculum is
easily adaptable for general use outside of Utah. Note: Utah Museum of
Natural History currently only source for CALEEP's Kamchatka Myth Posters.

Utah Geologic Hazards   For Grades 4 - Senior
Utah Museum of Natural History   High School

This includes a two-part slide presentation and a two foot square model of a
section of the Wasatch Front. Part I - mountain leveling processes of
rockfall, landslide, mudflow, flood, and lake level rises. Part II - mountain
building process-earthquake. It gives a general explanation of earthquakes,
reviews the situation in Utah and what could happen in a major earthquake.
This is followed by an earthquake safety session. Follow-up activities on
earthquake safety are left with the classroom teacher. These were adapted from
CALEEP materials to reflect the Utah scene.

CONCLUSION

Earthquakes are an international problem that require everyone's
attention. An on-going earthquake education program incorporated into all
grade levels will provide a continually developing foundation of science and
safety information for students and staff tailored to their learning and
emotional needs. Students of all ages must be able to take self-protective
actions during an earthquake. Factual information on the science of
earthquakes will help place the need for learning safety actions within the
context of naturally occurring phenomena like weather, will help dispel common
misperceptions that could inadvertently result in physical and emotional harm,
and will help build a future population of knowledgeable adults capable of
making decisions concerning appropriate policies needed to reduce earthquake
hazards.

References

Ross, K.E.K. and Shuell, T.J. 1989, October. <u>Children's beliefs about</u> <u>earthquakes</u>. Paper presented at meeting of the Northeastern Educational Research Association, Ellenville, NY.

Ross, K.E.K. and Shuell, T.J. 1990a, November. <u>The earthquake information</u> <u>test: Validating an instrument for determining student misconceptions</u>. Paper presented at meeting of the Northeastern Educational Research Association, Ellenville, NY.

Ross, K.E.K. and Shuell, T.J. 1990b. <u>After Loma Prieta--what do children</u> <u>outside of California believe about earthquakes</u>? Unpublished manuscript.

* Notes principle author

An example of aseismic measures for water supply systems: the case of Yokosuka City

Hiroshi Sato[I], Akinobu Watanabe[II], and Seiyu Kamata[II]

ABSTRACT: In this paper, as an example of systematic anti-earthquake measures for water supply system, we will discuss about the case of Yokosuka City. Most important feature of water supply system in Yokosuka City is that 99 % of water resources depends on the long distant places. Then, we must consider not only how to prevent the damage to the facilities during earthquakes, but also how to secure the drinking water as much as possible in the city area. This report is to demonstrate the fundamental ideas in earthquake preparedness of water supply facilities and executed counter-measure acts as an example of synthetic earthquake counter-measure acts for lifeline facilities by a local government.

1. INTRODUCTION

In general, water supply facilities are consisted of such three basic elements as a point, line and plane element. For example, a filtration plant and a water distribution basin may be defined as a point element, and all kinds of pipeline and a network of pipelines may be defined as line and plane elements, respectively. These facilities are affected by the conditions of the change of ground conditions and the passage of time. In the earthquake counter-measure acts for water supply facilities, therefore, the empirical and local regulations may have been employed.

This report is to demonstrate the fundamental ideas in earthquake preparedness of water supply facilities and the executed counter-measure acts in Yokosuka city, which is an example of synthetic earthquake counter-measure acts for the lifeline facilities executed by a local government.

2. SPECIFIC FEATURES OF WATER SUPPLY NETWORK IN YOKOSUKA CITY

Yokosuka city is located in the center part of Miura Peninsula and faces to Tokyo Bay in a east direction and to Sagami Bay in a west direction as shown in Fig. 1. This city has the population of 430,000 and the capability of 320,000m^3 per day for water supply. The water supply network in Yokosuka city is a kind of closed system due to the geographical situation. That is, since there is a few water sources within the city, more than 99% of water should be secure from the neighboring cities. The extended distance of traveling pipelines becomes about 30 to 70 km. And there are many ground faults and narrow valleys within the city, because it is closed to the trough of Sagami Bay called the source of earthquakes. Thus this city is under the geological condition that the ground has a weakness for earthquakes. From the

I National Defense Academy, Yokosuka, Japan
II Yokosuka City Waterworks Bureau, Yokosuka, Japan

view point of water operation, due to the geographical restriction, there is only one intake of water within the city and the water is distributed by the natural flow after pumping water up from 19 pump stations to 27 water distribution basins. Accordingly the water supply act itself depends vitally on the other lifeline as electricity.

Water systems

─────⊗───── Only one intra-city source of water	Hashirimizu water system (2,000 m³/day): Completed in 1876
─────●───── 53 km to Yokosuka city	Hanbara water system (10,000 m³/day): Completed in 1921
─── ─●─ ─── 29 km to Yokosuka city	Arima water system (79,000 m³/day): Completed in 1945
═════○═════ 28 km to Yokosuka city	Kosuzume water system (231,900 m³/day): Completed in 1965
▬▬▬●▬▬▬ 63 km to Yokosuka city	Sakawa river water system (20,000 m³/day): Completed in 1978

Major facilities from water sources to the city

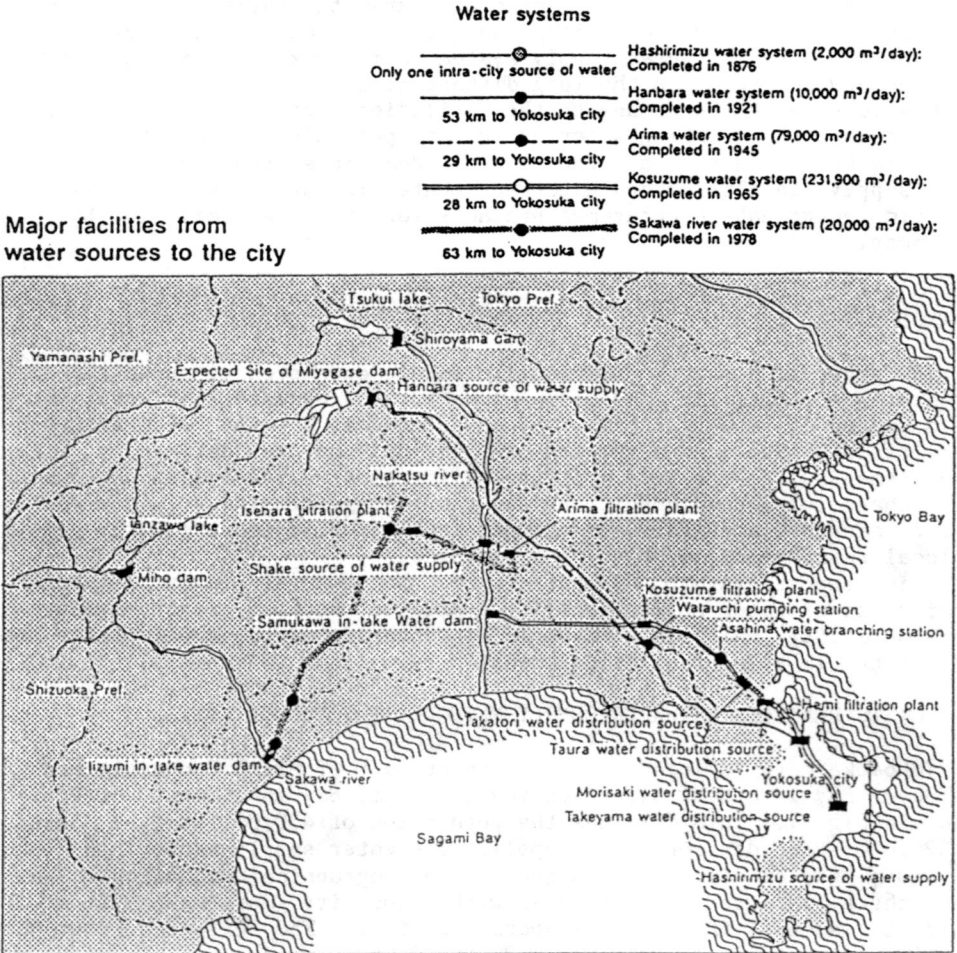

Fig. 1 Major facilities from water sources to the city

3. FUNDAMENTAL POLICIES FOR SEISMIC PROTECTION OF WATER SUPPLY FACILITIES[1]

With these circumstances as back ground, the diagnosis of existing facilities has been done regularly by Yokosuka City Waterworks Bureau to secure the drinking water for citizen. Deciding the following five fundamental policies:(1) to reinforce the trunk of transmission and distribution pipelines, (2) to develop the alternative water sources in emergency, (3) to secure the drinking water, (4) to establish the water supply system and (5) to define the role to each personnel, the acts of seismic protection have been proceeded seriously for seven years between 1979 and 1985. Some examples of the acts of seismic protection are discussed in the following chapter.

4. EXECUTED EXAMPLES OF SEISMIC PROTECTION

4.1 Aseismic diagnosis of existing facilities[2]

In order to understand the present situation of water supply facilities, the key facilities are selected from the existing one and then the aseismicity of them are diagnosed in accordance with the Guideline for Seismic Protection Works of Water Facilities(1979). 6 pump stations, 10 distribution basins and 47 distribution pipelines with the extended length of 82km are chosen as the object facilities of aseismic diagnosis.

Since the distribution pipelines are constructed in the underground, they are constituted as the complicated network. Even when a part of network system is damaged, the total system of pipelines may loose its function. Thus the aseismicity of key pipelines which cover the whole city should be diagnosed. The procedure of diagnosis has been done with two stages. In the first stage of diagnosis, the geological map of traverse along pipelines, the geological condition learned from the field survey, the informations on faults, the results obtained by the response analysis based on the above mentioned guideline and the considerations on liquefaction are used for evaluating the aseismicity of total pipelines and for predicting the behaviors of their pipelines in earthquake. In the second stage, based on the results by the first stage of diagnosis, the pipeline running along the basin of Hirasaku River, where the weak ground mostly distributes within the city, was picked up. Then the dynamic analysis was carried out for estimating the behaviors of this pipeline with the modeled ground.

It is found from the diagnosis that the expected damage in pipeline comes up to the extension distance of 19km and then the prescribed damage rate to the length of 82km for the whole pipelines becomes 23%. Based on the analytical results obtained from the diagnosis, the renewing plan for distribution pipelines are scheduled, and the replacing of existing pipelines to the aseismic one and the saving for emergency restoration resources have been proceeded.

4.2 Examples of countermeasures

(1) Reinforcement of the trunk pipelines for the transmission and distribution

The pipeline structure may be mostly damaged in every water supply facilities. 18 tunnels which amount to the total length of 10,500m are existed in the pipeline of Arima and Kosuzume water systems. Since these tunnels were made by handwork and were severely weathered, all parts of tunnel without lining were completely fixed up and the bridges of aqueduct were renewed as shown in Photo-1.

Photo 1　Renewed water piping brige

Photo 2　Well for emergency use

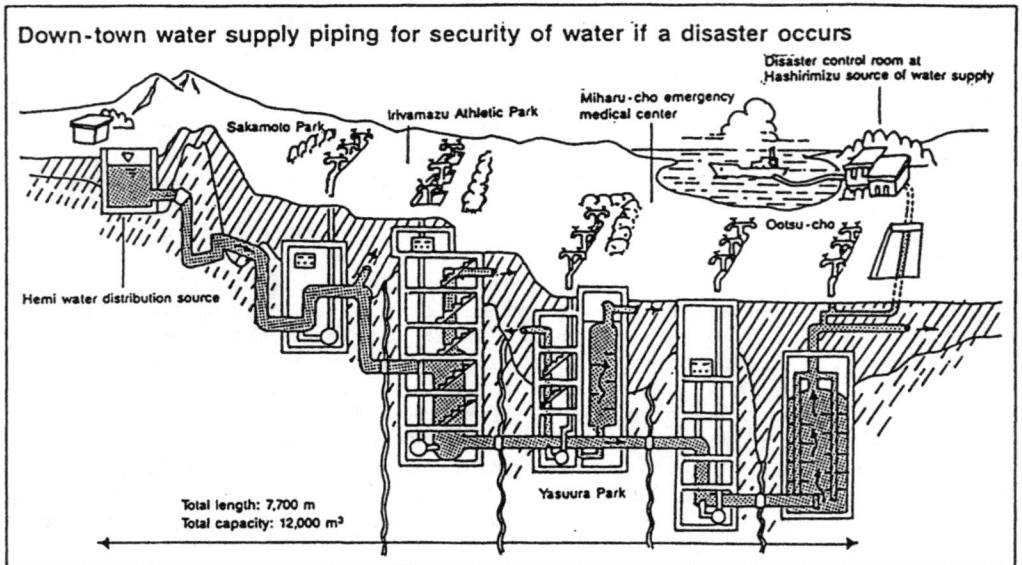

Fig. 2　Down-town water supply piping for security of
water if a disaster occurs

(2) Exploitation of new water resources in emergency

Since it is not enough to secure the drinking water only from Hashirimizu source of water supply which has a capacity of 2,000m^3 per day, the emergency wells for gathering water having a capacity of 3,000m^3 per day were founded on the basis of the measured data of natural radioactivity in this area as shown in Photo-2. Also in the west district of the city, the gathering water wells having a capacity of 2,000m^3 per day were founded. Then the emergency water sources in earthquake run up to 7,000m^3 per day in amount.

(3) Security of drinking water
   In order to secure immediately after disaster, the drinking water of 3 days (that is, 3 liters per day for a person) for the citizen, the following counter-measures were being put into practice.
 i)To prevent from flowing due to the breaking of pipe, the water distribution basin was equipped with the emergency shut-off valve apparatus.
 ii)To secure the drinking water in emergency for a wide spread refuge, the trunk line of water distribution was provided with the function of storage as shown in Fig. 2.

Photo 3 Steam-locomotive shaped emergency reservoir directly connected to water supply piping

Photo 4 A joint with simple mechanism for temporary piping

iii)The water storage tanks, which connected directly to the water pipe, was founded in a refuge within the city as shown in Photo-3.

(4) Establishment of water supply system
   Since 7,000m^3 per day of underground water and 85,600m^3 of storage water are secured as the water source for emergency use, the system which enables to supply 20 liters per day to one person for 45 days is arranged. As for the measures of water transportation from Hashirimizu source, the use of aseismic pipeline, the transportation

Photo 5 Motorcycle reconnaissance

by ship and that by land can be employed. To cope with any of these measures, a joint with simple mechanism shown in Photo-4 was developed for connecting the pipe and they have been stored for emergency use.

(5) Definition of official duties

　　To save the life of citizen in earthquake disaster, securing the drinking water, supplying it immediately and recovering the water supply facilities as soon as possible are the official duties of waterworks. Thus the system for mobilizing officials, which enables to make them understood completely their duties and to do the efficient acts, and the system for supporting of related institutions were constituted. And the motorcycle reconnaissance were organized to grasp immediately the disaster situation as shown in Photo-5.

4.3　Standardization of select works of pipes[3]

　　After completed both the replacement of normal pipes with aseismic one in the trunk pipeline and the reinforcement of water supply points, the guideline on selecting pipes for water supply was standardized in 1985 on the basis of the results by surveying the local works of pipe selection. The objective of standardization is to arrange effectively the network of water supply pipelines which has the total extension of 1,200km within the city. In this process, first of all, the geological informations on Yokosuka city, which are possessed individually by some institutions, were combined and rearranged. Then

Table 1　Rank of seismic risk

Kind of Ground	Risk of liquefaction		
	15<PL	5<PL≤15	PL≤5
30<Hs	A	A	B
10<Hs≤30	A	B	C
Hs≤10	A	C	C
Land slide Fault	A		

Hs:Thickness of surface stratum(m)
PL:Liquefaction index

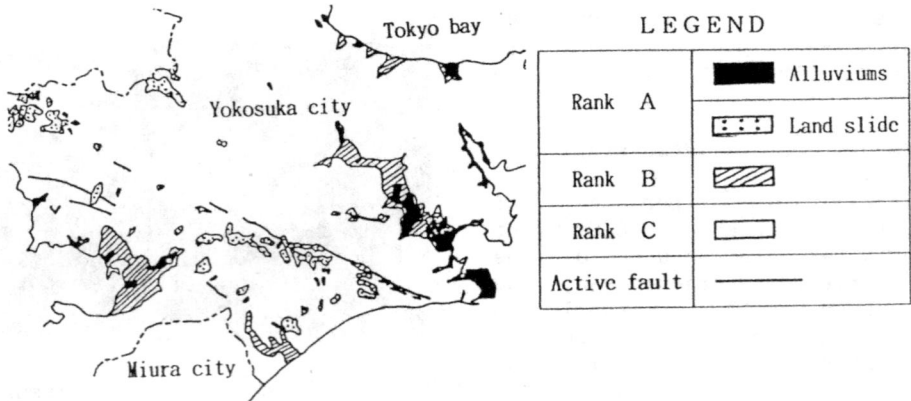

Fig. 3　Zoning in accordance with the classification of seismic risks

the rearranged informations were computerized. Considering both the strain distribution of ground and the occurrence probability of liquefaction, the ground within the city was classified into three ranks of seismic risks as shown in Table-1. Also the whole area of city was zoned in accordance with the classification of seismic risks as shown in Fig. 3. As the indication of renewing the pipeline in each rank of ground, the standard of the usable type of pipes was framed as given in Table-2.

5. CONCLUDING REMARKS

As for the seismic protection works of water facilities in Yokosuka city, some features of Yokosuka city and a few examples of aseismic counter-measures for water supply pipelines are presented in this report.

Table 2   Standard of the usable type of pipes

Rank	Steel pipe	Ductile iron pipe
A	Special thickness	S Ⅱ, S
B	Ordinary thickness	A, K, T
C	Ordinary thickness	A, K, T

The counter-measures of water supply facilities against earthquake in Yokosuka city has been improved gradually from a monistic treatment to a dualistic one. That is, at the present, the two approaches of improving aseismicity in the whole facilities by a renewal of old pipes etc. and establishing the decision-making system which can respond immediately to the unexpected situation in the future have been pursued simultaneously. It is considered that constructing the mapping system for making maximum use of the all kinds of data and dividing the water supply network into small blocks, which is the fundamental counter-measure for making efficient operate of water supply, will contribute toward improving the total system of water supply works.

REFERENCES

1)Yokosuka City Waterworks Bureau:Vital Water for Disaster,1982.2(in Japanese)
2)Y.Saito et.al: Survey of Aseismicity of the Water Supply Facilities in Yokosuka City I,II,III, Journal of Water Supply Association, 1981.9.10.12 (in Japanese).
3)Yokosuka City Waterworks Bureau: Report for Surveying the Local Works of Pipe Selection, 1986.2 (in Japanese).

Damage associated with three early Eastern North American earthquakes

Anne E. Stevens[1]

ABSTRACT

Earthquakes in June 1638, February 1663 and September 1732 were felt widely in northeastern North America. Primary documents are analyzed for reports of their effects on and within structures. In 1638 damage in the Boston—Plymouth area was insignificant in kind and extent. Damage in 1663 consisted of a few fallen chimneys at Trois-Rivières and a few damaged chimneys plus a few examples of displaced furniture and light damage at Québec. In 1732 earthquake damage was confined to Montréal. Three large buildings were seriously damaged, one of which had been poorly maintained. While a number of houses were damaged, they may have been less numerous than previously thought. In all three earthquakes, population living near the epicentral area was large enough to be certain that all significant damage had been reported and that such information is still available today in archives or in printed documents.

INTRODUCTION

Engineering assessment of seismic risk for a particular project begins with evaluation of the seismic hazard at the site and in its vicinity. Such evaluation depends on two rather different types of information about earthquake activity, instrumental and historical. Since most of the larger earthquakes in eastern North America occurred before suitable instrumentation had been developed, historical records of earthquakes are very important in the evaluation of seismic hazard. Written accounts of earthquakes that occurred in previous centuries help us estimate the rate of earthquake activity and the maximum size of earthquake in each earthquake zone. The older earthquakes are seldom useful in defining zone boundaries, as their locations may be quite uncertain, sometimes by 50 or 100 km, or more.

Lacking instrumental data, estimates of earthquake location and size depend upon an interpretation of the effects of such earthquakes, principally on people and on structures. This paper treats three earthquakes that were reported felt in settled areas in the 1600s and 1700s and which caused some damage. These reported effects are examined in terms of population and types of construction at the time in order to determine what was typical damage.

HISTORICAL EXAMPLES

This paper focuses on the damage associated with the earthquakes of June 1638, February 1663 and September 1732; it will not discuss determination of their locations and magnitudes. All three earthquakes have been catalogued

---

[1] Geological Survey of Canada, 1 Observatory Crescent, Ottawa, Canada K1A 0Y3
GSC Contribution Number 51990.

with maximum intensities of IX to X on the Modified Mercalli (MM) intensity scale, which were converted to magnitudes of the order of 7 (Smith 1962; Basham et al. 1979). These estimates of intensity and magnitude were based largely on non-representative damage, on accounts of landslides or on descriptive adjectives such as "great", "fearful" and "terrible" used in some of the original reports. When the damage caused by these earthquakes is isolated from these factors, then the three earthquakes would appear to have been smaller, considerably smaller in maximum intensity, perhaps somewhat smaller in magnitude. Aftershocks were reported felt in the weeks following each of these three earthquakes, but caused no further damage. Damage reports analyzed in this paper are drawn from primary documents written by persons having some first-hand knowledge of these earthquakes.

June 1638

The first earthquake in eastern North America for which a specific date can be unambiguously assigned occurred on Friday afternoon, 11 June 1638, somewhere in the vicinity of Boston, Massachusetts. About half a dozen documents still exist that were written by persons with first-hand knowledge of this event (Stevens and Gouin 1991), which was reported felt in New England in communities as widely separated as 150 km. It was also felt in the St. Lawrence valley, probably at one location near the settlement at Québec, at a distance of 500 km north of Boston. Based on recent experience, these distances suggest a magnitude not less than 5. Very minor damage was reported in the Plymouth colony, located on the Atlantic coast about 50 km southeast of Boston. In other communities, the historical documents describe the vibrations and the reactions of the populace, but say nothing about damage, from which we may assume that none occurred anywhere but Plymouth.

At Plymouth the following damage was reported (Bradford 1952 edition, page 302): " ... the earth began to shake and came at length with that violence as caused platters, dishes and suchlike things as stood upon shelves, to clatter and fall down." That is the total extent of the reported damage. In fact, "damage" is probably too strong a word; nevertheless it will be retained. We can accept as reliable this account of damage caused by the 1638 earthquake, since it forms part of a history written by William Bradford, who was governor of Plymouth colony for most of the years between 1620 and 1647. In another report written about 120 years later, but based on documents of the mid-1600s, some no longer available, Hutchinson (1760, page 90) added chimney tops to the short list of damage: " ... and the tops of chimnies [sic] in some places shook down ...." Note that no specific communities were named in this latter report. Comparing these effects with those of the Modified Mercalli intensity scale, a maximum intensity of MM V or VI could be assigned. This would be consistent with a magnitude of 5, although there is no satisfactory way to convert maximum intensity to magnitude. As already noted, this paper does not attempt to estimate magnitude for the June 1638 earthquake, but simply notes that its magnitude could not have been less than 5.

To place this reported damage in perspective, consider the size of the settlements at Boston and Plymouth and the type of house construction. A traveller from England arriving in Boston in July 1638 noted that it "was rather a village, than a town, there being not above twenty or thirty houses" (Josselyn 1674, pages 19-20). Later authors estimated the population of Boston in mid-1638 at about 1,500 persons, the entire European population of Massachusetts Bay at about 10,000 persons (Rutman 1965, page 179; Langdon 1966, page 44). Damage occurring in any of these communities would certainly have been noted and recorded, even if it had been as minor as fallen dishes. Plymouth and its surrounding communities were much smaller, with a population of several hundreds, rather than several thousands (Langdon 1966, pages 82-83,87).

Houses in the Plymouth and Massachusetts colonies in the mid-1600s were usually a simple 2-storey "frame structure covered with boards or shingles, with a central chimney rising in its center" (Langdon 1966, page 146). Roofs were thatched, chimneys of wood daubed with clay (Rutman 1965, pages 31, 203). Several decades later, the 1692 Building Code for Boston (1699, pages 3–4) required new buildings more than eight feet in length or width and more than seven feet high to be constructed of stone or brick instead of timber and to be roofed with slate or tile instead of shingle to minimize the spread of fire. In addition, non-conforming wooden buildings constructed since 1688 were to be roofed with slate or tile. While these new measures reduced the fire hazard, they may not have improved earthquake resistance.

In considering the damage, or lack thereof, reported for the earthquake of June 1638, it is recognized that small wooden houses, including wooden chimneys, are inherently earthquake-resistant. However, widespread reports of fallen dishes throughout the Boston area would have been possible if the shaking had been strong enough. Since such reports were confined to the Plymouth colony, two explanations may be suggested for Boston — neither of which can be proved or disproved. First, the strong ground motion was not particularly strong — hence no damage, no overturned objects. Or, secondly, Boston was close enough to the epicentre that the strong ground motion was essentially high-frequency vertical motion, with weak horizontal components. A modern analogue might be the Miramichi, New Brunswick, earthquake of 09 January 1982, magnitude 5.7, which, although widely felt and causing some minor cracks inside buildings at about 100 km distance, did not disturb even dishes in cabins in the epicentral area (Stevens 1983). While historical documents of the 1600s contain sufficient information to permit estimates of location and magnitude of the June 1638 earthquake, as reported elsewhere (Stevens and Gouin 1991), the damage reports are too limited to draw any firm conclusions. Was the damage negligible because the horizontal components of strong ground motion were not very strong or because there was little to disturb in early New England buildings? This remains an open question.

February 1663

Twenty-five years later, a strong earthquake in the St. Lawrence valley early Monday evening, 05 February 1663, was widely felt in eastern Canada and the northeastern United States, to distances of over 500 km. For example, the main shock and several aftershocks were reported felt in Montréal (Mondoux 1942, pages 184–185) and in Boston. The earthquake and its long series of aftershocks, reported felt mainly at Québec and at Tadoussac, have been described by many writers in the intervening 300 years in both popular and scientific articles. The original and derivative accounts of the effects in New France were devoted largely to numerous examples of landslides and slumps discovered for several months thereafter, and to the superstitions of the frightened population. Sifting fact from fiction is not straightforward. The present paper focuses narrowly on damage reports and does not discuss location, except to note that the Charlevoix–Kamouraska area has generally been designated as the epicentral area, which is about 100 km and 200 km, respectively, northeast of Québec and Trois-Rivières.

The following information on damage emerges from the three primary documents, the annual report from the Jesuit missions in the colony of New France, the residence journal of the Jesuits at Québec, and a letter from the head of the Ursuline community at Québec. These documents included some first-hand experiences plus many second- and third-hand reports gathered from European and Indian travellers passing through Québec during the seven months from early February to the end of August 1663. In his covering letter dated 04 September 1663, Lalemant (1663) wrote that not one life had been lost, nor any earthly goods. In Chapter 2 of his annual report, he noted that they had not lost one child, not even one hair from the head. In his single reference to damage, he expressed amazement that only several chimneys had been demolished, but he did not indicate their location. His statement would imply that if other damage had occurred to buildings or contents anywhere in New

France, it must have been of much less importance than chimney damage. The Jesuit residence journal (Laverdière and Casgrain 1871), with monthly entries in 1663 also written by Lalemant, noted, in its February paragraph, that certain chimneys had been damaged and that there had been other light losses and damage. Since entries in this journal generally concerned Québec or Jesuit activities there, unless otherwise stated, we may assume that the journal entry referred to damage at Québec. These two damage accounts by Lalemant contrast sharply with the more dramatic description in the following document.

In a lengthy letter about the earthquake written to her son in France and dated August—September 1663, Marie de l'Incarnation (1663) emphasized the supernatural and unusual; in the final paragraph she noted that no one had perished, nor even been injured. Her letter included a few personal observations — overturned furniture, fallen stones, floors that separated, walls that were split. From the context, these observations were probably made at the Ursuline residence at Québec. Nothing was said subsequently about repair or the need for repair. On the other hand, the stone building must have been vigorously shaken, as thick dust spread out on all sides. From the Jesuit journal we know that the Jesuit fathers visited the Ursuline community regularly to say mass; they would certainly have known about any damage there. Movement of the specific items mentioned by Marie de l'Incarnation must thus have been confined to a few examples, and must not have been a general fall of stones and cracking of floors and walls, as might be concluded from her letter alone. Marie de l'Incarnation also quoted directly from a letter written from Trois-Rivières which said, in part, that as houses there were all of wood, damage had been confined to the fall of a few chimneys.

The total damage attributed to the February 1663 earthquake thus consisted of a few fallen chimneys in Trois-Rivières, a few damaged chimneys at Québec, and some overturned furniture and minor cracking at Québec. No damage occurred at Montréal. Lalemant's description in the Jesuit journal was brief and factual. The annual report of Lalemant and the letter of Marie de l'Incarnation devoted most of their space to landslides beyond the settlements and stories told by frightened persons, which implies that not much actual damage to buildings or contents had occurred; otherwise more than a few short sentences would have been devoted to it. As explained later, archival records of building contracts in 1663 should be examined for any evidence of earthquake-related repairs.

In 1663, the population in the region of Québec totalled 1,976 persons; settlement extended only about 40 km downriver (i.e. northeast) from Québec; the mission post at Tadoussac at the mouth of the Saguenay was occupied for only part of the year. The Montréal and Trois-Rivières regions had 597 and 462 inhabitants, respectively, for a total of about 3,000 persons in the St. Lawrence Valley (Trudel 1973, Chapter 18; Harris and Matthews 1987, Plate 46). Recall that the Boston area had achieved a population of 10,000 twenty-five years earlier. Despite the small population, there was regular correspondence within New France and beyond to France; any significant building damage would have been noted and reported. In the mid-1600s most houses had wooden roofs and stone chimneys and more than 10% of the houses in Québec and Montréal had stone walls. At Québec, settlement almost from the beginning had occurred both along the river in lower town and above the cliff in upper town, where both the Jesuit and Ursuline buildings were located (Harris and Matthews 1987, Plates 49 and 55; Trudel 1973, Chapter 18). It is not possible to say from the limited reports whether the effects in upper and lower town were markedly different.

As with the New England earthquake of 1638, the St. Lawrence valley earthquake of 1663 caused no significant damage, in part due to the small population, hence relatively few houses, and to the predominance of small wooden houses. A maximum intensity of MM VI would satisfy the damage observations at Trois-Rivières and Québec, which however are generally thought to have been some distance from the epicentre. Aftershocks were frequent during the first night and continued off and on for at least seven months, which suggests that the main shock must have been fairly large in order to have spawned so many aftershocks, and hence its peak ground motions must have been fairly strong. We then can again ask why there were not more reports of fallen dishes and displaced furniture. The frequency range of the strong motions in the settlements was obviously not tuned to that necessary to displace household objects.

September 1732

The next widely-felt earthquake in eastern North America that caused some damage occurred in the late autumn of 1727, again in the Boston area. Since a similar shock occurred in Canada just a few years later, the Canadian earthquake is chosen for examination. The earthquake of Tuesday morning, 16 September 1732, near Montréal, Québec, had been catalogued for many decades as a major earthquake, which had killed one person and injured others. A re-examination by G. Leblanc (1981) proved from primary documents that previous interpretations had been greatly exaggerated. In particular, no one had been killed or seriously injured. He found a maximum intensity of MM VII to VIII compatible with damage reported in Montréal; he had available however only some of the damage information to be presented below. This new information would support a maximum intensity of not more than MM VII. Basing his magnitude estimate on the area and distances for which the earthquake had been reported felt, Leblanc concluded that a magnitude in the range 5.6 to 6.0 would satisfy these data. Damage was confined to the town of Montréal, if we overlook a report in two Boston newspapers that several articles there had fallen from shelves. No damage whatsoever was reported at Québec (Chaussegros de Lery, 1732) and there is no archival evidence of damage at Trois-Rivières.

Chaussegros de Lery (1732), chief engineer for the colony of New France, in an official report one month after the main shock, described the damage to houses and to the wall around the town, based on information that he had received from persons arriving at Québec from Montréal. He (1733a; 1734; 1733b) subsequently prepared itemized statements of repairs made to two large stone buildings and noted work done on a third. These last three documents have not been previously presented in connection with the 1732 earthquake.

For houses, he reported a good proportion of chimneys fallen, other chimneys split [cracked] and many walls of houses opened [cracked], but he gave no statistics. Two other reports cited figures of 567 chimneys (Lahaise 1973, page 38) and more than 300 houses (Duplessis 1732). These figures are compatible, since urban stone houses had at least two chimneys each. On the other hand, these figures, as well as some other details in these two documents, should be used with caution as they were mentioned, respectively, only in a history of Hôtel-Dieu of Montréal written some years later by Soeur Cuillérier, who had herself experienced the earthquake, and in personal correspondence written at Québec from second-hand reports. If these figures were to be accepted, then comparison with an estimated population of 3,000 in Montréal in 1732 (Miquelon 1987, page 145) would suggest that almost half of the houses did suffer some damage. In reviewing the exchange of official government correspondence between Québec and France [not all referenced herein], the effects of the earthquake seemed to be of little relative importance, which would suggest that very much less than 50% of the houses had been damaged. As thousands of building contracts in New France between 1640 and 1760 are still available in various archives (Moogk 1977, page 11), it should be possible to extract and analyze information on work done in Montréal between mid-September 1732 and the spring of 1733 and thus to determine the true extent of the damage.

In 1727, a comprehensive building code was issued for all the towns in the St. Lawrence valley, with the basic principle that construction was to be durable and safe (Moogk 1977, Chapter 3). Many of these regulations, including the requirement for urban stone construction, had already been issued for Montréal in 1721. By 1732, just over 50% of the houses in Montréal were made of stone; virtually all chimneys were stone; roofing materials were planks or shingles (Harris and Matthews 1987, Plate 55; Moogk 1977, Chapter 3). Whether houses built in Montréal between 1721 and 1732 resisted earthquake damage better than earlier houses might make an interesting study.

The town walls in Montréal were under reconstruction in 1732, wood being replaced by stone. The town itself occupied a rectangular area about 1½ km by ½ km parallel to the St. Lawrence River (Harris and Matthews 1987, Plate 49). Chaussegros de Lery (1732) reported that alignment of the walls and their embankments had not been changed by the earthquake, which proves that no slumping nor liquefaction had occurred there. He noted the following minor damage, which had already been repaired for "60 livres". At the just completed Saint-Laurent gate several stones had shifted. In the masonry section of the wall, several parapet tops had shifted and several stones from the top of the gun slits had fallen. At the northern bastion a crack had developed, but closed again during an aftershock. Without estimating the purchasing power of "60 livres" in 1732, the context of this and other reports shows that this sum was not, in fact, considered very important.

Two of the large stone buildings in Montréal are known to have been damaged sufficiently to warrant requests to the King of France to pay for repairs, since they served the community. The first of these was the hospital complex, Hôtel-Dieu, which included an interconnected hospital, residence and church, plus some smaller outbuildings. The main buildings, each several hundred feet in length, were two to three stories in height (Mondoux 1942, Chapter 21). In the request for financial assistance, the head of Hôtel-Dieu noted that the monastery had been nearly completely ruined and she feared that its walls might completely collapse during the winter frosts (Le Vasseur 1732). All the chimneys had been toppled, walls were split [cracked] so as to show through daylight and part of the frame had moved out half a foot. The letter did acknowledge, however, that two previous fires [1695 and 1721] had weakened the walls. Extensive damage in the 1721 fire had caused the entire hospital complex to be temporarily relocated for 3½ years, lacking funds for immediate repairs (Mondoux 1942, Chapter 21). The requested assistance of "640 livres" was granted (Beauharnois and Hocquart 1734; Ducharme 1973), which was ten times the amount spent to fix the town walls. Repairs to the frame of the residence, reported above to have moved out half a foot, were itemized as only "60 livres", which makes that particular damage seem less severe than the written description alone would suggest. The repair costs were divided as follows (Chaussegros de Lery 1733a): rebuild or repair five chimneys, 36%; repairs to hospital, church and residence, 34%; repairs to other small buildings, 30%. The actual damage thus seems less dramatic than would be inferred from the partly objective, partly emotional descriptions by hospital staff who were reacting also to a further disruption to their hospital duties and to their living quarters.

The second damaged stone building in Montréal was the Récollets building. Their request, although acknowledged to be valid, was refused since the French government was short of funds (Beauharnois et Hocquart 1735). The total repairs estimated at "1884 livres", three times that for Hôtel-Dieu, were divided as follows (Chaussegros de Lery 1734): rebuild one chimney and parts of two walls, 53%; repair many ceilings where pieces had fallen, plus cracks in one chimney, 15%; general repairs to frame of building, 13%; other repairs, 19%. Chaussegros de Lery (1733b, folio 228v) entered additional information regarding earthquake damage in his 1733 financial statements for the colony of New France. Three masons and two labourers were paid a total of "120.75 livres" for 49½ person-days of work at the Montréal residence of "M. le Général", which suggests damage to a third large stone building. Some other details in this entry require clarification. Again, to put the damage to large stone buildings in perspective, town maps (Harris and Matthews 1987, Plate 49) show Hôtel-Dieu close to buildings such as Église Notre-Dame and Séminaire Saint-Sulpice, which were apparently undamaged or else suffered only minor damage, as no details are recorded (Maurault 1929, Chapters 1 and 5). The Récollets were located several blocks further away. More archival research is necessary to determine the structural characteristics of these buildings and to find evidence for repairs.

Historians have paid very little attention to the 1732 Montréal-area earthquake, probably since its economic impact was considerably less than that of the large fires of 1695, 1721 and 1734 (Mondoux 1942, Chapters 19, 21, 22; Moogk 1977, Chapter 3). No comprehensive report of earthquake damage was published at the time. Summaries by Chaussegros de Lery (1732; 1733a; 1733b; 1734) and related correspondence provide the only objective information to date. With some effort, further information on damage or lack thereof could be extracted from various archives.

CONCLUSIONS

Most previous analyses of these three early earthquakes have concentrated on their more dramatic effects, assigning high maximum intensities and, by implication, large magnitudes. This paper has preferred to document the effects to structures and to note relevant population size and building types. When viewed in this context, the maximum intensities typical of a particular settlement are much reduced. Location and magnitude of these or any other earthquakes should not however be based on damage or maximum intensities, but on more representative parameters such as size of the felt area. As might have been expected, the damage recorded from the New England earthquake of 1638 and from the earthquakes near Québec and Montréal in 1663 and 1732, respectively, increased as the population of the settlements increased, being insignificant in 1638 and minimal in 1663. Inadequate maintenance was probably a contributing factor to the damage at the Hôtel-Dieu hospital complex in Montréal in 1732. Statements of repairs to this and two other large stone buildings in Montréal show what was repaired, but provide no details on why damage occurred. Chimney and other house damage in 1732 may have been less important than currently believed; an archival search for relevant building contracts in Montréal could document the true extent of such damage. A similar search might aid interpretation of the 1663 earthquake. Relevant details for all three earthquakes are too sketchy at present to identify any of the building damage with poor soils. While damage to structures was not in any sense catastrophic in these early earthquakes and while no one was even injured, one must not conclude that earthquakes of similar size near the same urban areas today would be of little consequence. Efforts must continue to improve the earthquake resistance of new and existing structures.

REFERENCES

Basham, P.W., Weichert, D.H. and Berry, M.J. 1979. Regional assessment of seismic risk in eastern Canada. Bull. Seis. Soc. Am. 69, 1567–1602.

Beauharnois, C. and Hocquart, G. 1734. Letter dated 20 April 1734 from Versailles, France. Public Archives of Canada, MG-1, Archives des Colonies, série B, volume 61, folios 91–92.

Beauharnois, C. and Hocquart, G. 1735. Letter dated 12 April 1735 from the Président du Conseil de Marine. In: Inventaire des papiers de Lery conservés aux Archives de la province de Québec, volume 1, edited by P.-G. Roy, Québec, 1939, page 267.

Bradford, W. 1952 edition. Of Plymouth Plantation, 1620–1647. Edited by S.E. Morison, Alfred A. Knopf, New York, 448 pp. [Note that Bradford's manuscript was completed by 1650, but not printed for the first time until 1856.]

Building Code for Boston, 1692. 1699. Acts and Laws, of His Majesties Province of the Massachusetts-Bay, in New-England, [first chapter: An Act for building with stone or brick in the town of Boston, and preventing fire]. Green and Allen, Boston.

Chaussegros de Lery, G.-C. 1732. Letter dated Québec, 20 October 1732. In: Inventaire des papiers de Lery conservés aux Archives de·la province de

Québec, volume 1, edited by P.-G. Roy, Québec, 1939, 258–260.

Chaussegros de Lery, G.-C. 1733a. Statement of damage to Hôtel-Dieu buildings, dated Montréal, 04 August 1733. Public Archives of Canada, MG-1, Canada–Correspondance générale, série C-11-A, volume 60, folio 285.

Chaussegros de Lery, G.-C. 1733b. Financial statements of repairs to government property in 1733, dated at Québec, 18 October 1733. Public Archives of Canada, MG-1, Canada–Correspondance générale, série C-11-A, volume 60, folios 211–232.

Chaussegros de Lery, G.-C. 1734. Statement of repairs to Récollet building, dated Québec, 30 September 1734. Public Archives of Canada, MG-1, Canada–Correspondance générale, série C-11-A, volume 58, folios 187–187v.

Ducharme, J. 1973. Les revenus des Hospitalières de Montréal au XVIIIe siècle. In: L'Hôtel-Dieu de Montréal (1642–1973), edited by M. Allard, Les Cahiers du Québec, Éditions Hurtubise HMH Ltée, Montréal, 209–244.

Duplessis de Ste-Hélène, Mère Marie-Andrée. 1732. Letter dated Québec, October 1732. In: Leblanc, G. 1981, page 547 [see below].

Harris, R.C. (editor) and Matthews, G.J. (cartographer/designer) 1987. Historical Atlas of Canada: volume I, From the Beginning to 1800. University of Toronto Press, Toronto, 198 pp.

Hutchinson, T. 1760. The History of the Colony of Massachusets-Bay, from the First Settlement thereof in 1628, until its Incorporation with the Colony of Plimouth, Province of Main, &tc. by the Charter of King William and Queen Mary, in 1691. volume 1, M. Richardson, London, second edition, 566 pp.

Josselyn, J. 1674. An Account of Two Voyages to New-England. Giles Widdows, London, 279 pp.

Lahaise, R. 1973. L'Hôtel-Dieu du Vieux-Montréal (1642–1861). In: L'Hôtel-Dieu de Montréal (1642–1973), edited by M. Allard, Les Cahiers du Québec, Éditions Hurtubise HMH Ltée, Montréal, 11–56.

Lalemant, H. 1663. Annual report for 1662–1663, dated at Québec, 04 September 1663 (Relation de ce qui s'est passé de plus remarquable aux missions des Pères de la Compagnie de Jésus en la nouvelle France, ès années 1662 et 1663). In: Relations des Jésuites, volume 5, 1656–1665, Éditions du Jour, Montréal, 1972.

Langdon, G.D. 1966. Pilgrim Colony: A History of New Plymouth, 1620–1691. Yale University Press, New Haven, 257 pp.

Laverdière, C.-H. and Casgrain, H.-R. (editors) 1871. Le Journal des Jésuites. L. Brousseau, Québec, 403 pp.

Leblanc, G. 1981. A closer look at the September 16, 1732, Montreal earthquake. Can. J. Earth Sci., 18, 539–550.

Le Vasseur, Soeur. 1732. Letter dated Montréal, October 1732. In: Leblanc, G. 1981, page 548–549 [see above].

Marie de l'Incarnation, Rév. Mère. 1663. Letter dated August–September 1663. In: Marie de l'Incarnation Ursuline (1599–1672), Correspondance. Edited by Dom Guy Oury, Abbaye Saint-Pierre, Solesmes, 1075 pp., 1971.

Maurault, O. 1929. La Paroisse: Histoire de l'Église Notre-Dame de Montréal. Louis Carrier & Cie, Les Éditions du Mercure, Montréal, 334 pp.

Miquelon, D. 1987. New France 1701–1744, "A Supplement to Europe". McClelland and Stewart, Toronto, 345 pp.

Mondoux, Soeur Maria. 1942. L'Hôtel-Dieu, Premier Hôpital de Montréal, 1642–1763. Thérien Frères Limitée, Montréal, 419 pp.

Moogk, P.N. 1977. Building a House in New France: An Account of the Perplexities of Client and Craftsmen in Early Canada. McClelland and Stewart, Toronto, 144 pp.

Rutman, D.B. 1965. Winthrop's Boston: Portrait of a Puritan Town, 1630–1649. University of North Carolina Press, Chapel Hill, 324 pp.

Smith, W.E.T. 1962. Earthquakes of eastern Canada and adjacent areas, 1534–1927. Pub. Dom. Obs., Ottawa, 26, 271–301.

Stevens, A.E. (editor) 1983. Miramichi, New Brunswick, Canada, earthquake sequence of 1982: A preliminary report. Earthquake Engineering Research Institute, Berkeley, EP–53, March 1983, 99 pp.

Stevens, A.E. and Gouin, P. 1991. The New England earthquake of June 1638. (in preparation).

Trudel, M. 1973. The Beginnings of New France, 1524–1663. McClelland and Stewart, Toronto, 323 pp.

Index of authors
Index des auteurs

www.ingramcontent.com/pod-product-compliance
Lightning Source LLC
Chambersburg PA
CBHW060925210326
41597CB00042B/4465